THIRD EDITION

Protocols for Secure Electronic Commerce

THIRD EDITION

Protocols for Secure Electronic Commerce

Mostafa Hashem Sherif

CRC Press is an imprint of the
Taylor & Francis Group, an **informa** business

CRC Press
Taylor & Francis Group
6000 Broken Sound Parkway NW, Suite 300
Boca Raton, FL 33487-2742

First issued in paperback 2018

ISBN-13: 978-1-4822-0374-5 (hbk)
ISBN-13: 978-1-138-58605-5 (pbk)

Library of Congress Cataloging-in-Publication Data

Names: Sherif, Mostafa Hashem, author.
Title: Protocols for secure electronic commerce / Mostafa Hashem Sherif.
Description: Third edition. | Boca Raton : Taylor & Francis, a CRC title, part of the Taylor & Francis imprint, a member of the Taylor & Francis Group, the Academic Division of T&F Informa, plc, [2016] | Includes bibliographical references and index.
Identifiers: LCCN 2015040118 | ISBN 9781482203745 (alk. paper)
Subjects: LCSH: Electronic commerce. | Bank credit cards. | Computer networks--Security measures.
Classification: LCC HF5548.32 .S5213 2016 | DDC 658.8/72028558--dc23
LC record available at http://lccn.loc.gov/2015040118

Visit the Taylor & Francis Web site at
http://www.taylorandfrancis.com

and the CRC Press Web site at
http://www.crcpress.com

Contents

Preface to the Third Edition

Electronic commerce has become part of the daily routine of consumers and businesses alike on a global level. Millions of secured transactions are made over the Internet every day; new intermediaries have established themselves in both the consumer and business spaces; mobile communication offers an important channel for transactions and payments; cryptocurrencies are pushing the boundary of digital money while web services are gradually replacing EDI (electronic data interface) in business-to-business transactions.

Past editions of the book provided a comprehensive and readable survey of the protocols for securing e-commerce and e-payments in consumer and business applications. The current edition preserves this spirit while tracking the changes since September 2003, when the text of the second edition was finalized. Despite the changes, those familiar with the previous editions should still be able to easily navigate the book. First-time readers will find a clear guide to current issues concerning the security of e-commerce, with pointers to more specialized texts listed in the bibliography.

The treatment of bank cards is now divided between two chapters, one for magnetic stripe cards and another for the chip-and-PIN technology. The material on mobile commerce covers the different service architectures, technologies, and security concerns specific to mobile terminals. All aspects of micropayment systems, whether as electronic purses, online systems, or research prototypes, have been consolidated into one chapter. The discussion on PayPal is commensurate with PayPal's increased role for consumers and businesses. Bitcoin and other cryptocurrencies that did not exist back in 2003 are treated in a separate chapter. I have retained, however, shortened versions of the chapters on WTLS (Wireless Transport Layer Security) and SET (Secure Electronic Transaction) so that the lessons learned from their failures are not forgotten. Furthermore, some of the innovations that the SET protocol introduced may be useful in the future.

For academic use, suggested answers to the end-of-chapter review questions and PowerPoint© presentations are available from the CRC website.

Several chapters, in particular Chapters 5, 8, 9, 10, 12, and 14, benefited from the thoughtful reviews and suggestions generously provided by Fred Burg (AT&T, retired), Dr. Pita Jarupunphol (Phuket Rajabhat University, Thailand), Fahad Najam (Cowen Group), and Waleed Saad (IBM), as well as two anonymous reviewers. Jonathan Warren kindly provided guidance on his paper "Bitmessage: A Peer-to-Peer Message Authentication and Delivery System" and Wei Dai graciously provided his correspondence with the mysterious Satoshi Nakamoto concerning B-money. I acknowledge conversations with Dr. Isam Habbab on the applications of the birthday problem in cryptography and e-mail exchanges with Dr. Guenther Starnberger from the Vienna University of Technology on the use of QR codes in mobile commerce. Of course, I have not necessarily followed all their suggestions and any remaining deficiencies are entirely my responsibility.

Thanks are also due to Mrs. Margery Ashmun, Reference Librarian, Drew University, New Jersey, for her help in locating some references.

Despite a delay of about 2 years, the publisher from Taylor & Francis/CRC Press, Nora Konopka, did not waiver in her support and understanding. I also acknowledge the excellent help from Laurie Oknowsky, production assistant, and the rest of the CRC team including Florence Kizza, Linda Leggio, and Kyra Lintholm. The meticulous copy editing, managed by Vijay Bose, from SPi Global, uncovered many inaccuracies and inconsistencies that slipped through during the long gestation of the manuscript.

Finally, the support and encouragement of family and friends were, as usual, indispensable.

Mostafa Hashem Sherif
Tinton Falls, New Jersey

Preface to the Second Edition

The presence and influence of online commerce are growing steadily despite, if not because of, the dot.com frenzy. With the speculators gone and in the absence of unsubstantiated claims, it is now possible to face the real problems of the information society in a rational and systematic manner. As more virtual services are offered to the general public or among businesses, the security of the networked economy will need to consider many factors. Potential solutions can go along so many directions, as additional parties with different priorities and requirements are brought online. The interconnection and fusion of local spaces can only mean that electronic commerce (e-commerce) security will require global actions, including global technical standards and organizational agreements. These activities, however, do not occur in vacuum; compromises will have to be made to cope with existing infrastructures, processes, laws, or social organizations that were not designed for online activity.

The aim of this book is to help the reader address these challenges. Its intended audience ranges from readers of the periodic "IT-Review" of the *Financial Times*, who may want to understand the technical reasons behind the analysis, to graduate students in technical and informational domains who would like to understand the context in which technology operates. In updating the text, I strove to maintain the goals of the first edition of providing a comprehensive, though readable, compendium to the protocols for securing e-commerce and electronic payments. I tried to provide enough technical details so that readers could gain a good grasp of the concepts, while leaving the details to more specialized works listed in the bibliography. Chapters were revised or completely rewritten to reflect technical advances and continuous developments and to include new topics such as mobile commerce (m-commerce). In doing so, I have benefited from the experience gained in teaching the material to improve the presentation and correct errors. In some cases, such as for SET, I decided to maintain topics that did not succeed in the marketplace because of the many innovative ideas that were involved.

For academic use, I followed the suggestions of several instructors and added review questions at the end of each chapter. In addition, PowerPoint® presentations will be available from the CRC website (www.crcpress.com) on the topics discussed in each chapter.

My French editor, Eric Sulpice, generously supplied me with information on the development of smart cards in Europe. Kazuo Imai, vice president and general manager of the network laboratories of NTT DoCoMo, provided me with technical information on i-Mode®. Professors Manu Malek of the Stevens Institute of Technology (Hoboken, New Jersey) and Mehmet Ulema from Manhattan College, New York, gave me useful comments on the content and its presentation.

Once again, I thank CRC Press, in particular Dr. Saba Zamir, editor-in-chief of the series, for her confidence, and the editorial team of Nora Konopka, Samar Haddad, and Jamie Sigal for their assistance and Lori Eby for her excellent copy editing.

Finally, the trust and encouragement of relatives and friends were, as usual, indispensable.

Mostafa Hashem Sherif
Tinton Falls, New Jersey
July 2002–September 2003

Preface to the First Edition

The purpose of this book is to present a synthesis of the protocols used today to secure electronic commerce. The book addresses several categories of readers: engineers, computer scientists, consultants, managers, and bankers. Students interested in the computer applications in the area of payments will find this book a useful introduction that will guide them toward more detailed references.

The book is divided into three parts. The first consists of Chapters 1 through 3 and is a general introduction to the multiple aspects of electronic commerce. The second part is formed by Chapters 4 through 12 and details the various aspects of electronic money: Electronic Data Interchange (EDI), payments with bank cards, and micropayments with electronic purses, digital money, and virtual checks. The third and final part comprises Chapters 13 through 15 and presents smart cards, efforts for converging heterogeneous payment systems, and some thoughts on the future of electronic commerce.

The field of electronic commerce covers several topics that are evolving continuously, and it is not possible to cover all of them here. We would be grateful to feedback from readers regarding errors, omissions, or additional material for consideration.

This book appears in a French version titled *La Monnaie Électronique: Systèmes de Paiement Sécurisé* (coauthored with Professor Ahmed Sehrouchni from the École Nationale Supérieure des Télécommunications [ENST], Paris, France) published by Eyrolles.

The discussions that I had with participants in the project "PECUNIA" of the now defunct AT&T-Unisource helped clarify many details concerning the payment systems. I thank, in particular, Maria Christensen, Greger S. Isaksson, and Lennart E. Isaksson, all three from the research unit of the Swedish operator Telia, and Philip Andreae (consultant) and Patrick Scherrer, who led the project. Aimé Fay, my former colleague at AT&T France and author of the dictionary on banking technology, *Dico Banque*, graciously guided my first steps in the field of payment systems. The research conducted with Luis Lucena while he was a graduate student at ENST-Paris and with my colleagues at the National Technical University of Athens, Greece—Maria Markakis, Georges Mamais, and Georges Stassinoupoulos—helped me evaluate the effect of computer telephony integration (CTI) on electronic commerce. Chapters 6 and 7 were influenced profoundly by contributions from A. Yassin Gaid and Farshid Farazmandnia during the course of their internship at AT&T, Paris, France in 1997 as part of their graduation project from ENST-Paris. The results of their work have been published in French and in English.

CRC Press have been very patient throughout the long gestation of this project. The project would not have started without Saba Zamir, editor-in-chief of the series Advanced and Emerging Communications Technologies, and Gerald T. Papke, senior editor with CRC Press.

My thanks also extend to Donna Coggshall who reviewed and edited the first English version of the manuscript. Fred Burg, my colleague at AT&T, reviewed the first two chapters and suggested some stylistic improvements. Andrea Tarr introduced me to Bert V. Burke, the founder and CEO of Every Penny Counts Inc. (EPC), who provided information included in Chapter 14.

Finally, I am grateful to friends and relatives who greatly supported me throughout the preparation of this book.

Mostafa Hashem Sherif
Neuilly-sur-Seine, France
October 1997

Tinton Falls, New Jersey
October 1999

Author

Mostafa Hashem Sherif, PhD, is a principal member of the technical staff at AT&T in Middletown, New Jersey. He received a BSc in electronics and communications and an MSc in electrical engineering from Cairo University, Egypt, in 1972 and 1975, respectively; and a PhD in engineering from the University of California, Los Angeles, in 1980. In 1996, he earned a master's of science in management of technology from Stevens Institute of Technology (Hoboken, New Jersey). Dr. Sherif is a senior member of the IEEE.

His book publications include:

1. *Protocols for Secure Electronic Commerce*, CRC Press, Boca Raton, FL, 2000, 2004. (A French version of the first edition was coauthored with A. Sehrouchni and published by Eyrolles, Paris, France, under the title *La Monnaie électronique*.)
2. *Managing Projects in Telecommunication Services*, John Wiley & Sons, Hoboken, NJ, 2006.
3. *Paiements électroniques sécurisés*, Presses polytechniques et universitaires romandes, Lausanne, Switzerland, 2007.

He is the editor or coeditor of the following publications:

1. *Handbook of Enterprise Integration*, CRC Press, Boca Raton, FL, 2009.
2. Coeditor with T. M. Khalil of *Management of Technology, Vol. 2. Management of Technology Innovation and Value Creation, Selected Papers from the 16th International Conference on Management of Technology*, World Scientific, Singapore, 2008.
3. Coeditor with T. M. Khalil of and contributor to *New Directions in Technology Management, Selected Papers from the 13th International Conference on Management of Technology*, Elsevier Science, Ltd., Oxford, UK, 2006.

He has contributed to these publications:

1. *The Encyclopedia of Life Support Systems*, UNESCO-EOLSS, Paris, France, 2007.
2. *Advanced Topics in Information Technology Standards and Standardization Research*, Vol. 1, Kai Jakobs, editor, Idea Group Publishing, Hershey, PA, 2006.
3. *Electrical Engineering Handbook*, 3rd ed., Richard C. Dorf, editor, CRC Press, Boca Raton, FL, 2006.
4. *Managing IP Networks, Challenges and Opportunities*, Salah Aidarous and Thomas Plevyak, editors, IEEE Press, Piscataway, NJ, 2003.

He is involved with the following scientific journals:

1. From January 2016, he was the area editor for the management of industry section, *Technovation*.
2. Associate editor of the *International Journal of IT Standards and Standardization Research*.
3. Standards coeditor of the *IEEE Communications Magazine* since 1993.

His other activities include:

1. Member of the Steering Committee of the Kaleidoscope Series of Conferences organized by the International Telecommunication Union (ITU).
2. Member of the Steering Committee of the *IEEE Symposium on Computers and Communication* from 1995 to 2006.
3. Participation in activities on innovation and technology management (1987, 1989, 1996, 1998, 1999, 2000, and 2002) sponsored by the National Science Foundation.
4. Member of the Evaluation Committee for the Commission on Science and Technology, State of New Jersey, in 2000–2001 and 2001–2002, reviewing preproposals for the R&D Excellence Program.

Acronyms

3D SET	3-Domain SET
3-D SSL	3-Domain SSL
6TOC	6C Toll Operators Committee
AAC	Application Authentication Cryptogram
ACH	Automated Clearing House
ACS	Access Control Server
ADS	Activation During Shopping
AEAD	Authenticated Encryption with Associated Data
AECE	Asociación Española de Comercio Electrónico (Association for Electronic Commerce)
AES	Advanced Encryption Standard
AFNOR	Association Française de Normalisation (French Association for Standardization)
AH	Authentication Header
AIAG	Automotive Industry Action Group
AID	Application Identifier
AIR-IMP	AIR Interline Message Procedures
Ajax	Asynchronous JavaScript and XML
ANSI	American National Standards Institute
ANX®	Automotive Network eXchange
APDU	Application Protocol Data Unit
API	Application Programming Interface
ARD	Authentication-Related Data
ARQC	Authorization Request Cryptogram
ASC	Accredited Standards Committee
ASCII	American Standard Code for Information Interchange
ASIC	Application-Specific Integrated Circuit
ASN.1	Abstract Syntax Notation 1
ATM	Automated Teller Machine
ATQA	Answer to Request A
ATR	Answer to Reset
BACS	Banker's Automated Clearing Service
BC	Biometric Consortium
BCBP	Bar-Coded Boarding Pass
BEAST	Browser Exploit Against SSL/TLS
BER	Basic Encoding Rules
BIN	Bank Identification Number
BIP	BioAPI Interworking Protocol
BIP	Bitcoin Improvement Proposal
BIS	Bank for International Settlements
BLE	Bluetooth Low Energy
BPEL	Business Process Execution Language
BPSS	Business Process Specification Schema
BREACH	Browser Reconnaissance and Exfiltration via Adaptive Compression of Hypertext
BSI	Business Service Interfaces
BSP	Bank Settlement Payment
BTX	Bildschirmtext
C-SET	Chip-Secured Electronic Transaction
CA	Certification Authority
CAFE	Conditional Access for Europe
CALS	Computer-aided Acquisition and Logistics Support became Continuous Acquisition and Life-cycle Support, then Commerce at Light Speed
CAPI	Cryptographic Application Programming Interface
CAPTCHA	Completely Automated Public Turing Test to Tell Computers and Humans Apart
CAR	Confirmation and Authentication Response
CARGO-IMP	CARGO Interchange Message Procedures
CAS	Channel-Associated Signaling
CAVV	Cardholder Authentication Verification Value
CBC	Cipher Block Chaining
CBEFF	Common Biometric Exchange File Format
CCA2	Adaptive Chosen Ciphertext Attack
CCD	Cash Concentration and Disbursement
CCD	Charge-Coupled Device
CCITT	Comité Consultatif International Télégraphique et Téléphonique
CCM	Counter with Cipher Block Chaining—Message Authentication Code Mode
CCS	Common Channel Signaling
CEFIC	Conseil Européen des Fédérations de l'Industrie Chimique (European Council of Industrial Chemistry Federations)
CEN	Comité Européen de Normalisation (European Committee for Standardization)
CEPS	Common Electronic Purse Specifications
CFB	Cipher Feedback
CFONB	Comité Français d'Organisation et de Normalisation Bancaires
CHAPS	Clearing House Automated Payment System
CHIPS	Clearing House Interbank Payment System
CID	Card Identification Number

CIDX	Chemical Industry Document Exchange
CIE	Customer-Initiated Entry
CMC7	Caractères Magnétiques Codés à 7 Bâtonnets (Magnetic Characters Coded with 7 Sticks)
CMP	Certificate Management Protocol
CMS	Cryptographic Message Syntax
CMVP	Cryptographic Module Validation Program
COPPA	Children Online Privacy Protection Act
CORBA	Common Object Request Broker Architecture
CORDIS	Community Research and Development Information Service
CPP	Collaboration Protocol Profile
CPPA	Collaboration Protocol Profile Agreement
CPS	Certification Practice Statement
CPU	Central Processing Unit
CREIC	Centres Régionaux d'Échange d'Images Chèques (Regional Chambers for the Exchange of Check Images)
CRIME	Compression Ratio Info-leak Made Easy
CRL	Certification Revocation List
CSS	Cascading Style Sheets
CTI	Computer Telephony Integration
CTP	Corporate Trade Payments
CTX	Corporate Trade Exchange
CVC	Card Verification Code
CVD	Card Verification Digit
CVM	Cardholder Verification Method
CVN	Card Verification Number
CVV	Card Verification Value
CWA	CEN Working Agreement
DAC	Decentralized Autonomous Corporation
DAO	Decentralized Autonomous Organization
DAP	Directory Access Protocol
DCOM	Distributed Component Object Model
DDOL	Dynamic Data Authentication data Object List
DDOS	Distributed Denial of Service
DEA	Data Encryption Algorithm
DEC	Digital Equipment Corporation
DEDICA	Directory-based EDI Certificate Access and Management
DER	Distinguished Encoding Rules
DES	Data Encryption Standard
DF	Dedicated File
DISP	Directory Information Shadowing Protocol
DKIM	DomainKeys Identified Mail
DNA	Deoxyribonucleic Acid
DNS	Domain Name System
DOD	Department of Defense
DOP	Directory Operational Binding Management Protocol
DPA	Differential Power Analysis
DPA	Dynamic Passcode Authentication
DRAM	Dynamic Random Access Memory
DSA	Digital Signature Algorithm
DSA	Directory System Agent
DSL	Digital Subscriber Line
DSP	Directory System Protocol
DSS	Digital Signature Standard
DTD	Document Type Definition
DTLS	Datagram Transport Layer Security
DTMF	Dual Tone Modulation Frequency
DTSU	Draft Standard for Trial Use
DUA	Directory User Agent
EAI	Enterprise application integration
EAN	European Article Number
EBA	European Banking Authority
ebCPPA	Electronic Business Collaboration Profile and Agreement
EBPP	Electronic Bill Payment and Presentment
EBT	Electronic Benefit Transfer
ebMS	Electronic Business Messaging Service
ebXML	Electronic Business XML
ECB	Electronic Codebook
ECC	Elliptic Curve Cryptography
ECCHO	Electronic Check Clearing House Organization
ECDH	Elliptic Curve Diffie–Hellman
ECDSA	Elliptic Curve Digital Signature Algorithm
ECI	Échange d'Images Chèques (Exchange of Check Images)
ECMA	European Computer Manufacturers Association
ECML	Electronic Commerce Modeling Language
ECP	Electronic Check Presentment
EDE	Encryption–Decryption–Encryption
EDGE	Exchange Data rates for GSM Evolution
EDI	Electronic Data Interchange
EDICC	EDI Canadian Council
EDIFACT	Electronic Data Interchange for Administration, Commerce and Transport
EDIINT	EDI Internet Integration
EEPROM	Electrically Erasable Programmable Read-Only Memory
EF	Elementary File
EFF	Electronic Frontier Foundation
EFT	Electronic Funds Transfer
EIA	Electronics Industry Association

EIC	Échange d'Images Chèques (Exchange of Truncated Checks)
EID	Electronic Identity Card (in Sweden)
EIPP	Electronic Invoice Payment and Presentation
EJB	Enterprise Java Beans
EMV	EuroPay, MasterCard, Visa
EPN	Electronic Payments Network
EPO	Electronic Payment Order
EPOID	Electronic Payment Order Identifier
EPROM	Electrically Programmable Read-Only Memory
ERP	Enterprise Resource Planning
ESB	Enterprise Service Bus
ESP	Encapsulating Security Payload
ESPRIT	European Strategic Program on Research in Information Technology
ETEBAC	Échange Télématique Entre les Banques et leurs Clients (Telematic Exchange among Banks and Their Clients)
ETSI	European Telecommunications Standards Institute
eUICC	embedded Universal Integrated Circuit Card
FACNET	Federal Acquisition Computer Network
FAQ	Frequently Asked Questions
FBI	Federal Bureau of Investigation
FCBA	Fair Credit Billing Act
FeRAM	Ferrite Random Access Memory
FERET	Face Recognition Technology
FIDO	Fast Identity Online Alliance
FinCEN	Financial Crimes Enforcement Network
FINREAD	Financial Transactional IC Card Reader
FinXML	Fixed Income Markup Language
FIPS	Federal Information Processing Standard
FIXML	Financial Information Exchange Markup Language
FPGA	Field Programmable Gate Array
FpML	Financial Products Markup Language
FSML	Financial Services Markup Language
FSR	Financial Services Roundtable
FSSB	Forensic Sciences Standards Board
FSTC	Financial Services Technology Consortium
FTC	Federal Trade Commission
FTP	File Transfer Protocol
FWT	Frame Waiting Time
FXML	Financial Exchange Markup Language
GCM	Galois/Counter Mode
GDP	Gross Domestic Product
Gie	Groupement d'intérêt économique
GMT	Greenwich Mean Time
GPL	General Public License
GPON	Gigabit Passive Optical Network
GPRS	General Packet Radio Service
GSA	General Services Administration
GSM	Groupe Spécial Mobile (Global System for Mobile Communication)
GSA	General Services Administration
GTA	Global Trust Authority
GTDI	General-purpose Trade Data Interchange
GTIN	Global Trade Item Number
HA-API	Human Authentication Application Program Interface
HEDIC	Healthcare EDI Coalition
HHA	Handheld Authenticator
HIBCC	Health Industry Business Communications Council
HMAC	Hashed Message Authentication Code
HSM	Hardware Security Module
HSPA+	Evolved High-Speed Packet Access
HTML	HyperText Markup Language
HTTP	HyperText Transfer Protocol
IAD	Issuer Authentication Data
IANA	Internet Assigned Numbers Authority
IATA	International Air Transport Association
IBIA	International Biometrics & Identification Association (formerly, International Biometric Industry Association)
ICAO	International Civil Aviation Organization
ICBA	International Conference on Biometric Authentication
ICC	Integrated Circuit Card
ICL	Image Cash Letter
ICMP	Internet Control Message Protocol
ICS	Issuers' Clearinghouse Service
IDEA	International Data Encryption Algorithm
IEC	International Electrotechnical Commission
IETF	Internet Engineering Task Force
IFTM	International Forwarding and Transport Message
IFX	Interactive Financial Exchange
IKE	Internet Key Exchange
ILO	International Labour Organization
IMA	Internet Merchant Account
IMAP	Internet Message Access Protocol
IMOTO	Internet, Mail Order, and Telephone Order
IMSI	International Mobile Subscriber Interface
INCITS	International Committee for Information Technology Standards
InterNIC	Internet Network Information Center
IoT	Internet of Things

IP	Internet Protocol
IPSec	Internet Protocol Security
IQA	Image Quality Analysis
IRC	Internet Relay Chat
IRD	Image Replacement Document
IRML	Investment Research Markup Language
IRS	Internal Revenue Service
ISAKMP	Internet Security Association and Key Management Protocol
ISI	Information Science Institute
ISO	International Organization for Standardization
ISP	Internet Service Provider
ISTH	International Standards Team Harmonization
ITAR	International Traffic in Arms Regulation
ITF	Interrogator Talks First
ITLS	Integrated Transport Layer Security
ITU	International Telecommunication Union
ITU-T	International Telecommunication Union—Telecommunication Standardization Sector
IVR	Interactive Voice Response
J2ME™	Java™ 2 Platform Micro Edition
J2SE	Java™ 2 Platform Standard Edition
JCB	Japan Credit Bureau
JIT	Just-in-Time
JPEG	Joint Photographic Expert Group
JVM	Java Virtual Machine
KEA	Key Exchange Algorithm
L2TP	Layer 2 Tunneling Protocol
LACES	London Airport Cargo EDP Scheme
LAI	Location Area Identity
LDAP	X.500 Lightweight Directory Access Protocol
LICRA	Ligue internationale contre le racisme et l'Antisémitisme (International League against Racism and Anti-Semitism)
LLCP	Logical Link Control Protocol
LSAM	Loading Secure Application Module
LTE	Long-Term Evolution
LZS	Lempel–Ziv–Stac
M2M	Machine to Machine
MAC	Media Access Control
MAC	Message Authentication Code
MaidSafe	Massive Array of Internet Disks—Secure Access For Everyone
MD	Message Digest
MDDL	Market Data Definition Language
MEL	MULTOS Executable Language
MF	Master File
MIC	Message Integrity Check
MICR	Magnetic Ink Character Recognition
MIME	Multipurpose Internet Mail Extensions
MISPC	Minimum Interoperability Specification for PKI Components
MIT	Massachusetts Institute of Technology
MITL	Multi Industry Transport Label
MMS	Multimedia Messaging
MMSC	Multimedia Messaging Center
MPOST	Mobile Point-of-Sale Terminal
MRO	Maintenance, Repair and Operations
MSB	Money Services Business
MSB	Most Significant Bit
MTA	Message Transfer Agent
mTAN	Mobile Transaction Authentication Number
MUA	Mail User Agent
MULTOS	Multi-application Operating System
MUSCLE	Movement for the Use of Smart Cards in a Linux Environment
NACHA	National Automated Clearing House Association
NAETEA	Network Assisted End-To-End Authentication
NBIS	NIST Biometric Image Software
NCFTA	National Cyber-Forensics and Training Alliance
NDEF	NFC Data Exchange Format
NESSIE	New European Schemes for Signature, Integrity, and Encryption
NFC	Near-Field Communication
NFIS	NIST Fingerprint Image Software
NFS	Network File System
NIST	National Institute of Standards and Technology
NMAC	Nested Message Authentication Code
NNTP	Network News Transfer Protocol
NSA	National Security Agency
NTM	Network Trade Model
NVM	Nonvolatile Memory
NWDA	National Wholesale Druggists Association
NYCE	New York Currency Exchange
NYCH	New York Clearing House
OAEP	Optimal Asymmetric Encryption Padding
OAGi	Open Applications Group
OASIS	Organization for the Advancement of Structured Information Standards
OCF	Open Card Framework
OCSP	Online Certificate Status Protocol
ODETTE	Organisation des données échangées par télétransmission en Europe (Organization for Data Exchange by Tele Transmission in Europe)
OECD	Organisation for Economic Co-operation and Development
OFB	Output Feedback

OFTP	ODETTE File Transfer Protocol
OFX	Open Financial Exchange
OI	Order Information
OMA	Open Mobile Alliance
OMG	Object Management Group
OSI	Open System Interconnection
OTA	Over the Air
OTP	One-Time Password
P2P	Peer-to-Peer or Person-to-Person
P2PE	Point-to-Point Encryption
P3P	Platform for Privacy Preference
PABP	Payment Application Best Practices
PAN	Primary Account Number
PBOC	People's Bank of China
PC	Personal Computer
PCI DSS	Payment Card Industry Data Security Standard
PCMCIA	Personal Computer Memory Card International Association
PEM	Privacy Enhanced Mail
PFS	Perfect Forward Secrecy
PGP	Pretty Good Privacy
PI	Payment Instructions
PIN	Personal Identification Number
PKCS	Public Key Cryptography Standards
PKI/PKIX	Public Key Infrastructure
PMI	Privilege Management Infrastructure
POP	Point of Purchase
POP	Post Office Protocol
POSA/R	Point-of-Sale Activation and Recharge
POZ	POS ohne Zahlungsgarantie
PPD	Prearranged Payment and Deposit
PPP	Point-to-Point Protocol
PPS	Protocol Parameter Selection
PROM	Programmable Read-Only Memory
PSAM	Purchase Secure Application Module
PSP	Payment Service Provider
PTS	PIN Transaction Security
QR	Quick Response
RA	Root Authority
RADIUS	Remote Authentication Dial-In User Service
RAM	Random Access Memory
RATS	Request Answer to Select
RBAC	Role-Based Access Control
RCP	Reference Control Parameter
RFC	Request for Comment
RFID	Radio-Frequency Identification
RIE	Renegotiation Information Extension
RIM	Registry Information Model
RIXML	Research Information Exchange Markup Language
ROM	Read-Only Memory
RPC	Remote Procedure Call
RPPS	Remote Payment and Presentment Service
RTGS	Real-Time Gross Settlement
S-HTTP	Secure HyperText Transfer Protocol
S/MIME	Secure Multipurpose Internet Mail Extensions (Secure MIME)
SAFER	Secure and Fast Encryption Routine
SAIC	Science Applications International Corporation
SAK	Select Acknowledge
SAM	Security Application Module
SAML	Security Assertion Markup Language
SASL	Simple Authentication and Security Layer
SCQL	Structured Card Query Language
SCSV	Signaling Cipher Suite Value
SD	Secure Digital
SDAD	Signed Dynamic Application Data
SDMI	Secure Digital Music Initiative
SDML	Signed Document Markup Language
SE	Secure Element
SEIS	Secured Electronic Information in Society
SEPA	Single Euro Payments Area
SERMEPA	Servicios para medios de pago (Services for Payment Instruments)
SET	Secure Electronic Transaction
SET SCCA	SET Compliance Certification Authority
SETREF	SET Reference Implementation
SGML	Standard Generalized Markup Language
SHA	Secure Hash Algorithm
SIM	Subscriber Identity Module
SIT	Système Interbancaire de Télécompensation (Interbanking Clearance and Settlement System)
SITA	Société Internationale de Télécommunications Aéronautiques (International Society for Aeronautical Telecommunications)
SITPRO	Simplification of International Trade Procedures
SMS	Short Message Service
SMTP	Simple Mail Transfer Protocol
SNEP	Simple NDEF Exchange Protocol
SNMP	Simple Network Management Protocol
SNNTP	Secure Network News Transfer Protocol
SOA	Service-Oriented Architecture
SOA	Source of Authority
SOAP	Simple Object Access Protocol
SPF	Sender Policy Framework
SQL	Structured Query Language
SRAM	Static Random Access Memory

SSC	Serial Shipment Container Code
SSH	Secure Shell
SSL	Secure Sockets Layer
SSO	Single Sign-On
STIP	Small Terminal Interoperability Platform
STK	SIM Application Tool Kit
STP	Straight-Through Processing
STPML	Straight-Through Processing Extensible Markup Language
Suica	Super Urban Intelligent IC Card
SVPo	Small Value Payment Co.
SwA	SOAP with Attachment
SWIFT	Society for Worldwide Interbank Financial Telecommunication
SwiftML	Society for Worldwide Interbank Financial Telecommunications Markup Language
TACACS	Terminal Access Controller Access System
TARGET	Trans-European Automated Real-time Gross settlement Express Transfer System
TC	Transaction Certificate
TCP	Transmission Control Protocol
TD	Transaction Data
TDCC	Transportation Data Coordinating Committee
TDEA	Triple Data Encryption Algorithm
TDI	Trade Data Interchange
TEDIS	Trade Electronic Data Interchange System
TEK	Token Encryption Key
TGS	Ticket Granting Server
TID	Transaction ID
TIFF	Tagged Image File Format
Tip	Titre Interbancaire de Paiement (Interbank Payment Title)
TLS	Transport Layer Security
TMSI	Temporary Mobile Subscriber Identity
TOTAL	Tag Talks Only After Listening
TRSM	Tamper-Resistant Security Module
TSM	Trusted Service Manager
TTC	Terminal Transaction Counter
TWIST	Transaction Workflow Innovation Standard Team
UBL	Universal Business Language
UCC	Uniform Commercial Code
UCD	Universal Companion Document
UCS	Uniform Communication Standards
UDDI	Universal Description, Discovery, and Integration
UDP	User Datagram Protocol
UEJF	Union des Étudiants Juifs de France (Jewish Student Union of France)
UHF	Ultra High Frequency
UICC	Universal Integrated Circuit Card
UID	Unique Identifier
UML	Unified Modeling Language
UMM	UN/CEFACT Modelling Methodology
UMTS	Universal Mobile Telecommunications System
UN-JEDI	United Nations Joint Electronic Data Interchange
UN-TDI	United Nations Trade Data Interchange
UN/CEFACT	Centre for Trade Facilitation and Electronic Business
UN/ECE	United Nations Economic Commission for Europe
UNCID	United Nations Rules of Conduct for Interchange of Trade Data by Teletransmission
UNCITRAL	United Nations Commission on International Trade Law
UNCL	United Nations Code List
UNI	User Network Interface
UNIFI	Universal Financial Industry
UPC	Universal Product Code
URL	Uniform Resource Locator
USC	University of Southern California
USAT	USIM Application Toolkit
USIM	Universal Subscriber Identity Module
VAN	Value-Added Network
VAT	Value-Added Tax
VDSL	Very High Bit Rate Digital Subscriber Line
VICS	Voluntary Interindustry Commerce Standards
VLSI	Very Large-Scale Integration
VPN	Virtual Private Network
W3C	World Wide Web Consortium
WAE	Wireless Application Environment
WAN	Wide Area Network
WAP	Wireless Application Protocol
WBMP	Wireless Application Protocol Bitmap
WCT	WIPO Copyright Treaty
WDP	Wireless Datagram Protocol
WEP	Wired Equivalent Privacy
WIF	Wallet Import Format
WIM	Wireless Identification Module
WINS	Warehouse Information Network Standard
WIPO	World Intellectual Property Organization
WML	Wireless Markup Language

WOIP World Organization for Intellectual Property
WPA Wi-Fi Protected Access
WPTT WIPO Performance and Phonogram Treaty
WS-BPEL Web Services Business Process Execution Language
WS-CDL Web Services Choreography Description Language
WSDL Web Services Description Language
WSP Wireless Session Protocol
WTLS Wireless Transport Layer Security
WTO World Trade Organization
WTP Wireless Transaction Protocol
X-KISS XML Key Information Service Specification
X-KRSS XML Key Registration Service Specification
XACML Extensible Access Control Markup Language
XARA Cross-App Resource Access
XBRL Extensible Business Reporting Language
xCBL XML Common Business Library
XDR External Data Representation
XFRML Extensible Financial Reporting Markup Language
XHTML Extensible HyperText Markup Language
XHTML MP XHTML Mobile Profile
XKMS XML Key Management Specification
XML Extensible Markup Language
XOR Exclusive OR
ZKA Zentraler Kreditausschuß

1

Overview of Electronic Commerce

Electronic commerce is at the conjunction of the advances in microelectronics, information processing, and telecommunication that have redefined the role of computers, first in enterprises and now in daily life, beyond that of process and production control. In the early phase, business supply networks or distribution channels were automated for optimal scheduling of production based on feedback from markets. Since the 1980s, a series of innovations opened new vistas to electronic commerce in consumer applications through automatic cash dispensers, bank cards, and Internet and wireless transactions. Simultaneously, money took the form of bits moving around the world, including the form of cryptocurrencies. Electronic commerce is also evolving in a *virtual* economy, focusing on services with looser temporal, geographic, or organizational obligations and with less consumption of natural resources and pollution hazards (Haesler, 1995).

This chapter presents a general introduction to various aspects of electronic commerce: its definition, various categories, its effects on society, its infrastructure, and what fraud means for individuals.

1.1 Electronic Commerce and Mobile Commerce

Electronic commerce (e-commerce), as defined by the French Association for Commerce and Electronic Interchange*—a nonprofit industry association created in 1996—is "the set of totally dematerialized relations that economic agents have with each other." Thus, e-commerce encompasses physical or virtual goods (software, information, music, books, etc.), as well as the establishment of users' profiles based on demographic and behavioral data collected during online transactions.

The Open Mobile Alliance (OMA) defines mobile commerce (m-commerce) as "the exchange or buying or selling of services and goods, both physical and digital, from a mobile device" (Open Mobile Alliance, 2005, p. 8). Typically, the buyer and the seller interact over a mobile network before the customer can engage the financial transaction. This initial phase includes advertising, discovery and negotiation of the price, and the terms and conditions.

Thus, both electronic commerce and mobile commerce blend existing technologies to create new financial services accessible from desktops, mobile terminals, phones, or pads. The ubiquity of mobile phones outside the industrialized countries has opened access to financial services, with telephone companies establishing payment networks through cash stored or transferred by phone. In recent years, mobile commerce services were extended to many financial services such as bill presentment and payment, loans, salary payment, and life insurance policies (the telecommunication company has all information that an underwriter needs from its subscribers, such as name, birth date, and address).

In this book, unless explicitly mentioned, the term *electronic commerce* will encompass all transactions irrespective of the access method, wireline or wireless, as well as the type of money used. There are instances, however, where the term *mobile commerce* implies specific characteristics. First, mobile transactions can be location based, that is, commercial offers can be modulated according to the current location of the mobile terminal. (In all cases, subscribers can receive tailored offers based on their demographic profile, preferences, and transaction history.) Second, in some countries, the limit for the amount that a user can transfer at any one time or during a specified time interval is lower when the transaction is conducted over a mobile network. Also, in some countries, mobile financial services require a partnership between telecommunication companies and banks or that telecommunication companies have a banking license. Other countries such as Kenya allow mobile money accounts to be unattached to any financial account, and mobile banking is exempt from the regulation of typical banking institutions (Bird, 2012; Crabtree, 2012; Demirguc-Kunt and Klapper, 2012).

Table 1.1 summarizes the main differences between Internet commerce and mobile commerce.

It should be noted that more than 90% of retail purchases in the United States are still conducted offline (Mishkin and Ahmed, 2014), but by 2016, 9% of the retail sales in the United States would be online. Of this, 8% would be mobile commerce transactions, which correspond in value to $90 billion in 2017 (Huynh, 2012).

* Association Française pour le Commerce et les Échanges Électroniques (AFCEE).

TABLE 1.1
Comparison of Electronic Commerce and Mobile Commerce

Characteristics	Internet Commerce	Mobile Commerce
User terminal	Computer or workstation	End device is a smart card with additional capabilities. With the advent of smartphones, mobile devices have increased their computational capabilities and storage capacity.
Ubiquity	End device is tethered	The end-user device is mobile.
Access	Wireline and wireless access	Wireless and cellular access.
Security	The end terminal can implement strong cryptography Firewalls may be added to filter the incoming traffic	End devices may have limited computational capabilities. End devices are more prone to theft and destruction. The radio interface between the terminals and the network access points adds additional threats.
Application development	A limited number of known companies	Application development is open to many designers with little experience in security and privacy.
Application environment	Depends on network technology; the choices of operating systems are limited.	Heterogeneous; depends on network technology, operating system, and security requirements.
Focus	Business-to-business and business-to-consumer applications	Geolocalized services and personal exchanges.
Personalization	Computer or workstation could be shared among users	Mobile devices are typically individualized.

Furthermore, with the increase in the number of mobile broadband subscribers worldwide to around seven billion, mobile commerce may become the dominant channel by 2018 (Taylor, 2013).

Depending on the nature of the economic agents and the type of relations among them, the applications of e-commerce fall within one of four main categories:

1. Business to business (B2B), where the customer is another enterprise or another department within the same enterprise. A characteristic of these types of relations is their long-term stability. This stability justifies the use of costly data processing systems, the installation of which is a major project. This is particularly true in information technology systems linking the major financial institutions. It should be noted that currently, mobile commerce does not include business-to-business transactions.
2. Business to consumer (B2C) at a distance through a telecommunication network, whether fixed or mobile.
3. Proximity or face-to-face commerce includes face-to-face interactions between the buyer and the seller as in supermarkets, drugstores, coffee shops, and so on. These interactions can be mediated through machines using contactless payment cards or mobile phones.
4. Peer-to-peer or person-to-person commerce (P2P) takes place without intermediaries, such as the transfer of money from one individual to another.

1.1.1 Examples of Business-to-Business Commerce

Business-to-business e-commerce was established long before the Internet. Some of the pre-Internet networks are as follows:

1. Société Internationale de Télécommunications Aéronautiques (SITA—International Society for Aeronautical Telecommunications), the world's leading service provider of IT business solutions and communications services to the air transport industry. Today, SITA links 600 airline companies and around 2000 organizations that are tied to them.
2. SABRE, an airline reservation system that was formerly owned by American Airlines, while in 1987, Air France, Iberia, and Lufthansa, established a centralized interactive system for reservations of air transport (Amadeus) to link travel agents, airline companies, hotel chains, and car rental companies. The settlement of travel documents among airline companies (changing airline companies after the ticket had been issued, trips of several legs on different airlines) is done through the Bank Settlement Payment (BSP) system.
3. Society for Worldwide Interbank Financial Telecommunications (SWIFT), whose network was established in 1977 to exchange standardized messages for the international transfer of funds among banks.
4. Banking clearance and settlement systems as discussed in Chapter 2.

Standardization of business-to-business e-commerce networks started with the X12 standard in North America and Electronic Data Interchange for Administration, Commerce and Transport (EDIFACT) in Europe. In the early 1980s, the U.S. Department of Defense (DOD) launched the Continuous Acquisition and Life-cycle Support (CALS) to improve the flow of information with its suppliers. In 1993, President Bill (William) Clinton initiated the exchange of commercial and technical data electronically within all branches of the federal government (Presidential Executive Memorandum, 1993). The Federal Acquisition Streamlining Act of October 1994 required the use of EDI in all federal acquisitions. A taxonomy was later developed to describe various entities and assign them a unique identifier within the Universal Data Element Framework (UDEF). With the installation of the Federal Acquisition Computer Network (FACNET) in July 1997, federal transactions can be completed through electronic means from the initial request for proposal to the final payment to the supplier.

Today, the adoption of the Internet as the worldwide network for data exchange is encouraging the migration toward open protocols and new standards, some of which will be presented in Chapter 4.

1.1.2 Examples of Business-to-Consumer Commerce

Interest in business-to-consumer e-commerce started to grow in the 1980s, although this interest varied across different countries. In Germany, remote banking services were conducted through the *Bildschirmtext* (BTX) system. The users of BTX were identified with a personal identification code and a six-digit transaction number (Turner, 1998).

In France, the Minitel service was undoubtedly one of the most successful pre–World Wide Web online business-to-consumer systems, lasting more than 40 years from the 1980s till its retirement in June 2012. Access was through a special terminal connected to an X.25 data network called Transpac via the public switched telephone network (PSTN). Until 1994, the rate of penetration of the Minitel in French homes exceeded that of personal computers in the United States (France Télécom, 1995; Hill, 1996). The crossover took place in 2002, when Internet users in France exceeded those using the Minitel (41%–32%), while turnover dropped from €700 billion in 2000 to €485 billion in 2002 (Berber, 2003; Selignan, 2003).

The Minitel uses the kiosk mode of operation. According to this model, the provider of an online service delegates the billing and the collection to the telephone operator for a percentage of the amounts collected. After collection of the funds, the operator compensates the content providers. Collection of small amounts by a nonbank could be justified because banks could not propose a competing solution for consolidating, billing, and collecting small amounts from many subscribers. At the same time, the financial institutions benefit from having a unique interface to consolidate individual transactions. However, because of the 30-day billing cycle, the telephone company was in effect granting an interest-free loan to its subscribers, a task that is usually associated with banks.

The role of a carrier as a payment intermediary has been carried over into many of the models for mobile commerce or commerce. One of the first examples was the i-mode® service from the Japanese mobile telephony operator NTT DoCoMo (Enoki, 1999; Matsunaga, 1999).

It is worth noting that originally communications were not encrypted but users did not mind giving their banking coordinates over the links. This shows that the sense of security is not merely a question of sophisticated technical means but that of a *trust* between the user and the operator.

Let us now look closer at three business-to-consumer applications over the web. These are the site auction eBay®, Amazon, and Stamps.com™ or Neopost.

1.1.2.1 eBay

The auction site eBay illustrates a successful pure player that established a virtual marketplace. The site supplies a space for exhibiting merchandises that overcome the geographic dispersion of the potential buyers and the fragmentation of the supply. In this regard, eBay provides a space to exhibit merchandises and to negotiate selling conditions; in particular, it provides a platform that links the participants in return of a commission on the selling price. The setup is characterized by the following properties:

- Participants can join from anywhere they may be, and the site is open to all categories of merchandises or services. The market is thus fragmented geographically or according to the commercial offers.
- Buyers have to subscribe and establish an account on eBay to obtain a login and define their password.
- The operator depends on the evaluation of each participant by its correspondents to assign them a grade. The operators preserve the right to eliminate those that do not meet their obligations.
- The operator does not intervene in the payment and does not keep records of the account information of the buyers.

These conditions have allowed eBay to be profitable, which is exceptional in consumer-oriented sites.

1.1.2.2 Amazon

Amazon started in 1995 as an online bookseller but also established a structure for consumers with the following components:

- Electronic catalogues
- Powerful search engines
- Rapid processing of searches and orders
- A capacity of real-time queries of inventories
- Methods of correct and quick identification of buyers
- Techniques for securing online payments
- Logistics for tracking and delivery

This infrastructure required large investments. For more than 6 years, Amazon remained in the red and its first profitable quarter was the last trimester of 2001 for a total yearly loss of $567 million (Edgecliffe-Johnson, 2002). With this infrastructure in place, Amazon evolved into a portal for other retailers looking to outsource their web operations or sell their products through Amazon.com. Another business is Amazon Web Services, which is a complete set of infrastructure and application services in the cloud, from enterprise applications and big data projects to social games and mobile apps. In 2013, it had more than 2 million third-party sellers, many of them small businesses (Bridges, 2013).

1.1.2.3 Stamps.com and Neopost

Electronic stamping systems Stamps.com and Neopost allow the printing of postage stamps with a simple printer instead of postage meters, thereby avoiding going to the post office. They are addressed to individual consumers, home workers, and small enterprises to save them the hassle of going to post offices. A 2D barcode contains, in addition to the stamp, the destination address and a unique number that allows fir tracking the letter.

Stamps.com operates online and with the query of an authorization center each time a stamp is printed. It uses a 2D barcode for postage and additional information related to the shipment, such as the sender's identity to assist in its tracking. It has established key relationships with the U.S. Postal Service, United Parcel Service (UPS), Federal Express (FedEx), Airborne Express, DHL, and other carriers. In contrast, Neopost is a semi-online system where stamping of envelopes continues without central intervention, as long as the total value of the stamps does not exceed the amount authorized by the authorization server. Neopost ranks number one in Europe and number two worldwide in mailroom equipment and logistics systems.

Both systems faced significant operational difficulties. These operational difficulties arose from the tight specifications of postal authorities for the positioning of the impressions, which are, in turn, a consequence of the requirements of automatic mail sorters. There is also a need to adapt to users' software and to all printer models. The total cost for the operator includes that of running a call center to assist users in debugging their problems. Finally, users must pay a surcharge of about 10% to the operator. From an economic point, the offer may not be attractive for all cases. However, the decision by the U.S. Postal Rate Commission in December 2000 to reduce the rate for Internet postage for some shipments (express mail, priority mail, first-class package service) was a significant victory.

1.1.3 Examples of Proximity Commerce

Proximity commerce or face-to-face commerce is using several methods to replace cash.

Prepaid cards can be one of two types: closed-loop and open-loop cards. The first kind is used only at the card sponsor's store or stores (the so-called gift cards), so the monetary value does not represent a legal tender. The second kind carries a network brand (e.g., Visa or MasterCard) and contains a record for a legal tender so that it can be used wherever the brand is accepted, including purchases, bill payments, and cash withdrawals (Coye Benson and Loftesness, 2010, pp. 72, 96–98).

Prepaid cards are used in many applications, such as transportation, mobile communication, and to play games. In Japan, it is used to play *pachinko,* a popular game of pinballs propelled with the objective of producing a winning combination of numbers. In France, prepaid telephone cards have been widely used since the 1980s (Adams, 1998). Prepaid cards have been used in South Africa to pay for electricity in rural communities and remote areas (Anderson and Bezuidenhoudt, 1996). In the United States, prepaid card transactions are considered a special type of debit card transactions.

Mobile Point-of-Sale Terminals (MPOSTs) are suitable for small, low-ticket businesses, such as street vendors, and all other merchants that do not qualify for enrolment in bank card networks. Payments can be accepted from terminals with near-field communication technology (NFC) capabilities by attaching a dongle to an audio port or a docking port or a wireless card reader. Many solution providers enhance their packages with marketing features such as the analysis of the collected sales data and the management of commercial offers.

1.1.4 Examples of Person-to-Person (Peer-to-Peer) Commerce

The ubiquity of network connectivity is driving the growth of the economy of sharing and collaboration.

The principle is to share excess and idle capacity on a very short-term basis. This is the principle of public bicycle systems, car rentals (Zipcar), taxi services (Uber), or holiday rentals (Airbnb).

Person-to-person lending—also known as marketplace lending—is built around the Internet or mobile networks. Its scope includes international transfer of worker remittances, student loans, hedge funds, wealth managers, pension funds, and university endowments. The payment can be directly from a bank account or a payment card (credit or debit). The transfer is credited to the recipient's bank or mobile account, and the recipient is notified of the deposit with the Short Message Service (SMS). For example, Square Cash users can open an account with their debit card and send money to a recipient using an e-mail address or a mobile phone number. Venmo, which is now part of PayPal, allows groups to split bills easily.

In general, handling small transfers between individuals is not a big or lucrative part of the financial system. Person-to-person transactions increased in importance, however, following the large banking fines and charges imposed on international banks for breaching U.S. sanctions, the highest being $8.9 billion onto BNP Paribas in 2014. Faced with the cost of meeting regulatory demands, many global banks have severed their links with correspondent institutions in several developing countries, particularly in Africa. This pullback has opened the way to new approaches to the transfer of workers' remittances. For example, TransferWise, a UK company, operates in the following manner. The customer selects a recipient and a currency and the amount to be transferred. The recipient receives the payment in the chosen currency withdrawn from the account of the initiator, piggybacking on another transfer going in the opposite direction. Thus, money is transferred without physically crossing borders so that the commission can be much lower than typical bank charges for international transfers. Obviously, this is possible only if the monetary flows in both directions balance each other. If this is not the case, then a system of brokers will be needed. Other person-to-person lenders in the United Kingdom include Funding Circle, Zopa, and WorldRemit.

WebMoney is another global transfer service based in Moscow, Russia. It provides online financial services, person-to-person payment solutions, Internet-based trading platforms, merchant services, and online billing systems. In the United States, the main operators include Prosper and Lending Club and SoFi, the latter specializing in student loans.

In a sense, this is a modern form of *hawala* transactions, a traditional person-to-person money transfers based on an honor system, with trust built on family or regional connections. It is common in the Arab World, the Horn of Africa, and the Indian subcontinent. In a typical hawala transaction, a person approaches a broker with a sum of money to be transferred to a recipient in another city or country. Along with the money, he or she specifies some secret information that the recipient will have to reveal before the money is paid. Independently, the transfer initiator informs the recipient of that secret code. Next, the hawala broker contacts a correspondent in the recipient's city with the transaction details including the agreed secret code. When the intended recipient approaches the broker's agent and reveals the agreed password, the agent delivers the amount in the local currency minus a small commission. The settlement among brokers is handled by whatever means they have agreed to (goods, services, properties, cash, etc.). Thus, the transaction takes place entirely outside the banking system, and there is no legal enforceability of claims.

Some banks have responded with their own mobile applications that allow their customers to send money to another person using the recipient's phone number, instead of specifying the routing of the recipient's bank and an account number. Examples of these applications are Pingit from Barclays Bank, or Paym from the Payments Council, the industry group responsible for defining payment mechanisms in the United Kingdom. In other cases, partnerships have been established among banks and peer-to-peer companies, such as Santander's collaboration with Funding Circle or Union Bank with Lending Club.

Peer-to-peer lending and online crowdfunding platforms are raising questions on how to adapt existing regulations to the new phenomena in terms of prudential requirements, protections in case of firm failures, disclosures, and dispute resolutions (Mariotto and Verdier, 2014).

1.2 Effects of the Internet and Mobile Networks

Originally, the Internet was an experimental network subsidized by public funds in the United States and by the large telecommunication companies for exchanges within collaborating communities. The community spirit was translated into an economy of free advice or software freely shared. The Free Software Foundation, for example, introduced a new type of software licensing, called *General Public License*, to prevent the takeover of free software by commercial parties and to further its diffusion, utilization, or modification. Even today, despite the domination of financial interests, many innovations on the Internet technologies depend to a large extent on volunteers who put their efforts at the disposal of everyone else.

The U.S. decision to privatize the backbone of the Internet from 1991 redirected the Internet to the market economy, in particular through the information highway project of the e Clinton-Gore Administration (Sherif, 1997). Furthermore, the invention of the World Wide Web, with its visual and user-friendly interface, stimulated the development of virtual storefronts. Similarly, the introduction of Extensible Markup Language (XML) and its specialized derivatives improved the ease with which business data are exchanged.

Nevertheless, the transformation of the *county fair* into a *supermarket* took more effort than originally anticipated. For one, the utilization of the Internet for economic exchanges clashed with the culture of availability of information free of charge. Other impediments relate to concerns regarding the security of information, because security on the public Internet is afterthought. Finally, from an operational viewpoint, the integration of electronic commerce with the embedded payment systems required significant investments in time, equipment, and training.

The security of transactions in transit, as well as the associated stored data was and remains to be a challenge. Despite many efforts, fraud for online transactions remains higher than for offline transactions. Periodically, the private data for millions of individuals stored by reputable banking institutions and online retailers are stolen. Some of the world's largest corporations, major retailers, financial institutions, and payment processors along with the U.S. largest electronic stock market are periodically penetrated. Another plague that is poisoning the life of many users is unsolicited electronic advertisement (spam).

At the same time, there are legitimate concerns regarding the collection and the reuse of personal data from the web. The consolidation of information tying buyers and products, which allows the constitution of individualized portfolios corresponding to consumer profiles, could be a threat to the individual's privacy.

In this regard, the nonlocalization of the participants in a commercial transaction introduces completely new aspects, such as the conflict of jurisdictions on the validity of contracts, the standing of electronic signatures, consumer protection, and the taxation of *virtual* products. Finally, new approaches are needed to address *virtual* products such as information, images, or software—products that pose major challenges to the concepts of intellectual property and copyrights.

As a social activity, the penetration of the Internet is also affected by the culture and social environment. Figure 1.1 depicts the percentage of individuals using

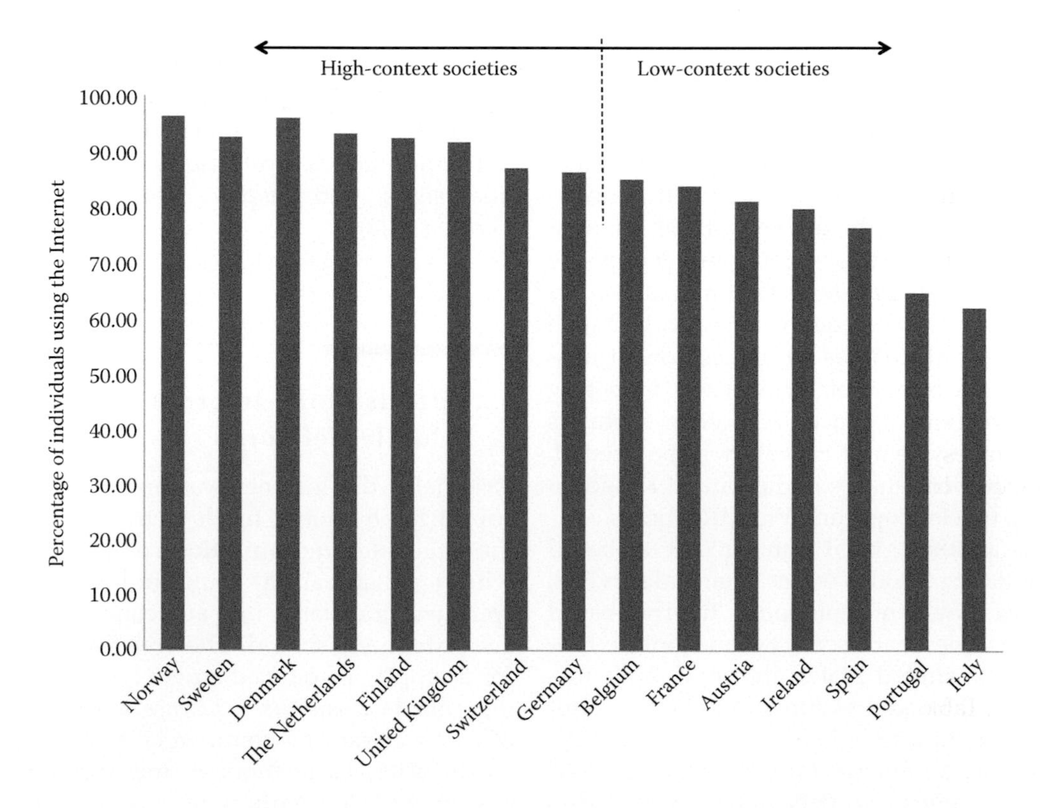

FIGURE 1.1
Percentage of individuals using the Internet in Western Europe in 2014. (From the International Telecommunication Union—Telecommunication Development Sector, Percentage of individuals using the Internet (Excel), July 8, 2015, available at http://www.itu.int/en/ITU-D/Statistics/Pages/stat/default.aspx, last accessed January 28, 2016.)

TABLE 1.2

Percentage of Individuals Using the Internet (2000–2014)

	Percentage of Individuals Using the Internet														
Country	**2000**	**2001**	**2002**	**2003**	**2004**	**2005**	**2006**	**2007**	**2008**	**2009**	**2010**	**2011**	**2012**	**2013**	**2014**
Austria	33.73	39.19	36.56	42.70	54.28	58.00	63.60	69.37	72.87	73.45	75.17	78.74	80.03	80.62	81.00
Belgium	29.43	31.29	46.33	49.97	53.86	55.82	59.72	64.44	66.00	70.00	75.00	81.61	80.72	82.17	85.00
Denmark	39.17	42.96	64.25	76.26	80.93	82.74	86.65	85.03	85.02	86.84	88.72	89.81	92.26	94.63	95.99
Finland	37.25	43.11	62.43	69.22	72.39	74.48	79.66	80.78	83.67	82.49	86.89	88.71	89.88	91.51	92.38
France	14.31	26.33	30.18	36.14	39.15	42.87	46.87	66.09	70.68	71.58	77.28	77.82	81.44	81.92	83.75
Germany	30.22	31.65	48.82	55.90	64.73	68.71	72.16	75.16	78.00	79.00	82.00	81.27	82.35	84.17	86.19
Ireland	17.85	23.14	25.85	34.31	36.99	41.61	54.82	61.16	65.34	67.38	69.85	74.89	76.92	78.25	79.69
Italy	23.11	27.22	28.04	29.04	33.24	35.00	37.99	40.79	44.53	48.83	53.68	54.39	55.83	58.46	61.96
The Netherlands	43.98	49.37	61.29	64.35	68.52	81.00	83.70	85.82	87.42	89.63	90.72	91.42	92.86	93.96	93.17
Norway	52.00	64.00	72.84	78.13	77.69	81.99	82.55	86.93	90.57	92.08	93.39	93.49	94.65	95.05	96.30
Portugal	16.43	18.09	19.37	29.67	31.78	34.99	38.01	42.09	44.13	48.27	53.30	55.25	60.34	62.10	64.59
Spain	13.62	18.15	20.39	39.93	44.01	47.88	50.37	55.11	59.60	62.40	65.80	67.60	69.81	71.64	76.19
Sweden	45.69	51.77	70.57	79.13	83.89	84.83	87.76	82.01	90.00	91.00	90.00	92.77	93.18	94.78	92.52
Switzerland	47.10	55.10	61.40	65.10	67.80	70.10	75.70	77.20	79.20	81.30	83.90	85.19	85.20	86.34	87.00
United Kingdom	26.82	33.48	56.48	64.82	65.61	70.00	68.82	75.09	78.39	83.56	85.00	85.38	87.48	89.84	91.61

Source: International Telecommunication Union—Telecommunication Development Sector, Percentage of individuals using the Internet (Excel), July 8, 2015, available at http://www.itu.int/en/ITU-D/Statistics/Pages/stat/default.aspx, last accessed January 28, 2016.

the Internet in Western European countries as of 2014, available from the International Telecommunication Union (ITU). Figure 1.1 shows that there are two groups of countries: those where, in 2014, more than 85% of the population use the Internet and those where the percentage of individuals using the Internet lies between 60% and 85%. As can be seen from the data in Table 1.2 and illustrated in Figure 1.2 that the difference is persistent over the period of 2000–2014.

Due to Hall, these numbers can be explained by taking into account the classification of societies into *low-context* and *high-context* societies (Hall and Hall, 1990). In high-context societies, interpersonal relations and oral networks have a much more important place than in the low-context societies of northern Europe where communication takes explicit and direct means, such as the written word. In contrast, high-context societies, those of central and southern Europe, are less receptive, particularly because the Internet has to compete with traditional social networks. In Figure 1.2, the dotted lines are reserved to countries with high-context cultures.

The situation is substantially different regarding mobile commerce. Table 1.3 reproduces the top 20 countries in terms of the number of mobile subscriptions per 100 inhabitants as recorded by the ITU for 2012. It is noted that 19 countries in the top 20 list are from the developing world. In many emerging economies, the financial infrastructure is not developed or widely distributed, which shows the importance of mobile commerce in meeting the needs for a banking system, particularly in rural areas.

In particular, there is a growing list of countries where more than 20% of adults have adopted some form of mobile money, as shown in Table 1.4 (Demirguc-Kunt and Klapper, 2012). As will be seen in Chapter 10, mobile phones can be used to emulate contactless cards used for transportation and access control. This explains why payment applications using mobile phones (mobile money) have been deployed in many developing countries to offer financial services to the unbanked, even though M-PESA in Kenya remains the lightning rod for mobile payment applications.

The World Bank, the Melinda and Bill Gates Foundation, and others have funded mobile programs, some of which are listed in Table 1.5 (International Telecommunication Union, 2013). In the developed world, many urban areas with high commuter traffic presents a great market potential for mobile payments with contactless chip cards at vending machines or parking kiosks.

Smartphones offer many additional features for mobile commerce. Although the term appeared in 1997 to describe mobile terminals with advanced computing capabilities, the term took hold in January 2007, when Steve Jobs, then Apple's chief executive, brandished an iPhone front of a rapt audience. The launch of Apple Pay in 2014, which allows iPhone 6 users to pay for goods with a tap of their handset, has been a catalyst for the broader adoption of mobile payments in

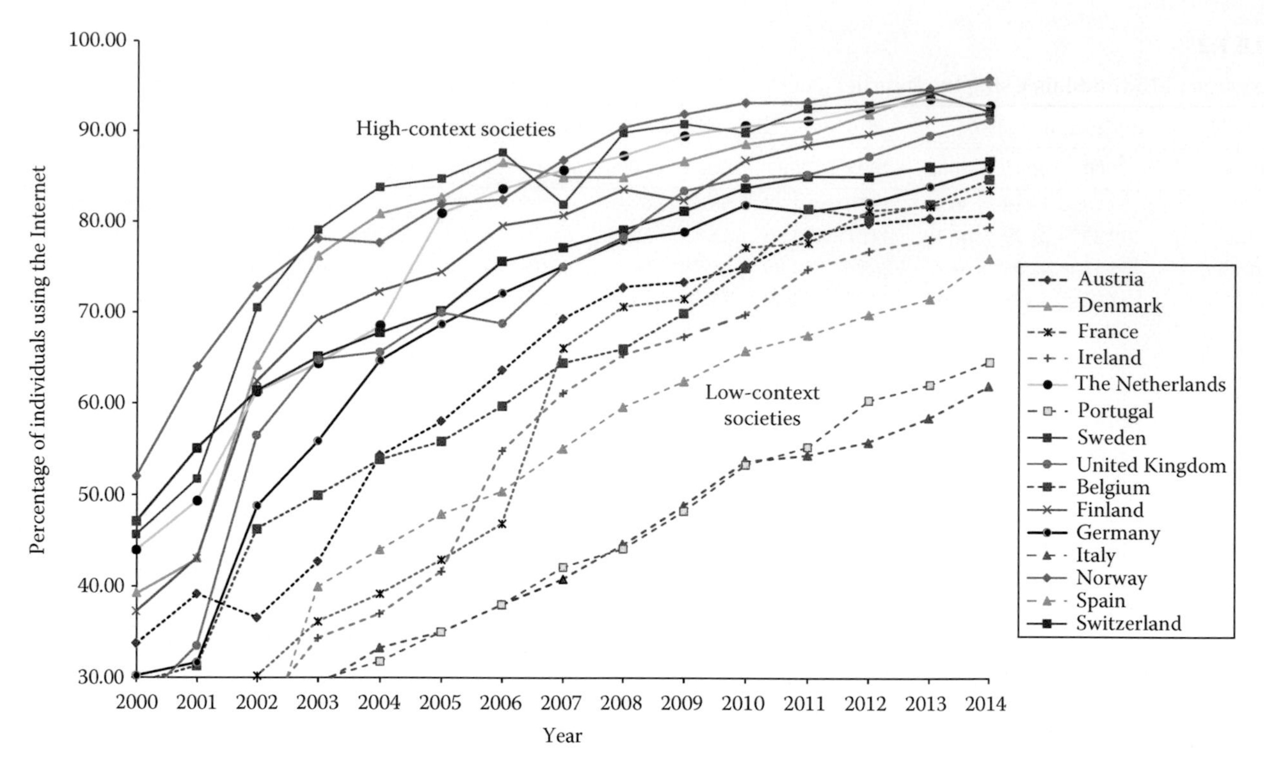

FIGURE 1.2
Evolution of the percentage of individuals using the Internet in Western Europe from 2000 to 2014. (From International Telecommunication Union—Telecommunication Development Sector, Percentage of individuals using the Internet (excel), July 8, 2015, available at http://www.itu.int/en/ITU-D/Statistics/Pages/stat/default.aspx, last accessed January 28, 2016.)

TABLE 1.3

Top 20 Countries with the Number of Mobile/Cellular Telephone Subscriptions per 100 Inhabitants in 2012

Rank	Country	No. of Subscriptions
1	Macao, China	284.34
2	Hong Kong, China	227.93
3	British Virgin Islands	205.45
4	Antigua and Barbuda	198.62
5	Kuwait	191.11
6	Gabon	187.36
7	Panama	186.73
8	Saudi Arabia	184.68
9	Russia	183.52
10	Suriname	182.90
11	Oman	181.73
12	Montenegro	177.94
13	Kazakhstan	175.39
14	Maldives	172.84
15	Finland	172.51
16	United Arab Emirates	169.94
17	Cayman Islands	168.27
18	Anguilla	163.41
19	Dominica	161.53
20	Austria	161.21

TABLE 1.4

Countries Where the Percentage of Adults Using Mobile Money in 2011 Equaled or Exceeds 20%

Country	Percentage of Adults Using Mobile Money in 2011
Albania	31
Algeria	44
Angola	26
Congo, Rep	37
Gabon	50
Kenya	68
Somalia	34
Sudan	52
Swaziland	20
Tajikistan	29
Tunisia	25
Uganda	27

the United States. In January 2015, it was estimated that $2 of every $3 spent via contactless cards on the three largest U.S. card networks originated from Apple Pay (Mishkin and Fontanella-Khan, 2015).

The seductive power of smartphones comes from their size, their computational power, their

TABLE 1.5

Examples of Mobile Money Applications

Mobile Money Application	Countries of Use
Airtel Money	India, Uganda, Tanzania, Kenya
EasyPaisa	Pakistan
EKO	India
GCash	The Philippines
M-PESA	Kenya, Tanzania, South Africa, and Afghanistan
MTN Mobile Money	Uganda, Ghana, Cameroon, Ivory Coast, Rwanda, and Benin
TCash	Haiti
WIZZIT	South Africa

connectivity, and their openness to mobile applications (apps) that supplement the capabilities of the device. This has security implications as well. Unlike applications designed for laptop and desktop computers, which are produced by identifiable parties, there are millions of applications for smartphones and tablets that have been designed with minimal attention to security or data protection. Furthermore, these downloaded applications could have access to personal information on the mobile device, which makes them potential conduits for attacks. Furthermore, mobile devices have limited options for running protection software.

1.3 Network Access

Network access for remote communication can be through either a wireline or a wireless infrastructure. The main properties of the access are the available bandwidth in bits per second (bits/s), the reliability in terms of downtime or time to repair, as well as the availability of resources to ensure access in case of increased load.

1.3.1 Wireline Access

The physical medium for wireline access can be copper cables, coaxial cables, and optical fibers. The bit rates depend on the access technology. Today, it is used to transmit the so-called triple play services (voice, data, high-definition television) to subscribers on the access side.

Cable companies use the DOCSIS 3.0 technology to support rates up to 100 Mbit/s. Various flavors of digital subscriber line (DSL) achieve high bit rates on

TABLE 1.6

Comparison of the Maximum Bit Rates for Selected Wireline Access Technologies

Technology	ITU-T Recommendation	Downstream (Max)	Upstream (Max)
VDSL2	G.993.2	100 Mbit/s	40 Mbit/s
GPON	G.984	2488 Mbit/s	1244 Mbit/s
XGPON	G.987	10 Gbit/s	2.5 Gbit/s
G.FAST	G.9701	700 Mbit/s	200 Mbit/s

traditional copper lines. Vectored very high bit rate digital subscriber line (VDSL) or VDSL2 of ITU-T G.993.2 provide high transmission speeds on short loops at the 8, 12, 17, or 30 MHz bands, with bit rates up to 100 Mbit/s downstream and 40 Mbit/s upstream.

Gigabit passive optical networks (GPONs) carry all service types in addition to analog telephony. The GPON technology described in ITU-T G.984.X (X varies from 1 to 7) offers up to 2488 Mbit/s in the downstream direction and 1244 Mbit/s upstream. The traffic of any service type can be encrypted using Advanced Encryption Standard (AES). The deployment configuration depends on the reach of the optical fiber. A further development is the XGPON of the ITU-T G.987 series of recommendation, with the maximum bit rates of 10 Gbit/s downstream and 2.5 Gbit/s upstream.

The so-called G.FAST (ITU-T G.9701) technology maximizes the use of the copper link by transmitting in frequency bands (106 or 212 MHz) and exploiting cross-talk cancellation techniques (denoted as frequency division vectoring). In theory, the bit rates over a twisted copper pair can reach 1 Gbit/s in both directions, but for longer copper lines (66 m), the maximum total bit rate is around 900 Mbit/s: 700 Mbit/s downstream and 200 Mbit/s upstream. Table 1.6 summarizes the maximum transmission speeds of all these technologies.

1.3.2 Wireless Access

Several protocols are available for wireless access. With Groupe Spécial Mobile (GSM—Global System for Mobile Communication), the bit rate with SMS does not exceed 9.6 kbit/s. To reach 28 or 56 kbit/s (with a maximum bit rate of 114 kbit/s), General Packet Radio Service (GPRS) was used. Exchange Data rates for GSM Evolution (EDGE) increased the rate to 473.6 kbit/s.

Newer generations of technologies have increased the rates, as shown in Table 1.7. These technologies use innovative spectrum allocation and management techniques required to utilize the radio spectrum availability in an

TABLE 1.7

Comparison of Wireless Access Technologies

	Technology			
Feature	**HSPA (3G) (Mbit/s)**	**HSPA+ (3G+) (Mbit/s)**	**LTE (Mbit/s)**	**LTE Advanced (4G)**
Downlink peak rate	14	168	299.6	1 Gbit/s
Uplink peak rate	5.76	22	75.4	
Typical user downlink rate	1–4	1–10	10–100	To be defined
Typical user uplink rate	0.5–2	0.5–4.5	5–50	To be defined

Note: HSPA, High-speed packet access; HSPA+, evolved high-speed packet access; LTE, long-term evolution.

optimal fashion and to enhance the information carrying capacity of the radio channels.

The Mobile WiMAX (IEEE 802.16e) mobile wireless broadband access (MWBA) standard (marketed as WiBro in South Korea) is sometimes branded as 4G and offers peak data rates of 128 Mbit/s downlink and 56 Mbit/s uplink over 20 MHz wide channels. The peak download is 128 Mbit/s, and the peak upload is 56 Mbit/s. Wireless local area networks can offer access points, in particular, WiFi (IEEE 802.11x), WiMax (802.16x), and Mob-Fi (IEEE 802.20) technologies. These operate respectively at the various frequencies and theoretical bit rates (maximum throughputs). Finally, to connect islands together, such as Indonesia, satellites are the only economical way to establish reliable communication (Bland, 2014).

1.4 Barcodes

Barcodes provide a simple and inexpensive way to encode textual information about items or objects in a form that machines can read, validate, retrieve, and process. The set of encoding rules is called *barcode symbology*. Barcodes can be read with smartphones equipped with a digital camera. A software client installed on the smartphone controls the digital camera to scan and interpret the coded information, allowing mobile users to connect to the web with a simple point and click of their phones, thereby making mobile surfing easier.

Linear barcodes consist of encoded numbers and letters (alphanumeric ASCII characters) as a sequence of parallel black and white lines of different widths. The specifications for 1D barcodes define the encoding of the message, the size of quiet zones before and after the barcode, and the start and stop patterns. One-dimensional codes have limited storage capacity and provide indices to backend databases. Most consumer products today are equipped with black-and-white stripes or Global Trade Item Number (GTIN). GTIN provides a faster and accurate way to track items as it progresses through the various distribution networks from producer to consumer so that the related information can be updated automatically.

GTIN is a linear or 1D barcode that follows a specific standard. The main standards are the Universal Product Code (UPC) or the European Article Number (EAN) (EAN is a superset of UPC). The UPC barcodes were defined by the Uniform Code Council (UCC) to track items at U.S. retail stores to speed up the checkout process and to keep track of inventory. The 12-digit UPC code comprises two 6-digit fields: the first reserved for a manufacturer identification number, while the second contains the item number. The whole UPC symbol adds to the UPC barcode its human-readable representation in the form of digits printed below the code. In a similar fashion, the EAN-13 barcodes are used worldwide to tag products sold at retail points with International Article Numbers as defined by the GS1 organization.

The United States Post Office, for example, uses several codes to track the mail:

- *Postnet* is used to encode zip codes and delivery points.
- *Planet* is used to track both inbound and outbound letter mail.
- The *Intelligent Mail Barcode* (IM barcode) is a 65 barcode to replace both Postnet and Planet.
- *USPS GS1-128* (*EAN-128*) is used for special services such as delivery confirmation.

2D codes store more information than 1D codes, including links to websites without any manual intervention.

The symbol elements in 2D barcodes consist of dark and light squares. The 2D specifications define the encoding of the message, the size of quiet zones before and after the barcode, the finder or position detection patterns, and error detection and/or correction of information. The finder pattern is used to detect and locate the 2D barcode symbol and to compute the characteristics of a symbol (e.g., size, location, orientation).

With error detection and correction, the original data can be retrieved when the symbol is partially damaged, provided that the finder pattern is not damaged.

There are two categories of 2D barcodes: stacked codes and matrix codes. Stacked codes consist of a series of rows one on top of the other, with each row represented as a 1D barcode. Data representation in matrix codes consists of a matrix of black and white squares or cells. Most stacked 2D barcodes require specialized laser scanning devices (raster scanners) capable of reading in two dimensions simultaneously. Charge coupled device (CCD) imagers can read these codes, which is why camera phones can be used for the scanning and decoding of matrix codes. Matrix codes have larger storage capacity than stacked codes. Some of the 2D barcodes that can be read from mobile phones are presented next.

Data Matrix was developed in 1987 and standardized finally in 2006 as ISO/IEC 16022:2006. Its main characteristic is its small symbol size to fit on tiny drug packages. Figure 1.3a shows a Data Matrix symbol with its finder pattern on the perimeter: two solid L-shaped solid lines and two opposite broken lines. The finder pattern defines the physical size of the symbol, its orientation and distortion, and its cell structure. Furthermore, the level of error correction with the Reed–Solomon algorithm is automatically determined in accordance with the symbol size (Kato et al., 2010, pp. 60–66). The 2D Data Matrix barcode symbology has been adopted in specific standards to identify parts by the International Air Transport Association (IATA), the Automotive Industry Action Group (AIAG), the Electronics Industry Association (EIA), and the U.S. Department of Defense.

PDF417 is a stacked code, developed in 1991 and standardized in 2006 as ISO/IEC 15438. PDF stands for Portable Data File while 417 signifies that a codeword consists of up to 17 units, each formed by a sequence of 4 bars and 4 spaces of varying widths. The code represents American Standard Code for Information Interchange (ASCII) characters only and, therefore, is used mainly in Europe and the United States. Figure 1.3b shows the structure of a printed PDF417 barcode symbol. It comprises 3 to 90 rows, each row consisting of piling up to 30 codewords on top of one another. A codeword represents 1 of 929 possible patterns from 1 of 3 different sets of symbols, patterns, or clusters used alternatively on every third row (Kato et al., 2010, pp. 32–35). Error correction is based on the Reed–Solomon algorithm. The U.S. Department of Homeland Security has selected PDF417 for RealID-compliant driver licenses and state-issued identification cards. PDF417 and Data Matrix can be used to print postages accepted by the U.S. Postal Service. PDF417 is also used by FedEx on package labels and by the International Air Transport Association (IATA) for the bar-coded boarding pass (BCBP) standard for airline paper boarding passes. Note that BCBP also restricts the barcodes that could be used on mobile phones for boarding passes to the Data Matrix, QR (Quick Response), and Aztec barcodes.

A derivative barcode, MicroPDF417, also adopted as ISO/IEC 24728:2006, is for applications where space is restricted.

MaxiCode was developed by UPS in 1992 for the high-speed sorting and tracking of units. It has a fixed size and a limited data capacity. One of its characteristic features is a circular finder pattern made up of three concentric rings. It was standardized in ISO/IEC 16023:2000. Figure 1.3c shows a MaxiCode symbol.

Quick Response (QR) Code was developed in Japan in 1994 by Denso Wave Incorporated to store Japanese ideograms (Kanji characters) in addition to other types of data such as ASCII characters, binary characters, and image data. Model 1 QR Code is the original version of the specification. An improved symbology with alignment patterns to protect against distortion, known as QR Code Model 2, has been approved as ISO/IEC 18004:2006. There are 40 symbol versions of the QR Code model (from 21×21 cells to 177×177 cells), each with a different size and data capacity (Kato et al., 2010, pp. 51–60, 226–230). Versions differ by the amount of data that can be stored and by the size of the 2D barcode. The resolution of the camera phone puts a practical limit on the versions that mobile applications can use. As a result, only versions 1–10 are available in mobile applications so that the maximum data capacity corresponds to 57×57 cells (Kato and Tan, 2007).

As shown in Figure 1.3d, one visual characteristic of the QR Code is the finder pattern, which consists of three separate parts located at three corners of the symbol. During scanning, these pattern blocks are the first to be detected. Micro QR Code is a small-sized QR Code for applications where less space and data storage capacity are required. In this code, the finder pattern consists of a block located at one corner, as shown in Figure 1.3e.

The QR Code is used to represent a user's Bitcoin address as discussed in Chapter 14. It can be used at a point-of-sale terminal to represent a payment request with a destination address, a payment amount, and a description so that a Bitcoin wallet application in a smartphone can prefill the information and display it in a human-readable format to the purchaser.

The Aztec Code is another 2D matrix symbology defined in ISO/IEC 24778:2008, as shown in Figure 1.3f. It was developed in 1995 to use less space than other matrix barcodes. Many transportation companies in Europe, such as Eurostar and Deutsche Bahn, use the Aztec barcode for tickets sold online. This is one of the matrix codes that BCBP standard defines for use on mobile phones together with Data Matrix and QR.

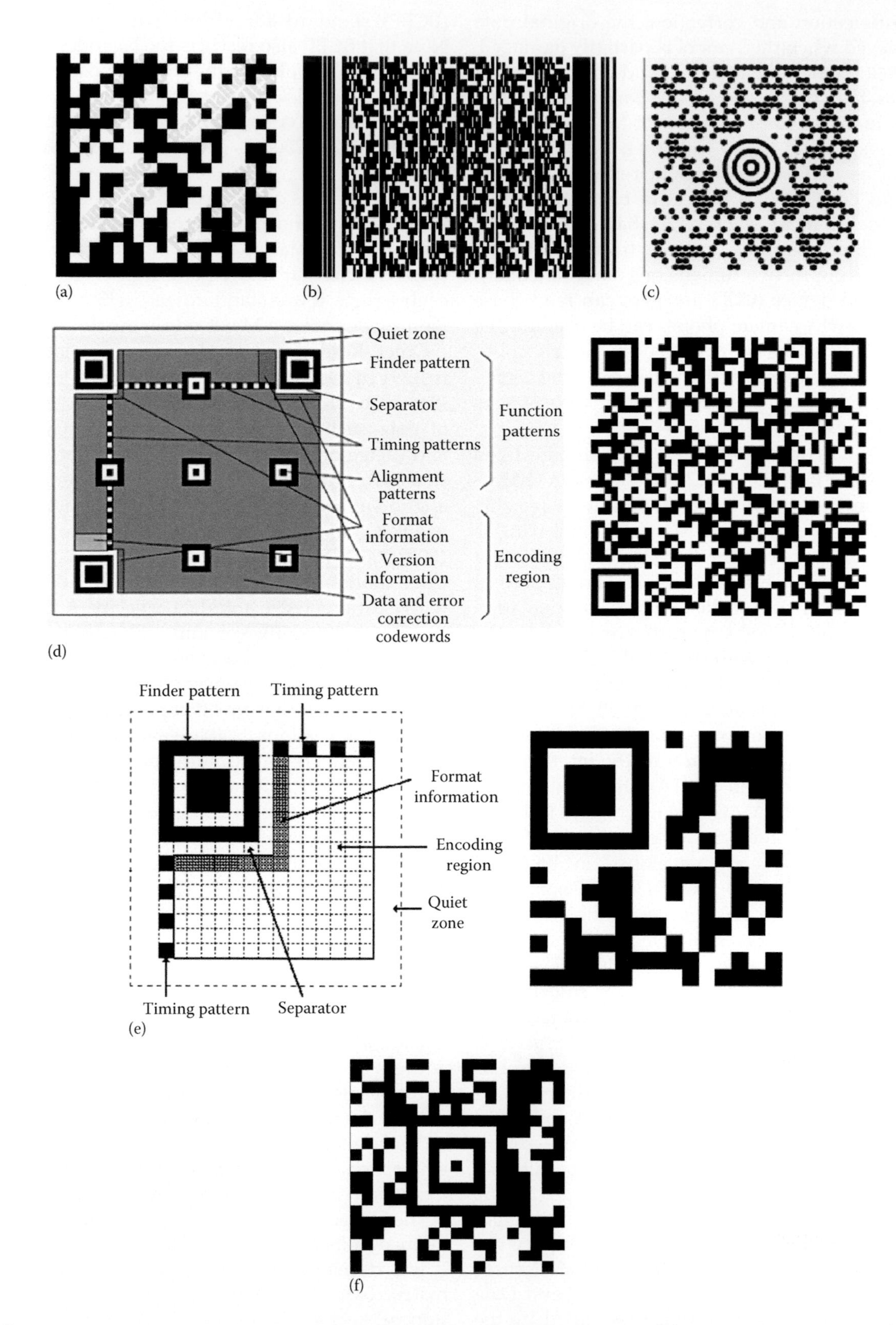

FIGURE 1.3
Structure of some 2D barcodes: (a) Data Matrix, (b) PDF417, (c) MaxiCode, (d) QR Code, (e) Micro QR Code, and (f) Aztec Code.

TABLE 1.8

Comparison of the Major 2D Barcodes

Characteristics	Data Matrix	PDF417	MaxiCode	QR Code[a]	Aztec
Year of development	1987	1991	1992	1994	1995
Category	Matrix	Stacked	Matrix	Matrix	Matrix
Maximum Theoretical Storage Capacity[a]					
Numeric characters	3116	2710	138	7089	3832
Alphanumeric characters	2335	1850	93	4296	3067
Binary in octets	1556	1108	—	2953	1914
Kanji characters	—	—	—	1817	—
Error correction algorithm	Reed–Solomon	Reed–Solomon	Reed–Solomon	Reed–Solomon	Reed–Solomon
Level of error correction	Automatically determined according to the symbol size	9 levels	2 levels	4 levels	19 levels
Standard	ISO/IEC 16022	ISO/IEC 15438	ISO/IEC 16023	ISO/IEC 18004	ISO/IEC 24778

Source: Kato, H. and Tan, K.T., *IEEE Pervasive Comput. Mobile Ubiquit. Syst.*, 6(4), 76, 2007, available at http://ro.ecu.edu.au/do/search/?q=Pervasive%202D%20barcodes%20for%20camera%20phone%20applications&start=0&context=302996, last accessed January 26, 2014.

[a] The respective maximums in mobile applications are 2061 (numeric), 1249 (alphanumeric), 856 (binary), and 538 (Kanji).

A symbol is made of square modules on a square grid, with a square bull's-eye pattern at their center. There are 36 predefined symbol sizes to choose from, and the user can also select from 19 levels of error correction (http://www.barcode-soft.com/aztec.aspx, last accessed January 14, 2016).

Table 1.8 presents a comparison of the major 2D barcodes (Gao et al., 2009; Grover et al., 2010; Kato et al., 2010). It should be noted that the exact data capacity depends on the structure of the data to be encoded.

Smart posters are equipped with 2D barcodes that can be read with a phone camera. The code represents the Uniform Resource Locator (URL) of a website that the mobile application can access for additional information without entering the URL manually. There are clear security implications to this capability because the URL can be that of a nefarious site that inserts malware or virus software. One simple check is to verify visually that the tag is printed with the poster and not inserted afterward.

The performance characteristics of mobile barcode scanners are evaluated by the following parameters (Von Reischach et al., 2010):

- Accuracy of the measurement, that is, its correctness. Accuracy is affected by the ambient light, reflections from the surface on which the barcode is displayed, error correction capabilities of the code, and so on.
- Reliability refers to the consistency of the measurement across repetitions. This parameter depends on the visual user interface and the ease of placement of the mobile phone in front of the code to be scanned.
- Speed, typically less than 5 seconds for user acceptance.
- The quality of the user interface.

As of now, there is no standardized procedure to assess the performance of scanners. So even though new versions of scanners are being continuously released, the diversity of platforms and the large variety of cameras make it difficult to evaluate the suitability of the scanner/device combination beforehand.

1.5 Smart Cards

The smart card (or integrated circuit card) is the culmination of a technological evolution starting with cards with barcodes and magnetic stripe cards. Smart cards or microprocessor cards contain within the thickness of their plastic (0.76 mm) a miniature computer (operating system, microprocessor, memories, and integrated circuits). These cards can carry out advanced methods of encryption to authenticate participants, to guarantee the integrity of data, and to ensure their confidentiality.

The first patents for integrated circuit cards were awarded in the 1970s in the United States, Japan, and France, but large-scale commercial development started in Japan in the 1970s and in France in 1980s. In most markets, microprocessor cards are gradually replacing magnetic stripe cards in electronic commerce (e-commerce) applications that require large capacity for storage, for the processing of information, and for security (Dreifus and Monk, 1998).

TABLE 1.9

Worldwide Shipment of Smart Secure Devices in 2014 (Millions of Units)

Area	Integrated Circuit Cards with Contacts	Contactless Integrated Circuit Cards
Telecom	5200	—
Banking and financial services	2050	880
Government—Health care	380	260
Transport/device manufacturers	190	210
Others	400	70
Total	8220	1360

Source: Eurosmart, May 6, 2015, http://www.eurosmart.com/facts-figure.html, last accessed January 28, 2016.

The figures for worldwide shipment of smart secure devices in 2014 are shown in Table 1.9.

Microprocessor cards were first applied in the 1980s for telephone payment cards and then in the Subscriber Identity Module (SIM) of GSM handsets. In the 1990s, they started to be used in health-care systems in Belgium, France, and Germany. The cards Sésame-Vitale in France, VersichertenKarte in Germany, and Hemacard in Belgium provide personalized interfaces to national informatics system for the acquisition and storage of health data of individual cardholders (McCrindle, 1990, pp. 143–146). In these cards, several applications may coexist within the same card but in isolation (*sandboxing*). Access to the card depends on a personal identification number (PIN), but communication with each application is controlled by a set of secret keys defined by the application providers.

Some specialty cards are related to client fidelity programs, as well as to physical access control (enterprises, hotels, etc.), or logical security (software, confidential data repositories, etc.). These cards are often associated with a display to show dynamic passwords. They can also be used in combination with one or more authentication techniques to provide multifactor authentication, including with biometrics. The use of smart cards for access control and identification started in the 1980s in many European universities (Martres and Sabatier, 1987, pp. 105–106; Lindley, 1997, pp. 36–37, 95–111).

Smart cards can be classified according to several criteria, for example:

- Usage in a closed or open system, that is, whether the monetary value is for a specific application or as a total and immediate legal tender. A transit card is used in a closed system, while credit and debit cards have a backend connection to the banking system.
- Duration of card usage, which distinguishes disposable cards from rechargeable cards.
- Intelligence of the card, which ranges from a simple memory to store information to wired logic and programmable microprocessor.
- Necessity for direct contact with the card reader to power up the card, which distinguishes contact cards from contactless. The technology of contactless integrated circuit cards is also known as radio-frequency identification (RFID).
- Characteristic application of the card, which differentiates monoapplication cards from multiapplication cards. If the card is integrated with a mobile phone, for example, multiple independent applications are expected to run for debit, credit, transportation access, or loyalty programs.

A contact smart card must be inserted into a smart card reader to establish electrical connections with gold-plated points on the surface of the card. In contrast, a contactless card requires only close proximity to a reader. Both the reader and the card are equipped with antennae and establish communication using radio frequencies. Contactless cards derive their power from this electromagnetic link through induction. The contactless chip and its antenna can be embedded in mobile devices such as smartphones. In general, the data transfer rate with contact cards is 9.6 kbit/s, whereas the rates for contactless cards can reach 106 kbit/s; but the actual rate depends on the applied clock. With the near-field communication (NFC) technology used with many mobile terminals, the bit rates are 106, 212, and 424 kbit/s.

RFID cards are used for tracking and access control. For example, electronic passports (e-passports) embedded with RFID chips (or contactless integrated circuit cards) operate at 13.56 MHz RFID and are based on the specifications of the International Civil Aviation Organization (ICAO). These comply with the ISO/IEC 14443 standards. Furthermore, ISO/IEC 7501-1:2008 defines the specifications of electronic passports including globally interoperable biometric data.

To improve the management of physical objects along the supply chain, an RFID tag containing a unique identifier is attached to objects during manufacturing. This tag would then be read at various intervals up to its point of sale. The major target benefits are a better control of inventories, faster material handling time, safer and more secure supply chains, and potential post-sale applications.

Some airport baggage handling systems use RFID tags to improve the accuracy and speed of automatic baggage routing. A growing number of new cars sold today contain an RFID-based vehicle immobilization system. It includes an RFID reader in the steering column and a tag in the ignition key. The computer in the car prevents the ignition from firing unless the reader detects the proper signal.

In general, RFID tags often complement barcodes with product codes as described in §§ 1.4 in supply chain management.

Urban transportation systems with high commuter traffic constitute a large market for contactless payments. The value is preloaded value, and the deduction per ticket is recorded on the card itself. Table 1.10 lists other transportation systems that have already adopted contactless cards (Tan and Tan, 2009; Bank for International Settlements, 2012b, p. 31).

In transportation systems, a microchip is embedded inside the card to store the payment information in a secured microcontroller and internal memory. An antenna is integrated in the surface of the card and operates at the carrier frequency of 13.56 MHz. This antenna enables card readers to track the card within 10 cm of the gates as well as data transmission. The card reader at the gate (called the validator) transfers energy to the microchip in the card to establish communication. The data protocols must resolve contentions among several cards vying for access simultaneously. Typically, a server at each station supervises all the gates and maintains various access control lists, such as blacklisted cards (for buses, the validation lists could be distributed to each bus of the fleet). For security reasons, the amount is capped: it is £20 in the United Kingdom and €30 in continental Europe (Schäfer and Bradshaw, 2014). In the United States, a typical cap is $100 and the minimum is $5 (e.g., for the ORCA card).

TABLE 1.10

Contactless Payment Cards in Public Transport Systems

Year of Introduction	City/Country	Scheme Name
1996	Singapore	CashCard
1997	Hong Kong	Octopus
1999	Baltimore–Washington Metropolitan Area/United States	SmarTrip
2001	Tokyo/Japan	Suica
	Singapore	EZ-Link
	Paris/France	Navigo
2002	Taipei City/Taiwan	EasyCard
2003	London/United Kingdom	Oyster Card
2004	Seoul/Korea	T-Money
2006	Rotterdam/The Netherlands	OV-Chipkaart
2009	Seattle, Washington	ORCA

The size of the investment necessary for converting from paper tickets to electronic tickets is still a deterrent from small operators. Standardization of contactless smart cards and NFC, however, has progressed enough to allow benefits from economy of scale, particularly in the back-office systems. Furthermore, location-based services can generate additional revenues to transportation operators. The obvious drawback is that electronic tickets can be used to monitor movement, which would not be possible with conventional paper tickets.

1.6 Parties in Electronic Commerce

The exchanges in a simple transaction electronic commerce transaction cover at least four categories of information:

1. Documentation on the merchandise
2. Agreement on the terms and conditions of the sale
3. Payment instructions
4. Information on the shipment and delivery of the purchases

The documentation relates to the description of the goods and services offered for sale, the terms and conditions of their acquisition, the guarantees that each party has to offer, and so on. These details can be presented online or offline, in catalogs recorded on paper, or on electronic media.

The agreement between the client and the merchant can be implicit or explicit. It is translated into an order with the required object, the price, the expected date of delivery and acceptable delays, and the means and conditions for payment.

The payment method varies according to the amount in question, the distance or proximity of the merchant and the client, and the available instruments. Regardless of the method used, payment instructions have a different path than that for the exchange of exchange of monetary value through specific interbanking networks.

Finally, the means of delivery depends on the nature of the purchased and the terms of the sale; it can precede, follow, or accompany the payment. The delivery of electronic or digital objects such as files, images, or software can be achieved online. In contrast, the processing, delivery, and guarantees on physical goods or services

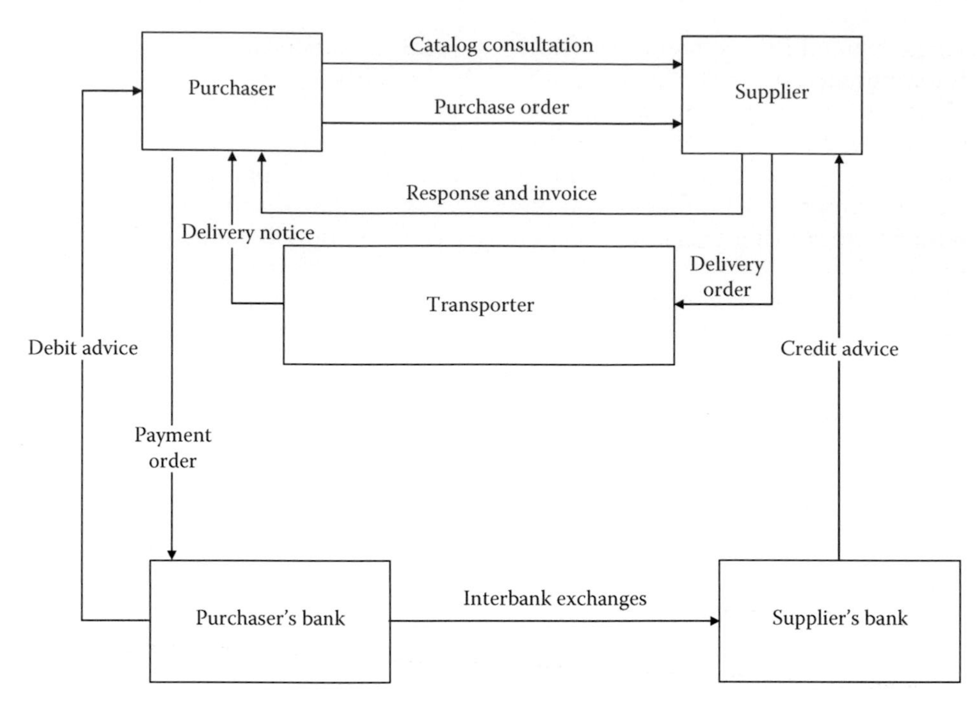

FIGURE 1.4
Typical exchanges and actors in an acquisition transaction.

require a detailed knowledge of insurance procedures and, in international trade, of customs regulations.

Figure 1.4 illustrates the various exchanges that come into play in the acquisition of a physical good and its delivery to the purchaser.

Partial or complete dematerialization of the steps in commercial transactions introduces additional security concerns. These requirements relate to the authentication of parties in the transaction and guarantees for the integrity and confidentiality of the exchanges and the proofs in case of disagreements. These functions are generally carried by intermediaries that handle the heterogeneity of interfaces to other parties.

We now discuss the main parties involved in electronic commerce.

1.6.1 Banks

To respond to the growth of electronic commerce and online operation, banks have been upgrading their infrastructure to accommodate new access methods and to improve security. In doing so, most banks have had to manage several generations of hardware and software, particularly when their expansion has been through acquisitions. Generally speaking, the banking IT infrastructure continues to include subsystems that are no longer manufactured, and it is estimated that global banks spend about 70%–87% of their IT expenditure on maintenance and not on new systems (Arnold and Ahmed, 2014). Interconnecting elements or *middleware* may mask this heterogeneity at the cost of complexity and inefficiency. In the meantime, start-ups unencumbered by legacy systems are free to explore new ways and experiment with innovative idea, putting pressure on traditional organizations.

Online access has increased the traffic intensity to the banks' back office. For example, people check their accounts more regularly from tablets and smartphones, which could strain systems designed for less frequent requests. Outages (such as a shutdown of the Royal Bank of Scotland in 2012) could prevent millions of customers from accessing their accounts for extended periods, denting the banks' reputations (Arnold and Ahmed, 2014). Also, the role of banks may be radically changed if the use of cryptocurrencies such as Bitcoin, becomes commonplace.

In response, some established banks have started offering cloud storage services to corporate clients. Others have partnered with technology companies to improve their service and/or to get market leads in return for promotional work.

1.6.2 Payment Intermediaries

Payment intermediaries convert purchase requests into financial instructions to banks and card schemes. They provide several functions such as payment aggregation, gateway functionalities, and payment processing. Some also intervene to manage the security of the payment infrastructure.

New intermediaries such as PayPal are at the same time partners or competitors of banks in accessing the end users. In the case of cryptocurrencies (such as Bitcoin), the intermediaries create the currencies and update the common financial ledger with the transactions in a process described in Chapter 14. Another role they could play is that of matching buyers of bitcoins with sellers using central bank currencies as a payment instrument and/or holding funds in bitcoins.

The main functions that intermediaries play are aggregation, gateway operation, payment processing, and certification.

1.6.2.1 Aggregators

Aggregators are companies that specialize in the collection, integration, synthesis, and online presentation of consumer data obtained from several sources. The goal is to save end customers the headache of managing multiple passwords of all the websites that have their financial accounts by replacing them with a single password to a site from which they can recover all their statements at once: bank statement, fidelity programs, investment accounts, money market accounts, mortgage accounts, and so on. Ultimately, these aggregators may be able to perform some banking functions.

1.6.2.2 Gateways

Payment gateway service providers allow merchant sites to communicate with multiple acquiring or issuing banks and accept different card schemes, including multipurpose cards, as well as electronic purses. Thus, it relieves the merchant for integrating various payment instruments to their website and automating payment operations. Some of the payment gateways assist the merchant in identifying risk factors to resolve customer inquiries about declined payments.

1.6.2.3 Payment Processors

Processors are usually sponsored by a bank, which retains the financial responsibility. Front-end processors handle the information for the merchant by verifying the card details, authenticating the cardholder, and requesting authorization for the transaction from the issuing banks and/or the card associations. A payment processor relays the issuer bank's decision back to the merchant. At the end of each day, payment processors initiate the settlements among the banks. Payment processors may also act as a notary by recording of exchanges to ensure nonrepudiation and, as a trusted third party, the formation and distribution of revocation lists.

1.6.2.4 Certification Authorities and Trusted Service Managers

These intermediairies are part of the security infrastructure of online payments. They include the certification of merchants and clients, the production and escrow of keys, fabrication and issuance of smart cards, and constitution and management of electronic and virtual purses. These functions could lead to new legal roles, such as electronic notaries, trusted third parties, and certification authorities (Lorentz, 1998), whose exact responsibilities remain to be defined.

Trusted Service Managers (TSMs) are third parties that distribute the secure information (cryptographic keys and certificates) for services and manage the life cycle of a payment application, particularly in mobile applications. The operations they manage include secure downloading of the applications to mobile phones; personalization, locking, unlocking, and deleting applications according to requests from a user or a service provider; and so on. Depending on the operational model, they may also be used to authenticate the transaction parties and authorize the payment.

1.6.3 Providers and Manufacturers

The efficiency and security of electronic transactions may require specialized hardware. For example, with cryptocurrencies, application-specific integrated circuits (ASICs) have been designed to speed up the computations. NFC-enabled terminals, such as mobile phones or tablets, play a role in the security arrangements as discussed in Chapter 10.

Smart card readers must resist physical intrusions and include security modules. These card readers can be with contacts or contactless using a variety of technologies.

Specialized software, such as shopping card software at merchant sites, provides capabilities to administer merchant sites and to manage payments and their security as well as the interfaces to payment gateways. Additional capabilities include the management of shipment such as real-time shipping calculations and shipment tracking. Additional function includes management of discounts in the form of coupons, gift certificates, membership discounts, and reward points. The software may also be managed through third parties (i.e., as through cloud services).

Other specialized functions may be needed in back-office operations. Back-office processing relates to accounting, inventory management, client relations, supplier management, logistical support, analysis of customers' profiles, marketing, as well as the relations with government entities such as online submission of tax reports. Data storage and protection, both online and offline, is an important aspect of that overall performance of electronic commerce systems.

1.7 Security

The maintenance of secure digital commerce channels is a complex enterprise. The pressures for speed and to cut cost are driving reliance on third parties that may compromise security. In addition, new fraudulent techniques relay on deception through phone calls or e-mail to uncover confidential information (phishing or social engineering).

We discuss here three aspects of security from a user's point of view: individual loss of control of their own data, loss of confidentiality, and service disruption.

1.7.1 Loss of Control

Loss of control is the result of usurping the identity of authorized users or subverting the authentication systems used for access control. As will be presented in Chapter 8, it is relatively easy to clone payment cards with magnetic stripes, and when the merchant neither has access to the physical card nor can authenticate the cardholder, such as for Internet payments, transactions with magnetic stripe cards are not easily protected. Some websites, such as fakeplastic.net, offered one-stop shopping to counterfeiters, with embossed counterfeit payment cards fetching as much as $15. On these sites, customers can browse through holographic overlays that can be used to make fake driver's licenses for many U.S. states. They could also select the design and look of payment cards using templates bearing the logos of major credit card companies. The cards would be then embossed with stolen payment card numbers and related information gleaned from the magnetic stripe on the back of legitimate payment cards. Purchases could be made using bitcoins for partial anonymity or U.S. dollars (Zambito, 2014).

Furthermore, mobile devices travel with people so their movements can be tracked, including what websites they consult, or correspondence by phone, e-mail, or various text messaging applications. With new applications, financial information or health information can be tracked, which may be used for industrial espionage or market manipulation. At a minimum, the knowledge can be shared and aggregated and sold without proper disclosure.

Finally, intermediaries in electronic commerce have access to a large amount of data on shopkeepers and their customers. For example, they can track sales and customer retention data and provide shopkeepers the means to compare their performances to anonymized list of their rivals. Small- and medium-size enterprises typically outsource the management of these data, which means that they do not control them.

1.7.2 Loss of Confidentiality

Electronic commerce provides many opportunities to profit from malfeasance. Stealing and selling credit card data and clearing out bank accounts have come easier with the spread of electronic and mobile commerce schemes. This often happens after breaking into a company's computer systems.

Malware can be installed on the victims' terminals by enticing them to click on links or to download files through carefully crafted e-mails, often based on very specialized knowledge on individuals and companies. In business environment, the malware can be used to search for trading accounts on all the machines in a network and then to issue automatic trading instructions, if the account is taken over.

The data gathered, on executives, for example, can be used for market manipulations such as shorting stocks before attacking listed companies, buying commodity futures before taking down the website of a large company, or using confidential information on products, mergers, and acquisitions before playing the markets.

1.7.3 Loss of Service

Service disruption can be used to prevent the digital commerce transactions by attacks on the servers used by any of the participants or the networking infrastructure, for example, taking down the website of a large company.

1.8 Summary

Electronic commerce covers a wide gamut of exchanges with many parties, whether it is monetary or nonmonetary exchanges. Its foundations are simultaneously commercial, socioeconomic, industrial, and civilizational. The initial applications of e-commerce in the 1980s were stimulated by the desire of the economic agents, such as banks and merchants, to reduce the cost of data processing. With the Internet and mobile networks in place, electronic commerce targets a wider audience. Online inventories expand the possible audience of any given shopkeeper. Parts of the Internet economy are nonmonetary in the form of open source or free software. Also, some of the value of electronic commerce comes from the value of learning by searching.

By increasing the speed and the quantity as well as the quality of business exchanges, e-commerce rearranges the internal organizations of enterprises and modifies the configurations of the various players. Innovative ways of operation eventually emerge with new intermediaries, suppliers, or marketplaces.

Security was added as afterthought to the Internet Protocols; transactions on the Internet are inherently not safe, which poses the problem of protection of the associated private information. The commercialization of cryptography, which until the 1990s was strictly for military applications, offers a partial remedy but it cannot overcome insecure system designs.

Electronic and mobile commerce require a reliable and available telecommunication infrastructure. Mobile commerce, as a *green field* technology, took hold much faster because it is cheaper to deploy mobile networks, and the rapid succession of cellular or wireless technologies improved the performance substantially.

Questions

1. Comment on the following definitions of e-commerce adapted from the September 1999 issue of the *IEEE Communications Magazine*:
    - It is the trade of goods and services in which the final order is placed over the Internet (John C. McCarthy).
    - It is the sharing and maintaining of business information and conduction of business transactions by means of a telecommunication network (Vladimir Zwass).
    - It consists of web-based applications that enable online transactions with business partners, customers, and distribution channels (Sephen Cho).
2. What are the goods sold in digital commerce (electronic or mobile)?
3. Name the 2D barcodes used for (a) Bitcoin addresses, (b) train tickets on Eurostar, (c) UPS, (4) a paper boarding pass on an airline, (e) EIA, (f) U.S. DoD, and (g) U.S. IDs.
4. Compare the roles of barcode technology with RFID technology in supply chain management.
5. What are the specific security challenges for mobile commerce compared to electronic commerce from laptop or desktop devices?
6. Discuss the role of standards in electronic commerce. Consider standards at the terminals, networks, and applications. Provide advantages and disadvantages of standards.
7. Compare at least three characteristics of business-to-business commerce with business-to-consumer commerce.
8. Describe the kiosk business model.
9. What are the main benefits and main costs related to business-to-business e-commerce?
10. List two anticompetitive issues that may arise in electronic procurement.
11. What are the three aspects of security from a user's viewpoint?

2

Money and Payment Systems

In this chapter, we describe the financial context within which the dematerialization of the means of payment takes place. First, the *classical* forms of money are introduced, and their evolution is shown in several industrialized countries based on comparative data collated by the Bank for International Settlements (BIS). When the focus is on the United States, data from the Federal Reserve will also be used. Next, *emerging* monies in either *electronic* or *virtual* forms are presented. The clearance and settlement systems are discussed, using examples from the United States, the United Kingdom, and France. Finally, the main drivers for innovations in banking and payment systems are highlighted.

2.1 Mechanisms of Classical Money

The term *money* designates a medium that can be used to certify the value of the items exchanged with respect to a reference system common to all parties of the transaction (Berget and Icard, 1997; Dragon et al., 1997, p. 17; Fay, 1997, p. 112; Redish, 2003). Thus, money represents the purchasing power for goods and services and has three functions:

1. It serves as a standard of value to compare different goods and services. These values are subjective and are affected, among other things, by currency fluctuations.
2. It serves as a medium of exchange, as an intermediary in the process of selling one good for money, thereby replacing barter. To fulfill that function, individuals that trade goods with money must expect that the value of money will remain stable.
3. It serves as a store of value and of purchasing power. Money permits postponement of the utilization of the product of the sales of goods or services. This saving function is maintained on the condition that the general level of prices remains stable or increases only slightly.

These functions of money may be considerate as the hierarchy illustrated in Figure 2.1 (Ali et al., 2014b). Clearly, a medium of exchange must also be a store of value.

Many assets can store value (e.g., houses) without being used as media of exchange. For an asset to be a medium of exchange, the two parties of a transaction must agree to treat it as a way to store value. Finally, for an asset to be considered a unit of account, it must be accepted as a medium of exchange in a significant portion of the economy. This implies that there is a secure system for transferring it from an entity to another, that is, a secure payment network. In addition, there must be a way for recording the values and their owners, that is, a ledger. Currently, financial institutions are organized in a hierarchical network, with the central bank of a given economy at the top, charging with orchestrating the clearance and settlement procedures among the other banks in the hierarchy. Payments within the system use identifiers such as the participant's bank routing number and account number.

Money is also a tool used to coordinate economic and social activities. As a consequence, its role as a means of exchange may at times conflict with its role as a store of value. The first role contributes to speculative demands in the financial sphere while the other facilitates transactions in the real economy (Varoufakis, 2013). The practical terms of money depend on theoretical considerations on its nature and its intrinsic value. Primitive forms of money corresponded to needs for storage and exchange on the basis of valued objects. Accordingly, money first took a materialistic nature in the form of a coin with a specific weight and minted from a precious metal. Today, the value of money corresponds to a denomination that is independent of the material support medium.

Thus, a monetary unit is a sign with a real discharging power that an economic agent would accept as payment in a specific geographic region. This discharging power is based on a legal notion (i.e., a decision of the political power) accompanied by a social phenomenon (acceptance by the public). This sign must satisfy specific conditions:

- It must be divisible to cover a wide range of small, medium, and large amounts.
- It must be convertible to other means of payment.

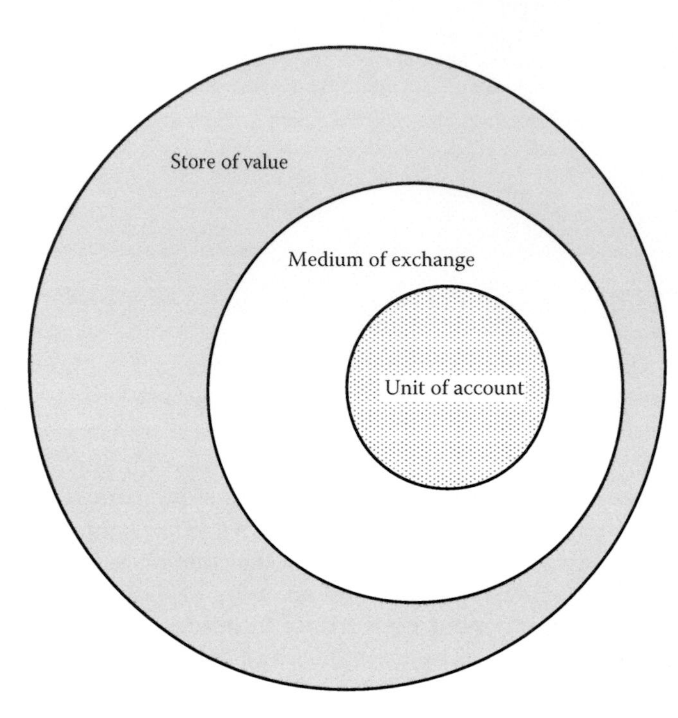

FIGURE 2.1
The hierarchical functions of money.

- It must be recognized in an open community of users. This is because money exists only inasmuch as its issuer enjoys the trust of other economic agents.
- It must be protected by the coercive power of a state.

As a consequence, the only monetary sign that has a real discharging power is the set of notes issued by a central bank or the coins minted by a government mint (central bank or fiat currency). This set, which is called "fiduciary money," is total and immediate legal tender within a specific territory, usually a national boundary, with two important exceptions. On the one hand, 10 countries have dollarized their economy by adopting the U.S. dollar as currency, while 34 others have indexed their currency to its value. On the other hand, the European Union has adopted the Euro as currency without a political union. Note, however, that payment by coins can be restricted by legislation.

While the nominal power corresponds to the face value imprinted on the note or the coin, the real value resides in the trust in the issuer. This is the same for the money that a bank, or generally a credit institution, creates by making available to a nonfinancial agent a certain quantity of means of payment to be used in exchange of an interest proportionate to the risks and the duration of the operation. This money is called "scriptural money" and is a monetary sign tied to the trust that the issuer enjoys in the economic sphere. For example, when Bank A creates scriptural money, the discharging power of that scriptural money depends on the confidence that this bank enjoys and on the system of guarantees that surround its utilization, under the supervision of political authorities (e.g., a central bank).

It should be noted that a merchant is free to accept or reject payments with scriptural money but not with fiduciary money. Note also that scriptural money is traceable, while fiduciary money is not.

Local currencies are complementary currencies used to stimulate the economic activity within a specific geography. For example, the Bristol Pound in the United Kingdom works alongside the sterling for participants based in or around Bristol. Bristol Pounds are printed by specialist printers and incorporate security features, but their acceptance is not obligatory because the Bristol Pound is not a legal tender. Alternative currencies were initiated by cooperatives at the local level in Greece in response to the austerity measures. Working on the principle of exchanges, they provide a directory of offers (and wants) that members can satisfy. Each member has an online account that starts at zero when they join. Members earn credits or IOUs by using their skills to meet specific needs for a member of the community, such as preparing food, carpentry, gardening, childcare, and medical care. These IOUs are recorded in a centralized accounting system. Members can spend their currencies to buy a specific product or service or can pay instead with their expertise. According to some estimates, in August 2015, there were about 80 such currencies varying in size from dozens of members to thousands (Roberts, 2015). TEM, whose name is the Greek acronym for Local Alternative Unit, is one of the largest networks, which was founded in 2010 early in the debt crisis.

Returning to classical money, its material support must meet the following requirements (Camp et al., 1995; Kelly, 1997):

- Be easily recognizable
- Have a relatively stable value across transactions
- Be durable
- Be easy to transport and use
- Have negligible production cost compared with the values exchanged in the transactions

The power of money can be transferred from one economic agent to another with the help of a means of payment or a *payment instrument.* Let us have a brief review of these instruments.

2.2 Payment Instruments

Instruments of payment facilitate the exchange of goods and services and respond to specific needs. Each instrument has its own social and technological history that orients its usage in specific areas. Today, banks offer a large number of means tied to the automatic processing of transactions and to the progressive dematerialization of monetary supports. The means utilized vary from one country to another. A general inventory of the means of payment takes any of the following forms:

- Cash (in the form of metallic coins or paper notes)
- Checks
- Credit transfers
- Direct debits
- Interbank transfers
- Bills of exchange or negotiable instruments
- Payment cards (debit or credit)

Only the central banks have the monopoly to generate fiduciary money, whether in print or stored in smart (chip) cards or in electronic or virtual (cloud-based) purses. Interbanking networks are thus strictly regulated and monitored by the monetary authorities in each country. The case of digital monies (such as Bitcoin) is an exception that is causing a rethinking of nature of money in digital economies.

Float is the elapsed time difference between the initiation of the payment and the availability of funds. The monetary value of the execution time depends on the interest rate and is a cost for the party that cannot access the funds and a gain to the intermediaries. Individuals are most sensitive to cost and convenience. Small and medium enterprises are most sensitive to cost and float. Large corporations prefer automated procedures that allow them to optimize cash flow and are best positioned to negotiate fees and execution times. Banks are most sensitive to risk minimization (including security) and maximizing revenues. The central bank's primary concern is the systemic risk that could destabilize the entire financial system. Thus, many of these scriptural instruments result from innovations in banking to increase the overall efficiency of the payment systems and not by legislation. For example, in the United States, electronic funds transfer (EFT) was developed by the National Automated Clearing House Association (NACHA), now called NACHA—the Electronic Payment Association, an industry association. Similarly in France, credit transfers and the Interbank Payment Title (Tip) are regulated by the Centre Français d'Organisation et de Normalisation Bancaire (CFONB—French Center for Banking Organization and Standardization) and interbank organizations.

The use of payment instruments is measured in two ways: the number of transactions and the total value of these transactions. Table 2.1 reproduces data from the Bank for International Settlements (BIS) regarding the number of transactions using various instruments of payment by nonbanks in selected countries in 2012* (Bank for International Settlements, January 2012).

The data show that payment systems are still a transitional state following the introduction of more efficient electronic methods. For example, while in 2000, checks were still the mostly used scriptural money among nonbanks in the United States (58.3%) and in France (37.9%), their utilization in these two countries dropped to around 15% in 2012. The Netherlands stopped the use of checks in 2003, and they are almost extinct in Sweden and Switzerland.

According to the BIS data, payment card transactions are now dominant in many places, with the majority being debit transactions as shown in Table 2.2.

The latest survey from the Federal Reserve in the United States confirms that debit card transactions constitute about 64% of the total volume of card transactions in the United States (Federal Reserve, 2013, 2014).

Direct debits account for around 50% of the transactions by nonbanks in Germany and about 25% in the Netherlands, a legacy of the Postal Giro system. The Giro systems were established in the late nineteenth century to provide banking services to rural populations† by preauthorizing the post office bank to debit their accounts regularly to pay their creditors. In Germany, it has evolved into a direct debit method called *POS ohne Zahlungsgarantie* (POZ), where the merchant can check that there is sufficient balance in the buyer's bank account to cover the purchase, but without a guarantee from the acquirer (the merchant's bank) (Rankl and Effing, 2010, p. 783).

Credit transfers constitute the most used instrument of scriptural money in Switzerland (51.8%). They are still important in Belgium (37.7%), Germany (33.8%), Italy (30.0%), the Netherlands (28.9%), and Sweden (25.7%).

Finally, Singapore stands out because e-money transactions form about 88% of the volume of nonbank transactions.

* The Bank for International Settlements (BIS) was established in 1930 and is owned by 60 of the world's largest central banks.

† Transfers can also be initiated by paying in cash at the post office counter.

TABLE 2.1

Utilization of Scriptural Money by Nonbanks in Selected Countries in 2012

Millions of Transactions						
Country	**Checks**	**Credit/Debit Cards**	**Credit Transfers**	**Direct Debit**	**E-Money**	**Total**
Belgium	5.4	1,226.9	946.8	285.6	46.2	2,510.91
Canada	748.0	7,484.8	986.9	699.3	—	9,918.98
France[a]	2,805.6	8,475.0	3097.2	3,543.4	52.2	18,068.32
Germany	34.4	3,182.2	6151.0	8,809.5	33.6	18,210.69
Italy[b]	275.7	1,629.0	1261.3	602.3	191.2	4,263.02
The Netherlands	—	2,642.9	1694.2	1,368.6	148.2	5,853.92
Singapore	74.6	235.3	39.9	56.4	3015.1	3,421.32
Sweden	0.2	2,190.0	859.0	297.0	—	3,346.20
Switzerland	0.3	671.8	776.5	47.6	2.8	1,499.00
United Kingdom	848.0	10,546.0	3693.1	3,416.7	—	18,503.80
United States	18,334.5	77,938.6	8493.6	12,821.7	—	117,588.00
Percentage Utilization						
Country	**Checks**	**Credit/Debit Cards**	**Credit Transfers**	**Direct Debit**	**E-Money**	**Total (%)**
Belgium	0.2	48.9	37.7	11.4	1.84	100
Canada	7.5	75.5	9.9	7.1	—	100
France[a]	15.5	47.0	17.1	19.6	0.29	99.5
Germany	0.2	17.5	33.8	48.4	0.18	100
Italy[b]	6.5	38.2	30.0	14.1	4.49	92.8
The Netherlands	—	45.1	28.9	23.4	2.53	100
Singapore	2.2	6.9	1.2	1.6	88.13	100
Sweden	0.0	65.4	25.7	8.9	—	100
Switzerland	0.0	44.8	51.8	3.2	0.19	100
United Kingdom	4.6	57.0	20.0	18.5	—	100
United States	15.6	66.3	7.2	10.9	—	100

[a] Other payment instruments (usually bills of exchange) accounted for 94.89 million transactions in 2012.
[b] Other payment instruments accounted for 303.58 million transactions in 2012.

TABLE 2.2

Credit versus Debit Card Transactions in Selected Countries in 2012

Millions of Transactions			
Country	**Total Payment Cards**	**Debit Cards Only**	**Debit Card (%)**
Belgium	1,226.9	1,087.6	88.7
Canada	7,484.8	4,357.3	58.2
Germany	3,182.2	1,091.5	34.3
The Netherlands	2,642.9	2,530.7	95.8
Singapore	235.3	235.3	100.0
Sweden	2,190.0	1,810.0	82.7
Switzerland	671.8	456.0	67.9
United Kingdom	10,546.0	8,155.0	77.3
United States	77,938.6	51,717.2	66.4

2.2.1 Cash

In each country, cash constitutes the fiduciary money that the central bank and the public treasury issue in the form of notes and coins (Fay, 1997, p. 83). This instrument of payment is available free of charge to individuals. Banks cover the costs for managing the payments, the withdrawals from branches or teller machines, and the costs of locking up the money. In retail commerce, banks usually charge their customers for their services if they have to process large amounts of notes or coins and perform the counting and the sorting of bills and coins.

Cash remains the preferred means of payment for proximity (or face-to-face) commerce (Bank for International Settlements, 2012b, p. 5). The current trend in Western countries is to use cash for relatively small amounts, while medium and large amounts are handled with scriptural instruments. On the basis of this suggestion, the French Comité des Usagers (Users Committee) defined a micropayment as a "payment, particularly in the case of a face-to-face payment, where, given the absence of any specific constraint, cash is the preferred instrument" (Sabatier, 1997, p. 22).

The cost of using cash includes printing new bills, transporting them in armored trucks to distribute them to banks and to replenish automatic cash dispensers and, finally, to retire and recycle old notes. To this cost, one must add the cost of fraud detection equipment at

TABLE 2.3

Percentage of Cash in the Narrow Money in Selected Countries (1985–2013)

Year	1985	1986	1987	1988	1989	1990	1991	1992	1993	1994	1995	1996	1997	1998	1999	2000	2001
Belgium	36.6	35.3	34	34.8	32.5	31.3	31.5	31.5	29.6	27.1	27.2	27.5	26.5	23.8	20.4	19.9	20.4
Canada	44.5	45.7	43.6	44.6	43.8	46.1	47.0	47.1	44.0	44.2	17.1	14.3	14.2	14.5	15.6	15.8	15.6
France	16.0	15.3	15.2	15.2	15.2	15.1	15.8	15.9	15.3	15.1	14.2	13.3	13.1	11.0	12.6	12.8	8.4
Germany	31.1	31.3	32.2	33.4	32.6	27.1	29.9	29.9	29.6	29.6	29.1	27.6	27.2	24.1	23.4	21.8	13.8
Hong Kong	—	—	—	—	—	—	—	—	—	—	—	38.6	42.8	45.5	48.5	45.0	49.4
Italy	14.2	14.1	13.8	14.3	15	14.4	14.2	15.7	15.5	16.0	16.3	16.1	16.1	16.1	14.4	14.3	12.6
Japan	28.6	27.4	28.3	31.5	35.3	36	33.1	31.2	31.1	30.7	26.9	26.1	25.8	25.3	24.8	25.0	26.0
The Netherlands	32.1	31.2	32.4	31.3	30.3	29.5*	28.6*	27.2	25.1	25.0	22.1	18.0	15.7	14.1	12.8	11.4	7.1
Singapore	—	—	—	—	—	—	—	—	—	—	—	38.1	38.9	37.3	36.4	33.9	36.3
Sweden[a]	—	—	—	—	—	—	11.5	10.8	10.7	10.7	10.5	9.9	10.0	10.2	10.6	13.1	13.0
Switzerland	—	—	—	—	—	—	21.8	21.6	19.7	30.6	18	17.3	15.6	15.5	15.3	15.8	19.1
United Kingdom[b]	19.5	17.0	14.5	14.1	6.5	6	5.6	4.8	4.5	4.6	4.6	4.9	5.0	5.0	5.0	5.0	5.3
United States	31.0	29.4	28.7	27.2	28.2	29.2	28.5	28.5	28.5	30.7	32.6	36.0	39.0	41.4	45.4	48.1	53.2

	2002	2003	2004	2005	2006	2007	2008	2009	2010	2011	2012	2013
Canada	14.7	14.22	13.66	13.33	13.24	12.83	12.06	10.99	10.50	10.18	9.85	9.74
Euro Area	15.70	16.83	17.52	16.71	17.22	17.87	19.44	18.20	18.17	18.77	18.15	18.04
Hong Kong	48.0	39.68	37.17	44.87	42.41	37.69	37.83	31.12	32.31	33.97	32.79	—
Japan	22.9	19.90	17.51	16.92	17.03	17.22	17.43	17.15	16.86	16.36	16.28	16.01
Korea	—	—	—	7.84	7.48	9.24	9.27	9.56	10.10	10.98	11.54	12.27
Singapore	37.8	36.27	34.18	34.57	32.43	28.84	27.39	23.79	21.79	21.04	20.69	20.47
Sweden[a]	12.5	12.05	11.47	10.22	0.01	0.01	0.01	7.34	6.60	6.22	5.66	4.61
Switzerland	17.4	14.78	15.25	15.10	16.42	17.28	15.76	13.28	12.54	12.14	12.05	12.39
United Kingdom[b]	5.2	4.95	4.86	4.60	4.48	4.48	4.37	4.47	4.44	4.54	4.50	4.39
United States	55.20	54.37	53.87	56.85	59.15	59.50	54.57	53.85	52.54	48.74	46.66	45.82

[a] As a percentage of the M3 monetary aggregate.
[b] As a percentage of the M2 aggregate starting from 1989.

retailers and banks. It is estimated to cost between 0.4% and 1% of the gross domestic product (GDP) (Rambure and Nacamuli, 2008, p. 25; Birch, 2012, p. 34). Other studies show that the marginal cost of a cash transaction (i.e., the cost of an additional transaction) is around double that of a debit card transaction. At the same time, the cost of manufacture of currency is much less than the face value of the money. This difference represents a loan without interest to the central banks (seignorage), which are now the monopoly suppliers of banknotes. This profit can be estimated by multiplying the notes and coin in circulation by the long-term rate of interest on government securities (Rambure and Nacamuli, 2008, p. 25). Some argue, however, that because there is zero interest rate on cash, central banks have less leeway to stimulate depressed economies, because people will change their deposits for cash should interest rates become negative (*Financial Times*, 2015b).

The use of cash relies on the trust of the various parties in the validity of bank notes and coins. To sustain this trust, responsible authorities introduce security measures on a continuous basis using advanced technologies. Notes are made from plastic or polymers to improve the resistance to humidity, wear, and tear. Authentication depends on hard to replicate features that protect against counterfeiting, such as the following:

- Deep stripes and patterns that appear to move.
- Changing images: the first is displayed at first glance and the second when held up to a light.
- Wide silver threads on the reverse side of the note that reveal text when held up to a light.
- Drawings on the reverse side of the note that change color when exposed to ultraviolet light.

Depicted in Table 2.3 is the part of cash in the narrow money for selected countries between 1985 and 2013 (Bank for International Settlements, 1996, 1997, 2000, 2001, 2002, 2003, 2004a,b, 2005, 2006, 2007, 2009a,b, 2011, 2012, 2013, 2014; Dragon et al., 1997). In most of the countries, narrow money is measured using the M1*

* M1 is the total amount of currency in circulation as well as monies in checking accounts. M2 is M1 plus monies in saving accounts and money market funds. M3 is M2 plus bank certificates of deposits and other institutional accounts, such as accounts in foreign currencies and, for the United States, Eurodollar deposits in foreign branches of U.S. banks.

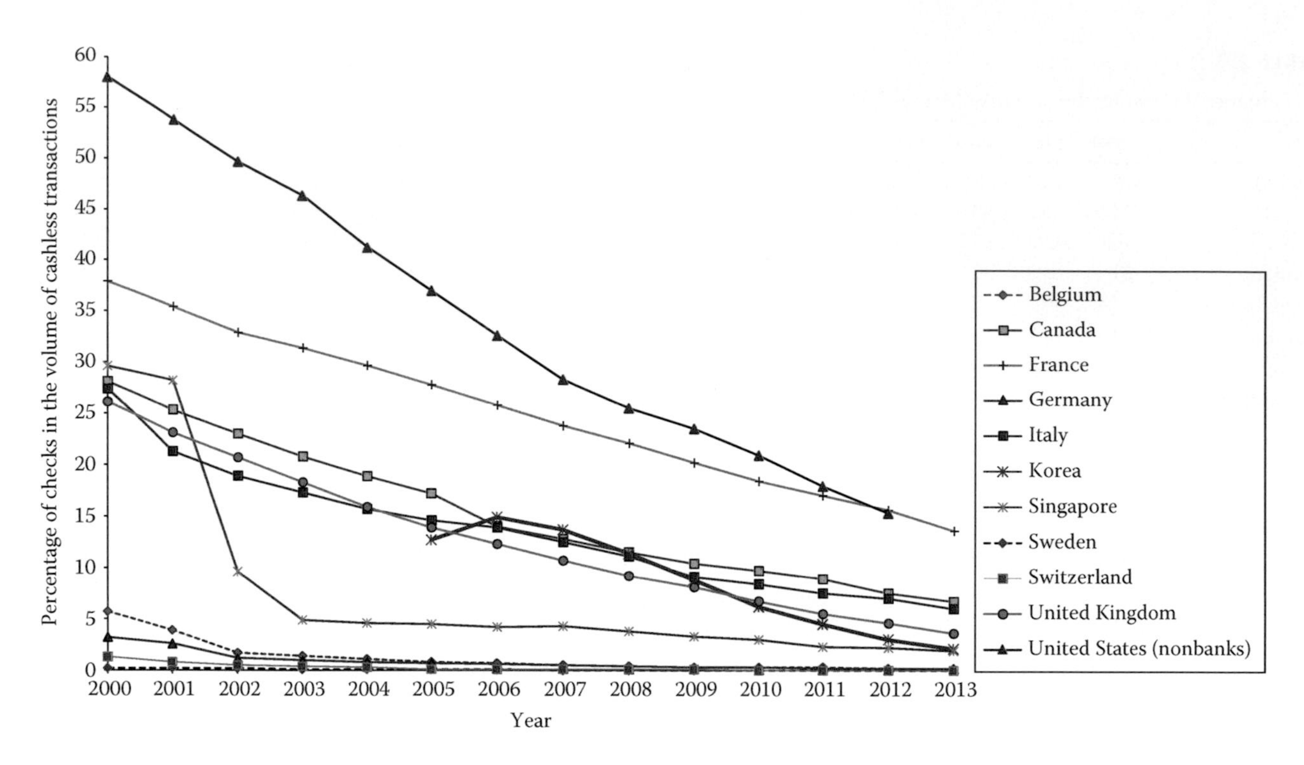

FIGURE 2.2
The contribution of cash to the narrow money for selected countries.

monetary, with the exception of Sweden, which uses the M3 aggregate and the United Kingdom, which has been using the M2 aggregate since 1989. The data are presented in graphical form in Figure 2.2.

Clearly, the contribution of fiduciary money varies tremendously among countries, particularly that the definition of narrow money is not uniform across countries. Within the Euro zone, cash payments in Germany are common for large items such as cars and it was under pressure from Germany that a €500 bill was issued (Rambure and Nacamuli, 2008, p. 39). The big drop in Canada around 1994 reflected a change in the ways narrow money is measured. Nevertheless, the general trend is a decreasing contribution of cash with the exception of Korea, where cash constitutes about 12% of the narrow money. As of 2011, the countries with the highest ratio are the United States and Hong Kong, respectively, with 53.2% and 49.4%. The countries with the lowest use of cash are the United Kingdom and Sweden and the United Kingdom at around 4.5%.

Between 2000 and 2007, the usage of cash increased in the United States. One possibility is that many foreigners hold dollar notes and some countries use the U.S. dollar as their official currency. Another is that the cost of banking services had put them out of reach of a large sector of the population (about 25%–30%, i.e., more than 40 million people whose income is less than $25,000) (Hawke, 1997; Mayer, 1997, p. 451; Demirguc-Kunt and Klapper, 2012, p. 15). It is estimated that the number of U.S. households without a bank account range from 18% to 25% (Handa et al., 2011; Paglieri, 2014, pp. 94–96). New programs by the federal government addressing the *unbanked* offer some solutions to the problem as will be explained later. Technology enthusiasts, however, are using these statistics to predict the demise of cash if not calling for phasing it out (Zorpette, 2012). Other arguments against cash include tracking the underground economy, enhancing tax collection, and improving monetary policy (*Economist*, 2014b). Another proposal is to have cash available at a cost so that users would pay for the privilege of cash (*Financial Times*, 2015b). Yet, all these academic arguments overlook the significant portion of people who are using cash out of necessity because there is no other instrument available to them. In addition, tax evasion is also available to those who can afford complicated schemes and legions of accountants and lawyers to hide huge sums of money in tax havens and evade taxation.

2.2.2 Checks

Check processing includes the fabrication, printing, distribution, sorting, examination to identify the signature and capture the written data, decision on acceptance or rejection, reconciliation, archiving, and so on. Checks can be lost, stolen, delayed, or damaged, and they can be counterfeit. The expense for stolen and/or fraudulent checks is estimated to be about 1% of the total amount. A working figure for the total charge for check processing is 50 cents to a U.S. dollar (Dragon et al., 1997, pp. 110–126). This means that checks are an expensive instrument of payment not only for banks but also for heavy users.

Another reason that corporations dislike checks is that they require manual handling and reconciliation. Nevertheless, checks remained popular with small businesses because of their familiarity, their convenience, and the interest they earn on the float, that is, the calendar difference between when the check is issued and when the funds are actually withdrawn. However, as will be seen in Chapter 15, electronic check presentment (ECP) has reduced the float. This may encourage the migration of small- and medium-sized companies to other means of payments. In the long term, the dematerialization of checks or their replacement with other electronic payment instruments will have consequences on the labor market and the thousands of jobs involved directly or indirectly with checks processing.

Table 2.4 shows the relative weight of checks in the total volume of scriptural transactions by nonbanks in selected countries from 2000 to 2013 using data from the BIS. These data are represented in graphical form in Figure 2.3. Clearly, the contribution of checks to the volume of scriptural transactions by nonbanks is

TABLE 2.4

Percentage of Checks in the Volume of Scriptural Transactions by Nonbanks in Selected Countries (2000–2013)

Year	2000	2001	2002	2003	2004	2005	2006	2007	2008	2009	2010	2011	2012	2013
Belgium	5.7	3.9	1.7	1.4	1.1	0.8	0.7	0.5	0.4	0.3	0.3	0.3	0.2	0.2
Canada	28.1	25.3	23.0	20.8	18.9	17.2	14.0	12.8	11.5	10.4	9.7	8.9	7.5	6.7
France	37.9	35.4	32.9	31.4	29.7	27.8	25.8	23.8	22.1	20.2	18.4	17.0	15.6	13.6
Germany	3.2	2.6	1.2	1.0	0.8	0.7	0.6	0.5	0.4	0.3	0.3	0.2	0.2	0.2
Italy	27.4	21.3	18.9	17.3	15.7	14.6	13.9	12.5	11.1	9.1	8.4	7.5	7.0	6.0
Korea	—	—	—	—	—	12.7	14.9	13.7	11.4	8.8	6.2	4.5	3.0	2.1
Singapore	29.6	28.2	9.6	4.9	4.6	4.5	4.2	4.3	3.8	3.3	3.0	2.3	2.2	1.9
Sweden	0.2	0.2	0.2	0.1	0.1	4.5	0.1	0.0	0.0	0.0	0.0	0.0	0.0	0.0
Switzerland	1.3	0.8	0.5	0.4	0.3	0.1	0.1	0.1	0.1	0.1	0.0	0.0	0.0	0.0
United Kingdom	26.1	23.1	20.7	18.3	15.9	13.9	12.3	10.7	9.2	8.1	6.7	5.5	4.6	3.6
United States (nonbanks)	58.0	53.8	49.7	46.4	41.3	37.0	32.6	28.3	25.5	23.5	20.9	17.9	15.6	—

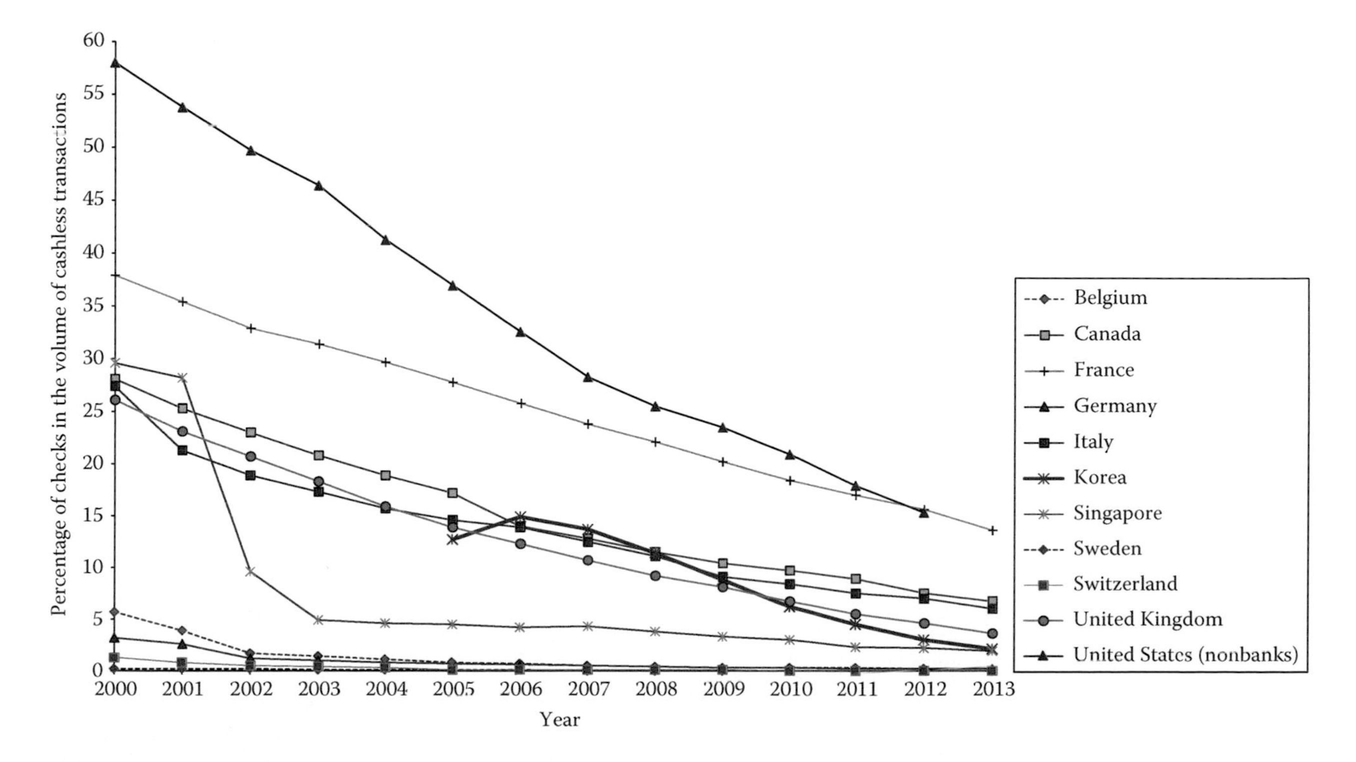

FIGURE 2.3
Evolution of check usage by nonbanks in the volume of scriptural transactions in selected countries.

decreasing uniformly across all countries. It should be noted that the data represent the checks paid and not the checks written. Because, once the check is converted into an electronic record it becomes a source document for the automatic clearing systems and is no longer treated as a traditional check.

Based on the data, it is possible to classify the various check usage patterns into four classes:

1. The United States and France constitute the first group where the volume of transactions by checks in 2012 remains substantial, respectively, 16.2% and 15.2% of the total number of scriptural transactions by nonbanks. In the United States, the number of checks declined by more than 50% from 2000 (from 41.9 to 183 billion in 2013), replaced by other payment instruments, including Internet and mobile transactions. More specifically, the rate of decrease was around 10% in the 2012–2013 interval. Should this rate remain constant, checks will constitute a minor percentage in the volume of transactions in both countries by around 2025. Checks will continue to be used in the United States for values exceeding $25,000 or if the payer is a federal entity (Federal Reserve, 2014, p. 192).
2. In the second group of countries, in particular Canada, Italy, Korea, Singapore, and the United Kingdom, the contribution from checks in 2013 varied between 6.7% and 1.9% of the volume of scriptural transactions by nonbanks. In these countries, the annual rate of usage decreases by around 15%–20%.
3. In a third group, checks play an insignificant role, such as in Belgium, Germany, Sweden, and Switzerland.
4. A fourth group of countries have withdrawn checks from usage; this is the case in the Netherlands in 2003.

For the United States, the volume of checks for nonbank exchanges includes U.S. Treasury checks and postal money orders. Interbank transfers for the settlement of large values and automated clearing of direct debit and credit transfers among banks were excluded.

The relative importance of the values exchanged by checks in the same countries in the period from 2000 to 2013 is given in Table 2.5 and in graphical form in Figure 2.4.

In Singapore, the contribution of checks to the total value of transactions was approximately 65% in 2013. In Canada, their proportion in the total exchange of value dropped from 96.5% in the early 1990s to approximately 47% in 2013. In the United States and Korea, the contribution of checks is around 30%. For all the other countries in the sample, the contribution is less than 10%. In particular, checks have insignificant role in the transfer of values in Sweden and Switzerland.

The time series for Canada shows several discontinuities due to changes in the reporting methodologies for the periods 2001–2005 and 2006 onward. In addition, with the introduction of the Large Value Transfer System (LVTS) in February 1995, many large payments moved from checks to credit transfer payments (Bank for International Settlements, 2001, note 5, Table 13, p. 23; 2007, note 2, Table 8, p. 22; 2010, note 1, Table 8, p. 62).

In the United States, the values exchanged among nonbanks with checks declined from 86.7% to 33.6% in 2012. One contributing factor is a federal law that took effect on January 1, 1999, that mandates the use of credit transfers for all payments by the federal government, with the exception of tax refunds. A large percentage of the *unbanked* receive federal payments such as social

TABLE 2.5

Percentage of Contribution of Checks in the Value Exchanged with Scriptural Transactions by Nonbanks in Selected Countries (2000–2013)

Year	2000	2001	2002	2003	2004	2005	2006	2007	2008	2009	2010	2011	2012	2013
Belgium	0.5	0.6	0.7	2.7	2.5	1.9	1.7	1.5	1.1	1.1	1.1	1.1	1.0	0.8
Canada	84.1	52.9	50.2	41.2	38.2	36.8	63.3	61.2	57.8	54	52.4	50.5	49.0	46.8
France	2.8	2.3	2.4	2.4	1.9	12.9	10.4	9.9	9.0	7.7	7.4	6.4	5.9	4.8
Germany	3.2	2.7	2.3	2.1	1.7	1.5	1.7	0.6	0.6	0.5	0.4	0.4	0.3	0.3
Italy	17.9	18.5	17.7	16.7	15.7	14.7	14.0	12.2	12.1	10.4	9.1	8.5	7.5	6.4
Korea	—	—	—	—	—	40.2	47.4	46.5	46.9	44.9	43.1	38.6	34.3	30.4
Singapore	4.7	3.6	3.8	75.2	74.2	73.3	73.3	74.8	70.8	68.6	68.6	67.1	65.7	64.8
Sweden	0.2	0.2	0.3	0.6	0.7	0.6	0.5	0.5	0.6	0.4	0.2	0.2	0.2	0.0
Switzerland	0.1	0.0	0.0	0.0	0.2	0.1	0.1	0.1	0.1	0.0	0.0	0.0	0.0	0.0
United Kingdom	2.9	2.6	2.6	2.4	2.2	2.0	1.8	1.5	1.8	1.8	1.6	1.4	1.1	0.0
United States	66.8	64.7	62.5	61.1	60.0	57.5	55.1	49.9	46.5	44.6	41.6	37.6	33.6	—

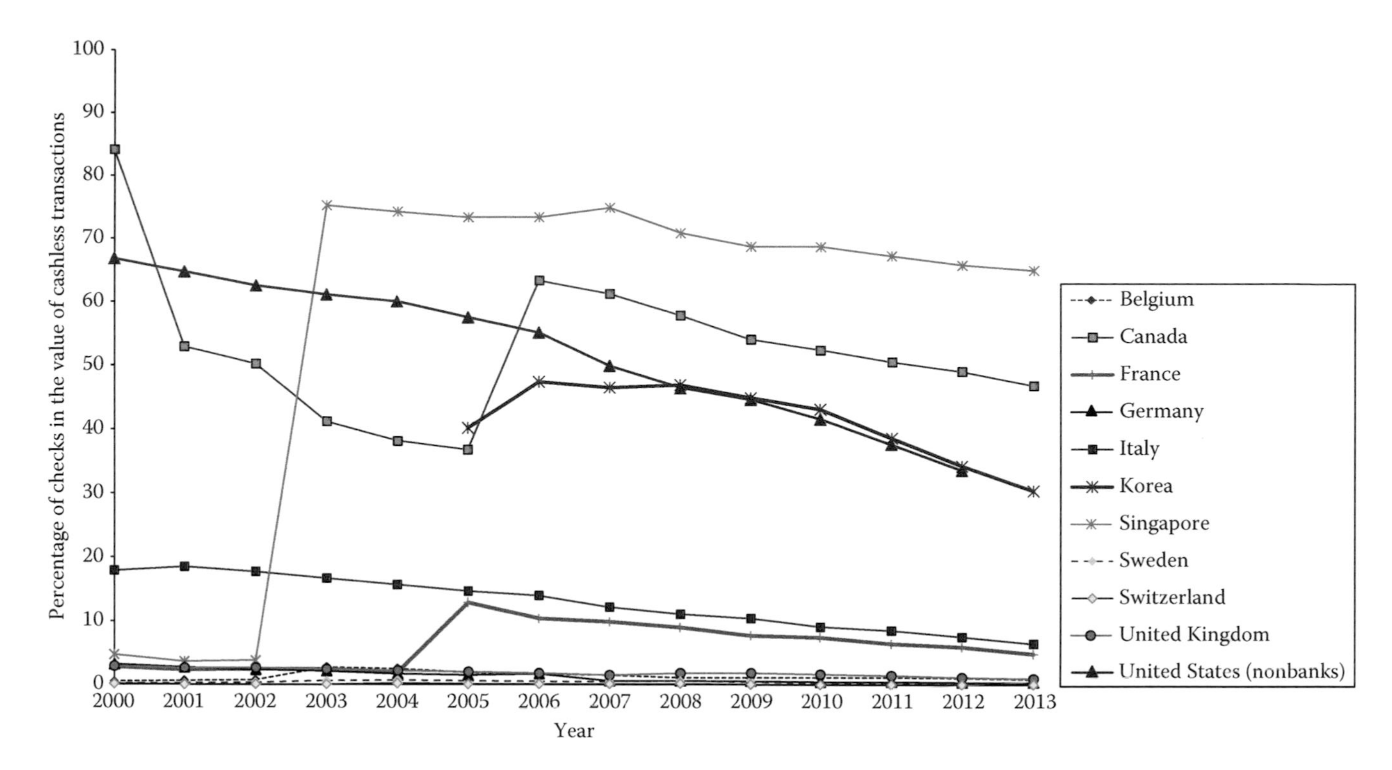

FIGURE 2.4
Evolution of check usage by nonbanks in terms of the value of scriptural transactions in selected countries.

security checks, veteran benefits, and other supplemental support. Prior to 2011, they could receive their benefits by checks sent by postal mail or through an Electronic Transfer Account (ETA). When checks were used, the recipients would cash their checks through banks or check-cashing outlets for a set fee. In 1999, ETAs were designed to be low-cost direct deposit accounts for federal benefit payments (social security, veteran's benefits, railroad retirements, etc.), but the take-up rates were very low. On March 2013, the U.S. Treasury Department started to distribute the benefits to the unbanked using a prepaid debit card, the electronic benefit transfer (EBT) card, such as the Direct Express® Debit MasterCard® card, with the Comerica Bank as the financial agent. EBT cards are restricted to certain types of purchases from any merchant that applies to, and is approved in the program. This migration is reflected in the reduction of the volume of checks, as well as in the increased use of direct debits and debit cards in the United States.

However, the move to electronic networks increases the vulnerability of beneficiaries to outages and connectivity issues, exposing them to large-scale disruptions. For example, on October 12, 2013, beneficiaries in 17 states (e.g., Ohio and Michigan) could not use their stamped debit cards following routine tests by the vendor (Xerox) that resulted in the system's failure (*Sunday Star-Ledger*, 2013).

Another problem is that each time welfare recipients withdraw cash from their EBT accounts, they have to pay fees to the banks (typically $3). In 2013, it was estimated that welfare recipients in California were charged $19 million in cashing fees (Paglieri, 2014, p. 93).

2.2.3 Credit Transfers

Credit (or wire) transfers (also called direct credits) are a means to transfer funds between accounts at the initiative of the debtor. The initiator instructs his/her bank to debit his account; the bank verifies the instructions and the availability funds. It then debits the originator's account and sends a credit transfer message to the beneficiary's bank or to a clearinghouse to clear and settle the payment. The beneficiary's bank credits the receiver's account after verification. This instrument requires the debtor to know the beneficiary's bank and bank accounts and is considered a *push* payment. This is the reason it is usually used in bulk transfers and prearranged payment deposits such as for salaries and pensions.

Wire transfers are used for very large interbank payments, as well as for many payments of small values by nonbanks. The statistics presented are for nonbanks only, that is, interbank fund transfers are not included. The majority of nonbanking credit transfers today are initiated electronically from Internet banking portals, accounting software packages or Enterprise Resource Planning (ERP) systems. The data available from the BIS, reproduced in part in Table 2.6 and illustrated in

TABLE 2.6

Percentage of Credit Transfers by Nonbanks in the Volume of Scriptural Transactions in Selected Countries (2000–2013)

Year	2000	2001	2002	2003	2004	2005	2006	2007	2008	2009	2010	2011	2012	2013
Belgium	41.0	48.2	47.4	43.8	43.9	43.2	42.5	42.8	42.1	41.6	42.1	41.0	37.7	37.9
Canada	9.6	10.2	9.9	10.4	10.8	10.9	11.2	10.3	10.2	10.4	10.6	10.6	9.9	9.6
France	17.7	17.8	19.4	19.1	18.6	17.1	17.7	17.0	17.1	17.1	17.6	17.1	17.2	17.2
Germany	45.7	45.4	45.7	43.1	42.2	42.2	42.2	36.0	35.5	35.1	33.9	34.2	33.8	31.5
Italy	36.6	38.2	34.8	34.8	33.8	32.8	32.6	32.1	30.6	32.9	32.7	32.6	31.9	30.1
Korea	—	—	—	—	—	25.4	23.2	23.7	24.0	22.9	22.0	21.6	20.8	20.2
The Netherlands	40.4	39.1	37.0	35.5	33.8	32.5	32.5	32.7	32.0	30.9	30.3	29.4	28.9	21.3
Singapore	4.7	5.1	1.9	1.0	1.0	1.1	1.2	1.3	1.4	1.3	1.3	1.1	1.2	1.1
Sweden	65.4	60.3	38.0	32.0	35.3	34.9	34.9	29.5	27.1	26.5	25.8	26.8	25.7	24.8
Switzerland	61.8	58.5	57.3	55.0	56.4	56.5	56.5	56.4	55.3	54.6	54.2	53.0	51.8	51.1
United Kingdom	17.8	17.4	17.4	18.0	19.7	21.4	21.2	21.0	20.9	20.6	20.5	20.2	20.0	19.6
United States	5.2	5.7	5.8	5.7	5.9	6.2	6.4	6.5	6.8	7.0	7.2	7.0	7.2	—

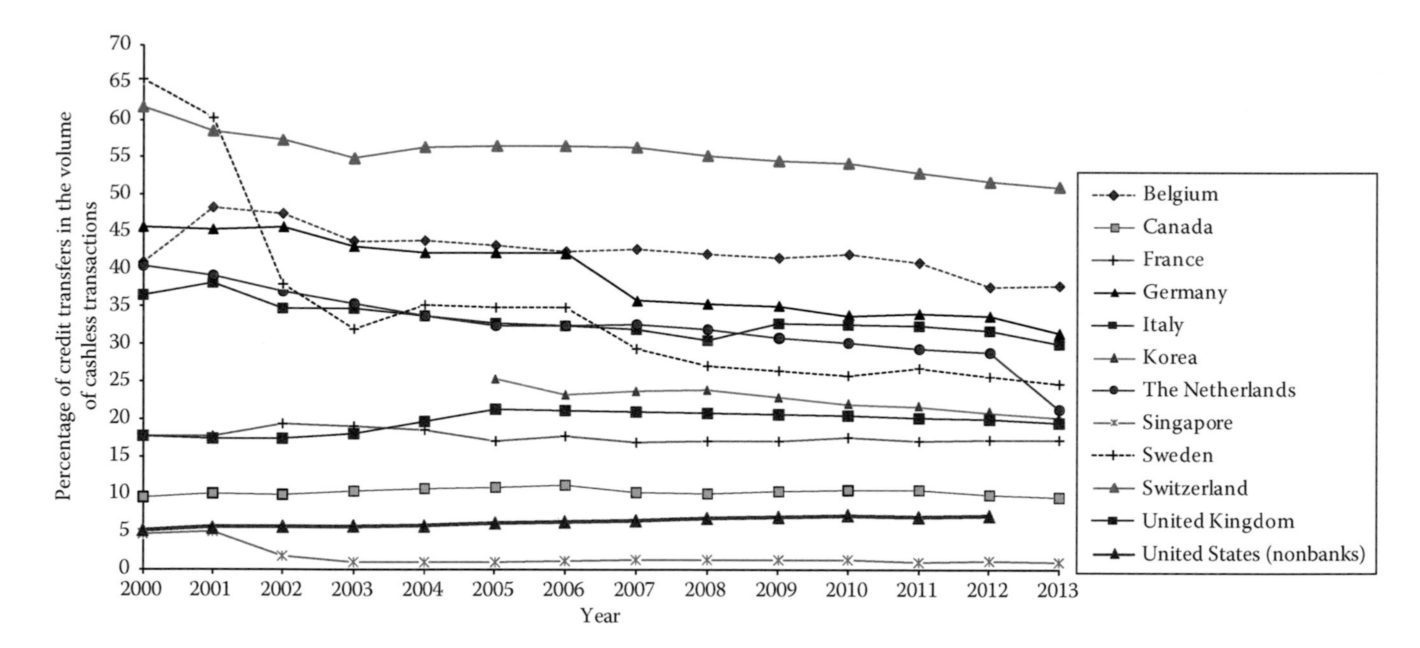

FIGURE 2.5
Evolution of the use of credit transfers by nonbanks in selected countries (in volume).

Figure 2.5, underline the evolution of the contribution of credit transfers in the volume of scriptural transactions between 2000 and 2013.

The data show that there are four distinct groups of countries in terms of the proportion of credit transfer payments by nonbanks in the total number of transactions:

1. Countries where credit transfers constituted more than 25% of the volume of scriptural transactions in 2013: Switzerland (51.1%), Belgium (37.9%), Germany (31.5%), and Italy (30.1%).
2. Countries where the volume of credit transfers is stable: the United Kingdom and Korea are both at around 20% and France is at around 17%.
3. Countries where the volume of credit transfers is decreasing: Sweden (from 34.9% in 2006 to 24.8% in 2013) and the Netherlands (from 32.5% in 2006 to 21.3% in 2013).
4. Countries where credit transfers constitute less than 10% of the volume of transactions: Canada (9.6%), the United States (7.2%), and Singapore (1.1%).

TABLE 2.7

Percentage of Credit Transfers by Nonbanks in the Transfer of Value with Scriptural Transactions in Selected Countries (2000–2013)

Year	2000	2001	2002	2003	2004	2005	2006	2007	2008	2009	2010	2011	2012	2013
Belgium	99.0	99.0	98.7	94.6	94.9	95.5	95.6	96.0	96.3	96.1	96.0	95.7	95.5	95.6
Canada	84.4	42.1	44.0	51.8	54.2	55.4	57.3	22.9	25.0	27.5	28.9	30.6	32.2	33.8
France	96.1	96.8	96.7	96.7	97.2	80.1	83.6	84.0	84.9	86.3	86.6	87.5	87.8	88.4
Germany	85.6	84.2	85.4	87.3	87.4	88.5	87.7	84.1	84.1	82.6	80.6	80.2	80.8	80.9
Italy	70.4	75.2	75.8	76.7	77.3	78.4	79.2	81.4	82.7	84.1	85.6	86.3	86.7	88.1
Korea	—	—	—	—	—	55.8	49.3	50.2	49.8	52.0	53.8	58.1	62.2	65.9
The Netherlands	93.3	93.6	93.2	93.2	93.9	94.6	94.6	94.4	93.8	93.8	93.8	93.1	92.9	92.0
Singapore	95.0	96.1	95.8	14.5	15.0	15.5	15.5	14.7	17.1	18.6	18.5	19.4	20.2	20.7
Sweden	94.8	94.1	90.8	90.3	89.4	89.5	89.8	90.1	89.7	89.4	89.8	90.2	90.5	90.7
Switzerland	99.7	99.7	99.7	96.1	95.9	95.8	96.1	96.6	96.6	96.4	96.3	96.4	96.3	96.3
United Kingdom	96.1	96.4	96.3	96.4	96.4	96.6	96.9	97.3	96.6	96.3	96.3	96.4	96.8	96.9
United States	14.9	16.2	17.7	18.2	19.9	21.5	23.4	26.4	28.7	29.9	31.9	34.4	36.5	—

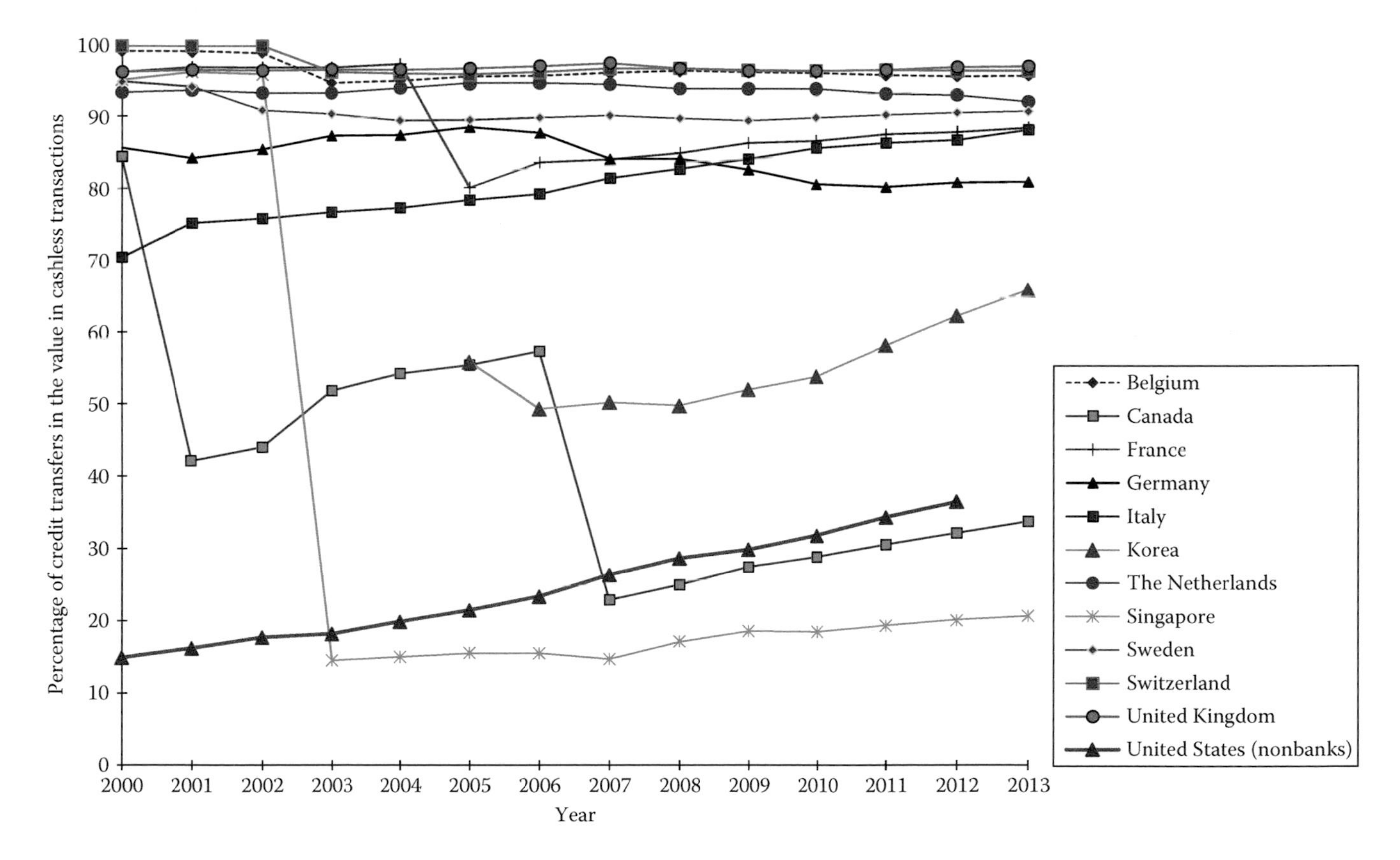

FIGURE 2.6
Evolution of the use of credit transfers by nonbanks in the transfer of value in selected countries.

The data reproduced in Table 2.7 and represented graphically in Figure 2.6 illustrate the evolution of the exchange of value by credit transfers for nonbanks between 2000 and 2013. The contribution of credit transfers exceeds 80% and may reach 96% in the following countries: the United Kingdom (96.9% in 2013), Belgium (96.5%), Switzerland (96.5%), the Netherlands (92%), Sweden (90.7%), France (88.4%), Italy (88.1%), and Germany (80.9%). In 2013, their use in Korea stood at 65.9% of the total value exchanged while their contribution in Singapore was at 20.7%.

The value exchanged in the United States increased steadily from 14.9% in 2000 to 36.5% in 2012. A major trigger was the federal mandate that all government payments be made with credit transfers, with the exception of tax returns starting from January 1, 1999. This encouraged "all depository financial institutions" (banks, thrifts, credit unions, savings, and loans) to join the ACH network as well (Coye Benson and Loftesness, 2010, p. 50). Direct deposits of payroll, interest, pensions, royalties, and dividends use the ACH code "prearranged payment and deposit" (PPD). The transfer of funds (both debit and credit) using EDI (Electronic Data Interchange) with ANSI ASC X12 or UN/EDIFACT messages also increased in corporate applications.

As mentioned earlier in the discussion of value exchanged by checks, changes in the reporting methodologies for the periods 2001–2005 and 2006 onward explain the peculiar shape of the time series for Canada.

2.2.4 Direct Debit

Direct debit is used for recurrent payments (e.g., utility bills, subscriptions), that is, it is a *pull* payment. To initiate a direct debit, the debtor signs on paper or electronically an agreement to pay future amounts. These payment instruments allow large bill producers to collect their dues at a predetermined date, thereby optimizing cash flow and treasury management. Direct debits save time and some cost for debtors, but they require some guarantee that the payer will be able to contest the amounts charged easily and then to recover the funds collected in error. The contribution of direct debits to the volume of scriptural payments in some countries is shown in Table 2.8. These data are presented graphically in Figure 2.7.

The consequence of Giro transfers appears one more time—the countries where direct debit is most frequently used are Germany and the Netherlands. Direct debit has increased in the United States in parallel with the development of preauthorized bill payments for consumer as well as business applications. This has encouraged some start-ups such as GoCardless to focus on setting up direct debit payment schemes for small businesses and individual-owned businesses. Other than for the United States and Germany, however, the value of the payments conveyed is less than 10% of the total value of cashless payments among nonbanks as indicated in Table 2.9 and the corresponding graphs depicted in Figure 2.8.

2.2.5 Interbank Transfers

The Titre Interbancaire de Paiement (Tip—Interbank Payment Title) is a specific instrument introduced in France in 1988. It is different from typical debit transfers in that a signature is required for each payment on a special form that the creditor supplies. Its main advantage is that it can be easily adapted to electronic payment to allow remote payments by telephones or through data networks. In this way, the creditor still sends the Tip by postal mail, while the client signs electronically. The signature can be sent on the Internet (previously on the Minitel) or by entering a special code over the phone.

2.2.6 Bills of Exchange

A bill of exchange (or a negotiated instrument) is a remote payment reserved for professional relations, giving either the debtor or the creditor the initiative of the payment. If the debtor is the initiator, the instrument is called a "promissory note," whereas if the creditor has the initiative, it is a "bill of exchange" proper. In either case, creditors give the documents they possess to their banks that then send the bill of exchange to the debtor banks. The promissory note resembles a check drawn on a checking account with the assurance of payment and the possibility of a discount fee for the beneficiary.

TABLE 2.8

Percentage of Direct Debits in the Volume of Scriptural Transactions by Nonbanks in Selected Countries (2000–2013)

Year	2000	2001	2002	2003	2004	2005	2006	2007	2008	2009	2010	2011	2012	2013
Belgium	11.8	11.5	9.9	11.5	11.8	11.6	11.7	11.4	11.3	11.4	10.3	10.6	11.4	11.6
Canada	7.5	7.8	7.9	8.0	8.0	8.0	7.0	7.0	7.0	7.0	6.9	6.9	7.0	7.1
France	16.6	16.8	16.5	17.4	18.2	17.8	18.5	19.0	19.2	20.0	20.1	20.3	19.7	19.1
Germany	38.1	38.6	37.5	40.8	41.4	41.9	42.8	49.4	49.4	49.4	50.1	48.8	48.4	49.8
Italy	12.6	13.2	14.1	14.2	14.6	14.5	14.7	14.9	16.0	15.7	15.8	15.5	15.2	14.9
Korea	—	—	—	—	—	17.8	16.9	16.5	14.2	12.6	11.8	11.1	10.8	10.0
The Netherlands	29.1	27.9	27.8	27.9	28.1	27.2	27.0	26.4	25.4	25.0	24.1	23.9	23.4	24.6
Singapore	5.6	5.8	3.2	2.5	2.5	2.7	2.7	0.1	2.3	2.2	2.2	1.7	1.6	1.5
Sweden	8.0	7.7	10.4	9.8	8.6	8.5	9.2	9.5	8.9	8.8	9.1	9.3	8.9	8.7
Switzerland	5.3	5.4	5.3	5.5	5.1	5.0	4.5	3.8	3.6	3.5	3.3	3.2	3.2	3.1
United Kingdom	19.4	19.4	19.8	19.7	19.7	19.6	19.8	19.9	20.2	19.8	19.5	18.7	18.5	17.9
United States	3.3	3.9	4.5	5.2	6.8	8.1	9.3	10.4	11.0	10.9	11.0	10.8	11.1	—

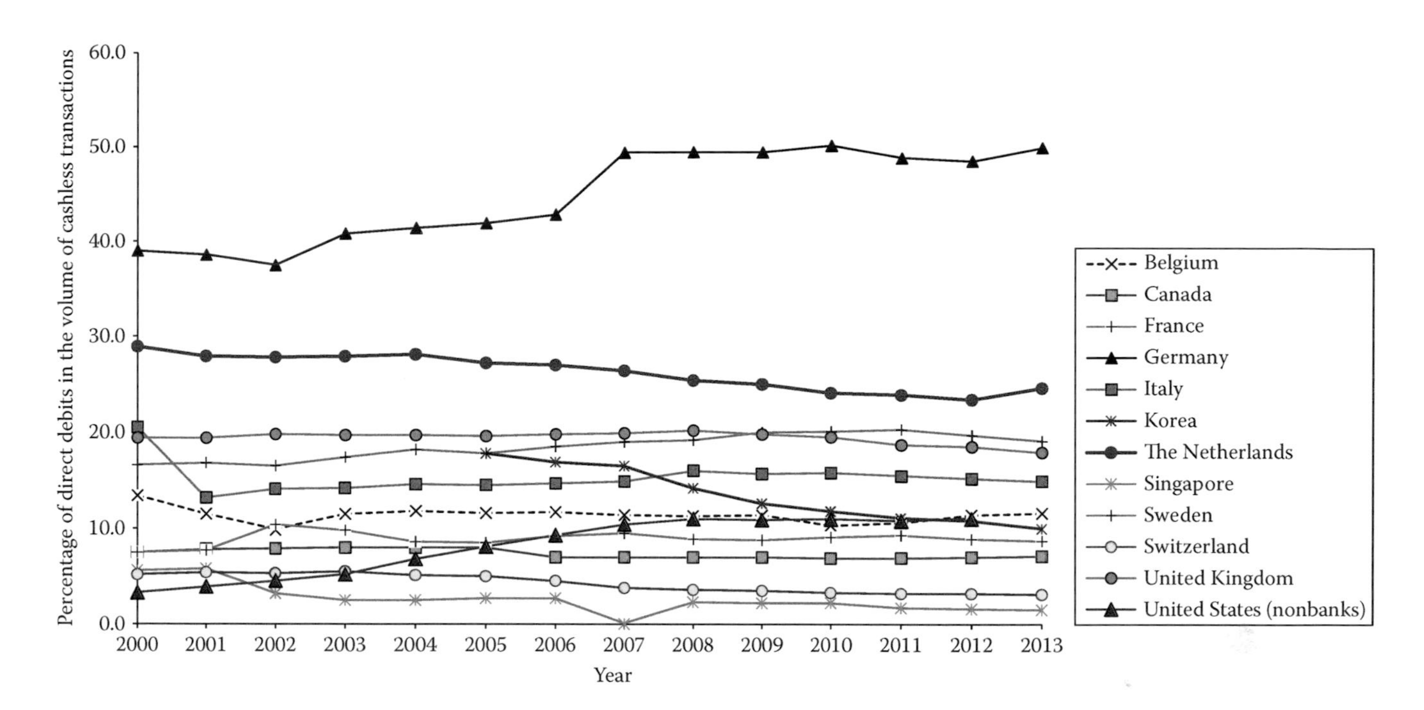

FIGURE 2.7
Evolution of the use of direct debits by nonbanks in selected countries (in volume).

TABLE 2.9

Percentage of Direct Debits in the Transfer of Value with Scriptural Transactions by Nonbanks in Selected Countries (2000–2013)

Year	2000	2001	2002	2003	2004	2005	2006	2007	2008	2009	2010	2011	2012	2013
Belgium	0.3	0.3	0.3	1.5	1.5	1.5	1.5	1.4	1.4	1.5	1.4	1.6	1.9	1.9
Canada	3.2	2.7	0.7	3.9	4.2	4.3	7.7	8.2	8.9	9.7	9.8	10.0	9.6	10.0
France	0.8	0.7	0.6	0.7	0.8	5.4	4.6	4.7	4.6	4.5	4.6	4.7	4.8	5.1
Germany	10.8	12.1	14.4	10.2	10.8	9.6	10.1	15.1	15.1	16.4	18.3	18.8	18.4	18.5
Italy	3.2	3.4	2.1	3.6	2.9	4.0	3.8	3.8	3.8	4.1	3.9	3.8	4.3	3.9
Korea	—	—	—	—	—	2.5	2.2	1.6	0.8	0.8	0.7	0.8	0.8	0.9
The Netherlands	5.3	5.0	4.9	5.3	5.3	4.2	4.2	4.3	4.8	4.7	4.7	5.2	5.3	6.0
Singapore	0.2	0.2	0.2	6.2	6.5	6.5	6.3	5.9	6.7	7.0	7.0	7.2	7.5	7.6
Sweden	2.7	2.9	2.7	3.9	2.7	3.9	3.9	3.8	3.8	3.9	3.9	3.9	3.6	3.5
Switzerland	0.2	0.2	0.1	2.0	2.3	2.2	2.0	1.5	1.5	1.5	1.6	1.6	1.6	1.6
United Kingdom	0.8	0.7	1.0	0.9	0.8	1.0	0.9	0.8	1.1	1.3	1.4	1.5	1.4	1.4
United States	15.7	16.3	17.7	17.6	16.7	17.3	17.6	19.3	20.0	20.7	21.4	22.5	24.1	—

2.2.7 Payment Cards

There are several categories of payment cards, depending on the type of the payment function, the functionalities supported and their use in open or closed systems. These are as follows (Slawsky and Zafar, 2005, pp. 4–6):

- Check guarantee cards. This type is now deprecated.
- Cash withdrawal (cash advance) cards, for use at automated teller machines (ATMs). A PIN is needed for cash withdrawals from automated tellers.
- Payment cards, which can take several forms:
  - Debit cards are typically linked to a bank account, combining the functions of cash withdrawal and check guarantee. Debit cards are further subdivided as follows:
    - Immediate debit cards, in which case the amount of the transaction is withdrawn from the debtor's account immediately.
    - Deferred debit cards, in which case the debtor's account is debited at a fixed date of the monthly payment cycle such as

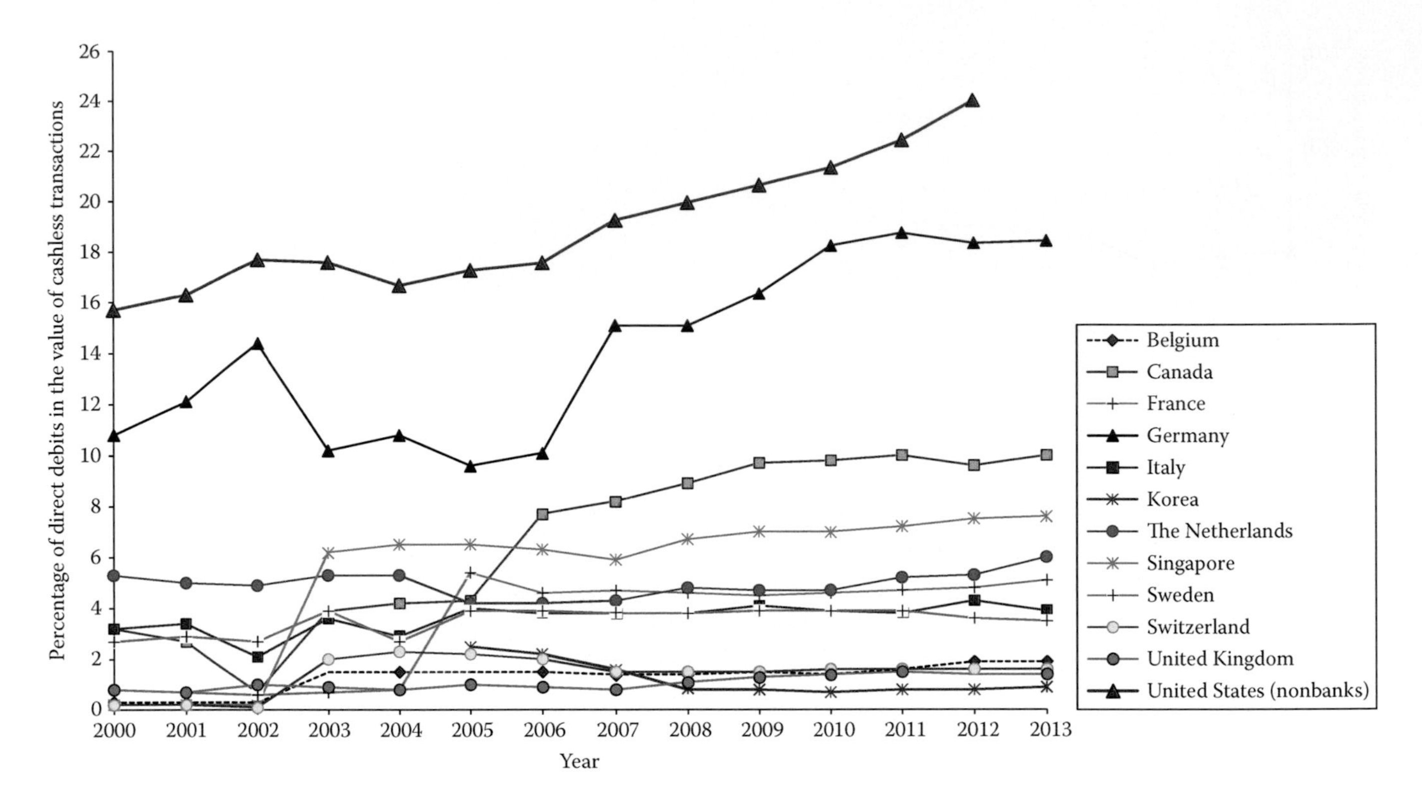

FIGURE 2.8
Evolution of the use of direct debits by nonbanks in the transfer of value in selected countries.

the end of the month, for the total purchases since the last statement through a direct debit. The amounts are accumulated up to an agreed ceiling.

– Prepaid debit cards such as NetSpend are not necessarily tied to a bank account. The monetary value is added through the direct deposits by reloading in Reload centers or through PayPal®. As mentioned earlier, the U.S. Treasury Department is issuing the Direct Express Debit MasterCard card as a prepaid debit card for federal benefits to the unbanked. They are prepaid in the sense that it is the federal government that loads the value and not the individual owner, but they function like any other debit card.

- Credit cards, which offer a revolving credit facility. The holder, upon receipt of the monthly statement, can settle the full amount or pays only a part (subject to a minimum amount), in which case the issuing institution will charge interest on the outstanding balance.
- Charge cards such as American Express or Diner's Card. They are nonrevolving credit cards because the cardholder has to pay all the charges incurred in full at the end of the billing period.
- Prepaid or stored-value cards; they can be disposable, that is, for a set amount fixed at purchase or reloadable. These cards can also be associated with electronic purses. Disposable prepaid cards are not typically PIN protected and, in that case, are vulnerable to loss and/or theft, while rechargeable cards typically require authentication of the cardholder when reloading or making payments. Prepaid cards dedicated to a single application (telecommunications, public transport, or for a specific merchant) are private label cards denoted also as closed-loop cards.
- Cards for corporate use are as follows:
    - Corporate cards for the expenses incurred by the employees during the course of their work-related activities.
    - Purchasing or procurement cards, which are deferred debit cards to cover the payments for nonrecurrent charges and small amounts.

They are linked to an enterprise account allowing the cardholder, as a representative of the enterprise, to make purchases. The service that the card scheme offers includes the generation of management reports and accounting and fiscal reports.

- Affinity cards are cards linked to charities that collect a percentage of the purchases.
- Private label cards issued by merchants, retail chains, transport companies or businesses for use within their own system of stores, outlets to pay for goods or services. They can be prepaid cards or, with the backing of a credit institution, a payment card. Often, the data collected help the provider to understand customer behavior to improve marketing and sales campaigns. Gift cards are often tied to merchant loyalty programs and are dealt by specialized payment processors with product offerings to meet the needs of the particular merchant.
- General-purpose prepaid debit cards, such as Green Dot or Net Speed, are offered by a non-depository institution with the cooperation of a depository institution to sponsor access to the financial network.
- A specific card type is being used in the United States, denoted as electronic benefit transfer (EBT) card to disburse benefits to specific individuals but is used for a defined set of purchases at selected stores.

In a typical card payment, merchants query authorization servers of the card networks before approving the expense. The authorization servers filter the query according to specific criteria to detect abusive or unauthorized transactions. The filtering process utilizes preestablished criteria, for example, whether a spending ceiling has been reached, or if a large number of transactions took place in a specific interval. After the delivery of the purchased goods or services, the merchant requests the acquirer bank to recover its dues; this request is then forwarded to the issuer bank. Both authorization and settlement requests can be combined in the case of virtual goods delivered over the Internet. The grouping of transactions is also possible. Similarly, in the case of reimbursement, either because of a returned or defective product, the merchant gives its bank instructions to credit the client's account. Finally, the microprocessor cards (smart cards) reduce the possibility of counterfeit as shown in Chapter 9.

Thus, payment card purchases involve in addition to the buyer and the seller, the banks of each of the parties and the credit card scheme. The merchant's bank is called the acquirer bank because it acquires the credits, and the buyer's bank is called the issuer bank, since it has issued the cards to members that it has authenticated. The merchant bank (acquirer) generally pays interchange fees to the issuer (the cardholder's bank) separately from any fee that the merchant and/or the buyer pay to the card scheme. Interchange fees create an incentive for issuers to encourage cardholders to use their card, for example, through reward points. When the transaction is cleared, settlements are made among the banks by using national and international circuits for interbank exchanges. Figure 2.9 depicts the interactions among these four entities: the issuer, the acquirer, the client, and the merchant. This arrangement is called a "four-party system."

Open-loop (i.e., general-purpose) prepaid cards may use established schemes such as Visa or MasterCard for disbursement. In the United States, they ride on the same ACH infrastructure for the clearing and settlement of debit cards.

General-purpose prepaid cards that do not go through this network provide less security. For example, a Green Dot card is associated with the so-called MoneyPak account. A 14- or 16-digit serial number on the back of a Green Dot card is the password to withdraw value from that account. This serial number is exposed after scratching the back of the card. The value in the MoneyPak account can then be used to load prepaid debit cards, make deposits into PayPal accounts, and pay participating merchants in anonymous transactions. Merchants typically limit how much value can be loaded into a single Green Dot card, so several purchases must be made from several outlets to accumulate important value. However, the MoneyPak accounts are not insured because they are not bank accounts.

Scammers have used this feature to target homes and businesses by impersonating as Internal Revenue Service agents and convincing the victim that they owe back taxes. Following the threat of a fine, arrest, or even deportation, the victim is instructed to buy sufficient Green Dot cards corresponding to the amount allegedly owed and then hand over the cards' serial numbers. In a more elaborate scheme, the caller phone number that appears on the victim's phone would be that of a law enforcement agency to convince the victim of an impending arrest unless they pay a fine with Green Dot cards. In response to all these reports, Green Dot announced in December 2014 that they will be pulling their cards from the market.

Not shown in Figure 2.9 are card processors, which play the role of a trusted third party and a notary. They perform some initial screening of the transaction on behalf of the merchant before querying the authorization server. Based on the response, they perform some

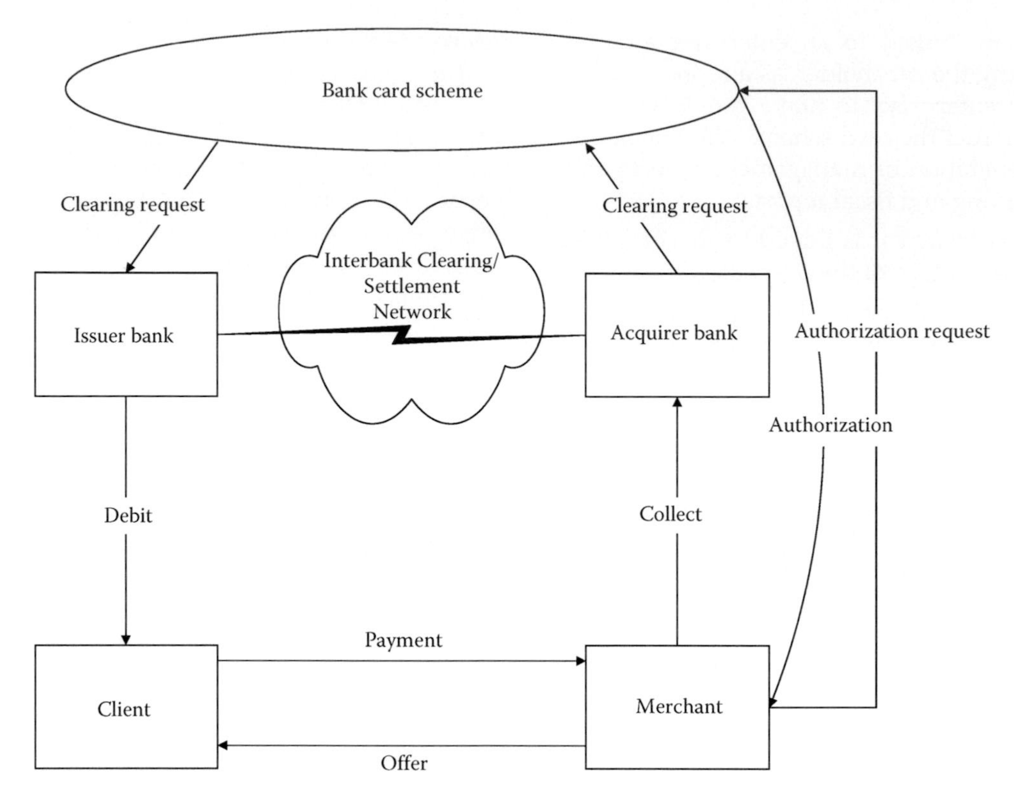

FIGURE 2.9
Message exchanges during a card transaction.

additional verification and then relay the information back to the merchant to complete the transaction. At the end of each day, processors reconcile the transactions to initiate the settlements among the banks involved.

In the 2010s, a new service entity, the token service provider, appeared in the processing of payment by cards. The role of the token service provider is to issue payment tokens that act as surrogate for a payment card in online or wireless transactions. The token can be restricted to a specific communication technology, or a merchant, or a mobile application (mobile wallet) (EMV, 2014).

Until the full introduction of chip cards using Europay, MasterCard, and Visa (EMV), debit cards in the United States will continue to be of two kinds: signature debit cards and personal identification number (PIN)-authenticated debit cards. With signature debit cards, cardholder authentication relies on the visual comparison of the signature at the point of sale with the signature on the back of the card, which gives a weak protection against counterfeiting or fraud (as explained in Chapter 8). As the name implies, the user must enter a PIN to authorize the transaction with PIN-authenticated debit cards.

Signature debit transactions are routed through the Visa or MasterCard networks, whereas PIN-authenticated debit transactions may also be routed over ATM networks (also known as debit networks) such as Interlink, STAR, Maestro, or New York Currency Exchange (NYCE) (Coye Benson and Loftesness, 2010, pp. 67–69). PIN debit payment networks require PIN encryption in the authorization request but have a lower transaction cost than the typical signature/credit network.

In the United States, the rules for allocating the liability for credit card fraud in face-to-face contact purchases and online purchases are different. In the first case, it is the issuer who is liable for the fraud, while in the case of fraudulent online purchases, the card issuer charges back the amount to the acquirer bank (the merchant's banks). Moreover, under the Fair Credit Billing Act (FCBA) of 1974, the maximum liability for unauthorized use of a credit card is $50. Moreover, if the credit card holder reports a theft or a loss of a card before it is used, they are not responsible for any unauthorized charges. If a credit card number is stolen, but not the card, the cardholder is also not liable for unauthorized use. This regulation does not apply for debit card transactions.

Table 2.10 provides the proportion (in volume) of scriptural payments made by payment cards in selected countries. The data are depicted in graphical form in Figure 2.10.

The proportion of transactions with payment cards is increasing in all countries except Singapore, where it is

TABLE 2.10

Percentage in the Volume of Payments by Nonbanks of Card Transactions in Selected Countries (2000–2013)

Year	2000	2001	2002	2003	2004	2005	2006	2007	2008	2009	2010	2011	2012	2013
Belgium	35.8	32.6	34.6	36.8	37.4	39.0	40.3	41.0	42.5	43.6	44.7	46.1	48.9	49.3
Canada	54.8	56.7	56.7	60.7	62.2	63.9	69.1	69.8	71.3	72.2	72.8	73.6	75.5	76.7
France	27.8	30.0	31.4	32.0	33.4	37.2	37.9	40.0	41.5	42.4	43.6	45.4	47.2	49.8
Germany	11.8	13.2	15.3	15.0	15.3	14.9	14.2	13.8	14.4	14.8	15.5	16.6	17.5	18.4
Italy	32.8	27.3	29.1	33.7	35.6	37.5	37.8	39.0	40.2	39.9	40.0	40.5	41.1	43.2
Korea	—	—	—	—	41.7	42.7	44.4	49.0	54.5	58.7	62.0	64.9	67.4	41.7
The Netherlands	29.3	31.9	32.7	34.7	36.5	36.7	38.0	38.9	40.7	42.4	43.6	45.1	51.8	36.5
Singapore	27.7	29.6	11.2	6.5	7.2	7.7	8.5	8.2	8.5	7.9	6.8	6.9	13.5	7.2
Sweden	26.7	31.7	51.4	56.1	56.5	55.8	61.0	64.0	64.7	65.1	63.9	65.4	66.5	56.5
Switzerland	29.7	34.1	33.5	36.3	36.5	37.1	38.2	39.6	40.7	41.3	43.0	44.8	45.7	36.5
United Kingdom	36.6	40.1	41.2	44.8	45.1	46.6	48.4	49.8	51.5	53.2	55.6	57.0	58.9	45.1
United States	33.5	36.6	42.5	45.9	45.9	48.6	51.7	54.7	56.8	58.5	60.9	64.4	66.3	—

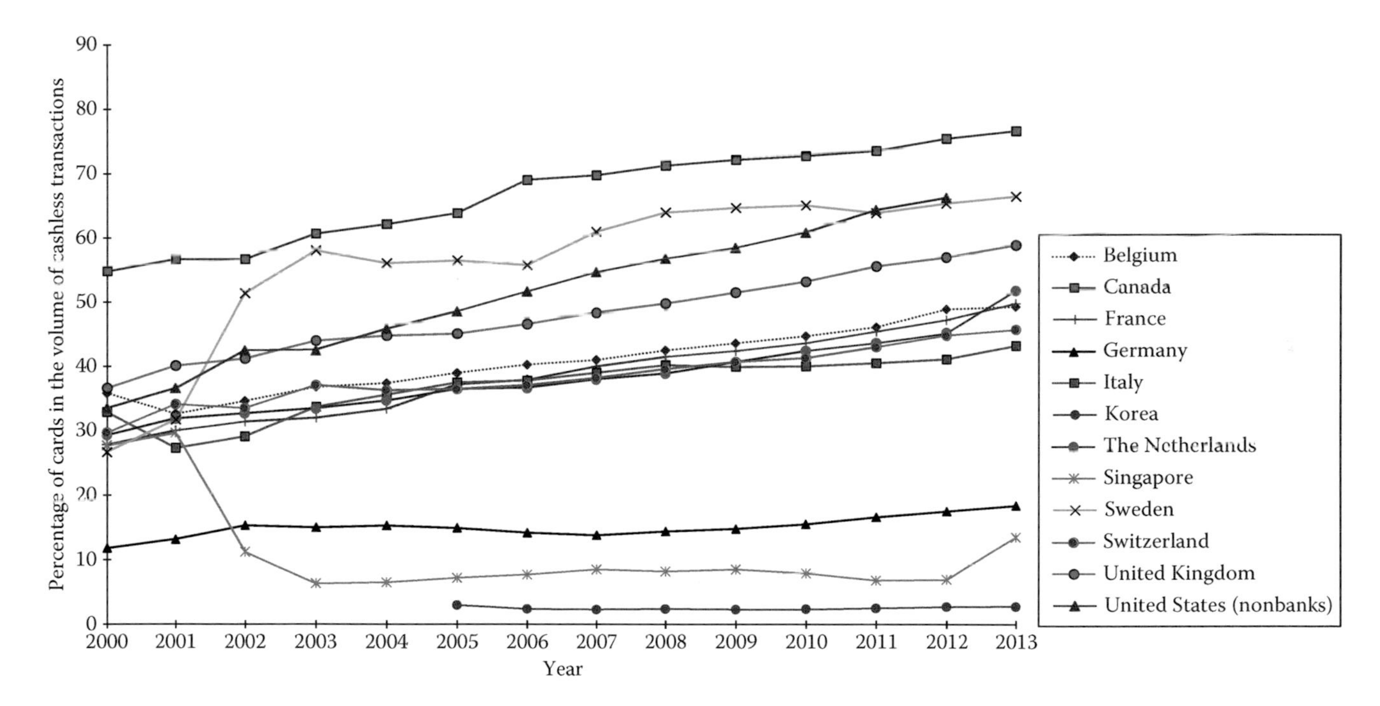

FIGURE 2.10
Evolution of the percentage of payment with cards by nonbanks in selected countries (in volume).

decreasing due to the rapid development of electronic purses, as will be discussed later in the book. The increase can be explained by the diffusion of cards in the population as well as a good geographic coverage of ATMs. The greatest percentage is found in Canada, where card transactions constitute more than ¾ of the total volume of transactions (76.7%), the majority (56.1%) by credit cards. The United States is next with 64.2% of the volume of transactions by nonbanks, with 56.2% by credit cards. The lowest usage is in Singapore and Germany, respectively, at 7.2% and 18.4%. In the remaining countries of the sample, the percentage in volume of scriptural payments made by cards is between 30% and 60%.

Table 2.11 gives the proportion of value exchanged by payment cards in the same countries from 2000 to 2013. The data are depicted in graphical form in Figure 2.11. These results confirm that card payment is actually a mass instrument for small amounts (less than 10% of the total value exchanged by all scriptural instruments). In the United States, for example, payment cards represent 67% of the volume of all cashless transactions among nonbanks in the United States and about 4% in value.

TABLE 2.11

Percentage in the Value of Payments by Nonbanks of Card Transactions in Selected Countries (2000–2013)

Year	2000	2001	2002	2003	2004	2005	2006	2007	2008	2009	2010	2011	2012	2013
Belgium	0.2	0.2	0.2	1.1	1.1	1.1	1.2	1.1	1.2	1.3	1.5	1.6	1.7	1.7
Canada	2.1	2.2	2.3	3.1	3.3	3.5	7.5	7.7	8.3	8.8	8.8	8.9	9.2	9.4
France	0.2	0.2	0.2	0.2	0.2	1.5	1.3	1.4	1.4	1.4	1.5	1.4	1.5	1.7
Germany	0.4	0.4	0.4	0.4	0.5	0.5	0.5	0.2	0.2	0.2	0.3	0.3	0.3	0.3
Italy	0.8	1.0	1.2	1.2	1.3	1.4	1.4	1.3	1.3	1.3	1.3	1.3	1.4	1.4
Korea	—	—	—	—	—	3.0	2.4	2.3	2.4	2.3	2.3	2.5	2.7	2.8
The Netherlands	1.4	1.4	1.5	1.5	1.3	1.2	1.2	1.3	1.4	1.5	1.5	1.7	1.8	2.0
Singapore	0.2	0.2	0.2	4.0	4.1	4.5	4.7	4.4	5.2	5.7	5.7	6.1	6.4	6.6
Sweden	2.2	2.9	5.3	5.1	6.1	5.9	5.5	5.5	6.0	6.3	6.1	5.7	5.6	5.8
Switzerland	0.1	0.1	0.1	1.6	1.6	1.8	1.8	1.9	1.9	2.0	2.1	2.0	2.1	2.1
United Kingdom	0.2	0.3	0.3	0.3	0.4	0.4	0.4	0.4	0.5	0.6	0.7	0.7	0.7	0.7
United States	2.6	2.7	2.9	3.1	3.4	3.6	3.9	4.4	4.7	4.8	5.0	5.5	5.7	—

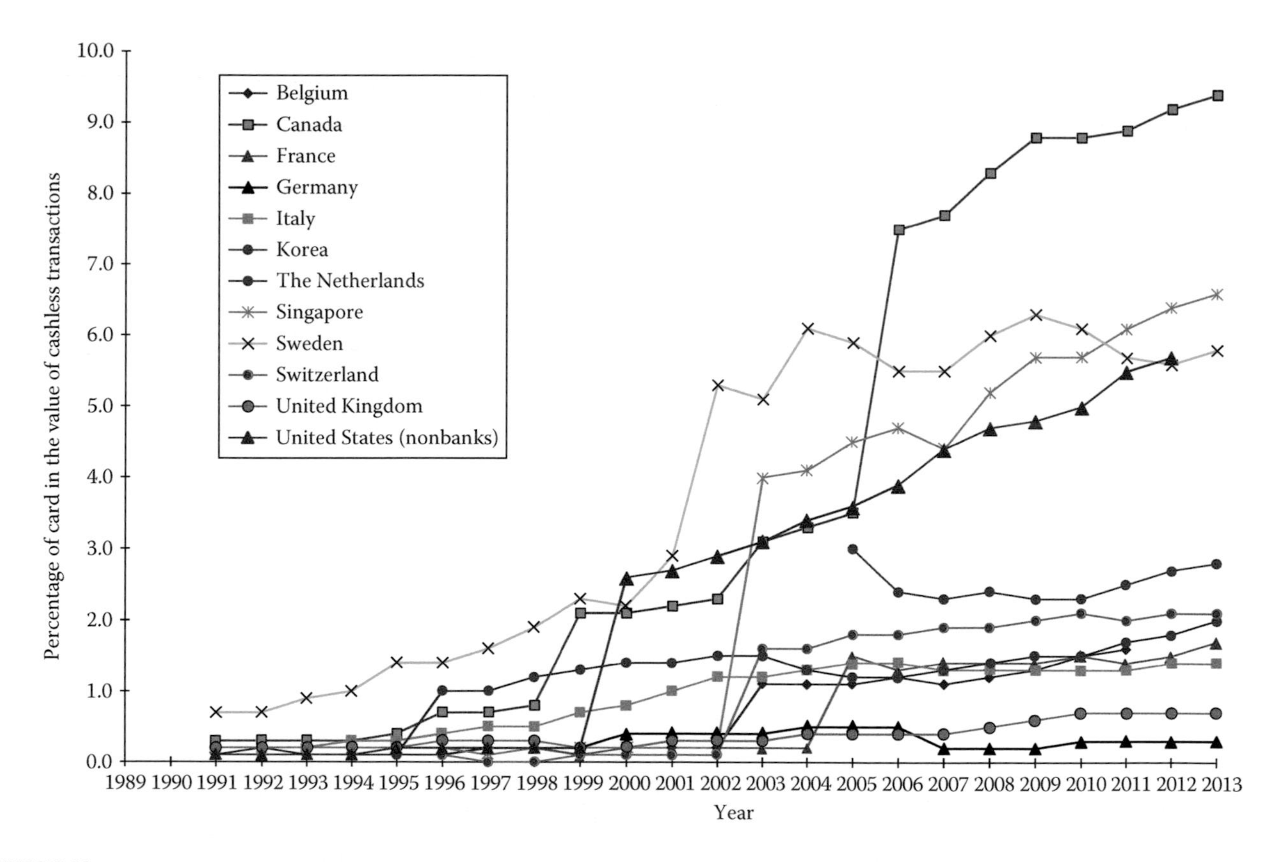

FIGURE 2.11
Evolution of the percentage of payment with cards by nonbanks in selected countries (in volume).

This explains why electronic purses have the potential of displacing them.

For a long time, debit cards were more common than credit cards in many European countries. In the late 1990s, U.S. banks started to push debit cards to their customers when they realized the potential for high profit margins. Today, transactions with debit cards form a substantial percentage (44%) of card transactions among nonbanks in the United States. However, prepaid debit cards do not earn interest on the stored amount. Also, 62% of the volume of transaction in the United States is from signature debit cards (Federal Reserve System, 2011, p. 59). As explained in Chapter 8, these cards can be easily cloned while the legislation does not protect debit cardholders from fraudulent use. Decoupled debit cards are another controversial mutant that Capital

One introduced in 2007 (before the financial crisis). Here, cardholders do not have an account with the issuer bank, so the transaction is authorized before verifying the availability of funds. The assumption is that the revenues from the interchange fees that the acquirer pays to the issuer are significantly higher than the risks of nonavailability of funds as well as fraud (Coye Benson and Loftesness, 2010, p. 58).

The growth of the volume of noncash payments in the U.S. population is illustrated in Table 2.12 from the Federal Reserve (2013, 2014).

The Federal Reserve study indicates that the decline in the volume of checks is most significant for values of $500 or less and that checks written by individuals or to individuals declined must faster than business-to-business checks. In fact, the average value of consumer checks increased from $1103 in 2003 to $1420 in 2012, indicating that they are less used for low-value point-of sale transactions (Federal Reserve, 2013, p. 9). Checks are also used among banks for the transfer of large amounts. Figure 2.12 depicts the shift to credit transfers and direct debits for nonbank payments. It is not possible to estimate accurate transaction values however, because a detailed study that the Federal Reserve conducted in 2012 showed that some of the previously reported ACH values are not valid and the graphs are discontinuous around 2011, when the error was detected, which is shown a sudden break in the time

TABLE 2.12

Evolution of the Volume of Selected Noncash Instruments in the United States (2000–2012)

	Volume in Billions of Transactions				
Instrument	**2000**	**2003**	**2006**	**2009**	**2012**
Checks (paid)	4.1				
General-purpose cards	20.6	30.8	44.3	58.4	73.9
Credit	12.3	15.2	19.0	19.5	23.8
Debit	8.3	15.6	25.0	37.5	47.0
Prepaid	—	—	0.3	1.3	3.1
Private label cards	3.3	3.8	4.6	5.1	6.0
Credit	3.3	3.8	2.7	1.5	2.4
Prepaid	—	—	1.9	3.6	3.6
EBT cards	0.5	0.8	1.1	2.0	2.5
ACH (credit transfers and direct debits)	0.5	0.8	1.1	2.0	2.5
Total	72.4	81.4	95.2	108.1	122.4

Sources: Federal Reserve, The 2013 Federal Reserve Payments Study, Recent and long-term trends in the United States: 2003–2012, Summary report and initial data release, December 19, 2013, available at https://www.frbservices.org/files/communications/pdf/research/2013_payments_study_summary.pdf, last accessed August 16, 2015; Federal Reserve, The 2013 Federal Reserve Payments Study, Recent and long-term trends in the United States: 2000–2012, Detailed report and updated data release, July 2014, available at https://www.frbservices.org/files/communications/pdf/general/2013_fed_res_paymt_study_detailed_rpt.pdf, last accessed August 16, 2015.

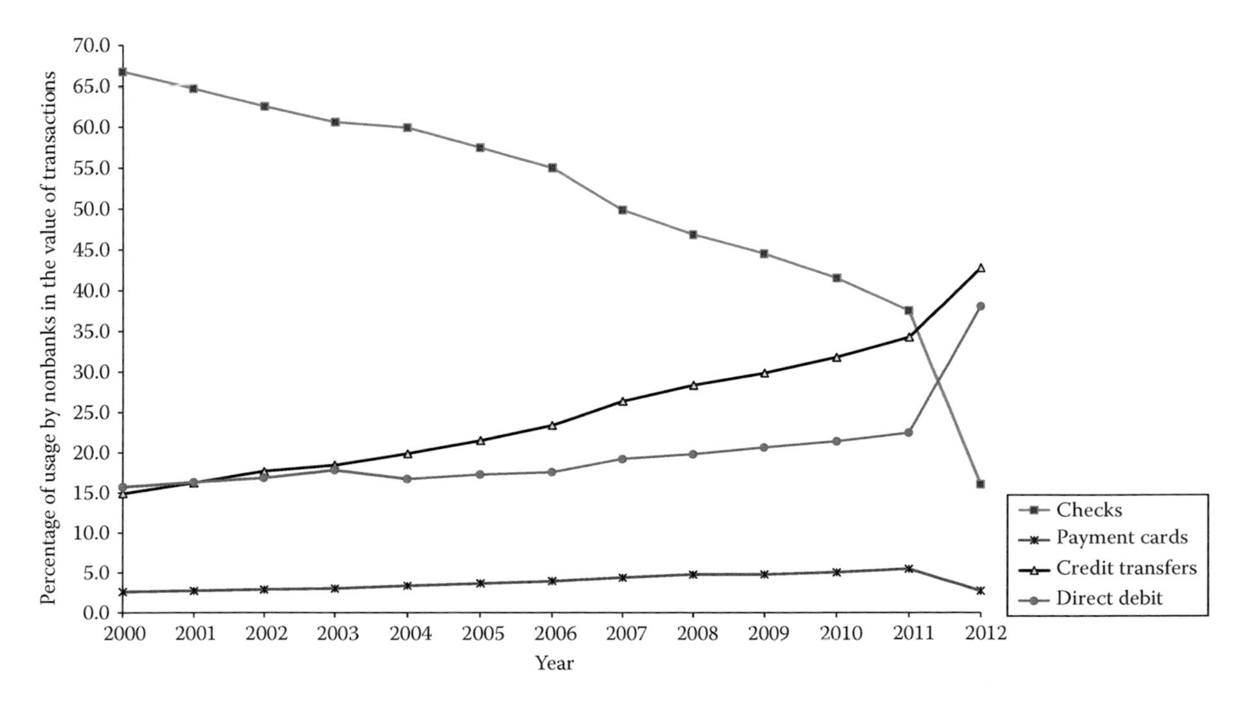

FIGURE 2.12
The shift from checks to credit transfers and direct debit for the transfer of value among nonbanks in the United States between 2000 and 2013.

series (Bank for International Settlements, 2014, p. 426, note 2; Federal Reserve System, 2014, pp. 72–73). As a consequence, the trend can be tracked accurately for the volume of transactions and not for their value.

It should be noted here that there are differences in the data collection methodologies of the Federal Reserve System and the Bank for International Settlements in relation to card data. For consistency across countries, BIS does not distinguish between general-purpose cards and private label cards and does not consider prepaid cards, including EBT cards, as a separate category. Finally, BIS uses the tri-annual studies from the Federal Reserve as the basis to interpolate the missing data points of the time series.

2.3 Types of Dematerialized Monies

With the increased use of prepaid cards, such as telephone cards in the 1980s (Martres and Sabatier, 1987, pp. 85–87) in France, three forms of dematerialized currencies have been promoted: electronic money, virtual money, and digital money. In general, the use of dematerialize money appears in the following areas (Bank for International Settlements, 2012b): (1) new forms of card payments, (2) Internet payments, (3) mobile payments, and (4) electronic bill presentment and payments.

It should also be noted that the definition of electronic money varies across countries, which in turn affect the reported statistics on its usage (Bank for International Settlements, 2012b, p. 5).

2.3.1 Electronic Money

The Bank for International Settlements defines electronic money "as a monetary value measured in fiduciary units that is stored in an electronic device owned or available to the consumer" (Bank for International Settlements, 1996, p. 13). This definition corresponds to money stored in a smart card or on the server of a payment intermediary. Its scriptural character is related to the status of the issuer (since it is not issued by the central bank) and to the traceability of the transactions and the movement of money.

The units of payment contained in the cards or in the software are bought either with fiduciary money or by charging to a bank account. The discharging power of these units is restricted to those merchants who accept them. This is the reason certain experts consider that electronic money does not exist in a strict sense, because it is neither legal tender nor does it have discharging power (Fay, 1997, p. 113).

In most countries, only banks are allowed to pay interest. Therefore, if the payment intermediary is a nonbank, the user will not earn interest.

2.3.2 Virtual Money

Virtual money differs from electronic money in that its support, its representation, and its mode of payment do not take tangible forms. Virtual money can be contained in software programs that allow payments to be carried out on open networks, such as the Internet, using cloud-based virtual purses dedicated to the transactions with a given service provider.

Starting with the definition of the BIS for electronic money, one can consider virtual money as a referent (or a pointer) to a bank account. The scriptural character of the virtual money is also tied to the status of the issuer (it is not issued by the central bank) and to the traceability of the transactions.

In the limiting case, virtual money may also be a virtual token (or *Jeton*) issued by a trusted issuer for a unique usage in a closed circuit.* The notion of *openness* versus *closure* relates to the final utilization of the means of payment to make purchases without any *a priori* restrictions and independently of the issuer. The purchasing power of virtual money is described with these Jetons, which are recognized only in a restricted commercial circuit for the exchange of goods and services. Electronic money, in contrast, is a multipurpose payment mechanism that is recognized in general commercial circuits because it is the electronic version of legal tenders. Restaurant or manufacturer coupons or airline miles and the virtual currencies used for Internet games such as the QQ coins issued by Tencent or Facebook Credits are representative of closed virtual currency schemes. In these schemes, there is no fixed relation between the Jeton and the legal currency; the service provider, for example, can arbitrarily change the rate. For telecommunication, transport, or Internet gaming, the flow is from legal currencies to the virtual currency (Jeton). If the flow is bidirectional, such as in the case of credit card reward points, there are two exchange rates: a buy rate and a sell rate. Virtual money systems may also be a route for trading fiat currencies, for example, by buying with dollars the Jetons for *World of Warcraft* and then selling them for yens.

* The dictionary (*Webster's New Collegiate Dictionary*, 1975) definition of token is "a piece resembling a coin issued as money by some person or a body other than a de jure government." Despite this clear statement, there has been a tendency to mix legal coins with tokens (see, e.g., Camp et al., 1995). To avoid the potential confusion, this book will use the French word *Jeton* to mean a coin issued by a nongovernmental body.

The experience with telephone cards showed that a *nonbank* can effectively help in collecting amounts that are individually marginal by attaching them to the telephone bill. The use of phone ticks for micropayments was even considered within the European project Conditional Access for Europe (CAFE) that ran from 1992 to 1996 (Pedersen, 1997). The basic idea was to use the agreements among telephone companies to handle and settle small payments and construct a virtual currency between suppliers and consumers, even when the parties are not using the same legal tender. PayPal allowed its subscribers to translate their Jetons stored with PayPal into debit transactions to fund their purchases over the Internet.

Governments and judiciaries everywhere are facing the issue of how to regulate online economies, because the entrance of huge sums of virtual money from Internet games into the real economy can threaten the financial stability of the legal monetary system. QQ coins, for example, were accepted as payment and sold for legal tender until the Chinese government issued a formal notice to curb this practice in 2007 (Dickie, 2007). In 2009, the Chinese Ministry of Culture and Ministry of Commerce issued a rule banning the exchange of virtual Jetons for real goods and services (Peck, 2012).

2.3.3 Digital Money

Digital money is an ambitious project not only to solve the problem of online payment but also to replace the central authority for issuing legal tenders. Like regular money, each piece has a unique serial number. However, the support for this money is *virtual*, the value being stored in the form of algorithms in the memory of the user's computer, on a hard disk, in a smart card, or of a smartphone. It is worth noting that the U.S. Department of Treasury's definition of "virtual currency" includes both digital money as well as Jetons in electronic form: "a medium of exchange that operates like a currency in some environments (...but) does not have legal tender status in real currency" (Financial Crimes Enforcement Network, 2013).

Digital money can be centralized or decentralized. The first kind, for example, DigiCash, has a centralized repository managed by a central administrator. The second kind are decentralized currencies that have neither a central repository nor a central administration. They are also known as cryptocurrencies because they use advanced cryptographic algorithms to prevent fraudulent transactions. An example of such currencies is bitcoin.

Promoters of digital money would like it to have a real discharging power so that the economic agents in as large an area as possible would be able to accept it in return for payments. One of the characteristics of digital money compared with other electronic payment systems is the possibility of making the transactions completely anonymous, that is, of dissociating the instrument of payment from the identity of the holder, just as in the case of fiduciary money.

Another salient characteristic of the digital money is that it is minted by a client software on a network node using computing resources available to individuals or groups of individuals. In the DigiCash scheme, the minted digital coin is also sealed by the bank. The creditor that receives the digital money in exchange of a product or a service verifies the authenticity using the public key of the issuer bank. Anonymity is thus guaranteed, but it is not easy to transfer the value between two individuals without the intervention of the bank of the issuer. Furthermore, as each algorithmic step is associated with a fixed value, the problem of change causes some complications. The main actors are users, exchanges, trading platforms, inventors, and wallet developers. The Bitcoin scheme, in contrast, removes the role of the bank totally and is totally peer to peer.

One of the destabilizing aspects of this digital money is that it could lead to forming new universal monies independent of the current monetary system. In fact, more than 200 different cryptocurrency schemes were said to be in operation by mid-2014. Furthermore, a digital currency that is indeed international would collide with the various regional and local currencies and would disturb the existing economies. The question is no longer exclusively technological and touches upon the aspects of national sovereignty and foreign intervention. The economic and political stakes of such a proposition are enormous and may lurk behind the screen of juridical disputes.

2.4 Purses, Holders, and Wallets

2.4.1 Electronic Purses and Electronic Token (Jeton) Holders

According to the BIS, an *electronic purse* is "a reloadable multipurpose prepaid card that can be used for retail or other face-to-face payments." This means of payment can substitute, if the holder wishes, for other forms of monies. Electronic purses are thus portable electronic registers for the funds that the cardholder possesses stored in a physical card. These registers contain a precharged value that can serve as an instrument of exchange in open monetary circuits. The protection that affords the stored value of money is based on the difficulty (if not the impossibility) of fabricating a fake card or manipulating the registers.

Where an electronic purse is used depends on the identity of the issuer (merchant, bank, merchants association, etc.) and its prerogatives under the law. Banking networks are by definition open wherever the electronic money corresponds to a legal currency. In contrast, a purse from a nonbank is restricted because it can only contain Jetons and, in principle, can only be used in closed circles and for predefined transactions involving the issuer.

Electronic purses are attractive to banks because they permit a reduction in the transaction cost and can replace coins, notes, or checks for small amounts. They can be considered a cybernetic form of the traveler checks that were first introduced by American Express in 1890.

Electronic purses and electronic Jeton holders have already proved their economic effectiveness in proximity commerce and payment through automatic machines. Recently, they have been extended for debits authorized or initiated over wireless networks. They have an advantage in micropayments because the transaction cost of other payment instruments may exceed the amounts involved. It is possible, however, to combine electronic purses and Jeton holders in a multiapplication card. A merchant may be associated with a bank to issue a fidelity card while offering credit facilities (as managed by the bank).

A common Jeton holder is found in the form of service cards for communication or transport services, or stored in a mobile/smartphone where the units of payment give the right to access the associated service. Some of the cards used in transportation systems were expanded to retail payments. In particular, the Octopus card, a single-purpose transport card owned by major transport operators of Hong Kong, obtained in 1997 the authorization to expand to retail payments, thereby acting as a new intermediary. In 2010, nontransport-related payments represented about 40% of the value of all transactions (Adams, 2003; Bank for International Settlements, 2012b, p. 32).

Table 2.13 summarizes the financial and legal differences between electronic purses and electronic Jeton holders.

TABLE 2.13

Comparison between Electronic Purses and Electronic Jeton Holders

Characteristic	Electronic Purse	Electronic Jeton Holder
Expression of purchasing power	Legal tender	Consumption unit
Unit of payment	Universal: can settle any payment in a defined territory	Specific to transactions involving the issuer
Guarantor of purchasing power	Bank	Service provider
Charging of value	By a bank or its agent	Unregulated
Circuit of financial services	Open	Closed

2.4.2 Virtual Purses and Virtual Jeton Holders

A *virtual purse* is an account precharged with units of legal money and stored in the collection system of a nonbank (Remery, 1989; Bresse et al., 1997, p. 26; Sabatier, 1997, p. 94). Online access to this virtual purse is achieved with a software installed in the personal computer of the client or with a mobile application of a smartphone. In other words, virtual purses constitute a cloud-based payment system.

In a cloud-based payment system, the service providers open several subaccounts under their own bank accounts. Each subaccount is reserved for a specific subscriber to their cloud-based service, whether buyers or merchants. The client's subaccount is called a virtual purse, while the merchant subaccount is denoted as the virtual cash register. The client's purchasing power is indicated in the virtual purse and refers to the subaccount under the operator's account. The purse is called "virtual," because the value stored is not physically touchable, yet the units of payment correspond to legal tenders. What clients have on the hard disk of their personal computer or their mobile terminal is a copy of the balance of this subaccount. In addition, the terminal (computer or smartphone) may contain various files for the cryptographic security of the operation.

Each purchase debits the client's virtual purse and credits the merchant's virtual purse with the amount of the transaction minus the operator's commission. At predefined intervals, the operator makes an overall payment to each merchant corresponding to the amounts that have accumulated in their respective virtual cash registers.

Cloud-based services provide a way to overcome the processing limitations of mobile telephone terminals and protect clients' assets should their terminals fail or are lost. They are also suited to micropayments, because the grouping of payments before initiating the compensation makes this approach economical. One example is ExxonMobil's SpeedPass, a payment system linked to a payment card account and dedicated to gasoline purchases from ExxonMobil gas stations. Card numbers or other personal information are not stored on the SpeedPass device, so the user's information is protected from unauthorized access.

One limitation of cloud-based purses is that network availability and reliability can be a cause of concern, because the connection can be disrupted after a transaction has started and before it is completed.

WebMoney is another global online funds transfer and payment system established in 1998 and based in Moscow, Russia. WebMoney transactions require the installation of stand-alone application or can be initiated through the WebMoney website. After registration, the virtual purse is referred to as a WebMoney purse and is funded by wire transfer. WebMoney transactions cannot be reversed or canceled and can only be used for online purchases.

2.4.3 Digital Wallets

A digital wallet is a software that provides access to the payment infrastructure from a desktop, a mobile terminal, or a network server. It mimics the various functions of physical wallets, that is, to store various payment cards, merchant coupons, loyalty cards, receipts, and so on, including electronic or virtual purses. One component of the wallet is an authentication mechanism for users.

A mobile wallet is a special stand-alone application (*app*) on a mobile terminal to store, access, manage, and use identification and payment instruments in a secure way (European Payments Council, 2012, p. 86). It implements complex transactions and access multiple websites or payment instruments, such as credit and debit cards, loyalty accounts, merchant accounts, gift cards, coupons, and electronic money. The user scrolls through the instruments that the wallet contains to select the one to use when making the payment at the point of sale. Alternatively, the wallet may be linked through an account on a server, a solution that avoids storing the card information in the secure element of the mobile phone. Another approach to avoid storing the card number in the terminal is through tokenization.

An electronic purse is one of the payment applications in a wallet. In a mobile wallet, user authentication is performed either in the secure area of the mobile terminal or on a remote secure payment server (i.e., in the *cloud*). In this case, the service is under the control of the payment service provider (PSP) or a trusted service manager (TSM), and the mobile terminal is the access method to the payment server. In addition, the communication channel between the terminal and the remote server must be secure and the transaction protocol must be able to recover should the mobile link fade before the transaction is completed.

There are many mobile wallets (Google Wallet, the Softcard [formerly known as ISIS]—a mobile wallet provided by a joint venture of several U.S. mobile network operators—AT&T, T-Mobile and Verizon Wireless—Apple Passbook, LevelUp, GoPago, QuickTap, Osaifu-Keitai, etc.). The differences include aspects such as the applet implementation and whether the account information is stored in the mobile device or in a secure server.

2.4.4 Diffusion of Electronic Purses

Table 2.14 gives the portion of the volume of scriptural payments using electronic money in several countries. These data for countries where the volume exceed 3% are depicted in the graphs of Figure 2.13. Singapore is distinct from all other countries as being the first where payments from electronic purses form an important part of the volume of transactions (about the third). Italy is in a distant second place where the proportion of transactions using electronic purses does not even reach 4% of the total volume.

The growth of the electronic purse in Singapore is the fruit of a planned effort to replace coins with the CashCard, a reloadable contactless electronic purse introduced in 1996 for small amounts (tolls, vending machines, parking fees, etc.). It was launched by the Network for Electronic Transfers (NETS) that is owned by local banks and was made compulsory. Furthermore, starting from 2008, Singapore accorded electronic purses the same legal status as cash, that is, no merchant can refuse e-cash (*Banking Technology*, 2001). The numbers in Table 2.15 confirm that, as expected, the values settled with electronic purses form a tiny proportion of the total values exchanged.

TABLE 2.14

Percentage with Electronic Money in the Volume of Scriptural Transactions by Nonbanks in Selected Countries (2000–2013)

Year	2000	2001	2002	2003	2004	2005	2006	2007	2008	2009	2010	2011	2012	2013
Belgium	4.1	3.9	7.1	6.4	5.9	5.4	4.9	4.3	3.7	3.1	2.5	2.0	1.8	1.1
France	—	—	—	—	—	0.1	0.1	0.2	0.2	0.2	0.2	0.3	0.3	0.3
Germany	0.2	0.2	0.3	0.3	0.3	0.2	0.2	0.3	0.3	0.3	0.2	0.2	0.2	0.2
Italy	—	—	—	0.1	0.3	0.2	1.0	1.5	2.1	2.4	3.1	3.9	4.7	5.8
Korea						2.5	2.3	1.7	1.4	1.3	1.3	0.8	0.5	0.3
The Netherlands	0.9	1.0	2.6	3.0	3.4	0.6	3.9	3.9	3.7	3.5	3.3	3.1	2.5	2.2
Singapore	32.4	31.3	74.1	85.4	85.3	84.5	84.2	83.2	84.3	84.7	85.5	88.1	88.1	81.9
Switzerland	2.0	2.3	2.1	2.1	1.9	1.8	1.7	1.5	1.4	1.2	1.1	0.7	0.2	0.1

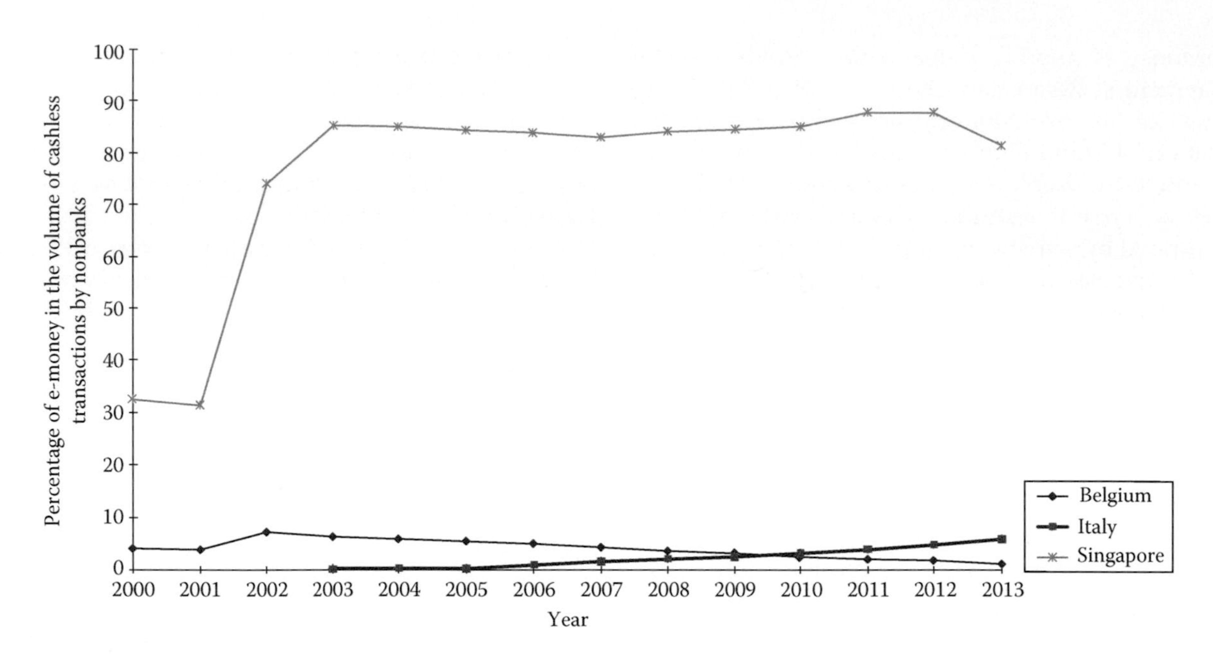

FIGURE 2.13
Evolution of payment by nonbanks with electronic money in selected countries (in volume).

TABLE 2.15

Percentage with Electronic Money in the Value of Scriptural Transactions by Nonbanks in Selected Countries (2000–2013)

	2000	2001	2002	2003	2004	2005	2006	2007	2008	2009	2010	2011	2012	2013
Belgium	0.00	0.00	0.00	0.00	0.00	0.01	0.01	0.01	0.01	0.01	0.01	0.01	0.01	0.00
Italy	—	—	—	—	—	0.02	0.03	0.03	0.05	0.06	0.08	0.10	0.14	0.13
Singapore	0.00	0.00	0.00	0.20	0.20	0.23	0.22	0.19	0.21	0.23	0.20	0.22	0.22	0.22

This type of transactions must be distinguished from unregulated ventures such as Liberty Reserve, which the U.S. authorities shut down in 2013 as the "biggest money laundering scheme ever." Liberty Reserve was operating from Costa Rica as a worldwide online bank without a banking license and without authenticating its customers. Accounts were created after providing an e-mail address and a date of birth without documentation. Deposits in U.S. dollars, euros, or gold were all labeled with a new currency affixed with the prefix "Liberty." In exchange, the company took 1% of each transaction and for an additional 75 cents hid a user's account number on a transfer.

2.5 Transactional Properties of Dematerialized Currencies

From an information technology viewpoint, computer monetary transactions must satisfy certain conditions that can be expressed in terms of the following properties (Camp et al., 1995):

- *Atomicity*: This is an all-or-none property. A transaction has to occur completely for its consequences to take place. Otherwise, the state anterior to the transaction must be restored.
- *Consistency*: All parties must agree on the critical fact of the exchange.
- *Isolation*: The absence of interference among transactions so that the final result of a set of transactions that may overlap will be equivalent to the result when the transactions are executed in a nonconcurrent serial order.
- *Durability*: This is the property that the system returns to its previous state following a breakdown during operation.

From an end-user viewpoint, the reliability of the system as a whole depends on the atomicity of the transactions; that is, a transaction must occur in its entirety or not occur at all. No buyer should be forced to pay for an interrupted transaction. Atomicity is the property of payments made by cash, by checks, by credit transfers, or by debit cards. In contrast, transactions by credit cards or by deferred credit are not always atomic, if the client can revoke the transaction during the time interval between the end of the transaction and the instant at which the amount is debited to the client's account. Although cash payments are isolated, check transactions do not have this characteristic since an overdraft may occur depending on the order of successive deposits and withdrawals.

2.5.1 Anonymity

Anonymity means that the identity of the buyer is not explicitly utilized to settle the obligations. Personalization, in contrast, establishes a direct or indirect relationship between the debtor and the means of payment. Cash in the form of notes and metallic coins is anonymous because it has no links to the nominal identity of holders and their banking references. In the case of remote financial transactions, anonymity raises two questions: the ability to communicate anonymously and the ability to make anonymous payments. Clearly, an anonymous communication is a necessary condition for anonymous payments because, once the source of a call has been identified, the most sophisticated strategies for masking the exchanges would not be able to hide the identity of the caller.

For bank cards and electronic or virtual purses and holders, there are four types of anonymity (Sabatier, 1997, pp. 52–61, 99):

1. The plastic support is anonymous if it does not contain any identifier that can establish a link with the holder. This is the case of gift cards. On the other side, the support of a bank card is not anonymous because it carries the card number as well as the cardholder's name and account.
2. Recharging an electronic purse with value is an anonymous transaction, if it does not establish a link with the identity of the holder, for example, charging of a smart card with the aid of cash. The transaction loses temporarily its anonymity if it is protected by a personal identification number (PIN) because the identity is taken into consideration. However, anonymity can be restored if the transaction is not archived.
3. A transaction is partially anonymous if the information collected during its progress does not establish a link with the holder's bank account. One such example is when payment transactions are grouped by accumulating the total sum of the transactions within a given period. In this case, however, it is possible to discover the identity of the cardholder because the grouped transactions must be tied with a bank account for clearance and settlement.
4. Anonymity for face-to-face transactions is different from anonymity for remote transactions. In proximity commerce, the utilization of a smart card with offline verification can protect the identity of the holder and the subject of the transaction. This is because the algorithms for authentication and identification, which are stored within the memory of the card, will operate independently of any management center. The case of remote commercial transactions whose smooth operation requires that both parties identify themselves without ambiguity to prevent any future contest of the authenticity of the exchange is different. In this case, complete anonymity is incompatible with nonrepudiation. The maximum that can be achieved is partial anonymity; for example, merchants would not have access to the references of the holder, and this information would be collected and stored by a trusted third party that will be an arbiter if a dispute arises.

2.5.2 Traceability

Scriptural money is tied to the status of the issuer and the user, which allows the monitoring of a transaction in all its steps; it is thus personalized and traceable. Nontraceability means that the buyer would not only be anonymous, but that two payments made by the same person could not be linked to each other, no matter what (Sabatier, 1997, p. 99). In smart cards, for example, a *protected zone* preserves an audit trail of the various operations executed. However, by ensuring the total confidentiality of the exchanges with the help of a powerful cryptographic algorithm, third parties external to the system would not be able to trace the payments and link two different payments made with the same card.

Any guarantee for merchandise delivery as well as ambitions to arbitrate disputes run counter to nontraceability of transactions. The question of proof becomes quickly complicated because the laws on *guarantees* and *confidentiality* vary widely among countries.

Table 2.16 shows a comparison of the different means of payments on the basis of the previous properties.

TABLE 2.16

Transactional Properties of Different Methods of Payments

	Atomicity	Consistency	Isolation	Durability	Anonymity	Traceability
Cash	Yes	Yes	Yes	Yes	Yes	*No*
Checks	Yes	Yes	*No*	Yes	*No*	Yes
Credit transfer	Yes	Yes	Yes	Yes	*No*	Yes
Direct debit	Yes	Yes	Yes	Yes	*No*	Yes
Debit card	Yes	Yes	Yes	Yes	*No*	Yes
Credit card	*No*	Yes	Yes	Yes	*No*	Yes
Electronic purse	Yes	Yes	Yes	Yes	*Maybe*	*Maybe*
Virtual purse	Yes	Yes	Yes	Yes	*Maybe*	*Maybe*

2.6 Overall Comparison of the Means of Payment

The diversity of payment instruments indicates that they are not perfect substitutes for one another in all contexts and applications. In addition, historical specificities and cultural influences are translated in significant differences among regions and countries. The choice of a given payment instrument depends on various factors such as the following (Rambure and Nacamuli, 2008, pp. 23–27):

- Ease of use and convenience to the debtor or the creditor
- Terms, conditions, and execution time
- Cost, in terms of fees charged on either the debtor or the creditor as well as processing costs
- Security including anonymity
- Traceability and auditability
- Ease of automation
- Type of commerce (proximity or remote)
- Frequency of payments

Among the classical means, the choice for face-to-face commerce is limited to cash, checks, and payment cards. The choice is much larger for remote payments, which indicates that the requirements differ according to applications and that there is not a uniformly optimal solution. For remote business payments, credit transfers, direct debit, and, when they are available, various types of interbank exchanges are more suitable.

While the main strength of cash is in the area of retail commerce, it is neither suitable for remote payments or for business-to-business payments. The check is the only means of payment that is adapted to most cases, which explains its persistence in some countries. However, the cost of transactions by checks or by payment cards does not make them suitable for micropayments. Stored-value systems, such as electronic or virtual purses, may be able to displace cash and checks in this area, because they can satisfy more or less the same need, while offering the possibility of making small payments in an economic manner.

Summarized in Table 2.17 is the previous discussion on the domain of utilization of the various means of payment.

Table 2.18 summarizes the various properties of money in terms of six criteria:

1. The nature of money
2. The support of money (the container)
3. The location of the value store
4. The representation of the value
5. The mode of payment
6. The means or instrument of payment

TABLE 2.17

Domains of Utilization of Means of Payment

Means of Payment	Face-to-Face Payments	Remote Payments	Business-to-Business Payments
Cash	Yes		
Check	Yes	Yes	Yes
Credit transfer		Yes	Yes
Direct debit		Yes	
Interbank transfer		Yes	
Bank card	Yes, with a reader		Yes
Electronic or virtual purse	Yes, with a reader	Possible	Possible

TABLE 2.18
Properties of Money Types

Type of Money	Nature of Money	Support (the Container)	Value Store	Value Representation	Mode of Payment	Means of Payments (Instrument)
Fiduciary	Concrete, material	Paper, piece of metal	Safe, wallet, purse	Bank notes, coins	Face-to-face transaction	Bank notes, coins
Scriptural	Immaterial (an account maintained by a credit institution)	Magnetic, optical, electronic	Account maintained by a credit institution	Numerical value	Remote, face-to-face (retail, automatic machines)	Check, debit card, credit card, credit transfer
		Integrated circuit card	Electronic purse			
		Computer/ smartphone	Virtual purse (memory allocated by an intermediary)			Electronic funds transfer

2.7 Practice of Dematerialized Money

2.7.1 Protocols of Systems of Dematerialized Money

Depicted in the block diagram of Figure 2.14 are the financial and control flows among participants in a system of dematerialized money (Sabatier, 1997, pp. 46–47):

- *Relation 1* defines the interface between the client (the purse holder) and the operator responsible for charging the purse with electronic monetary values. This operator verifies the financial solvency of the holder or the validity of the payment that the holder makes with the classical instruments of payment. After verification, the operator updates the value stored in the electronic or virtual purse.
- *Relation 2* controls the junction between the charging operator and the issuing bank, if the operator is a nonbank.
- *Relation 3* relates to the interbanking relations between the issuing bank and the acquiring bank (the bank of the merchant) and depends on the regulations at hand.
- *Relation 4* defines the interface of the acquiring operator and the acquiring bank to acquire the credits owed to the merchant.
- *Relation 5* describes the procedures for collection and compensation to credit the merchant's account with the values corresponding to the electronic values exchanged.
- *Relation 6* represents the purchase transaction and the transfer of electronic value from the client to the merchant simultaneously.

In a person-to-person transaction, the recipient of the transfer replaces the merchant.

The *charging protocol* of a system of dematerialized money specifies the procedures to request the authorization and the transfer of electronic value toward the

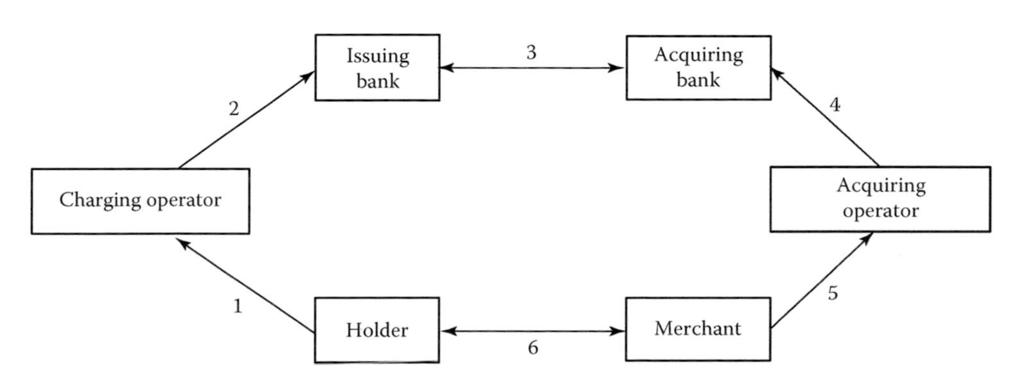

FIGURE 2.14
Block diagram of the flow in a transaction by dematerialized money.

holder's purse in exchange for a payment acceptable to the charging operator (e.g., cash, bank card, checks, or even another electronic purse). The protocol relates to Relations 1 and 2 when the charging operator is a nonbank, otherwise to Relation 1 only. In this latter case, Relation 2 falls within the realm of interbank relationships. Feeding an electronic or virtual purse is considered a collection of funds from the public, which in most countries is a banking monopoly. Only a credit institution is allowed to credit a purse whose units can be utilized for the purchase of products that have not been defined earlier. With the help of a system for point-of-sale activation and recharge (POSA/R of cards), the reloading of value can be done from points connected to the banking networks.

Relation 6 includes two distinct protocols: a *purchasing protocol* during the negotiation of the price and the purchase conditions and a *payment protocol*. The payment can be made directly to the merchant or through an intermediary. The corresponding architectures will be discussed in the following text. In general, the means used for the security of payments do not extend to the purchase protocol, even though the simple fact of knowing that a communication between the partners is taking place can be in itself interesting information. In some schemes, the negotiation that precedes the payment is also protected.

The query of the authorization server can be the responsibility of the merchant or supplier who directly queries the financial circuits. However, an intermediary can relieve merchants of this job and collect in their stead the necessary authorization in return for a negotiated fee.

In systems where the verification is online, the interrogation of the authorization server is systematic for each purchase irrespective of the amounts. These systems are predominant in the United States for credit cards because the cost of a telephone communication is negligible. Online verification was retained in the Secure Electronic Transaction (SET) protocol for remote payments by payment cards on the Internet.

Systems with semionline verification interrogate the authorization server only for certain situations; for example, when the amount of the transaction exceeds a critical threshold or when the transaction takes place with merchants who are more exposed to risk because of the nature of their activity (such as gas stations). An automatic connection is set up periodically to transmit the details of the transactions and to update the security parameters (blacklisted cards, authorization thresholds, etc.). The French system for bank cards is semionline.

Finally, the whole verification is done locally in the case of offline architectures that are based on secure payment modules incorporated in the merchant cash registers. Remote collection and update of the security parameters take place once every 24 hours usually at night.

The terminals used for electronic payment in semionline or offline payment systems are computationally more powerful than those for the online systems. Intelligent terminals have the following responsibilities: (1) reading and validating the parameters of the means of payment; (2) authenticating the holders; (3) controlling the ceiling expenditures allowed to the holder, calculating the proof of payment, generating the sales ticket, and recording of the acceptance parameters; and (4) periodically exchanging data and files with the collection and authorization centers. These terminals must therefore be equipped with an adequate security application module (SAM) to perform the operations of authentication and verification according to the protocols of the payment system used.

The security of online systems is theoretically higher, because they allow a continuous monitoring of the operating conditions and the real-time evaluation of the risks. This assumes that the telecommunications network is reliable and is available at all times. The choice of a semionline system can be justified if the cost of connection to the telecommunications network is important or if the cost of the computational load is too high for the amounts involved.

The protocols must be able to resist attacks from outside the system as well as from any misappropriation by one of the participants (Zaba, 1996). Thus, a third party that is not a participant must not be able to intercept the messages, to manipulate the content, to modify the order of the exchanges, or to resend valid but old messages (this type of attack is called the man-in-the-middle attack). Similarly, the protocols must resist false charges, for example,

- Attributing the recharge to a different purse than the one identified
- Debiting a purse by a false server
- Attributing a different amount than the amount requested
- Replaying a previously authenticated charge
- Repudiating a charge that has been correctly executed or recalling a payment that has been made

In general, the protocols must be sufficiently robust to return to the previous state following a transmission error, particularly if the recharging is done through the Internet.

Finally, the protocol for collection, acquisition, clearance, and settlement that Relations 4 and 5 describe varies depending on whether the acquiring operator is a bank. The purpose is to collect in a secure manner the electronic values stored in the merchants' terminals, to group these values according to the identity of each acquiring bank, and to inform the respective bank of the acquired amount. In the case where the acquiring operator is a bank, which is the most common situation envisaged, Relation 4 falls within the domain of the interbanking relations that are defined by the law.

It should be noted that the functioning of the system must include other protocols, which are not represented in Figure 2.14:

- An initialization protocol to allow the purse holder to subscribe to an account at the operator of the system of e-commerce
- A peer-to-peer transfer protocol to allow the transfer of the electronic monetary value from one purse to another, among holders equipped with compatible readers, and without the intervention of a third party
- A discharging protocol to control the inverse transfer of the electronic money in the purse to a bank account
- A shopping protocol, which is not treated in this book

Some systems of dematerialized money seem to be able to accept peer-to-peer transfers and discharging operations. For example, the electronic purse Mondex was designed to allow the transfer of value between two purses.

2.7.2 Direct Payments to the Merchant

In systems where the payment is directly given to the merchant, clients transmit the coordinates of their account to the merchant. In a classical configuration, the merchant may use one of the well-tested mechanisms such as direct debit or credit transfers.

To make a payment from the client terminal using a purse or a payment card, a payment gateway must intervene to guarantee the isolation of the banking network from the Internet traffic. It is the gateway that will receive the client's request before contacting the authentication and authorization servers to make the function completely transparent to the banking circuits. In this manner, the gateway operator is called upon to become a trusted third party and a notary. Thus, the payment gateway connects the merchant to multiple acquiring or issuing banks and card schemes as well as electronic purses. Thus, it relieves the merchant for integrating various payment instruments to their website. In principle, common standards are necessary to reduce implementation problems; however, the majority of innovative payments and methods have very limited interoperability.

The gateway operator cannot assume the role of charging operator unless it is a credit institution. In this case, the gateway can take on a supplementary role as a change agent and can accept payment in the currency of the client and pay the merchants the amount that is due in the currency of choice.

The location of the payment gateway with the payment architecture is illustrated in Figure 2.15. Although the diagram shows access to the authorization server through the Internet, an alternative configuration is to have the server connected directly to the secure financial network.

An example for an m-commerce application is the payment of parking fees with mobile phones. For the driver this means no need for small change or for a physical ticket, since wardens use GSM terminals. There is also the possibility to extend the parking time at a distance using the phone. The parking fees are aggregated, and the actual transaction is executed automatically at the end of each month.

The disadvantage of direct payments is that the cardholder and the merchant will have to agree on all the details of the protocol beforehand, which impedes open or spontaneous exchanges. The merchant site will have to be able to manage all payment schemes that could be potentially used. Finally, the buyer would have to own a purse for each currency that may be used that, due to cost of inconvenience, may hinder acceptance of the scheme.

Payments mediated by intermediaries can overcome some of these drawbacks. Mobile payments have the advantage of providing a uniform interface through the mobile network.

2.7.3 Payment via an Intermediary

Figure 2.16 shows the position of a payment intermediary or payment service provider in the circuits of e-commerce. The intermediary is generally a nonbank whose role is to hide from the participants the differences among the various payment instruments. This allows participants to avoid the hassle of having specific software for the various systems of payment. The payment service provider represents the issuer and acquirer banks on the Internet side, and on the secure financial networks, it acts on behalf of the client and merchant.

The function depends on prior subscription by clients to the intermediary to give them proxy power to act on their behalves. There are two possibilities:

1. For payments by bank card or by electronic purse, the intermediary usually will have the client's payment coordinates that were previously sent through a secure channel. The intermediary uses this information to debit the buyer's account for the purchases made and to credit the suppliers with the amounts due

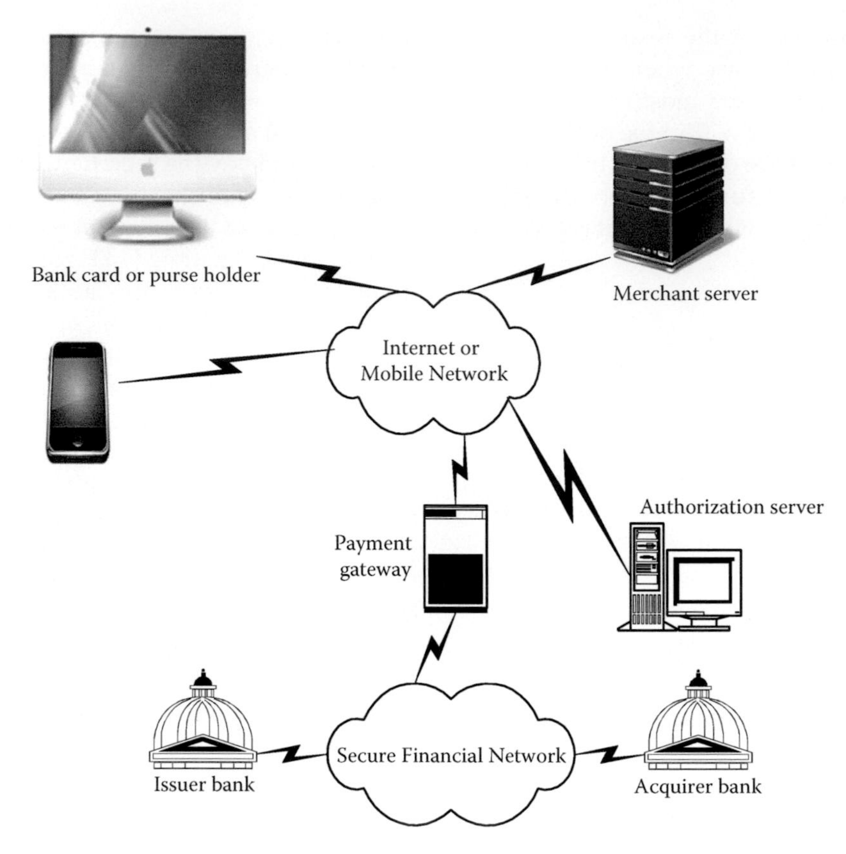

FIGURE 2.15
Position of the payment gateway in e-commerce.

to them. To establish a connection, the holder utilizes an identifier (that could be encrypted with a secret key) as an indication to the intermediary. Settlements can be made for each transaction individually or, in the case of small payments, by a periodic global invoice grouping the individual amounts.

2. For payments with a virtual purse, as was previously explained, the intermediary opens in its own bank subaccounts for the various users and merchants that subscribe to the intermediation service. Users prepay their subaccounts by direct credit or by a bank card or any other established instrument. Following each transaction, the intermediary debits the user's subaccount to the benefit of the *virtual cash register* (subaccount) of the merchant. The intermediary groups the transactions and periodically sends requests to the banking network to settle the account after withdrawing a commission on the turnover.

The same operator can add to the function of the intermediary other roles such as billing and collection for the suppliers, management of the payment instruments for the merchant, management of the cross-borders commerce (exchange rates, import and export taxes, shipping of physical goods, etc.). These roles are often complementary, especially for a worldwide operation.

This trilateral architecture calls for a trusted third party to (1) manage the encryption keys, their generation, distribution, archiving, and revocation; (2) manage the subscriptions of merchants and clients, their certification, and authentication; and (3) update the directories and the blacklists or revocation lists.

The electronic notary can put in place a nonrepudiation service to time stamp the exchanges, archive the transactions, and so on. Depending on the legislation, the intermediary may also act as a small claims judge to settle differences between the merchant and the client on faulty deliveries, defective or nondelivered items, incorrect deciphering keys, and so on. Thus, from a strictly financial viewpoint, the intermediary plays the role of an escrow service to ensure the delivery versus payment of the goods or services.

The software architecture of an electronic purse comprises software application on the consumer side called *digital wallet* managing the various fields of an online order (buyer's name, address, banking coordinates, delivery address, etc.) that connects to the server side

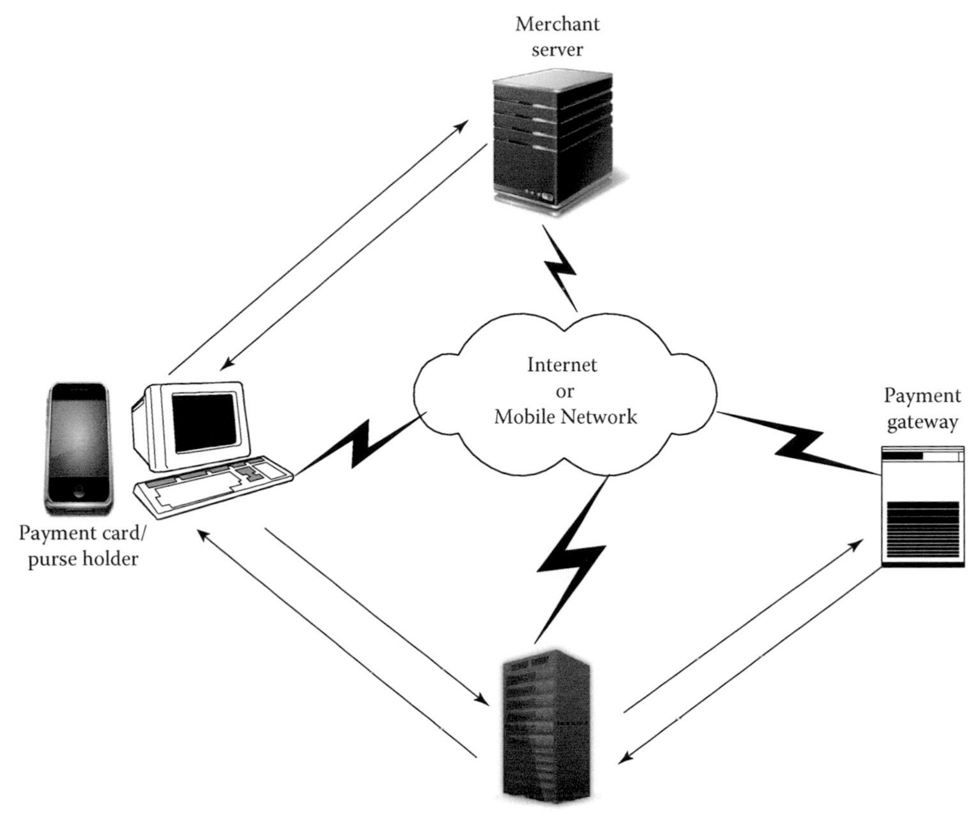

FIGURE 2.16
Position of the payment intermediary in e-commerce.

at the merchant or service provider side. On the server side of the service provider (i.e., in the *cloud*), it is called a *virtual wallet*. Some have introduced the expression *hot wallet* to designate a digital walled that is connected to the Internet, while a *cold wallet* is offline.

The role of nonbanks as a third party in retail payments has increased significantly in mobile and Internet payments. For example, NACHA approved the WEB code for ACH transactions to encourage bill payments over the Internet. Third-party service providers were allowed to use that code as well, which gave PayPal the opportunity to offer inexpensive debit transactions to its subscribers. In 2011, the WEB code was extended to ACH debits authorized or initiated from wireless terminals. Routing these transactions over the ACH system rather than the card networks reduced the cost of these transactions. It should be noted that the transactions are subject to NACHA rules and not U.S. law, which has implications on the risks that users or service providers take.

In Europe, the Single Euro Payments Area (SEPA) initiative was launched in 2008 to harmonize and standardize the procedures for noncash retail payments (credit transfers, direct debits, and card transactions) in 33 European countries: those in the euro countries, the European Union (EU) states with currencies other than the euro and non-EU countries.* The goal is to make payments across Europe as fast, safe, and efficient as national payments are today and stimulate cashless euro payments, including mobile money services, to anyone located anywhere in Europe. This program changed the domestic automatic clearing rules and required overcoming technical, legal, and market barriers among participating countries. Also, payment service providers were recognized as nonbank entities subject to lighter capital requirements and regulatory supervision than credit institutions.

2.8 Clearance and Settlement in Payment Systems

Historically, clearance and settlement took place when all bank representatives would meet every working day in a special house to compare their respective credits in

* The list of countries is available at http://www.ecb.europa.eu/paym/sepa/about/countries/html/index.en.html.

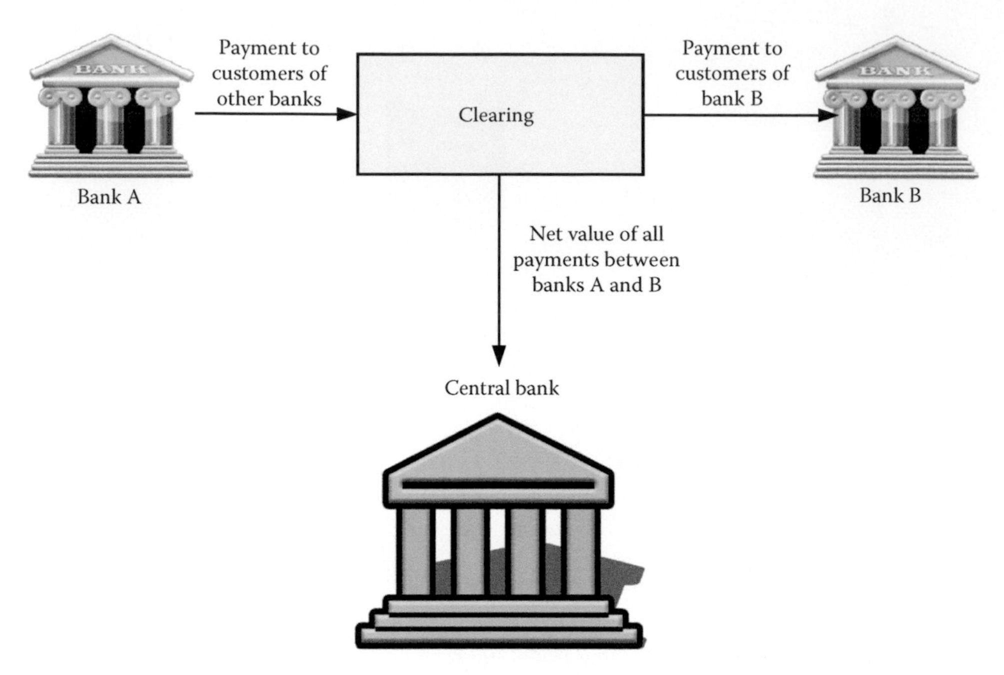

FIGURE 2.17
Clearing and settlement in payment systems.

the various financial instruments and then settle their accounts by exchanging money. Today, the clearance and settlement chain shown in Figure 2.17 is automated and depends on computer networks.

Clearing is the set of processes of transmitting, reconciling and, in some cases, confirming payment orders or security transfer instructions prior to settlement. Clearing may also include the netting of instructions and the establishment of final positions for settlement. This turns the promise of payment (e.g., in the form of checks or electronic payment request) into an actual flow of money from one bank to another. Settlement is an event by which the settlement bank (the central bank) discharges bank debts by transferring corresponding funds from the settlement account of the debtor bank to that of the creditor account. "Netting" is an operation by which all bank obligations are aggregated and consolidated into the net debit or net credit of each bank. This operation reduces the number of transactions affected during the settlement but increases the total risk to the system in the case that one of the participating banks fails. The risks are of two types: the transaction does not clear at all (credit or counterparty risk) or it is settled late (liquidity risk). Next, the central bank instructs each bank to fund or draw funds from its settlement account. In the United States, the settlement is handled by the Federal Reserve.

The net obligation is calculated periodically, most typically daily. Several countries such as India, the Netherlands, and the United Kingdom have introduced multiple settlements in the day (Bank for International Settlements, 2012b, pp. 29–30).

In a bilateral net settlement system, each bilateral combination of the participants is considered separately. In a multilateral net settlement system, each participant settles with several parties with a single transaction. Finally, with real-time gross settlement (RTGS), each transaction settles individually as it is being processed. This system is used for settling large-value payments and is necessary to avoid the credit/counterparty risk due to bank failures.

With SEPA, the European payment systems are converging onto a common architecture. Existing national schemes will operate in parallel with the SEPA scheme until national schemes are switched off.

Classification of the settlement networks can be based on several criteria, for example:

- The nature of the processing
  - Large-value systems
  - System for retail transactions that process many daily transactions of relatively small values
- The ownership and the management of the network
  - Network owned by the central bank
  - Private network owned by the members of a group of banks
  - Private network leased to the banks on a use basis
- The way the settlement is done
  - Netting, whether bilateral or multilateral

- Real-time gross settlement
- Grouping is used when the transfer occurs among different entities of the same group of companies to avoid paying settlement charges repeatedly

The following sections contain additional clarifications on the clearance and settlement systems in the United States, the United Kingdom, and Europe.

2.8.1 United States

The Federal Reserve System is the central bank. It represents what is called today a private–public partnership that is composed of a central governmental agency; the Board of Governors, in Washington, DC; and 12 regional Federal Reserve Banks located in Atlanta, Boston, Chicago, Cleveland, Dallas, Kansas City, Minneapolis, New York, Philadelphia, Richmond, St. Louis, and San Francisco. The Federal Reserve Banks play three roles: regulator, provider of settlement services to U.S. banks, and operator of the Fedwire service for large-value settlements and of one of the two large-scale Automated Clearing House (ACH) networks. The other network is called the Electronic Payments Network (EPN) of the Clearing House Payments Company, a company owned by 18 of the world's largest commercial banks.

As a regulator, the Federal Reserve defines the implementation details for the laws passed by Congress and specifies the binding requirements for banks to operate under.

There are two large-value settlement systems in the United States, Fedwire and CHIPS (Clearing House Interbank Payment System). Fedwire is a service of the Federal Reserve Banks and is available to all federal chartered banks. It is also available to state-chartered banks that have an account at one of the Federal Reserve Banks. Fedwire is used for the real-time transfer of funds.

CHIPS is owned by some large banks and is managed by the New York Clearing House (NYCH) (now called Clearing House Payments Company and owner of SPVCo). CHIPS does not use RTGS but utilizes a form of multilateral netting.

In the United States, the ACH network is a batch processing, store-and-forward network, governed by the NACHA operating rules. In the past, it provided for the interbank clearing of electronic payments for depository financial institutions as well as for large-value payments among businesses, mortgage payments, and payrolls. Recently, it has been used for web-based payments and other types of debit transactions, such as electronic conversion of consumer checks. This has lowered the average value of the amounts in ACH transactions.

As mentioned earlier, the Federal Reserve System and Electronics Payments Network (EPN) are the two ACH operators. Transactions received from financial institutions during the day are stored, processed, and sorted by destination for transmission later. EPN does not have same-day operations, but the Federal Reserve has a same-day treatment for debit transactions. Settlement is based on net positions, either bilateral or multilateral. The net positions are then transmitted to the central bank for settlement or to relevant RTGS, which will settle them with other high-value payment systems, such as Fedwire.

The ACH Network supports several payment applications and is both a credit and a debit payment system. Each ACH application is identified by a specific *standard entry class code*. The codes are defined by NACHA–the Electronic Payment Association, formerly known as National Automated Clearing House Association (NACHA) located in Washington, DC. The organization specifies the ACH rules that bind the financial institutions and define the network operations, all within the Federal Reserve regulations. Some standard entry class codes are shown in Table 2.19 (Thierauf, 1990, pp. 170–172; Emmelhainz, 1993; NACHA, 2003).

In terms of dollar values, CCD ACH transactions accounted for the largest share of all ACH values at 56.5%, 65.5% of the value of all ACH debits, and 50.5% of the value of ACH credits. In terms of the number of transactions, PPD code accounted for 48.1% of the volume of ACH transactions. For ACH credits, PPD accounted for 74.5% of the volume. In terms of ACH debits, PPD, ARC, and WEB transactions comprised the largest three volumes at 30.4%, 26.4%, and 25%, respectively (Federal Reserve, 2011, pp. 47–48). A consequence of the use of the WEB transaction code was an increase in the number of small-value payments, hence lowering of the average value of an ACH transaction.

Business-to-business transactions are the focus of financial Electronic Data Interchange (EDI) and subsequent evolutions as discussed in Chapter 4. Electronic check presentment (ECP) and point-of-sale check approval use the codes POP, ARC and BOC, and RC. eCheck transactions start as native ACH transactions and use the code WEB.

Payment service providers also use this code to provide person-to-person transactions. First, they collect funds from the sender with a WEB transaction and then use an ACH credit transfer to deliver the funds to the receiver.

Depending on the order of succession of the two transactions, there could be at least three possibilities:

1. The payment service provider waits until the debit clear before credit is sent. This is the good funds model.

TABLE 2.19

ACH Standard Entry Class Codes

Types of Applications	Code	Comments
Corporate applications	CCD+	Single or multiple invoices with addenda of up to 80 characters each. These addenda are not standardized, which makes it difficult to automate the processing of account receivables by the receiving company.
	CCD (cash concentration and disbursement/ corporate credit or debit)	Used for consolidating funds held by one corporation across multiple accounts.
	CTP (Corporate Trade Payment)	An older code that was used by corporate and governmental entities to pay their creditors. The CTP system was discontinued in 1996 and was replaced by the CTX system.
	CTX (corporate trade exchange)	CTX allows variable-length fields. CTX can carry addenda records up to 9999 records of up to 80 characters each, that is, remittance data can accompany the payment. In a business-to-business environment, most companies are paying multiple invoices at once, so the addenda include the associated references and adjustments for each invoice.
Consumer applications	CIE (customer-initiated entry)	This code is used for situations where the user initiates the transfer of funds to a merchant through some home banking product or a bill payment service provider.
	PPD (prearranged payment and deposit)	PPD debit transactions are used primarily for recurring bill payments. PPD credit transactions are used for direct deposit of payroll, pension, and benefit payments to consumers.
	RCK (Re-presentation Check Entry)	Code used to represent a returned check.
	TEL (telephone-initiated entry)	Code to authorize a transaction by phone.
	WEB (Internet-initiated entry)	Code to authorize a transaction over the Internet.
Consumer/corporate applications	ARC (Accounts Receivable Conversion)	This is a code to convert checks received via postal mail to an ACH debit. The paper check is used to collect the routing number, account number, the check serial number, and dollar amount. It is mostly used for checks drawn on personnel accounts for bill payments and under some conditions for corporate or business accounts.
	BOC (back-office conversion)	Transactions where the actual conversion occurs in the merchant's bank's (originator bank) back office.
	POP (point of purchase)	Used when a check is scanned at the point of sale to create a one-time debit transaction by capturing the check data and the payment amount. The register generates the ACH transaction to debit the customer's account. The check is returned to the buyer stamped as void.

2. The debit and credit are processed simultaneously. This is the risk model because the debit transaction may not go through.
3. The service provider has already the monetary value in its system if the two parties have accounts with that service provider. These accounts were replenished by debiting each party banking account through the ACH network. The person-to-person transaction transfers value from one account to the other. This is the stored value model.

Table 2.20 shows the comparison of the contribution from each of these clearance and settlement systems in 2006 and 2013 (Bank for International Settlements, 2012a, 2014). The table includes transactions that are not routed through the clearing and settlement system because they involve customers of the same bank ("on us").

The increase in the amounts and volumes of transactions automatically cleared is clearly visible. In addition to the growth in the traditional consumer and business categories such as payroll, prearranged bill payment, and cash concentration and disbursement, Internet-initiated (WEB) payments have expanded ACH applications from direct debits to online-initiated ACH credit payments, as well as person-to-person (P2P) payments.

TABLE 2.20

Contribution of Various Clearance and Settlement Systems in the United States in 2006 and 2013

Nature of the Contribution	2006 Volume (in Millions of Transactions)	2006 Value (in Billions of US $)	2013 Volume (in Millions of Transactions)	2013 Value (in Billions of US $)
Large-Value Systems				
CHIPS	77.9	394,567.3	103.1	379,984.8
Fedwire	133.6	572,645.8	134.2	713,310.4
Checks				
Private clearinghouses	12,992.1	12,953.2	6,806.2	8,500.8
Federal Reserve	11,476.0	16,740.0	6,171.0	8,136.9
Automated Clearinghouses				
Private (ACH)	4,726.7	12,362.9	8,070.4	18,744.4
Direct transfers	2,694.9	4,957.1	5,031.3	6,283.0
Credit transfers	2,031.8	7,405.8	3,039.1	12,461.4
Federal Reserve	7,596.6	13,976.4	9,481.4	19,953.9
Direct transfers	4,490.4	6,238.7	5,166.6	8,328.2
Credit transfers	3,106.2	7,737.7	4,314.8	11,625.7
Memorandum Items				
"On-us" wire	—	—	61.0	172,701.8
"On-us" checks	6,052.9	11,907.0	5,092.5	8,874.9
"On-us" ACH	2,322.4	4,632.7	5,049.3	98,824.6
Direct transfers	1,495.5	2,089.6	3,376.7	48,613.8
Credit transfers	826.9	2,543.1	1,672.6	50,210.8
Total	45,378.2	1,039,785.3	40,969.1	1,429,032.5

2.8.2 United Kingdom

The Clearing House Automated Payment System (CHAPS) is for large-value transfers (credit and direct debit). CHAPS was launched in 1984 as a decentralized system but operates now as a centralized RTGS system. It is operated by the Bank of England on behalf of the CHAPS Clearing Company Ltd., jointly owned by its member banks and the Bank of England. CHAPS has a Euro component within the Trans-European Automated Real-time Gross settlement Express Transfer (TARGET) that links participating national RTGS systems. From a historical viewpoint, the Town Clearing Company Ltd. was responsible for the same-day settlement of transactions of very large values (£500,000 or more) until it ceased operation on February 24, 1995 (Eaglen, 1988; Tyson-Davies, 1994).

There are three systems for retail payments. The services of the Cheque & Credit Clearing Company Ltd. concern a paper credit scheme where payments must be effected by checks if they are mailed or by check and cash at a bank or a Post Office. As its name indicates, the Cheque & Credit Clearing Company clear the paper credits and checks and the net amounts are settled via CHAPS (Rambure and Nacamuli, 2008, pp. 97–98). BACS (Banker's automated clearing service), founded in 1968, was the world's first and largest system dealing with credit transfers and direct debits (Fallon and Welch, 1994). Finally, the Faster Payment Service, which started in May 2008, clears Internet and telephone payments up to £10,000 and standing orders up to £100,000 in 2 hours at most (Rambure and Nacamuli, 2008, pp. 98–99).

Table 2.21 gives a breakdown of the transactions cleared through each of these systems in 2006 and 2008 (Bank for International Settlements, 2012a, 2014).

2.8.3 France

The structure of the French clearance system has changed profoundly between 1995 and 2008 as a consequence of efforts to streamline the process, to follow technological evolutions and due to the adoption of the Euro as a single European currency.

Since February 18, 2008, TARGET2-Banque de France is the single French large-value payment system. TARGET (Trans-European Automated Real-time Gross settlement Express Transfer) was the first real-time settlement system for Euro transactions (less than 2 minutes after debiting the issuer account). TARGET2 was introduced in 2007 and is based on a single shared platform with a standards interface with access to the SWIFT (Society for Worldwide Interbank Financial Telecommunications) network using SWIFT messages.

TABLE 2.21

Contribution of Various Clearance and Settlement Systems in the United Kingdom in 2006 and 2013

	2006		2013	
Year	Volume (in Millions of Transactions)	Value (in Billions of £)	Volume (in Millions of Transactions)	Value (in Billions of £)
Large-Value Payment Systems				
CHAPS Euro (TARGET Component)				
Credit transfers sent	5.57	42,203.1	—	—
Credit transfers sent within CHAPS Euro	1.44	10,768.0	—	—
Credit transfers sent to another TARGET component	4.13	31,435.1	—	—
Credit transfers received from another TARGET component	2.08	31,422.3		
CHAPS Sterling				
Credit transfers	33.02	58,321.6	34.98	70,138.9
Retail Payment Systems				
Check and Credit Clearings				
Total national transactions sent	1,381.76	1,266.6	587.26	576.1
Checks	1,271.05	1,205.1	47.37	24.6
Credit transfers	110.71	61.1	539.89	551.5
BACS				
Total national transactions sent	5,361.59	3,426.5	5,695.03	4,218.6
Credit transfers	2,503.83	2,581.7	2,170.12	3,103.2
Direct debits	2,857.76	844.8	3,524.91	1,115.1
Faster Payment Services				
Total national transactions sent			967.63	771.4
Credit transfers			967.63	771.4
Total	6,784.02	135,373.50	7,284.90	75,704.70

It operates out of two active regions, Germany and Italy, with a data warehouse located in France (Rambure and Nacamuli, 2008, pp. 81, 85).

Depicted in Table 2.22 are the contribution of each of these systems in 2006 and 2013 (Bank for International Settlements, 2012a, 2014).

2.9 Drivers of Innovation in Banking and Payment Systems

The creation of negotiable instruments was the result of many technical and legal innovations in Western Europe and North America starting with the seventeenth century. The frequency and intensity of changes have increased during the last three decades, particularly after the introduction of Internet banking and mobile banking. The main factors that influence innovations in banking and payment systems are as follows:

- Technical developments
- Business needs
- User preferences
- Legislation and regulation
- Standards
- Ideology

The weight that any of these factors take depend on the socioeconomic context, as the following discussion will attempt to clarify.

2.9.1 Technical Developments

Banking has relied on technologies since the weight of metals were used as units for the exchange of value. Technological changes within banking have contributed to the shaping, the size, scope, and type of financial intermediaries. Today, technology permeates all aspects of banking from means of payments and access terminals to the clearing and settlement architectures. Overall, the goals of technology in business applications are to enhance efficiency, lower processing costs, reduce time lags within the system, increase capacity, and improve security. In consumer applications, cost and convenience are the leading drivers.

TABLE 2.22

Contribution of Various Clearance and Settlement Transactions in France in 2006 and 2013

	2006		2013	
	Volume (in Millions of Transactions)	Value (in Billions of €)	Volume (in Millions of Transactions)	Value (in Billions of €)
Large-Value Systems				
TARGET2-BDF				
Credit transfers sent	4.58	135,189.1	9.12	87,565.1
Credit transfers sent within TARGET component	2.12	107,991.9	4.53	53,538.7
Credit transfers sent to another TARGET component	2.46	27,197.1	4.59	34,026.5
Credit transfers received from another TARGET component	2.21	27,196.3	5.36	32,798.2
PNS				
Credit transfers sent	6.614	14,862.0	—	—
Retail Payment Systems				
CORE				
Total transactions sent	12,181.53	5,030.5	13,635.44	5,376.7
Credit transfers	1,846.01	1,872.1	1,969.02	2,760.9
Direct debits	2,077.28	688.9	2,337.02	1,005.6
Card payments	4,421.53	208.9	6,596.33	306.0
ATM transactions	636.63	35.5	611.21	37.6
Checks	3,100.67	1,818.7	2,050.31	1,016.4
Promissory notes	99.41	406.3	71.56	250.1
Total	12,194.93	182,277.90	13,649.92	125,740.0

The technologies in electronic commerce cover a wide gamut. This can be seen clearly in the area of mobile payments where the enabling technologies include the networks, the devices, the microprocessor chips, the antennas, the operating systems, and the security protocols.

Any electronic money must meet the following conditions to replace the physical fiduciary money (Fay, 1997, pp. 113, 115):

1. It must be issued by a source that has the confidence of those that will hold that money.
2. Each monetary unit must have a unique number and must be unfalsifiable.
3. Clearly identified signs must guarantee the quantity represented.

A final technological factor is the switching costs and networking effects in the retail banking industry because the new currency must be accepted by a large number of actors.

2.9.2 Business Needs

Real-time payment processing has reduced the credit (i.e., no funds) and liquidity risks (i.e., late settlements) that import corporations. However, higher speeds and faster payments increase the risks of funding being diverted or stolen before the fraud is detected. Hence, improvements in security, including authentication and verification have been major concerns in a business environment.

In retail applications, the main factors that encourage merchants' adoption are the influence of large buyers, the payment fees, the switching cost to new payment methods, and their effect on decreasing incidents of fraud and theft.

2.9.3 User Preferences

User acceptance of new payment systems depends on many technical, political, and social factors. The propensity of individuals to adopt these innovations depends on the following factors (Bank for International Settlements, 2012b, p. 26):

- The peculiarities of specific payment instruments in terms of ease of use and security
- The demographics, for example, age, education, income level
- The characteristics of the transactions, that is, amount, type of goods or service or location
- The financial incentives, that is, charges, discounts or reward program

Users can be government institutions, enterprises corporations, small- and medium-sized businesses, and individual consumers. In all cases, a technological innovation has to fit or be adapted to the working environment. New means of payment should satisfy specific requirements on the cost of implementation and operation, security, and individual control of payment schedule. In particular, the confidentiality of the transaction must be protected, not only against surveillance but also against the abuses of unethical merchants.

2.9.4 Legislation and Regulation

Banking and payment services evolve and compete within a specific regulatory framework that can encourage or stifle innovation. Regulators have to balance the advantages of lowering entry barriers to increase competition with the concern for the stability of the financial sector. The U.S. banking system is unique compared to other countries due to the large number of depository financial institutions, including local banks and credit unions and the lack of nationwide branching in addition to the multiplicity for financial networks and regulatory and standardization entities.

Electronic invoicing was allowed and recognized in Italy in 1990 (Pasini and Chaloux, 1995) and electronic tax collection started in France in 1992. Similarly, in 1999, after the U.S. federal government initiated the electronic payment of Social Security benefits via ACH, banks, credit unions, savings and loans, and thrifts were stimulated to join the ACH network. The Check Clearing for the 21st Century Act (Check 21 Act), which became effective in 2004, stimulated the use of electronic checks when it accorded the scanned check (Image Replacement Document) the same legal status as the paper original. Remote deposit capture by using devices to create digital image of checks has increased steadily, including scanning them using the camera of a mobile telephone. In parallel, NACHA's WEB transaction code stimulated the migration of bill payments to web-based systems.

In Europe, the euro conversion had a major impact on the architecture, software and hardware of banks, and clearance and settlement systems. The Payment Services Directive (2007/64/EC) of the European Union allowed new institutions to offer payment services by becoming a payment service provider (PSP), with less restrictive licenses and a lower regulatory burden than those institutions with a full banking license, provided that they do not extend credit to its customers (Bank for International Settlements, 2012b, p. 37, n. 55; Mariotto and Verdier, 2014).

Fraud prevention, money laundering, and consumer protection are important issues for regulators and policymakers. In addition, the increased role of mobile operators in the transfer and/or storage of funds is blurring the boundary between regulation of telecommunications services and regulation of financial services. For example, when applied tax rates differ for revenue and financial services, mobile operators may be at a comparative disadvantage with respect to banks and other financial service providers.

Security of payments covers many aspects: certification of merchants and clients, the production and escrow of keys, fabrication and issuance of smart cards, and constitution and management of electronic and virtual purses. Other activities include fraud detection, the recording of exchanges to ensure nonrepudiation, the formation and distribution of revocation lists, and so on. These functions should lead to the birth of new legal roles, such as electronic notaries, trusted third parties, and certification authorities (Lorentz, 1998), whose exact responsibilities remain to be defined.

The security of payments is not sufficient to protect users. It is legislation that must prevent fraud and breaches of trust and protect the right to privacy. Public authorities are thus directly concerned with e-commerce and not only because of its potential effects on employment in the banking sector. Laws need to be written for monetary transactions and the purchase of nonmaterial goods online, especially on a worldwide basis. Most of the examples mentioned in this regard relate to taxation and the exploitation of personal data collected during transactions. Thus, the development of e-commerce, if not of the information society, requires the definition of new global rules, such as the legal recognition of electronic signatures, the uniform protection of individual and consumer rights, as well as the protections given to intellectual properties.

2.9.5 Standards

From an economic perspective, a monetary standard is the set of monetary arrangements and institutions governing the supply of money (Bordo, 2003). Here, the focus is on voluntary standards that create a stable group of operation and facilitate agreements among the participants. The actors include equipment manufacturers, network operators, service providers, application developers or content providers, and content managers (such as bill aggregators). The agreements are voluntary because contrary to regulations and legislations, there are no legal sanctions for not following them (even though the penalties may be economic). One such example is the EMV specifications for integrated circuit cards that define the manufacturing of these cards and the interfaces to payment as well as security applications. Cryptographic algorithms are also standardized so as to ensure end-to-end protection of the transaction irrespective of the parties involved.

From a strictly technical viewpoint, standardization covers the format and content of the exchanges, the protocols that manage the exchange of electronic files, including the security aspects, the nature of the telecommunication networks used, and the information technology platform of the exchanges. Within each enterprise, company standards cover internal policies, service offers, electronic documents archival, retrieval and backup, and management of legal responsibilities.

Mobile commerce, in particular, faces many challenges in terms of the diversity of mobile devices, applications, communication channels, banks, currencies, and service providers. Roaming adds to the financial risks that banks and mobile network operators could face. Interoperability is a key issue and becomes tougher with more complex banking transactions and the proliferation of smartphone apps for mobile money. ITU-T issued two recommendations related to secure mobile financial transactions. ITU-T Recommendation Y.2740 describes the principles of security system development for mobile commerce and mobile banking systems, the probable risks and means of risk reduction. Recommendation Y.2741 specifies the general architecture of a security solution for mobile commerce and mobile banking, describing the key participants, their roles, and the operational scenarios of the mobile commerce and mobile banking systems. Currently, however, there are no realistic assessments of the problems that could arise due mobile malware or technological failures.

2.9.6 Ideology

Cash was invented in the seventh century BC, and remains the most convenient way to pay for everyday purchases (that can be denoted as proximity or face-to-face commerce). Furthermore, coinage remains almost universally a state prerogative, while bank notes and other private instruments such as payment cards constitute a large percentage of money in circulation. Anarchism, as a political philosophy, advocates stateless societies based on nonhierarchical free associations. The influence of this ideology is visible in attempts to create alternative currencies that are not issued by government-endorsed central banks or backed by a national currency. Cryptocurrencies, such as bitcoin, rely on cryptographic algorithms to create the currencies on a peer-to-peer and a decentralized basis. These currencies could recover the anonymity lost by removing cash. Cryptocurrencies also fit with the paradigm of a virtual economy in which financial wealth can be created without physical production and where profits can be made without physical assets.

2.10 Summary

Different payment instruments have been developed through the ages to handle transactions between buyers and sellers of goods and services and the corresponding financial operations (transfer of funds, loan originations, and loan fulfillments). Emerging forms of money take into account new needs and exploit technological advances. Innovative noncash payments methods, collectively denoted as alternate payment initiation methods, are being proposed and tested over Internet and mobile networks. Simultaneously, there are experiments with currencies, such as the adoption of the Euro in Europe as well as the development of a stateless, algorithmic currency with Bitcoin. In short, the overall situation is dynamic and has not reached a new steady state.

With the exception of the use of integrated circuit cards with the EMV security infrastructure, most innovations have been for specific markets and only a few have had an international reach. In other words, the world of payments is responding much more rapidly to technological changes than to globalization. States have had a mixed role in initiating or steering these developments.

In the banking sector, technology has been used to speed up payment processing, either through faster settlement or through faster payment initiation, as well as to improve security. In addition, standardized protocols have been used in the financial networks so as to reduce the cost of equipment and operation.

Another characteristic of the landscape is the significant increase in the role of nonbanks in retail payments, owing to the deregulatory mood that is prevailing since the 1980s as well as the effect of new technologies that allows nonbanks to compete. Ultimately, innovations in retail payments would reshape the environment in terms of payment choices to the various users, individuals or corporations.

Financial inclusion has served as an important driving force for mobile payments in many countries, either under a government mandate or because of the new business opportunities opened up by an untapped market.

The new payment instruments are expected to have significant impact on employment. Policy makers will have to prepare the workforce whose jobs are on the wane (cash collection, check processing, etc.) to the new jobs that these innovations are introducing.

Questions

1. What factors affect consumer selection of a payment instrument?
2. What factors affect merchants' adoption of new payment instruments?
3. What were the intended and unintended consequences of the approval of the WEB transaction code in 2001?

4. Compare the use of money in open versus closed payment circuits.
5. What transactions are excluded in the estimates of nonbank payments in the United States?
6. What is the difference between the money stored in a transportation card and the money stored in an electronic purse?
7. What are the advantages of prepaid cards in public transportation systems?
8. What is the advantage of carrying addenda records with each financial transaction using the CTX credit transaction in ACH?
9. What are the risks in using debit cards in the United States (before the introduction of EMV)?
10. What are some of the necessary steps to ensure the success of a payment system based on smart cards?
11. Match items from column A with the corresponding item in column B.

Column A	Column B
CCD debit	Supplier payment
TEL	Recurring bills
PPD debit	Payroll, pension, and benefit payment
CIE	Transaction by phone
CCD credit	Government tax collection
PPD credit	Home banking

12. Compare client-based and server-based (virtual) wallets.
13. What are the factors that encourage the use of RTGS?
14. What are the advantages of RTGS for retail payments?
15. Compare credit risk with liquidity risk.
16. Match items from column A with the corresponding item in column B.

Column A	Column B
Fedwire	The risk that the inability of one of the participants to meet its obligations, or a disruption in the system itself, could affect the other participants to meet their obligations
Systemic risk	RTGS in Canada
Large-Value Transfer System (LVTS)	RTGS in the United States

17. What are the risks associated with contactless payments?
18. What are the risks associated with prepaid cards?
19. What is the Bank for International Settlements?
20. What is the difference between clearance and settlement?
21. What is SEPA?
22. In a push transaction, one party sends a payment to its correspondent. In a pull transaction, one party receives a payment from the other party. Indicate which of the following transactions is a push transaction:
    - Wire transfer
    - Credit transfer
    - Check
    - ACH debit
23. What is the main role of NACHA?
24. What is the role of the U.S. Federal Reserve System?
25. What is SWIFT?
26. Evaluate the evolution from magnetic strip cards to smart cards in payments from the point of view of the user, the merchant, the bank, the card manufacturer the chip manufacturer and the card manufacturer.
27. What are the arguments for and against cash?
28. Compare the technologies currently used for bank notes, bank cards, paper checks and electronic bill presentment, electronic bill presentment in terms of cost, security, and user's convenience. What are the core competencies for each technology? What possible technology transactions are possible?

3

Algorithms and Architectures for Security

The security of electronic commerce (e-commerce) transactions covers the security of access to the service, the correct identification and authentication of participants (to provide them the services that they have subscribed to), the integrity of the exchanges, and, if needed, their confidentiality. It may be necessary to preserve evidences to resolve disputes and litigation. All these protective measures may counter users' expectations regarding anonymity and nontraceability of transactions.

This chapter contains a short review of the architectures and algorithms used to secure electronic commerce. In particular, the chapter deals with the following themes: definition of security services in open networks, security functions and their possible location in the various layers of the distribution network, mechanisms to implement security services, certification of the participants, and the management of encryption keys. Some potential threats to security are highlighted particularly as they relate implementation flaws.

This chapter has three appendices. Appendices 3A and 3B contain a general overview of the symmetric and public key encryption algorithms, respectively. Described in Appendix 3C are the main operations of the Digital Signature Algorithm (DSA) of the American National Standards Institute (ANSI) X9.30:1 and the Elliptic Curve Digital Signature Algorithm (ECDSA) of ANSI X9.62, first published in 1998 and revised in 2005.

3.1 Security of Open Financial Networks

Commercial transactions depend on the participants' trust in their mutual integrity, in the quality of the exchanged goods, and in the systems for payment transfer or for purchase delivery. Because the exchanges associated with electronic commerce take place mostly at a distance, the climate of trust that is conducive must be established without the participants meeting in person and even if they use dematerialized forms of money or even digital currencies. The security of the communication networks involved is indispensable: those that link the merchant and the buyer, those that link the participants with their banks, and those linking the banks together.

The network architecture must be capable to withstand potential faults without important service degradation, and the physical protection of the network must be insured against fires, earthquakes, flooding, vandalism, or terrorism. This protection will primarily cover the network equipment (switches, trunks, information systems) but can also be extended to user-end terminals. However, the procedures to ensure such protection are beyond the scope of this chapter.

Recommendations X.800 (1991) from the International Telecommunication Union Telecommunication Standardization Sector (ITU-T) categorizes the specific informational threats into two main categories: passive and active attacks. Passive attacks consist in the following:

1. Interception of the identity of one or more of the participants by a third party with a mischievous intent.
2. Data interception through clandestine monitoring of the exchanges by an outsider or an unauthorized user.

Active attacks take several forms such as the following:

- Replay of a previous message in its entirety or in part.
- Accidental or criminal manipulation of the content of an exchange by a third party by substitution, insertion, deletion, or any unauthorized reorganization of the user's data *d*.
- Users' repudiation or denial of their participation in part or in all of a communication exchange.
- Misrouting of messages from one user to another (the objective of the security service would be to mitigate the consequences of such an error as well).
- Analysis of the traffic and examination of the parameters related to a communication among users (i.e., absence or presence, frequency, direction, sequence, type, volume), in which this analysis would be made more difficult by encryption.
- Masquerade, whereby one entity pretends to be another entity.

- Denial of service and the impossibility of accessing the resources usually available to authorized users following the breakdown of communication, link congestion, or the delay imposed on time-critical operations.

Based on the preceding threats, the objectives of security measures in electronic commerce are as follows:

- Prevent an outsider other than the participants from reading or manipulating the contents or the sequences of the exchanged messages without being detected. In particular, that third party must not be allowed to play back old messages, replace blocks of information, or insert messages from multiple distinct exchanges without detection.
- Impede the falsification of Payment Instructions or the generation of spurious messages by users with dubious intentions. For example, dishonest merchants or processing centers must not be capable of reutilizing information about their clients' bank accounts to generate fraudulent orders. They should not be able to initiate the processing of Payment Instructions without expediting the corresponding purchases. At the same time, the merchants will be protected from excessive revocation of payments or malicious denials of orders.
- Satisfy the legal requirements on, for example, payment revocation, conflict resolution, consumer protection, privacy protection, and the exploitation of data collected on clients for commercial purposes.
- Ensure reliable access to the e-commerce service, according to the terms of the contract.
- For a given service, provide the same level of service to all customers, irrespective of their location and the environmental variables.

The International Organization for Standardization (ISO) standard ISO 7498-2:1989 (ITU-T Recommendation X.800, 1991) describes a reference model for the service securities in open networks. This model will be the framework for the discussion in the next section.

3.2 OSI Model for Cryptographic Security

3.2.1 OSI Reference Model

It is well known that the Open Systems Interconnection (OSI) reference model of data networks establishes a structure for exchanges in seven layers (ISO/IEC 7498-1:1994):

1. The *physical layer* is where the electrical, mechanical, and functional properties of the interfaces are defined (signal levels, rates, structures, etc.).
2. The *link layer* defines the methods for orderly and error-free transmission between two network nodes.
3. The *network layer* is where the functions for routing, multiplexing of packets, flow control, and network supervision are defined.
4. The *transport layer* is responsible for the reliable transport of the traffic between the two network endpoints as well as the assembly and disassembly of the messages.
5. The *session layer* handles the conversation between the processes at the two endpoints.
6. The *presentation layer* manages the differences in syntax among the various representations of information at both endpoints by putting the data into a standardized format.
7. The *application layer* ensures that two application processes cooperate to carry out the desired information processing at the two endpoints.

The following section provides details about some cryptographic security functions that have been assigned to each layer.

3.2.2 Security Services: Definitions and Location

Security services for exchanges used in e-commerce employ mathematical functions to reshuffle the original message into an unreadable form before it is transmitted. After the message is received, the authenticated recipient must restore the text to its original status. The security consists of six services (Baldwin and Chang, 1997):

1. *Confidentiality*, that is, the exchanged messages are not divulged to a nonauthorized third party. In some applications, the confidentiality of addresses may be needed as well to prevent the analysis of traffic patterns and the derivation of side information that could be used.
2. *Integrity* of the data, that is, proof that the message was not altered after it was expedited and before the moment it was received. This service guarantees that the received data are exactly what were transmitted by the sender and that they were not corrupted, either intentionally or by error in transit in the network. Data integrity is also needed for network management data such as configuration files, accounting, and audit information.
3. *Identification*, that is, the verification of a pre-established relation between a characteristic

(e.g., a password or cryptographic key) and an entity. This allows control of access to the network resources or to the offered services based on the privileges associated with a given identity. One entity may possess several distinct identifiers. Furthermore, some protection against denial-of-service attacks can be achieved using access control.

4. *Authentication* of the participants (users, network elements, and network element systems), which is the corroboration of the identity that an entity claims with the guarantee of a trusted third party. Authentication is necessary to ensure nonrepudiation of users as well of network elements.
5. *Access control* to ensure that only the authorized participants whose identities have been duly authenticated can gain access to the protected resources.
6. *Nonrepudiation* is the service that offers an irrefutable proof of the integrity of the data and of their origin in a way that can be verified by a third party, for example, the nonrepudiation that the sender sent the message or that a receiver received the message. This service may also be called authentication of the origin of the data.

The implementation of the security services can be made over one or more layers of the OSI model. The choice of the layer depends on the several considerations as explained in the following text.

If the protection has to be accorded to all the traffic flow in a uniform manner, the intervention has to be at the physical or the link layers. The only cryptographic service that is available at this level is confidentiality by encrypting the data or similar means (frequency hopping, spread spectrum, etc.). The protection of the traffic at the physical layer covers all the flow, not only user data but also the information related to network administration: alarms, synchronization, update of routing table, and so on. The disadvantage of the protection at this level is that a successful attack will destabilize the whole security structure because the same key is utilized for all transmissions. At the link layer, encryption can be end to end, based on the source/destination, provided that the same technology is used all the way through.

Network layer encipherment achieves selective bulk protection that covers all the communications associated with a particular subnetwork from one end system to another end system. Security at the network layer is also needed to secure the communication among the network elements, particularly for link state protocols, where updates to the routing tables are automatically generated based on received information then flooded to the rest of the network.

For selective protection with recovery after a fault, or if the network is not reliable, the security services will be applied at the transport layer. The services of this layer apply end to end either singly or in combination. These services are authentication (whether simple by passwords or strong by signature mechanisms or certificates), access control, confidentiality, and integrity.

If more granular protection is required or if the nonrepudiation service has to be ensured, the encryption will be at the application layer. It is at this level that most of the security protocols for commercial systems operate, which frees them from a dependency on the lower layers. All security services are available.

It should be noted that there are no services at the session layer. In contrast, the services offered at the presentation layer are confidentiality, which can be selective such as by a given data field, authentication, integrity (in whole or in part), and nonrepudiation with a proof of origin or proof of delivery.

As an example, the Secure Sockets Layer (SSL)/Transport Layer Security (TLS) protocols are widely used to secure the connection between a client and a server. With respect to the OSI reference model, SSL/TLS lie between the transport layer and the application layer and will be presented in Chapter 5.

In some cases, it may be sufficient for an attacker to discover that a communication is taking place among partners and then attempt to guess, for example:

- The characteristics of the goods or services exchanged
- The conditions for acquisition such as delivery intervals, conditions, and means of settlement
- The financial settlement

The establishment of an enciphered channel or "tunnel" between two points at the network layer can constitute a shield against such types of attack. It should be noticed, however, that other clues, such as the relative time to execute the cryptographic operations, or the variations in the electric consumption or the electromagnetic radiation, can permit an analysis of the encrypted traffic and ultimately lead to breaking of the encryption algorithms (Messerges et al., 1999).

3.3 Security Services at the Link Layer

Internet Engineering Task Force (IETF) Request for Comment (RFC) 1661 (1994) defines the link-layer protocol Point-to-Point Protocol (PPP) to carry traffic between two entities identified with their respective IP addresses. The Layer 2 Tunneling Protocol (L2TP) defined in IETF

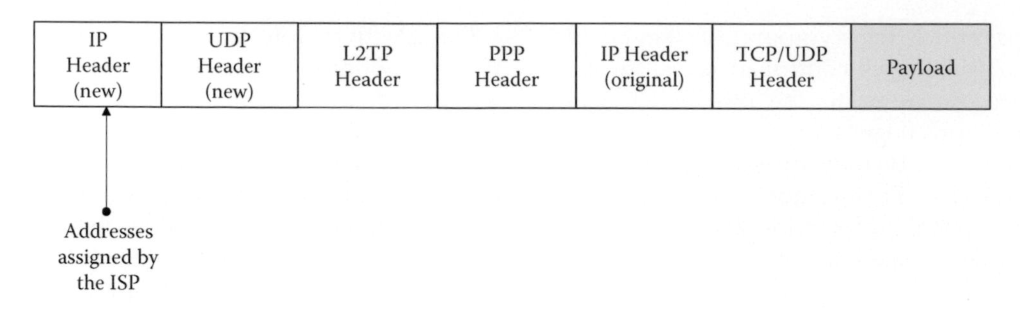

FIGURE 3.1
Layer 2 tunneling with L2TP.

RFC 2661 (1999) extends the PPP operation by separating the processing of Internet Protocol (IP) packets within the PPP frames from that of the traffic flowing between the two ends at the link layer. This distinction allows a remote client to connect to a network access server (NAS) in a private (corporate) network through the public Internet, as follows. The client encapsulates PPP frames in an L2TP tunnel, prepends the appropriate L2TP header, and then transports the new IP packet using the User Datagram Protocol (UDP). The IP addresses in the new IP header are assigned by the local Internet Service Provider (ISP) at the local access point. Figure 3.1 illustrates the arrangement where the size of the additional header ranges from 8 to 16 octets (1 to 2 octets for PPP, 8 to 16 octets for L2TP). Given that the overhead for UDP is 8 octets and that the IP header is 20 octets, the total additional overhead ranges from 37 to 46 octets.

Although L2TP does not provide any security services, it is possible to use Internet Protocol Security (IPSec) to secure the layer 2 tunnel because L2TP runs over IP. This is shown in the following section.

3.4 Security Services at the Network Layer

The security services at this layer are offered from one end of the network to the other end. They include network access control, authentication of the users and/or hosts, and authentication and integrity of the exchanges. These services are transparent to applications and end users, and their responsibilities fall on the administrators of network elements.

Authentication at the network layer can be simple or strong. *Simple* authentication uses a name and password pair (the password may be a one-time password), while *strong* authentication utilizes digital signatures or the exchange of certificates issued by a recognized certification authority (CA). The use of strong authentication requires the presence of encryption keys at all network nodes, which imposes the physical protection of all these nodes.

IPSec is a protocol suite defined to secure communications at the network layer between two peers. The most recent road map to the IPSec documentation is available in IETF RFC 6071 (2011). The overall security architecture of IPSec-v2 is described in IETF RFC 2401; the architecture of IPSec-v3 is described in RFC 4301 (2005).

IPSec offers authentication, confidentiality, and key management and is not tied to specific cryptographic algorithms. The Authentication Header (AH) protocol defined in IETF RFCs 4302 (2005) and 7321 (2014) provides authentication and integrity services for the payload as well as the routing information in the original IP header. The Encapsulating Security Payload (ESP) protocol is described in IETF RFCs 4303 (2005) and 7321 (2014) that define IPSec-v3. ESP focuses on the confidentiality of the original payload and the authentication of the encrypted data as well as the ESP header. Both IPSec protocols provide some protection against replay attacks with the help a monotonically increasing sequence number that is 64 bits long. The key exchange is performed with the Internet Key Exchange (IKE) version 2, the latest version of which is defined in RFCs 7296 (2014) and 7427 (2015).

IPSec operates in one of two modes: the transport mode and the tunnel mode. In the transport mode, the protection covers the payload and the transport header only, while the tunnel mode protects the whole packet, including the IP addresses. The transport mode secures the communication between two hosts, while the tunnel mode is useful when one or both ends of the connection are a trusted entity, such as a firewall, which provides the security services to an originating device. The tunnel mode is also employed when a router provides the security services to the traffic that it is forwarding (Doraswamy and Harkins, 1999). Both modes are used to secure virtual private networks with IPSec as shown in Figure 3.2. The AH protocol is used for the transport mode only, while the ESP is applicable to both modes.

Illustrated in Figure 3.3 is the encapsulation in both cases. In this figure, the IPSec header represents either

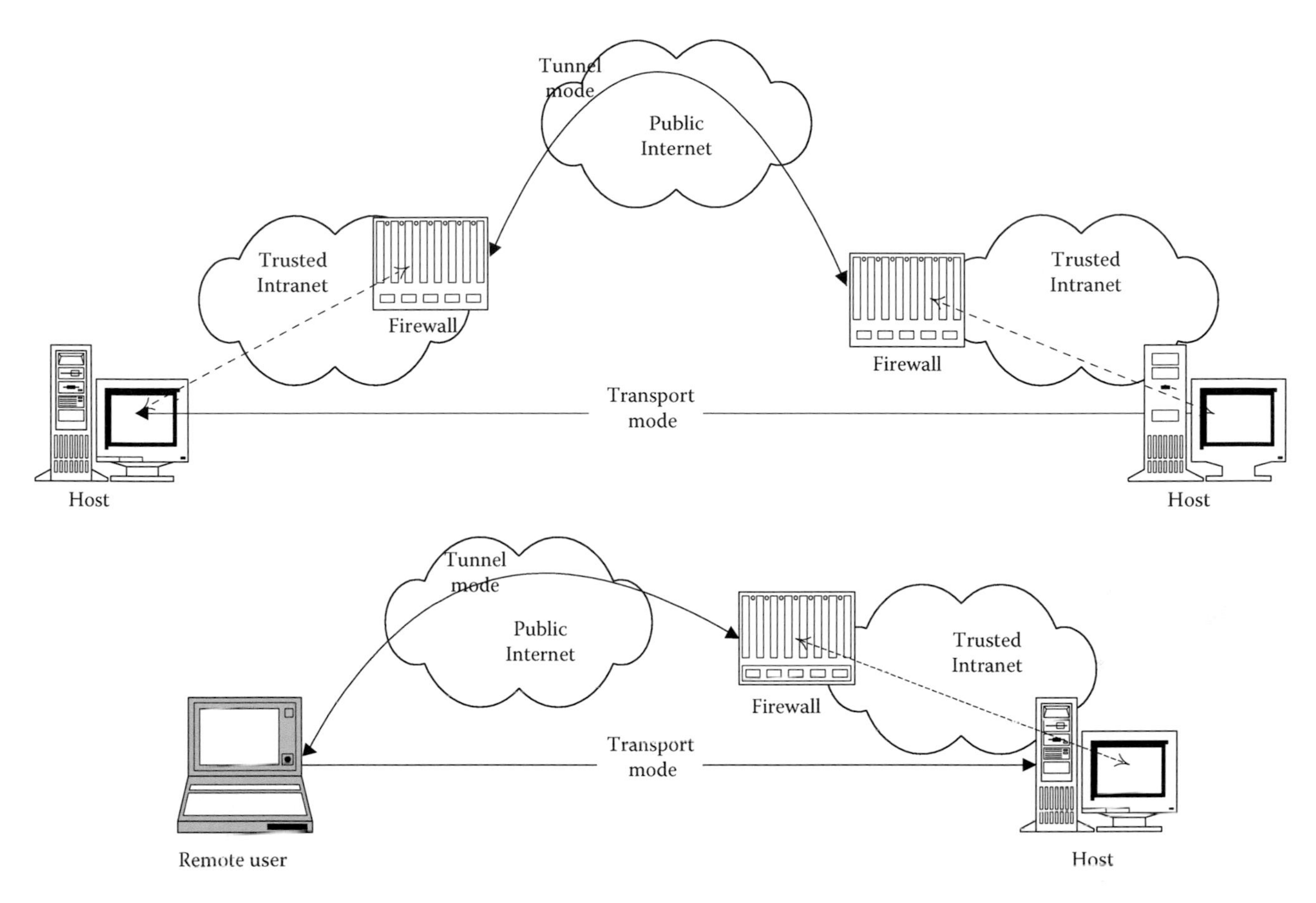

FIGURE 3.2
Securing virtual private networks with IPSec.

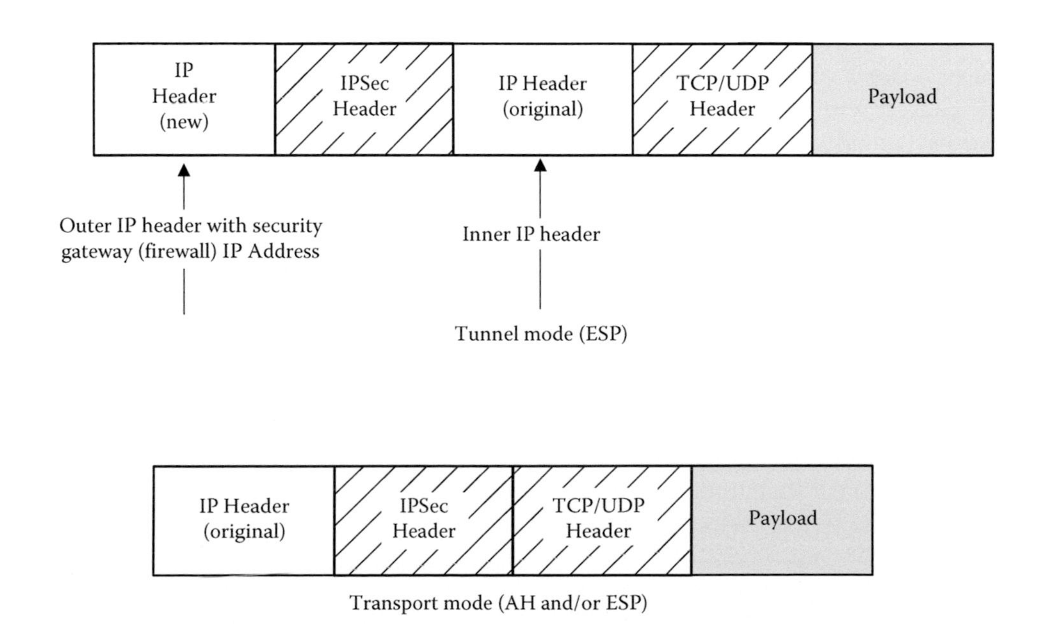

FIGURE 3.3
Encapsulation for IPSec modes.

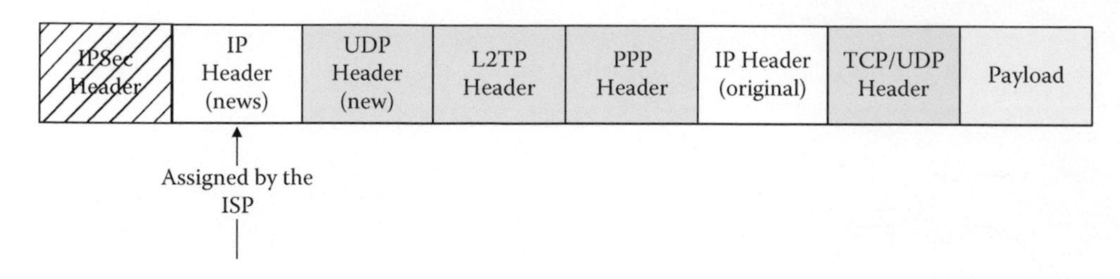

FIGURE 3.4
Encapsulation for secure network access with L2TP and IPSec.

the ESP header or both the ESP and the AH headers. Thus, routing information associated with the private or corporate network can be encrypted after the establishment of a Transmission Control Protocol (TCP) tunnel between the firewall at the originating side and the one at the destination side. Note that ESP with no encryption (i.e., with a NULL algorithm) is equivalent to the AH protocol.

In verifying the integrity, the contents of fields in the IP header that change in transit (e.g., the "time to live") are considered to be zero. With respect to transmission overheads, the length of the AH is at least 12 octets (a multiple of 4 octets for IPv4 and of 6 octets for IPv6). Similarly, the length of the ESP header is 8 octets. However, the total overhead for ESP includes 4 octets for the initialization vector (if it is included in the payload field), as well as an ESP trailer of at least 6 octets that comprise a padding and authentication data.

The protection of L2TP layer 2 tunnels with the IPSec protocol suite is described in IETF RFC 3193 (2001). When both IPSec and L2TP are used together, various headers are organized as shown in Figure 3.4.

IPSec AH (RFC 4302) and IPSec ESP (RFC 4303) define an antireplay mechanism using a sliding window that limits how far out of order a packet can be, relative to the authenticated packet with the highest sequence number. A received packet with a sequence number outside that window is dropped. In contrast, the window is advanced each time a packet is received with a sequence number within the acceptable range. RFCs 4302 and 4303 define minimum window sizes of 32 and 64 octets.

3.5 Security Services at the Application Layer

The majority of the security protocols for e-commerce operate at the application layer, which makes them independent of the lower layers. The whole gamut of security services is now available:

1. Confidentiality, total or selective by field or by traffic flow
2. Data integrity
3. Peer entity authentication
4. Peer entity authentication of the origin
5. Access control
6. Nonrepudiation of transmission with proof of origin
7. Nonrepudiation of reception with proof of reception

The Secure shell (SSH®*), for example, provides security at the application layer and allows a user to log on, execute commands, and transfer files securely.

Additional security mechanisms are specific to a particular usage or to the end-user application at hand. For example, several additional parameters are considered to secure electronic payments, such as the ceiling of allowed expenses or withdrawals within a predefined time interval. Fraud detection and management depends on the surveillance of the following (Sabatier, 1997, p. 85):

- Activities at the points of sale (merchant terminals, vending machines, etc.)
- Short-term events
- Long-term trends, such as the behavior of a subpopulation, within a geographic area and in a specific time interval

In these cases, audit management takes into account the choice of events to collect and/or register, the validation of an audit trail, the definition of the alarm thresholds for suspected security violations, and so on.

The rights of intellectual property to dematerialized articles sold online pose an intellectual and technical challenge. The aim is to prevent the illegal reproduction of what is easily reproducible using "watermarks" incorporated in the product. The means used differ depending on whether the products protected are ephemeral (such as news), consumer oriented (such as films, music, books, articles, or images) or for production (such as enterprise software).

* Secure Shell and SSH are registered trademarks of SSH Communications Security, Ltd. of Finland.

Next, we give an overview of the mechanisms used to implement security service. The objective is to present sufficient background for understanding the applications and not to give an exhaustive review. More comprehensive discussions of the mathematics of cryptography are available elsewhere (Schneier, 1996; Menezes et al., 1997; Ferguson et al., 2010; Paar and Pelzl, 2010).

3.6 Message Confidentiality

Confidentiality guarantees that information will be communicated solely to the parties that are authorized for its reception. Concealment is achieved with the help of encryption algorithms. There are two types of encryption: symmetric encryption, where the operations of message obfuscation and revelation use the same secret key, and public key encryption, where the encryption key is secret and the revelation key is public.

3.6.1 Symmetric Cryptography

Symmetric cryptography is the tool that is employed in classical systems. The key that the sender of a secret message utilizes to encrypt the message is the same as the one that the legitimate receiver uses to decrypt the message. Obviously, key exchange among the partners has to occur before the communication, and this exchange takes place through other secured channels. The operation is illustrated in Figure 3.5.

Let M be the message to be encrypted with a symmetric key K in the encryption process E. The result will be the ciphertext C such that

$$E[K(M)] = C$$

The decryption process D is the inverse function of E that restores the clear text:

$$D(C) = M$$

There are two main categories of symmetric encryption algorithms: block encryption algorithms and stream cipher algorithms. Block encryption acts by transforming a block of data of fixed size, generally 64 bits, in encrypted blocks of the same size. Stream ciphers convert the clear text one bit at a time by combining the stream of bits in the clear text with the stream of bits from the encryption key using an exclusive OR (XOR).

Table 3.1 presents in alphabetical order the main algorithms for symmetric encryption used in e-commerce applications.

Fortezza is the Cryptographic Application Programming Interface (CAPI) that the National Security Agency (NSA) defined for security applications running on PCMCIA (Personal Computer Memory Card International Association) cards. SKIPJACK algorithm is used for encryption, and the Key Exchange Algorithm (KEA) is the algorithm for key exchange. The experimental specifications of IETF RFC 2773 (2000) describe the use of SKIPJACK and KEA for securing file transfers.

The main drawback of symmetric cryptography systems is that both parties must obtain, one way or another, the unique encryption key. This is possible without too much trouble within a closed organization; on open networks, however, the exchange can be intercepted. Public key cryptography, proposed in 1976 by Diffie and Hellman, is one solution to the problem of key exchange.

3.6.2 Public Key Cryptography

Algorithms of public key cryptography introduce a pair of keys for each participant, a private key SK and a public key PK. The keys are constructed in such a way that it is practically impossible to reconstitute the private key with the knowledge of the public key.

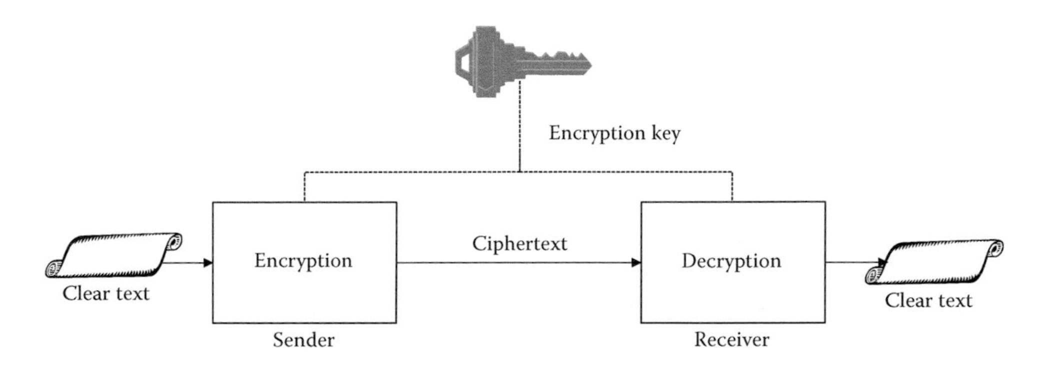

FIGURE 3.5
Symmetric encryption.

TABLE 3.1

Symmetric Encryption Algorithms in E-Commerce

Algorithms	Name and Description	Block Size in Bits	Key Length in Bits	Standard
AES	Advanced Encryption Standard	Blocks of 128, 192, or 256 bits	128, 192, or 256	FIPS 197
DES	Data Encryption Standard	Blocks of 64 bits	56	FIPS 81, ANSI X3.92, X3.105, X3.106, ISO 8372, ISO/IEC 10116
IDEA (Lai and Massey, 1991a,b)	International Data Encryption Algorithm	Blocks of 64 bits	128	—
RC2	Developed by Ronald Rivest (Schneier, 1996, pp. 319–320)	Blocks of 64 bits	Variable (previously limited to 40 bits for export from the United States)	No; proprietary
RC4	Developed by R. Rivest (Schneier, 1996, pp. 397–398)	Stream	40 or 128	No, but posted on the Internet in 1994
RC5	Developed by R. Rivest (1995)	Blocks of 32, 64, or 128 bits	Variable up to 2048 bits	No; proprietary
SKIPJACK	Developed for applications with the PCMCIA card Fortezza	Blocks of 64 bits	80	Declassified algorithm; version 2.0 available at http://csrc.nist.gov/groups/ST/toolkit/documents/skipjack/skipjack.pdf, last accessed January 25, 2016
Triple DES	Also called TDEA	Blocks of 64 bits	112	ANSI X9.52/NIST SP 800-67 (National Institute of Standards and Technology, 2012a)

Note: FIPS, Federal Information Processing Standard.

Consider two users, A and B, each having a pair of keys (PK_A, SK_A) and (PK_B, SK_B), respectively. Thus,

1. To send a secret message x to B, A encrypts it with B's public key and then transmits the encrypted message to B. This is represented by

$$e = PK_B(x)$$

2. B recovers the information using his or her private key SK_B. It should be noted that only B possesses SK_B, which can be used to identify B. The decryption operation can be represented by

$$x = SK_B(e) \quad \text{or} \quad x = SK_B\left[PK_B(x)\right]$$

3. B can respond to A by sending a new secret message x' encrypted with the public key PK_A of A:

$$e' = PK_A(x')$$

4. A obtains x' by decrypting e':

$$x' = SK_B(e') \quad \text{or} \quad x' = SK_A\left[PK_A(x')\right]$$

The diagram in Figure 3.6 summarizes these exchanges.

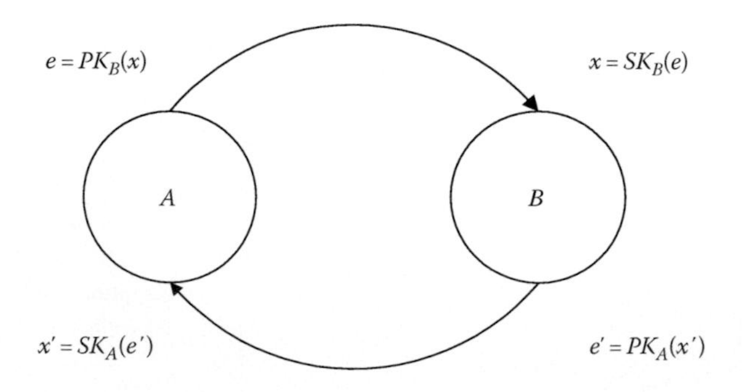

FIGURE 3.6
Confidentiality of messages with public key cryptography. (From ITU-T Recommendation X.509 (ISO/IEC 9594-8), Information technology—Open systems interconnection—The directory: Public-key and attribute certificate frameworks, 2012, 2000. With permission.)

It is worth noting that the preceding exchange can be used to verify the identity of each participant. More precisely, A and B are identified by the possession of the decryption key, SK_A or SK_B, respectively. A can determine if B possesses the private decryption key SK_B if the initial message x is included in the returned message x' that B sends. This indicates to A that the communication has been made with the entity that possesses SK_B. B can also confirm the identity of A in a similar way.

The *de facto* standard for public key encryption is the algorithm RSA invented by Rivest et al. (1978). In many new applications, however, elliptic curve cryptography (ECC) offers significant advantages as described in Appendix 3B.

3.7 Data Integrity

The objective of the integrity service is to eliminate all possibilities of nonauthorized modification of messages during their transit from the sender to the receiver. The traditional form to achieve this security is to stamp the letter envelope with the wax seal of the sender. Transposing this concept to electronic transactions, the seal will be a sequence of bits associated univocally with the document to be protected. This sequence of bits will constitute a unique and unfalsifiable "fingerprint" that will accompany the document sent to the destination. The receiver will then recalculate the value of the fingerprint from the received document and compare the value obtained with the value that was sent. Any difference will indicate that the message integrity has been violated.

The fingerprint can be made to depend on the message content only by applying a *hash function*. A hash function converts a sequence of characters of any length into a chain of characters of a fixed length, L, usually smaller than the original length, called hash value. However, if the hash algorithm is known, any entity can calculate the hash value from the message using the hash function. For security purposes, the hash value depends on the message content and the sender's private key in the case of a public key encryption algorithm or a secret key that only the sender and the receiver know in the case of a symmetric encryption algorithm. In the first case, anyone who knows the hash function can calculate the fingerprint with the public key of the sender; in the second case, only the intended receiver will be able to verify the integrity. It should be noted that lack of integrity can be used to break confidentiality. For example, the confidentiality of some algorithms may be broken through attacks on the initialization vectors.

The hash value has many names: compression, contraction, message digest, fingerprint, cryptographic checksum, message integrity check (MIC), and so on (Schneier, 1996, p. 31).

3.7.1 Verification of the Integrity with a One-Way Hash Function

A *one-way hash function* is a function that can be calculated relatively easily in one direction but with considerable difficulty in the inverse direction. A one-way hash function is sometimes called a compression function or a contraction function.

To verify the integrity of a message whose fingerprint has been calculated with the hash function $H()$, this function should also be a one-way function; that is, it should meet the following properties:

1. *Absence of collisions*: In other words, the probability of obtaining the same hash value with two different texts should be almost null. Thus, for a given message x_1, the probability of finding a different message x_2, such that $H(x_1) = H(x_2)$, is extremely small. For the collision probability to be negligible, the size of the hash value L should be sufficiently large.
2. *Impossibility of inversion*: Given the fingerprint h of a message x, it is practically impossible to calculate x such that $H(x) = h$.
3. *A widespread among the output values*: This is because a small difference between two messages should yield a large difference between their fingerprints. Thus, any slight modification in the original text should, on average, affect half of the bits of the fingerprint.

Consider the message X. It will have been divided into n blocks, each consisting of B bits. If needed, padding bits would be appended to the message, according to a defined scheme, so that the length of each block reaches the necessary B bits. The operations for cryptographic hashing are described using a compression function $f()$ according to the following recursive relationship:

$$h_i = f\left(h_{i-1}, x_i\right), \quad i=1,\ldots,n$$

In this equation, h_0 is the vector that contains an initial value of L bits and $x = \{x_1, x_2, \ldots, x_n\}$ is the message subdivided into n vectors of B bits each. The hash algorithms that are commonly used in e-commerce are listed in Table 3.2 in alphabetical order.

For the MD5 and SHA-1 hashing algorithms, the message is divided into blocks of 512 bits. The padding consists, in appending to the last block, a binary "1," then as many "0" bits as necessary for the size of the last block with padding to be 448 bits. Next, a suffix of 8 octets is added to contain the length of the initial message (before padding) coded over 64 bits,

TABLE 3.2

Hash Functions Utilized in E-Commerce Applications

Algorithm	Name	Signature Length (*L*) in Bits	Block Size (*B*) in Bits	Standardization
AR/DFP	Hashing algorithms of German banks			German Banking Standards
DSMR	Digital signature scheme giving message recovery			ISO/IEC 9796
MCCP	Banking key management by means of public key algorithms using the RSA cryptosystem; signature construction by means of a separate signature			ISO/IEC 1116-2
MD4	Message digest algorithm	128	512	No, but described in RFC 1320
MD5	Message digest algorithm	128	512	No, but described in RFC 1321
NVB7.1, NVBAK,	Hashing functions used by Dutch banks			Dutch Banking Standard, published in 1992
RIPEMD	Extension of MD4, developed during the European project RIPE (Menezes et al., 1997, p. 380)	128	512	
RIPEMD-128	Dedicated hash function #2	128	512	ISO/IEC 10118-3
RIPEMD-160	Improved version of RIPEMD (Dobbertin et al., 1996)	160	512	
SHA	Secure Hash Algorithm (replaced by SHA-1)	160	512	FIPS 180
SHA1 (SHA-1)	Dedicated Hash Function #3 (revision and correction of the Secure Hash Algorithm)	160	512	ISO/IEC 10118-3
				FIPS 180-1 (National Institute of Standards and Technology, 1995), FIPS 180-4 (National Institute of Standards and Technology, 2012b)
SHA-2		224, 256	512	FIPS 180-4
		384, 512	1024	

which brings the total size of the last block to 512 bits of 64 octets.

For a hash algorithm with an output of n bits, there are $N = 2^n$ hashes, which are assumed to be equally probable. Based on the generalization of the birthday problem, the event that there are no collisions after q hashes has a probability given by (Barthélemy et al., 2005, pp. 390–391; Paar and Pelzl, 2010, pp. 299–303):

$$\prod_{i=1}^{q-1}\left(1-\frac{i}{N}\right) \approx \prod_{i=1}^{q} e^{-\frac{i}{N}} = e^{-\sum_{i=1}^{q-1}\frac{1}{N}}$$

$$= e^{-\frac{q(q-1)}{2N}}$$

Therefore, the probability of at least one collision is given by

$$1-e^{-\frac{q(q-1)}{2N}}$$

Now, put the probability of at least one collision to be 50%. Solving for q by taking the natural logarithm of both sides, we get

$$q(q-1) = 2N\ln(2)$$

Since q is large, $(q-1) \approx q$, therefore,

$$q^2 = 2N\ln(2),$$

$$\text{i.e., } q = \sqrt{2N\ln(2)}$$

$$= 1.18\sqrt{N} \approx \sqrt{N} \text{ or } 2^{\frac{n}{2}}$$

Thus, the number of messages to hash to find a collision is roughly the square root of the number of possible output values. The so-called birthday attack on cryptographic hash functions of 256 bits indicates that the probability of a collision exceeds 50% after around 2^{128} hashes.

In 1994, two researchers, van Oorschot and Wiener, were able to detect collisions in the output of MD5 (van Oorschot and Wiener, 1994), which explains its gradual replacement with SHA-1. In 2007, Stevens et al. (2009) exploited the vulnerability of MD5 to collisions to construct two X.509 certificates for SSL/TLS traffic with the identical MD5 signature but different public keys and belonging to different entities without the knowledge and/or assistance of the relevant certification authority. Knowledge that rogue SSL/TLS certificates can be easily forged accelerated the migration of certification authorities away from MD5.

TABLE 3.3

Comparison of the Hash Algorithms

Algorithm		Output Size (Bits)	Block Size (Bits)
MD5		128	512
SHA-1		160	512
SHA-2	SHA-224	224	512
	SHA-256	256	
	SHA-384	384	1024
	SHA-512	512	
	SHA-512/224	224	
	SHA-512/256	256	
SHA-3	SHA3-224	224	1152
	SHA3-256	256	1088
	SHA3-384	384	832
	SHA3-512	512	576
	SHAKE128	Arbitrary	1344
	SHAKE256		1088

Source: http://en.wikipedia.org/wiki/SHA-3.

Federal Information Processing Standard (FIPS) 180-2 is the standard that contains the specifications of both SHA-1 and SHA-2. SHA-2 is a set of cryptographic hash functions (SHA-224, SHA-256, SHA-384, SHA-512, SHA 512/224, SHA-512/256) published in 2001 by the National Institute of Standards and Technology (NIST) as a U.S. Federal Information Processing Standard (FIPS). SHA-2 includes a significant number of changes from its predecessor, SHA-1.

In 2007, NIST announced a competition for a new hash standard, SHA-3, to which 200 cryptographers from around the world submitted 64 proposals. In December 2008, 51 candidates were retained in the first round. In July 2009, the candidate list was reduced to 14 entries. Five finalists were selected in December 2010. After a final round of evaluations, NIST selected Keccak as SHA-3 in October 2012. Keccak was developed by a team of four researchers from STMicroelectronics, a European semiconductor company, and is optimized for hardware but can be implemented in software as well. It uses an entirely different technique from previous cryptographic algorithms, which use a compression function to process fixed-length blocks of data. Keccak applies a permutation process to extract a fingerprint from the data, a technique given the name of a "sponge function" (Harbert, 2012).

Table 3.3 provides a comparison of the parameters of the various hash algorithms.

3.7.2 Verification of the Integrity with Public Key Cryptography

An encryption algorithm with a public key is called "permutable" if the decryption and encryption operations can be inverted, that is, if

$$M = PK_X\left(SK_X\left(M\right)\right)$$

In the case of encryption with a permutable public key algorithm, an information element M that is encrypted by the private key SK_X of an entity X can be read by any user possessing the corresponding public key PK_X. A sender can, therefore, sign a document by encrypting it with a private key reserved for the signature operation to produce the seal that accompanies the message. Any person who knows the corresponding public key will be able to decipher the seal and verify that it corresponds to the received message.

Another way of producing the signature with public key cryptography is to encrypt the fingerprint of the document. This is because the encryption of a long document using a public key algorithm imposes substantial computations and introduces excessive delays. Therefore, it is beneficial to use a digest of the initial message before applying the encryption. This digest is produced by applying a one-way hash function to calculate the fingerprint that is then encrypted with the sender's private key. At the destination, the receiver recomputes the fingerprint. With the public key of the sender, the receiver will be able to decrypt the fingerprint to verify if the received hash value is identical to the computed hash value. If both are identical, the signature is valid.

The block diagram in Figure 3.7 represents the verification of the integrity with public key encryption of the hash. In this figure, h represents the hash function, O represents the encryption with the public key, and O^{-1} represents the decryption function to extract the hash for comparison with the recalculated hash at the receiving end.

The public key algorithms, which are frequently used to calculate digital signatures, are listed in Table 3.4.

Even though this method allows the verification of the message integrity, it does not guarantee that the identity of the sender is authentic.

In the case of a public key, a signature produced from a message with the signer's private key and then verified with the signer's corresponding public key is sometimes called a "signature scheme with appendix" (IETF RFC 3447, 2003).

3.7.3 Blind Signature

A *blind signature* is a special procedure for a notary to sign a message using the RSA algorithm for public key cryptography without revealing the content (Chaum, 1983, 1989). One possible utilization of this technique is to time-stamp digital payments.

Consider a debtor who would like to have a payment blindly signed by a bank. The bank has a public key e, a private key d, and a public modulo N. The debtor

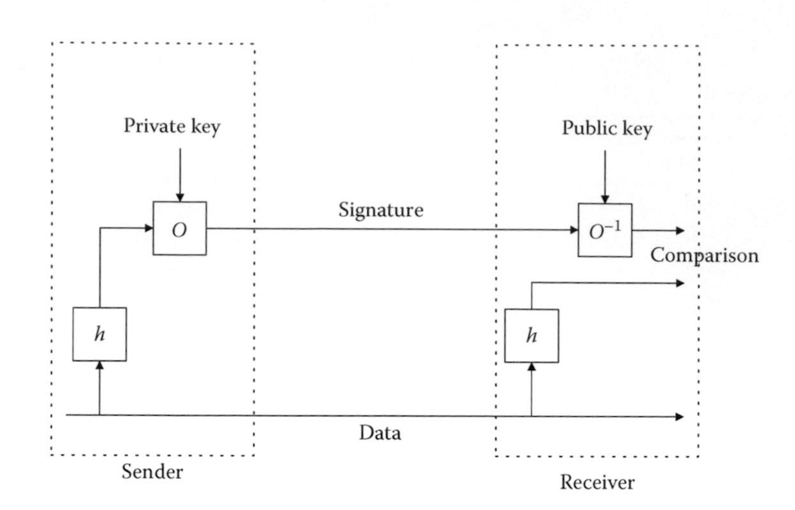

FIGURE 3.7
Computation of the digital signature using public key algorithms and hashing.

TABLE 3.4
Standard Public Key Algorithms for Digital Signatures

Algorithm	Comments	Signature Length in Bits	Standard
DSA	Digital Signature Algorithm is a variant of the ElGamal algorithm and published the Digital Signature Standard (DSS) proposed by NIST (National Institute of Standards and Technology) in 1994.	320, 448, 512	FIPS 186-4 (National Institute of Standards and Technology, 2013)
ECDSA	Elliptic Curve Digital Signature Algorithm, first standardized in 1998.	384, 488, 512, 768, 1024	ANSI X9.62:2005
RSA	This is the *de facto* standard algorithm for public key encryption; it can also be used to calculate signatures.	512–1024	ISO/IEC 9796

chooses a random number k between 1 and N and keeps this number secret.

The payment p is "enveloped" by applying

$$\left(pk^e\right)\text{mod } N$$

before sending the message to the bank. The bank signs it with its private key so that

$$(pk^e)^d \text{mod } N = p^d k \text{ mod } N$$

and returns the payment to the debtor. The debtor can now extract the signed note by dividing the number by k. To verify that the note received from the bank is the one that has been sent, the debtor can raise it to the e power because (as will be shown in Appendix 3B)

$$(p^d)^e \text{mod } N \equiv p \text{ mod } N$$

The various payment protocols for digital money take advantage of blind signatures to satisfy the conditions of anonymity.

3.7.4 Verification of the Integrity with Symmetric Cryptography

The message authentication code (MAC) is the result of a one-way hash function that depends on a secret key. This mechanism guarantees simultaneously the integrity of the message content and the authentication of the sender. (As previously mentioned, some authors call the MAC the "integrity check value" or the "cryptographic checksum.")

The most obvious way of constructing a MAC is to encrypt the hash value with a symmetric block encryption algorithm. The MAC is then affixed to the initial message, and the whole is sent to the receiver. The receiver recomputes the hash value by applying the same hash function on the received message and compares the result obtained with the decrypted MAC value. The equality of both results confirms the data integrity.

The block diagram in Figure 3.8 depicts the operations where h represents the hash function, C the encryption function, and D the decryption function.

Another variant of this method is to append the secret key to the message that will be condensed with the hash functions.

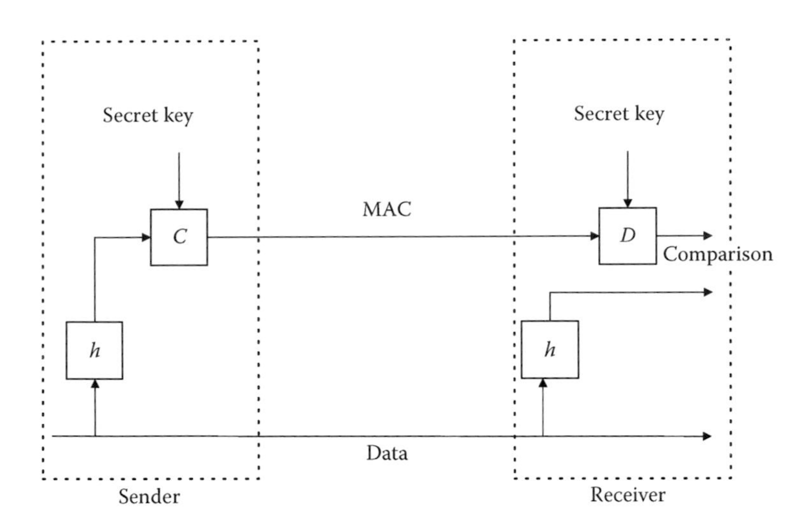

FIGURE 3.8
Digital signature with symmetric encryption algorithms.

It is also possible to perform the computations with the compression function $f()$ and use as an initial value the vector of the secret key, k, of length L bits in the following recursion:

$$k_i = f(k_{i-1}, x_i), \quad i = 1, \dots, n$$

where $x = \{x_1, x_2, \dots, x_n\}$ is the message subdivided into n vectors, each of B bits. The MAC is the value of the final output k_n.

The procedure that several U.S. and international standards advocate, for example, ANSI X9.9 (1986) for the authentication of banking messages and ISO 8731-1 (1987) and ISO/International Electrotechnical Commission (IEC) 9797-2 (2002) for implementing a one-way hash function, is to encrypt the message with a symmetric block encryption algorithm either in the cipher block chaining (CBC) or in the cipher feedback (CFB) modes. The MAC is the last encrypted block, which is encrypted one more time in the same CBC or CFB mode.

The following key hashing method augments the speed of computation in software implementation and increases the protection, even when the one-way hash algorithm experiences some rare collisions (Bellare et al., 1996).

Consider the message X subdivided into n vectors of B bits each and two keys (k_1 and k_2), each of L bits. The padding bits are added to the end of the initial message according to a determined pattern. The hashing operations can thus be described with the help of two compression functions $f_1()$ and $f_2()$:

$$k_i^1 = f_1(k_{i-1}^1, x_i)$$

$$k_i^2 = f_2(k_{i-1}^2, k_i^1)$$

where

k_0^1 and k_0^2 are the initial values of k_1 and k_2, respectively
$x = x_1, x_2, \dots, x_n$

The result that this method yields is denoted as the Nested Message Authentication Code (NMAC). It is, in effect, constructed by applying compression functions in sequence, the first on the padded initial message and the second on the product of the first operation after padding.

The disadvantage of this method is that it requires access to the source code of the compression functions to change the initial values. In addition, it requires the usage of two secret keys. This explains the current popularity of the keyed hashed message authentication code (HMAC), which is described in IETF RFC 2104 (1997) and NIST FIPS 198-1 (2008). This method uses one single key k of L bits.

Assuming that the function $H()$ represents the initial hash function, the value of the HMAC is computed in the following manner:

$$HMAC_k(x) = H\ [\bar{k} \oplus opad \,\|\, H(\bar{k} \oplus ipad, x0]$$

In this construction, $\bar{k}$ is the vector k of minimum length of L bits, which after padding with a series of 0 bits will reach a total length of B bits. The variables *opad* and *ipad* are constants for outer padding and inner padding, respectively. The variable *opad* is formed with the octet $0x36$ repeated as many times as needed to constitute a block of B bits. The variable *ipad* is the octet $0x5C$ repeated as many times. For MD5 and SHA-1, SHA-256, and SHA-512, the number of repetitions is 64. Finally, the symbols $\|$ and $\oplus$ in the previous equation denote, respectively, the concatenation and exclusive OR operations.

It should be noted that with the following representation

$$k^1 = f_1\left(\bar{k} \oplus ipad\right)$$

$$k^2 = f_2\left(\bar{k} \oplus opad\right)$$

the HMAC becomes the same as the nested MAC.

3.8 Identification of the Participants

Identification is the process of ascertaining the identity of a participant (whether a person or a machine) by relying on uniquely distinguishing feature. This contrasts with authentication, which is the confirmation that the distinctive identifier indeed corresponds to the declared user.

Authentication and identification of a communicating entity take place simultaneously when one party sends to the other in private a secret that is only shared between them, for example, a password or a secret encryption key. Another possibility is to pose a series of challenges that only the legitimate user is supposed to be capable of answering.

A digital signature is the usual means of identification because it associates a party (a user or a machine) with a shared secret. Other methods of simultaneous identification and authentication of human users exploit distinctive physiological and behavioral characteristics such as fingerprints, voiceprints, the shape of the retina, the form of the hand, signature, gait, and so on. The biometric identifiers used in electronic commerce are elaborated upon the following section.

3.9 Biometric Identification

Biometry systems use pattern recognition algorithms on specific physiological and/or behavior characteristics of individuals. Biometric identification systems recognize the identity of an individual by matching the features extracted from a biometric image with an entry in a database of templates. Identification is used mostly in forensic and law enforcement applications. Verification systems, in contrast, authenticate a person's identity by comparing a newly captured biometric image with that of person's biometric template stored in the system storage or in that user's credential (or identity card) (Maltoni et al., 2003, p. 3). Verification is the operation that controls access to resources such as a bank account or a payment system.

Biometrics was reserved until recently for forensic and military applications but is now used in many civilian applications. Biometric systems present several advantages over traditional security methods based on what the person knows (password or personal identification number [PIN]) or possesses (card, pass, or physical key). In traditional systems, the user must have several keys or cards, one for each facility, and remember different passwords to access each system. These keys or cards can be damaged, lost, or stolen, and long passwords can be easily forgotten, particularly if they are difficult and/or changed at regular intervals. When a physical card or key or a password is compromised, it is not possible to distinguish between a legitimate user and one who has acquired access illegally. Finally, the use of biological attributes for identification and authentication bypasses some of the problems associated with cryptography (e.g., key management).

There are two main categories of biometric features. The first category relates to behavioral patterns and acquired skills such as speech, handwriting, or keystroke patterns. In contrast, the second category comprises physiological characteristics such as facial features, iris morphology, retinal texture, hand geometry, or fingerprints. Methods based on gait, odor, or genetics using deoxyribonucleic acid (DNA) have limited applications for electronic payment systems. Methods using vascular patterns, palm prints, palm veins, or ear features have not been applied on a wide scale in electronic commerce.

Biometric verification consists of five steps: image acquisition during the registration phase, features extraction, storage, matching, and decision-making. The digital image of the person under examination originates from a sensor in the acquisition device (e.g., a scanner or a microphone). This image is processed to extract a compact profile that should be unique to that person. This profile or signature is then archived in a reference database that can be centralized or distributed according to the architecture of the system. In most cases, registration must be done in person in the presence of an operator to record the necessary biometric template.

The accuracy of a biometric system is typically measured in terms of a false acceptance rate (an impostor is accepted) and false rejection rate (a legitimate user is denied access). These rates are interdependent and are adjusted according to the required level of security. Other measures distinguish between the decision error rates and the matching errors, in the cases that the system allows multiple attempts or includes multiple templates for the same user (Faundez-Zanuy, 2006).

The choice of a particular system depends on several factors, such as the following:

1. The accuracy and reliability of the identification or verification. The result should not be affected by the environment or by aging.
2. Cost of installation, maintenance, and operation.
3. Scale of applicability of the technique; for example, handwriting recognition is not useful for illiterate people.
4. The ease of use.
5. The reproducibility of the results; in general, physiological characteristics are more reproducible than behavioral characteristics.
6. Resistance to counterfeit and attacks.

ISO/IEC 19794 consists of several parts specifying the data formats for biometric data as follows:

1. *Part 2*: Finger minutiae data
2. *Part 3*: Finger pattern spectral data
3. *Part 4*: Finger image data
4. *Part 5*: Face image data
5. *Part 6*: Iris image data
6. *Part 7*: Signature time series data
7. *Part 8*: Finger pattern skeletal data
8. *Part 9*: Vascular image data
9. *Part 10*: Hand geometry silhouette data

Part 11 concerns dynamic data collected during a manual signature and is currently being prepared.

3.9.1 Fingerprint Recognition

Fingerprinting is a method of identification and identity verification based on the ridge patterns of the fingerprints. The method is based on the belief that these ridge patterns are unique to each individual and that, in normal circumstances, they remain stable over an individual's life. In the past, fingerprints were collected by swiping the finger tips in a special ink and then pressing them over a paper to record a negative image. Fingerprint examiners would then look for specific features or *minutiae* in that image as distinguishing characteristics that identify an individual. Today, fingerprints are captured electronically by placing the finger on a small scanner using optical, electric, thermal, or optoelectronic transducers (Pierson, 2007, pp. 82–87).

Optical transducers are charge-coupled devices (CCDs) that measure the reflections of a light source on the finger and focused through a prism. Electric transducers measure the fluctuations in the capacitance between the user's fingers and sensors on the surface of a special mouse or home button. Another electric technique measures the changes in the electric field between a resin plate on which the finger rests and the derma with a low-tension alternating current injected into the finger pulps. Thermal techniques rely on a transducer tracking the temperature gradient between the ridges and the minutiae. Optoelectronic methods employ a layer of polymers to record the image of the fingerprint on a polymer layer that converts the image into a proportional electric current. Finally, ultrasound sensors are more capable of penetrating the epidermis to get an image of the pattern underneath it but are not yet ready for mass-market applications.

During the enrollment phase, the image of the user's fingerprint is recorded and then processed to extract the features that form the reference template that describes the user's minutiae. The template must consist of stable and reliable indices, insensitive to defects in the image that could have been introduced by dirty fingers, wounds, or deformities. Depending on the application, the template may be stored in a central database or recorded locally on a magnetic card or an integrated circuit card issued to the individual. ISO/IEC 19794-2 (2011) defines three standard data formats for the data elements of the templates.

During verification, fingerprints may be used alone or to supplement another identifier such as a password, a PIN, or a badge. The system then processes the new image to extract the relevant features for the matcher to conduct a one-to-one comparison with the user's template retrieved from storage.

There are two well-known categories of fingerprint recognition algorithms. Most algorithms are minutiae based and measure a variety of features, such as ridge endings and bifurcations (when two lines split), and compare them to the stored minutiae templates. Image-based methods use other characteristics such as orientation, ridge shape, and text texture. The latter approach is particularly useful when the establishment of a reliable minutiae map is made difficult, for example, because of a bad quality imprint or fingerprint damage due to a high degree of oxidative stress and/or immune activation.

The verification algorithm must be insensitive to potential translations, rotations, and distortions. The degree of similarity between the features extracted from the captured image and those in the stored template is described in terms of an index that varies from 0% to 100% or 1 to 5. The traditional measure was a count of the number of minutiae that corresponded, typically 12–17 (Ratha et al., 2001; Pierson, 2007, pp. 168–169).

NIST and the Federal Bureau of Investigation (FBI) collaborated to produce a large database of fingerprints gathered from crime scenes with their corresponding minutiae to train and evaluate new algorithms for automatic fingerprint recognition. NIST also provides a Biometric Image Software (NBIS) for biometric processing and analysis, free of charge, and with no licensing requirements (a previous version was called NIST Fingerprint Image Software or NFIS). It includes a fingerprint image quality algorithm, NFIQ, that analyzes that fingerprint image and assigns it a quality value of 1 (highest quality) to 5 (lowest quality).

NIST has conducted several evaluations of fingerprint technologies. In the latest evaluations in 2014, several databases were used from the federal government and law enforcement with different fingerprint combinations varying from a single index up to 10 fingers, impressions on flats, and plain and rolled impressions. (The rolled impressions are obtained during a roll from side to side to capture the full width.) It was found that, for a false-positive identification rate of 10^{-3}, the false-negative identification rates (i.e., the miss rates) were 1.9% for a single index finger to 0.10 for 10 fingers plain and rolled, that is, more fingers improved accuracy. The most accurate results were achieved with 10 fingers, with searches in datasets of 3 and 5 million subjects (Watson et al., 2014).

There are now several large-scale applications using fingerprints as biometrics. In 2002, the International Civil Aviation Organization (ICAO) adopted the following biometrics for machine-readable travel documents: facial recognition, fingerprint, and iris recognition. The templates are recorded as 2D PDF417 barcodes, or encoded in magnetic stripes, integrated circuits cards with contacts or contactless, or on optical memory (International Civil Aviation Organization, 2004, 2011). In 2003, the International Labor Organization (ILO) added two finger minutia–based biometric templates to the international identity document of seafarers encoded in the PDF417 stacked code format of ISO/IEC 15438 (International Labor Organization, 2006). In 2009, India launched the Aadhar national identity scheme using, the fingerprints of the 10 digits and iris biometrics to assign each citizen a unique 12-digit identification number (Unique Identification Authority of India, 2009).

Scanned fingerprints are also used in many payment applications. A typical approach used in payment is to use the fingerprint captured by placing the finger on the reader for identification then entering a PIN to authorize the payment. Citibank launched a biometric credit card in Singapore in 2006 using equipment from Pay By Touch. This technology was also used by several grocery chains in the United States and in the United Kingdom. In 2013, Apple incorporated a fingerprint scanner in its iPhone 5S under the brand name Touch ID. The percentage of false rejects in commercial systems reaches about 3%, and the rate of false acceptance is less than one per million.

Fingerprint spoofing has a long tradition but was more difficult with analog techniques. Since 2001, it was shown that automatic finger recognition with digital images is vulnerable to reconstructions of fake fingers made of gelatin ("gummy fingers") or of silicon rubber filled with conduct carbon ("conductive silicon fingers"). The patterns imprinted on the artificial fingers can be extracted from the negative imprints of real fingers, or from latent fingerprints lifted with a specialized toolkit, digitizing them with a scanner and then using the scanned image to imprint the artificial finger or from high-resolution photos of the true fingerprints (Matsumoto et al., 2002; Maltoni et al., 2003, pp. 286–291; Pierson, 2007, pp. 116–120; Galbally et al., 2010, 2011). A more recent technique is to use minutiae templates to reconstruct a digital image that is similar to the original fingerprint (Galbally et al., 2010). One advantage of thermal and capacitive sensors is that the quality of fake fingerprint images using gummy fingers is low, which could help in identifying spoofing scenarios (Galbally et al., 2011). However, the Chaos Computer Club in Germany announced less than a month after the introduction of Apple's Touch ID, which uses a capacitive transducer, that the system could be fooled with a fingerprint photographed on a glass surface (Chaos Computer Club, 2013; Musil, 2013).

It is worth noting that these results infirm the assumption made by the ILO in 2006 in allowing visible biometric data that "it is sufficiently difficult to reconstitute from the biometric data that will be stored in the barcode either an actual fingerprint (..) or a fraudulent device that could be used to misrepresent seafarer intent or presence" (International Labor Organization, 2006, § 5.1, p. 9).

It should be noted that some subjects do not have fingerprints or have badly damaged fingerprints.

Finally, it should be noted that the individuality of fingerprints remains an assumption. Most human experts and automatic fingerprint recognition algorithms declare that fingerprints have a common source if they are "sufficiently" similar in a given representation scheme based on a similarity metric (Maltoni et al., 2003, chapter 8). This has important implications in forensic investigation but probably will be less significant in payment applications.

3.9.2 Iris Recognition

The iris is the colored area between the white of the eye and the pupil, with a texture that is an individual characteristic that remains constant for many years. The digital image of the iris texture can be encoded as an *IrisCode* and used for the identification of individuals with accuracy and an error probability of the order of 1 in 1.2 million. It is even possible to distinguish among identical twins and to separate the two irises of the same person (Flom and Safir, 1987; Daugman, 1994, 1999; Wildes, 1997). The inspection is less invasive than a retinal scan.

This technique was developed by John Daugman and was patented by Iridian Technologies—previously known as IriScan, Inc, a company formed by two ophthalmologists and a computer scientist. This patent expired in 2004 in the United States and in 2007 in the rest of world, leaving the field open to new entrants. Since 2006, it is a division of Viisage Technology. A major supplier of the technology is currently the Lithuanian company Neurotechnology.

During image acquisition, the person merely faces a camera connected to computer at a distance of about 1 m. Iris scanning software can also be downloaded to mobile phones and smartphones. Some precautions need to be respected during image capture, particularly to avoid reflections by ensuring uniform lighting. Contact lenses produce a regular structure in the processed image.

Image acquisition can be accomplished in special kiosks or from desktop cameras. The latter are cheaper because they were proven difficult for some users that they must look into a hole to locate an illuminated ring. ISO/IEC 19794-6 (2011) defines the format of the captured data.

Iris recognition is typically used as a secondary identifier in addition to fingerprint imaging. Other potential applications include the identification of users of automatic bank teller machines, the control of access either to a physical building or equipment or to network resources. It is being evaluated to increase the speed of passenger processing at airports. As mentioned earlier, India is collecting iris images together with the 10 fingerprints and an image of the face for its new national identity scheme (Unique Identification Authority of India, 2009).

It was shown, however, that the accuracy of the current generation of commercial handheld iris recognition devices degrades significantly in tactical environments (Faddis et al., 2013). Furthermore, it was demonstrated that the templates or IrisCodes have sufficient information to allow the reconstruction of synthetic images that closely match the images collected from real subjects and then deceive commercial iris recognition systems. The success rate of the synthetic image ranges from 75% to 95%, depending on the level set for the false acceptance rate. The attack is also successful that even the reconstructed image of the iris does not use the stored templates that the system used for recognition (Galbally et al., 2012). This creates the possibility of stealing someone's identity through their iris image.

3.9.3 Face Recognition

Police and border control agencies have been using facial recognition to scan passports. This has been done on the basis of a template that encodes information derived from proprietary mathematical representations of features in the face image, such as the distance between the eyes, the gap between the nostrils, the dimension of the mouth. U.S. government initiatives, starting with the Face Recognition Technology (FERET) program in September 1993, have provided the necessary stimulus to improve the speed and the accuracy of the technology. ISO/IEC 19794-5 (2011) defines the format for the captured data.

Some consumer applications, such as Apple's iPhoto or Google's Picasa, have included facial recognition to identify previously defined faces in photo albums. Another consumer application is Churchix (http://churchix.com, last accessed January 26, 2016), designed to help church administrators and event managers track their members attendance by comparison with a database of reference photos. Nowadays, almost all mobile phones have a photographic camera and newer programs that allow people to take a picture of a person and then search the Internet for possible matches to that face. Polar Rose, a company from Malmö, Sweden, which produced such software, was acquired by Apple in 2010. There are also tools to help search for pictures on Facebook. Other applications allow users to identify a person in a photo and then search the web for other pictures in which that person appears (Palmer, 2010).

Facial recognition, however, is not very accurate in less controlled situations. The accuracy is sensitive to accessories such as sunglasses, beards or mustaches, grins, or head tilts (yaw) by 15°. It can be combined with other biometrics such as fingerprints (Yang, 2010; de Oliveir and Motta 2011) or keystrokes (Popovici et al., 2014) to increase the overall system accuracy.

An international contest for face verification algorithms was organized in conjunction with the 2004 International Conference on Biometric Authentication (ICBA). The National Institute of Standards and Technology (NIST) has tracked the progress of face recognition technologies in a series of benchmark evaluations of prototype algorithms from universities and the research industrial laboratories in 1997, 2002, 2006, 2010, and 2013. In recent years, many participants were major commercial suppliers from the United States, Japan, Germany, France, Lithuania, and Chinese academics. The 2010 results showed that an algorithm from NEC was the most accurate with chances of identifying an unknown subject from a database of 1.6 million records about 92%, the search lasting about 0.4 second. The search duration, however, increased linearly with the size of the enrolled population, while the speed of less accurate algorithms increased more slowly with the size of the database (Grother et al., 2010). In general, the differences between the best and worst performances can be significant.

The National Police Academy in Japan followed NIST study with a focused investigation of NEC algorithm to evaluate the influence of several factors such as (1) the effect of age, due to the time lapse between the recorded picture and the subject; (2) the shooting angle, that is, the yaw; (3) the effect of face expression; and (4) of accessories such as a cap, sunglasses, beards, and mustaches. The results confirmed that from 2001 to 2010, the top of the line face recognition algorithms improved significantly. The overall recognition rate was 95% (compared to 48% in 2001). In particular, 98% of the images without glasses were classified correctly (compared to 73% in 2001). Similar improvements were observed for the effects of the yaw (provided that the angle is less than 30°) of the expression, and of the various accessories, such as spectacles (Horiuchi and Hada, 2013). It should be noted that the results were obtained with pictures taken under good lighting conditions, which is not always the case for police investigations, which often start with low-quality images, taken at bad angles and under poor lights and/or with small face sizes.

NIST (2013) study used reasonable quality law enforcement mug shot images, poor-quality web cam images collected in similar detention operations and moderate quality visa application images. The range thus covered high quality in identity documents (passports, visa applications, driver licenses) to poorly controlled surveillance situations. The 2013 study confirmed that NEC had the most accurate set of algorithms, with an error rate of less than half of the second place algorithm. It was followed by those supplied by Morpho (a Safran company), which merged its algorithms with those acquired from L1 Identity Solutions in 2011. Other leading suppliers are Toshiba, Cognitec Systems, and 3M/Cogent. Algorithms with lesser accuracy (error rate higher than 12%) are those from Neurotechnology, Zhuhai Yisheng, HP, Decatur, and Ayonix. Some Chinese universities provided high-accuracy algorithms as well. The ranking of performance across algorithms, however, must be weighed by application-specific requirements; some algorithms are more suited to recognition of difficult webcam images (Grother and Ngan, 2014).

3.9.4 Voice Recognition

Depending on the context, voice recognition systems have one of two functions:

1. *Speaker identification*: This is a one-to-many comparison to establish the identity of the speaker. A digital vocal print, newly acquired from the end user, is compared with a set of stored acoustic references to determine the person from its utterances.
2. *Speaker verification*: This case consists in verifying that the voice imprint matches the acoustic references of the person that the speaker pretends to be.

The size of the voice template that characterizes an individual varies depending on the compression algorithm and the duration of the record.

Speaker verification is implemented in payment systems as follows. The voice template that characterizes a subject is extracted during registration through the pronunciation of one or several passwords or passphrases. During the verification phase, the speaker utters one of these passwords back. The system then attempts to match the features extracted from new voice sample with the previously recorded voice templates. After identity verification, the person is allowed access.

A bad sound quality can cause failures. In remote applications, this quality depends on several factors such as the type of the telephone handset, ambient noise (particularly in the case of hands-free telephony), and the type of connection (wire line or wireless). The voice quality is also affected by a person's health, stress, emotions, and so on.

Some actors have the capability of mimicking other voices. Furthermore, there are commercial personalized text-to-speech systems that produce voice automation prompts using familiar voices (Burg, 2014). Speech synthesis algorithms require about 10–40 hours of professionally recorded material with the voice to be mimicked. Using archival recordings, they could also bring voices back from the dead. An easier method to defraud the system would to be playback recordings of authentic commands. This is why automatic speaker recognition systems must be supplemented with other means of identification.

3.9.5 Signature Recognition

Handwritten signatures have long been established as the most widespread means of identity verification particularly for administrative and financial institutions. They are a behavioral biometrics that changes over time and is affected by many physical and emotional conditions. ISO/IEC 19794-7 specifies two formats for the time series describing the captured signatures. One format is for general use and the other is more compact for use with smart cards.

Systems for automatic signature recognition rely on the so-called permanent characteristics of an individual handwriting to match the acquired signature with a prerecorded sample of the handwriting of the person whose identity is to be verified. It goes without saying that handwritten signatures do not work in areas with high illiteracy and are not persistent before 16 years of age (Blanco-Gonzalo et al., 2013).

Signature recognition can be static or dynamic. In static verification, also called offline verification, the signature is compared with an archived image of the signature of the person to be authenticated. For dynamic signature recognition, also called online verification, signing can be with a stylus on a digitizing tablet or the fingertip on a pad connected to a computer, or on the screen of a tablet or a mobile phone. The algorithms are based on techniques such as dynamic time warping to compare two time series even with their different timescales. The parameters under consideration are derived from the written text and various movement descriptors such as the pressure exercised on the pad or the screen, the speed and direction of the movement, the accelerations and decelerations, the angles of the letters, and the azimuth and the altitude of the pen with respect to the plan of the pad. The static approach is more susceptible to forgeries because the programs have yet to include all the heuristics that graphologists take into consideration based on their experience. There are enormous differences in the way people from different countries sign, and there are specific approaches for non-Latin scripts (Impedovo and Pirlo, 2008).

Even though users prefer stylus-based devices to finger-based systems, an initial evaluation showed their performances are comparable and even that finger signing on an Apple iPad2 can exceed that of stylus-based devices, but there are no obvious explanations (Blanco-Gonzalo et al., 2013).

3.9.6 Keystroke Recognition

Keystroke recognition is a technique based on an individual's typing style in terms of rhythm, speed, duration and pressure of keystrokes, and so on. It is based on the premise that each person's typing style is distinct enough to verify a claimed identity. Keystroke measures are based on several repetitions of a known sequence of characters (e.g., the login and the password) (Obaidat and Sadoun, 1999; Dowland et al., 2002).

There are no standards for the data collected, and the user profile is based on a small dataset, which could result in high error rates. Typically, the samples are collected using structured texts at the initiation of a session to complement a classic login approach. The keystrokes are monitored unobtrusively as the person is keying in information. This approach, however, does not protect against session hijacking, that is, seizing control of the session of a legitimate user after a valid access. There is an active research to include continuous verification of the user's identity using free text, that is, what users type during their normal interaction with a computing device (Ahmed and Traore, 2014; Popovici et al., 2014). Another possibility to reduce error rate and to avoid session hijack is to combine keystroke dynamics with face recognition (Giot et al., 2010).

One important security risk of keystroke monitoring is that any software positioned at an input device can also leak sensitive information, such as passwords (Shah et al., 2006).

3.9.7 Hand Geometry

In the past several years, hand geometry recognition has been used in large-scale commercial applications to control access to enterprises, customs, hospitals, military bases, prisons, and so on.

The principle is that some features related to the shape of the human hand, such as the length of the fingers, are relatively stable and peculiar to an individual. The image acquisition requires the subject's cooperation. The user positions the hand with outstretched fingers flat on a plate facing the lens of a digital camera. The fingers are spread, resting against guiding pins soldered on the plate. This plate is surrounded by mirrors on three sides to capture the frontal and side-view images of the hand. The time for taking one picture is about 1.2 seconds. Several pictures (three to five) are taken, and the average is stored in memory as a reference to the individual. One commercial implementation uses a 3D model with 90 input parameters to describe the hand geometry with a template of 9 octets.

3.9.8 Retinal Recognition

The retina is a special tissue of the eye that responds to light pulses by generating proportional electrical discharges to the optical nerve. It is supplied by a network of blood vessels according to a configuration that is characteristic of each individual and that is stable throughout life. The retina can even distinguish among twins. A retinal map can be drawn by recording the reflections of a low-intensity infrared beam with the help of a charge-coupled device (CCD) to form a descriptor of 35 octets.

The equipment used is relatively large and costly, and image acquisition requires the cooperation of the subject. It entails looking through an eyepiece and concentration on an object while a low-power laser beam is injected into the eye. As a consequence, the technique is restricted to access control to high-security areas: military installations, nuclear plants, high-security prisons, bank vaults, and network operation centers. Currently, this technique is not suitable for remote payment systems or for large-scale deployment.

3.9.9 Additional Standards

FIPS 201 (National Institute of Standards and Technology, 2013) defines the procedures for personal identity verification for federal employees and contractors, including poof of identity, registration, identity card issuance and reissuance, chains of trust, and identity card usage. The biometrics used are fingerprints, iris images, and facial images. NIST Special Publication (SP) 800-76-2 contains the technical specifications for the biometric data used in the identification cards. In particular, the biometric data use the Common Biometric Exchange File Format (CBEFF) of ANSI INCITS 398 (Podio et al., 2014).

ISO/IEC 7816-11 (2004) specifies commands and data objects used in identity verification with a biometric means stored in integrated circuit cards (smart cards). ITU-T Recommendation X.1084 (2008) defines the Handshake protocol between the client (card reader or the terminal) and the verifier, which is an extension of TLS.

ANSI INCITS 358, published in 2002 and revised in 2013, contains BioAPI 2.0, a standard interface that allows a software application to communicate with and utilize the services of one or more biometric technologies. (ISO/IEC 19784-1 [2006] is the corresponding international standard). ITU-T Recommendation X.1083 (ISO/IEC 24708) specifies the BioAPI Interworking Protocol (BIP), that is, the syntax, semantics, and encodings of set of messages ("BIP messages") that enable a BioAPI-conforming application to request biometric operations in BioAPI-conforming biometric service providers. It also specifies extensions to the architecture and the BioAPI framework (specified in ISO/IEC 19784-1) that support the creation, processing, sending, and reception of BIP messages.

BioAPI 2.0 is the culmination of several U.S. initiatives starting in 1995 with the Biometric Consortium (BC). The BC was charted to be the focal point for the U.S. government on research, development, testing, evaluation, and application of biometric-based systems for personal identification/verification. In parallel, the U.S. Department of Defense (DOD) initiated a program to develop a Human Authentication–Application Program Interface (HA–API). Both activities were merged, and version 1.0 of the BioAPI was published in March 2000. The BioAPI specification was also the basis of the ANSI 358-2002[R2012], first published in 2002 and revised in 2007 and 2013.

ANSI X9.84 (2010) describes the security framework for using biometrics for the authentication of individuals in financial services. The scope of the specification is the management of the biometric data throughout the life cycle: collection, storage, retrieval, and destruction. Finally, the Fast Identity Online Alliance (FIDO) was established in July 2012 to develop technical specifications that define scalable and interoperable mechanisms to authenticate users by hand gesture, a biometric identifier, or pressing a button.

3.9.10 Summary and Evaluation

There is a distinction between the flawless image of biometric technology presented by promoters and futuristic authors and the actual performance in terms of robustness and reliability. For example, the hypothesis that biometric traits are individual and that they are sufficiently stable in the long term is just an empirical observation. Systems that perform well in many applications have yet to scale up to cases involving tens of millions of users. False match rates, which may be acceptable for specific tasks in small-scale applications, can be too costly or even dangerous in applications designed to provide high level of security. For some techniques, the biometrics images obtained during data acquisition vary significantly from one session to another.

Attacks on biometric systems can be grouped into two main categories: those that are common to all information systems and those that are specific to biometrics. The first category of attacks comprises the following:

1. Attacks on the communication channels to intercept the exchanges, for example, to snoop on the user image of the biometrics in question or the features extracted from that image. The stolen information can be replayed as if it came from a legitimate user.
2. Attacks of the various hardware and software modules, such as physical tampering with the sensors or inserting malware at the feature extraction module or the matcher module.
3. Attack on the database where the templates are stored.

Attacks specific to biometrics constitute the second category. The automatic recognition system is deceived with the use of fake fingers or reconstructed iris patterns.

In total, there are eight places in a generic biometric system where attacks may occur (Ratha et al., 2001), these are identified in the following list:

1. Fake biometrics are presented to the sensor such as a fake finger, a copy of a signature, or a face mask.
2. Replay attacks to bypass the sensor with previously digitized biometric signals.
3. Override of the feature extraction process with a malware attack to allow the attacker to control the features to be processed.
4. A more difficult attack involves tampering with the template extracted from the input signal before it reaches the matcher. This is the case if they are not collocated and are connected over a long-haul network. Another condition for this attack to succeed is that the method for coding the features into a template is known.
5. An attack on the matcher so that it produces preselected results.

6. Manipulation of the stored template, particularly if it is stored on a smart card to be presented to the authentication system.
7. Interception and modification of the template retrieved from the database on its way to the matcher.
8. Overriding the final decision by changing the display results.

Encryption solves the problem of data production, and digital signatures solve the problem of data integrity. Replay attacks can be thwarted by time-stamping the exchange or by including nonces. If the matcher and database are colocated in a secure location, many of these attacks can be thwarted. However, automated biometric authentication is still susceptible to attacks against the sensor by replaying previous signals or using fake samples.

Current research activities are related to solving these problems. For example, since 2011, the University of Cagliari in Italy and Clarkson University in the United States have hosted LivDet, an annual completion for fingerprint "liveness" (i.e., aliveness) detection. The goal of the competition is to compare different algorithms to separate "fake" from "live" fingerprint images. One of the criteria used in software-based techniques of liveness detection is the quality of the image, on the assumption that degree of sharpness, color and luminance levels and local artifacts, and other aspects of a fake image will have a lower quality than a real sample acquired under normal operating conditions (Galbally et al., 2014). Hardware-based techniques add some specific sensors to detect living characteristics such as sweat, book pressure, or specific reflection properties of the eye.

In 2014, to improve the scientific basis of forensic evidence in courts, NIST and the U.S. Department of Justice jointly established the Forensic Sciences Standards Board (FSSB) within NIST's structure to foster the development and forensic standards. The board comprises members from the U.S. universities and professional forensic associations.

One of the most problematic issues is that once a biometric image or template is compromised, it is stolen forever and cannot be reissued, updated, or destroyed. Reissuing a new PIN or password is possible, but having a new set of fingerprints is very difficult.

3.10 Authentication of the Participants

The purpose of authentication of participants is to reduce, if not eliminate, the risk that intruders might masquerade under legitimate appearances to pursue unauthorized operations.

As stated previously, when the participants utilize a symmetric encryption algorithm, they are the only ones who share a secret key. As a consequence, the utilization of this algorithm, in theory, guarantees the confidentiality of the messages, the correct identification of the correspondents, and their authentication. The key distribution servers also act as authentication servers (ASs), and the good functioning of the system depends on the capability of all participants to protect the encryption key.

In contrast, when the participants utilize a public key algorithm, a user is considered authentic when that user can prove that he or she holds the private key that corresponds with the public key that is attributed to the user. A certificate issued by a certification authority indicates that it certifies the association of the public key (and therefore the corresponding private key) with the recognized identity. In this manner, identification and authentication proceed in two different ways, the identity with the digital signature and the authentication with a certificate. Without such a guarantee, a hostile user could create a pair of private/public keys and then distribute the public key as if it was that of the legitimate user.

Although the same public key of a participant could equally serve to encrypt the message that is addressed to that participant (confidentiality service) and to verify the electronic signature of the documents that the participant transmits (integrity and identification services), in practice, a different public key is used for each set of services.

According to the authentication framework defined by ITU-T Recommendations X.500 (2001) and X.811 (1995), simple authentication may be achieved by one of several means:

1. Name and password in the clear
2. Name, password, and a random number or a time stamp, with integrity verification through a hash function
3. Name, password, a random number, and a time stamp, with integrity verification using a hash function

Strong authentication requires a certification infrastructure that includes the following entities:

1. *Certification authorities* to back the users' public keys with "sealed" certificates (i.e., signed with the private key of the certification authority) after verification of the physical identity of the owner of each public key.
2. A database of authentication data (*directory*) that contains all the data relative to the private encryption keys, such as their value, the

duration of validity, and the identity of the owners. Any user should be able to query such database to obtain the public key of the correspondent or to verify the validity of the certificate that the correspondent would present.

3. A *naming* or *registering authority* that may be distinct from the certification authority. Its principal role is to define and assign unique *distinguished names* to the different participants.

The certificate guarantees the correspondence between a given public key and the entity, the unique distinguished name of which is contained in the certificate. This certificate is sealed with the private key of the certification authority. When the certificate owner signs documents with the private signature key, the partners can verify the validity of the signature with the help of the corresponding public key that is contained in the certificate. Similarly, to send a confidential message to a certified entity, it is sufficient to query the directory for the public key of that entity and then use that key to encrypt messages that only the holder of the associated private key would be able to decipher.

3.11 Access Control

Access control is the process by which only authorized entities are allowed access to the resources as defined in the access control policy. It is used to counter the threat of unauthorized operations such as unauthorized use, disclosure, modification, destruction of protected data, or denial of service to legitimate users. ITU-T Recommendation X.812 (1995) defines the framework for access control in open networks. Accordingly, access control can be exercised with the help of a supporting authentication mechanism at one or more of the following layers: the network layer, the transport layer, or the application layer. Depending on the layer, the corresponding authentication credentials may be X.509 certificates, Kerberos tickets, simple identity, and password pairs.

There are two types of access control mechanisms: identity based and role based. Identity-based access control uses the authenticated identity of an entity to determine and enforce its access rights. In contrast, for role-based access control, access privileges depend on the job function and its context. Thus, additional factors may be considered in the definition of the access policy, for example, the strength of the encryption algorithm, the type of operation requested, or the time of day. Role-based access control provides an indirect means of bestowal of privileges through three distinct phases: the definition of roles, the assignment of privileges to roles, and the distribution of roles among users. This facilitates the maintenance of access control policies because it is sufficient to change the definition of roles to allow global updates without revising the distribution from top to bottom.

At the network layer, access control in IP networks is based on packet filtering using the protocol information in the packet header, specifically the source and destination IP addresses and the source and destination port numbers. Access control is achieved through "line interruption" by a certified intermediary or a firewall that intercepts and examines all exchanges before allowing them to proceed. The intermediary is thus located between the client and the server, as indicated in Figure 3.9. Furthermore, the firewall can be charged with other security services, such as encrypting the traffic for confidentiality at the network level or integrity verification using digital signatures. It can also inspect incoming and outgoing exchanges before forwarding them to enforce the security policies of a given administrative domain. However, the intervention of the trusted third party must be transparent to the client.

The success of packet filtering is vulnerable to packet spoofing if the address information is not protected and if individual packets are treated independently of the other packets of the same flow. As a remedy, the firewall can include a proxy server or an application-level gateway that implements a subset of application-specific functions. The proxy is capable of inspecting all packets in light of previous exchanges of the same flow before allowing their passage in accordance with

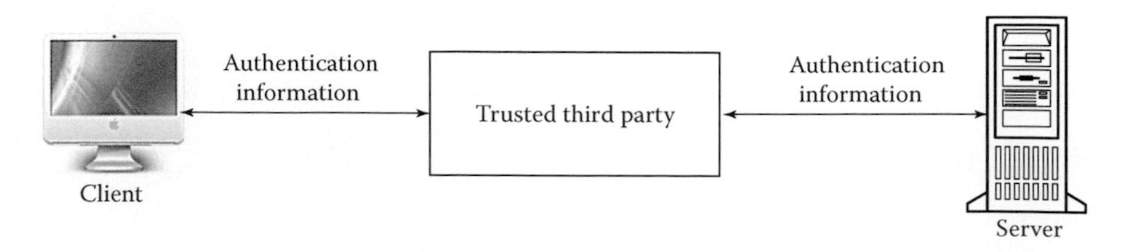

FIGURE 3.9
Authentication by line interruption at the network layer.

the security policy in place. Thus, by filtering incoming and outgoing electronic mail, file transfers, exchanges of web applications, and so on, application gateways can block nonauthorized operations and protect against malicious codes such as viruses. This is called a *stateful* inspection. The filter uses a list of keywords, the size and nature of the attachments, the message text, and so on. Configuring the gateway is a delicate undertaking because the intervention of the gateway should not prevent daily operation.

A third approach is to centralize the management of the access control for a large number of clients and users with different privileges with a dedicated server. Several protocols have been defined to regulate the exchanges among network elements and Access Control Servers. RFC 6929 (2013) specifies Remote Authentication Dial in User Service (RADIUS) for client authentication, authorization, and for collecting accounting information of the calls. In RFC 1492 (1993), Cisco described a protocol called Terminal Access Controller Access System (TACACS), which was later updated in TACACS+. Both RADIUS and TACACS+ require a secrete key between each network element and the server. Depicted in Figure 3.10 is the operation of RADIUS in terms of a client/server architecture. The RADIUS client resides within the Access Control Server while server relies on an X.509 directory through the protocol Lightweight Directory Access Protocol (LDAP). Both X.509 and LDAP will be presented later in this chapter.

Note that both server-to-client authentication and user-to-client authentication are outside the scope of RADIUS. Also, because RADIUS does not include provisions for congestion control, large networks may suffer degraded performance and data loss. It should be noted that there are some known vulnerabilities in RADIUS or in its implementations (Hill, 2001).

Commercial systems implement two basic approaches for end-user authentication: one-time password and challenge response (Forrester et al., 1998). In a typical one-time password system, each user has a device that generates a number periodically (usually every minute) using the current time, the card serial number, and a secret key held in the device. The generated number is the user's one-time password. This procedure requires that the time reference of the Access Control Server be synchronized with the card so that the server can regenerate an identical number.

In challenge–response systems, the user enters a personal identification number to activate handheld authenticators (HHA) and then to initiate a connection to an Access Control Server. The Access Control Server, in turn, provides the user with a random number (a challenge), and the user enters this number into a handheld device to generate a unique response. This response depends on both the challenge and some secret keys shared between the user's device and the server. It is returned to the Access Control Server to compare with the expected response and decide accordingly.

Two-factor authentication is also used in many services, particularly in mobile commerce domain. The site to be accessed sends a text to the user that is requesting access with a new code and a password each time the user attempts to login.

3.12 Denial of Service

Denial-of-service attacks prevent normal network usage by blocking the access of legitimate users to the network resources they are entitled to, by overwhelming

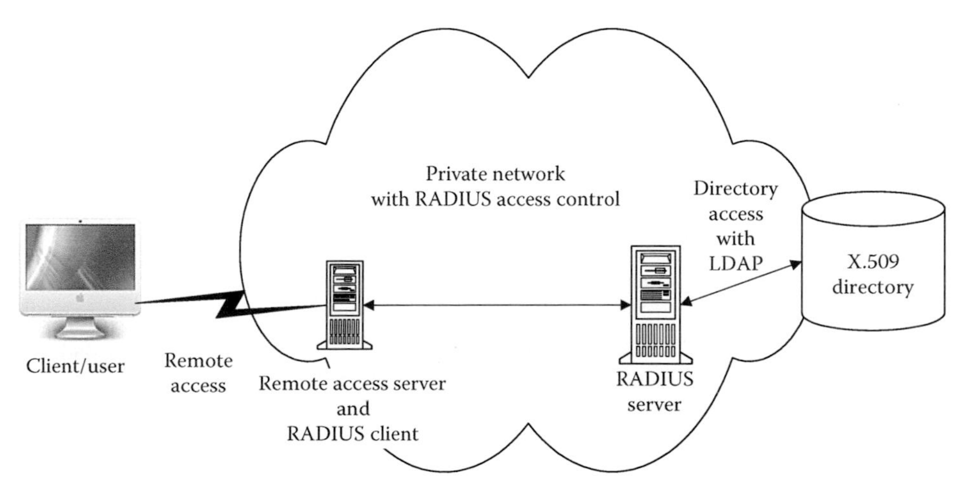

FIGURE 3.10
Remote access control with RADIUS.

the hosts with additional or superfluous tasks to prevent them from responding to legitimate requests or to slow their response time below satisfactory limits.

In a sense, denial of service results from the failure of access control. Nevertheless, these attacks are inherently associated with IP networks for two reasons: network control data and user data share the same physical and logical bandwidths; and IP is a connectionless protocol where the concept of admission control does not apply. As a consequence, when the network size exceeds a few hundred nodes, network control traffic (due, e.g., to the exchange of routing tables) may, under some circumstances, occupy a significant portion of the available bandwidth. Further, inopportune or ill-intentioned user packets may able to bring down a network element (e.g., a router), thereby affecting not only all endpoints that rely on this network element for connectivity but also all other network elements that depend on it to update their view of the network status. Finally, in distributed denial-of-service (DDOS) attacks, a sufficient number of compromised hosts may send useless packets toward a victim around the same time, thereby affecting the victim's resources or bandwidth or both (Moore et al., 2001; Chang, 2002).

As a point of comparison, the public switched telephone network uses an architecture called common channel signaling (CCS) whereby user data and network control data travel on totally separate networks and facilities. It is worth noting that CCS was introduced to protect against fraud. In the old architecture, called Channel-Associated Signaling (CAS), the network data and the user data used separate logical channels on the same physical support.

Denial-of-service attacks work in two principal ways: forcing a protocol state machine a deadlock situation and overwhelming the processing capacity of the receiving station.

One of the classical examples of protocol deadlocks is the SYN flooding attack that perturbs the functioning of the TCP protocol (Schuba et al., 1997). The handshake in TCP is a three-way exchange: a connection request with the SYN packet, an acknowledgment of that request with SYN/ACK packet, and finally, a confirmation from the first party with the ACK packet (Comer, 1995, p. 216). Unfortunately, the handshake imposes asymmetric memory and computational loads on the two endpoints, the destination being required to allocate large amounts of memory without authenticating the initial request. Thus, an attacker can paralyze the target machine by exhausting its available resources by sending a massive number of fake SYN packets. These packets will have spoofed source addresses so that the acknowledgments are sent to hosts that the victim cannot reach or that do not exist. Otherwise, the attack may fail because unsolicited SYN/ACK packets at accessible hosts provoke the transmission of RST packets, which, upon arrival, would allow the victim to release the resources allocated for a connection attempt.

Internet Control Message Protocol (ICMP) is a protocol for any arbitrary machine to return control and error information back to the presumed source. To flood the victim's machine with messages that overwhelm its capacities, an ICMP echo request ("ping") is sent to all the machines of a given network using the subnet broadcast address but with victim's address falsely indicated as the source.

The Code Red worm is an example of attacks that exploit defects in the software implementation of some web servers. In this case, Hypertext Transfer Protocol (HTTP) GET requests larger than the regular size (in particular, a payload of 62 octets instead of 60 octets) would cause, under specific conditions, a buffer overflow and an upsurge in HTTP traffic. Neighboring machines with the same defective software will also be infected, thereby increasing network traffic and causing a massive disruption (CERT/CC CA-2001-19, 2002).

Given that IP does not separate user traffic from that of the network, the best solution is to identify all with trusted certificates. However, authentication of all exchanges increases the computational load, which may be excessive in commercial applications. Short of this, defense mechanisms will be developed on a case-by-case basis to address specific problems as they arise. For example, resource exhaustion due to the SYN attack can be alleviated by limiting the number of concurrent pending TCP connections, reducing the time-out for the arrival of the ACK packet before calling off the connection establishment, and blocking packets to the outside that have source addresses from outside.

Another approach is to reequilibrate the computational load between the two parties by asking the requesting client to solve a *puzzle* in the form of simple cryptographic problems before the allocated resources needed to establish a connection. To avoid replay attacks, these problems are formulated using the current time, a server secret, and additional information from the client request (Juels and Brainard, 1999). This approach, however, requires programs for solving puzzles specific to each application that are incorporated in the client browser.

A similar approach is to require the sender to do some hash calculation before sending a message to make spamming uneconomic in the form of a proof of work (Warren, 2012). The difficulty of the proof of work is made proportional to the size of the message, and each message is time-stamped to protect the network against denial-of-service attacks by flooding old messages.

The aforementioned solutions require a complete overhaul of the Internet architecture. Yet, in their absence, electronic commerce is vulnerable to interferences from botnets that can be thwarted with *ad hoc* solutions only. A *botnet* is a virtual network constituted by millions of terminals infected with a specialized virus or worm called a *bot*. The bot does not destroy data on the infected terminal nor does it affect any of the typical services; all what it does is to give the bot master remote control. Each client of botnet can be instructed to send messages surreptitiously in spamming campaigns or in distributed denial-of-service attacks. The bot master is also responsible for the bot maintenance: fixing errors, ensuring that the bot remains undetected, and changing the IP address of the master server periodically to prevent tracing. Once all is in place, the bot master offers the botnet services in an online auction to the highest bidder, which can be a political opponent or a commercial competitor of the website under attack.

3.13 Nonrepudiation

Nonrepudiation is a service that prevents a person who has accomplished an act from denying it later, in part or as a whole. Nonrepudiation is a legal concept to be defined through legislation. The role of informatics is to supply the necessary technical means to support the service offer according to the law. The building blocks of nonrepudiation include the electronic signature of documents, the intervention of a third party as a witness, time stamping, and sequence numbers. Among the mechanisms for nonrepudiation are a security token sealed with the secret key of the verifier that accompanies the transaction record, time stamping, and sequence numbers. Depending on the system design, the security token sealed with the verifier's secret key can be stored in a tamper-resistant cryptographic module. The generation and verification of the evidence often require the intervention of one or more entities external to parties to the transaction such as a notary, a verifier, and an adjudicator of disputes.

ITU-T Recommendation X.813 (1996) defines a general framework for nonrepudiation in open systems. Accordingly, the service comprises the following measures:

- Generation of the evidence
- Recording of the evidence
- Verification of the evidence generated
- Retrieval and reverification of the evidence

There are two types of nonrepudiation services:

1. *Nonrepudiation at the origin.* This service protects the receiver by preventing the sender from denying having sent the message.
2. *Nonrepudiation at the destination.* This service plays the inverse role of the preceding function. It protects the sender by demonstrating that the addressee has received the message.

Threats to nonrepudiation include compromise of keys or unauthorized modification or destruction of evidence. In public key cryptography, each user is the sole and unique owner of the private key. Thus, unless the whole system has been penetrated, a given user cannot repudiate the messages that are accompanied by his or her electronic signature. In contrast, nonrepudiation is not readily achieved in systems that use symmetric cryptography. A user can deny having sent the message by alleging that the receiver has compromised the shared secret or that the key distribution server has been successfully attacked. A trusted third party would have to verify each transaction to be able to testify in cases of contention.

Nonrepudiation at the destination can be obtained using the same mechanisms but in the reverse direction.

3.13.1 Time Stamping and Sequence Numbers

Time stamping of messages establishes a link between each message and the date of its transmission. This permits the tracing of exchanges and prevents attacks by replaying old messages. If clock synchronization of both parties is difficult, a trusted third party can intervene as a notary and use its own clock as reference.

The intervention of the "notary" can be in either of the following mode:

- Offline to fulfill functions such as certification, key distribution, and verification if required, without intervening in the transaction.
- Online as an intermediary in the exchanges or as an observer collecting the proofs that might be required to resolve contentions. This is a similar role to that of a trusted third party of the network layer (firewall) or at the application layer (proxy) but with a different set of responsibilities.

Let's assume that a trusted third party combines the functions of the notary, the verifier, and the adjudicator. Each entity encrypts its messages with the secret key that has been established with the trusted third party before sending the message. The trusted third party decrypts the message with the help of this shared secret with the intervening party, time-stamps it, and then reencrypts it with the key shared with the other party. This approach requires the establishment of a secret key between each entity and the trusted third party that acts as a delivery messenger. Notice, however, that the time-stamping procedures have not been normalized and each system has its own protocol.

Detection of duplication, replay, as well as the addition, suppression, or loss of messages is achieved with the use of a sequence number before encryption. Another mechanism is to add a random number to the message before encryption. All these means give the addressee the ability to verify that the exchanges genuinely took place during the time interval that the time stamp defines.

3.14 Secure Management of Cryptographic Keys

Key management is a process that continues throughout the life cycle of the keys to thwart unauthorized disclosures, modifications, substitutions, reuse of revoked or expired keys, or unauthorized utilization. Security at this level is a recursive problem because the same security properties that are required in the cryptographic system must be satisfied in turn by the key management system.

The secure management of cryptographic keys relates to key production, storage, distribution, utilization, withdrawal from circulation, deletion, and archiving (Fumer and Landrock, 1993). These aspects are crucial to the security of any cryptographic system. SP 800-57 (National Institute of Standards and Technology, 2012c) is a three-part recommendation from NIST that provides guidance on the management. Each part is tailored to a specific audience: Part 1 is for system developers and system administrators, Part 2 is aimed at system or application owners, while Part 3 is more general and targets system installers, end users, as well as people making purchasing decisions.

3.14.1 Production and Storage

Key production must be done in a random manner and at regular intervals depending on the degree of security required.

Protection of the stored keys has a physical aspect and a logical aspect. Physical protection consists of storing the keys in safes or in secured buildings with controlled access, whereas logical protection is achieved with encryption.

In the case of symmetric encryption algorithms, only the secret key is stored. For public key algorithms, storage encompasses the user's private and public keys, the user's certificate, and a copy of the public key of the certification authority. The certificates and the keys may be stored on the hard disk of the certification authority, but there is some risk of possible attacks or of loss due to hardware failure. In cases of microprocessor cards, the information related to security, such as the certificate and the keys, are inserted during card personalization. Access to this information is then controlled with a confidential code.

3.14.2 Distribution

The security policy defines the manner in which keys are distributed to entitled entities. Manual distribution by mail or special dispatch (sealed envelopes, tamper-resistant module) is a slow and costly operation that should only be used for the distribution of the root key of the system. This is the key that the key distributor utilizes to send to each participant their keys.

An automatic key distribution system must satisfy all the following criteria of security:

- Confidentiality
- Identification of the participant
- Data integrity: by giving a proof that the key has not been altered during transmission or that it was not replaced by a fake key
- Authentication of the participants
- Nonrepudiation

Automatic distribution can be either point to point or point to multipoint. The Diffie–Hellman key exchange method (Diffie and Hellman, 1976) allows the two partners to construct a master key with elements that have been previously exchanged in the clear. A symmetric session key is formed next on the basis of the data encrypted with this master key or with a key derived from it and exchanged during the identification phase. IKE is a common automated key management mechanism designed specifically for use with IPSec.

To distribute keys to several customers, an authentication server can also play the role of a trusted third party and distribute the secret keys to the different parties. These keys will be used to protect the confidentiality of the messages carrying the information on the key pairs.

3.14.3 Utilization, Withdrawal, and Replacement

The unauthorized duplication of a legitimate key is a threat to the security of key distribution. To prevent this type of attack, a unique parameter can be concatenated to the key, such as a time stamp or a sequence number that increases monotonically (up to a certain modulo).

The risk that a key is compromised increases proportionately with time and with usage. Therefore, keys have to be replaced regularly without causing service interruption. A common solution that does not impose a significant load is to distribute the session keys on the same communication channels used for user data. For example, in the SSL/TLS protocol, the initial exchanges provide the necessary elements to form keys that would be valid throughout the session at hand. These elements flow encrypted with a secondary key, called a key encryption key, to keep their confidentiality.

Key distribution services have the authority to revoke a key before its date of expiration after a key loss or because of the user's misbehavior.

3.14.4 Key Revocation

If a user loses the right to employ a private key, if this key is accidentally revealed, or, more seriously, if the private key of a certification authority has been broken, all the associated certificates must be revoked without delay. Furthermore, these revocations have to be communicated to all the verifying entities in the shortest possible time. Similarly, the use of the revoked key by a hostile user should not be allowed. Nevertheless, the user will not be able to repudiate all the documents already signed and sent before the revocation of the key pair.

3.14.5 Deletion, Backup, and Archiving

Key deletion implies the destruction of all memory registers and magnetic or optical media that contain either the key or the elements needed for its reconstruction.

Backup applies only to encryption keys and not to signature keys; otherwise, the entire structure for nonrepudiation would be put into question.

The keys utilized for nonrepudiation services must be preserved in secure archives to accommodate legal delays that may extend for up to 30 years. These keys must be easily recoverable in case of need, for example, in response to a court order. This means that the storage applications must include mechanisms to prevent unrecoverable errors from affecting the ciphertext.

3.14.6 A Comparison between Symmetric and Public Key Cryptography

Systems based on symmetric key algorithms pose the problem of ensuring the confidentiality of key distribution. This translates into the use of a separate secure distribution channel that is preestablished between the participants. Furthermore, each entity must have as many keys as the number of participants with whom it will enter into contact. Clearly, the management of symmetric keys increases exponentially with the number of participants.

Public key algorithms avoid such difficulties because each entity owns only one pair of private and public keys. Unfortunately, the computations for public key procedures are more intense than those for symmetric cryptography. The use of public key cryptography to ensure confidentiality is only possible when the messages are short, even though data compression before encryption with the public key often succeeds in speeding up the computations. Thus, public key cryptography can complement symmetric cryptography to ensure the safe distribution of the secret key, particularly when safer means such as direct encounter of the participants, or the intervention of a trusted third party, are not feasible. Thus, a new symmetric key could be distributed at the start of each new session and, in extreme cases, at the start of each new exchange.

3.15 Exchange of Secret Keys: Kerberos

Kerberos is the most widely known system for the automatic exchange of keys using symmetric encryption. Its name is from the three-headed dogs that, according to Greek mythology, guarded the gates of Hell.

Kerberos comprises the services of online identification and authentication as well as access control using symmetric cryptography (Neuman and Ts'o, 1994). It allows management access to the resources of open network from nonsecure machines such as the management of student access to the resources of a university computing center (files, printers, etc.). Kerberos has been the default authentication option since Windows 2000. Since Windows Vista and Windows Server 2008, Microsoft's implementation of the Kerberos authentication protocol enables the use of Advanced Encryption Standard (AES) 128 and AES 256 encryption with the Kerberos authentication protocol.

The development of Kerberos started in 1978 within the Athena project at the Massachusetts Institute of Technology (MIT), financed by the Digital Equipment Corporation (DEC) and IBM. Version 5 of the Kerberos protocol was published in 1994 and remains in use. RFC 4120 (2005) provides an overview and the protocol specifications. Release 1.13.12 is the latest edition and was published in May 2015.

The system is built around a Kerberos key distribution center that enjoys the total trust of all participants with whom they all have already established symmetric encryption keys. Symmetric keys are attributed to individual users for each of their account when they register in person. Initially, the algorithm used for symmetric encryption is the Data Encryption Standard (DES), but AES was later added in 2005 as per RFC 3962. Finally, in 2014, DES and other weak cryptographic algorithms were deprecated (RFC 6649, 2012).

The key distribution center consists of an authentication server (AS) and a ticket-granting server (TGS). The AS controls access to the TGS, which in turn, controls access to specific resources. Every server shares a secret key with every other server. Finally, during the registration of the users in person, a secret key is established with the AS for each user's account. With this arrangement, a client has access to multiple resources during a session with one successful authentication, instead of repeating the authentication process for each resource. The operation is explained as follows.

After identifying the end user with the help of a login and password pair, the authentication server (AS) sends the client a session symmetric encryption key to encrypt data exchanges between the client and the TGS. The session key is encrypted with the symmetric encryption key shared between the user and the authentication server. The key is also contained in the session ticket that is encrypted with the key preestablished between the TGS and the AS.

The session ticket, also called a ticket-granting ticket, is valid for a short period, typically a few hours. During this time period, it can be used to request access to a specific service; which is why it is also called an initial ticket.

The client presents the TGS with two items of identification: the session ticket and an authentication title that are encrypted with the session key. The TGS compares the data in both items to verify the client authenticity and its access privileges before granting access to the specific server requested.

Figure 3.11 depicts the interactions among the four entities: the client, the *AS*, the *TGS*, and the desired merchant server or resource *S*.

The exchanges are now explained.

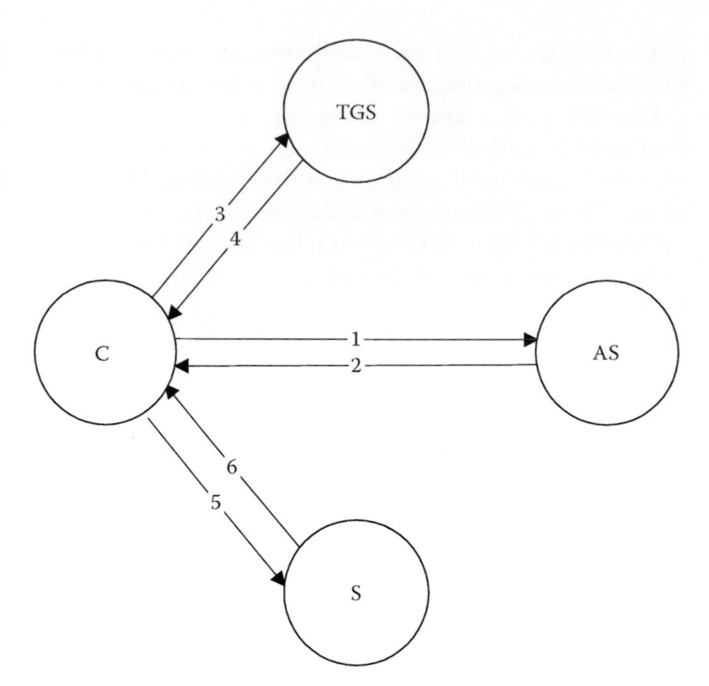

FIGURE 3.11
Authentication and access control in Kerberos.

3.15.1 Message (1): Request of a Session Ticket

A client *C* that desires to access a specific server *S* first requests an entrance ticket to the session from the Kerberos *AS*. To do so, the client sends message consisting of an identifier (e.g., a login and a password), the identifier of the server *S* to be addressed, a time stamp H_1 and a random number *Rnd*, both to prevent replay attacks.

3.15.2 Message (2): Acquisition of a Session Ticket

The Kerberos authentication server responds by sending a message formed of two parts: the first contains a session key K_{CTGS} and the number *Rnd* that were in the first message, both coded with the client's secret key K_C, and the second includes the session ticket T_{CTGS} destined for the TGS and encrypted by the latter's secret key between itself and the Kerberos authentication server.

The session (ticket-granting ticket) includes several pieces of information, such as the client name *C*, its network address Ad_C, the time stamp H_1, the period of validity of the ticket *Val*, and the session key K_{CTGS}. All these items, with the exception of the server identity TGS, are encrypted with the long-term key K_{TGS} that the TGS shares with the *AS*. Thus,

$$T_{CTGS} = TGS, K_{TGS}\{C, Ad_C, H_1, Val, K_{CTGS}\}$$

and the message sent to the client is

$$K_C\{K_{CTGS}, Rnd\}, T_{CTGS}$$

where $K\{x\}$ indicates encryption of the message x with the shared secret key K. The client decrypts the message with its secret key K_C to recover the session key K_{CTGS} and the random number. The client verifies that the random number received is the same as was sent as a protection from replay attacks. The time stamp H_1 is also used to protect from replay attacks. Although the client will not be able to read the session ticket because it is encrypted with K_{TGS}, it can extract it and relay it in the server.

By default, the session ticket T_{CTGS} is valid for 8 hours. During this time, the client can obtain several service tickets to different services without the need for a new authentication.

3.15.3 Message (3): Request of a Service Ticket

The client constructs an authentication title *Auth* that contains its identity *C*, its network address Ad_c, the service requested *S*, a new time stamp H_2, and another random number Rnd_2 and then encrypts it with the session key K_{CTGS}. The encrypted authentication title can be represented in the following form:

$$Auth = K_{CTGS}\{C, Ad_C, S, H_2, Rnd_2\}$$

The request of the service ticket consists of the encrypted authentication title and the session ticket T_{CTGS}:

$$Service\ request = Auth, T_{CTGS}$$

3.15.4 Message (4): Acquisition of the Service Ticket

The TGS decrypts the ticket content with its secret key K_{TGS}, deduces the shared session key K_{CTGS}, and extracts the data related to the client's service request. With knowledge of the session key, the server can decrypt the authentication title and compare the data in it with those that the client has supplied. This comparison gives formal proof that the client is the entity that was given with the session ticket by the server. The time stamps confirm that the message was not an old message that has been replayed. Next, the TGS returns a service ticket for accessing the specific server *S*.

The exchanges described by Messages (3) and (4) can be repeated for all other servers available to the user as long as the validity of the session ticket has not expired.

The message from the TGS has two parts: part one contains a service key K_{CS} between the client and the server *S* and the number Rnd_2 both coded with shared secret key K_{CTGS} and part two contains the service ticket T_{CS} destined to the server *S* and encrypted by secret key, K_{STGS}, shared between the server *S* and the *TGS*.

As earlier, the service ticket destined for the server *S* includes several pieces of information, such as the identity of the server *S*, the client name *C*, its network address Ad_C, a time stamp H_3, the period of validity of the ticket *Val*, and, if confidentiality is desired, a service key K_{CS}. All these items, with the exception of the server identity *S*, are encrypted with the long-term key K_{STGS} that the *TGS* shares with the specific server. Thus,

$$T_{CS} = S, K_{STGS}\{C, Ad_C, H_3, Val, K_{CS}\}$$

and the message sent to the client is

$$K_{CTGS}\{K_{CS}, Rnd_2\}, T_{CS}$$

The client decrypts the message with the shared secret key K_{CTGS} to recover the service key K_{CS} and the random number. The client verifies that the random number received is the same as was sent as a protection from replay attacks.

3.15.5 Message (5): Service Request

The client constructs a new authentication title $Auth_2$ that contains its identity *C*, its network address Ad_c, a new time stamp H_3, and another random number Rnd_3 and then encrypts it with the service key K_{CS}. The encrypted authentication title can be represented as follows:

$$Auth_2 = K_{CS}\{C, Ad_C, H_4, Rnd_3\}$$

The request of the service consists of the encrypted new authentication title and the service ticket T_{CS}:

$$Service\ request = Auth_2, T_{CS}$$

3.15.6 Message (6): Optional Response of the Server

The server decrypts the content of the service ticket with the key K_{STGS} it shares with the TGS to derive the service key K_{CS} and the data related to the client. With knowledge of the service key, the server can verify the authenticity of the client. The time stamps confirm that the message is not a replay of old messages. If the client has requested the server to authenticate itself, it will return the random number, Rnd_3, encrypted by the service key K_{CS}. Without knowledge of the secret key K_{CS}, the server would have not been able to extract the service key K_{CS}.

The preceding description shows that Kerberos is mostly suitable for networks administered by a single-administrative entity. In particular, the Kerberos key distribution center fulfills the following roles:

- It maintains a database of all secret keys (except of the key between the client and the server, K_{CS}). These keys have a long lifetime.
- It keeps a record of users' login identities, passwords and access privileges. To fulfill this role, it may need access to an X.509 directory.
- It produces and distributes encryption keys and ticket-granting tickets to be used for a session.

3.16 Public Key Kerberos

The utilization of a central depot for all symmetric keys increases the potential of traffic congestion due to the simultaneous arrival of many requests. In addition, centralization threatens the whole security infrastructure, because a successful penetration of the storage could put all the keys in danger (Sirbu and Chuang, 1997). Finally, the management of the symmetric keys (distribution and update) becomes a formidable task when the number of users increases.

The public key version of Kerberos simplifies key management, because the server authenticates the client directly using the session ticket and the client's certificate sealed by the Kerberos certification authority. The session ticket itself is sealed with the client's private key and then encrypted with the server public key. Thus, the service request to the server can be described as follows:

$$Service\ request = S, PK_S\{Tauth, Kr, Auth\}$$

with

$$Auth = C, certificate, [Kr, S, PK_C, Tauth]SK_C$$

where

Tauth is the initial time for authentication

Kr is a one-time random number that the server will use as a *symmetric* key to encrypt its answer

{…} represents encryption with the server public key, PK_S while […] represents the seal computed with the client's private key, SK_C

This architecture improves speed and security.

3.16.1 Where to Find Kerberos?

The official web page for Kerberos is located at http://web.mit.edu/kerberos/www/index.html. A Frequently Asked Questions (FAQ) file on Kerberos can be consulted at the following address: ftp://athena-dist.mit.edu/pub/kerberos/KERBEROS.FAQ. Tung (1999) is a good compendium of information on Kerberos.

The Swedish Institute of Computer Science is distributing a free version of Kerberos, called Heidmal. This version was written by Johan Danielsson and Assar Westerlund. The differences between Heidmal and the MIT APIs are listed at http://web.mit.edu/kerberos/krb5-1.13/doc/appdev/h5l_mit_apidiff.html.

In general, commercial vendors offer the same Kerberos code that is available at MIT for enterprise solutions. Their main function is to provide support for installation and maintenance of the code and may include administration tools, more frequent updates and bug fixes, and prebuilt (and guaranteed to work) binaries. They may also provide integration with various smart cards to provide more secure and movable user authentication. A partial list of commercial implementations is available at http://web.ornl.gov/~jar/commerce.htm.

3.17 Exchange of Public Keys

3.17.1 The Diffie–Hellman Exchange

The Diffie–Hellman algorithm is the first algorithm for the exchange of public keys. It exploits the difficulty in calculating discrete algorithms in a finite field, as compared with the calculation of exponentials in the same field. The technique was first published in 1976 and entered in the public domain in March 1997.

The key exchange comprises the following steps:

1. The two parties agree on two random large integers, p and g, such that g is a prime with respect to p. These two numbers do not have to be necessarily hidden, but their choice can have a substantial impact on the strength of the security achieved.
2. A chooses a large random integer x and sends to B the result of the computation:

$$X = g^x \bmod p$$

3. B chooses another large random integer y and sends to A the result of the computation:

$$Y = g^y \bmod p$$

4. A computes

$$k = Y^x \bmod p = g^{xy} \bmod p$$

5. Similarly, B computes

$$k = Y^x \bmod p = g^{xy} \bmod p$$

The value k is the secret key that both correspondents have exchanged.

The aforementioned protocol does not protect from man-in-the-middle attacks. A secure variant is as follows (Ferguson et al., 2010, pp. 183–193):

1. The two parties agree on (p, q, g) such that
   a. p and q are prime numbers, typically $2^{255} < q < 2^{256}$, that is, q is a 256-bit prime and $p = Nq + 1$, for some even N
   b. $g = \alpha^N \pmod{p}$, with $\alpha \in \mathbb{Z}_p^*$, $\mathbb{Z}_p^* = 1, \ldots, p-1$ is the finite field modulo p, such that $g \neq 1$ and $g^q = 1()\bmod p$
2. A chooses a large random integer $x \in [1, \ldots, q-1]$ and sends B the result of the computation:

$$X = g^x \bmod p$$

3. B verifies that
   a. $X \in [2, \ldots, p-1]$
   b. q is a divisor of $(p - 1)$
   c. $g \neq 1$ and $g^q = 1 \pmod{p}$
   d. $X^q = 1$
4. Next, it chooses a large random integer $y \in [1, \ldots, q-1]$ and sends A the result of the computation:

$$Y = g^y \bmod p$$

5. A verifies that $Y \in [2, \ldots, p-1]$ and that $Y^q = 1$ and then computes

$$k = Y^x \bmod p = g^{xy} \bmod p$$

6. Similarly, B computes

$$k = Y^x \bmod p = g^{xy} \bmod p$$

SSL/TLS uses the method called ephemeral Diffie–Hellman, where the exchange is short-lived, thereby achieving *perfect forward secrecy*, that is, that a key cannot be recovered after its deletion.

The Key Exchange Algorithm (KEA) was developed in the United States by the National Security Agency (NSA) based on the Diffie–Hellman scheme. All calculations in KEA are based on a prime modulus of 1024 bits with a key of 1024 bits and an exponent of 160 bits.

3.17.2 Internet Security Association and Key Management Protocol

IETF RFC 4306 (2005) combines contents that were described over several separate documents. One of these is Internet Security Association and Key Management Protocol (ISAKMP), a generic framework to negotiate point-to-point security associations and to exchange key and authentication data among two parties. In ISAKMP, the term *security association* has two meanings. It is used to describe the secure channel established between two communicating entities. It can also be used to define a specific instance of the secure channel, that is, the services, mechanisms, protocol and protocol-specific set of parameters associated with the encryption algorithms, the authentication mechanisms, the key establishment and exchange protocols, and the network addresses.

ISAKMP specifies the formats of messages to be exchanged and their building blocks (payloads). A fixed header precedes a variable number of payloads chained together to form a message. This provides a uniform management layer for security at all layers of the ISO protocol stack, thereby reducing the amount of duplication within each security protocol. This centralization of the management of security associations has several advantages. It reduces connect setup time, improves the reliability of software, and allows for future evolution when improved security mechanisms are developed, particularly if new attacks against current security associations are discovered.

To avoid subtle mistakes that can render a key exchange protocol vulnerable to attacks, ISAKMP includes five default exchange types. Each exchange specifies the content and the ordering of the messages during communications between the peers.

Although ISAKMP can run over TCP or UDP, many implementations use UDP on port 500. Because the transport with UDP is unreliable, reliability is built into ISAKMP.

The header includes, among other information, two 8 octet "cookies"—also called "syncookies"—which constitute an *anticlogging* mechanism, because of their role against TCP SYN flooding. Each side generates a cookie specific to the two parties and assigns it to the remote peer entity. The cookie is constructed, for example, by hashing the IP source and destination addresses, the UDP source and destination ports and a locally generated secret random value. ISAKMP recommends including the data and the time in this secret value. The concatenation of the two cookies identifies the security association and gives some protection against the replay of old packets or SYN flooding attacks. The protection against SYN flooding assumes that the attacker will not intercept the SYN/ACK packets sent to the spoofed addresses used in the attack. As was explained earlier, the arrival of unsolicited SYN/ACK packets at a host that is accessible to the victim will elicit the transmission of an RST packet, thereby telling the victim to free the allocated resources so that the host whose address has been spoofed will respond by resetting the connection (Juels and Brainard, 1999; Simpson, 1999).

The negotiation in ISAKMP comprises two phases: the establishment of a secure channel between the two communicating entities and the negotiation of security associations on the secure channel. For example, in the case of IPSec, Phase I negotiation is to define a key exchange protocol, such as the Internet Key Exchange (IKE) and its attributes. Phase II negotiation concerns the actual cryptographic algorithms to achieve IPSec functionality.

IKE is an authenticated exchange of keys consistent with ISAKMP. IKE is a hybrid protocol that combines the aspects of the Oakley Key Determination Protocol and of SKEME. Oakley utilizes the Diffie–Hellman key exchange mechanism with signed temporary keys to establish the session keys between the host machines and the network routers (Cheng, 2001). SKEME is an authenticated key exchange that uses public key encryption for anonymity and nonrepudiation and provides means for quick refreshment (Krawczyk, 1996). IKE is the default key exchange protocol for IPSec. The protocol was first specified in 1998, and the latest revision of IKEv2 is defined in RFC 7296 (2014) and RFC 7427 (2015).

None of the data used for key generation is stored, and a key cannot be recovered after deletion, thereby achieving perfect forward secrecy. The price is a heavy cryptographic load, which becomes more important the shorter the duration of the exchanges. Therefore, to minimize the risks from denial-of-service attacks, ISAKMP postpones the computationally intensive steps until authentication is established.

ISAKMP is implemented in OpenBSD, Solaris, Linux, and Microsoft Windows and in some IBM products. Cisco routers implement ISAKMP for VPN negotiation using the cryptographic library from Cylink Corporation. A denial-of-service vulnerability of this implementation was discovered and fixed in 2012 (Cisco, 2012).

3.18 Certificate Management

When a server receives a request signed with a public key algorithm, it must first authenticate the declared identity that is associated with the key. Next, it will verify if the authenticated entity is allowed to perform the requested action. Both verifications rely on a certificate that a certification authority has signed. As a consequence, certification and certificate management are the cornerstones of e-commerce on open networks.

Certification can be decentralized or centralized. Decentralized certification utilizes PGP (Pretty Good Privacy) (Garfinkel, 1995) or OpenPGP. Decentralization is popular among those concerned about privacy because each user determines the credence accorded to a public key and assigns a confidence level in the certificate that the owner of this public key has issued. Similarly, a user can recommend a new party to members of the same circle of trust. This mode of operation also eliminates the vulnerability to attacks on a central point and prevents the potential abuse of a single authority. However, the users have to manage the certificates by themselves (update, revocation, etc.). Because that load increases exponentially with the number of participants, this mode of operation is impractical for large-scale operations such as online commerce.

In a centralized certification, a *root* certification authority issues the certificates to *subordinate* or *intermediate* certification authorities, which in turn certify other secondary authorities or end entities. This is denoted as X.509 certification, using the name of the relevant recommendation from the ITU-T. ITU-T Recommendation X.509 is identical to ISO/IEC 9594-1, a joint standard from the ISO and the IEC. It was initially approved in November 1988, and its seventh edition was published in October 2012. It is one of a series of joint ITU-T and ISO/IEC specifications that describe the architecture and operations of public key infrastructures (PKI). Some wireless communication systems such as IEEE 802.16 use X.509 certificates and RSA public key encryption to perform key exchanges.

X.500 (ISO/IEC 9594-1) provides a general view of the architecture of the directory, its access capabilities, and the services it supports.

X.501 (ISO/IEC 9594-2) presents the different information and administrative models used in the directory.

X.509 (ISO/IEC 9594-8) defines the base specifications for public key certificates such as using identity certificates and attribute certificates.

X.511 (ISO/IEC 9594-3) defines the abstract services of the directory (search, creation, deletion, error messages, etc.).

X.518 (ISO/IEC 9594-4) for searches and referrals in a distributed directory system using the Directory System Protocol (DSP).

X.519 (ISO/IEC 9594-5) specifies four protocols. The Directory Access Protocol (DAP) provides a directory user agent (DUA) at the client side access to retrieve or modify information in the directory. The Directory System Protocol (DSP) provides for the chaining of requests to directory system agents (DSA) that constitute a distributed directory. The Directory Information Shadowing Protocol (DISP) provides for the shadowing of information held on DSA to another DSA. Finally, the Directory Operational Binding Management Protocol (DOP) provides for the establishment, modification, and termination of bindings between pairs of DSAs.

X.520 (ISO/IEC 9594-6) and X.521 (ISO/IEC 9594-7) specify selected attribute types (keywords) and selected object classes to ensure compatibility among implementations.

X.525 (ISO/IEC 9594-9) shares information through replication of the directory using DISP (Directory Information Shadowing Protocol).

The relationship among these different protocols is shown in Figure 3.12.

A simplified version of DAP, the Lightweight Directory Access Protocol (LDAP), is an application process that is part of the directory that responds to requests conforming to the LDAP protocol. The LDAP server may have the information stored in its local database or may forward the request to another DSA that understands the LDAP protocol. The latest specification of LDAP is defined in IETF RFCs 4511, 4512, and 4513 (2006). The main simplifications are as follows:

1. LDAP carried directly over the TCP/IP stack, thereby avoiding some of the OSI protocols at the application layer.
2. It uses simplified information models and object classes.
3. Being restricted to the client side, LDAP does not address what happens on the server side, for example, the duplication of the directory or the communication among servers.
4. Some directory queries are not supported.
5. Finally, Version 3 of LDAP (LDAPv3) does not mandate the strong authentication mechanisms of X.509. Strong authentication is achieved on a session basis for the TLS protocol.

IETF RFC 4513 (2006) specifies a minimum subset of security functions common to all implementations of LSAPv3 that use the SASL (Simple Authentication and Security Layer) mechanism defined in IETF RFC 4422 (2006). SASL adds authentication services and, optionally, integrity and confidentiality. Simple authentication is based on the name/password pair, concatenated with a random number and/or a time stamp with integrity protection using MD5.

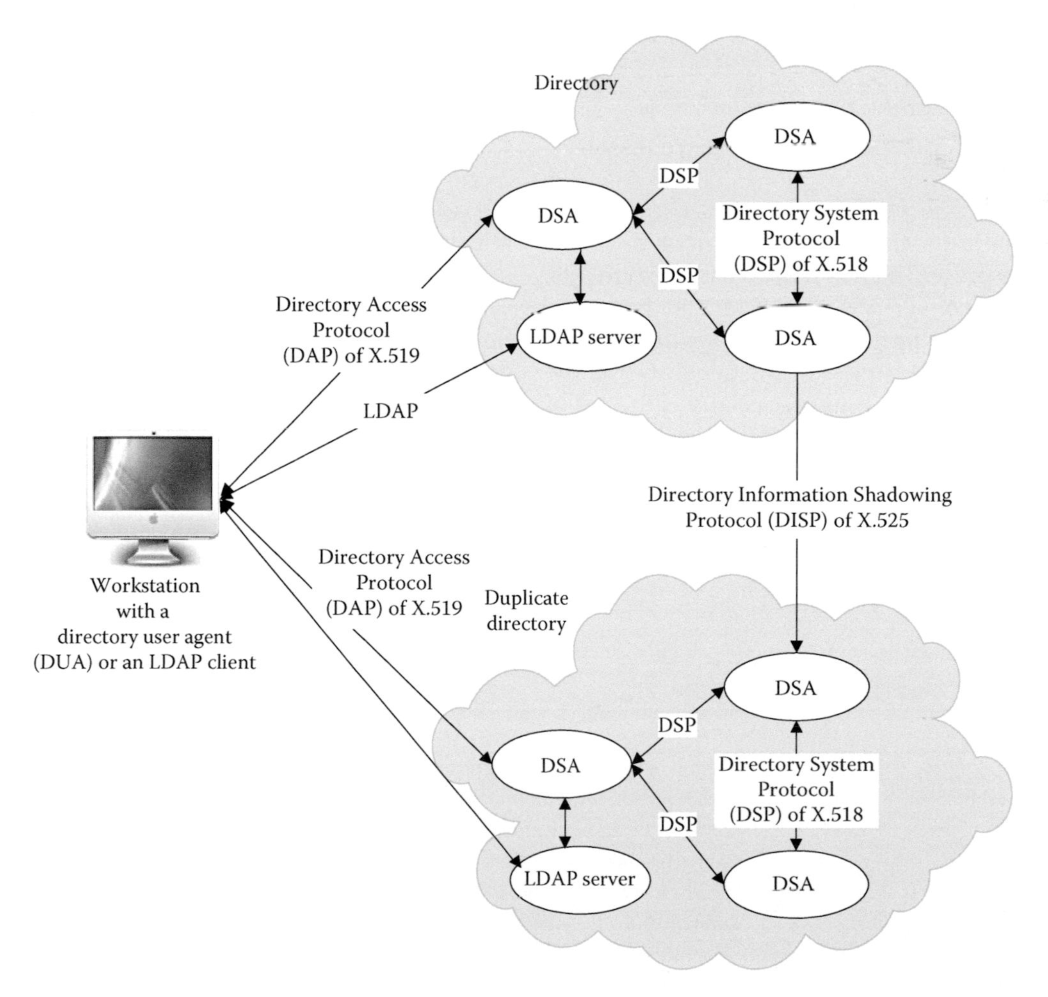

FIGURE 3.12
Communication protocols among the components of the X.500 directory system.

3.18.1 Basic Operation

After receiving over an open network a request encrypted using public key cryptography, a server has to accomplish the following tasks before answering the request:

1. Read the certificate presented.
2. Verify the signature by the certification authority.
3. Extract the requester public key from the certificate.
4. Verify the requester signature on the request message.
5. Verify the certificate validity by comparison with the Certificate Revocation List (CRL).
6. Establish a certification path between the public key certificate to be validated and an authority recognized by the relying party, for example, the root authority. That certification path—or chain of trust—starts from an end entity and ends at the authority that validates the path (the root certification authority or the trust anchor as explained later).
7. Extract the name of the requester.
8. Determine the privileges that the requester enjoys.

The certificate permits the accomplishment of Tasks 1 through 7 of the preceding list. In the case of payments, the last step consists of verifying the financial data relating to the requester, in particular, whether the account mentioned has sufficient funds. In the general case, the problem is much more complex, especially if the set of possible queries is large. The most direct method is to assign a key to each privilege, which increases the complexity of key management.

The Certificate Management Protocol (CMP) of IETF RFC 4210 (2005) specifies the interactions between the various components of a public key infrastructure for the management of X.509 certificates (request, creation, revocation, etc.). The Online Certificate Status Protocol (OCSP) of IETF RFC 6960 (2013) specifies the data exchanges between an application seeking the status of one or more certificates and the server providing the corresponding status. This functionality is desirable to sending a CRL to save bandwidth. IETF RFC 2585 (1999) describes how to use the File Transfer Protocol (ftp) and HTTP to obtain certificates and certification revocation lists from their respective repositories.

During online verification of the certificates, end entities or registration authorities submit their public key using the Certification Signing Request of PKCS #10 as defined in IETF RFC 2986 (2000).

3.18.2 Description of an X.509 Certificate

An X.509 certificate is a record of the information needed to verify the identity of an entity. This record includes the distinguished name of the user, which is a unique name that ties the certificate owner with its public key. The certificate contains additional fields to locate its owner's identity more precisely. The certificate is signed with the private key of the certification authority.

There are three versions of X.509 certificates, Versions 1, 2, and 3, the default being Version 1. X.509 Versions 2 and 3 certificates have an extension field that allows the addition of new fields while maintaining compatibility with the previous versions. Version 1 pieces of information are listed in Table 3.5.

Usually, a separate key is used for each security function (signature, identification, encryption, etc.) so that, depending on the function, the same entity may hold several certificates from the same authority.

There are two primary types of public key certificates: end entity certificates and certification authority (CA) certificates. The subject to an end entity public key certificate is not allowed to issue other public key certificates, while the subject of a CA certificate is another

TABLE 3.5

Content of a Version 1 X.509 Certificate

Field Name	Description
Version	Version of the X.509 (2001) certificate
Serial number	Certificate serial number assigned by the certification authority
Signature	Identifier of the algorithm and hash function used to sign the certificate
Issuer	Distinguished name of the certification authority
Validity	Duration of the validity of the certificate
Subject	References for the entity whose public is certified, such as the distinguished name, unique identifier (optional), and so on
Subject public key info	Information concerning the algorithm that this public key is an instance of and the public key itself

certification authority. CA certificates fall into the following categories:

- Self-issued certificates, where the issuer and the subject are the same certification authority.
- Self-signed certificates, where the private key that the certification authority used for signing is the same as the public key that the certificate certifies. This is typically used to advertise the public key or any other information that the certification authority wishes to make available.
- Cross-certificates, where the issuer is one certification authority and the subject is another certification authority. In a strict hierarchy, the issuer authorizes the subject certification authority to issue certificates, whereas in a distributed trust model, one certification authority recognizes the other.

Cross-certifications are essential for business partners to validate each other credentials. For cross-certification, policies and policy constraints must be similar or equivalent on both sides. This requires a mixture of technical, political, and managerial steps. On a technical level, NIST Special Publication 800-15 provides the minimum interoperability specifications for PKI components (MISPC) (Burr et al., 1997).

3.18.3 Attribute Certificates

X.509 defines a third type of public key certificates called *attribute* certificates that are digitally signed by an attribute authority to bind certain prerogatives, such as access control, to identity separately from the authentication of the identity information. More than one attribute certificate can be associated with the same identity.

Attribute certificates are managed by a *Privilege Management Infrastructure* (PMI). This is the infrastructure that supports a comprehensive authorization service in relation to a public key infrastructure. The PKI and PMI are separate logical and/or physical infrastructures and may be established independently but they are related. When a single entity acts as both a certification authority and an attribute authority, different keys are used for each kind of certificates. With a hierarchical role-based access control (RBAC), higher levels inherit the permissions accorded to their subordinates.

The Source of authority (SOA) is the ultimate authority to assign a set of privileges. It plays a role similar to the root certification authority and can be certified by that authority. The SOA may also authorize the further delegation of these privileges, in part or in full, along a delegation path. There may be restrictions on the power of delegation capability, for example, the length of the delegation path can be bonded, and the scope of privileges allowed can be restricted downstream. To validate the delegation path, each attribute authority along the path must be checked to verify that it was duly authorized to delegate its privileges.

Although it is quite possible to use public key identity certificates to define what the holder of the certificate may be entitled to, a separate attribute certificate may be useful in some cases, for example:

1. The authority for privilege assignment is distinct from the certification authority.
2. A variety of authorities will be defining access privileges to the same subject.
3. The same subject may have different access permissions depending on the role that individual plays.
4. There is the possibility of delegation of privileges, in full or in part.
5. The duration of validity of the privilege is shorter than that of the public key certificate.

Conversely, the public key identity certificate may suffice for assigning privileges, whenever the following occur:

1. The same physical entity combines the roles of certification authority and attribute authority.
2. The expiration of the privileges coincides with that of the public key certificate.
3. Delegation of privileges is not permitted or if permitted, all privileges are delegated at once.

3.18.4 Certification Path

The idea behind X.509 is to allow each user to retrieve the public key of certified correspondents so that they can proceed with the necessary verifications. It is sufficient therefore to request the closest certification authority to send the public key of the communicating entity in a certificate sealed with the digital signature of that authority. This authority, in turn, relays the request to its own certifying authority, and this permits an escalation through the chain of authorities, or certification path, until reaching the top of the certification pyramid (the root authority, RA). Figure 3.13 is a depiction of this recursive verification.

Armed with the public key of the destination entity, the sender can include a secret encrypted with the public key of the correspondent and corroborate that the partner is the one whose identity is declared. This is because, without the private key associated with the

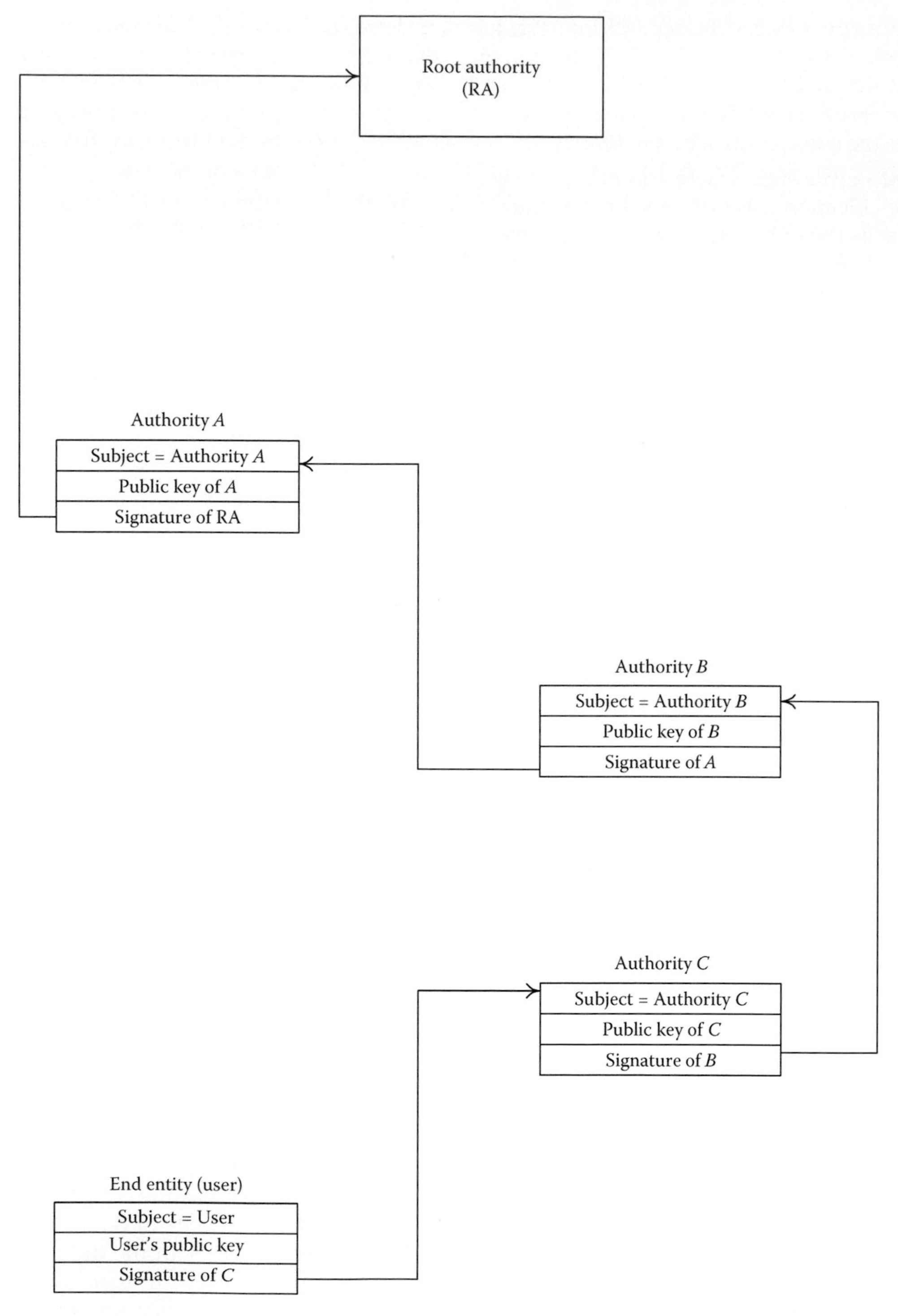

FIGURE 3.13
Recursive verification of certificates. (Adapted from Ford, W. and Baum, M.S., *Secure Electronic Commerce*, Pearson Education, Inc., Upper Saddle River, NJ, 1997. With permission.)

key used in the encryption, the destination will not be able to extract the secret. Obviously, for the two parties to authenticate themselves mutually, both users have to construct the certification path back to a common certification authority.

Thus, a certification path is formed by a continuous series of certification authorities between two users. This series is constructed with the help of the information contained in the directory by going back to a common point of confidence. In Figure 3.13, authorities *C*, *B*, and

A are the intermediate certification authorities. From the end entity's perspective, the root authority and an intermediate certification authority perform the same function, that is, they are functionally equivalent. However, the reliability of the system requires that each authority of the chain ensures that the information in the certification is correct. In other words, the security requires that none of the intermediate certification authorities are deficient or compromised.

It should be noted that the various intermediate authorities do not have to reside in the same country neither in the country of the root authority. This can be also different from the country where the data are stored. End users are therefore vulnerable to the different privacy laws in various jurisdictions. Moreover, governments in any country along the certification path have the possibility of initiating man-in-the-middle attacks by forcing the CAs in their jurisdiction to issue false certificates that can be used to intercept users data (Soghoian and Stamm, 2010).

Registration relates to the approval and rejection of certificate applications and the request of revocation or renewal of certificates and is different from key management and the certificate management. In some systems, the registration authority is different from the certification authority, for example, the human resources (HR) of a company while the certification is handled by the IT department. In such a case, the HR will have its own intermediate certification authority.

The tree structure of the certification path can be hierarchical or nonhierarchical as explained next.

3.18.5 Hierarchical Certification Path

According to a notational convention used in earlier versions of X.509, a certificate is denoted by

$$\text{authority}\langle\langle\text{entity}\rangle\rangle$$

Thus,

$$X_1\langle\langle X_2\rangle\rangle$$

indicates the certificate for entity X_2 that authority X_1 has issued, while

$$X_1\langle\langle X_2\rangle\rangle X_2\langle\langle X_3\rangle\rangle \ldots X_n\langle\langle X_{n+1}\rangle\rangle$$

represents the certification path connecting the end entity X_{n+1} to authority X_1. In other words, this notation is functionally equivalent to $X_1\langle\langle X_{n+1}\rangle\rangle$, which is the certificate that authority X_1 would have issued to the end entity X_{n+1}. By constructing this path, another end entity would be able to retrieve the public key of end entity X_{n+1}, if that other end entity knows X_{1p}, the public key of authority X_1. This operation is represented by

$$X_{1p} \cdot X_1\langle\langle X_2\rangle\rangle$$

where $\cdot$ is an infix operator, whose left operand is the public key, X_{1p}, of authority X_1, and whose right operand is the certificate $X_1\langle\langle X_2\rangle\rangle$ delivered to X_2 by that same certification authority. This result is the public key of entity X_2.

In the example depicted in Figure 3.14, assume that user A wants to construct the certification path toward another user B. A can retrieve the public key of authority W with the certificate signed by X. At the same time, with the help of the certificate of V that W has issued, it is possible to extract the public key of V. In this manner, A would be able to obtain the chain of certificates:

$$X\langle\langle W\rangle\rangle,\ W\langle\langle V\rangle\rangle,\ V\langle\langle Y\rangle\rangle,\ Y\langle\langle Z\rangle\rangle,\ Z\langle\langle B\rangle\rangle$$

This itinerary, represented by $A \to B$, is the forward certification path that allows A to extract the public key Bp of B, by an application of the operation $\cdot$ in the following manner:

$$\begin{aligned} Bp &= Xp \cdot (A \to B) \\ &= Xp \cdot X\langle\langle W\rangle\rangle W\langle\langle V\rangle\rangle V\langle\langle Y\rangle\rangle Y\langle\langle Z\rangle\rangle Z\langle\langle B\rangle\rangle \end{aligned}$$

In general, the end entity A also has to acquire the certificates for the return certification path $B \to A$, to send them to its partner:

$$Z\langle\langle Y\rangle\rangle,\ Y\langle\langle V\rangle\rangle,\ V\langle\langle W\rangle\rangle,\ W\langle\langle X\rangle\rangle,\ X\langle\langle A\rangle\rangle$$

When the end entity B receives these certificates from A, it can unwrap the certificates with its private key to extract the public key of A, Ap:

$$\begin{aligned} Ap &= Zp \cdot (B \to A) \\ &= Zp \cdot Z\langle\langle Y\rangle\rangle Y\langle\langle V\rangle\rangle V\langle\langle W\rangle\rangle W\langle\langle X\rangle\rangle X\langle\langle A\rangle\rangle \end{aligned}$$

As was previously mentioned, such a system does not necessarily impose a unique hierarchy worldwide. In the case of electronic payments, two banks or the fiscal authorities of two countries can mutually certify each other. In the preceding example, the intermediate certification authorities X and Z are cross-certified.

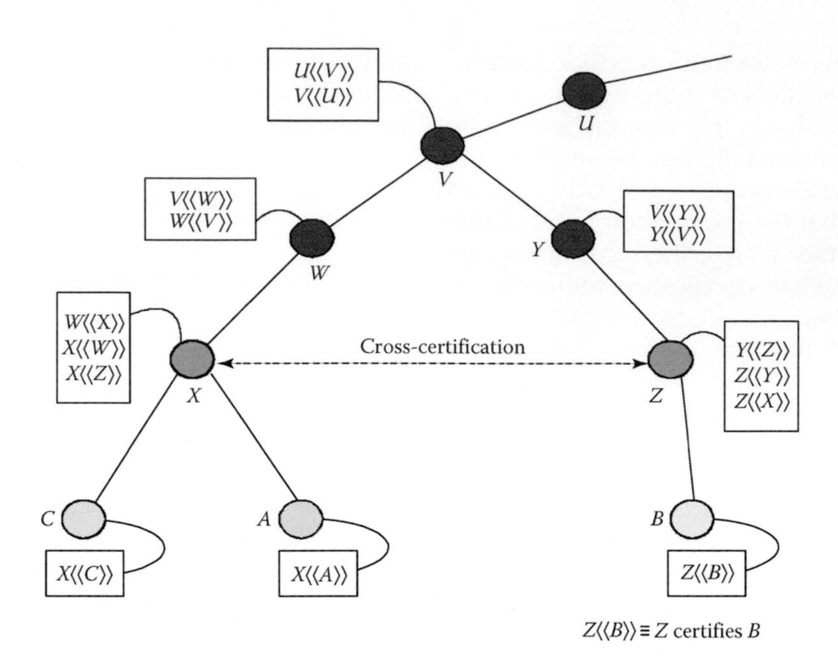

FIGURE 3.14
Hierarchical certification path according to X.509. (From ITU-T Recommendation X.509 (ISO/IEC 9594-8), Information technology—Open systems interconnection—The directory: Public key and attribute certificate frameworks, 2012, 2000. With permission.)

If *A* wants to verify the authenticity of *B*, it is sufficient to obtain

$$X\langle\langle Z\rangle\rangle, Z\langle\langle B\rangle\rangle$$

to form the forward certification path and

$$Z\langle\langle X\rangle\rangle$$

to construct the reverse certification path. This permits the clients of the two banks to be satisfied with the certificates supplied by their respective banks.

A root certification authority may require all its intermediary authorities to keep an audit trail of all exchanges and events, such as key generation, request for certification, validation, suspension, or revocation of certificates.

Finally, it should be noted that cross-certification applies to public key certificates and not to attribute certificates.

3.18.6 Distributed Trust Model

If certification authorities are not organized hierarchically, the end entities themselves would have to construct the certification path. In practice, the number of operations to be carried out can be reduced with various strategies, for example:

1. Two end entities served by the same certification authority have the same certification path, and can exchange their certificates directly. This is the case for the end entities *C* and *A* in Figure 3.15.
2. If one end entity is constantly in touch with users that a particular authority has certified, that end entity could store the forward and return certification paths in memory. This would reduce the effort for obtaining the other users' certificates to a query into the directory.
3. Two end entities that have each other's certificates mutually authenticate themselves without querying the directory. This reverse certification is based on the confidence that each end entity has in its certification authority.

Later revisions of X.509 have introduced a new entity called a Trust Anchor. This is an entity that a party relies on to validate certificates along the certification path. In complex environments, the trust anchor may be different from the root authority, for example, if the length of the certification path is restricted for efficiency.

3.18.7 Certificate Classes

In general, service providers offer several classes of certificates according to the strength of the link between the certificate and the owner's identity. Each class of certificates has its own root authority and possibly registration authorities.

Consider, for example, a three-class categorization. In this case, Class 1 certificates confirm that the distinguished name that the user presents is unique and unambiguous within the certification authority's domain and that it corresponds to valid e-mail address. They are typically used for domain registration. Class 1 certificates are used for modest enhancement of security through confidentiality and integrity verification. They cannot be used to verify an identity or to support nonrepudiation services.

Class 2 certificates are also restricted to individuals. They indicate that the information that the user has submitted during the registration process is consistent with the information available in business records or in "well-known" consumer databases. In North America, one such reference database is maintained by Equifax.

Class 3 certificates are given to individuals and to organizations. To obtain a certificate of this class, an individual has to be physically present with their public key in possession before an authority to confirm the identity of the applicant with a formal proof of identity (passport, identity card, electricity or telephone bill, etc.) and the association of that identity with the given public key. If the individual is to be certified as a duly authorized representative of an organization, then the necessary verifications have to be made. Similarly, an enterprise will have to prove its legal existence. The authorities will have to verify these documents by querying the databases for enterprises and by confirming the collected data by telephone or by mail. Class 3 certificates have many business applications.

3.18.8 Certificate Revocation

The correspondence between a public key and an identity lasts only for a period of time. Therefore, certification authorities must refer to revocation lists that contain certificates that have expired or have been revoked. These lists are continuously updated. Table 3.6 shows the format of the revocation list that Version 1 of X.509 has defined. In the third revision of X.509, other optional entries, such as the date of the certificate revocation and the reason for revocation, were added.

TABLE 3.6

Basic Format of the X.509 Revocation List

Field	Comment
Signature	Identifier of the algorithm used to sign the certificates and the parameters used
Issuer	Name of the certification authority
thisUpdate	Date of the current update of the revocation list
nextUpdate	Date of the next update of the revocation list
revokedCertificates	References of the revoked certificates including the revocation date

The CPS describes the circumstances under which certification of end users and the various intermediate authorities can be revoked and defines who can request that revocation. To inform all the entities of the PKI, CRLs are published at regular intervals—or when the certificate of an authority is revoked—with the digital signature of the certification authority to ensure their integrity. Among other information, the CRL indicates the issuer's name, the date of issue, the date of the next scheduled CRL, the revoked certificates serial numbers, and the specific times and reasons for revocation.

In principle, each certification authority has to maintain at least two revocation lists: a dated list of the certificates that it has issued and revoked and a dated list of all the certificates that the authorities that it recognizes have revoked. The root certification authority and each of its delegate authorities must be able to access these lists to verify the instantaneous state of all the certificates to be treated within the authentication system.

Revocation can be periodic or exceptional. When a certificate expires, the certification authority withdraws it from the directory (but retains a copy in a special directory, to be able to arbitrate any conflict that might arise in the future). Replacement certificates have to be ready and supplied to the owner to ensure the continuity of the service.

The root authority (or one of its delegated authorities) may cancel a certificate before its expiration date, for example, if the certificate owner's private key was compromised or if there was abuse in usage. In the case of secure payments, the notion of solvency, that is, that the user has available the necessary funds, is obviously one of the essential considerations.

Processing of the revocation lists must be done quickly to alert users and, in certain countries, the authorities, particularly if the revocation is before the expiration date. Perfect synchronization among the various authorities must be attained to avoid questioning the validity of documents signed or encrypted before the withdrawal of the corresponding certificates.

Users must also be able to access the various revocation lists; this is not always possible because current client programs do not query these lists.

3.18.9 Archival

Following certification expiration or its revocation, the records associated with a certificate are retained for at least the specific time periods. Table 3.7 shows the current intervals for Symatech. Thus, archival of Class 1

TABLE 3.7
Symantec Archival Period per Certificate Class

Certificate Class	Duration in Years
1	5
2	10.5
3	10.5

certificates lasts for at least 5 years after expiration of the certificate or its revocation, while the duration for Class 2 and 3 certificates are 10.5 years each (Symantec Corporation, 2015).

3.18.10 Recovery

Certification authorities implement procedures to recover from computing failures, corruption of data, such as when a user's private key is compromised, and natural or man-made disasters.

A disaster recovery plan addresses the gradual restoration of information services and business functions. Minimal operations can be recovered within 24 hours. They include certificate issuance or revocation, publication of revocation information, and recovery of key information for enterprises customers.

3.18.11 Banking Applications

A bank can certify its own clients to allow them access to their back account across the Internet. Once access has been given, the operation will continue as if the client were in front of an automatic teller machine. The interoperability of bank certificates can be achieved with interbank agreements, analogous to those that have permitted the interoperability of bank cards. Each financial institution certifies its own clients and is assured that the other institutions will honor that certificate.

As the main victims of fraud, financial institutions have partnered to establish their own certification infrastructures. In the United States, several banks including Bank of America, Chase Manhattan, Citigroup, and Deutsche Bank have formed IdenTrust (in 2000). The main purpose was to enable a trusted business-to-business e-commerce marketplace with financial institutions as the key trust providers. The institution changed its name into IdenTrust (http://www.identrust.com) in 2006 and was acquired by HID in 2014. It is one of two vendors accredited by the General Services Administration (GSA) to issue digital certificates.

At the same time, about 800 European institutions have joined forces to form a Global Trust Authority (GTA) as a nonprofit organization whose mission is to put in place an infrastructure of trust that can be used, by all sectors, to conduct cross-border e-business. In 2009, the project was suspended (IDABC, 2009).

3.19 Authentication

3.19.1 Procedures for Strong Authentication

Having obtained the certification path and the other side's authenticated public key, X.509 defines three procedures for authentication, one-way or unidirectional authentication, two-way or bidirectional authentication; and three-way or tridirectional authentication.

3.19.1.1 One-Way Authentication

One-way authentication takes place through the transfer of information from User A to User B according to the following steps:

- A generates a random number R^A used to detect replay attacks.
- A constructs an authentication token $M = (T^A, R^A, I_B, d)$ where T^A represents the time stamp of A (date and time) and I_B is the identity of B. T^A comprises two chronological indications, for example, the generation time of the token and its expiration date, and d is an arbitrary data. For additional security, the message can be encrypted with the public key of B.
- A sends to B the message:

$$B \rightarrow A, A\{(T^A, R^A, I_B, d)\}$$

where
$B \rightarrow A$ is the certification path
$A\{M\}$ represents the message M encrypted with the private key of A

B carries on the following operations:

- Obtain the public key of A, A_p, from $B \rightarrow A$, after verifying that the certificate of A has not expired.
- Recover the signature by decrypting the message $A\{M\}$ with Ap. B then verifies that this signature is identical to the message hash, thereby ascertaining simultaneously the signature and the integrity of the signed message.
- Verify that B is the intended recipient.
- Verify that the time stamp is current.

- Optionally, verify that R^A has not been previously used.

These exchanges prove the following:

- The authenticity of A, that is, the authentication token was generated by A
- The authenticity of B, that is, the authentication token was intended for B
- The integrity of the identification token
- The originality of the identification token, that is, it has not been previously utilized

3.19.1.2 Two-Way Authentication

The procedure for two-way authentication adds to the previous unidirectional exchanges similar exchanges but in the reverse direction. Thus, the following applies:

- B generates another random number R^B.
- B constructs the message $M' = (T^B, R^B, I_A, R^A, d)$, where T^B represents the time stamp of B (date and time), I_A is the identity of A, and R^A is the random number received from A. T^B consists of one or two chronological indications as previously described. For security, the message can be encrypted with the public key of A.
- B sends to A the message:

$$B\{(T^B, R^B, I_A, R^A, d)\}$$

where $B\{M'\}$ represents the message M' encrypted with the private key of B.

- A carries out the following operations:
  - Extracts the public key of B from the certification path and uses it to decrypt $B\{M'\}$ and recovers the signature of the message that B has produced; A verifies next that the signature is the same as the hashed message, thereby ascertaining the integrity of the signed information
  - Verifies that A is the intended recipient
  - Checks the time stamp to verify that the message is current
  - As an option, verifies that RB has not been previously used

3.19.1.3 Three-Way Authentication

Protocols for three-way authentication introduce a third exchange from A to B. The advantage is the avoidance of time stamping and, as a consequence, of a trusted third party. The steps are the same as for two-way identification but with $T^A = T^B = 0$. Then:

- A verifies that the value of the received R^A is the same that was sent to B.
- A sends to B the message:

$$A\{R^B, I_B\}$$

encrypted with the private key of A.

- B performs the following operations:
  - Verifies the signature and the integrity of the received information
  - Verifies that the received value of R^B is the same as was sent

3.20 Security Cracks

We have reviewed in this chapter the main concepts of how cryptographic algorithms may be used to support multiple security services. On a system level, however, the security of a system depends on many factors such as

1. The rigorousness of the authentication criteria
2. The degree of trust place in the root authority and/or intermediate certification authorities
3. The strength of the end entity's credentials (e.g., passport, birth certificate, driver's license)
4. The strength of the cryptographic algorithms used
5. The strength of the key establishment protocols
6. The care with which end entities (i.e., users) protect their keys

This is why the design of security systems that provide the desired degree of security requires high skill and expertise and attention to details. For example, if the workstations are not properly configured, users can be denied access or signed e-mails may appear invalid (Department of Defense Public Key Enabling and Public Key Infrastructure Program Management Office, 2010).

3.20.1 Problems with Certificates

Certification is essential for authenticating participants to prevent intruders from impersonating any side to spy on the encrypted exchanges. When an entity produces a certificate signed by a certification authority, this means that the CA attests that the information in the certificate is correct. As a corollary, the entry for that entity in the directory that the certification authority maintains has the following properties:

1. It establishes a relationship between the entity and a pair of public and private cryptographic keys.
2. It associates a unique distinguished name in the directory with the entity.
3. It establishes that at a certain time, the authority was able to guarantee the correspondence between that unique distinguished name and the pair of keys.

Each certification authority describes its practices and policies in a Certification Practice Statement (CPS). The CPS covers the obligations and liabilities, including liability caps, of various entities, their obligations. For example, one obligation is to maintain their cryptographic technology current and to protect the integrity of their physical and logical operations including key management.

There is no standard CPS, but IETF RFC 3647 (2003) offers guidance on how to write such a certification statement. The accreditation criteria of entities are also not standardized, and there is no code of conduct for certification authorities. Each operator defines its conduct, rights, and obligations and operates at its own discretion and is not obliged to justify its refusal to accredit an individual or an entity. There are also no standard criteria to evaluate the performance of a certification authority, so that browser vendors are left to their own discretion as to which certification authorities they should trust. There are almost no laws to prevent a PKI operator from cashing in on the data collected on individuals and their purchasing habits by passing the information to all those that might be interested (merchants, secret services, political adversaries, etc.). Thus, should the certification authorities fail to perform their role correctly, willingly, by accident or negligently, the whole security edifice is called into question.

The American Institute of Certified Public Accountants and the Canadian Institute of Chartered Accountants have developed a program to evaluate the risks of conducting commerce through electronic means. The CPA WebTrustSM is a seal that is supposed to indicate that a site is subject to quarterly audit on the procedures to protect the integrity of the transactions and the confidentiality of the information. However, there are limits to what audits can uncover as shown in the notorious case of DigiNotar.

DigiNotar was a certification authority agreed and audited by the Dutch government as part of its PKI program. In August 2011, however, following an investigation by the Dutch government, DigiNotar was forced to admit that more than 500 SSL/TLS certificates were stolen from compromised DigiNotar's servers and used to create fake DigiNotar certificates. In particular, the fake certificate for google.com was used to spy on to the TLS/SSL sessions of some 300,00 Iranians using gmail accounts. DigiNotar detected and revoked some of the fraudulent certificates without notifying the browsers' manufacturers such as Apple, Google, Microsoft, Mozilla, and Opera forcing them to issue updates to block access to sites secured with DigiNotar certificates. Finally, in September 2011, The Dutch Independent Post and Telecommunications Authority (OPTA) revoked DigiNotar as a certification authority (Schoen, 2010; Keizer, 2011; Nightingale, 2011).

In the case of DigiNotar, a fundamental problem was shown in the way certificates are used in consumer applications. Browsers maintain a list of trusted certification authorities or rely on the list that the operating system provides. Each of these authorities has the power to certify additional certification authorities. Thus, a browser ends up trusting hundreds of certificate authorities some which do not necessarily have the same policies. Because the authorities on the chain of trust may be distributed over several countries, different government agencies may legally compel any of these certification authorities to issue false certificates to intercept and hijack individuals' secure channels of communication. In fact, there are some commercial covert surveillance devices that operate on that principle (Soghoian and Stamm, 2010).

Browsers currently contact the CAs to verify that a certificate that a server presents has not been revoked, but they currently do not track changes in the certificate, for example, by comparing hashes (but there are Firefox add-ons that perform this function). In principle, browser suppliers work with certification authorities to respond and contain breaches by blocking fraudulent certificates. Experience has shown, however, that certification authorities do not always notify the various parties of security breaches promptly.

3.20.2 Underground Markets for Passwords

The spread of online electronic commerce has been correlated to the number of times individuals have to login to different systems and applications. In many cases, the back-end authentication systems are different so that users have to manage an increasing number of passwords, particularly that many systems force the users to replace their passwords periodically. In a typical large application, such as for a bank, the password database may include around 100 or 200 m password. Increasingly, users face the problem of creating and remembering multiple user names and passwords and often end up reusing passwords.

One of the consequences of all these factors is the rise in cybercrime, fuelled by underground markets of stolen passwords and of tools ("bots") to attempt automatic logins in many websites trying the passwords in a file until access is achieved.

3.20.3 Encryption Loopholes

Encryption is a tool to prevent undesirable access to a secret message. While the theoretical properties of the cryptographic algorithms are important, how are the fundamentals of cryptography implemented and used are essential. Brute-force attacks where the assailant systematically tries all possible encryption keys until getting the one that will reveal the plaintext. Table 3.8 provides the estimated time for successful brute-force attacks with exhaustive searches on symmetric encryption algorithms with different key lengths for the *current* state of technology (Paar and Pelzl, 2010, p. 12).

As a consequence, a long key is a necessary but not sufficient condition for secure symmetric encryption. Cryptanalysis focuses on finding design errors, implementation flaws or operational deficiencies to break the encryption and retrieve the messages, even without knowledge of the encryption key. GSM, IEEE 802.11b, IS-41, and so on are known to have faulty or deliberately weakened protection schemes.

The most common types of cryptological attacks include the following (Ferguson et al., 2010, pp. 31–36):

1. Attacks on the encrypted text assuming that the clear text has a known given structure, for example, the systematic presence of a header with a known format (this is the case of e-mail messages) or the repetition of known keywords.
2. Attacks starting with chosen plaintexts that are encrypted with the unknown key so as to deduce the key itself.
3. Attacks by replaying old legitimate messages to evade the defense mechanisms and to short-circuit the encryption.
4. Attacks by interception of the messages (man-in-the-middle attacks) where the interceptor inserts its eavesdrop at an intermediate point between the two parties. After intercepting an exchange of a secret key, for example, the interceptor will be able to decipher the exchanged messages while the participants think they are communicating in complete security. The attacker may also be able to inject fake messages that would be treated as legitimate by the two parties.
5. Attacks by measuring the length of encryption times, of electromagnetic emissions, and so on to deduce the complexity of the operations, and hence their form.
6. Attacks on the network itself, for example, corruption of the DNS, to direct the traffic to a spurious site.

TABLE 3.8

Estimated Time for Successful Brute-Force Attacks on Symmetric Encryption for Different Key Lengths with Current Technology

Key Length in Bits	Estimated Time with Current Technology
56–64	A few days
112–128	Several years
256	Several decades

In some cases, the physical protection of the whole cryptographic system (cables, computers, smart cards, etc.) may be needed. For example, bending of an optical fiber results in the dispersion of 1%–10% of the signal power; therefore, well-placed acoustic-optic devices can capture the diffraction pattern for later analysis.

A catalog of the causes of vulnerability includes the following (Schneier, 1997, 1998a; Fu et al., 2001):

1. Nonverification of partial computations.
2. Use of defective random number generators, because the keys and the session variables depend on a good supply source for nonpredictable bits.
3. Improper reutilization of random parameters.
4. Misuse of a hash function in a way that increases the chances for collisions.
5. Structural weakness of the telecommunications network.
6. Nonsystematic destruction of the clear text after encryption and the keys used in encryption.
7. Retention of the password or the keys in the virtual memory.
8. No checking of correct range of operation; this is particularly the case when buffer overflows can cause security flaws.

9. Misuse of a protocol can lead to an authenticator traveling in plaintext. For example, IETF RFC 2109 (1997)—now obsolete—specified that when the authenticator is stored in a cookie, the server has to set the Secure flag in the cookie header so that the client waits before returning the cookie until a secure connection has been established with SSL/TLS. It was found that some web servers neglected to set this flag thereby negating that protection. The authenticator can also leak if the client software continues to use even after the authentication is successful.

A new type of vulnerability was recently discovered and given the name of cross-app resource access attacks (XARA). Operating systems such as Apple's OS X and iOS attempt to isolate applications from each other even when the same user is running ("sandboxing") to prevent a malicious or compromised program from harming the others. To talk to each other, applications use interprocess communication channels. Apple's implementation of sandboxing has several flaws that can be exploited to enable a malicious app, though vetted by the Apple Store, to gain unauthorized access to the sensitive data of other apps (e.g., passwords) and utilize their resources surreptitiously (Xing et al., 2015). For example, Apple's credential management service, called the keychain, allows each application to record the user's credentials (the user's passwords, secret keys, and certificates) but does not offer an easy way to determine the owner of an existing keychain item and/or to authenticate that owner before granting access to that item. Taking advantage of this flaw, a malicious app can secretly allocate keychain attributes for apps that have not yet been installed in the hope that the user will install them later. In that case, the malicious app will have full access to the credentials. If the target application is already installed, the malicious app can delete the corresponding item of keychain and replace it with a new access control list under the attacker's control. So, when the target application updates the user's credentials, they will be divulged to the attacker. In June 2015, Apple was still working on a fix for these security flaws (Goodin, 2015).

It is also possible to take advantage of other implementation details that are not directly related to encryption. For example, when a program deletes a file, most commercial operating systems merely eliminate the corresponding entry in the index file. This allows recovery of the file, at least partially, with off-the-shelf software. The only means by which to guarantee total elimination of the data is to rewrite systematically each of the bits that the deleted file was using. Similarly, the use of the virtual memory in commercial systems exposes another vulnerability because the secret document may be momentarily in the clear on the disk.

Systems for e-commerce that are for the general public must be easily accessible and affordably priced. As a consequence, many compromises will be made to improve response time and the ease of use. However, if one starts from the principle that, sooner or later, any system is susceptible to unexpected attacks with unanticipated consequences, it is important the system make it possible to detect attacks and to accumulate proofs that are accepted by law enforcement personnel and the courts. The main point is to have an accurate definition of the type of expected threats and possible attacks. Such a realistic evaluation of threats and risks permits a precise understanding of what should be protected, against whom, and for how long.

3.20.4 Phishing, Spoofing, and Pharming

Phishing, *spoofing*, and *pharming* have been used to trick users to voluntarily reveal their credentials. These terms are neologisms that describe various tricks played on users. Although they are often used interchangeably, we attempt here to distinguish among them for the sake of clarity.

Phishing is a deceitful message to trick users into revealing their credentials (bank account numbers, passwords, payment card details, etc.) to unauthorized entities. The term was coined to indicate that fraudsters are "fishing" for online banking details of customers through e-mail. The sender can impersonate reputable companies such as banks, financial intermediaries or banks and exploits the whole gamut of human emotions and credulity, ranging from obligation toward friends in distress in foreign lands, fear of arrest warrants to greed and ignorance. For example, messages purportedly from a bank would alert the recipient that there is a security problem with the recipient's account that requires immediate attention and then ask the recipient to click on a link that seems legitimate under any false pretense (to restore a compromised account, to verify identity, etc.). However, by using HTML Forms in the body of the text or JavaScript event handler, the link seen in the e-mail is different from the actual link destination. Once the link is clicked, a window opens that contain the real site or a fake copy of the bank's website would ask for their account details or authentication data. When the user types the credential, they are captured and used to siphon the user's account (Drake et al., 2004).

After convincing the recipient that the e-mail is credible and originated from a trusted institution, phishing exploits the whole gamut of human emotions and credulity to persuade the recipient to divulge personal and financial information under the guise of verifying account, paying taxes, assisting a friend mugged in a foreign land, and so on.

Alternatively, a malware (malicious software) may be injected into the target device to record all keyboard inputs and e-mail the collected data periodically to what is called a mail drop (Schneier, 2004; Levin et al., 2012). An extreme form of this malware is called *ransomware*, where hackers seize control of the target system unless a ransom is paid. This happened in June 2015 to the police force of Tewksbury, Massachusetts (a suburb of Boston), which was only able to access its data back after paying a ransom in Bitcoin.

In February 2015, the Russian antivirus firm Kaspersky Lab issued a report detailing how a group calling themselves Carbanak was able to penetrate bank networks using phishing e-mails to steal over $500 million from banks in several countries and their private customers. In some instances, cash machines were instructed to dispense their contents to associates. In other cases, they altered databases to increase fraudulently balances on existing accounts and then pocket the difference without the knowledge of the account owner (*Economist*, 2015b).

Finally, pharming is the systematic exploitation of a vulnerability of the DNS servers due, for instance, to a coding error or to an implanted malware. As a result, instead of translating web names into their corresponding IP address, spurious IP addresses associated with a fake site are used. In other words, the effects of spoofing and pharming are the same, but in spoofing the victim is a willingly albeit gullible participant while in pharming the victim is unaware of the redirection.

To conduct such a campaign, an attacker registers a new domain with an Internet domain registrar using contact details similar to those of the target company as obtained with a *whois* lookup on the real domain. The attacker then creates a DNS record that points to the newly registered domain that will post a fake landing page to give the e-mail recipients the impression of genuineness so that they enter willingly their credentials. The deceptive site may even be a compromised website of another victim to which the attacker has uploaded the phishing kit. Some kits are freely available on the Internet, while others are distributed within closed communities in a hacking forum.

The root cause of all these problems is the lack of authentication in electronic mail exchanges. E-mail was originally intended for the communication among collaborators who knew each other and not for the users distributed among more than 70 million domains to communicate without any prior screening. Only partial solutions have been offered thus far.

RFC 4871 (2007) defines an authentication framework, called *DomainKeys Identified Mail* (DKIM), which is a synthesis of two proposals from Cisco and Yahoo! to associate a domain name identity to a message through public key cryptography. This can be applied at the message transfer agents (MTA) located in the various network nodes or at the e-mail clients, that is, readers or more formally mail user agents (MUA). The sending MTA checks if the source is authorized to send mail using its domain. If the message is authorized, it signs the body of the message and selected header fields and inserts both signatures in the new DKIM-Signature header field. The MTA of the receiving domain verifies the integrity of the message and the message headers.

The signatures are calculated through hashing with SHA-1 or SHA-256 and the encryption with RSA, with key sizes ranging from 512 to 2048 bits. RFC 4871 avoids a public key certification infrastructure and uses the DNS to maintain the public keys of the claimed senders. Accordingly, verifies have to request the sender public key from the DNS.

The scheme in RFC 4871 does not protect from relay attacks. For example, a spammer can send a message to an accomplice while the original message satisfies all the criteria set in RFC 4871. The accomplice inserts the extra fields needed to forward the message, but the signatures will remain valid, fooling the DKIM tool at the receiving end (see § 8.5 of RFC 4871). The end user will still be unable to determine if the message is a phishing attack or that it contains a malicious attachment. The information, however, may be useful for the receiving MTA to establish that a specific e-mail address is being used for spamming (Selzer, 2013). Also, because the DNS is not fully equipped for key management, an attacker can publish key records in DNS that are intentionally malformed.

The Sender Policy Framework (SPF) of RFC 7208 (2014) is another approach based on IP addresses. Work on the specifications took more than 10 years, but the concept was implemented and deployed before the RFC was published. Here, a domain administrator populates the SPF record of that domain's entry in the DNS with the IP addresses that are allowed to send e-mail from that domain and publish that SPF data as DNS TXT (type 16) Resource Record per RFC 1035 (1987). Any mail server or spam filter on the user's machine query the DNS for the domain's SPF record to verify the source of e-mail messages as they arrive. Only messages whose IP addresses, as specified in the MAIL FROM field, match one of the authorized IP addresses will be accepted. Thus, other mail servers can stop fake mail claiming to originate from that domain. The only benefit for the domain is to protect its reputation. Furthermore, there is no obligation for receiving mail servers to filter their incoming mail based on SPF, which they may avoid to avoid going through multiple DNS lookups.

There are attempts at understanding the problem and tracking the fraudsters down. For example, the

University of Alabama at Birmingham has established in 2007 the Computer Forensics Research Lab to store spam mails in a Spam Data Mine available for investigator. The lab has developed automated and manual techniques to analyze the mail identity, the phishing website, and determine the tool kits used to creating the phishing website. The university's Phishing Operations work with law enforcement and corporate investigators to identify groups or related phishing sites, particularly when the financial loss is significant (Levin et al., 2012).

Finally, there are websites that help distinguish between legitimate e-mails and traps designed to steal personal information and/or money. PayPal offers a specific address (spoof@paypal.com) where users can inquire if the message purportedly originating from PayPal is authentic.

3.21 Summary

There are two types of attacks: passive and active. Protection can be achieved with suitable mechanisms and appropriate policies. Recently, security has leaped to the forefront in priority because of changes in the regulatory environment and in technology. The fragmentation of operations that were once vertically integrated increased the number of participants in end-to-end information transfer. In virtual private networks, customers are allowed some control of their part of the public infrastructure. Finally, security must be retrofitted in IP networks to protect from the inherent difficulties of having user traffic and network control traffic within the same pipe.

Security mechanisms can be implemented in one or more layers of the OSI model. The choice of the layer depends on the security services to be offered and the coverage of protection.

Confidentiality guarantees that only the authorized parties can read the information transmitted. This is achieved by cryptography, whether symmetric or asymmetric. Symmetric cryptography is faster than asymmetric cryptography but has a limitation in terms of the secure distribution of the shared secret. Asymmetric (or public key) cryptography overcomes this problem; this is why both can be combined. In online systems, public key cryptography is used for sending the shared secret that can be used later for symmetric encryption. Two public key schemes used for sharing the secrets are Diffie–Hellman and RSA. As mentioned earlier, ISAKMP is a generic framework to negotiate point-to-point security and to exchange key and authentication data among two parties.

Data integrity is the service for preventing nonauthorized changes to the message content during transmission. A one-way hash function is used to produce a signature of the message that can be verified to ascertain integrity. Blind signature is a special procedure for signing a message without revealing its content.

The identification of participants depends on whether cryptography is symmetric or asymmetric. In asymmetric schemes, there is a need for authentication using certificates. In the case of human users, biometric features can be used for identification in specific situations. Kerberos is an example of a distributed system for online identification and authentication using symmetric cryptography.

Access control is used to counter the threats of unauthorized operations. There are two types of access control mechanisms: identity based and role based. Both can be managed through certificates defined by ITU-T Recommendation X.509. Denial of service is the consequence of failure of access control. These attacks are inherently associated with IP networks, where network control data and user data share the same physical and logical bandwidths. The best solution is to authenticate all communications by means of trusted certificates. Short of this, defense mechanisms will be specific to the problem at hand.

Nonrepudiation is a service that prevents a person who has accomplished an act from denying it later. This is a legal concept that is defined through legislation. The service comprises the generation of evidence and its recording and subsequent verification. The technical means to ensure nonrepudiation include electronic signature of documents, the intervention of third parties as witnesses, time stamping, and sequence numbering of the transactions.

3A Appendix: Principles of Symmetric Encryption

3A.1 Block Encryption Modes of Operation

The principal modes of operation of block ciphers are: electronic codebook (ECB) mode, cipher block chaining (CBC) mode, cipher feedback (CFB) mode, output feedback (OFB) mode, and counter (CTR) mode (National Institute of Standards and Technology, SP 800-38A, 2001).

The ECB mode is the most obvious, because each clear block is encrypted independently of the other blocks. However, this mode is susceptible to attacks by replay or reordering blocks, without detection. This is the reason this mode is only used to encrypt random data, such as the encryption of keys during authentication.

Some recent examples of the incorrect use of ECB is the Version 1 of Bitmessage, a peer-to-peer messaging system built on Bitcoin, which made it vulnerable (Buterin, 2012; Lerner, 2012; Warren, 2012).

The other three modes use a feedback loop protect against such types of attacks. They also have the additional property that they need an initialization vector to start the computations. This is a dummy initial ciphertext block whose value must be shared with the receiver. It is typically a random number or generated by encrypting a nonce (Ferguson et al., 2010, pp. 66–67). The difference among the three feedback modes resides in the way the clear text is mixed, partially or in its entirety, with the preceding encrypted block.

In the CBC mode, the input to the encryption algorithm is the exclusive OR of next block of plain test and the preceding block of the ciphertext. This is called "chaining" the plaintext blocks, as shown in Figure 3.15. Figure 3.16 depicts the decryption operation. In these figures, P_i represents the ith block of the clear message, while C_i is the corresponding encrypted block. Thus, the encrypted block C_i is given by

$$C_i = E_K\left(P_i \oplus C_{i-1}\right), \quad i = 0,1,\ldots$$

where

$E_K()$ represents the encryption with the secret key K

$\oplus$ is the exclusive OR operation

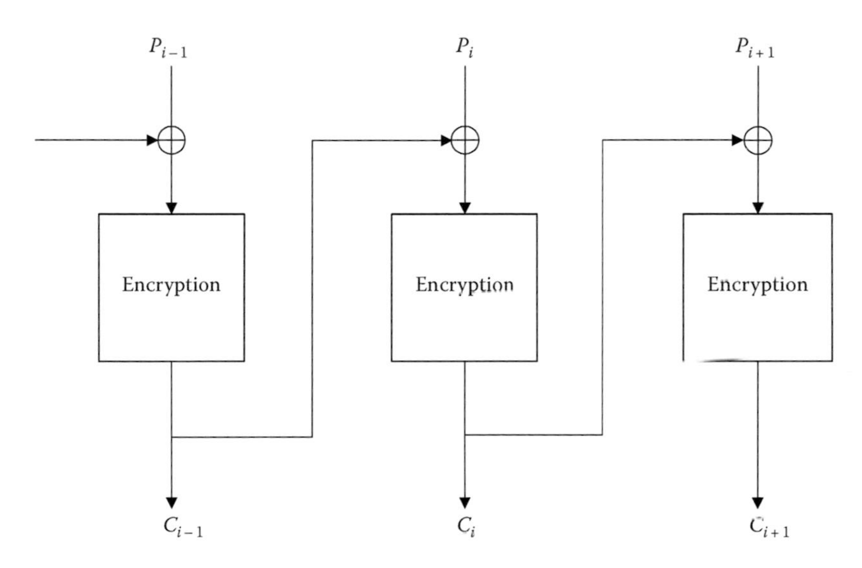

FIGURE 3.15
Encryption in the CBC mode.

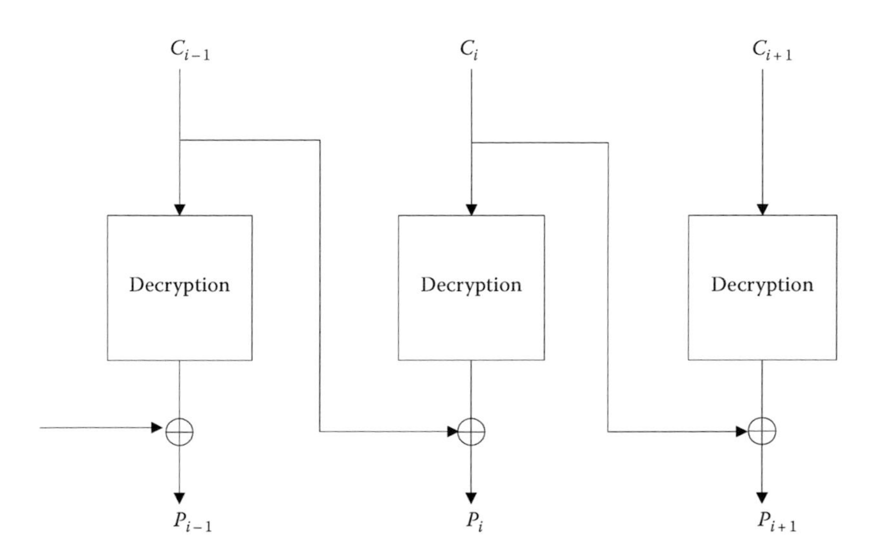

FIGURE 3.16
Decryption in the CBC mode.

The starting value C_0 is the initialization vector. The initialization vector does not need to be secret but it must be unpredictable and its integrity protected. For example, it can be generated by applying $E_K()$ to a nonce (a contraction of "*n*umber used *once*") or by using a random number generator (National Institute of Standards and Technology, 2001, p. 20). Also, the CBC mode requires the input to be a multiple of the cipher's block size, so padding may be needed.

The decryption operation, shown in Figure 3.16, is described by

$$P_i = C_{i-1} \oplus D_K(C_i)$$

Any subset of a CBC message will be decrypted correctly. The CBC mode is efficient in the sense that a stream of infinite length can be processed in a constant memory in linear time. It is useful for non-real-time encryption of files and to calculate the signature of a message (or its MAC) as specified for financial and banking transactions in ANSI X9.9 (1986), ANSI X9.19 (1986), ISO 8731-1 (1987), and ISO/IEC 9797-1 (1999). Transport Layer Security (TLS) also uses the CBC mode but, in general, the CFB and OFB modes are often used for the real-time encryption of a character stream, such as in the case of a client connected to a server.

In the CFB mode, the input is processed s bits at a time. A clear text block of b bits is encrypted in units of s bits (s = 1, 8, or 64 bits), with $s \le b$, that is, in $n = [s/b]$ cycles, where $[x]$ is the smallest integer less than x. The clear message block, P, is divided into n segments, $\{P_1, P_2, \ldots, P_n\}$ of s bits each; with extra bits padded to the trailing end of the data string if needed. In each cycle, an input vector I_i is encrypted into an output O_i. The segment P_i is combined through an exclusive OR function, with the most significant s bits of that output O_i to yield the ciphertext C_i of s bits. The ciphertext C_i is fed back and concatenated to the previous input I_i in a shift register, and all the bits of this register are shifted s positions of the left. The s left-most bits of the register are ignored, while the remainder of the register content becomes the new input vector I_{i+1} to be encrypted in the next round. The CFB encryption is described as follows:

$$I_i = LSB_{b-s}(I_{i-1}) \| C_{i-1}$$

$$O_i = E_K(I_i)$$

$$C_i = P_i \oplus MSB_s(O_i)$$

where

I_0 is the initialization vector
$LSB_j(x)$ is the least significant j bits of x
$MSB_j(x)$ is the most significant j bits
$E_K()$ represents the encryption with the secret key K

The decryption operation is identical to the roles of P_i and E_i transposed, that is,

$$I_i = LSB_{b-s}(I_{i-1}) \| C_{i-1}$$

$$O_i = E_K(I_i)$$

$$P_i = C_i \oplus MSB_s(O_i)$$

Depicted in Figure 3.17 is the encryption and illustrated in Figure 3.18 is the decryption.

Similar to CBC encryption, the initialization vector need not be secret but is preferably unpredictable and needs to be changed for each message. However, the chaining mechanism causes the ciphertext block C_i to depend on both P_i and the preceding O_i. The decryption operation is sensitive to bit errors, because one bit error in the encrypted text affects the decryption of n blocks.

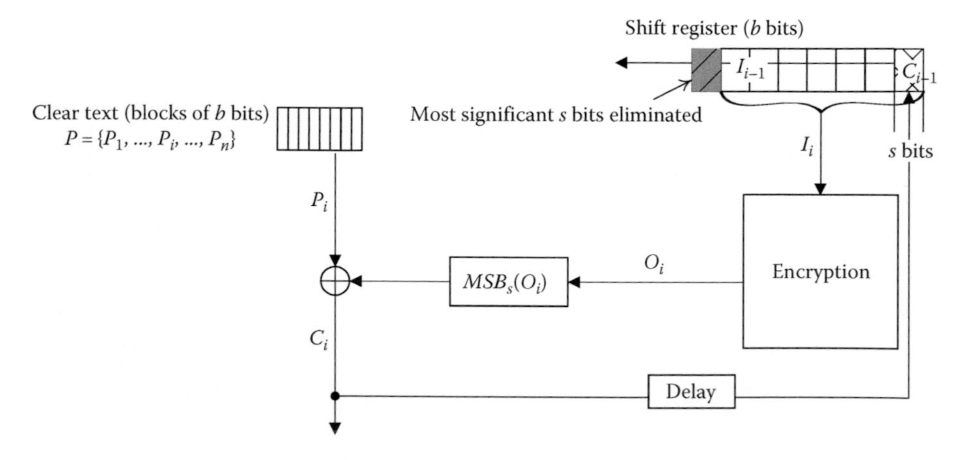

FIGURE 3.17
Encryption in the CFB mode of a block of b bits and s bits of feedback.

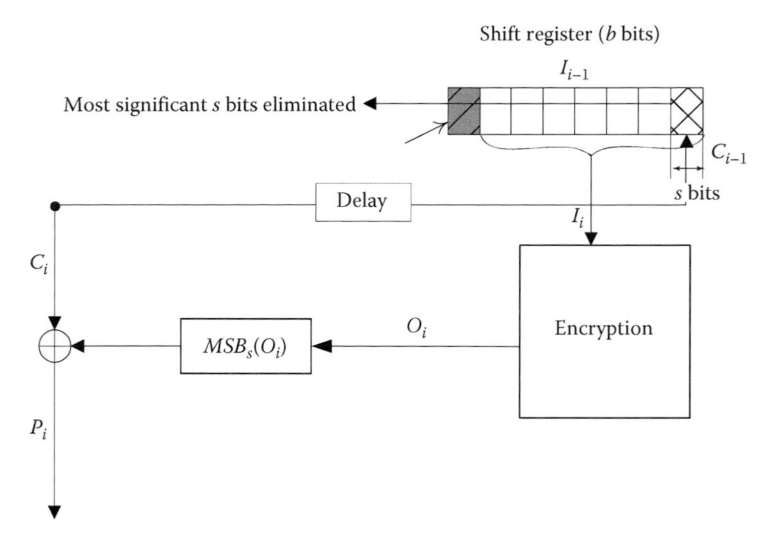

FIGURE 3.18
Decryption in the CFB mode of a block of b bits with s bits in the feedback loop.

If $s = b$, the shift register can be eliminated and the encryption is done as illustrated in Figure 3.19. Thus, the encrypted block C_i is given by

$$C_i = P_i \oplus E_K(C_{i-1})$$

where $E_K()$ represents the encryption with the secret key K.

The decryption is obtained with another exclusive OR operation as follows:

$$P_i = C_i \oplus E_K(C_{i-1})$$

which is shown in Figure 3.20.

The CFB mode can be used to calculate the MAC of a message. This method is also indicated in ANSI X9.9 (1986) for the authentication of banking messages, as well as ANSI X9.19 (1986), ISO 8731-1 (1987), and ISO/IEC 9797-1 (1999).

For the ECB, CBC, and CFB modes, the plaintext must be a sequence of one or more complete data blocks. If the data string to be encrypted does not satisfy this property, some extra bits, called padding, are appended to the plaintext. The padding bits must be selected so that they can be removed unambiguously at the receiving end. For example, it can consist of an octet with value 128 and then as many octets as required to complete the block (Ferguson et al., 2010, p. 64).

In the OFB mode, the input to the block cipher is not the clear text but a pseudo-random stream, called the key stream, which is derived from the ciphertext. The clear text message itself is not an input to the block cipher. This encryption scheme is called a "stream cipher." The encryption of the OFB mode is described as follows:

$$I_i = O_{i-1}$$

$$O_i = E_K(I_i)$$

$$C_i = P_i \oplus MSB_s(O_i)$$

where I_0 is the initialization vector. Each initialization vector must be used once, otherwise confidentiality may be compromised. In a stream cipher, there is no padding so if the last output block is a partial block of u bits, only the most significant u bits of that block are used in the exclusive OR operation. The lack of padding reduces the overhead, which is especially important with small messages.

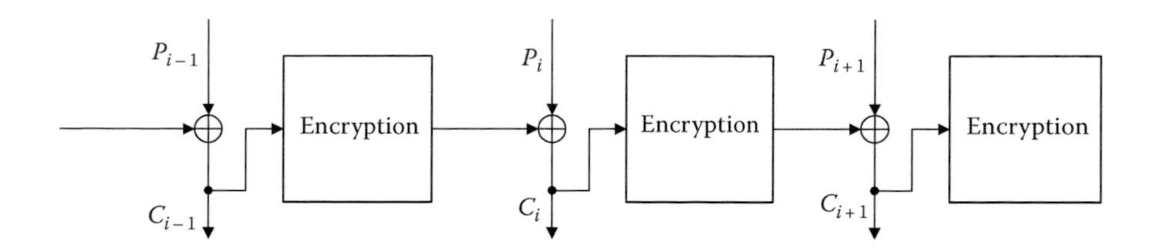

FIGURE 3.19
Encryption in the CFB mode for a block of b bits with a feedback of b bits.

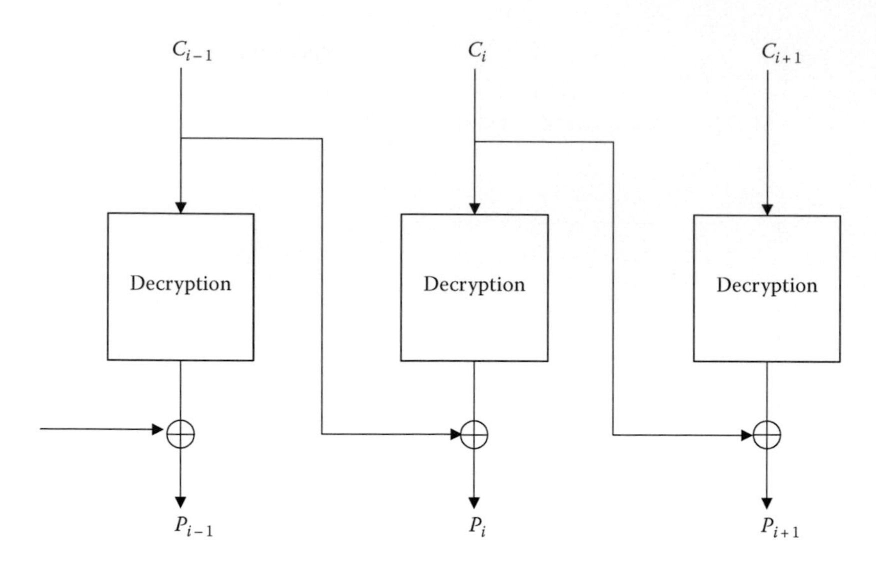

FIGURE 3.20
Decryption in the CFB mode for a block of b bits with a feedback of b bits.

The decryption process is exactly the same operation as encryption and is described as

$$I_i = O_{i-1}$$

$$O_i = E_K(I)$$

$$P_i = C_i \oplus MSB_s(O_i)$$

This is illustrated in Figures 3.21 and 3.22 for the encryption and decryption, respectively.

In the OFB mode, the input to the decryption depends on the preceding output only (i.e., it does not include the previous ciphertext), so errors are not propagated. This makes it suitable to situations where transmission is noisy. In this case, a single bit error in the ciphertext affects only one bit in the recovered text, provided that the values in the shift registers at both ends remain identical to maintain synchronization. Thus, any system that incorporates the OFB mode must be able to detect synchronization loss and have a mechanism to reinitialize the shift registers on both sides with the same value.

In the case where $s = b$, the encryption is illustrated in Figure 3.23 and is described by

$$O_i = E_k(O_{i-1})$$

$$C_i = P_i \oplus O_i$$

The decryption is described by

$$O_i = E_k(O_{i-1})$$

$$P_i = C_i \oplus O_i$$

and is depicted in Figure 3.24.

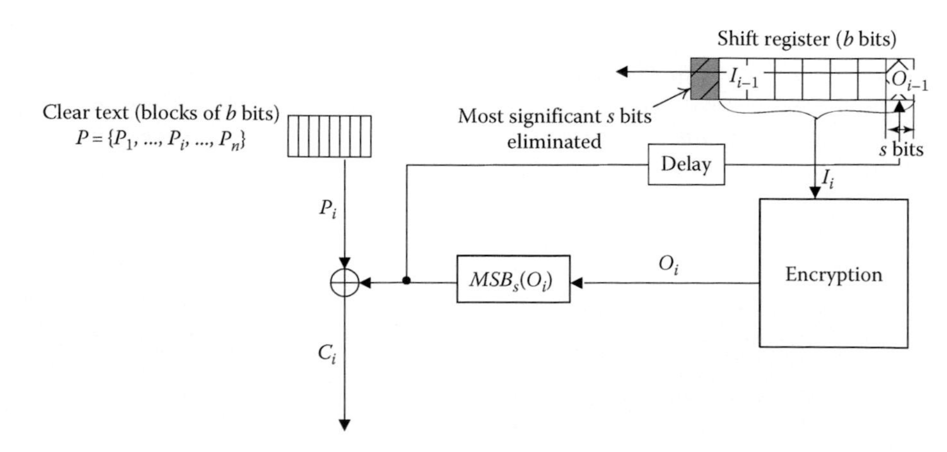

FIGURE 3.21
Encryption in OFB mode of a block of b bits with a feedback of s bits.

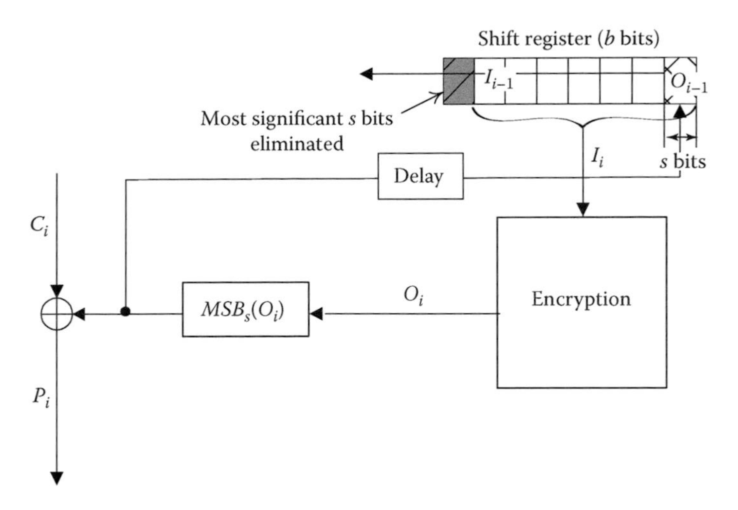

FIGURE 3.22
Decryption in OFB mode of a block of b bits with a feedback of s bits.

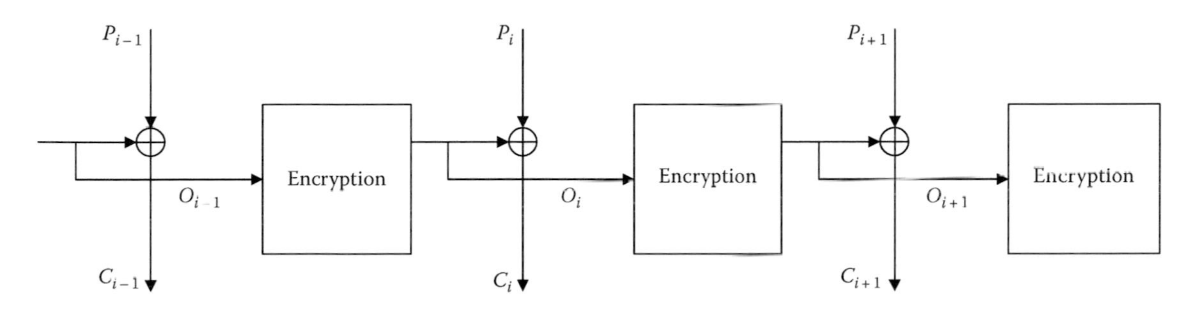

FIGURE 3.23
Encryption in OFB mode with a block of b bits and a feedback of b bits.

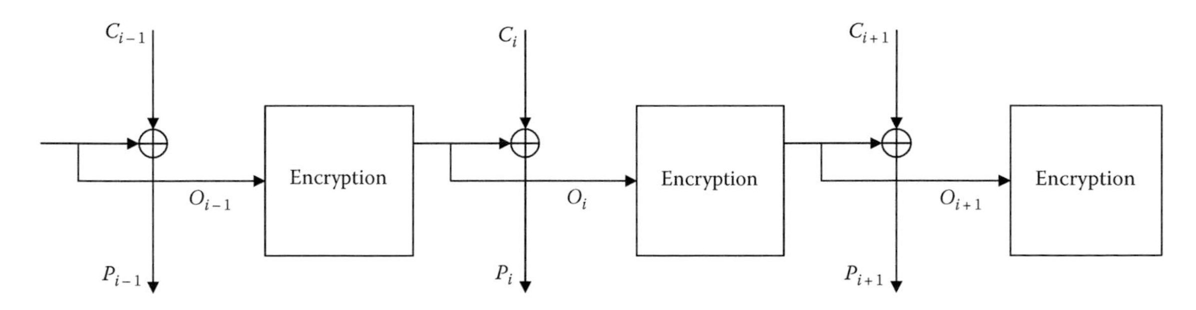

FIGURE 3.24
Decryption in OFB mode for a block of b bits with a feedback of b bits.

For security reasons, the OFB mode with $s = b$, that is, with the feedback size equal to the block size, is recommended (National Institute of Standards and Technology, 2001b; Barthélemy et al., 2005, p. 98).

Finally, the Counter (CTR) mode uses a set of input blocks, called counters, instead of the initialization vector. A cipher is applied to the input blocks to produce a sequence of output blocks that are used to produce the ciphertext. Given a sequence of counters, $T_1, T_2, \ldots, T_n$, the encryption operation is defined in SP 800-38A as follows (National Institute of Standards and Technology, 2001b):

$$O_i = E_K(T_i)$$

$$C_i = P_i \oplus O_i$$

Each counter in the sequence must be different from every other, that is, never repeat, for any given key. According to Appendix B of SP 800-38A, the counter block can be either a simple sequential counter that is incremented started from an initial string (i.e., $T_i = T_1 + j \bmod 2^b$) or a combination of a nonce for each message and a counter. For a 128-bit block size, a typical approach is to use T_i as the concatenation of a 48-bit message number, a 16-bit of additional nonce data and 64 bits for the block counter i so that one key can be used for encrypting at most 2^{48} different messages (Ferguson et al., 2010, p. 70). Because the nonce must be changed for every message, the plain text message is limited to $(2^{64} \times 128)/8 = 2^{68}$ octets. Should the last output block be a partial block of u bits, only the most significant u bits of that block are used in the exclusive OR operation, that is, there is no need for padding.

The decryption operation is as follows:

$$O_i = E_K(T_i)$$

$$P_i = C_i \oplus O_i$$

In both CTR encryption and CTR decryption, the output blocks O_i can be calculated in parallel and before the plaintext or the ciphertext are available. Also, any message block P_i can be recovered independently from any other message blocks, provided that the corresponding counter block is known; this allows parallel encryption, which makes suitable for high-speed data transmission. Another advantage is that the last block can be arbitrary and no padding is needed.

The CCM mode (Counter with CBC-MAC) is a derived mode of operation that provides both authentication and confidentiality for cryptographic block ciphers with a block length of 128 bits, such as the Advanced Encryption Standard (AES). In this way, the CCM mode avoids the need to use two systems: a MAC for authentication and a block cipher encryption for privacy. This results in a lower computational cost compared to the application of separate algorithm for both.

The inputs to the encryption are a nonce N, some additional data A, for example, network packet header, and the plain text P of length m. The length of N, $k_n <$ the length of the block k_b and is a multiple of 8 between 56 and 112 bits while the tag length k_t is a multiple of 16 between 32 and 128 bits. The data and the plaintext are authenticated while the CTR mode encryption is applied to the plain text only. First, a tag T of length $k_t \leq k_b$ is calculated as

$$T = CBC - MAC\left(N \| A \| P\right)$$

CBC-MAC acts on blocks of bit length k_b (typically 128 bits) and so if the last block has j bits, after the last full block has been encrypted, the ciphertext is encrypted again and the left-most j bits of the encrypted ciphertext and is exclusive ORed with the short block of the message to generate the tag T. Next, the tag T and the message P are encrypted with the CTR mode to form a ciphertext c of length $m + k_t$. The nonce N must not have been used in a previous CCM encryption during the lifetime of the key. The counters blocks CTR_i are generated from the nonce N using a function π as follows:

$$CTR_i = \pi(i, N, P, A)$$

This mode was developed to avoid the intellectual property issues around the use of another mode, the Offset Codebook (OCB) mode proposed for the IEEE 802.11i standard. The OCB mode, however, is still optional in that standard. The CCM mode is described in RFC 3610 (2003).

3A.2 Examples of Symmetric Block Encryption Algorithms

3A.2.1 DES and Triple DES

DES, also known as the Data Encryption Algorithm (DEA), was widely used in the commercial world for applications such as the encryption of financial documents, the management of cryptographic keys, and the authentication of electronic transactions. The algorithm was developed by IBM and then adopted as a U.S. standard in 1977. It was published in FIPS 81, then adopted by ANSI in ANSI X3.92 under the name of Data Encryption Algorithm (DEA). However, by 2008, commercial hardware costing less than $15,000 could break DES keys in less than a day on average.

DES operates by encrypting blocs of 64 bits of clear text to produce blocks of 64 bits of ciphertext. The key length is 64 bits, with 8 bits for parity control, which gives an effective length of 56 bits. The encryption and decryption are based on the same algorithm with some minor differences in the generation of subkeys.

In 2005, the National Institute of Standards and Technology (NIST) finally withdrew the DES standard following the development of the AES (Advanced Encryption Standard).

The vulnerability of DES to an exhaustive attack forced the search for an interim solution until a replacement algorithm could be developed and deployed. Given the considerable investment in the software and hardware implementations of DES, triple DES, also known as TDEA (Triple Data Encryption

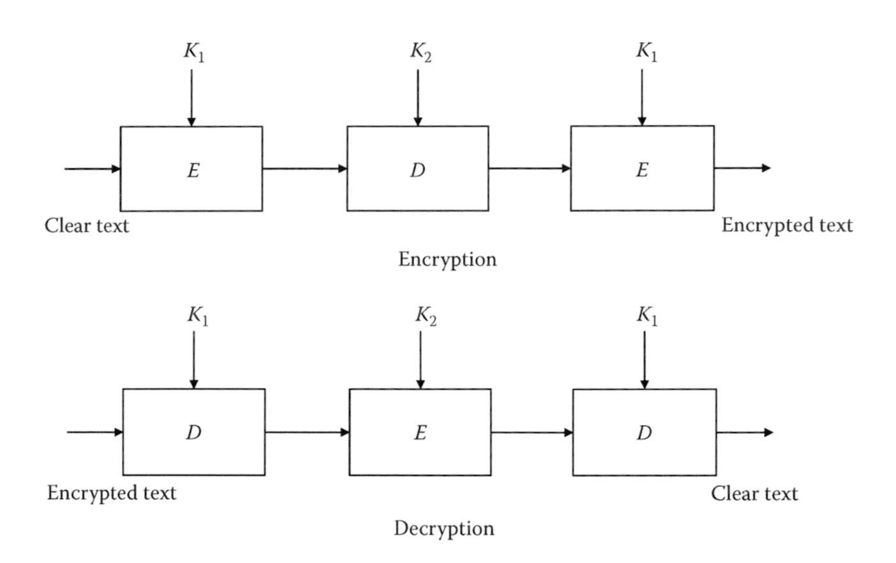

FIGURE 3.25
Operation of triple DES (TDEA) with two keys.

Algorithm), is based on the use of DES three successive times (Schneier, 1996, pp. 359–360). The operation of Triple DES with two different 56-bit keys is represented in Figure 3.25.

The use of three stages doubles the effective length of the key to 112 bits. The operations "encryption–decryption–encryption" aim at preserving compatibility with DES, because if the same key is used in all operations, the first two cancel each other.

Three independent 56-bit keys are highly recommended in federal applications (National Institute of Standards and Technology, 2012c, p. 37). In this case, the operation becomes as illustrated in Figure 3.26.

3A.2.2 AES

The Advanced Encryption Standard (AES) is the symmetric encryption algorithm that replaced DES. It is published by NIST as FIPS 197 (National Institute of Standards and Technology, 2001a) and is based on the algorithm Rijndael as developed by Joan Daemen of Proton World International and Vincent Rijmen from the

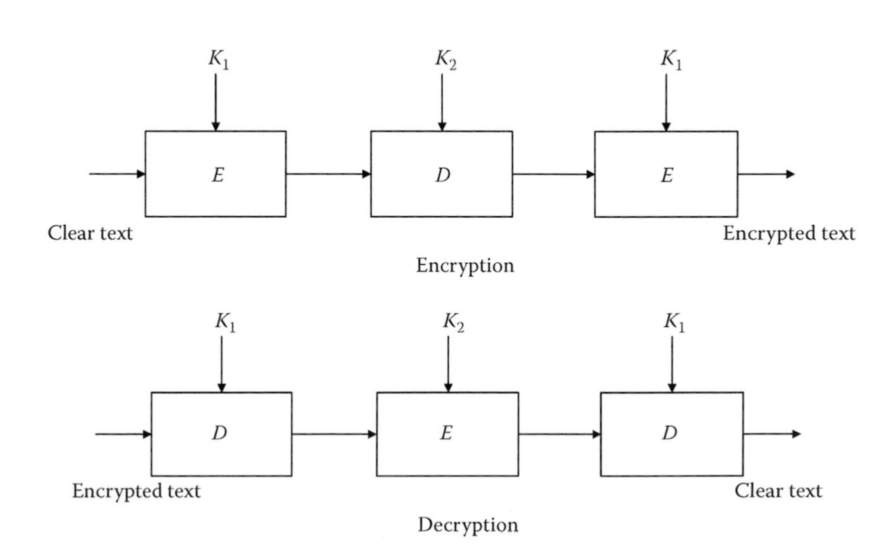

FIGURE 3.26
Operation of triple DES (TDEA) with three keys.

Catholic University of Leuven (Katholieke Universiteit Leuven). It is a block code with blocks of 128, 192, or 256 bits. The corresponding key lengths are 128, 192, and 256 bits, respectively. It is based on finding a solution of an algebraic equation of the form

$$y^2 = (x^3 + ax + b) \bmod n$$

The selection in October 2000 came after two rounds of testing following NIST's invitation for submission to cryptographers from around the world. In the first round, 15 algorithms were retained for evaluation. In the second round of evaluation, five finalists were retained: RC6, MARS, Rijndael, Serpent, and Twofish. All the second round algorithms showed a good margin of security. The criteria used to separate them are related to algorithmic performance: speed of computation in software and hardware implementations (including specialized chips), suitability to smart cards (low memory requirements), and so on. Results from the evaluation and the rational for the selection have been documented in a public report by NIST (Nechvatal et al., 2000).

SAFER+ (Secure And Fast Encryption Routine) was one of the candidate block ciphers, based on a nonproprietary algorithm developed by James Massey for Cylink Corporation (Schneier, 1996, pp. 339–341). An enhanced version of that algorithm SAFER-128 is used in Bluetooth networks for confidentiality and mutual authentication.

AES CGM (Galois/Counter Mode) and AES CCM (Counter with Cipher Block Chaining—Message Authentication Code Mode) are two modes of operation of AES with key lengths of 128, 192, or 256 bits (National Institute of Standards and Technology, 2001b, 2007). These modes provide Authenticated Encryption with Associated Data (AEAD), where the plaintext is simultaneously encrypted and its integrity protected. The input may be of any length but the ciphered output is generally larger than the input because of the integrity check value. CCM is described in RFC 5116 (2008). AES-GCM provides an effective defense against side-channel attacks but at the expense of performance.

As of now, there are no known attacks that would break a correct implementation of encryption with AES. AES is free of royalties and highly secure, and today, it is often the first choice for symmetric encryptions.

The AES competition is generally viewed as having provided a tremendous boost to the cryptographic research community's understanding of block ciphers, and a tremendous increase in confidence in the security of some block ciphers.

3A.2.3 RC4

The stream cipher RC4 was originally designed by Ron Rivest and became public in 1994. It uses a variable key length ranging from 8 to 2048 bits. It has the advantage of being extremely fast when implemented in software; as a consequence, it is commonly used in cryptosystems such as SSL, TLS, the wireless security protocols WEP (Wired Equivalent Privacy) of IEEE 802.11, WPA (Wi-Fi Protected Access) of IEEE 802.11i and some Kerberos encryption modes used in Microsoft Windows.

The statistical distribution of single octets as well as in the entire RC4-encrypted stream shows deviations from a uniform distribution. In other words, there are strong biases toward particular values at specific positions and some bit patterns occur in the output stream more frequently than others. In 2013, it was shown how to use the 65,536 single octet statistical biases in the initial 256 octets of the RC4 ciphertext to mount an attack on SSL/TLS via Bayesian analysis (AlFardan et al., 2013c). The attack, however, requires access to between 2^{28} and 2^{32} copies of the same data encrypted using different keys. This could be done with a browser infected with a JavaScript malware to make these many connections to a server and, accordingly, give the attacker enough data. This attack may not be always practical, but it is quite possible that other attacks are known but not revealed, which may explain why RC4 was never approved as a Federal Information Processing Standard (FIPS).

3A.2.4 New European Schemes for Signature, Integrity, and Encryption

The following algorithms have been selected by the New European Schemes for Signature, Integrity, and Encryption (NESSIE) project:

1. *MISTY*: This is a block encryption algorithm. The block size is 64 bits and the key is 128 bits.
2. AES with 128 bits.
3. *Camellia*: The block length is fixed at 128 bits but the key size can be 128, 192, or 256 bits.
4. *SHACAL-2*: The block size is 256 bits and the key is 512 bits.

Camellia is also used in Japanese government systems and is included in the specifications of the TV-Anytime Forum for high-capacity storage in consumer systems.

3A.2.5 eSTREAM

In 2004, ECRYPT, a Network of Excellence funded by the European Union, announced the eSTREAM competition to select new stream ciphers suitable for widespread adoption. This call attracted 34 submissions. Following hundreds of security and performance evaluations, the eSTREAM committee selected a portfolio containing several stream ciphers.

3A.2.6 IDEA

IDEA was invented by Xuejia Lai and James Massey circa 1991. The algorithm takes blocks of 64 bits of the clear text, divides them into subblocks of 16 bits each, and encrypts them with a key 128 bits long. The same algorithm is used for encryption and decryption. IDEA is clearly superior to DES but has not been a commercial success. The patent is held by a Swiss company Ascom-Tech AG and is not subject to U.S. export control.

3A.2.7 SKIPJACK

SKIPJACK is an algorithm developed by the NSA for several single chip processors such as Clipper, Capstone, and Fortezza. Clipper is a tamper-resistant very large-scale integration (VLSI) chip used to encrypt voice conversation. Capstone provides the cryptographic functions needed for secure e-commerce and is used in Fortezza applications. SKIPJACK is an iterative block cipher with a block size of 64 bits and a key of 80 bits. It can be used in any of the four modes ECB, CBC, CFB (with a feedback of 8, 16, 32, or 64 bits), and OFB with a feedback of 64 bits.

3B Appendix: Principles of Public Key Encryption

The most popular algorithms for public cryptography are those of Rivest, Shamir, and Adleman (RSA) (1978), Rabin (1979), and ElGamal (1985). Nevertheless, the overwhelming majority of proposed systems in commercial systems are based on the RSA algorithm.

It should be noted that RSADSI was founded in 1982 to commercialize the RSA algorithm for public key cryptography. However, its exclusive rights ended with the expiration of the patent on September 20, 2000.

3B.1 RSA

Consider two odd prime numbers **p** and **q** whose product $\mathbf{N} = \mathbf{p} \times \mathbf{q}$. **N** is the modulus used in the computation that is public, while the values p and q are kept secret.

Let $\varphi(\mathbf{N})$ be the Euler totient function of **N**. By definition, $\varphi(\mathbf{N})$ is the number of elements formed by the complete set of residues that are relatively prime to **N**. This set is called the reduced set of residues modulo **N**.

If **N** is a prime, $\varphi(\mathbf{N}) = \mathbf{N} - 1$. However, because $\mathbf{N} = \mathbf{p} \times \mathbf{q}$ by construction, while **p** and **q** are primes, then

$$\varphi(\mathrm{N}) = (\mathbf{p}-1)(\mathbf{q}-1)$$

According to Fermat's little theorem, if **m** is a prime and **a** is not a multiple of **m**, the

$$\mathbf{a}^{\mathrm{m}-1} \equiv 1(\mathrm{mod}\,\mathbf{m})$$

Euler generalized this theorem in the form

$$\mathbf{a}^{\varphi(\mathrm{N})} \equiv 1(\mathrm{mod}\,\mathbf{N})$$

Choose the integers **e**, **d** both less than $\varphi(\mathbf{N})$ such that the greatest common divisor of $(\mathbf{e}, \varphi(\mathbf{N})) = 1$ and $\mathbf{e} \times \mathbf{d} \equiv 1 \bmod (\varphi(\mathbf{N})) = 1 \bmod ((\mathbf{p} - 1)(\mathbf{q} - 1))$.

Let **X**, **Y** be two numbers less than **N**

$$\mathbf{Y} = \mathbf{X}^{\mathrm{e}} \bmod \mathbf{N} \quad \text{with } 0 \le \mathbf{X} < \mathbf{N}$$
$$\mathbf{X} = \mathbf{Y}^{\mathrm{d}} \bmod \mathbf{N} \quad \text{with } 0 \le \mathbf{Y} < \mathbf{N}$$

because, by applying Fermat's little theorem,

$$\mathbf{Y}^{\mathrm{d}} \bmod \mathbf{N} = (\mathbf{X}^{\mathrm{e}})^{\mathrm{d}} \bmod \mathbf{N} = \mathbf{X}^{\mathrm{ed}} \bmod \mathbf{N} = \mathbf{X}^{\varphi(\mathrm{N})}$$
$$\equiv 1(\mathrm{mod}\,\mathbf{N}) = 1 \bmod \mathbf{N}$$

To start the process, a block of data is interpreted as an integer. To do so, the total block is considered as an ordered sequence of bits (of length, say, λ). The integer is considered to be the sum of the bits by giving the first bit the weight of $2^{\lambda-1}$, the second bit the weight of $2^{\lambda-2}$, and so on until the last bit that will have the weight of $2^0 = 1$.

The block size must be such that the largest number does not exceed the modulo **N**. Incomplete blocks must be completed by padding bits with either 1 or 0 bits. Further padding blocks may also be added.

The public key of the algorithm **PK** is the number **e**, along with **n**, while the secret key **SK** is the number **d**. RSA achieves its security from the difficulty of factoring **N**. The number of bits of **N** is considered to be the key size of the RSA algorithm. The selection of the primes **p** and **q** must make this factorization as difficult as possible.

Once the keys have been generated, it is preferred that, for reasons of security, the values of p and q and all intermediate values such as the product $(\mathbf{p} - 1)(\mathbf{q} - 1)$ be deleted. Nevertheless, the preservation of the values of **p** and **q** locally can double or even quadruple the speed of decryption.

3B.1.1 Chosen-Ciphertext Attacks

It is known that plain RSA is susceptible to a chosen-ciphertext attack (Davida, 1982). An attacker who wishes to find the decryption of a message $m \equiv c^d \bmod \mathbf{N}$ of a ciphertext c chooses a random integer s and asks for the decryption of $c' \equiv s^e c \bmod \mathbf{N}$. The answer is $(s^e c)^d \bmod \mathbf{N} \equiv s^{ed} c^d \bmod \mathbf{N} = 1 \times c^d \bmod \mathbf{N} \equiv c^d \bmod \mathbf{N}$

It is also known that a number of attacks are based on the fact that around ¼ of the bits of the private key are leaked, such as by timing and power analysis attacks. Also, the strength of cryptosystem depends on the choice of the primes (Pellegrini et al., 2010).

3B.1.2 Practical Considerations

To increase the speed of signature verification, suggested values for the exponent e of the public key are 3 or $2^{16} + 1$ (65,537) (Menezes et al., 1997, p. 437). Other variants designed to speed up decryption and signing are discussed in Boneh and Shacham (2002).

For short-term confidentiality, the modulus **N** should be at least 768 bits. For long-term confidentiality (5–10 years), at least 1024 bits should be used. Currently, it is believed that confidentiality with a key of 2048 bits would last about 15 years.

In practice, RSA is used with padding schemes to avoid some of the weaknesses of the RSA algorithm as described earlier. Padding techniques such as Optimal Asymmetric Encryption Padding (OAEP) are standardized as discussed in the next section.

3B.2 Public Key Cryptography Standards

The public key cryptography standards (PKCS) are business standards developed by RSA Laboratories in collaboration with many other companies working in the area of cryptography. They have been used in many aspects of public key cryptography, which are based on the RSA algorithm. At the time of writing this section, their number has reached 15.

PKCS #1 (RFC 2437 [1998]; RFC 3447 [2003]) defines the mechanisms for data encryption and signature using the RSA algorithm. These procedures are then utilized for constructing the signatures and electronic envelopes described in PKCS #7. Following the presentation of an adaptive chosen-ciphertext attack on PKCS #1 (Bleichenbacher, 1998), more secure encodings are available with PKCS #1v1.5 and PKSC #1v2.1, the latter of which is defined in RFC 3447. In particular, PKCS #1v2.1 defines an encryption scheme based on the Optimal Asymmetric Encryption Padding (OAEP) of Bellare and Rogaway (1994). PKCS #2 and #4 are incorporated in PKCS #1.

PKCS #3 defines the key exchange protocol using the Diffie–Hellman algorithm.

PKCS #5 describes a method for encrypting an information using a secret key derived from a password. For hashing, the method utilizes either MD2 or MD5 to compute the key starting with the password and then encrypts the key with DES in the CBC mode.

PKCS #6 defines a syntax for X.509 certificates.

PKCS #7 (RFC 2315, 1998 defines the syntax of a message encrypted using the basic encoding rules [BER] of ASN.1 [Abstract Syntax Notation 1]) (Steedman, 1993) of ITU-T Recommendation X.209 (1988). These messages are formed with the help of six content types:

1. *Data*, for clear data
2. *SignedData*, for signed data
3. *EnvelopedData*, for clear data with numeric envelopes
4. *SignedAndEnvelopedData*, for data that are signed and enveloped
5. *DigestedData*, for digests
6. *EncryptedData*, for encrypted data

The secure messaging protocol, S/MIME (Secure Multipurpose Internet Mail Extensions) and the messages of the SET protocol, designed to secure bank card payments over the Internet, utilize the PKSC #7 specifications.

PKCS #8 describes a format for sending information related to private keys.

PKCS #9 defines the optional attributes that could be added to other protocols of the series. The following items are considered: the certificates of PKCS #6, the electronically signed messages of PKCS #7, and the information on private keys as defined in PKCS #8.

PKCS #10 (RFC 2986, 2000) describes the syntax for certification requests to a certification authority. The certification request must contain details on the identity of the candidate for certification, the distinguished name of the candidate, his or her public key, and optionally, a list of supplementary attributes, a signature of the preceding information to verify the public key, and an identifier of the algorithm used for the signature so that the authority could proceed with the necessary verifications. The version adopted by the IETF is called CMS (Cryptographic message syntax).

PKCS #11 defines a cryptographic interface called *Cryptoki* (Cryptographic Token Interface Standard) between portable devices such as smart cards or PCMCIA cards and the security layers.

PKCS #12 describes a syntax for the storage and transport of public keys, certificates, and other users' secrets. In enterprise networks, key pairs are transmitted to individuals via password protected PKCS #12 files.

PKCS #13 describes a cryptographic system using elliptic curves.

PKCS #15 describes a format to allow the portability of cryptographic credentials such as keys, certificates, passwords, PINs among application and among portable devices such as smart cards.

The specifications of PKCS #1, #7, and #10 are not IETF standards because they mandate the utilization of algorithms that RSADSI does not offer free of charge. Also note that in PKCS #11 and #15, the word *token* is used to indicate a *portable device capable of storing persistent data.*

3B.3 PGP and OpenPGP

Pretty Good Privacy is considered to be the commercial system whose security is closest to the military grade. It is described in one of the IETF documents, namely, RFC 1991 (1996). PGP consists of six functions:

1. Public key exchange using RSA with MD5 hashing
2. Data compression with ZIP, which reduces the file size and redundancies before encryption (Reduction of the size augments the speed for both processing and transmission, while reduction of the redundancies makes cryptanalysis more difficult.)
3. Message encryption with IDEA
4. Encryption of the user's secret key using the digest of a sentence instead of a password
5. ASCII "armor" to protect the binary message for any mutilations that might be caused by Internet messaging systems (This armor is constructed by dividing the bits of three consecutive octets into four groups of 6 bits each and then by coding each group using a 7-bit character according to a given table. A checksum is then added to detect potential errors.)
6. Message segmentation

The IETF did not adopt PGP as a standard because it incorporates proprietary protocols. OpenPGP is based on PGP but avoids these intellectual property issues. It is specified in RFCs 4880 (2007). RFC 6637 (2012) describes how to use elliptic curve cryptography (ECC) with OpenPGP.

OpenPGP is currently the most widely used e-mail encryption standard. Companies and organizations that implement OpenPGP have formed the OpenPGP Alliance to promote it and to ensure interoperability. The Free Software Foundation has developed its own OpenPGP conformant program called GNU Privacy Guard (abbreviated GnuPG or GPG). GnuPG is freely available together with all source code under the GNU General Public License (GPL).

3B.4 Elliptic Curve Cryptography

Elliptic curve cryptography (ECC) is a public key cryptosystem where the computations take place on an elliptic curve. Elliptic curves have been applied in factoring integers, in proving primality, in coding theory and in cryptography (Menezes, 1993). Variants of the Diffie–Hellman and DSA algorithms on elliptic curves are the Elliptic Curve Diffie–Hellman algorithm (ECDH) and the Elliptic Curve Digital Signature Algorithm (ECDSA), respectively. They are used to create digital signatures and to establish keys for symmetric cryptography. Diffie–Hellman and ECDH are comparable in speed, but RSA is much slower. The advantage of elliptic curve cryptography is that key lengths are shorter than for existing public key schemes that provide equivalent security. For example, the level of security of 1024-bit RSA can be achieved will elliptic curves with a key size in the range of 171–180 bits (Wiener, 1998). This is an important factor in wireless communications and whenever bandwidth is a scarce resource.

Elliptic curves are defined over the finite field of the integers modulo a primary number p [the Galois field GF(p)] or that of binary polynomials [GF(2^m)]. The key size is the size of the prime number or the binary polynomial in bits. Cryptosystems over GF(2^m) appear to be slower than over GF(p) but there is no consensus on that point. Their main advantage, however, is that addition over GF(2^m) does not require integer multiplications, which reduces the cost of the integrated circuits implementing the computations.

NIST has standardized a list of 15 elliptic curves of varying key lengths (NIST, 1999). Ten of these curves are for binary fields and five are for prime fields. These curves provide confidentiality equivalent to symmetric encryption with keys of length 80, 112, 128, 192, and 256 bits and beyond.

Table 3.9 shows the comparison of the key lengths of RSA and elliptic cryptography for the same level of

TABLE 3.9

Comparison of Public Key Systems in Terms of Key Length in Bits for the Same Security Level

RSA	Elliptic Curve	Reduction Factor RSA/ECC
512	106	5:1
1,024	160	7:1
2,048	211	10:1
5,120	320	16:1
21,000	600	35:1

Source: Menezes, A., *Elliptic Curve Public Key Cryptosystems*, Kluwer, Dordrecht, the Netherlands, 1993.

TABLE 3.10

NIST Recommended Key Lengths in Bits

Symmetric	RSA and Diffie–Hellman	Elliptic Curve
80	1,024	160–112
112	2,048	224–255
128	3,072	256–283
192	7,680	384–511
256	15,360	512+

Source: Barker, E. et al., National Institute of Standards and Technology (NIST), Recommendation for Key Management – Part 1: General, (Revision 3), NIST Special Publication SP 800-57, July 2012c.

security measured in terms of effort to break the system (Menezes, 1993).

Table 3.10 gives the key sizes recommended by the National Institute of Standards and Technology for equivalent security using symmetric encryption algorithms (e.g., AES, DES or SKIPJACK) or public key encryption with RSA, Diffie–Hellman, and elliptic curves for equivalent security.

Thus, for the same level of security per the currently known attacks, elliptic curve–based systems can be implemented with smaller parameters. This computational efficiency is compared to Diffie–Hellman in Table 3.11 for several key sizes.

Another related aspect is the channel overhead for key exchanges and digital signatures on a communications link. In channel-constrained environments or when computing power, memory and battery life of devices are critical such as in wireless communication, elliptic curves offer a much better solution than RSA or Diffie–Hellman. As a result, many companies in wireless communications have embraced elliptic curve cryptography. Furthermore, many nations (e.g., the United States, the United Kingdom, and Canada) have adopted elliptic curve cryptography in the new generation of equipment to protect classified information.

TABLE 3.11

Relative Computation Costs of Diffie–Hellman and Elliptic Curves

Symmetric Key Size in Bits	Diffie–Hellman Cost: Elliptic Curve Cost
80	3:1
112	6:1
128	10:1
192	32:1
256	64:1

Source: National Security Agency (NSA), The case for elliptic curve cryptography, Central Security Service, January 15, 2009, https://www.nsa.gov/business/programs/elliptic_curve.shtml, last accessed June 20, 2015, no longer accessible.

One disadvantage of ECC is that it increases the size of the encrypted message significantly more than RSA encryption. Furthermore, the ECC algorithm is more complex and more difficult to implement than RSA, which increases the likelihood of implementation errors, thereby reducing the security of the algorithm. Another problem is that various aspects of elliptic curve cryptography have been patented notable the Canadian company Certicom, which holds over 130 patents related to elliptic curves and public key cryptography in general.

3C Appendix: Principles of the Digital Signature Algorithm and the Elliptic Curve Digital Signature Algorithm

According to the Digital Signature Algorithm (DSA) defined in ANSI X9.30:1 (1997), the signature of a message M is the pair of numbers r and s computed as follows:

$$r = (g^k \bmod p)$$

and

$$s = \{k^{-1}\ [SHA(M) + x\,r]\} \bmod q$$

The following are given:

- p and q are primes such that $2^{511} < p < 2^{1024}$, $2^{159} < q < 2^{160}$, and q is a prime divisor of $(p-1)$, that is, $(p-1) = mq$ for some integer m.
- $g^q = h^{(p-1)/q} \bmod p$ is a generator polynomial modulo p of order q. The variable h is an integer $1 < h < (p-1)$ such that $h^{(p-1)/q} \bmod p > 1$. By Fermat's little theorem, $g^q = h^{(p-1)} \bmod p$ because $g < p$. Thus, each time the exponent is a multiple of q, the result will be equal to 1 (mod p).
- x is a random integer such that $0 < x < q$.
- k is a random integer in the interval $0 < k < q$ and is different for each signature.
- k^{-1} is the multiplicative inverse of k mod q, that is, $(k^{-1} \times k) \bmod q = 1$.
- $SHA()$ is the SHA-1 hash function.
- x is the private key of the sender, while the public key is (p, q, g, y) with $y = g^x \bmod p$.

The DSA signature consists of the pair of integers (r, s). To verify the signature, the verifier computes

$$w = s^{-1} \bmod q$$
$$u_1 = SHA(M)w \bmod q$$
$$u_2 = r \cdot w \bmod q$$
$$v = (g^{u_1} \cdot y^{u_2} \bmod p) \bmod q$$

The signature is valid if $v = r$. To show this, we have

$$v = \left\{\left[g^{[SHA(M)\cdot w \bmod q]}\, y^{r\cdot w \bmod q}\right] \bmod p\right\} \bmod q$$
$$= \left\{\left[g^{[SHA(M)\cdot w \bmod q]}\, g^{r\cdot x\cdot w \bmod q}\right] \bmod p\right\} \bmod q$$
$$= \left\{\left[g^{[SHA(M)+x\cdot r]w \bmod q}\right] \bmod p\right\} \bmod q$$
$$= \{g^{(k\cdot s\cdot w) \bmod q} \bmod p\} \bmod q$$
$$= (g^{k \bmod q} \bmod p) \bmod q$$
$$= (g^{k} \bmod p) \bmod q,$$

since the generator is of order q by construction

$$= r$$

The strength of the algorithm is heavily dependent on the choice of the random numbers.

The random variable k is also transmitted with the signature. This means that if verifiers also know the signer's private key, they will be able to pass additional information through the channel established through the value of k.

The Elliptic Curve Digital Signature Algorithm (ECDSA) of ANSI X9.62:2005 is used for digital signing, while ECDH can be used to secure online key exchange. Typical key sizes are in the range of 160–200 bits. The ECDSA keys are generated as follows (Paar and Pelzl, 2010, pp. 283–284):

- Use an elliptic curve E with modulus p, coefficients a and b, and a point A that generates a cyclic group of prime order q, that is, $A^q = 1 \bmod p$ and all the elements of this cyclic group can be generated with A.
- Choose a random integer y such that $0<y<q$ and compute $B = y\,A$.
- The private key is y and the public key is (p, a, b, q, A, B).

The ECDSA signature for a message M is computed as follows:

- Choose an integer k randomly with $0<k<q$.
- Compute $R = kA$ and assign the x-coordinate of R to the variable r.
- Compute $s = (H(M) + y\cdot r)k^{-1} \bmod q$, with $H()$ a hash function.

The signature verification proceeds as follows:

- Compute $w = s^{-1} \bmod q$.
- Compute $u_1 = w\cdot H(x) \bmod q$ and $u_2 = w\cdot r \bmod q$.
- Compute $P = u_1 A + u_2 B$.
- The signature is valid if and only if x_p the x-coordinate of point P satisfies the condition.

$$x_p = r \bmod q.$$

Questions

1. What are the major security vulnerabilities in a client/server communication?
2. What are the services needed to secure data exchanges in e-commerce?
3. Compare the tunnel mode and transport mode of IPSec.
4. Why is the Authentication header (AH) less used with IPSec than the Encapsulating Security Payload (ESP) protocol?
5. What factors affect the strength of encryption?
6. Discuss some potential applications for blind signatures.
7. Discuss the difference between biometric identification and biometric verification.
8. Compare and contrast the following biometrics: fingerprint, face recognition, iris recognition, voice recognition, and hand geometry.
9. Discuss some of the vulnerabilities of biometric identification systems.
10. What is needed to offer nonrepudiation services?
11. What conditions favor denial-of-service attacks?
12. Which of the following items is not in a digital public key certificate?
    (a) the subject's public key
    (b) the digital signature of the certification authority
    (c) the subject's private key
    (d) the digital certificate serial number

13. What is cross-certification and why is it used? What caveats need to be taken considered for it to work correctly?
14. What is the Sender Policy Framework (SPF)? Describe its advantages and drawbacks.
15. What is the DomainKeys Identified Mail (DKIM)? Describe its advantages and drawbacks.
16. Using the case of AES as a starting point, define a process to select a new encryption algorithm.
17. Compare public key encryption and symmetric encryption in terms of advantages and disadvantages. How to combine the strong points of both?
18. What are the reasons of the current interest in elliptic curve cryptography (ECC)?
19. Explain why the length of the encryption key cannot be used as the sole measure for the strength of an encryption.
20. Speculate on the reasons that led to the declassification of the SKIPJACK algorithm.
21. What are the problems facing cross-certification? How are financial institutions attempting to solve them?
22. What is the counter mode of block cipher encryption? What are its advantages?
23. With the symbols as explained in Appendix 3C, show that a valid ECDSA signature satisfies the condition $r = x_p \bmod q$.
24. List the advantages and disadvantages of controlling the export of strong encryption algorithms from the United States.

4

Business-to-Business Commerce

The electronic exchange of commercial data was a direct consequence of the computerization of businesses as a way to increase operational efficiencies through automation. The need for standardization became acutely apparent after the introduction of just-in-time (JIT) production in global enterprises and increased focus on the efficiency of the procurement process to improve the response to market stimuli as well as control operational costs. Specifically, supply networks are constituted by firms that purchase goods, their suppliers, and their suppliers' suppliers. Increasing the efficiency of the supply networks is a key objective of supply chain management by linking the activities of buying, manufacturing, and shipping products from suppliers to purchasing firms to improve response to market conditions. More recently, the adoption of the Internet in commercial applications has generated new possibilities in the fragmented markets and for small and medium enterprises.

This chapter summarizes the current status of business-to-business electronic commerce, often known as B2B e-commerce or e-business. The focus is on the technical aspects of business-to-business commerce and its security. We start by reviewing the legacy protocols used in Electronic Data Interchange (EDI) and the various ways to manage their coexistence with new frameworks based on distributed processing and object-oriented designs. The focus is on communications based on the Extensible Markup Language (XML), in particular, the electronic business (using) XML* (ebXML) framework, web services, and the various initiatives for financial service–oriented XML. We also review attempts at standardization and harmonization, particularly for the security of the exchanges for credit transfers or electronic billing.

4.1 Drivers for Business-to-Business Electronic Commerce

Several factors have stimulated and facilitated business-to-business electronic commerce (e-commerce) and the associated integration of systems. These include advances in computer networks and information processing, globalization, the need for organizational agility to cope with competition and rapid development, and market positioning through the customization of products and services, as well as regulatory compliance.

4.1.1 Progress in Telecommunications and Information Processing

Computers were first introduced as stand-alone systems to improve data processing functions in selected applications, such as process control, financial transaction processing, and business and administrative automation. At first, computerization accelerated work procedures without modifying the established ways of operation. With experience, it became apparent that individual computer systems could be tied together to avoid the cost and delays due to rekeying the same data repeatedly and incurring unnecessary transcription errors. The availability of secure and reliable telecommunication infrastructure offering high transmission speeds stimulated the growth of enterprise computer networks, while advances in the microprocessor technology and software engineering transformed computers into productivity tools for workers and individuals.

For example, to optimize the supply chain by coordinating planning, scheduling, and execution, Toyota introduced just-in-time (JIT) techniques to connect shop floors with the back office. JIT was used to provide suppliers and partners with advanced visibility into parts designs, engineering measures, and inventory levels. Similarly, with the use of barcodes and later radio-frequency identification (RFID) tags, the distribution channels and the retail environment became another source of feedback to the supply chain, thereby improving inventory management and production scheduling.

4.1.2 Globalization

Exchanges among partner enterprises have long relied on proprietary electronic networks for business-to-business transactions. Initially, each industrial sector devised its own rules for automated and structured exchange independently. The explanation is quite

* The word "using" is sometimes used to distinguish the suite of modular specifications for the conduct of business on the Internet from the specific dialect of the XML that is used in electronic business exchanges. See www.ebxml.org/geninfo.htm.

simple: in networked services, rivalry does not prevent the parties from cooperating on reasonable terms and conditions to take advantage of the network externalities (i.e., the value of the service offered increases with the number of participants to that network).

New communication and information technologies have allowed companies to split up tasks and to disperse the locations of their execution. More recently, enterprises in advanced countries have concentrated on benefiting from wage differentials to move some of their production facilities and to use service desks in various places around the world to offer continuous 24 hour customer care. Outsourcing has been used to off-load supplementary activities to specialized providers, to reduce labor costs, and to be utilized as an antiunion strategy. Each phase of the outsourcing movement depended on the availability of integrated enterprise systems worldwide. First, manufacturers delocalized their factories or sourced their components from a variety of global producers. Next, the supply chains were restructured as enterprises focused on their core competencies and outsourced nonessential activities and even some internal operations such as payroll and pension plan administration, human resources administration, and information technology (IT) support. Nowadays, service industries are splitting functions, such as accounting or customer interface and distributing them to providers on a worldwide basis. This distribution ensures round-the-clock operation and increases the performance by seeking expertise wherever it may be. The downside, however, is the increased vulnerability of their supply chains and communication networks to risks outside their control.

One effect of outsourcing is that it increases the need for interoperable and flexible IT systems. Integration of proprietary, project-specific solutions is costly and adds delays and risks of lock-in to a given manufacturer or technology. This has stimulated the search for new ways to put together enterprise systems that are more flexible and more responsive to changes in the business needs.

4.1.3 Quest for Organizational Agility

The combined effects of deregulation, globalization, and new technologies have changed the landscape in many industries and highlighted the need for organizational agility. The need to shorten product development times has stimulated concurrent engineering as a method to facilitate the rapid development of integrated solutions without excessive costs or delays by engaging the various functional disciplines in tandem rather than in sequence. In the service industries, particularly in networked services such as telecommunications, airlines, or banking, new services from large companies depend on hundreds of computer support systems, many of which with different architectures and operating under different models, to handle the aspects of order entry, provisioning, installation, quality management, and so on.

The first important consequence is that constant communications must be maintained among the various functional departments, irrespective of the information architecture within each. With mobility, the strict boundaries among the three service spheres of information processing—the home, the enterprise, and the road—are becoming more and more blurred. The second is that this communication and information processing architecture, including its operations support systems, needs to be quickly reconfigurable to adapt to changes in the environment or whenever a new service is introduced. This would reduce the time and effort needed to consolidate the distinct systems into a flexible and an "evolvable" structure.

4.1.4 Personalization of Products and Services

One way in which firms have been competing is by devoting more resources to offer a personalized attention to their customers by exploiting what information technology can offer. This driver for systems integration concerns the capability of tailoring a specific product or a service offer to an individual's profile.

Traditionally, marketing was used to identify the characteristics and the needs of a certain segment of potential users to tailor the product or service to them. This new personalization focuses on individuals' expectations, needs, and wants based on their locations or their environment. As a strategy, many organizations store the integrated information on their customers in data warehouses culled from any number and variety of data sources. Yet, these databases may be distributed among several entities and have different formats so that a service provider would have to negotiate access with each entity separately and then convert the data to a common format. The promise of the service-oriented architecture (SOA) and web services is that most—if not all—of these negotiations would be automatic and online and that very little data conversion would be needed by using a common format.

4.1.5 The Legal Environment and Regulatory Compliance

Governments and regulations have played a significant role in encouraging systems integration. The European Commission spurred European organizations and businesses to use electronic exchanges in the course of their commercial activities, and various European customs authorities have harmonized and automated their procedures. Similarly, in the United States, the Federal Acquisition Streamlining Act of October 1994 required the use of Electronic Data Interchange (EDI) in all federal acquisitions. The Expedited Funds Availability Act of 1990 obliged U.S. banks to make the funds of deposited checks available within a certain interval and, indirectly, forced banks to establish the necessary mechanisms to exchange information and reduce the chances of fraud.

4.2 Four Stages of Systems Integration

The basis for business-to-business electronic commerce is systems integration at different levels, which are labeled as follows:

1. Interconnectivity
2. Functional interoperability
3. Semantic interoperability
4. Optimization and innovation

Typically, these levels come sequentially, in the sense that, for example, functional interoperability is not possible without interconnectivity. Similarly, semantic consistency and interoperability is a prerequisite to optimizing the current system or thinking about improved and innovative ones.

4.2.1 Interconnectivity

This is the most elementary state of integration. It relies on a telecommunication infrastructure to bring the disparate equipment or applications together so that they could coexist and exchange information through gateways, adaptors, and/or transformers (the term used depends on the discipline involved).

This level of integration can be labeled "loose integration" because the basic applications, functionalities, and use of the original equipment or application remain unaffected. To a large extent, the ways in which the business is performed are not modified. A key advantage of advanced telecommunications capacity is to eliminate the transport of the physical storage media. As a reminder, in early EDI systems, messengers would carry magnetic tapes with the data from one location to another to share the necessary data.

4.2.2 Functional Interoperability

Interoperability refers to the ability to make one equipment or application work with another directly, without requiring special effort from the end user. It requires functional and technical compatibility among the protocol interfaces of the network elements, the applications, and the data formats. From a strictly technical viewpoint, the successful exchange of data involves the following components (O'Callaghan and Turner, 1995):

- The structure of the exchanged data and the message formats
- The protocols for exchanging data
- The telecommunication networks
- The security procedures

A significant limitation of the traditional representation of data in EDI is its reliance on alphanumeric characters, even though in many sectors, such as automobile or public works, there is a need to include other types of data such as drawings. Network convergence onto the Transmission Control Protocol/Internet Protocol (TCP/IP) introduced the need to secure the transport of legacy EDI messages on the Internet and gave the opportunity to include all the objects of commercial transactions (text, graphics, images, sound, audio, video) by using structured documents. The various initiatives are described later in this chapter.

Functional interoperability implies coordination across organizational lines. The higher the number of interested parties and their dependencies, the stronger is the need to define common interface specifications. Enterprise resource planning (ERP) systems were introduced later to integrate the back office and production planning systems with incoming orders. Enterprise application integration (EAI) was one of the first architectural concepts to bring together the various heterogeneous platforms, applications, and information systems spread across various departments of an enterprise and separated by organizational boundaries so that they could access the same data and communicate using a common protocol. As a term coined by industry consultants, the definition of EAI remains fuzzy. For some, it is a way to achieve *ex post* interoperability of proprietary applications developed at different terms and with a variety of technologies. For others, the term covers an integration *ex ante* by defining common standards for the design of flexible distributed applications.

Interoperability can be achieved if both sides perform similar functions or if their interfaces are compatible. In this case, all parties have to agree to implement a common profile or interface template, while each may add optional features that should not affect interoperability. This depends on the standards defined either by a recognized standardization committee, through market dominance, or by regulation. In some circles, this level of interoperability is called "syntactic interoperability."

In addition, the end-to-end management aspects cover the following:

- Service offers (cataloging, order taking, payment, billing, logistics, etc.).
- Policies for flow control (purchase policies, traceability of orders, merchandise reception, security).
- Management of electronic documents (archival, retrieval, backup).
- Management of legal responsibilities. In some professions (notaries, bailiffs, etc.), it is essential to ensure integrity and to preserve the archives for a duration defined by law. Data directly or indirectly related to financial results must likewise be retrieved in case of financial or legal audits.

Private contracts (called interchange agreements) define the framework for bilateral electronic transactions: the technical and legal responsibilities of each party, the rules for authentication and identification of the various entities, and the ways to preserve and archive the electronic documents.

4.2.3 Semantic Interoperability

Semantic consistency is achieved when data elements and their meanings are shared (semantic unification). All parties could share the same model or, at a minimum, one model could be mapped without ambiguity into the other. Otherwise, there is the risk that each party would make incorrect assumptions about the other parties leading to semantic interoperability problems.

Standards for semantic consistency are those that define the "what" of the exchange, that is, a shared set of rules to interpret the exchanges and the subsequent actions (the "how" aspects are typically related to the syntax). In this regard, the term *ontology* was borrowed from the realm of metaphysics and epistemology to represent the agreed semantics (i.e., terms and their significance) for a domain of interest. The emphasis at the semantic level is on providing access to data and on minimizing potential errors due to human interpretation through the creation of standards for data definitions and formats. This means that the data elements are uniformly defined and that their significance is consistent from system to system and application to application. For example, the sharing of medical documents requires that the medical vocabularies among the various health-care professions have the same meaning and lead to the same actions. In other words, all parties have to share the same expectation about the effect of the exchange messages.

4.2.4 Optimization and Innovation

In this phase, systems integration becomes an enabler for systematic changes in the way technology is used or in the organizations or both. The treatment of information can be optimized through statistical controls and enhanced reporting activities such as improved customer care (retention, collection, etc.), increased capability of fraud detection, the ability to measure and predict returns on investment (ROI), to improve the efficiency of day-to-day operations, or the ability to refocus the mix of production in response to changes to the economic configuration.

4.3 Overview of Business-to-Business Commerce

An enterprise is the focus of convergent relationships with suppliers, partners, clients, and banks that result in information exchanges. However, a large number of data are reproduced from one form to another inside a given enterprise. When these data reentries are done manually, they can be a source of errors that must be later detected and corrected. In addition, paper documents must be organized and archived for legal and fiscal reasons. As a result, the cost of processing contract documents in paper can reach 7% of the total transaction cost (Breton, 1994; Dupoirier, 1995). The first objective of the dematerialization of business-to-business commerce is to eliminate this additional cost through the exchange of structured and predefined data among the information systems involved in the conduct of businesses, thereby streamlining the tasks of billing, account management, inventory management, and so on (Sandoval, 1990; Kimberley, 1991; Charmot, 1997b; Troulet-Lambert, 1997). A new need has emerged as a consequence of restructuring the supply chain as enterprises focused on their core competences and outsourced nonessential activities. Just-in-time management, in particular, requires unimpeded and continuous circulation of the information to coordinate in real-time production planning with the product delivery and market predictions.

Figure 4.1 depicts exchanges related to catalog consultation, offer and purchasing transactions, shipment notices, and merchandise reception as well as the financial data flowing within the banking network.

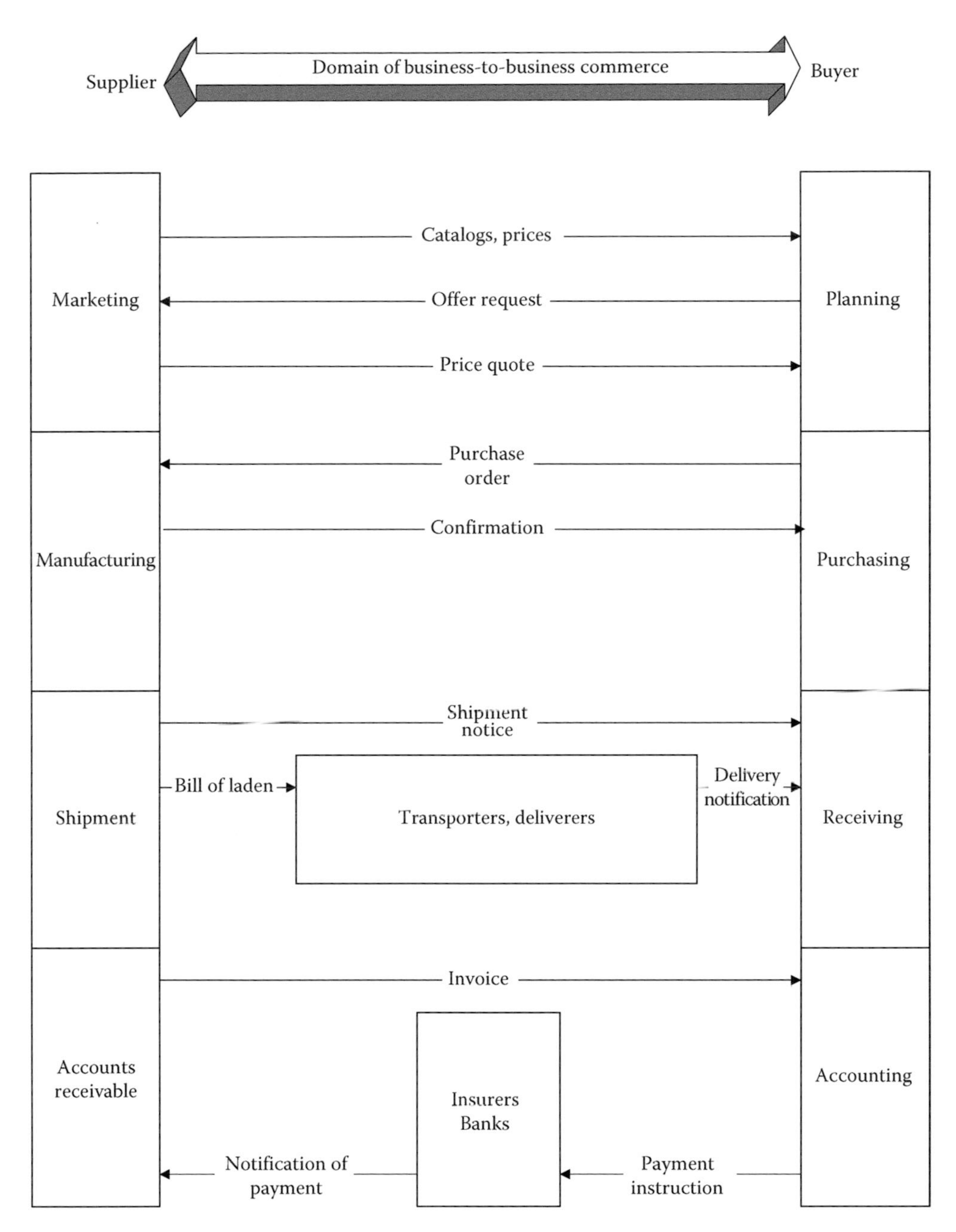

FIGURE 4.1
Message exchange in business-to-business commerce.

Assume that a purchaser has identified an item on the basis of the information available online. The purchaser then puts an electronic order into the supplier's ordering system, which responds to the purchaser with a quote. If that quote is accepted, the purchaser's software composes a purchase order. At the supplier's side, the purchase order is translated into the enterprise format and then routed to the various departments, such as accounting, the factory, or the warehouse. An electronic receipt is also sent to the purchaser to confirm the order. Once the order is filled, a shipment notification is constructed and transmitted to the purchaser as well as to the accounting department of the supplier. Reception of the shipment notification or the arrival of the invoice triggers the purchaser software to create a receiving file.

If the information system of the shipping company is also integrated in the same information system, the documents originating from the transporter (freight letter, delivery notification, etc.) will be composed in an automatic manner on the basis of the data from the supplier. Reception of the delivery notification automatically starts the accounting procedures of the buyer. The receipt notification is reconciled electronically with the initial purchase order and the invoice so as to prepare the payment instructions for the purchaser bank.

If the banking settlement is done electronically as well, the banks are responsible for carrying the payment notification to the supplier within the time frame that is specified by the terms and conditions of the purchase. The information system of the supplier must, in turn, reconcile the payment with the invoice so as to keep the accounting information up to date. Other back-office services include the preparation of the accounting and tax packages and archiving the data associated with the transaction.

From this presentation, it is clear that the development of standards for messages relative to the exchange of goods and services must be coordinated with the development of standards for the messages among the banks and other financial institutions that support the purchases and suppliers.

4.4 Short History of Business-to-Business Electronic Commerce

The first attempts for business-to-business electronic commerce (e-commerce) took place in the United States in the 1960s with the aim of improving the military supply logistics. Civilian applications soon followed in railroad, truck transportation, civil aviation, and international payments (credit transfers, credit cards, and the management of customs). As each industrial group was devising its rules for structuring data without consultation with the others, the U.S. Transportation Data Coordinating Committee (TDCC) was formed to work on the convergence of various specifications. Its first document, published in 1975, covered transport by air, by road, by railroads, and by maritime or river transport. A little bit later, the food and warehouse industries in the United States issued their respective standards, Uniform Communication Standard (UCS) and Warehouse Information Network Standard (WINS). Finally, large automobile manufacturers, such as General Motors, and large-scale retailers, such as the National Wholesale Druggists Association (NWDA), imposed their own rules to their subcontractors and their billing agents. To avoid the proliferation of sector or proprietary rules, the American National Standards Institute (ANSI) established in 1982 a syntax common to the different business sectors. This syntax is known as the ANSI X12 standard, which is widely followed in North America.

In the United Kingdom, the Department of Customs and Excise developed the first EDI for customs known as the London Airport Cargo EDP Scheme (LACES) at Heathrow airport in 1971. The objective was to speed the processing of documents used in the international trade (Tweddle, 1988; Walker, 1988). This activity, known as the Simplification of International Trade Procedures (SITPRO), produced the Trade Data Interchange (TDI), which was then submitted to the United Nations Economic Commission for Europe (UN/ECE) to facilitate international trade. It adopted this document as the UN trade data interchange (UN-TDI), which became the General-purpose Trade Data Interchange (GTDI) in 1981.

In Canada, large business companies established a standard system for messages and communications, and 1984, they formed the EDI Canadian Council (EDICC) that united the large distributors (stores, drugstores, warehouses, retailers). Telecom Canada offered in 1986 a translation service between the internal messaging format of each organization and the X12 format as well as secured electronic mailboxes.

Under the aegis of the United Nations Joint Electronic Data Interchange (UN-JEDI) Initiative, experts worked to harmonize the syntaxes developed in North America and Europe. This endeavor produced what is known as the EDI for Administration, Commerce, and Transport (EDIFACT) language. This agreement was adopted by the United Nations in 1987 and then by the International Organization for Standardization (ISO) as ISO 9735. Customs forms using EDIFACT are regularly used within the European Union since January 1993, when the borders among members states were opened (Granet, 1997).

The European Commission issued a model EDI contract to guide European organizations and businesses using electronic exchanges in the course of their commercial activities (European Commission, 1994). Finally, the United Nations Commission on International Trade Law (UNCITRAL) proposed a model law for the commercial use of international contracts in e-commerce that national legislation could use as a reference (UNCITRAL, 1996).

With the introduction of the Internet, the focus shifted to XML-based business-to-business communication with web services using the service-oriented architecture (SOA).

4.5 Examples of Business-to-Business Electronic Commerce

4.5.1 Banking Applications

In parallel with the aforementioned activities, the Society for Worldwide Interbank Financial Telecommunications

(SWIFT) was established in 1987 by 239 banks in 15 countries with the objective of relaying the interbank messages related to international fund transfers. The aim was to replace paper and telex communications with electronic messaging. The SWIFT standard contains 200 messages that cover all aspects of international finance: cash, retail, large amounts, settlement of real estate transactions, currency operations, treasury, derivatives, international trade, and so on. SWIFT syntax contains codes to identify the parties and the processing of the payment instructions in each country (Remacle, 1996). Bringing the system into full operation, however, required considerable debugging efforts that lasted until the late 1990s (initially, about half of the messages required manual correction).

4.5.2 Aeronautical Applications

Société Internationale de Télécommunications Aéronautiques (SITA—International Society for Aeronautical Telecommunications) was established in 1949 to serve the airline industry. Later, it established a network for the exchange of data concerning reservations, tariffs, passenger boarding, and so on, according to the standards of the International Air Transport Association (IATA). These are CARGO Interchange Message Procedures (CARGO-IMP) for freight and AIR Interline Message Procedures (AIR-IMP) for passengers. Another SITA service allows the selection, purchasing, and localization of spare parts used in aviation. Since March 1994, SITA started to use EDIFACT and ANSI X12 to structure the data carried by the International Forwarding and Transport Message (IFTM) exchanges with the X.400 messaging protocol and the ODETTE* File Transfer Protocol (OFTP). Later, it migrated to the Internet protocols: Transmission Control Protocol/Internet Protocol (TCP/IP) and File Transfer Protocol (FTP).

4.5.3 Applications in the Automotive Industry

The worldwide automotive industry is organized around a small number of manufacturers (General Motors, Ford, Daimler, Toyota, Renault, etc.) that obtain automotive components from several thousands of suppliers organized in a three-tiered pyramid as shown in Figure 4.2. The first tier is formed by around 1000 entities that are supplied by the second tier of about 5000 firms. Last, the third tier comprises about 50,000 suppliers, generally small or medium enterprises, which work simultaneously with several car manufacturers. Without standardization of the tools, the third-tier suppliers have to invest in training and maintenance

* ODETTE is the Organization for Data Exchange by Tele Transmission in Europe.

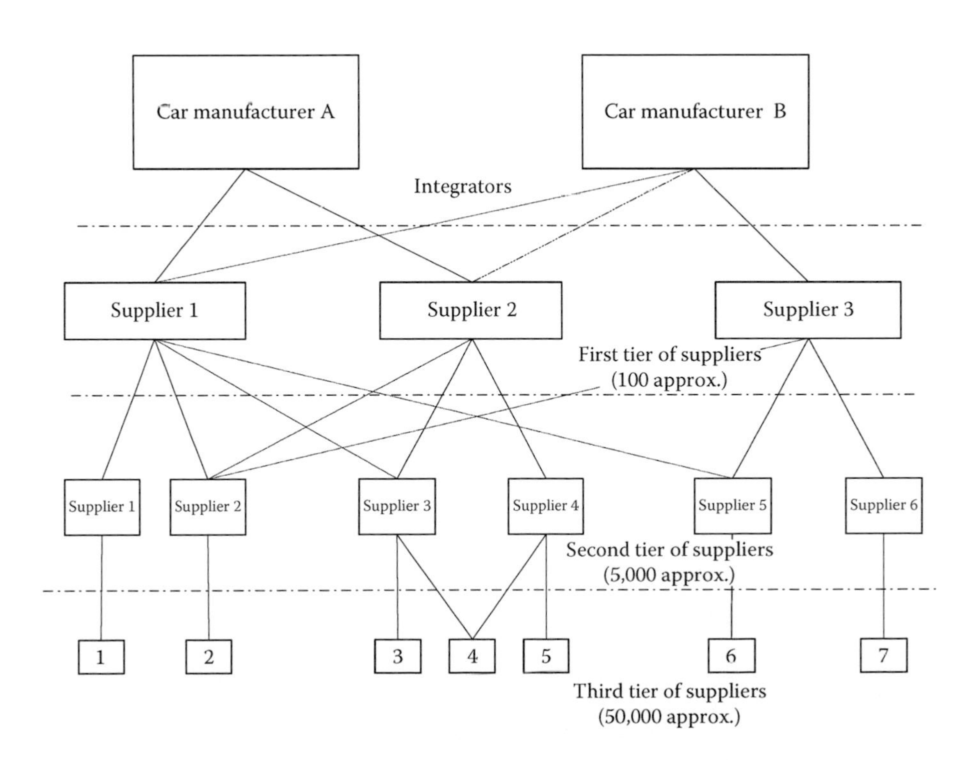

FIGURE 4.2
Pyramidal structure of the automobile industry as integrators and three tiers of suppliers.

of multiple programs for computer-aided design and communication to be able to work with the different automobile manufacturers they partner.

In 1984, the European automobile manufacturers formed ODETTE for the exchange of information between suppliers and car manufacturers. Among its activities was the standardization of a common transmission protocol (OFTP), as well as the content and the structure of the documents according to the syntax rules of EDIFACT in ISO 7372 (1993) (de Galzain, 1989). The North American equivalent is Automotive Network eXchange (ANX®), the network of the Automotive Industry Action Group (AIAG), which is based on the TCP/IP stack. AIAG was formed in 1982 to define rules for exchanging information among partners in the North American car industry.

Experience has underlined the cost of encryption for small and medium enterprises. Other difficulties arose from the incompatibility of certificates or the implementations of the Internet Protocol Security (IPSec) protocol. Other operational difficulties concern end-to-end quality of service guarantees and trouble localization and repair (Borchers and Demski, 2000).

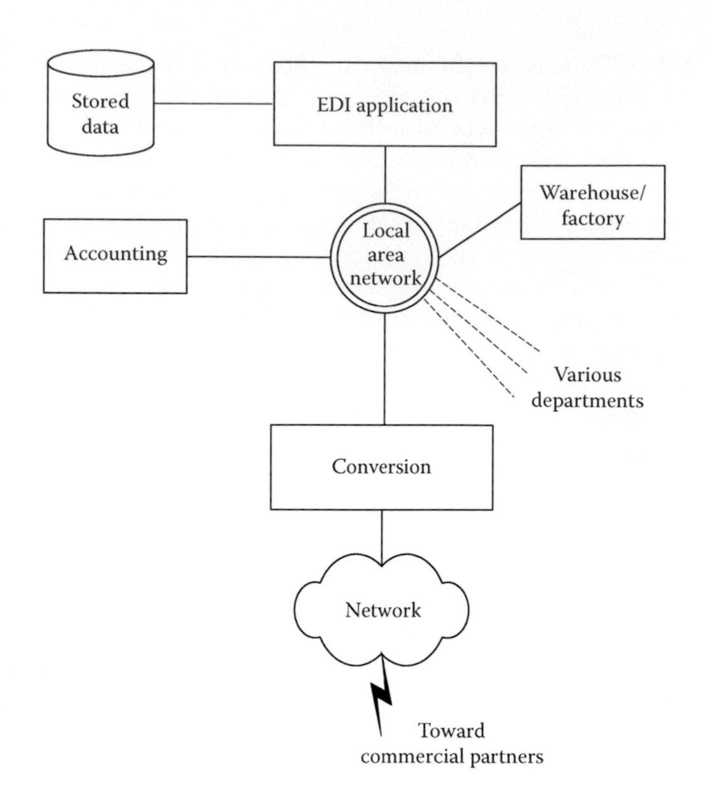

FIGURE 4.3
Components of business-to-business e-commerce systems.

4.5.4 Other Industries

The multinationals in the area of industrial chemistry employ the procedures of the Conseil Européen des Fédérations de l'Industrie Chimique (CEFIC—European Council of Industrial Chemistry Federations) in Europe and the Chemical Industry Document Exchange (CIDX) in the United States.

In the United States, the Health Industry Business Communications Council (HIBCC) groups various institutions are interested in the use of EDI for health care, in particular, the Healthcare EDI Coalition (HEDIC) and the National Wholesale Druggists Association (NWDA).

4.6 Evolution of Business-to-Business Electronic Commerce

Figure 4.3 represents the various components of a business-to-business e-commerce implementation.

The type of exchanges depends on the nature of the supply chain. This chain is called vertical if the procured goods intervene directly into the production and horizontal if they cover several industries, in which case the goods are called indirect. In the first case, the purchases are called strategic, while the purchases in the second case are for maintenance, repair, and operations (MRO). Another criterion of distinction is the duration of the contracts among commercial partners: long duration for daily production or temporary for emergencies. Thus, by taking into account the two criteria of urgency of need and the strategic aspects of goods and exchanges services, there are four types of platforms for business-to-business electronic commerce: the exchanges for excess inventory, EDI systems, MRO hubs, and the generalist catalogs.

The focus of traditional EDI was the purchase of strategic goods that are directly related to the chain of production or to service creation in a specific sector (automotive, chemistry, steel). However, nonstrategic purchases (equipment and office furniture, travels, etc.) that often represent the large bulk of the volume purchases remained managed in a traditional way. Thus, in the 1980s, the focus of EDI was on the format and content of messages exchanged for the purchase of direct goods related to the production chain or to the service creation in a specific sector. Specialized software at each end performed the necessary format conversions, while providers of value-added networks (VANs) were responsible for the details of the communication and its security, including the protocol conversion to and from the networks used within each trading organization. Thus, data were extracted from the enterprise databases, converted to a mutually agreed format, and secured before transmission to the VAN that is connecting the business partners.

At the destination, the received data are converted to the "in-house" format and then directed to the specific application that can process the data. The software that interfaces with both the transmission network and the internal systems and carries out the necessary format conversion and reorganization data is called "translator" or "converter."

Despite the expected benefits of automated procedures, the rate of penetration of the traditional EDI in industrialized countries varied between 1% in the United States and 5% in some European countries. Within the countries of the European Union, it was mostly used for intranational business, with applications directed toward intrasector transactions, in which the risk of litigation is at a minimum such as banking, transportation, retail distribution, aeronautical, or automobile manufacturing (del Pilar Barea Martinez, 1997; Landais, 1997).

The Internet stimulated new forms of business-to-business e-commerce, especially in fragmented markets, through the integration of supplier management with enterprise information systems related to order management, payments, inventory controls, and reporting capabilities. Some examples are e-procurement with online exchanges, MRO hubs, or yield managers (Kaplan and Sawhney, 2000; Phillips and Meeker, 2000). In these new platforms, all participants (suppliers and consumers) agree to open their information systems and enhance their security infrastructure. Access to the applications is through a web client.

Figure 4.4 depicts the evolution of the traditional EDI systems into these new forms that give different types of service. Thus, exchanges constitute a neutral platform that does not favor either the buyer or the vendor; in contrast, MRO hubs select their suppliers according to criteria that the buyer specifies, while yield managers bring together buyers and sellers through a catalog or auctions. The latter mechanism constitutes a prized method to link offer and demand and to determine the price. Intermediation sites have had mixed success depending on the industrial sector.

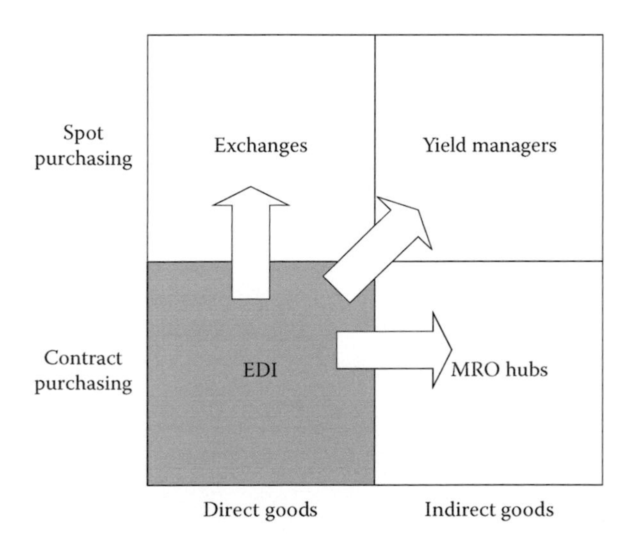

FIGURE 4.4
Platform versus type of business-to-business electronic commerce.

4.7 Implementation of Business-to-Business Electronic Commerce

Successful implementation and management of EDI require significant investment in project management, software maintenance, user training, and so on (Jackson, 1988). Furthermore, EDI procedures bring about major changes in the organization of work and in the power relationships within the enterprise. The magnitude and the cost of the task have often discouraged small and medium enterprises. The problems are well documented in the literature on enterprise resource planning (ERP) systems (Dey et al., 2010). They can be summarized as follows:

1. Unproven technology or supplier failures.
2. Unrealistic targets in terms of date or cost or scope.
3. Lack of standards and/or exclusive reliance on proprietary interfaces and protocols, which cause a manufacturer lock-in. Alternatively, the lack of interoperability of "standard-compliant" implementations may arise from deficiencies in the scope of the standards, ambiguities in the specifications, or flaws in the implementations themselves.
4. Lack of quality control, in the sense of specifications defining what is an acceptable performance level, either for the system as a whole or for individual components.
5. Inadequate customer care. The real customer here is the end user and not the department responsible for the integration project. In fact, in many cases, the system's requirements are defined without adequate study of the need of these end users.
6. Inadequate risk management plan, particularly technology risks.

Another type of difficulties relates to the legal status of contracts over electronic media and on the evidentiary value of dematerialized documents. The existing judicial systems were developed in a context where business practice was relying on paper documents (e.g., the legal requirement for a handwritten signature). Large corporations use private contracts among commercial partners (or interchange agreements) to define the framework for bilateral electronic transactions: the responsibilities of each party, the rules for identification and authentication of the various entities, and the ways to preserve and archive the electronic documents. This approach may be too burdensome for small and medium enterprises. The dematerialization of the support raises the question of the admissibility of electronic documents as evidence, their evidentiary value, and their long-term readability and preservation. Other aspects relate to the identification of the contracting parties, the validity of electronic signatures, the time stamping of the operations, and the authentication of the origin. This is why the adoption of laws regulating the use of electronic documents and accepting the use of digital signatures in recent years opens evidentiary law and commercial laws to technical advances (Bresse et al., 1997, pp. 162–166).

Figure 4.5 summarizes the evolution of the collaborative application within and outside enterprises, first from proprietary systems, then to the traditional EDI, and finally with the inclusion of the Extensible Markup Language (XML) and its variants in the information infrastructure of e-commerce and the establishment of enterprise portals.

The service-oriented architecture (SOA) is a new blueprint for systems integration starting from the design phase by moving away from monolithic applications, with their own embedded data tied to specific processes and business rules. SOA is based on the experience gained from distributed computing and object- and component-based designs. Here, a service is defined as a logical representation of a repeatable business activity that has a specified outcome and does not depend on other services, unless it reuses them or it is composed of other services. To maintain this independence, each service is responsible for updating the data it uses. Furthermore, the services are not necessarily under the control of a single administrative entity and they communicate using an Enterprise Service Bus (ESB), rather than through function calls from the body of the programs. By design, the applications are made to be modular and independent of the input data; the data they need are verified and authenticated separately and are accessed through standardized interfaces. Similarly, the business rules are defined and processed outside the applications. With dynamic service composition, services and workflows do not have to be defined at design time but can be adapted later to fit the context of the service consumer. As a result, the binding of an application with the data it processes and the business rules it applies is made at run time under the control of an external entity supervising the workflow. Finally, legacy software programs, computing devices, and networking resources that were not designed as loosely coupled components are encapsulated via standardized common interfaces.

4.8 X12 and EDIFACT

The objective here is not to give an exhaustive treatment of all alphanumeric EDI systems but rather to present the essential notions to understand the means

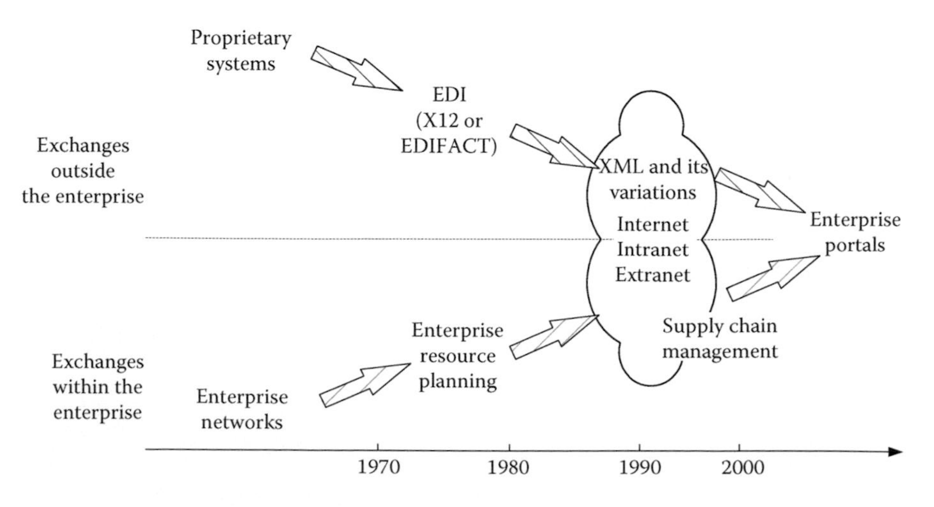

FIGURE 4.5
Portals as the convergence point of collaborative applications within and outside the enterprise.

to secure the EDI and the problems of integration with the Internet.

Each of these two systems defines a representation of the contents of administrative, commercial, or transport documents as alphanumeric language strings with the help of standardized conventions. The items mentioned in the documents are described with the help of elementary data common to all sectors (purchase order, credit note, instruction for payment, freight reservation, customs form, etc.). By combining these elementary data, it is possible to form composite data and data segments that can be organized according to precise rules to constitute the canonical messages. The differences between X12 and EDIFACT reside in the definitions of the data elements and the syntax rules, as well as the procedures employed to secure the exchanges.

4.8.1 Definitions

The basic units of an alphanumeric EDI exchange are *data elements* that are defined in a dictionary of elementary data. From a functional viewpoint, the data element is either a *service element* or an *application element*. Service elements contain the information that structures the transmission and are utilized in service segments. In contrast, application elements relate to the heart of the end-to-end transaction itself, that is, the data defined and agreed upon by the two parties of the transaction.

A *segment* is a logical set that includes a series of elements, simple or compound, and may include other segments. The order, content, the maximum number of repetitions of the constituents, and the way these repetitions should be organized are defined in the segment dictionary. To express a precise functionality, for example, a purchase order or a payment instruction, the segments are combined and organized in a *group of segments*.

There are two types of segments: segments denoted as "control segments" in X12 or "service segments" in EDIFACT and *application data segments*. Control (or service) segments are used to structure the content and to distinguish the various parties. The application data segments contain the application data organized by function. It is the entity in charge of managing the application that is responsible for specifying the coding and the organization of the application data segments.

A *transaction set* (X12) or *message* (EDIFACT) is the set of structured segments in the order defined in the directory of standard messages. These messages represent functions that are common to all activity sectors. For example, billing is based on a universal commercial practice and does not depend on the type of activity. There are two classes of messages:

1. *Service messages*: Formed with service segments and whose role is to correct syntax errors or application errors
2. *Application messages*: Formed with application segments

Messages may be mandatory or optional and may be repeated.

In general, messages consist of three distinct zones: header, detail (body), and trailer. The segment groups forming the "detail" concern different transactions that may be included in the same message.

The *functional group* gives the possibility of putting several messages within the same structure. Finally, the *interchange* is the external envelope of all messages originating from the same application, although they may relate to independent transactions.

4.8.2 ANSI X12

ANSI ASC X12 was defined by the ANSI Accredited Standards Committee (ASC) X12. The standardization of ANSI X12 is more advanced than that of EDIFACT; unfortunately, the syntax of X12 is positional, which makes it incompatible with EDIFACT.

Within the X12 terminology, a transaction set corresponds to the useful transmission of the *useful* content of the paper document (purchase order, invoice, etc.) between the computers of two organizations. Each transaction set consists of three parts: a header, a detail (body), and a trailer. The header announces the characteristics of the transmission, while the trailer contains control elements to verify the integrity of the information transmitted.

The body is organized in lines or segments that describe one particular aspect of the total action. In turn, a segment is composed of data elements and codes associated with the function to be performed. The order in which the elements are arranged, their composition, and the significance of the codes are all defined in the data dictionary. For example, an asterisk (*) separates two consecutive elements, and two asterisks indicate that an optional element has been eliminated. If the omitted element (or elements) was at the end of the segment, it would be replaced by the end of segment (N/L) to indicate the return to a new line.

As mentioned earlier, there are two types of segments: control segments and data segments. The segment headers and trailers, as well as the repetition loops of segments form the control segments. Table 4.1 gives some of the most frequently used transaction sets.

In general, the software for translating to the EDI format ("the EDI document") does not support all

TABLE 4.1

Examples of X12 Transaction Sets

Code	Meaning
810	Invoice
820	Payment order/remittance advice
824	Application advice (results for an attempt to modify a transaction)
827	Financial return notice (impossibility of carrying an 820 transaction)
850	Purchase order
855	Purchase order acknowledgment
856	Ship notice/manifest
997	Functional acknowledgment (to indicate that the received message is syntactically correct)

transaction sets and is limited to those used in the target domain of activity.

Several transaction sets can be combined in an X12 interchange in the following manner:

```
ISA* interchange header
     GS* functional group header
              ST* transaction set header
                  Segments (i.e., purchase
                    order no. 1)
              SE* end of transaction set
              ST* transaction set header
                  Segments (i.e., purchase
                    order no. 2)
              SE* end of transaction set
     GE* end of functional group
     GS* functional group header
              ST* transaction set header
                  Segments (i.e., purchase
                    order no. 3)
              SE* end of transaction set
              ST* transaction set header
                  Segments (i.e., purchase
                    order no. 4)
              SE* end of transaction set
     GE* end of functional group
IEA* interchange trailer
```

4.8.3 EDIFACT

EDIFACT is described in the following documents (Charmot, 1997a):

- The vocabulary or the data elements of ISO 7372 (2005)
- Directives for the composition of messages (document Trade/WP 4/R.840/Rev2)
- Syntax rules for structuring the canonical messages as per ISO 9735
- Directives for using the syntax (document Trade/WP 4/R.530)
- UN Rules of Conduct for Interchange of Trade Data by Teletransmission (UNCID)
- Dictionary of data elements, a subset of ISO 7372
- Dictionary of composite data elements
- Dictionary of canonical messages
- UN Code List (UNCL), a dictionary of recognized codes

Taking into account the individual practices of each country, it is obvious that the design of standardized messages for a given sector is a long-term effort. Messages are classified according to the progress in standardization as follows: status "0" is for messages that are still being defined; status "1" indicates that the message is undergoing trials; stable and standardized messages have the status of "2."

ISO 9735 recognizes several character sets defined in ISO/IEC 8859 for Latin and non-Latin alphabets including the right-to-left writing movements for Arabic and Hebrew.

Service segments allow structuring the content and distinguishing its many parts. They start with a *tag*. The service elements are identified with the letters *UN*; for example, *UNH* designates the header of a message, while *UNT* designates its trailer. The tags are defined by the standard EDIFACT in the directory entry of the corresponding segment. Qualifiers can be used together with tags to give a specific meaning to the functions of other data elements or segments, which is useful if several functions are represented within the same message. If optional elements are omitted, they will be replaced by the corresponding separator (data element or component of a composite data element) defined for the segment at hand.

The following sections give the details necessary to understand the security mechanisms of EDIFACT that will be presented later in this chapter.

4.8.3.1 UNB/UNZ and UIB/UIZ Segments

The pair of service segments (control segments) UNB/UNZ define, respectively, the beginning and the end of a *batch* exchange, that is, they bracket the envelope. The segments UIB/UIZ are the header and the trailer segments that have been proposed to envelop the interactive exchanges. An optional segment, service string advice UNA, is used if it is needed to redefine the component data element separator or the data element separator or to change the notational sign of the decimal, which is the "," by default. UNA consists of a sequence of nine characters (UNA followed by six characters according to a specific format that ISO 9735 [1998] has specified).

The UNB defines the transmission characteristics using the following elements:

- Syntax identifier
- Address of the sender
- Address of the recipient
- Date and time of preparation
- Unique interchange control reference
- Recipient reference or password
- Reference to the application used or, in the case of a single message, the message type (e.g., an invoice)
- Processing priority
- Whether an acknowledgment is requested
- An identifier of the communication agreement controlling the interchange
- A test indicator in the case that the interchange is a probe sent to verify the connection between the two ends

The UNZ segment is the interchange trailer that signals the end of the exchange. It includes two elements: the total number of messages of functional groups in the interchange and the same reference to the interchange as in the UNB.

These elements help check that the delivery of the interchange has not encountered problems.

4.8.3.2 UNH/UNT Segments

The service (control) segments UNH/UNT play the equivalent role of the preceding doubles but for the messages instead of the interchanges. Thus, each message begins with a header segment UNH and ends with a trailer segment UNT. The structure of an EDIFACT message is given in Figure 4.6.

The message header UNH contains control data such as the following:

- The sender's unique message reference number
- The message type, the version number of the standard used to compose the message
- The agency that controls the message specification, maintenance, and publication
- A common access reference, for example, the name of the message file, which will be used to cross-reference subsequent transfers of data

The UNT segment indicates the end of the message and contains the two elements for verification: the number of segments in the message and the message reference number present in the UNH segment.

4.8.3.3 The UNS Segment

The service segment UNS is a special segment used to separate the body or detail of a message from the header on one side and from the trailer on the other side.

4.8.3.4 UNG/UNE Segments

The header of a functional group UNG is a service segment that contains the following elements:

- Identifier of the message type in the functional group
- Identifier of the sender's application (e.g., the division or department of the enterprise that has prepared the functional group)
- Identifier of the recipient's application
- Date and type of preparation
- Functional group reference number
- The controlling agency that has specified the message
- The version of the standard that was used to compose the message
- Application password according to the recipient's request (optional)

The UNE service segment closes a functional group. It contains two verification elements: the number of messages in the functional group and the reference number for the functional group that is included in the header.

4.8.3.5 UNO/UNP Segments

Although it is possible to encapsulate a multiformat object, difficulties may arise at the level of handling the various objects and synchronizing them at the

Header	Data segments	Data segment	...	Message trailer

FIGURE 4.6
Structure of an EDIFACT message.

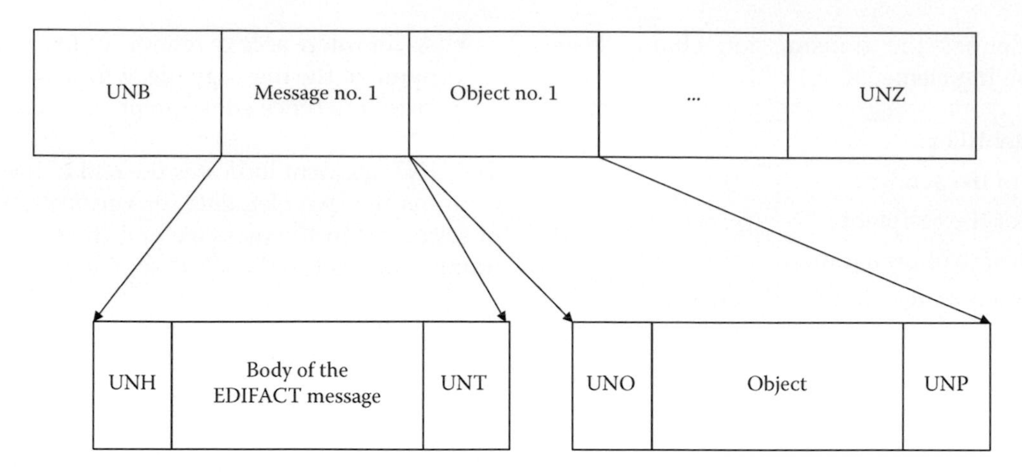

FIGURE 4.7
Encapsulation of a non-EDIFACT object within an interchange.

destination. To resolve this problem, two new EDIFACT service segments have been defined (Charmot, 1997b).

Depicted in Figure 4.7 is an object surrounded by the two segments, UNO and UNP, which bound the object within an interchange. UNO identifies the start of the object and can be used to specify its nature. UNP also gives the size of the objects in octets to verify the quality at reception.

4.8.3.6 *Structure of an Interchange*

An interchange is structured in the following manner:

Segment		Type
Service string advice *UNA*		Optional
Interchange header *UNB*		Mandatory
Functional group header *UNG*		Optional
	Message header *UNH*	Mandatory
	Message data	
	Message trailer *UNT*	Mandatory
Functional group trailer *UNE*		Optional
Interchange trailer *UNZ*		Mandatory

The structure of an interchange without functional groups is given in Figure 4.8. Figure 4.9 depicts the structure of an interchange with functional groups.

4.8.3.7 *A Partial List of EDIFACT Messages*

Table 4.2 contains some EDIFACT messages used in commercial transactions. These examples are given for illustrative purposes and are arranged in alphabetical order.

4.8.3.8 *Interactive EDIFACT*

The discussion thus far has essentially related to batch EDI, which is encountered in asynchronous and non-real-time transactions. Yet, online transactions are characterized by shorter response times than in traditional applications. The reference model for Open-EDI defined in ISO 14662 (2010) was first adopted in October 1996 (Troulet-Lambert, 1997). ISO 9735-3 (2002) introduced a syntax for interactive EDIFACT exchanges in 1998. This syntax was harmonized with Electronic Business XML to be presented later in this chapter.

Interchange header	Message no. 1	Message no. 2	...	Interchange trailer

FIGURE 4.8
Structure of an EDIFACT interchange without functional groups.

Interchange header	Functional group header	EDIFACT message	...	Functional group header	Functional group header	EDIFACT message	...	Functional group trailer	Interchange trailer

FIGURE 4.9
Structure of an EDIFACT interchange with functional groups.

TABLE 4.2

EDIFACT Messages (in Alphabetic Order)

Message Name	Function
CREADV	Credit advice
CREEXT (CREADV + REMADV)	Extended credit advice
DEBADV	Debit advice
DESADV	Dispatch advice
DOCADV	Documentary credit advice
DOCINF	Documentary credit issuance information
IFTMAN	Arrival notice
INVOIC	Invoice
ORDCHG	Purchase order change request
ORDERS	Purchase order
PAYEXT (PAYORD + REMADV)	Extended payment order
PAYMUL	Multiple payment order
PAYORD	Payment order
REMADV	Remittance advice

4.8.4 Structural Comparison between X12 and EDIFACT

Figure 4.10 depicts the correspondence of the control elements of both EDIFACT and ANSI X12 (Kimberley, 1991, p. 127).

Table 4.3 shows the correspondence between the terms used in header segments of the EDI interchanges in EDIFACT with those of ANSI X12, using Table K-1/X.435 of ITU-T Recommendation X.435.

4.9 EDI Messaging

In the past, EDI messaging used proprietary messaging systems such as Microsoft Exchange and Lotus cc:Mail. For the Internet, Simple Mail Transfer Protocol (SMTP) of the Internet Engineering Task Force Request for Comments (IETF RFCs 821, 1982) and its Multipurpose Internet Mail Extensions (MIME) described in the RFC 2045 (1996) define the standard mail protocol. Both messaging systems follow the same model of a user agent and a message transfer agent, even though the connection protocols between the components are not identical.

4.9.1 X.400

The X.400 messaging system was one of the first alternatives to proprietary messaging systems and to

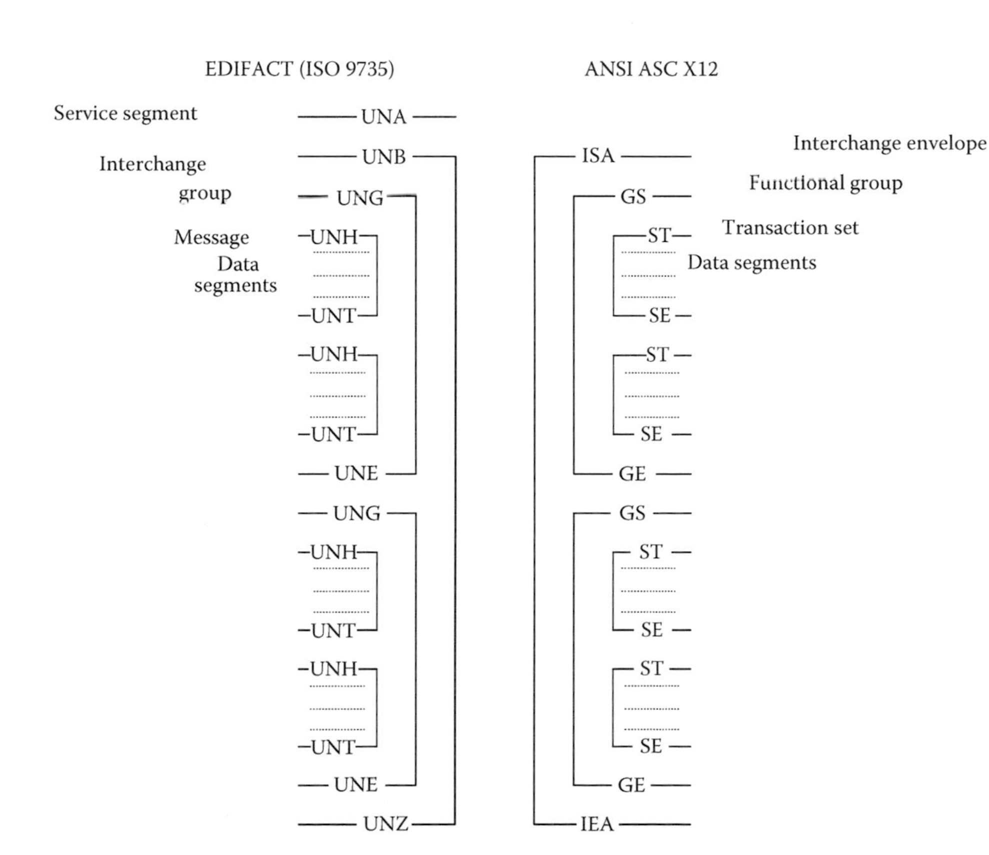

FIGURE 4.10
Comparison of the structures of EDIFACT and ANSI X12.

TABLE 4.3

Comparison of the Terms Utilized in the Headers of the EDI Interchanges

EDIFACT	ANSI X12
Interchange header (UNA and UNB)	Interchange header (ISA)
Functional group header (UNG)	Functional group header (GS)
Message header (UNH)	Transaction set header (ST)
Service string advice	1. Data element separator 2. Segment terminator 3. Subelement separator
Syntax identifier	1. Interchange standard identifier 2. Interchange version ID
Interchange sender	Interchange sender ID
Interchange recipient	Interchange receiver ID
Date/time of preparation	1. Interchange date 2. Interchange time
Interchange control reference	Interchange control number
Recipients reference, password	Security information
Application reference	—
Processing priority code	—
Acknowledgment request	Acknowledgment requested
Communications agreement ID	—
Test indicator	Test indicator
—	Authorization information

TABLE 4.4

Transfer Delay Objectives in X.400 Messaging Systems

Grade of Delivery	Time for Delivery of 95% of the Messages	Time for Forced Nondelivery (Hours)
Urgent	45 minutes	4
Normal	4 hours	24
Nonurgent	24 hours	36

third-party networks. In the past, X.400 was used as a generic name to include a whole series of International Telecommunication Union Telecommunication Standardization Sector (ITU-T) recommendations, such as X.420 for the encapsulation of EDI messages and X.402 for security services. The wide use of the Internet has made this approach obsolete.

Message labeling was use to treat each message according to the required degree of security. Furthermore, X.400 messaging systems were designed to meet some contractual levels of service quality, according to the degree of urgency of the traffic. These objectives are defined in Recommendation F.410 and are reproduced in Table 4.4.

4.9.2 The Internet (SMTP/MIME)

RFCs 2045 through 2049 (1996) define how MIME implementations can encapsulate EDI messages. RFC 2046 is compatible with the basic messaging protocol SMTP that is on top of the TCP/IP layers. This encapsulation allows the inclusion of different object types, as well as the transmission of non-ASCII text, which makes it suitable for EDI. MIME allows the separation of the body of a message into distinct parts, separated by delimiters. The delimiter is a demarcation line that can be defined as a sequence of characters that does not appear anywhere else in the message. The lines that follow the delimiter define the properties of the object for the recipient applications.

An EDI message that will use MIME will include, first, the SMTP message header that is defined in RFC 822 (1982) (the sequence of fields *From, To, Date, Subject,* etc.). The header precedes a series of declarations that indicate the content type, the initial representation of the characters, the coding used to protect the text from being mutilated in the Internet, and finally, the succession of body parts, each separated from the other with delimiters.

The coding type for EDI files has to be exclusively *base64* to oblige the sender's user agent to convert the text to the 7-bit ASCII code and append a "carriage return" and a "new line" <CR><LF> at the end of the file. The presence of the field "Content-Transfer-Encoding: base64" in the header allows the destination user agent to perform the inverse processing and to recover the initial content.

To illustrate with an example, and without attempting to specify all the details of MIME, a skeleton message can take the following form:

```
To: < RECIPIENT'S ELECTRONIC ADDRESS>
SUBJECT:
FROM: <SENDER'S ELECTRONIC ADDRESS>
DATE:
MIME-VERSION: 1.0
CONTENT-TYPE: MULTIPART/MIXED; BOUNDARY="ABXYXMSON"

-- ABXYXMSON
CONTENT-TYPE: TEXT/PLAIN;CHARSET="ISO/IEC-8859-1²

<THIS IS PREAMBLE TO AN EDI MESSAGE THAT can contain
accents>
-- ABXYXMSON
CONTENT-TYPE: APPLICATION/<EDI STANDARD>
CONTENT-TRANSFER-ENCODING: BASE64

<THIS IS THE EDI interchange coded according to
the standard indicated in the Content-Type>
-- ABXYXMSON
CONTENT-TYPE: APPLICATION/DRAWING; ID=260; NAME="COST"
CONTENT-TRANSFER-ENCODING: BASE64

<HERE IS THE GRAPHICAL FILE COST>
-- ABXYXMSON --
```

Three content types are recognized for EDI applications: EDI-X12, EDIFACT, and EDI-CONSENT. This last category is used for proprietary EDI applications. These content types are registered by the Internet Assigned Numbers Authority (IANA), the Internet registration authority.

It should be noted that, contrary to X.400 messaging, the messages included with this basic format are clear, thus easily readable. Also, there is no guarantee that the message will reach its destination.

4.10 The Security of EDI

The security of EDI includes both technical and managerial aspects. The technical part has many aspects related to the security of the exchanges as well as the network elements and network element management systems. In the following, we will restrict ourselves to the security of the exchange, the other aspects being outside the scope of this book.

4.10.1 X12 Security

Secure X12 transmissions use the security structures defined in X12.58 issued in December 1997. Before then, security consisted in using a password before each transmission and in each direction. Should a recipient want to respond, a new connection would have to be established in the inverse direction. Value-added networks provided security services by establishing the two unidirectional circuits instead of the two parties transparently.

Authentication uses the Data Encryption Standard (DES) algorithm to calculate a message authentication code (MAC). For protection against replay attacks, the standard recommends inserting in the message to be authenticated a combination of a date of composition and a unique identifier (such as a purchase order number). Nonrepudiation employs a public key encryption along with time stamping.

The standard generalizes the concept of digital signature in the form of assurances that express the *business intent*. An assurance is contained in newly defined segments: the S3A or S4A segments and the SVA segments. These segments are added before calculating the MAC or signing the message. The combination S4A/SVA frames the unsecured transaction before encryption or authentication and offers a first level of protection for the functional group. The combination S3A/SVA offers a second level. Each level has its own keys. Furthermore, at each security level, the keys utilized for a service (e.g., authentication or encryption) should be different.

The standard allows an optional compression of the message before encryption in addition to an optional filtering of the encrypted or compressed data. Filtering prevents the occurrence of binary sequences that may activate incorrectly the control functions of the transmission systems. Three types of filters are recognized: the conversion of each binary into two hexadecimal characters; the conversion of the binary data into a string of printable ASCII characters, which produces an expansion of the required bandwidth by about 23%; and, finally, an ASCII/Baudot filter to convert the binary data to a string of characters that belong to both ASCII and Baudot character sets, a procedure that results in an expansion of about 86%.

A transmission secured by these segments is illustrated as follows:

```
ISA* interchange header
     GS* functional group header
     S3S* security header—level 1
     S3A* assurance header—level 1
          ST* transaction set header
          S4S* security header—level 2
          S4A* assurance header—level 2
                 Segment details (e.g.,
                  purchase order)
           SVA* security value—level 2
           S4E* security trailer—level 2
           SE* transaction set trailer
     SVA* security value—level 1
     S3E* security trailer—level 1
     GE* functional group trailer
IEA* interchange trailer
```

Note, however, that the security mechanism gives the same protection to all parts of a transaction. This is a difference with EDIFACT services that offer finer resolution and permit the possibility of different types of protection according to the different fields of a transaction.

X12 can directly utilize the X.509 certificates delivered by a certification authority. Level 3 is considered to be sufficient for EDI applications (CommerceNet, 1997).

4.10.2 EDIFACT Security

The security of EDIFACT follows ISO 7498-2 (1989). This standard is the result of the European research program Trade Electronic Data Interchange System (TEDIS), which lasted from 1988 to 1994. The services offered are message integrity, authentication of the origin, and nonrepudiation (at the origin and at the destination). Confidentiality is not offered explicitly but may be constructed with the other services.

EDIFACT security services can be offered in two ways: by sending security segments "in-band" or "out-of-band." In the in-band approach, the security segments flow jointly with the messages to be protected, whereas in the out-of-band approach, separate security messages are used.

ISO 9735-5 (2002) considers the first case and defines the segments that must be inserted, including the means to distinguish them from user traffic.

The out-of-band security measures rely on the AUTACK message, defined in ISO 9735-6 (2002). This message contains either the hash of the EDIFACT structure that is mentioned or the signature of this same structure. The AUTACK message can also be used as an acknowledgment when it is sent by the recipient or by an entity that has the authority to act on behalf of the recipient. Used in this manner, the message confirms reception of the EDIFACT structure that was sent, the integrity of its content at destination, and helps establish nonrepudiation at the destination.

Finally, the management of EDIFACT certificates (inscription, renewal, replacement, revocation, delivery) and the generation, distribution, and management of keys are performed with the KEYMAN message defined in ISO 9735-9 (2002). The KEYMAN messages can also refer to the certification path and the revocation list defined according to X.509. These references are transported in a binary format between the segments UNO and UNP.

It should be noted that EDIFACT certificates are different from X.509 certificates both in format and in their method of management. To resolve this issue, the European Commission sponsored the Directory-based EDI Certificate Access and Management (DEDICA) project to allow access to secure EDI with X.509 certificates. A DEDICA gateway performs the necessary conversions, thereby saving the users from having to obtain and maintain two sets of certificates.

4.10.2.1 Security of EDIFACT Documents Using In-Band Segments

Figure 4.11 shows how security segments are inserted in the initial structure. Each USH/UST corresponds to a given security service. This permits the possibility of varying the offered service for distinct parts of the interchange; thus, a given interchange can include data from several transactions that do not require the same degree

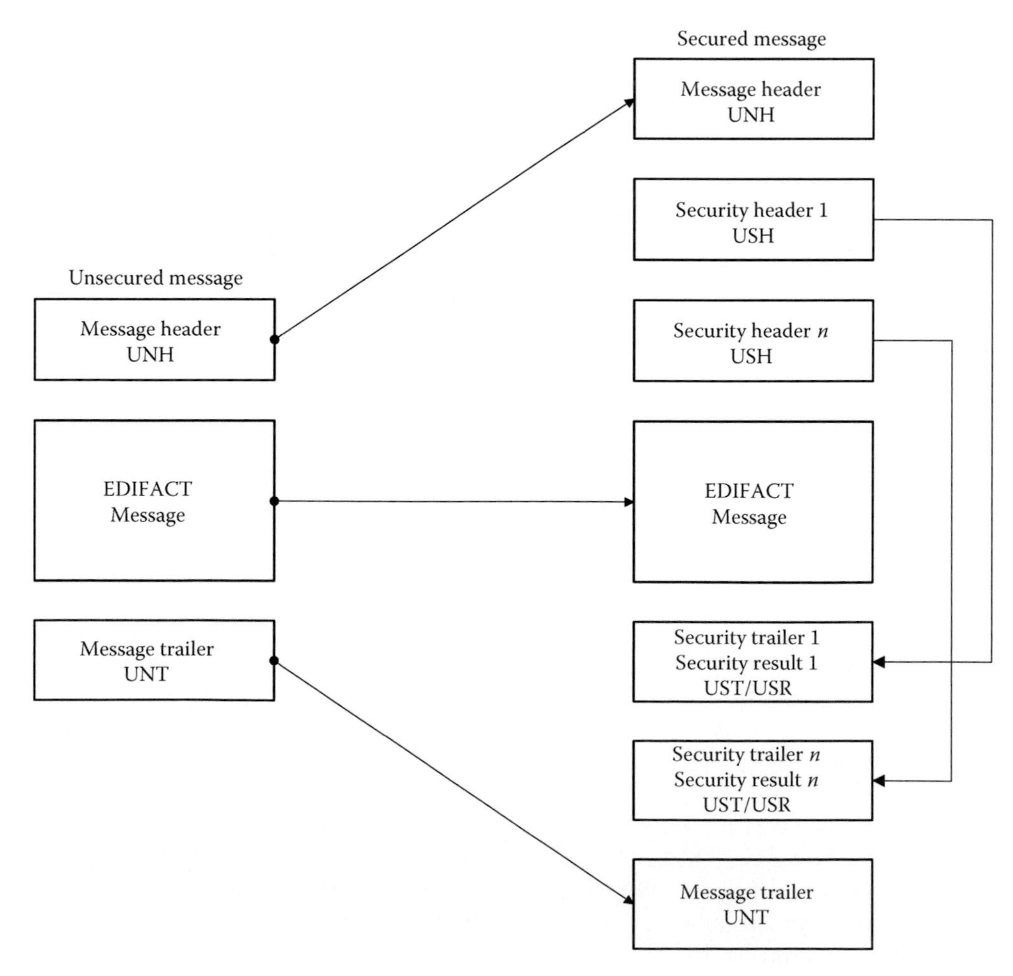

FIGURE 4.11
Security of EDIFACT messages with in-band segments.

TABLE 4.5
Security Structure for "In-Band" Segments

Label	Segment Name	Type	Maximum Number of Repetitions
Segment group 1		O	99
USH	Security header	M	1
USA	Security algorithm	O	3
Segment group 2		O	2
USC	Certificate	M	1
USA	Security algorithm	O	3
USR	Result to be validated	O	1
...		...	...
...		...	...
Segment group *n*		O	99
UST	Security trailer	M	1
USR	Result to be validated	O	1

Note: M, mandatory; O, optional.

of protection. The price of this flexibility is a further complication of the protocol, a complication that may seem excessive because the same effect could have been obtained by sending consecutive messages instead of a single complex message.

Each security structure begins with a header USH and ends with a trailer UST. It contains the following segments: USA, USC, USH, and USR according to the structure shown in Table 4.5. These segments inform the recipient of the invoked level of security and of the security mechanisms utilized and their parameters as well as the extent of the protected domain. These various segments will be recognized and processed by applications that demand security functions and will be safely ignored by other applications. They will be inserted in the interchange between the UNH and UNT segments for a message or between the UNO and UNP segments in the case of a non-EDIFACT object.

The advantage of this selective protection is that the consequences of a security breach would be limited to the concerned party without compromising the whole interchange. Another advantage is to allow selective access to the different data included within the interchange. For example, in a payment instruction, the intermediary banks do not need to know the identities of the creditor and the debtor.

Protection against loss or duplication of messages relies on optional fields that contain a sequential number counter and a time stamp.

The following sections give a more precise definition of the content of these various segment groups.

4.10.2.1.1 Segment Group No. 1

This is a group of segments (called the security header) that defines the security services, the security mechanism, and the elements necessary to carry out the validation calculations. The group can be repeated for each service offered (e.g., integrity, authentication of the origin, nonrepudiation of the origin) to the same message or if the same security service is benefiting several messages.

The security header segment USH contains the following:

- A mandatory security reference number
- The scope of the security application
- An indication whether an acknowledgment is requested
- The role of the intermediary providing the security services (e.g., document issuer, notary, witness, contracting party)
- Details concerning the sender and the receiver: name of the symmetrical encryption key or of the certificate in case of a public key encryption
- A time stamp (the date and time when the security services were applied)

To each security service invoked, there is a corresponding USA segment that defines the algorithm utilized and its parameters.

4.10.2.1.2 Segment Group No. 2

This group contains the elements that are pertinent to public key encryption, in particular, the certificates. Two repetitions are allowed to communicate both the sender and the recipient certificates. The latter case occurs if the sender utilizes the public key of the recipient instead of his/her own private key.

If the recipient does not have the public key of the certification authority, the group will contain references to the sender's public key through the certificate of authenticity, the hashing function that the certification authority employs for producing the certificate, and the public key algorithm that the certification authority uses to sign as well as its public key. These references will be contained in three USA segments. The segment USR will contain the signature of the public authority that will be used to verify the transmission integrity.

If the recipient knows the public key of the certification authority, it will be communicated by reference (e.g., the name of the file where it resides) so that the recipient could extract it from a secure database.

4.10.2.1.3 Segment Group n

The segment UST is the security trailer that separates the message from the security structure. It is followed by the segment USR that contains the outcome that the validation results should match when the computations are done according to the specifications of the header segment USH.

The following two examples illustrate how the mechanisms indicated earlier can be used to provide security services for the protection of EDIFACT messages.

Example 4.1: Nonrepudiation of the Origin and Message Integrity

There are several ways to ensure nonrepudiation of the origin. Assume that the mechanism used is to condense the message with a one-way hashing function and then encrypt the digest with a public key algorithm. Assume also that the certification authority utilizes a different hash function to sign the certificate of authenticity accorded to the sender. Finally, assume that the hash function used by the sender has already been the subject of a bilateral agreement with the recipient.

There are now four algorithms: the hash function and the public key algorithm that the sender utilizes to condense the messages and encrypt the digest and the hash function and the public key algorithm that the certification authority utilizes to calculate the digest of the certificate and to sign it, respectively. Because the sender's hash algorithm is the subject of a prior bilateral agreement, only three USA segments will be necessary. The sender's public key is obviously contained in the certificate that the public authority has signed.

Two USR segments contain the data needed to verify the integrity of the transmission, namely, the encrypted digest of the EDIFACT message and the signature of the certification authority.

Example 4.2: Nonrepudiation of the Origin and Confidentiality

For confidentiality, the EDIFACT message can be encrypted using a symmetrical key that will then be encrypted with the public key of the recipient. This is, in fact, the procedure specified by the protocol ETEBAC5. This is the protocol that French banks use to secure file exchanges with their client businesses.

Two groups of number 2 segments are needed: the first relates to the sender's public key certificate and the second to the recipient's public key that the sender will use to encrypt the secret symmetrical key. Each group corresponds a USR segment that contains the result to be validated to verify the integrity of the received message.

Four USA segments are involved. The first segment, which is within the segment group number 1, contains the parameters of the symmetrical algorithm. The other three segments, respectively, identify the public key algorithm of the sender, the hash function utilized by the certification authority to condense the sender's certificate before signing it, and, finally, the public key algorithm and the public key that the authority uses to sign the sender's certificate, respectively.

4.10.2.2 Security of EDIFACT Documents with Out-of-Band Segments: The AUTACK Message

The AUTACK message is used to provide the same levels of security as with in-band segments, including acknowledgment of reception. Authentication is ensured either by symmetrical encryption of the digest

TABLE 4.6

Security Structure of the AUTACK Message

Tag	Segment Name	Type	Maximum Number of Repetitions
UNH	Message header	M	1
Segment group 1		O	99
USH	Security header	M	1
USA	Security algorithm	O	3
Segment group 2		O	2
USC	Certificate	M	1
USA	Security algorithm	O	3
USR	Security result to confirm the validity of security as certified by the certification authority	O	1
USB	Identification of AUTACK (type, function, time stamp, sender, recipient)	M	1
Segment group 3		M	9999
USX	Security references to what is being secured	M	1
USY	Security references (result to be verified)	M	9
Segment group 4		M	99
UST	Security trailer	M	1
USR	Security result that verifies the security of the AUTACK message	O	1
UNT	Message trailer	M	1

Note: M, mandatory; O, optional.

of the structure to be protected or by exchange of certificates. For nonrepudiation of the origin, the digest is encrypted with a public key. Table 4.6 depicts the security structure of an AUTACK message.

Segment USX points to the EDIFACT structure to be secured and records the date and time of its creation. Each USX segment corresponds to one or several USY segments that contain the results to be matched to verify the validity of the message to which USX refers.

Segment USR of segment group number 4 is optional; it is only necessary when the AUTACK message itself is to be verified. For example, if USY contains the encrypted digest, USR may be omitted. However, if the AUTACK message is used as a secure acknowledgment, USR will contain the encrypted digest of the message. The whole structure will be enclosed in a pair of segments USH/UST (see Example 4.5).

Segment USB identifies the sender and the recipient of the interchange to be secured. It also contains a time stamp and indicates whether the sender has requested an acknowledgment.

Example 4.3: Authentication of the Origin of the EDIFACT Message and the Origin of the AUTACK Message as Well as the Integrity of the EDIFACT Message

In this example, a symmetrical algorithm is used with a secret key that both parties already have.

Segment USH contains, among other details, references to the symmetrical key, segment USY contains the digest of the EDIFACT message, and segment USR in group 4 contains the encrypted digest of the AUTACK message. When the recipient verifies that the value of the computed encrypted digest is identical to the value in the message, this will be a verification of the authenticity of the origin of both EDIFACT and AUTACK messages. The integrity of the EDIFACT message is verified if the value of its digest computed after reception is the same as the value indicated in the USY segment.

Example 4.4: Nonrepudiation of the Origin

The segment USC holds the sender certificate and the public key of the certification authority. Three USA segments are needed to identify, respectively, the public key algorithm that the sender uses for signature, the hashing function, and the public key algorithm that the certification authority employs to compute the digest of the sender certificate and to sign the certificate.

The USR segment of group 2 contains the signature of the certification authority that should be verified at reception. The USB has the security parameters and indicates whether an acknowledgment or the security date and time is required, as well as the identity of the sender and recipient.

Segment USX points to the secured EDIFACT message, and the USY segment has the sender signatures that the recipient will have to verify. The segment USR of group 4 has the signature of the sender as applied to the AUTACK message.

Nonrepudiation of the origin is the consequence of authentication of the origin of the EDIFACT and AUTACK messages, as verified with the sender's signature on both. Note that the sender's signature is used twice; accordingly, the USR segment of group 4 is redundant and can be eliminated at the origin.

Example 4.5: Interchange Comprises Two Messages: Nonrepudiation of the Origin of the First and Authentication of the Origin of the Second

The transmission will contain two sets of segment group number 1, one for each required service. The first service is offered in the same manner as for Example 4.4. Authentication of the origin of the second message is the other service. A USA segment will define the symmetrical algorithm used, its parameters, and its mode of operation. Because the sender's signature is verified for each message, securing the AUTACK message itself is not necessary.

Example 4.6: Nonrepudiation of the Origin with a Request of Acknowledgment

The transmission will contain two sets of segment group number 1, one for each required service. The first will relate to nonrepudiation of the origin and the second to the request of a secure acknowledgment.

Segment USC of the segment group number 2 contains the sender certificate and the public key of the certification authority. Three consecutive USA segments identify, respectively, the public key algorithm that the sender uses, the hash function, and the public key algorithm, both used by the certification authority to generate the digest and sign the sender's certificate.

Segment USB contains the time stamp of the AUTACK message, the credentials of both parties, and the references to the message.

Segment USX includes references to the message to be secured; segment USY has the digest of the message encrypted with the sender's private key.

Two sets of segment group number 4 are used, one for the EDIFACT message and the other for the AUTACK message. The USR segment of the last segment group number 4 will contain the signature of the AUTACK message to verify that it has arrived safely at the destination.

4.10.3 Protection of EDI Messages in Internet Mail

The need for transporting EDI messages with Simple Mail Transfer Protocol (SMTP) and Multipurpose Internet Mail Extensions (MIME) became more pressing as enterprises started to use the Internet protocols in their networks. Various specifications defined in IETF Requests for Comments (RFCs) 2045 through 2049 show how to encapsulate these messages to include different object types. In addition, RFC 3335 describes how EDI messages could be protected with encryption and digital signatures using Pretty Good Privacy (PGP)/MIME, Secure Multipurpose Internet Mail Extensions (S/MIME), or Secure HyperText Transfer Protocol (S-HTTP). In August 1997, the Electronic Data Interchange Internet Integration (EDIINT) group of the IETF conducted tests to verify the interoperability of the EDI translation software and S/MIME messaging systems.

Internet messaging systems have most of the functions in X.400 but do not authenticate the route and do not guarantee delivery times. Another limitation is that the structure of MIME precludes cross-referencing of segments because it does not accommodate referrals to different segments.

The outlines of the various IETF methods are as follows (RFC 3335, 2002):

- The sending organization sends the encrypted data and the signature of the message (message digest encrypted with the sender's private key) in an envelope constructed according to PGP/MIME, S/MIME, or S-HTTP and requests and acknowledgment (RFCs 1767, 2015, 2311).
- The recipient organization recovers the symmetrical encryption key and its parameters (e.g., the initialization vector) with its private key, decrypts the message, and checks the signature with the public key of the sender, thereby verifying simultaneously the integrity of the data and the authenticity of the sender.
- The recipient sends back a signed acknowledgment by encrypting its digest with the recipient's private key; the acknowledgment, in turn, contains the digest and the identifier of the received message.

In the IETF jargon, the digest is called "message integrity check" (MIC) and the acknowledgment is the "message disposition notification." If the acknowledgment is signed, it will be in the multipart format of MIME to accommodate the acknowledgment and the signature. The sender can then verify the integrity of the acknowledgment, thereby proving that

TABLE 4.7

Reference Documents for S/MIME

Document No.	Title
RFC 2632	S/MIME version 3 certificate handling
RFC 2633	S/MIME version 3 message specification
RFC 2634*	Enhanced security services for S/MIME
RFC 2437	PKCS #1: RSA Encryption Version 2.0
RFC 2315	PKCS #7: Cryptographic Message Syntax Version 1.5

* Updated by RFC 5035 (2007).

- The recipient has authenticated the sender
- The recipient recognizes having received the message that corresponds to the mentioned identifier
- The message has been received, with its integrity intact
- The recipient cannot deny having sent the acknowledgment

For encryption, PGP/MIME uses either PGP or OpenPGP, the first being described in IETF RFC 1991 (1996), while the second is in IETF RFC 2440 (1998). PGP is considered to be the closest commercial algorithm to military-grade performance (Garfinkel, 1995); its adoption as an IETF standard floundered because of its use of patented techniques, particularly the RSA key exchange and International Data Encryption Algorithm (IDEA), something that the IETF avoids by principle. OpenPGP has been able to go around this difficulty.

S/MIME is now in its third version, and its use is described by several RFCs, the more important are given in Table 4.7. Its second version has not been standardized by the IETF because it requires the usage of a patented key exchange mechanism that is not freely available.

S/MIME allows a choice between two symmetrical encryption algorithms DEC in the CBC mode or triple DES. The use of RC2 and RC5 is optional, but usage of RC2 with a key of 40 bits is not recommended. Hashing is done with either MD5 or SHA-1. The digest, called, MIC, and the symmetrical session keys are encrypted using the RSA algorithm with keys of 512 or 1024 bits. Note that in a multipart MIME message, computation of the digest takes into account the content of all parts, including the headers (RFC 2045, 1996).

4.10.4 Protocol Stacks for EDI Messaging

Figure 4.12 summarizes the different protocol used for EDI messaging without security, while Figure 4.13 depicts the protocol stack for secure EDI. A more rigorous comparison of the various protocols can be found in

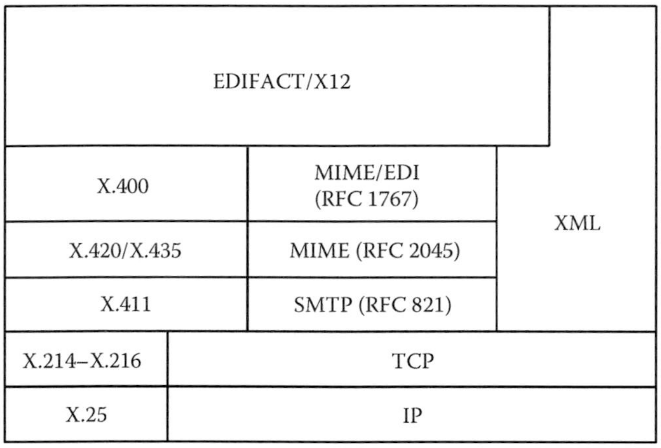

<table>
<tr><td colspan="3">EDIFACT/X12</td><td rowspan="4">XML</td></tr>
<tr><td colspan="2">X.400</td><td>MIME/EDI
(RFC 1767)</td></tr>
<tr><td colspan="2">X.420/X.435</td><td>MIME (RFC 2045)</td></tr>
<tr><td colspan="2">X.411</td><td>SMTP (RFC 821)</td></tr>
<tr><td>X.214–X.216</td><td colspan="3">TCP</td></tr>
<tr><td>X.25</td><td colspan="3">IP</td></tr>
</table>

FIGURE 4.12
Protocol stack for EDI messaging (without security).

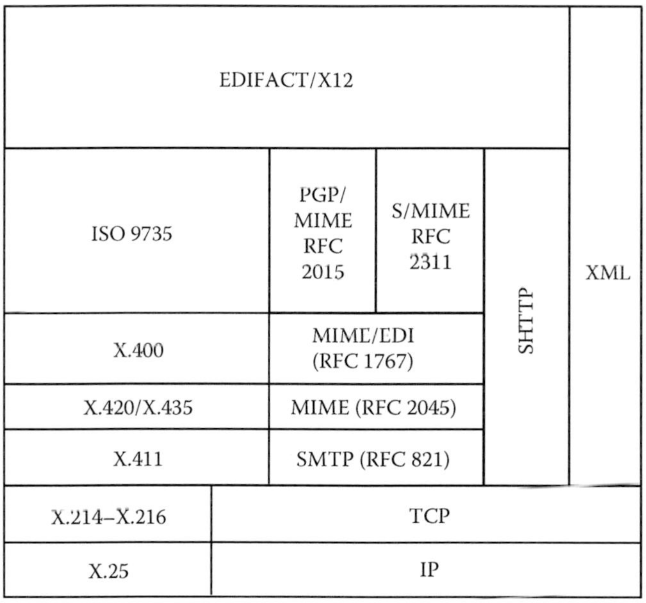

<table>
<tr><td colspan="5">EDIFACT/X12</td><td rowspan="5">XML</td></tr>
<tr><td colspan="2">ISO 9735</td><td>PGP/
MIME
RFC
2015</td><td>S/MIME
RFC
2311</td><td rowspan="4">SHTTP</td></tr>
<tr><td colspan="2">X.400</td><td colspan="2">MIME/EDI
(RFC 1767)</td></tr>
<tr><td colspan="2">X.420/X.435</td><td colspan="2">MIME (RFC 2045)</td></tr>
<tr><td colspan="2">X.411</td><td colspan="2">SMTP (RFC 821)</td></tr>
<tr><td>X.214–X.216</td><td colspan="5">TCP</td></tr>
<tr><td>X.25</td><td colspan="5">IP</td></tr>
</table>

FIGURE 4.13
Protocol stack for secure EDI.

specialized references on electronic messaging, such as Palme (1995) or Bouillant (1998). In these figures, X.25, X.400, and so on refer to the ITU-T—formerly known as the Comité Consultatif International Téléphonique et Télégraphique (CCITT)—recommendations that are part of the EDIFACT specifications.

4.11 Integration of XML and Traditional EDI

The integration of XML and traditional EDI using X12 or EDIFACT is needed for industries with considerable prior investment in EDI. Aggregate elements (or "classes") are combined to form messages, segments, or structured documents sent to the destination with the necessary instructions for interpreting them. Structuring the dialogs as documents allows consideration of all the data exchanged in the context of a commercial transaction, independent of their format (text, graphics, image, sound, audio, video), and not only those that can be represented by alphanumeric characters.

The main structured language used today is XML, a direct descendent of Standard Generalized Markup Language (SGML). SGML is a declarative metalanguage that describes the logical structure of a document using a model called the document type definition (DTD) (Dupoirier, 1995, pp. 107–120). The syntactic analysis of the SGML document in light of the associated DTD is independent of the information processing platform, which means that SGML defines simultaneously a methodological framework for the exchange of documents.

SGML was adopted in 1986 as ISO 8879. It inspired several efforts for automatic documentation, particularly HTML. A key concept of HTML is the utilization of a Uniform Resource Locator (URL) to locate document sources accessible on the Internet.

In 1994, Contribution N1737 was presented to ISO/IEC JTC1/SC18/WG8 to replace alphanumeric EDI with SGML (ISO, 1994), but this proposal was judged to be too radical. It was up to the World Wide Web Consortium (W3C) to define XML on the basis of SGML.

XML is used to define files of different types so that they can be shared among applications using the broadcast possibilities of the Internet. XML can enhance EDI solutions in several ways. It allows the attachment of multimedia objects to the exchanged messages and adds interactivity to EDI. It gives recipients the possibility to interpret and process of nonstandardized segments or messages by attaching their DTDs. It also increases the possibility of automatic translation of the data into the recipient language. It provides a mechanism to recall documents using their URLs, thereby avoiding their attachment in the exchanges. Finally, it can be the starting point of the definition of specialized markup languages for any type of documents (Bryan, 1998; Michard, 1999).

The architecture for integration is shown in Figure 4.14. An intermediary encapsulates or translates the X12/EDIFACT messages into XML exchanges to be displayed by the client on the user station.

We now present several approaches to express EDI exchanges into an XML syntax.

4.11.1 BizTalk®

Shepherded by Microsoft, BizTalk defines a framework called "BizTalk Framework" to use XML in

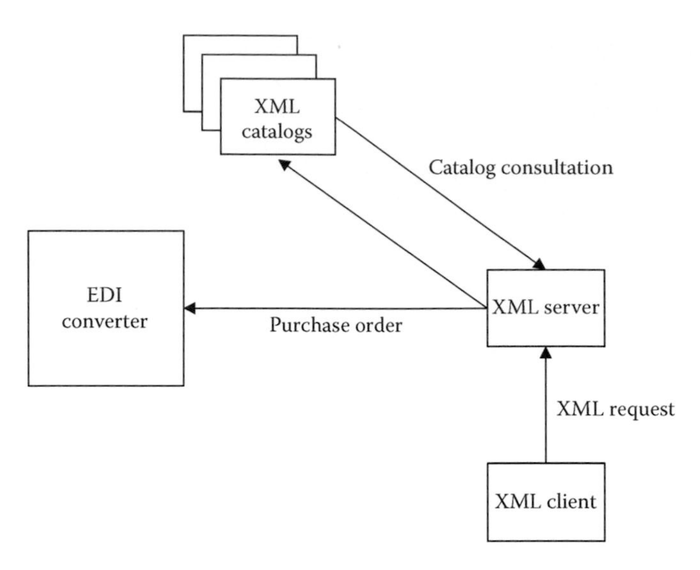

FIGURE 4.14
EDI/XML Integration.

e-commerce. The rules for BizTalk concern the definition and the publications of XML schemas and the use of XML messages for existing applications to communicate.

4.11.2 xCBL

The XML Common Business Library (xCBL) initiative was started in 1997 to express EDI messages (X12 and EDIFACT) in the XML syntax. The documents covered include cost estimates, invoicing, payment, order tracking, shipment of goods, delivery, and so on. Version 3.0, approved in November 2001, was the starting point to the development of UBL.

4.11.3 UBL

The Universal Business Language (UBL) was developed by the Organization for the Advancement of Structured Information Standards (OASIS) consortium to facilitate the production of XML schemas for business operations starting with a library of EDI templates. This language was initially promoted by CommerceNet, a nonprofit consortium of U.S. companies formed to promote global solutions for e-commerce (Tenenbaum et al., 1995). The first version of the specification codifies the transposition of the X12 and EDIFACT messages associated with the purchase order/invoice cycle into their XML equivalents schemas (Bosak, 2005). A version of UBL, called OIOXML, is used in Denmark for communication among government agencies, for public procurement, and in government ERP systems.

4.12 New Architectures for Business-to-Business Electronic Commerce

The switch to TCP/IP networks in the mid-1990s initiated a wave of migration from proprietary to standard solutions. This led to a major redesign of the architectures of business-to-business electronic. Semantic interoperability was added to the list of requirements in the face of the heterogeneity of business models and processes (Hasselbring, 2000).

This new architecture is represented in Figure 4.15. A global repository contains the structure of the documents, their logical components, and the business processes by referring to the dictionaries of each data category. This global repository or library of objects contains the core components of the messages sent (data structures or objects). The sender queries the repository to define the partner's profile and the scenarios of the exchanges. The requests are routed to specialized libraries that contain EDIFACT/X12 standards, the appropriate industry agreements or standards, or the elementary data and the collaborative agreements between the two organizations involved.

Reusable components are built using domain information that represents the common requirements within a specific industrial or economic sector. These components are then stored in global depositories and referred to in a directory accessible on the World Wide Web. The directory points to repositories storing the potential partners' profiles, the processes, the exchange scenarios, and the data dictionaries. The dynamic discovery of the services, service providers, and agreements and the potential partners are mediated through standardized protocols. Finally, at run time, that is, when a specific transaction takes place, the exchanges follow a predefined "choreography" specified in the business process model. The application programming interfaces (APIs) are responsible for the appropriate conversions.

Depending on the starting point, we can distinguish two major ways to implement this architecture. Starting with the traditional EDI, the electronic business Extensible Markup Language (ebXML) framework takes advantages of the capabilities of the World Wide Web and the use of XML in the message exchange. The other approach is the use of web services to integrate applications within and without the enterprise boundaries without manual intervention.

Both approaches use process modeling to cover the business processes, the data information models, the formats of the exchange, and the exchange protocols. Formal modeling also captures the implicit and explicit

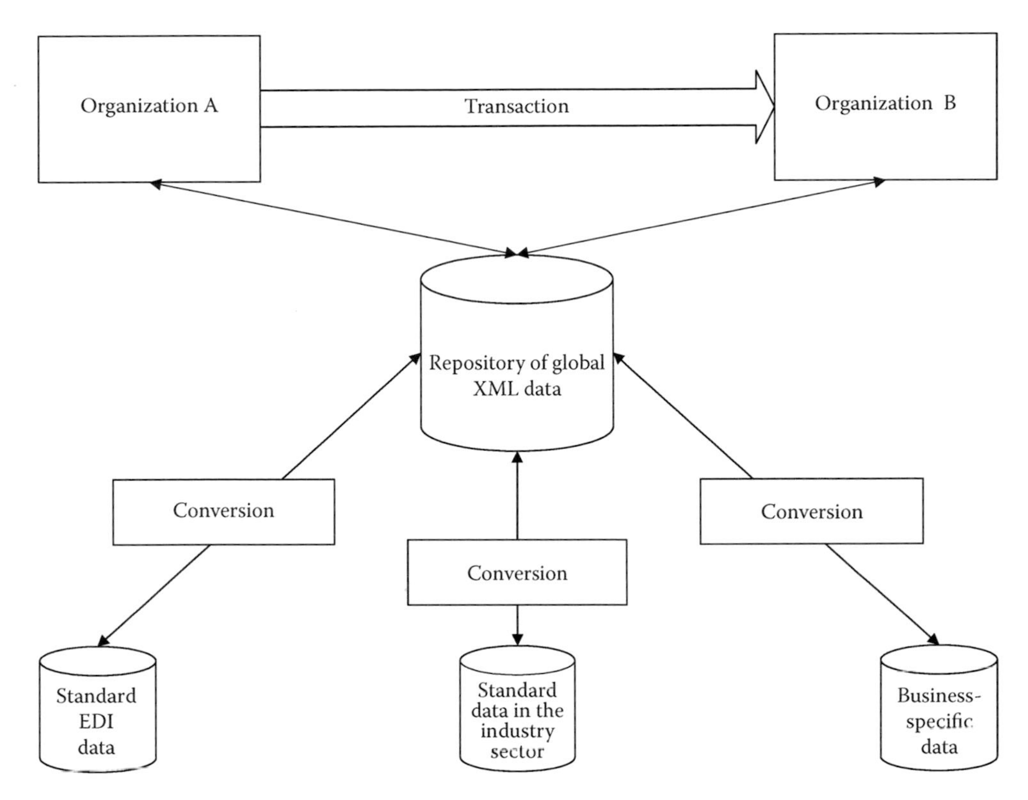

FIGURE 4.15
New architecture for the business-to-business electronic commerce.

knowledge about these business processes (what, how, who, where, how much) in a form that can be shared. For example, in the management of the supply chain for physical goods, the internal processes of the corporation (the "back office") must be integrated with external processes to identify and track parcels or crates in automated warehouses, following standards such as the Multi Industry Transport Label (MITL) and the serial shipment container code (SSCC). Thus, business process models represent the information flow, the relations between the objects from various perspectives (e.g., a data view, a functional view, a process view) and between the objects of different views. Finally, an agreement has to be drawn to define the specific formats of the business interchanges and the parameters of the business interfaces that each party has to conform to.

Process models are defined at an abstract level, that is, they should avoid implementation details. This separation increases the flexibility of the design and the implementation. Furthermore, model-driven development facilitates the evaluation of design alternatives and allows the automatic propagation of modifications top down from the requirements to the final realization.

The representation of business exchanges in the ebXML framework is inherited from the experience with the traditional EDI, while web services are intended for a much wider range of organizations with dissimilar business processes and scenarios. Thus, ebXML and web services provide distinct approaches for service discovery (Al-Masri and Mahmoud, 2007). Other differences concern the structure of the directories or registries; the nature of the contents of the specialized libraries in terms of standards, logical components, or business process; and the access protocols to the registries. Finally, because web services are more general in nature, it is important to establish a set of common vocabulary and concepts that describe the business logic (Dorn et al., 2007).

The formal model used to describe the business process in ebXML is the United Nations Modeling Methodology (UMM). UMM describes Open-EDI scenarios by adapting the Unified Modeling Language (UML) developed by the Object Management Group (OMG) by specifying new elements to model the content of exchanged messages (Langlois et al., 2005). Languages used for web services modeling include Business Process Execution Language (BPEL) also known as WS-BPEL and Web Services Description Language (WSDL).

UMM describes the processes associated with the information flow and the relevant actions and objects associated with the exchange of goods and services but does not provide specific mechanisms for message exchange nor the data to be exchanged. Thus, the focus is shifted from modeling business documents to the representation of the pieces of information needed in the management of the supply chain. This allows an automated analysis of risks in a global supply chain. With this arrangement, several implementation technologies could correspond to the same specification. Huemer et al. (2010) present a summary of business process modeling with UMM.

More generally, in SOA, existing software components are listed in a directory (also called registry or service broker) that can be queried in a standard manner (Erl, 2004, pp. 60, 324–364; Manouvier and Ménard, 2007; Bonnet et al., 2008). A precise contract defines the access conditions to a service, the rules to launch it, and the end conditions that must be validated for the result to conform to the service user's expectations. The contract and its functional clauses are described in standardized languages. To increase flexibility, rule definition is separated from service definition, so the same set of services can be used with changed rules. In this manner, the workflow can adapt quickly to changes without affecting the core elements of a given service. The sequencing of the workflow (choreography) and coordination of message exchanges between the services (orchestration) are under the control of an external entity, the orchestration engine, which executes the rules describing the business logic and/or composes them on the fly using exchanges with a rule engine.

SOA is related to web services, in the sense that both use XML-based protocols, but opinions differ on the exact nature of their relationship. Some authors consider that SOA has to rely on web services, while others tend to decouple XML specifications from SOA pointing to the extreme verbosity of XML-based documents, which could be a drawback when efficiency is needed.

4.13 Electronic Business (Using) Extensible Markup Language

As stated earlier, ebXML is a model-driven open framework to ensure the interoperability of transactions among business entities based on the object-oriented methodology and consistent with the Open-EDI reference model of ISO/IEC 14662. The specifications of ebXML were started jointly in September 1999 by the OASIS consortium* and the United Nations Centre for Trade Facilitation and Electronic Business (UN/CEFACT). The purpose was to take advantage of new technologies, such as XML (Extensible Markup Language and the World Wide Web), to increase the richness of business-to-business exchanges. In general, the business specifications for ebXML are done within UN/CEFACT and the information technology specifications are done by OASIS. Typically, they are then submitted to ISO to become an international standard.

The ebXML captures knowledge about the business under consideration in a technology-independent way to provide an operational view of the business rules called the "business-oriented view" (BOV). The technology-dependent way is denoted as the "functional service view" (FSV).

4.13.1 Architecture of ebXML

The ebXML framework is built on a directed peer-to-peer architecture with five major components and two perspectives. The business-oriented view comprises the business scenarios and the core components. The business scenarios capture the business knowledge in the form of process models for the interaction between the parties of a transaction or an interaction. The core components are the basic building blocks used to construct all the interactions. The functional view consists of the following:

- The specific profile of the Collaboration protocol as well as the agreements between the two parties on the various aspects of the transaction that are needed for its execution
- A registry of previously agreed information on objects, exchanges, and processes that could be reused
- A messaging protocol for communicating and exchanging DATA through standardized interfaces

These five parts are specified in a series of standards defined by UN/CEFACT, OASIS, and ISO, as listed in Table 4.8.

4.13.2 Business Scenarios

As mentioned earlier, typical business scenarios, such as catalog consultation, purchasing, payment, and delivery, are modeled using UMM. The corresponding business processes are described using an XML-based language called the Business Process Specification Schema (BPSS). Each process is described from three views:

* This consortium, founded in 1993, groups the main software developers and system integrators and focuses on standards for web services.

TABLE 4.8

Standards and Specifications of ebXML

Document Title	Source	URL	ISO
ebXML Technical Architecture Specification v1.0.4, February 16, 2001	UN/CEFACT and OASIS	http://www.ebxml.org/specs/ebTA.pdf	—
ebXML Business Process Specification Schema Technical Specification v2.0.4, December 21, 2006	OASIS	http://docs.oasis-open.org/ebxml-bp/2.0.4/OS/spec/ebxmlbp-v2.0.4-Spec-os-en.pdf	—
ebXML Collaboration Protocol Profile and Agreement Specification, Version 2.0, September 23, 2002	OASIS	http://www.oasis-open.org/committees/ebxml-cppa/documents/ebcpp-2.0.pdf	—
ebXML Messaging Services version 3.0: Part 1, Core Features	OASIS	http://docs.oasis-open.org/ebxml-msg/ebms/v3.0/core/os/ebms_core-3.0-spec-os.pdf	—
ebXML Message Service Specification Version 2.0, 1 April 2002	OASIS	http://www.ebxml.org/specs/ebMS2.pdf	—
ebXML Registry Information Model (RIM) v3.0, May 2005	OASIS	Not available	—
ebXML Registry Information Model (RIM) v2.0, April 2002	OASIS	http://www.oasis-open.org/committees/regrep/documents/2.0/specs/ebrim.pdf	—
ebXML Registry Services Specification (RS) v3.0, May 2005	OASIS	Not available	—
ebXML Registry Services Specification (RS) v2.0, April 2002	OASIS	http://www.oasis-open.org/committees/regrep/documents/2.0/specs/ebrs.pdf	—
Core Components Technical Specification –Part 8 of the ebXML Framework, November 15, 2003, Version 2.01	UN/CEFACT	http://www.unece.org/cefact/ebxml/CCTS_V2-01_Final.pdf	ISO 15000-5:2005, Electronic Business Extensible Markup Language (ebXML)—Part 5: ebXML Core Components Specification (CCS)
UML Profile for Core Components (UPCC), Version 1.0, Final Specification, January 16, 2008	UN/CEFACT	http://www.unece.org/cefact/codesfortrade/UPCC_UML-CoreComponent.pdf	—
UN/CEFACT's Modeling Methodology (UMM): UMM Meta Model—Base Module Version 1.0 Technical Specification, 2006-10-06	UN/CEFACT	http://www.unece.org/cefact/umm/UMM_Base_Module.pdf	—
XML Naming and Design Rules, Version 2.0, February 17, 2006	UN/CEFACT	http://www.unece.org/cefact/xml/XML-Naming-and-Design-Rules-V2.0.pdf	—

Note: All URLs last accessed on January 17, 2016.

1. A business domain view to describe the context and the business sector where the process applies
2. A business requirements view for the preconditions to execute the process
3. A business transaction view reserved for the exchanged data at the semantic level

4.13.3 Core Components

The core components are the objects, messages, and data elements that a given business sector uses as the building block for all processes, messages, and data exchanges. They are specified in UML and stored in a core components library. Agreements on these building blocks are reached at the international level under the auspices of the UN/CEFACT. They are then harmonized with those of other sectors to avoid duplication and encourage reuse.

The exchanged documents are formed from the core components, taking into account the business rules governing the exchange and the context of use. A business information entity is made from the core components arranged in a specific business context. Changes to the core components do not affect the choreography. Similarly, changes in the preconditions that affect the choreography, that is, the interchange agreement, do not ripple through to the data. With the separation of the business from the functional aspects, the information model is valid irrespective of the syntax used, for example, ANSI X12 or EDIFACT messages or XML documents.

4.13.4 Registry and Repository

The registry contains information about companies or organizations as well as definitions of the processes already defined for business collaborations and the metainformation concerning the objects in a business collaboration, such as the messages to be exchanged, the data definitions, and formats. The Registry Information Model (RIM) describes the internal data structure of the registry. The Registry Services (RS) describe the methods for accessing the registry via a browser over the World Wide Web or via queries using Structured Query Language (SQL).

The repository is the actual database that stores the entries. It can be centralized or distributed on a geographic basis or per business sector or using both criteria. The RS specifications explain how to synchronize the various distributed data stores and how to retrieve the information.

4.13.5 CPPA

The protocol profiles for collaboration describe the technical capabilities of the partners in a business transaction in terms of the services offered; the parameters of the connection, for example, the configuration and security parameters; the maximum delay before an acknowledgment is to be sent; the maximum number of attempts at a communication; and so on. For any given party, its business profile is based on the core components used in the interaction, the common business processes for the particular business sector, and the instant of the process model described in BPSS that is appropriate to the firm. Each company defines its profile in XML and stores it in the ebXML registry as a Collaboration Protocol Profile (CPP) as defined in the electronic business Collaboration Protocol Profile and Agreement (ebCPPA).

Companies find each other's profiles by querying the registry or by communicating directly. The technical agreements among the parties can be made automatically using an automatic matching process to determine if the two profiles can be matched, and if so, create a technical contract that binds the two parties. The Collaboration Protocol Agreement (CPA) defines the technical parameters of the interface and the messaging services as specified in the CPPA. The CPA does not cover the legal aspects of the collaboration.

4.13.6 Message Service Specification

The electronic business message service (ebMS) concerns the messaging protocol and the structure of the messages to be sent securely over the World Wide Web. The ebMS has a payload that is included over the Simple Object Access Protocol (SOAP) body and is called SOAP with Attachment (SwA). The SOAP protocol structures the exchanges in the form of requests and responses.

SOAP is a recommendation from the World Wide Web Consortium (W3C). SOAP messages are XML documents, typically transported using HyperText Transfer Protocol (HTTP) (Box et al., 2000; Erl, 2004, pp. 72–77). They structure communication in a decentralized, distributed environment by generalizing the remote procedure call (RPC) to start the execution of a procedure on a remote server. SOAP is independent of information technology platform but suffers from a lack of service guarantees concerning the transport of exchanges.

SOAP messages are XML documents formed of three components: an envelope, coding rules and conventions to make remote procedure calls and interpret their responses, and a payload. The envelope consists of a header and a body. The header supplies information for authenticating the exchange, including the routing information. The body contains the elements that are needed to interpret the exchanges as well as encryption and signature information. The presence of the body is mandatory, while the header is optional—because the two parties in communication can agree beforehand on the elements they supply. The ebMS adds security mechanism so that messages can be partially or totally encrypted and/or electronically signed to verify their integrity.

The Transport protocol is usually Hypertext Transfer Protocol (HTTP), but ebMS can also be used on top of Simple Mail Transfer Protocol (SMTP) or File Transfer Protocol (FTP).

The structure of the MIME ebMS message in a SOAP with Attachment (SwA) is shown in Figure 4.16, while Figure 4.17 illustrates the protocol stack of ebXML.

4.13.7 ebXML Operations

Each business defines its business processes following the steps specified in ebXML and requests them to be registered in the ebXML global registry. It may have to modify some of its proposed definitions to conform to the rules already established or reuse an already existing process specification and modify it according to its proper usage. Alternatively, it may start by querying the registry for an existing specification. In any

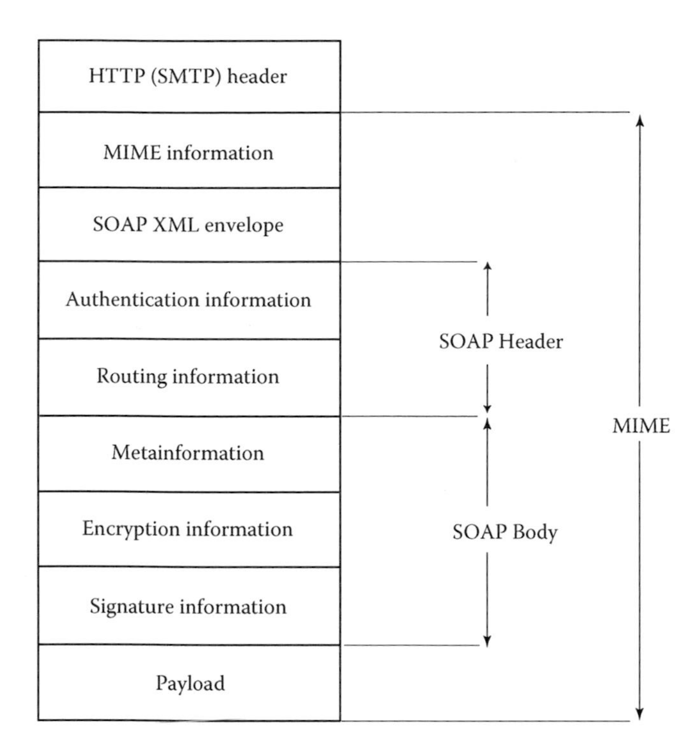

FIGURE 4.16
Structure of ebXML message specification (ebMS).

SOAP with Attachments
MIME
HTTP/SMTP
TLS/SSL
TCP/IP

FIGURE 4.17
Protocol stack for ebXML.

event, it submits a conforming CPP to be registered and stored.

To establish collaboration, a business queries the registry to find a match to a specific profile or a specific scenario of interest. For each match, it contacts the associated party to propose an arrangement using the ebXML procedures. The proposed arrangement outlines specific scenarios and messaging and security requirements. Once both parties settle on an agreement, they document it in a CPA.

Based on the details of the agreement, each party configures its interfaces, called Business Service Interfaces (BSI). The core components are selected according to the context of the specific transactions and the relevant business information entities.

If intermediaries are involved, such as for transportation and shipping, each party conducts similar negotiations with these intermediaries independently, because ebXML is a peer-to-peer arrangement.

In summary, ebXML operations comprise the following steps:

- Registration of each party and their profiles
- Discovery, where each party discovers the other's technical capabilities and negotiates an agreement
- Implementation, during which each party implements and configures an ebXML compliant application capable of carrying out the business collaboration described in the agreement
- Run time, where the business transaction takes place according to the choreography of the business process specification

The first three steps occur in the "design time" as depicted in Figure 4.18 (Sherif, 2010).

For a fully operational ebXML system, some additional architectural decisions will have to be made, such as

- The placement of the registries, directories, and repositories
- The procedures to verify the conformance of the stored components to the specifications
- The operational criteria and the guarantees on the service quality
- The maintenance and the management of faults

The ebXML registry, directory, and repository are in the process of development based on national registries. Japan and Canada, for example, are already implementing national ebXML central registries. Global electronic trade will then relay on the federation of these national ebXML registries.

Table 4.9 compares the main properties of the traditional EDI with ebXML.

Today, ebXML does not offer yet the same coverage as traditional EDI and the tools for development are not always available. ebXML will require some time to mature, because it depends on the involvement of enterprises in supplying all the necessary blocks.

4.14 Web Services

Web services are software applications or application components that are self-contained and can establish real-time communications with other applications or components on the World Wide Web, irrespective of the underlying physical or logical platforms, including

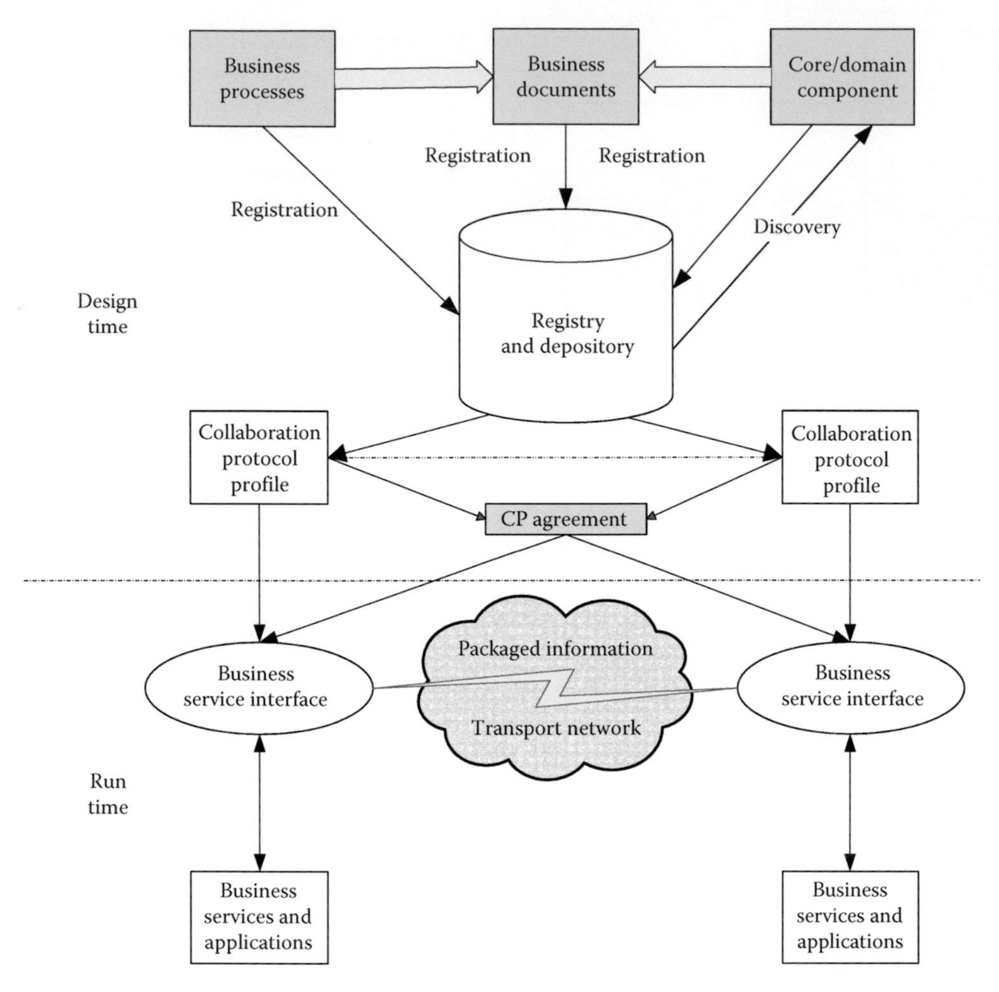

FIGURE 4.18
Technical operations of ebXML.

TABLE 4.9
Comparison of Traditional EDI and ebXML

Characteristic	Traditional EDI	ebXML
Design orientation	Documents	Processes
Nature of the links	Point to point	Peer to peer but allows multilinks
Interface configuration	Static and manual	Dynamic and automatic
Subject of standardization	Structured messages and data elements and segments	Methods, architectures, processes, and objects
Technical negotiation	Manual	Automatic
Operation	Batch	Transactional
Orchestration of the exchange kinetics	Managed by each application	Managed by a choreography process

operating systems or programming languages. In this manner, they support more flexible collaborative relations with other web services within the enterprise as well as among different enterprises. The idea behind web services was to offer building blocks for use in many applications starting from the ground up. Web services comprise networked software components and independent machines that communicate to form one entity that is distributed and decentralized. Thus, they have inherited many of the concepts, models, and specifications from other object-oriented architectures: Common Object Request Broker Architecture (CORBA), distributed component object model (DCOM), or Enterprise Java Beans (EJB). BizTalk from Microsoft was one of the first commercial offers (http://www.microsoft.com/biztalk).

4.14.1 Web Services Standards

The focus here is on standards that have been approved by the main standard setting organizations involved, OASIS and the World Wide Web Consortium (W3C).

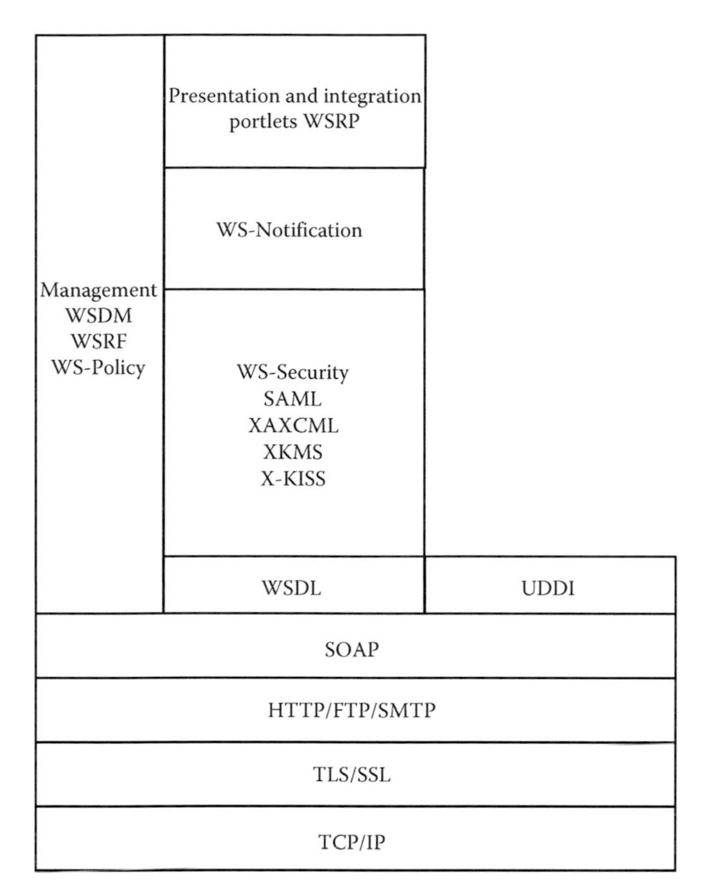

FIGURE 4.19
Protocol stack for web services.

The basic protocols of the web services are

- Web Services Description Language (WSDL)
- Universal Description, Discovery and Integration (UDDI)
- Simple Object Access Protocol (SOAP)

Additional protocols concern security, management, presentation, orchestration, routing and delivery of messages, and so on. Figure 4.19 is an attempt at presenting the protocol stack for web services.

We now present these various protocols.

4.14.2 Web Services Description Language

WSDL is used to describe the mechanisms for invoking web services according to a client–server model. The description includes the operation that the service does and the protocols (addresses, parameters, messages, and transport) needed to access it on the World Wide Web. This way, WSDL offers a uniform method for calling services (Christensen et al., 2001).

The description of the workflow internal to the state machine that represents the service (orchestration or service composition) is done using WS-BPEL. The exchange of messages among the various BPEL machines according to the rules agreed by the various parties (choreography) is described with the Web Services Choreography Description Language (WS-CDL).

4.14.3 Universal Description, Discovery, and Integration

UDDI describes a universal registry for the services, the businesses that offer them and their processes, organized in multiple ways, such as by name, by category, and by geographic location. The aim is to have a uniform interface so that enterprises can discover potential trading partners and basic information about them irrespective of the technologies used to construct the subsuming directories and their geographic location. Thus, the WSDL files could be distributed over many sites and/or administrative domains, with the registry acting as a logical structure pointing to specific directories where the information resides.

4.14.4 Simple Object Access Protocol

WS-Addressing provides a mechanism for exchanging endpoint logical addresses to be inserted in the SOAP envelopes. WS-Eventing specifies ways to provide asynchronous notification of events (e.g., the events associated with shipping an order) and subscribe to an event notification service (e.g., when a process has completed), which are needed for event-driven service-oriented architectures.

4.14.5 Security

Security Assertion Markup Language (SAML) is a protocol from OASIS that describes, using XML documents, user profiles and the requests for authentication before authorizing the access of an individual or an object to a given service (OASIS, 2005). It establishes equivalencies among administrative domains, each with its own policy for managing rights.

WS-Security specifications, also from OASIS, describe the conceptual and technical basis of integrity and authentication (OASIS, 2006). When a user identifies itself to a SAML server, the latter includes a description of the user's access rights within a SOAP envelope to be forwarded to the servers of the remaining web services within the same administrative domain. Thus, following the initial authentication, the user's access profile is propagated to avoid a second authentication in a process called single sign-on (SSO)—this is also called the federated identity management of users. The SAML structures are transported either using the POST method of HTTP or in SOAP messages. The elements used for authentication are defined using the SAML protocol. WS-Security provides support for multiple security token formats, multiple trust domains, multiple signature formats, and multiple encryption technologies. Among these tokens are X.509 certificates and Kerberos tickets.

The security of XML exchanges themselves is the subject of several recommendations and technical reports from the W3C. XML Encryption W3C, 2002) allows selective encryption by identifying the sections to be encrypted with specialized tags while the processing of signatures is specified in XML Signature Syntax and Processing (W3C, 2008).

The management of encryption keys has its own set of specifications. XML Key Management Specification (XKMS), now in its second version (W3C, 2005), defines the public key infrastructure for web services in two ways: XML Key Information Service Specification (X-KISS) and XML Key Registration Service Specification (X-KRSS). X-KISS provides a client with the possibility of avoiding the computational load of the security tasks such as encryption, signature, and authentication, by delegating them to specialized servers with larger computational power. X-KRSS specifies the protocol to register data about public keys so that they can be made available to all secured services on the web. XACML (Extensible Access Control Markup Language) defines the access rights to a web service and the ways to manage them (OASIS, 2013). It is worth noting that this specification overlaps with another non-W3C specification (WS-Authorization).

TABLE 4.10

Standardization of Web Services

Technical Service	Protocol	Function	Standards Organization
Presentation	WSRP (Web Services for Remote Portlets)	Aggregation of contents among enterprise portals	OASIS
Process orchestration	WS-BPEL (Web Service Business Process Execution Language)	Description and modeling of business processes	OASIS
Choreography	WS-CDL (Web Services Choreography Description Language)	To specify the rules of the message exchanges based on a common agreement of the parties involved	W3C
Transaction	WSTF (Web Services Transaction) that comprise the following: WS-Coordination WS-AtomicTransaction WS-BusinessActivity	The mechanisms for transaction processing and management	OASIS
Messaging	WS-Addressing	Mechanism to find and exchange the logical addresses	W3C
	WS-ReliableMessaging	To provide guarantees for the message delivery in the presence of network failures	OASIS
Distributed management of services	WSDM (Web Services Distributed Management)	End-to-end management of web services	OASIS
Security	SAML (Security Assertion Markup Language) WS-Security WS-Trust Web Services Secure Conversation Language (WS-SecureConversation) Web Services Security Policy Language (WS-SecurityPolicy)	See text	OASIS
	XML Encryption XML Signature Syntax and Processing XML Key Management Specification (XKMS) XML Key Information Service Specification (X-KISS) XML Key Registration Service Specification (X-KRSS) XACML (Extensible Access Control Markup Language) WS-Policy WS-PolicyAssertions WS-PolicyAttachment		W3C

Additional OASIS security specifications include the following:

1. WS-Trust to describe the trust model of web services
2. Web Services Secure Conversation Language (WS-Secure Conversation), which builds on the WS-Security and WS-Trust models to secure the communication among web services
3. Web Services Security Policy Language (WS-SecurityPolicy), which defines how to represent the security capabilities and requirements of a web service in machine-readable form so that the compatibilities of the policies could be determined automatically

The mainW3C security specifications are as follows:

1. The Web Services Policy 1.5—Framework (WS-Policy), which defines a framework for expressing the security constraints and requirements of web services as policy assertions.
2. Web Services Policy 1.5—Attachment (WS-PolicyAttachment). A policy attachment is a mechanism for associating a policy with one or more policy scopes, a policy scope being a collection of policy subjects to which a policy applies.
3. Web Services Policy Assertions Language (WS-PolicyAssertions).

4.14.6 Standardization of Web Services

Table 4.10 summarizes the partition of standardization activities between OASIS and the W3C.

Finally, the Web Services Interoperability Organization (WS-I) (http://www.ws-i.org) is an industry consortium that IBM and Microsoft initiated in 2003. It is focused on providing guidance, recommended best practices, and supporting test tools to ensure the interoperability of web services. It is now part of OASIS.

4.15 Relation of EDI with Electronic Funds Transfer

An enterprise conducting electronic commerce must track its cash flow among its various departments as well as its accounts with its bank. The activities correspond to the following:

1. Internal exchanges within the enterprise from the accounting department to the finance department, concerning, for example, notification of the due date for payment to suppliers and notification of salary payment due date and files for customers invoices.
2. Exchanges between the enterprise finance department and its bank concerning the various payment instruments, such as bills of exchange, check remittances, and credit transfers.
3. Exchanges between the bank and the finance department concerning the account statement, unpaid invoices, and statements of drafts (bills of exchange) and promissory notes.
4. Internal exchanges within the enterprise from the finance department to the accounting department, about notification of payment to suppliers, notification of salary settlement, and reconciliation of internal accounting ledgers and with the bank statement.

Figure 4.20 shows the aforementioned exchanges, with the number referring to the numbered list.

An enterprise carries out three types of accounts reconciliation used to harmonize the records from production and distribution with those of the accounting department (Dragon et al., 1997, pp. 378–380):

- Business reconciliation of the settlements with the invoices
- Financial reconciliation to adjust the actual cash flow with the forecast
- Account reconciliation between the bank statement and the enterprise records

The migration for paper-based to electronic support offers the possibility of joining these loops to facilitate tracking and reduce delays in settlements. Figure 4.21

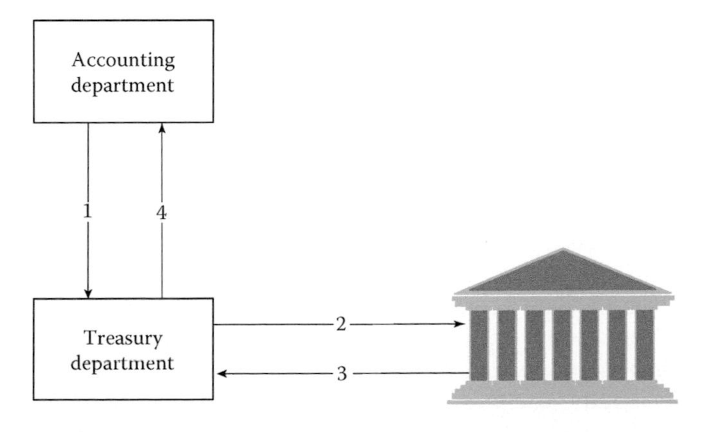

FIGURE 4.20
Exchanges within an enterprise as well as between the enterprise and its bank.

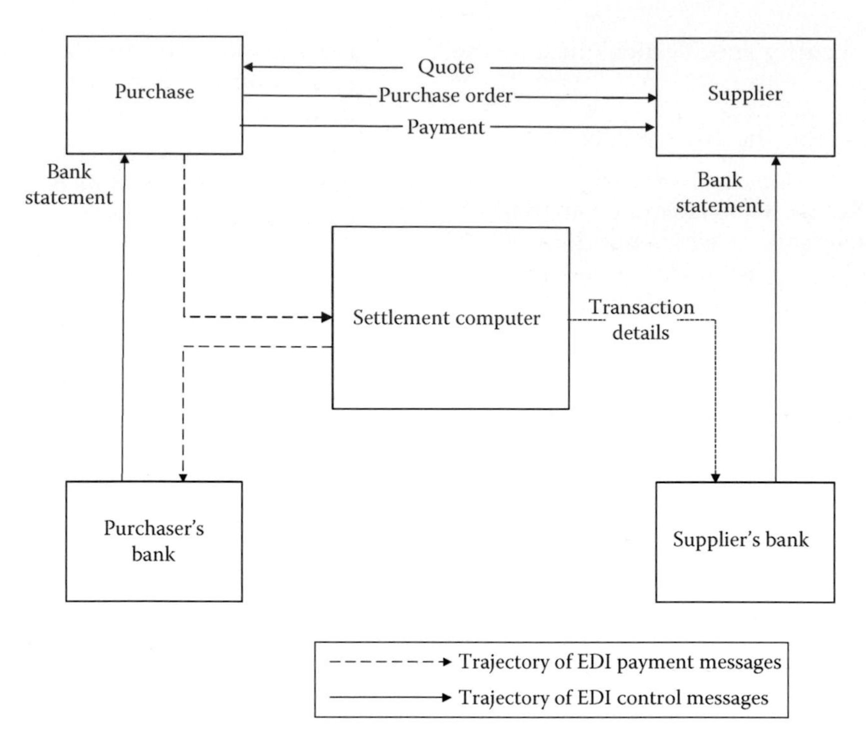

FIGURE 4.21
Trajectories of the EDI financial and control messages.

shows the trajectories of the financial EDI messages as well as those of the EDI control messages.

Each enterprise assembles its transaction records and presents them, in a common format, to the settlement computer through the telecommunication network (in early systems, messengers would carry magnetic tapes). The computer groups all valid requests, sorts them according to the drawer bank, and then sends them to their respective banks to debit the drawer's account. Because all parties are now sharing the same reference numbers, all types of reconciliation will be easier, particularly if they are automated. These exchanges are sketched in Figure 4.22.

Financial EDI has opened the way for automatic processing of fund transfers between an enterprise and its bank. Today, many enterprises utilize credit transfers for paying salaries, compensation, pensions, and benefits, as well as for direct debit notification, because the integration of the bank settlement system with the business circuits allows for better management of the settlement speed.

Manual settlements introduce three types of delays. The first type is due to the postal service; the second is the result of the manual treatment of the check by the beneficiary before depositing it in a bank account. Finally, the third delay comes from the bank settlement. If D is the day that the beneficiary deposits the payment instrument into a bank account, the settlement takes place on day $D + 1$, but the beneficiary's account is only credited on day $D + n$, where n varies according to the banking system and whether the banks are in the same state but is usually between 2 and 4. During this time, the drawee's bank has the amount working for its own account for $(n - 1)$ days.

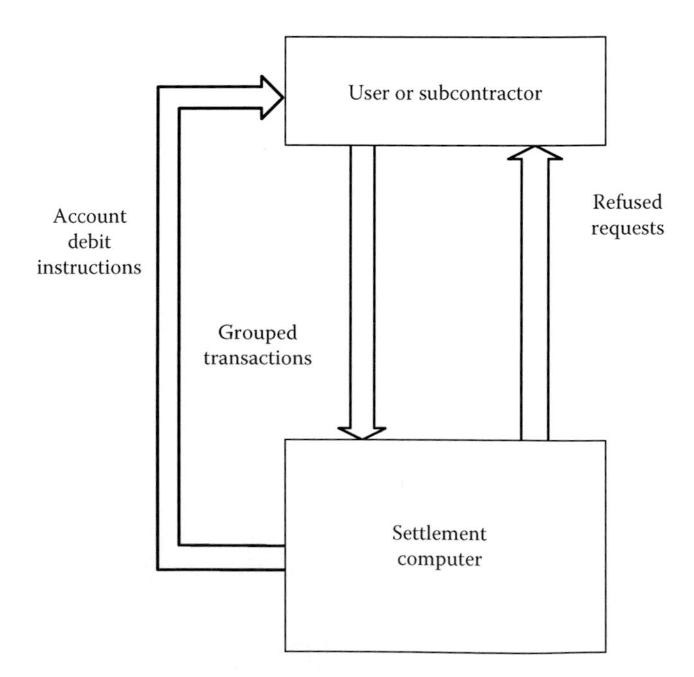

FIGURE 4.22
Exchanges involved in bank settlements.

Although the manual treatment allowed the payer to keep the funds for a longer period, the advantages of the electronic funds transfer would be shared among both the payer and the beneficiary. Each could take the reduction of the float and the faster notifications into account in the management of cash flow, either by modifying the payment date or by negotiating the terms for payment.

4.15.1 Funds Transfer with EDIFACT

Many financial EDIFACT messages have been standardized. These messages exclude interbank exchanges and focus on the exchanges between banks and their clients as well as the exchanges among businesses. This is because settlement architectures differ from one country to another and there is no justification to reconsider them just to conform to EDIFACT specifications. However, the normalized messages can encapsulate proprietary messages to give a common interface among the systems of the various units.

There are three categories of messages for financial EDIFACT:

1. Simple messages such as PAYORD and REMADV that describe a single operation with abundant comments
2. Detailed or "extended" messages such as PAYEXT that allow the juxtaposition of the details to the basic operation, for example, identifications of the invoices, the nature of settlements, and the reasons for nonpayment
3. Multiple messages, for example, PAYMUL, that include several financial transactions of the same kind

The diagram in Figure 4.23 represents the way that these messages can be utilized to effect a credit transfer in response to an invoice. The two Internet "clouds" could be either the same public network (if the exchanges are secured) or one or several private networks or value-added networks (Cafiero, 1991; Hendry, 1993, pp. 125–131).

In this transaction, the information that was initially in PAYORD (or PAYEXT) is copied in CREADV (or CREEXT). The client can also send REMADV directly to the supplier with the content of the CREEXT message.

The FINPAY message is for interbanking credit transfer through the SWIFT network and is used for international payments.

The REMADV message is only needed when the remittance and the payment instructions are separated

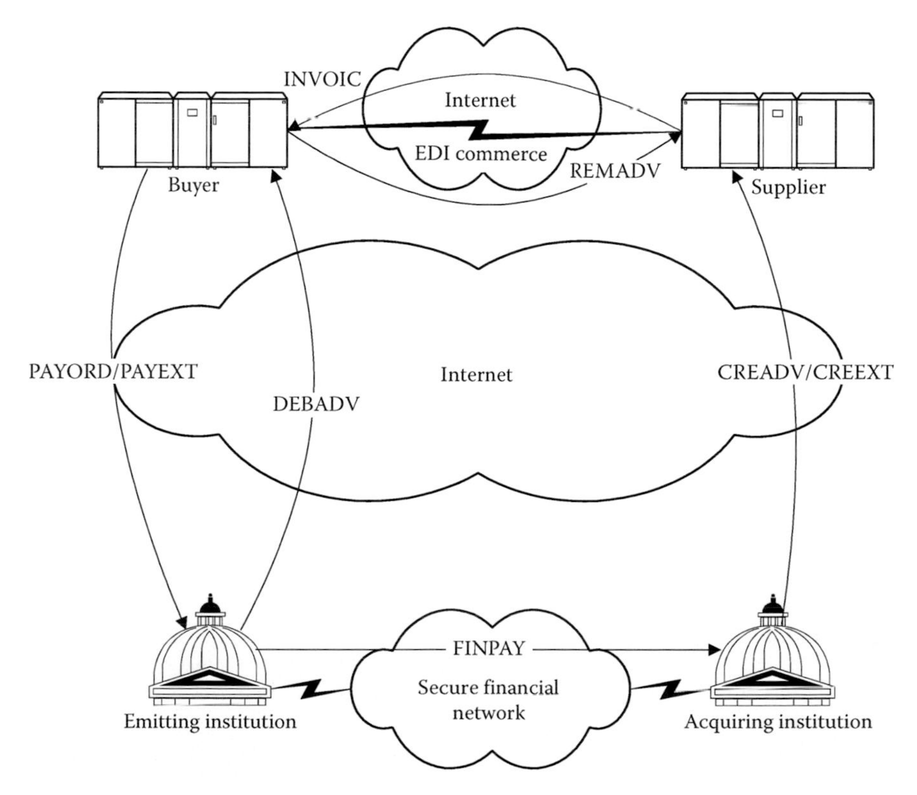

FIGURE 4.23
Credit transfer using the messages of financial EDIFACT.

in time. In such a case, the remittance advice informs the supplier that a payment will be made at the date that was defined in the supply contract. The figure sketches a bare-bones illustration of the operation and does not take into account all potential problems, such as the lack of funds and identification errors, which will trigger the exchange of the appropriate messages.

4.15.2 Fund Transfers with X12

Fund transfers with X12 follow the preceding outline with the Transaction Set 820 replacing the messages PAYORD/PAYEXT and DEBADV. Following each exchange, the recipient may use the Transaction Set 997 to inform the sender that the received message is syntactically correct, thereby ensuring the regular progress of each stage of the transaction. If the credit transfer cannot be done, the Transaction Set 827 is sent. Of course, all these messages can be secured with the procedures defined in X12.58 or those that are based on S/MIME. In a pilot experiment, the time needed for the transmission and the processing of these exchanges, assuming error-free transmissions, varied between 12 minutes per 100 instructions and 58 minutes for 1000 instructions instead of the average 2–4 days for manual settlement (Segev et al., 1996).

The delay between the order of a stock trade and its settlement increases the transaction costs and the financial risks. The Security Industry Association (SIA) in the United States set June 2005 as the date when all securities trades must be completed by the end of the day following the trade, down from the previous 3 days (the so-called $T + 1$ settlement). This concept of *Straight-Through Processing* (STP) covers activities from the front office, including gathering intelligence, till the final settlement passing by order taking and execution.

4.15.3 Financial Dialects of XML

Market pressures, competition, and in some cases regulatory demands have combined to create a situation where independent efforts to derive XML-based descriptions of financial transactions have multiplied leading to a large number of languages, some of which are listed in Table 4.11.

Companies report their financials in a standard form with XBRL, which is the mandatory format in the United Kingdom since 2010. One of the widely used XML derivatives is XMLPay developed by OASIS. The typical user of XMLPay is an Internet merchant or merchant aggregator who wants to dispatch credit card, corporate purchase card, Automated Clearing House (ACH), or other payment requests to a financial processing network. Using the data-type definitions specified by XMLPay, an XMLPay client sends requests to

TABLE 14.11

Some XML Derivatives for Financial Services

Acronym	Title	Remarks
FinXML	Fixed Income Markup Language	—
FIXML	Financial Information Exchange Markup Language	—
FpML	Financial products Markup Language	Over-the-counter (OTC) derivatives
FXML	Financial Exchange Markup Language	—
IFX	Interactive Financial Exchange	Specifications for online presentation of financial information, electronic bill presentment, electronic banking, and so on
IRML	Investment Research Markup Language	—
MDDL	Market Data Definition Language	Description of marketing research
NewsML	Electronic News Markup Language	Multimedia financial news
NTM	Network Trade Model	Stock and risk evaluation
OFX	Open Financial Exchange	Online exchange of payment messages among financial institution; used for electronic billing
RIXML	Research Information Exchange Markup Language	Document indexing
STPML	Straight-Through Processing Extensible Markup Language	—
TWIST	Transaction Workflow Innovation Standard Team	Wholesale financial transaction and account administration; financial supply chain (ordering, invoicing, financing, etc.)
XBRL	Extensible Business Reporting Language	Reporting of companies financials
XFRML	Extensible Financial Reporting Markup Language	—
XMLPay		Used for processing a broad range of web-based payment types

an associated XMLPay-compliant server. The specifications relate to merchant registration and configuration, merchant transaction, and report generation. PayPal, for example, uses that standard as an interface to web merchant systems.

It goes without saying that the presence of many specifications in the same sector is a major cause of concern because it increases the cost and difficulties of interoperability to the point of putting the benefits of electronic integration into question. This, in turn, has spurred initiatives to bring some order to the situation by paring the number of specifications. In 1999, the technology group for the Financial Services Roundtable, the OFX (Open Financial Exchange) and Gold specifications were merged into the IFX (Interactive Financial Exchange) to provide a unique way for describing online presentations of financial information, electronic bill presentment, electronic banking, and so on. Banks have adopted IFX to allow the communication among their ATMs irrespective of the hardware.

In 2003, IFX, the Open Applications Group (OAGi), SWIFT, and TWIST (Transaction Workflow Innovation Standards Team) collaborated to launch the International Standards Team Harmonization (ISTH) initiative to develop and promote a common "core payment XML kernel" that banks could use globally. In 2004, ISO 20022 was agreed to define the Universal Financial Industry (UNIFI) message scheme.

ISO 20022 provides a framework for standard protocols supporting financial services. It provides a common development platform based on UML modeling to define the business processes, the business transactions, and the associated information flows in a syntax-independent manner. The methodology also provides a set of XML design rules to convert UML messages into XML schemas. This focus is in contrast with the focus of UN/CEFACT, which includes the physical supply chain and the associated information flows. ISO 20022 is also being used for transaction in the SEPA (Single Euro Payments Area) program.

Unfortunately, the financial data elements defined in the repository of ISO 20022 are not necessarily consistent with the core components used in ebXML. This is because standardization of the messages exchanges for the purchase of goods and services is typically done independently from that of the underlying financial activities that support the physical supply chain. Also, ISP 20022 messages are wrapped in an XML wrapper that increases the total length of the message and, as a consequence, the bandwidth needed. This may form a bottleneck with millions of real-time transactions. Furthermore, the harmonization activities of ISTH covered only the aspects of the financial supply related to payments. To resolve this potential inconsistency, UN/CEFACT and ISO have defined a path for convergence under the UNIFI initiative (Potgieser, 2010). Accordingly, the financial core components in the UN/CEFACT registry will be derived from the data dictionary determined through UNIFI according to ISO 20022.

4.15.4 Electronic Billing

Electronic billing is called electronic bill payment and presentment (EBPP) or Electronic Invoice Payment and Presentation (EIPP). This is a logical step after the dematerialization of commercial exchanges and has been around since the 1980s through the Minitel. While this feature appeals to large bill producers (utilities and telecommunication companies), it may also interest the public at large, because it gives each subscriber the capability of viewing their bill on the screen and of paying it online. From a business-to-business perspective, the electronic links among buyers, suppliers, and their respective bankers provide firms with easier tools for financial management and inventory management.

The establishment of electronic billing on the Internet requires a larger opening of the information systems of the parties and uniformity in the message formats used. This is to group bills from different suppliers and consolidate them in a single bill that will be presented to the firm in a form that is compatible with its information system. These steps are shown in Figure 4.24.

The payment phase depicted in Figure 4.25 relies on collectors associated with credit institutions to collect the payments (with credit transfers, electronic checks, etc.) and deposit them in acquirer banks.

CheckFree is an aggregator that has captured a significant share in the online billing market of the United States. Billserv is an intermediary between billers and bill presentment aggregators, electronic banking systems and Internet portals. Billserv consolidates customer billing information from multiple billers and delivers it to aggregators for presentment to consumers. eServ, a service provided it provides, allows billers to outsource their EBPP services and rely on Billserv to manage billing information and deliver it to the many online front ends that consumers may choose to view and pay their bills. In the consumer space, MasterCard uses OFX in its Remote Payment and Presentment Service (RPPS) offer to banks. Finally, the operators of the settlement networks ACH and CHIPS compete to deliver electronic bills and payments on their respective networks.

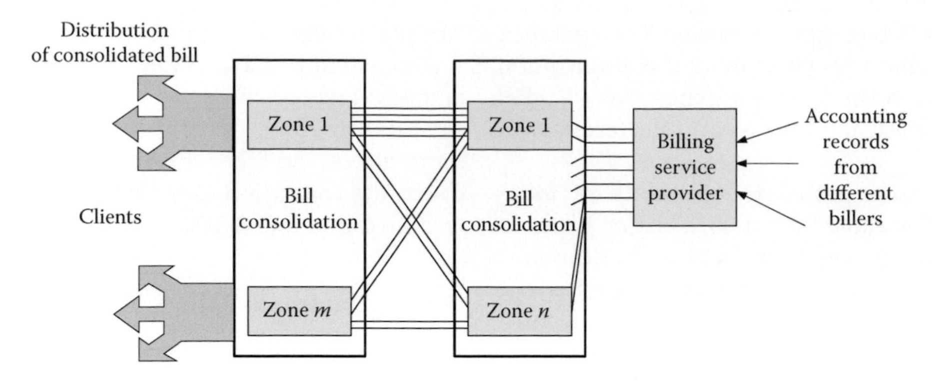

FIGURE 4.24
Grouping and consolidation of electronic bills.

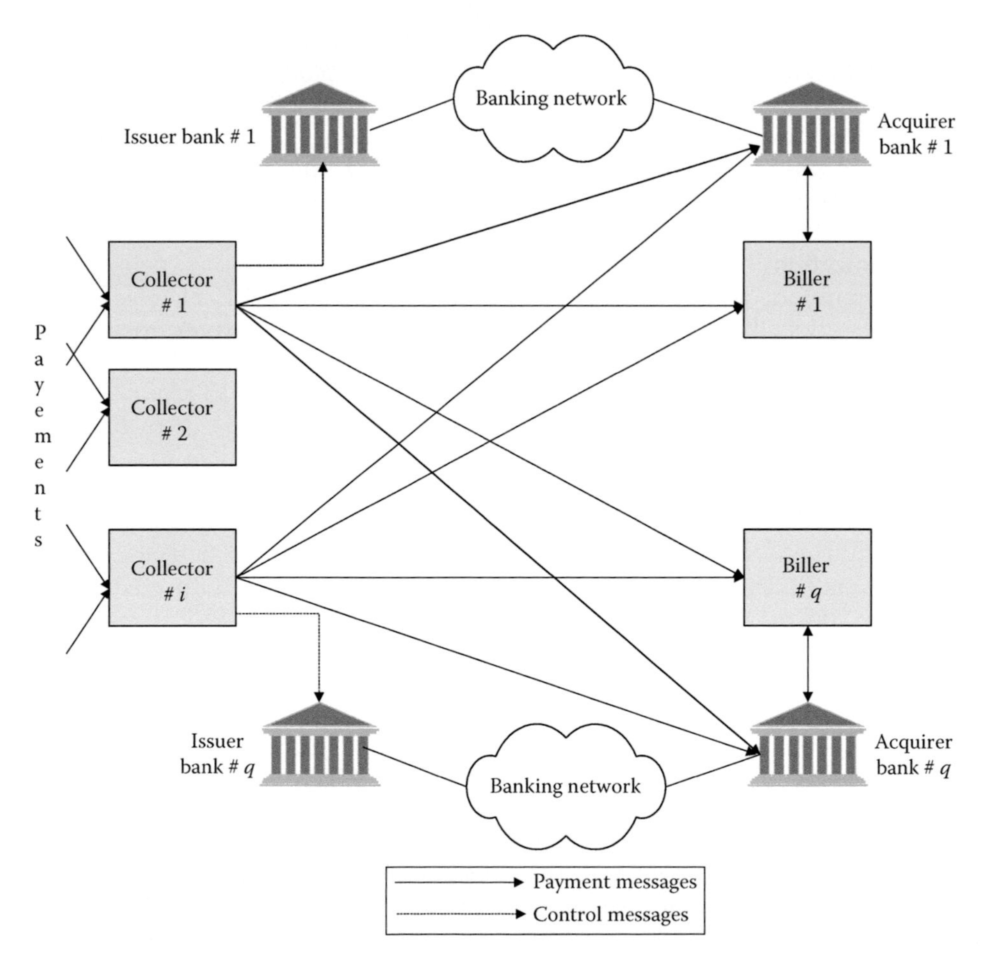

FIGURE 4.25
Online payment of electronic bills.

4.15.5 An Example for EDI Integration with Business Processes

Figure 4.26 depicts a simplified synthesis of all the EDI elements in a commercial transaction. The basis is the real-life efforts of a telephone operator (Bell Atlantic, today it is Verizon) to streamline its internal process and reduce processing cost of commercial invoices by at least 20% (Sivori, 1996). Of course, in this simplified view, many elements of a real transaction have been discarded for the sake of clarity.

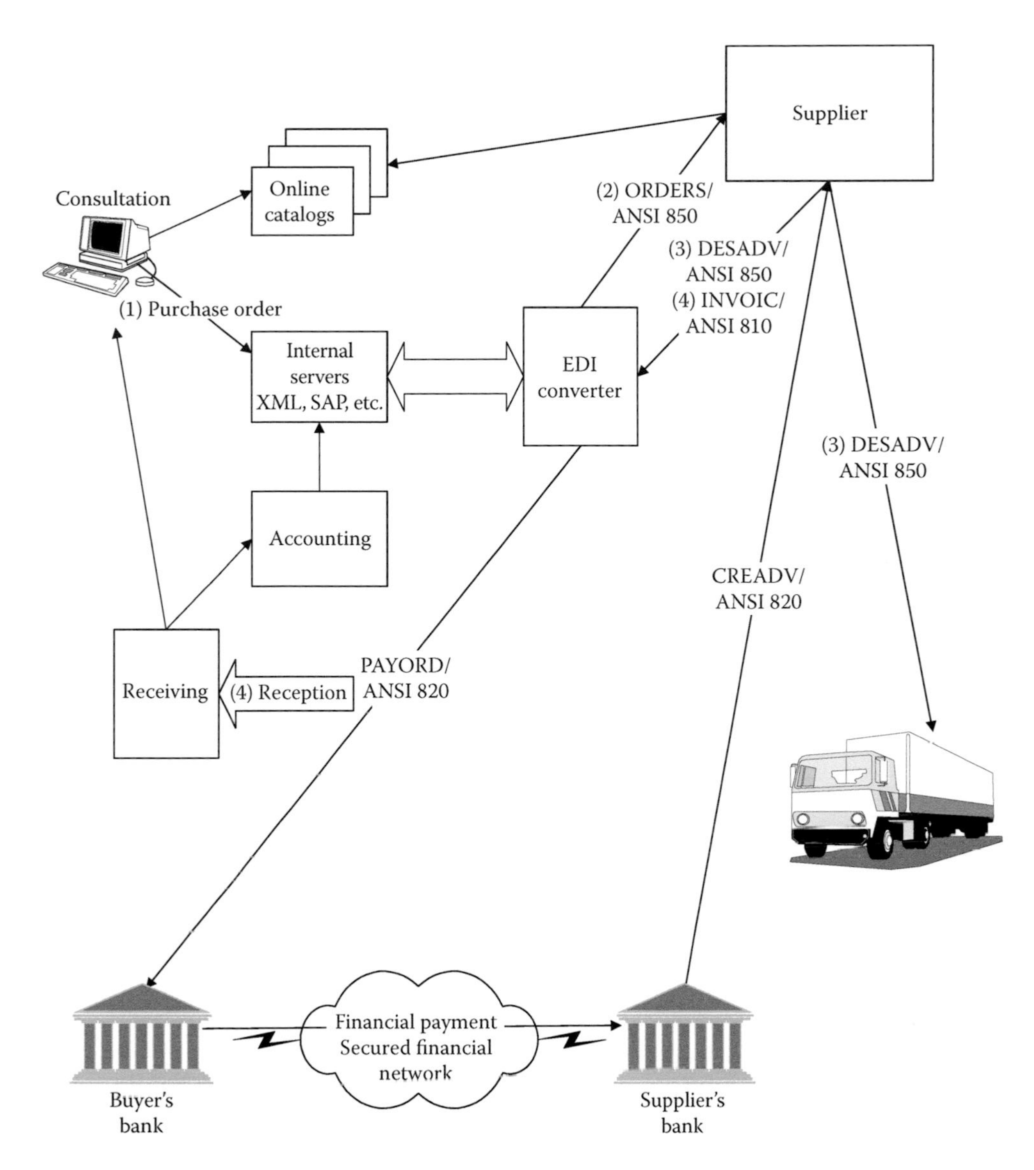

FIGURE 4.26
Synthesis for EDIFACT and X12. The numbers in parentheses show the sequence of the messages.

4.16 Summary

The dematerialization of business-to-business traffic started with tasks related to procurement. With the advancements in information technology (hardware and software) and in telecommunications, the focus expanded gradually to other technical or managerial areas to coordinate the separate elements of data or applications that were assembled as need arose to satisfy specific customer requests.

Business-to-business electronic commerce ensures that transaction records are exchanged among computer applications and processed according to business rules and regulations without manual intervention. This process eliminates the overhead and errors from repeated manual data entry at each step, thus increasing the speed and the efficiency of operations. Thus, the success of business-to-business electronic commerce depends on adherence to common conventions and agreements to insure the flow of information and its correct interpretation throughout the supply chain. This includes the integration of external inputs (orders, delivery notices, payments, etc.) with the internal information of the enterprise such as inventory controls or other management reports. Agreements among parties must ensure that the various business models are consistent to facilitate end-to-end Straight-Through Processing, that is, automated processing without manual intervention.

To achieve this goal, the message formats and the data they enclose must be structured according to common rules that are technology neutral and implementation independent. However, by modifying the course of the information flows within the enterprise and by establishing collaborative relations among separate organizations, business-to-business e-commerce causes a major reorientation in the ways information is managed.

Enterprise systems integration started as a vehicle and a methodology to bring disparate systems together through a common front end and mask the underlying computer and communication infrastructure. It has evolved into a systematic redesign of the information architecture, within enterprises and across enterprises, to ensure the flexibility and extensibility of the applications by design, in addition to their interoperability. Both aspects coexist in initiatives for systems integration, even when they are not explicitly stated. Initially, the focus was on data structuring to tie individual computer systems and avoid the cost and delays due to rekeying the same data repeatedly—with a chance of introducing errors.

At each technology transition, the established conventions are subverted, and economic chains are thrown into turbulence until new rules settle. As computers became essential to the daily routines of enterprises, message standardization leads to the traditional EDI protocols of X12 and EDIFACT. Likewise, the transition to open networks and distributed computing is underlining the need for new designs and operational techniques to take advantage of the distributed and open nature of the web. It takes time for the new ways to mature sufficiently so as to be deployed securely and reliably in real situations.

The magnitude of the investments that large enterprises had already committed to the legacy EDI solutions makes the coexistence of several generations unavoidable. No enterprise can afford to rethink its mode of operation every few months to follow the latest fashion. Without stability, it is highly unlikely that a commercial network would be able to survive. Solutions that use XML will gradually replace existing solutions as the infrastructure is renewed and as new requirements arise that the legacy systems cannot meet. Standardization of the information flow is essential to avoid ending up with a complicated architecture.

Currently, there are many approaches, such as ebXML, web services, or the various XML solutions in the financial section, that combine the same building blocks to end with similar but different answers. Just like in the case of traditional EDI, several years of experimentation are needed to stabilize the situation. The integration of the internal processes of several enterprises requires substantial prior thinking, particularly because the question of security over open networks like the Internet is not completely resolved. Based on past experience, we can expect that pressures to rationalize business operations will drive the various interoperability agreements that are essential for any new model for universal business-to-business commerce.

Questions

1. What is the main difference between EDIFACT and X12?
2. How can mobile commerce be integrated into the supply chain?
3. What are the main benefits and main costs related to business-to-business e-commerce?
4. How can business-to-business electronic commerce simplify the activities of the supply network?
5. Describe how business-to-business e-commerce is applied in the automotive industry.
6. Why are standards needed for business-to-business electronic commerce?
7. Describe the main standardization organizations in business-to-business electronic commerce.
8. What factors can affect order fulfillment in business-to-business e-commerce?
9. Compare and contrast traditional EDI and ebXML.
10. Compare the limitations and advantages of the platforms for traditional EDI as compared to web-based platforms.
11. Compare and contrast business-to-business and business-to-consumer e-commerce in terms of procurement processes, order sizes, market sizes, and workflow.

5

Transport Layer Security and Secure Sockets Layer

Secure Sockets Layer (SSL) and Transport Layer Security (TLS) are two widely used protocols to secure exchanges at the transport layer between a client and a server. SSL Version 1.0 was used internally within Netscape. Version 2.0 was released to the public in 1994 and integrated into the Netscape Navigator. Version 3.0 corrected deficiencies found in Version 2.0 and was the basis for RFC 2246 that defined TLS 1.0 in 1999 (Freier et al., 1996; Rescorla, 2001). TLS 1.0 was allowed for use to protect U.S. federal data; in contrast, SSL v3 was tolerated in limited, low-risk circumstances, such as to access vendor sites that did not support TLS (Chernick et al., 2005, p. 21, n. 20). TLS 1.0 was next adapted to wireless communication as Wireless TLS (WTLS). Two updates of TLS in 2006 and 2008, respectively, TLS 1.1 and 1.2, are defined in RFCs 4346 and 5246; these updates included a variety of countermeasures and workarounds to reported security threats. Finally, in 2011, RFC 6151 officially withdrew SSL Version 2.0. In parallel, the Datagram Transport Layer Security (DTLS) of RFC 4347 was defined to run on top of unreliable transport protocols; DTLS was later updated in RFC 6347. WTLS is discussed in Chapter 6.

This chapter focuses on the exchanges and during session and connection establishment for both SSL and TLS, explaining the main differences between the successive versions. Some (but not all) of the attacks against TLS/SSL are discussed. In fact, abundant information concerning possible attacks and how to defend against them is available. As usual, the primary references remain the standard documents mentioned in the text.

5.1 Architecture of SSL/TLS

Figure 5.1 shows the location of SSL/TLS in the TCP/IP protocol stack. It is seen that they operate above the transport layer and below the application layer of the OSI (Open systems interconnection) reference model. The handshake for session and connection establishment is at the session layer. During that handshake, the various cryptographic parameters are established to be used in the presentation layer to support various application protocols, such as S-HTTP (Secure HyperText Transfer Protocol) of RFC 2660 (1999). Because SSL/TLS are readily incorporated in browsers, they have totally eclipsed the SET (Secure Electronic Transaction) protocol, which was specifically designed to secure bank card transactions. SET is presented in Chapter 7.

Figure 5.2 presents the correspondence between SSL/TLS and other Internet protocols. Any transport protocol that, like TCP, offers reliable transmission can take advantage of the security services of SSL/TLS. In contrast, SSL/TLS do not protect UDP (User Datagram Protocol) exchanges because UDP does not offer a reliable transport service. In this case, the flow interruptions due to IP packet losses may be incorrectly interpreted as security breaks that would force disconnection of the communication. As a consequence, SSL/TLS do not protect protocols that ride on UDP such as Simple Network Management Protocol (SNMP), network file system (NFS), Domain Name Service (DNS), and voice-over IP (VoIP). As mentioned earlier, the DTLS of RFC 4347 can secure the traffic of these applications with mechanisms such as explicit sequence numbers, retransmission timers, and replay detection.

The Internet Assigned Numbers Authority (IANA) has allotted specific IP ports to some applications for communicating with SSL. These ports are given in Table 5.1. The applications listed in Table 5.2 use the shown port by widespread convention, that is, without an official attribution.

5.2 SSL/TLS Security Services

SSL/TLS define a framework to use the encryption and hashing algorithms that have been negotiated between the two parties to offer three security services: authentication, integrity, and confidentiality. With the help of a digital signature, it is possible to provide the elements necessary for a nonrepudiation service. This flexible structure is open to the integration of new algorithms as they are adopted.

SSL/TLS combine the operations of key establishment, confidentiality, signature, and hashing into one package denoted as cipher suite. There are now over 300 standardized cipher suites defined in various documents. In TLS 1.2, the mandatory cipher suite consists of

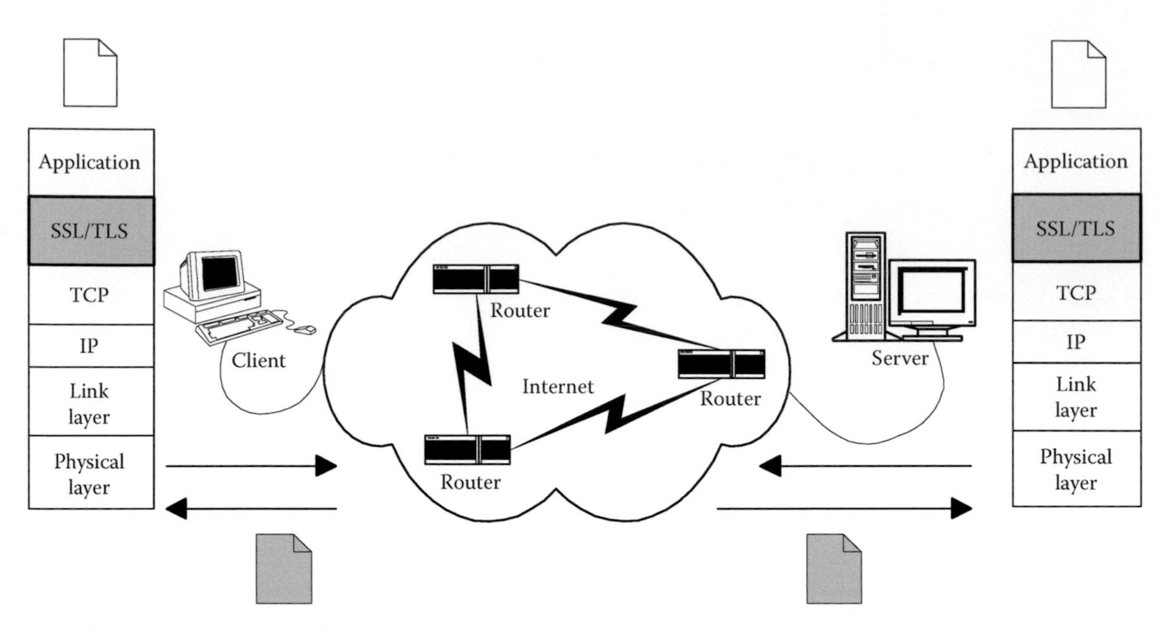

FIGURE 5.1
Position of SSL/TLS in the TCP/IP stack.

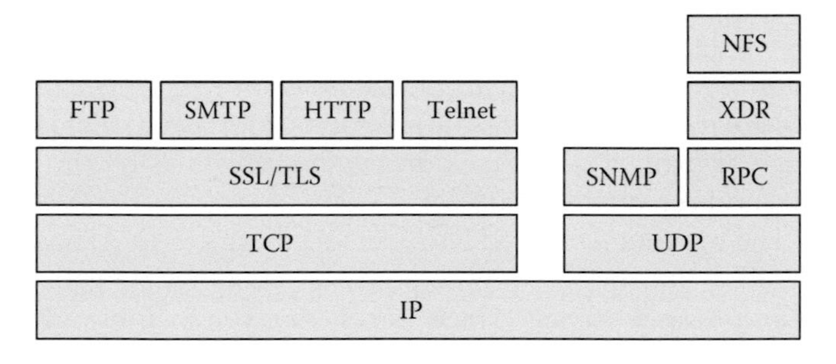

FIGURE 5.2
Relation of SSL/TLS with other Internet protocols (RPC, Remote Procedure Call; XDR, External Data Representation).

TABLE 5.1
IP Ports Assigned to Secure Applications with SSL/TLS

Secure Protocol	Port	Nonsecure Protocol	Application
S-HTTP	443	HTTP	Secure request–response transactions
SSMTP	465	SMTP	E-mail
SNNTP	563	NNTP	Network news
SSL-LDAP	636	LDAP	Light version of X.500
SPOP3	995	POP3	Remote access of mailbox with message download

Note: S-HTTP, Secure HyperText Transfer Protocol; SSMTP, Secure Simple Mail Transfer Protocol; SNNTP, Secure Network News Transfer Protocol; LDAP, Lightweight Directory Access; POP, Post Office Protocol.

TABLE 5.2
IP Ports Used without Formal Attribution to Secure Applications with SSL/TLS

Secure Protocol	Port	Nonsecure Protocol	Application
ftp-DATA	889	ftp	File transfer
ftps	990	ftp	Control of file transfer
IMAPS	991	IMAP4	Remote access to mail box with or without downloading of messages
TELNETS	992	Telnet	Remote access to a computer
IRCS	993	IRC	Internet chat; text conferencing

Note: IMAP, Internet Message Access Protocol; IRC, Internet Relay Chat.

RSA for certificate signature, AES for encryption with a key of 128 bits in the cipher block chaining (CBC) mode, and SHA-1 for integrity verification.

5.2.1 Authentication

Authentication uses a certificate that conforms to ITU-T Recommendation X.509 Version 3. It takes place only at the session establishment and before the first set of data has been transmitted. This service was optional in Version 2 of SSL and became mandatory for the server in Version 3.0 and for TLS. The server can require the client to authenticate itself and may refuse to establish the session because the certificate is lacking.

Key establishment or key exchange is the process of establishing a shared secret key to be used for encryption. During key establishment, authentication can be static or dynamic. In static authentication, the public encryption key is extracted from the certificate of the other side, typically the server's certificate. In dynamic authentication, the server certificate contains a signature key used to sign a temporary encryption key to protect its integrity. The pair of temporary keys are used to protect the data exchanged to construct the symmetric encryption key of the session.

Dynamic authentication was first used to avoid the limit of the RSA encryption keys to 512 bits, as was required by the U.S. export restrictions as defined in the International Traffic in Arms Regulation (ITAR). With dynamic authentication, a longer (1024-bit) key could be used to sign the temporary 512-bit key. Another advantage of dynamic authentication is that it avoids the use of a static key that, if compromised, would expose every SSL/TLS session that was ever established with that key.

SSL/TLS use RSA and the Diffie–Hellman algorithm for key exchange. There are three variants of Diffie–Hellman communications—static (or fixed), ephemeral, or anonymous:

1. In the fixed Diffie–Hellman, the public parameters are contained in the server's certificate. The client provides its corresponding public parameters either in a certificate or through message exchange, if it does not possess. In that case, the server provides a static Diffie–Hellman key while the client's key is ephemeral or temporary.
2. The ephemeral Diffie–Hellman is a technique to create one time secret keys. Each side sends its Diffie–Hellman public keys signed with its private RSA key. The recipient uses the corresponding public key to verify the certificate. Ephemeral keys achieve *perfect forward secrecy* (PFS) because the temporary keys on both sides cannot be recovered once they are destroyed (provided that the numbers used are unpredictable so that none of the generating parameters for one key are used to generate another key). Thus, the compromise of a session key or long-term private key does not expose any earlier session key.
3. In anonymous Diffie–Hellman, the public Diffie–Hellman parameters are exchanged without authentication. This method, however, is susceptible to man-in-the-middle attacks.

In 1999, RFC 2712 added the Kerberos (V5) authentication protocol to the TLS protocol.

SSL and TLS 1.0 used previously classified algorithms approved for cryptographic applications on the Fortezza PCMCIA card (National Institute of Standards and Technology, 1994). The Fortezza card uses a variant of the Diffie–Hellman algorithm called the Fortezza Key Exchange Algorithm (KEA) for key agreement and a block cipher algorithm called SKIPJACK for encryption. Signatures are computed with DSA, and message digests are calculated with the SHA-1 hash function. The integration of SSL/TLS with Fortezza caused problems (Rescorla, 2001, pp. 126–128), and this led to the support of Fortezza in TLS 1.1.

In TLS 1.2, elliptic curve cryptography was introduced for key exchange and signature.

5.2.2 Confidentiality

Message confidentiality is based on the utilization of the symmetric encryption algorithms, whether stream encryption or block encryption. The same algorithm is used on both sides, but each side uses its own key, sharing it with the other party. On the client side, the key is called the `client _ write _ key`, and it is called the `server _ write _ key` on the server side. Initially, the algorithms that could be used with SSL were DES (Data Encryption Standard), triple DES (3DES), DES40 (which is the same as DES but with a key size of 40 bits), RC2, RC4 with a key of 128 bits (RC4-128) or of 40 bits (RC4-40) for export outside the United States, IDEA (International Data Encryption Standard), and the algorithm SKIPJACK of Fortezza. As a historical note, DES40 and RC4-40 used keys of 40 bits to comply with the restrictions that the U.S. federal government had imposed on the export of cryptographic software; these restrictions were lifted in 1999.

RC4 was not approved for use in U.S. federal applications, and the minimum key length for all federal server certificates with RSA keys is 1024 bits (Chernick et al., 2005, pp. 22–23). However, many browsers use cipher suites with RC4 because it is extremely fast when implemented in software. Also, from 2010 onward, key sizes larger than 1024 bits are required in applications approved by the U.S. federal authorities (Chernick et al., 2005, p. 26).

RFC 4132 added the Camellia cipher suites to TLS 1.0. As mentioned in Chapter 3, Camellia is a block symmetric encryption with a block size of 128 and 128, 192, and 256-bit keys.

In 2006, TLS 1.1 added the Advanced Encryption Standard (AES) in its cipher suites and, as mentioned earlier, dropped Fortezza (RFC 4346).

TLS 1.2 supports elliptic curve cryptography (ECC) defined in ANSI X9.62 (2005) and Authenticated Encryption with Associated Data (AEAD) but removed both DES and IDEA. In AEAD, the plaintext is simultaneously encrypted and integrity protected. The input may be of any length; the ciphered output is generally larger than the input because of the integrity check

TABLE 5.3

Summary of Security Algorithms of SSL and TLS

Function	Algorithms		
	SSL/TLS 1.0	TLS 1.1	TLS 1.2
Key exchange	RSA, Diffie–Hellman, Fortezza, Kerberos	RSA, Diffie–Hellman	RSA, Diffie–Hellman, Elliptic Curve Diffie–Hellman
Signature	Digital Signature Algorithm (DSA), RSA	Diffie–Hellman	RSA, Digital Signature Algorithm (DSA), Elliptic Curve RSA. Elliptic Curve Digital Signature Algorithm (ECDSA)
Stream symmetric encryption	RC4 with keys of 40 or 128 bits	RC4 with 128 bits, Camellia	RC4 with 128 bits
Block symmetric encryption	DES, DES40, 3DES RC2 with 128 bits, IDEA, SKIPJACK	DES, 3DES, RC2, IDEA, AES with 128 and 256 bits, Camellia	3DES, AES with 128 and 256 bits
Authenticate Encryption with Associated Data (AEAD)	—	—	AES CGM, AES CCM
Hashing	MD5, SHA-1	MD5, SHA-1	MD5, SHA-1 with keys of 64, and 256

value. Backward compatibility with SSL Version 2.0 was removed as a requirement.

Thus, TLS 1.2 offers three encryption methods:

1. CBC mode encryption using the block ciphers 3DES or AES.
2. Stream encryption using the RC4 stream cipher with a 128-bit key.
3. AEAD in one of two modes of operation: the Galois/Counter Mode (GCM) or the counter with cipher block chaining and message authentication code (CCM) as described in RFC 5116. CCM and GCM are based on the AES algorithm (Dworkin, 2004, 2007) with keys of 128 or 256 bits. AES-CGM, however, reduces the performance.

Later in this chapter, we will see that the CBC mode of encryption in TLS and stream encryption with RC4 are both vulnerable to attacks.

5.2.3 Integrity

The integrity of the data is assured with hash functions that employ the HMAC procedure to ensure stronger protection against attacks (Bellare et al., 1996). The hash functions utilized can be either Secure Hash Algorithm (SHA) or MD5. The digest is treated by a series of operations that depend on a secret key, and the result is called the message authentication code (MAC).* This operation also serves as authentication because knowledge of the secrets utilized in the encryption of the digest is restricted to the two parties. Up to TLS 1.2, HMAC-MD5 and HMAC-SHA1 were used, and the hash function and the encryption functions were defined simultaneously according to the cipher suite. With TLS 1.2, the client can specify the hash function to be used in the digital signature independently from the encryption algorithm. In addition, the use of a 256-bit key for hashing calculations became the default (HMAC-SHA256), yielding a MAC of 32 octets. Thus, with TLS1.2, SHA can be used with key lengths of 224, 256, 384, and 512 bits.

In 2011, RFC 6151 indicated that MD5 is no longer acceptable where collision resistance is required, such as in digital signatures, but legacy applications can still use HMAC-MD5.

In TLS 1.2, a certificate containing the key for one signature algorithm may be signed using a different algorithm. The algorithm had to be the same in earlier specifications. Also, integrity verification may use SHA with a key of 256 bits in relevant cipher suites.

5.2.4 Summary of Security Algorithms

Table 5.3 summarizes the various securities supported by SSL/TLS.

Note that following the revelation in 2013 of the massive surveillance of Internet traffic by various governments, many companies and organizations responded by increasing the size of their TLS/SSL encryption keys, typically from 1024 to 2048 bits.

5.2.5 TLS Cryptographic Vulnerabilities

Even though TLS 1.2 specifies two general classes of encryption methods, recent investigations have shown that some have vulnerabilities that affect the security of the transaction. This section describes two categories of attacks. The first exploits the initialization vector used in the CBC mode of block encryption and is given the name of BEAST (browser exploit against SSL/TLS). The second vulnerabilities affect cipher suites using RC4 due to statistical biases in the encryption.

* Note that SSL/TLS denote two different computations as MAC: the digest and the HMAC.

5.2.5.1 Initialization Vector Attack (BEAST Attack)

In the CBC mode of block encryption, SSL 3.0 and TLS 1.0 generate the initialization vector of the first block. For messages longer than one block, the subsequent initialization vectors are chained, that is, each new vector is the last ciphertext from the previous message.

The chaining of initialization vectors across messages is considered a weak variant of the CBC mode of encryption, which typically requires a fresh random initialization vector for each block. Chaining initialization vectors allow the full plaintext to be recovered, under the following conditions (Bard, 2004; Duong and Rizzo, 2011):

1. The attacker knows which plaintext block contains the information, such a password or an HTTP session cookie.
2. The ciphertext blocks are known.
3. The attacker knows the values of the initialization vector for the next block.
4. The attacker is able to choose the next plaintext to be encrypted in the long message. This is possible through a malicious plug-in or a JavaScript in the user's browser.

Under the aforementioned conditions, the attacker can use the chosen plaintext to validate repeated guesses as to the value of a particular plaintext block until the correct one is identified.

A workaround is to use a dummy zero-length message (i.e., with only the HMAC and the padding) just to generate an initialization vector that an attacker may not know in advance. Many servers, however, consider an empty message as a signal for an end of file, so this workaround is not full proof.

TLS 1.1 and TLS 1.2 are not vulnerable to this attack, because an initialization vector is generated anew for each block.

5.2.5.2 The RC4 Statistical Bias Attack

The use of RC4 increased following the disclosure of the BEAST (browser exploit against SSL/TLS) attack in 2011 as a protection against the vulnerable CBC suites in SSL/TLS 1.0. As a consequence, it was estimated that about 50% of all TLS traffic was using RC4 in 2012.

Unfortunately, in 2013, it was shown that, under certain conditions, the known statistical biases in the RC4 encrypted text could be used to recover the first 220 octets after the Finished message of the Handshake protocol (which is 36 octets long), even without prior knowledge of the plaintext. First, the same plaintext must always be in the same position of the original text. In particular, the encrypted scream may include the password to access a mail server or encoded HTTP cookie headers, which always start with the string "Cookie": and end with a new line character. Also, the same plaintext must be encrypted with different keys: it is estimated that between 2^{28} and 2^{32} independent encryptions must be available to perform the statistical analysis (AlFardan et al., 2013).

To generate such a large number of independent encryptions, a browser could be infected with a JavaScript malware to send HTTPS requests repeatedly to a remote server. Because the corresponding cookies are automatically included in each of these requests in a predefined location, these requests can be targeted for analysis. Another possibility is to terminate TLS session after the encrypted cookie has been transmitted, thus forcing the establishment of a new TLS session for the next HTTPS request. In the case of an e-mail server, resetting the TCP connection between the e-mail client and the server after authentication may trigger some client configurations to resume the TLS session and retransmit the encrypted password.

While at the moment the attack may not yet be practical in all situations, other unknown vulnerabilities may exist. In fact, as mentioned earlier, the U.S. federal authorities have never approved RC4 for cryptographic applications and allowed its use in very limited circumstances, such as ordering supplies from vendors that support nothing stronger than RC4 (Chernick et al., 2005, p. 19, n. 17, p. 22, n. 22). As a consequence, there is a general consensus to avoid RC4 completely in TLS.

5.2.5.3 Forging X.509 Certificates

Some companies, such as Packet Forensics, market to law enforcement and intelligence agencies the capability to use forged certificates issued from any one of more than the 100 trusted certificate authorities in operation. This would lull the user in trusting the fake server in a classic man-in-the-middle attack. Nongovernment attackers could also use this technique to give the victim a false sense of security at will, particularly for electronic commerce transactions (Single, 2010).

5.3 SSL/TLS Subprotocols

The initial SSL/TLS protocols are four subprotocols: Handshake, Record, ChangeCipherSpec (CCS), and Alert; in 2012, the Heartbeat subprotocol was added. Figure 5.3 depicts the arrangement of the various components. It shows that the protocol Record is on top of the transport layer, while the other three protocols are between the application and the Record layer.

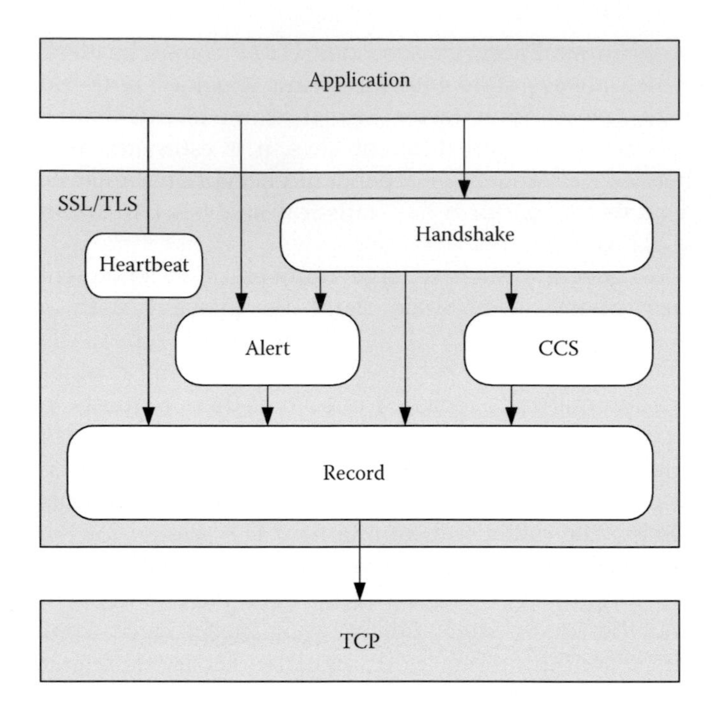

FIGURE 5.3
Protocol stack for the SSL/TLS subprotocols.

The Handshake protocol uses asymmetric cryptography for the authentication of the communicating parties, for the negotiation of the encryption and hash algorithms, and for the exchange of a secret, the PreMasterSecret. The function of the CCS protocol is to signal to the Record layer any changes in the security parameters. The Alert protocol indicates errors encountered during message verification as well as any incompatibility that may arise during the Handshake. The Record protocol applies symmetric cryptography with all the negotiated security parameters to protect the application data as well as the messages originating from the Handshake, the CCS, or Alert protocols. Finally, the Heartbeat protocol is used to maintain a connection even in the absence of data transfer.

5.3.1 SSL/TLS Exchanges

The exchanges described in SSL/TLS happen in two phases:

1. During the preliminary phase, identification of the parties, negotiation of the cryptographic attributes, and the generation and sharing of keys take place.
2. During the exchange of data, the security depends on the algorithms and parameters negotiated in the preliminary phase.

At any moment, it is possible to signal an intrusion or an error.

For SSL/TLS, a session is an association between two entities that have a common set of parameters and of cryptographic attributes. Whenever a client connects to a server, an SSL or a TLS session is launched. If the client connects to another process on the server, a new session is started without interrupting the initial session. If the client returns later to the first server and wishes to conserve the cryptographic choices already made, the client will ask the first process to resume the old session instead of starting a new one. To limit the risks of attacks through message interception, SSL/TLS recommend limiting the age of the session identifier to a maximum of 24 hours, but the exact duration is at the discretion of the server. In addition, the interrupted session can be resumed only if the proper suspension procedures have been used.

The concept of a connection in SSL/TLS has been introduced to allow an application to refresh (i.e., modify) certain security attributes (e.g., the encryption key) without affecting all the other attributes that have been negotiated at the start of a session. A session can contain several connections under the control of the applications. SSL/TSP sessions and connections can be illustrated with the help of their state variables and the associated security parameters.

5.3.2 State Variables of an SSL/TLS Session

An SSL/TLS session is uniquely identified with the following six state variables:

1. A *session ID*, which is an arbitrary sequence of 32 octets that the server selects to identify an active session or a session that can be reactivated.
2. The *peer certificate*, which is that of the correspondent and it conforms to Version 3 of X.509.
3. The *compression method*—SSL/TLS allow the possibility of negotiating a data compression method.
4. The *cipher spec*, which defines the encryption and hash algorithms used out of a preestablished list.
5. The *MasterSecret*, with a size is 48 octets, is shared between the client and the server. This parameter is used to generate all other secrets; therefore, it remains valid for the whole session.
6. A flag denoted *is resumable* to describe whether the session can be used to open new connections.

The cipher suite parameters are negotiated in the clear during session establishment by defining five elements (in SSL and TLS 1.0):

1. The type of encryption, whether it is a stream code or a block code.
2. The algorithm of encryption.
3. The hashing algorithm.
4. The size of the digest.
5. A binary value indicating the permissibility to export the encryption algorithm according to U.S. law on export of cryptography. This element was removed in TLS 1.1 and 1.2.

5.3.3 State Variables for an SSL/TLS Connection

The parameters that define the state of a connection during an SSL/TLS session are those that will be "refreshed" when a new connection is established. These parameters are as follows:

- Two random numbers (`server _ random` and `client _ random`) of 32 octets each. These numbers are generated by the server and the client, respectively, at the establishment of a session and for each new connection. The secret key will be derived using these random numbers, which means that these numbers are exchanged in the clear at the opening of a session. In contrast, during the establishment of an additional connection when the session is active, the process of encryption is fully functioning and the numbers are transmitted encrypted. The use of these numbers protects against replay attacks using ancient messages.
- Two secret keys, `server _ MAC _ write _ secret` and `client _ MAC _ write _ secret`. These keys will be employed within the hashing functions to calculate the message authentication code (MAC). The size of the MAC depends on the hashing algorithm used, for example, 16 octets for SHA-1 or 20 octets for MD5.
- Two keys for the symmetric encryption of data, one for the server side and the other for the client side—While the same algorithm is used by both parties, each can use its own key, `server _ write _ key` or `client _ write _ key`, respectively, provided that they share it with the other side. The key size depends on the encryption algorithm selected and legal stipulations on cryptography.
- Two initialization vectors (IV) for symmetric encryption in the CBC mode—One vector at the server side and the other at the client side. Each is a random string that is "exclusively ORed" with the plain text message before encryption. Their size depends on the selected algorithm. The IVs add randomness to the message and are sent along the message in the clear.
- Two sequence numbers, one for the server and the other for the client, each coded over 8 octets. These sequence numbers are now maintained separately for each connection and are incremented whenever a message is sent on this connection. This mechanism offers some protection against replay attacks since it prevents the reuse of already emitted messages.

Each connection has its own cryptographic parameters (keys and initialization vectors), but all connections of the same session share the *MasterSecret*. In addition, the confidentiality keys for each direction remain independent of each other.

The generation of quality random numbers is of concern to both the client and the server. The Cryptographic Module Validation Program (CMVP) is a collaboration of the NIST and the Communications Security Establishment of Canada to validate commercial products that purportedly generate random numbers.

5.3.4 Synopsis of Parameters Computation

Figure 5.4 illustrates the computation of MasterSecret starting from the PreMasterSecret and the parameters `client _ random` and `server _ random`. The value of the MasterSecret will remain constant throughout the session. The parameters `client _ random` and `server _ random` are exchanged in the clear while the PreMasterSecret is exchanged confidentially with the help of the Key Exchange Algorithm. The computation of the key from encrypting the hash key and the initialization vectors begins from the variables MasterSecret, `client _ random`, and `server _ random` in the manner depicted in Figure 5.5.

The opening of a new connection will lead to the recalculation of the variables `client _ random` and `server _ random`, although the value of the MasterSecret remains unchanged. As a consequence, the variables `client _ write _ key` (respectively, `server _ write _ key`) will be recomputed at the opening of a connection.

Note that for each new connection, each of the communicating entities uses a symmetric encryption key

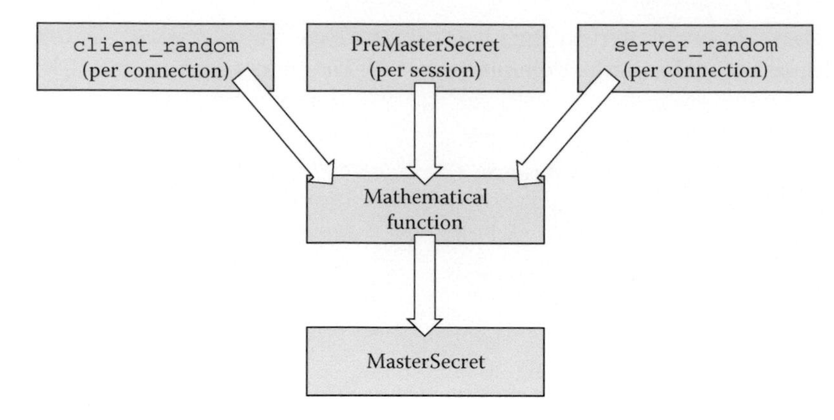

FIGURE 5.4
Construction of the MasterSecret at the start-up of a session.

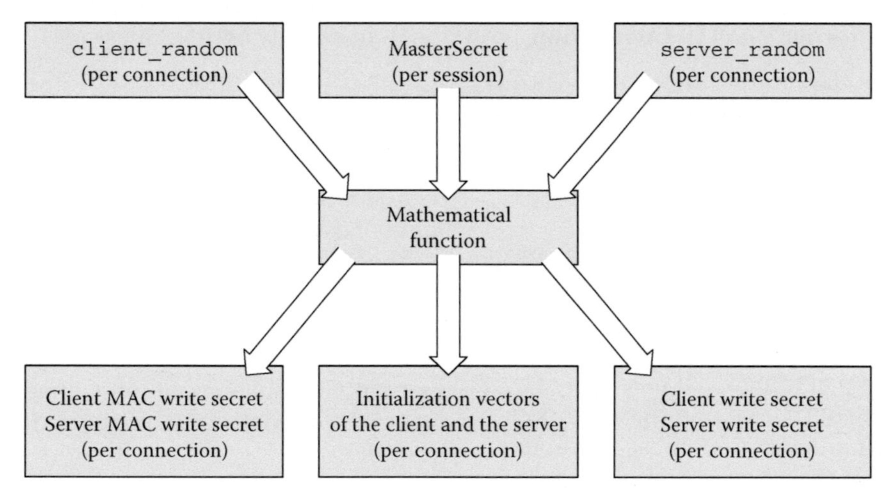

FIGURE 5.5
Generation of the secrets and the initialization vectors at the start-up of a session or a connection.

different from its partner. Thus, a flow in one direction is encrypted with a different key from corresponding flow in the opposite direction.

5.3.5 The Handshake Protocol

The Handshake protocol describes a series of message exchanges between the client and the server to establish the SSL/TLS channel using a set of negotiated security algorithms and parameters.

5.3.5.1 General Operation

The Handshake protocol begins with the mandatory authentication of the server, authentication of the client being optional. Once the authentication is established, both parties move to the negotiation phase to select the cipher suite that will be used throughout the session. Thus, the Handshake protocol conditions the whole process of secure data transfer, which makes it a prime target for potential aggressors.

Shown in Table 5.4 is a chronological list of the messages of the Handshake protocol and their significance. Figure 5.6 illustrates the exchanges during the establishment of a session.

5.3.5.2 Opening a New Session

A session that the client initiates begins with the transmission of the *ClientHello* message to the server. The server can also take the initiative by sending the *HelloRequest*. This message does not contain any information and is only used to alert the client that the server is ready. The subsequent exchanges take place in the same manner irrespective of the way the session has started.

In SSL Version 3.0 and TLS, the exchanges that take place during the opening of a new session comprise the following four stages:

TABLE 5.4

Chronological List of the Messages of the Handshake Protocol

Message	Message Type[a]	Direction of Transmission	Meaning
HelloRequest	O	Server → Client	Notice to the client to begin the Handshake.
ClientHello	M	Client → Server	This message contains the following: The version of the SSL protocol The random number `client_random` The session identifier: session_ID The list of cipher suites that the client selects The list of compression methods that the client selects
ServerHello	M	Server → Client	This message contains the following: The version of the SSL protocol The random number: `server_random` The session identifier: session_ID A cipher suite A compression method
Certificate	O	Server → Client Client → Server	This message contains the server's certificate or the client's certificate if the client has one and the server has requested it.
ServerKeyExchange	O	Server → Client	The server sends this message if it does not have a certificate or owns only a signature certificate.
CertificateRequest	O	Server → Client	The server sends this message to request the client's certificate.
ServerHelloDone	M	Server → Client	This message informs the client that the transmission of the ServerHello and subsequent messages has ended.
ClientKeyExchange	M	Client → Server	Message containing the PreMasterSecret encrypted with the server's public key.
CertificateVerify	O	Client → Server	Message allowing the explicit verification of the client's certificate.
Finished	M	Server → Client Client → Server	This message indicates the end of the Handshake and the beginning of data transmission, protected with the newly negotiated parameters.

Note: M, mandatory; O, optional.

1. Identification of the cipher suites available at each site
2. Authentication of the server
3. The exchange of secrets
4. The verification and confirmation of the exchanged messages

In 2011, RFC 6176 prohibited the support of SSL Version 2.0 in new implementations. TLS servers may continue to accept Version 2.0 ClientHello messages but would abort the connection if the client does not accept a higher version protocol.

The first messages exchanged between the client and the server, *ClientHello* and *ServerHello*, allow the negotiation of the session parameters, the encryption algorithms, and the secrets of the session. At the end of the negotiation, the client and the server would have chosen the version of the common protocol between the two sides.

All messages of the Handshake protocol have a header of 5 octets:

- The message type is coded over 1 octet.
- The version of the protocol is coded over 2 octets.
- The message length is coded over 2 octets.

The message data include the following items:

- The random numbers `client _ random` and `server _ random`, each of 32 octets, consisting each of two fields: the size of the first is 4 octets and comprises the time indicated by the client's internal clocks expressed in universal time; the size of the second is 28 octets and includes a random number generated by a random number generator.
- A session identifier of 0–32 octets, preceded by a field indicating its length. When a session is first opened, the length of the identifier is 1 octet, because it has not yet been assigned. When a connection is opened, the length is 32 octets.
- The list of cipher suites that the client supports. Each suite is a combination of various cryptographic algorithms and is represented by a code of 2 octets. Given that SSL/TLS specify that the length of that field ranges between 2 and $2^{16} - 1$ octets, the maximum number of distinct combinations is $2^{15} - 1$, which can also be represented by 2 octets. This value is transmitted before the list of proposed cipher suites. Once a cipher suite

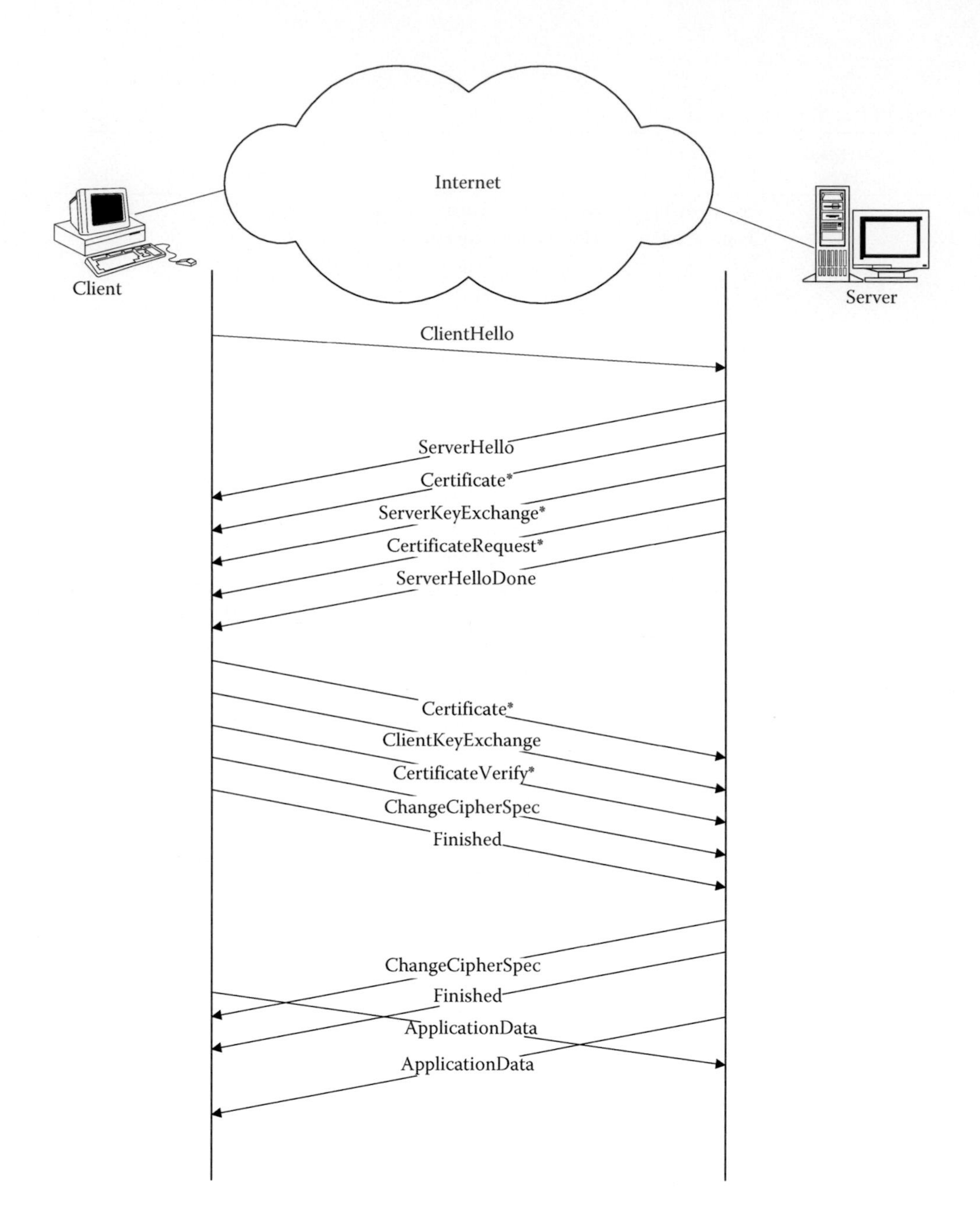

FIGURE 5.6
Messages exchanged during the establishment of a new session. * Optional message. (From Sherif, M.H. et al., Electronic payments on the Internet, *Proceedings of the ISCC98*, Athens, Greece, IEEE Computer Society Press, Washington, DC, 1998, pp. 353–358. Copyright 1998 IEEE.)

is selected, it will be applied to all connections of the same session.

- The list of compression methods that the client is able to support. The method is coded over 1 octet. SSL/TLS support up to 255 distinct methods, but three have been standardized: without compression identified by the code "0", the algorithm DEFLATE in RFC 1951 with the identifier "1" (RFC 3749) and the algorithm Lempel–Ziv–Stac (LZS) defined in RFC 1967 with the identifier "64" (RFC 3943).
- Starting with TLS 1.2, the ClientHello includes TLS extensions by which the client may request the server to provide additional functionalities. Servers indicate in the ServerHello message the extensions they support; however, they are not allowed to initiate an extension request.

Table 5.5 indicates the sizes of the variable fields in the ClientHello and ServerHello messages.

Thus, TLS 1.2 allows the client and the server to negotiate the following parameters: the protocol version,

TABLE 5.5

The Size of the Variable Fields in ClientHello and ServerHello Messages

Field	Maximum Length in Octets	Number of Octets to Code the Length
Session identifier	32	1
List of cipher suites	65,534	2
List of compression methods	255	1
List of extensions	65,534	2

the cipher suite, the compression algorithm, and the list of extensions. It is worth noting that compression allows some attacks whereby the attacker can discover secret information, such as the session token based on the size of the compressed HTTP requests. One attack is called Compression Ratio Info-Leak Made Easy (CRIME) and was demonstrated by Juliano Rizzo and Thai Duong in 2012. Yoel Gluck, Neal Harris, and Angelo Prado demonstrated another attack, called the Browser Reconnaissance and Exfiltration via Adaptive Compression of Hypertext (BREACH) attack, in 2013. As a consequence in September 2012, major vendors disabled TLS compression in the client and at the server side (Sarkar and Fitzgerald, 2013).

TLS 1.2 extensions include the support of GCM, CCM, and the Camellia cipher suites, as well as the Heartbeat Extension of RFC 6520 to support the Heartbeat subprotocol. When a Heartbeat Request message is received, the receiving side responds with a Heartbeat Response message carrying an exact copy of the incoming payload in the request.

Interoperation with SSL Version 2.0 requires that the ClientHello message be encapsulated in a Record message conforming to the SSL 2.0 specifications. However, this version has well-known security vulnerabilities, and specific measures had to be taken to prevent rollback attacks where a malicious party would be forcing its use to take advantage of its weaknesses. In the past, the minimum version allowed for clients on U.S. federal computers was Version 3.0 (Chernick et al., 2005, p. 21). As mentioned earlier, SSL 2.0 was officially withdrawn through RFC 6151 in 2011.

After sending the ClientHello message, the client waits for the arrival of the ServerHello message. This message must indicate the version of the protocol and the unique session identifier that the server has selected. The session identifier will be used to resume an already established session, as will be shown later. The message also contains the random number `server _ random`. By exchanging the random numbers, `client _ random` and `server _ random`, each party will be able to reproduce the secrets of its correspondent, thereby sharing the secrets.

The message must include the cipher suite that the server has retained from among the choices that the client had suggested. This suite will be used to ensure the confidentiality and the integrity of the data. If no common suite is available, the server (or the client, according to the case) will generate the error message `close _ notify` as specified by the Alert protocol and the session will be dropped. Note, however, that some servers do not send the message `close _ notify` as required but close the TCP session without warning.

It is important to remember that this negotiation takes place in the clear. An intruder may attempt to intercept the ClientHello message and replace it by a spurious message to request a less robust algorithm. Protection against this attack can be achieved with the Finished message that is exchanged at the end of the Handshake. This message is the first to be protected with the just negotiated algorithms, keys, and secrets. Recipients of Finished messages must verify that the contents are correct. Once a side has sent its Finished message and received and validated the Finished message from its peer, it is ready to exchange application data over the connection.

5.3.5.3 Authentication of the Server

In this second phase, the server authenticates itself by sending the *Certificate* message with the following:

1. The Version 3.0 X.509 certificate of the server, which includes the public key of the cipher suite previously selected. The protocol Alert will indicate an error if the certificate is not included.
2. The certification path and the certificate of the certification authority.

As a reminder, an X.509 certificate has the following form:

$$\text{Certificate} = \text{Name}_{\text{Server}} \| \text{Name}_{\text{CA}} \| \| \text{PK}_{\text{Server}} \| \text{SK}_{\text{CA}} \times \left\{ \text{H}\left(\text{Name}_{\text{Server}} \| \text{Name}_{\text{CA}} \| \text{PK}_{\text{Server}} \right) \right\}$$

where

$\text{Name}_{\text{Server}}$ is the server name
Name_{CA} is the name of the certification authority
$\text{PK}_{\text{Server}}$ is the server public key
$\text{SK}_{\text{CA}}\{x\}$ is the signature of x with the private key of the certification authority
$\text{H}(y)$ is the digest of y

If the server does not have a certificate, then instead of the Certificate message, it transmits the ServerKeyExchange message. Obviously, in e-commerce applications, servers must be authenticated.

Even if a server presents a certificate, it may be obliged to transmit the ServerKeyExchange message under the following conditions:

- With the Fortezza Key Exchange Algorithm, the ServerKeyExchange message is sent without a signed key, because it is provided in the certificate. The message contains a random value that is used as part of the KEA agreement process.
- If the RSA key exchanged is ephemeral, the X.509 certificate is for signature and the ServerKeyExchange message contains the ephemeral public key of the server and its signature to verify the integrity of the exchanged key.
- If the ephemeral Diffie–Hellman method is used for key exchange, a signed key (usually with the DSA algorithm) is included in the ServerKeyExchange message to verify its integrity before establishing the joint key.

Thus, for dynamic authentication (i.e., a temporary key), the server must have a signature certificate and the server transmits two consecutive messages: the Certificate message, then the ServerKeyExchange message. The Certificate message includes the public key for signature of the server. The client verifies the certificate with the public key of the certification authority. The ServerKeyExchange message contains the public parameters of the algorithm used to exchange the secret key for symmetric encryption. The server signs these parameters using the private key that corresponds to the public key already recovered in the first step. Upon receiving this second message, the client ensures that the public parameters of the Key Exchange Algorithm are those of the server by verifying the signature using the public key already received. This message contains the digest of the concatenation of the variables `client _ random` and `server _ random`, as well as the other server parameters.

After authenticating itself, the server can ask the client to do the same. The server next sends the CertificateRequest message that contains a list of the types of certificates requested, arranged in the order of the server's preference for the certification authorities.

Following these messages, the server sends the ServerHelloDone message that signifies to the client that it is finished and that it is waiting for a response.

5.3.5.4 Exchange of Secrets

If the server asks the client to authenticate itself using the CertificateRequest message, the client must respond by including its certificate, if it has one, in the Certificate message. If the client is not able to give a certificate, the answer will be no_certificate. This is merely a warning and not a fatal error, unless the server requires client authentication, in which case the connection will be interrupted. Otherwise, user authentication will take place at the application level with a login and a password.

The client sends next the ClientKeyExchange message with the content depending on the algorithm used for key exchange and the type of certificate according to the following:

- If the RSA algorithm is used for key exchange, the PreMasterSecret is encrypted, either with the public key of the server or with a temporary key contained in the ServerKeyExchange message as explained earlier.
- If the key exchange is according to the Diffie–Hellman algorithm, the message ClientKeyExchange contains the public key that the client sends to the server. Each site can perform separately the necessary computations that will yield a shared secret, whose value will be the PreMasterSecret.
- If the key exchange is based on the Fortezza algorithm, a TEK is calculated from the parameters that the server has sent in the ServerKeyExhange message. This key will then be used to encrypt the client key, client_write_key, the secret PreMasterSecret, and the initialization vectors.

5.3.5.5 Key Derivation for SSL

Once the MasterSecret has been obtained, the encryption keys are derived from the encryption block as follows:

$$\begin{aligned}\text{key\_block} = {} & \text{MD5}\left(\text{MasterSecret} \,\|\, \text{SHA}\left('\text{A}' \,\|\, \text{MasterSecret} \,\|\, \text{server\_random} \,\|\, \text{client\_random}\right)\right) \,\| \\ & \text{MD5}\left(\text{MasterSecret} \,\|\, \text{SHA}\left('\text{BB}' \,\|\, \text{MasterSecret} \,\|\, \text{server\_random} \,\|\, \text{client\_random}\right)\right) \,\| \\ & \text{MD5}\left(\text{MasterSecret} \,\|\, \text{SHA}\left('\text{CCC}' \,\|\, \text{MasterSecret} \,\|\, \text{server\_random} \,\|\, \text{client\_random}\right)\right) \,\| \\ & [\ldots]\end{aligned}$$

where
|| denotes the concatenation operator
'A' represents the character A, and so on

The computation is repeated as many times as necessary so that the encryption block can be of sufficient length to extract the following components:

```
client_MAC write_secret[hash_size]
server_MAC write_ secret[hash_size]
client_write_key[key_length]
server_write_key[key_length]
client_write_IV[size_of_the_initialization_
  vector] /* for Fortezza only, not for
  export from the U.S. */
server_write_IV[size_of_the_initialization_
  vector] /* for Fortezza seulement, not for
  export from the U.S. */
```

SSL Version 3.0 specified MD5 as the hash function used for the formation of the key block. In 1994, however, collisions were detected in the output of MD5 (van Oorschot and Wiener, 1994). Also, MD5 is not approved as a FIPS hash algorithm. As a consequence, the key derivation in TLS was changed.

5.3.5.6 Key Derivation for TLS

The derivation of the secrets in TLS is similar to that of SSL Version 3.0, but both MD5 and SHA-1 are used.

More precisely, MasterSecret is obtained from the PreMasterSecret through the formula:

$$\text{MasterSecret} = \text{PRF}\begin{pmatrix}\text{PreMasterSecret,"master secret"} \\ \|\text{client\_random}\|\text{server\_random}\end{pmatrix}$$

where
|| indicates concatenation

PRF is a pseudo-random function of the form PRF (secret, label || seed). It has two arguments, the first is the PreMasterSecret and the second is concatenation of a label and a seed, where label is a series of characters, in this case the string "master secret," and seed is a random number formed by the concatenation of `client _ random` and `server _ random`.

As explained earlier, the values of `client _ random` and `server _ random` are generated by the client and server, respectively, and exchanged through the ClientHello and ServerHello messages. The MasterSecret will be included in the MAC calculation of the CertificateVerify, ChangeCipherSpec, and Finished messages.

The pseudo-random function (PRF) is built as follows:

$$\text{PRF(secret, label} \| \text{seed)} = \text{P\_MD5(S1, label} \| \text{seed)} \oplus \text{P\_SHA}-1\text{(S2, label} \| \text{seed)}$$

where ⊕ is the Exclusive OR operation,

$$\begin{aligned}&\text{P\_Hash(secret, seed)}\\ &= \text{HMAC\_Hash[secret, A(1)} \| \text{seed]}\\ &\quad \| \text{HMAC\_Hash[secret, A(2)} \| \text{seed]}\\ &\quad \| \text{HMAC\_Hash[secret, A(3)} \| \text{seed]} \|\end{aligned}$$

P_Hash represents P_MD5 or P_SHA-1 depending on the algorithm used (MD5 or SHA-1). The values A(1), A(2), A(3),... are computed recursively from the value of the seed and the label as follows:

$$\begin{aligned}&A(0) = \text{seed}\\ &A(i) = \text{HMAC\_Hash}(\text{secret}, A(i-1))\end{aligned}$$

The secrets S1 and S2 are derived by dividing the *PreMasterSecret* into two parts of equal size. S1 is taken from the first half of the secret, and S2 from the second half. Their length is created by rounding up the length of the overall secret divided by two:

$$\text{LS1} = \text{LS2} = \text{ceil}(\text{LS}/2)$$

The function ceil(x) gives the smallest integer equal or larger than x. LS, LS1, and LS2 are, respectively, the lengths of the initial secret and of S1 and S2. If LS is odd, the two sizes are made equal by adding an octet to the beginning of the second half, where the last octet of the first half is copied.

By processing each half with a different algorithm, the vulnerability of one algorithm (MD5 in particular) does not weaken the overall protection.

Thus, in TLS 1.0 and 1.1, the PRF is created by splitting the secret into two halves and using one half to generate data with MD5 and the other half to generate data with P_SHA-1, then Exclusive ORing the outputs of these two expansion functions together.

Four iterations of the function PRF are sufficient to get 64 octets. In this case, all the octets of the output from P_MD5 will be used but only 64 out of the 80 octets that P_SHA-1 produces. In contrast, to get 80 octets, there is a need to have five iterations of P_MD5 and four iterations of P_SHA-1.

In TLS 1.2, a single hash function (such as SHA-256) replaces the MD5/SHA-1 combination, that is,

$$\text{PRF (secret, label, seed)} = \text{P\_Hash (secret, label} \parallel \text{seed)}$$

With SHA-256, 32 octets are produced with a 256-bit key.

For a secret k and a message m, HMAC_hash is computed for a given hash function H() with the formula

$$\text{HMAC\_hash } (k, m) = \text{H}\left[\bar{k} \oplus \text{opad} \parallel \text{H}(\bar{k} \oplus \text{ipad}, m)\right]$$

As explained in Chapter 3, the secret $\bar{k}$ is derived from the initial k of L bits by padding a sequence of "0"s to form the block size of the relevant hashing algorithm. The variables *opad* and *ipad* are constants for outer padding and inner padding, respectively, $0x36$ and $0x5C$.

Having obtained the MasterSecret, the PRF is reapplied to form the block from which the keys will be extracted, that is, the symmetric encryption keys, the hash keys, and the initialization vectors for the client and the server:

$$\text{key\_block} = \text{PRF (MasterSecret,"key expansion",}$$
$$\text{server\_random} \parallel \text{client\_random)}$$

Here, the label is the character string "key expansion" while the seed is formed of the concatenation of server_random and client_random. Note that their order is the inverse of their order in the computation of the MasterSecret. The computation is repeated a sufficient number of times to produce all the parameters of the selected cipher suite, that is,

```
client_write_MAC_key [MAC_key_length]
server_write_MAC_key [MAC_key_length]
client_write_key[length_of_the_encryption_
  key]
server_write_key[length_of_the_encryption_
  key]
/* For non exportable algorithms only/
client_write_IV[size_of_the_initialization_
  vector]
client_write_IV[size_of_the_initialization_
  vector]
```

5.3.5.7 Exchange Verification

If the client has a certificate for digital signature, it will then send an explicit confirmation with the CertificateVerify message. This message contains the digest of all previous messages starting from ClientHello, with the exception of the container message itself. This digest is encrypted with the client's private key. The purpose of this message is to give the server the ability to authenticate the client by verifying its signature. It is not sent in Fortezza's Handshake.

Accordingly, the content of the CertificateVerify message of SSL Version 3 will be

$$\text{Hash\{MasterSecret} \parallel \text{pad\_2} \parallel \text{Hash}$$
$$\times\left(\text{handshake\_messages}\right) \parallel \text{MasterSecret} \parallel \text{pad\_1)\}}$$

where Hash represents either MD5 or SHA-1, depending on the case. The pad_1 and pad_2 fields contain repetitions of the octets 0x36 and 0x5c, respectively; the number of repetitions is 48 in the case of MD5 and 40 for SHA-1. The hash is calculated over all previous handshake messages, that is, the current message is excluded.

In TLS 1.0 and 1.1, the CertificateVerify contains the hash of the Handshake messages, that is, either MD5 (handshake_messages) or SHA-1 (handshake_messages). For TLS 1.2, the hash is computed with whatever hashing algorithm in the selected cipher suite.

The client then invokes the protocol ChangeCipherSpec to start the encryption of the exchanges with the choices made in the previous two phases.

The client sends immediately the Finished message. This the first message encrypted with the just selected cryptographic parameters. For security reasons, a received Finished message is processed only if it follows the associated ChangeCipherSpec message (Wagner and Schneier, 1996).

5.3.5.8 Verification and Confirmation by the Server

The Finished message contains the elements that allow the server to verify the integrity of all the Handshake messages from the client starting with ClientHello, thereby foiling any man-in-the-middle attack.

For SSL Version 3, the verification data are the hash calculated with the following formula:

$$\text{Hash\{MasterSecret} \parallel \text{pad\_2} \parallel \text{Hash}$$
$$\times \text{(handshake\_messages} \parallel \text{Sender} \parallel\parallel \text{MasterSecret} \parallel$$
$$\text{pad\_1)\}}$$

where Hash is either MD5 or SHA-1, depending on the hashing algorithm used. The field Sender contains the string "client" or "server" depending on the case at hand.

For TLS 1.0 and 1.1, the verification data are defined as follows:

$$\text{PRF(MasterSecret, finished\_label, MD5}$$
$$\times \text{(handshake\_messages} \parallel \text{SHA}$$
$$-1\text{(handshake\_messages))}$$

The finished_label is the string "client_finished" on the client side or "server_finished" on the server side. The data are enclosed in the verify_data field and are denoted as client_verify_data on the client side and server_verify_data on the server side.

For TLS 1.2, the verify_data field is calculated as follows:

$$\text{PRF (MasterSecret, finished\_label,}$$
$$\text{Hash (handshake\_messages))}$$

where Hash is the hashing algorithm used. In all cases, the field handshake_messages contains the exchanged messages with the exception of the ChangeCipherSpec and the current message. The ChangeCipherSpec message is not included in the digest because it is not part of the Handshake protocol.

In previous versions of TLS, the length of client_verify_data was always 12 octets. In TLS 1.2, the length of the client_verify_data field is 12 octets by default, but the actual length depends on the cipher suite. Similarly, the length of the server_verify_data is 24 octets by default. The corresponding lengths for SSL Version 3 are 36 and 72 octets, respectively.

Upon receipt of the Finished message, the server attempts to reproduce the same hash with the messages that it has received previously concatenated to the MasterSecret. It compares the result with the content of the Finished message that just arrived from the client. This step will allow detection of any intruder that would have intercepted and modified the messages.

Once the verification has been done, the server sends in turn the message ChangeCipherSpec and Finished. Here, again, the Finished message is generated from all the messages that have been sent and transmitted encrypted. This allows the client to perform the verification as well. The server starts to send the application data.

It should be noted that the total message length exchanged and considered in the Finished message is different from the client side than from the server side. This difference arises because the second Finished message includes the first Finished message. For session establishment, it is the client that sends the first message (see Figure 5.6) while, as we will see shortly, the server sends the first Finished during connection establishment (see Figure 5.9).

In SSL, data transmission begins just after sending the Finished message. This gives an intruder the opportunity to modify either the ClientHello message or the ServerHello message to force the use of a weak cipher suite, then recover the PreMasterSecret, and divert the Finished message to impersonate one or both parties. To defend against this attack, called rollback attack, the Finished messages must be exchanged from both sides before transmission starts; a requirement in TLS 1.0 (Mitchell et al., 1998; RFC 2246, 1999). Thus, both the client and the server have the hash of all the exchanged handshake messages. Furthermore, in TLS 1.2, RFC 5246 provides another defense by specifying that the version indicated in the ClientKeyExchange message must be identical to the version in the ClientHello message.

5.3.6 The ChangeCipherSpec Protocol

The ChangeCipherSpec (CCS) protocol consists of a single 1-octet message with the same name as the protocol. The ChangeCipherSpec message signals to the protocol Record that encryption can start with the negotiated cryptographic algorithms. Before this message, the encryption was the task of the Handshake protocol. Following receipt of this message, the protocol Record layer on the transmit side will have its method of encrypting the messages to be sent. On the receive side of the far-end entity, the protocol Record layer will have to modify its read attributes to be able to decipher the received messages.

5.3.7 Record Protocol

As seen earlier, the Record protocol intervenes only after the transmission of the ChangeCipherSpec message. During the session establishment, the role of the protocol Record layer is to encapsulate the Handshake data and to transmit them without modification toward the TCP layer.

During data encryption, the Record protocol receives data from the upper layers (Handshake, Alert, CCS, HTTP, ftp, etc.) and transmits them to the TCP after performing in order the following functions:

1. Fragmentation of the data in blocks of maximum size of 2^{14} octets
2. Data compression, a function considered but not supported in the current version of the specifications
3. Generation of the digest to ensure the integrity service
4. Data encryption to ensure the confidentiality service

The sequence of operations is MAC computation and then encryption. The Record protocol fragments the message received from the upper layers into payloads and prepends 13 octets to each payload as follows: a sequence number of 8 octets and a header of 5 octets. This header indicates the message type according to its origin subprotocol (Handshake, Alert, CCS, or application data). It identifies the version of the SSL/TLS protocol used and the length of the encapsulated data blocks. This is different from the actual length of the record because the latter includes the MAC and possible padding. Not that this sequence number is not included in Record message itself.

For each fragment of data (the payload), the MAC is computed on the 8-octet sequence number, the 5-octet header, and the payload. The sequence number is used to detect potential replay or reordering attacks. It is not possible, however, to correct the effects of these attacks because the Record protocol does not have a field dedicated to transport these numbers.

The SSL Version 3 MAC is computed as follows:

$$\text{Hash}(\text{MAC\_write\_secret} \,\|\, \text{pad\_2} \,\|\, \text{Hash}$$
$$\times(\text{MAC\_write\_secret} \,\|\, \text{pad\_1} \,\|\, \text{seq\_number}$$
$$\|\, \text{length} \,\|\, \text{payload})$$

For TLS, the MAC is computed as follows:

$$\text{HMAC}\,(\text{MAC\_write\_key}, \text{seq\_num} \,\|\, \text{type} \,\|\, \text{version}$$
$$\|\, \text{length} \,\|\, \text{payload})$$

where

- || denotes concatenation
- seq_num is the 8 octets sequence number for this fragment of data

In SSL Version 3, the hash is an early version of HMAC, pad_1 is the octet 0x36 repeated 38 times for MD5 and 40 times for SHA-1, and pad_2 is the octet 0x5c repeated the same number of times. In TLS 1.2, the algorithms SHA-256, SHA-385, and SHA-512 are also used. Note that MAC_write_key is the new name for MAC_write_secret in TLS 1.2.

Padding is used with block symmetric encryption (e.g., DES, 3DES, or AES) in the CBC mode to ensure that the size of the plaintext to be encrypted is a multiple of the block size of the selected cipher, for example, 8 octets for DES/3DES and 16 octets for AES. The padding can range in size from 0 to 255 octets and is preceded by a field of 1 octet that indicates the length. Each octet in the padding data is filled with the padding length value. For example, "0x00," "0x01||0x01," "0x02||0x02||0x02,"... are examples of valid pad sequences. Notice that the padding is added after the HMAC has been computed so that its integrity is not protected nor it is authenticated.

Padding, however, makes the decryption process sensitive to errors in the coding logic and inadequate parsing. Furthermore, the total processing time can depend on the peculiarities of the underlying plaintext and the padding and not only on the size of the ciphertext, as it should be for security reasons. An attacker on the same local area network segment as the targeted TLS client can exploit the timing differences to recover a block of plaintext using about 2^{23} TLS sessions. A variant of this attack, called Lucky 13 attack, allows the partial recovery of the plaintext in 2^{16} sessions (AlFardan and Paterson, 2003a,b).

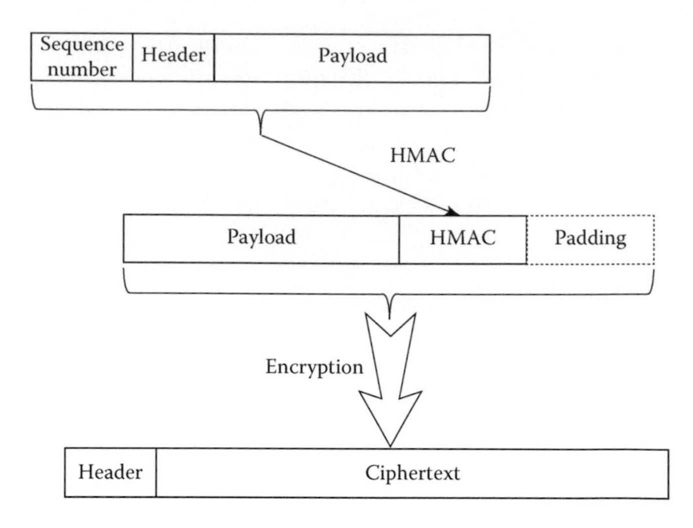

FIGURE 5.7
Exchanges during a connection establishment.

Figure 5.7 represents the formation of an SSL/TLS record. The padding is shown as a dotted box because it is only added with the CBC mode of block symmetric encryption.

The inverse tasks are performed on the receive side: decryption, integrity verification, decompression, and reassembly. If the computed digest is not identical to the one that was received, the Record protocol invokes the Alert protocol to relay the error message to the transmit unit.

Figure 5.8 illustrates the various tasks for the Record protocol.

5.3.8 Connection Establishment

The establishment of the first SSL/TLS connection is the same as the establishment of a session as explained earlier. If an SSL/TLS has already been established, TCP flows can transit in both directions. Thus, the establishment of a new connection consists in refreshing the parameters `client _ random` and `server _ random` with the ClientHello and ServerHello messages while preserving the encryption and hashing algorithms already selected. A new authentication is avoided and, contrary to what happens for session establishment, the ClientHello and ServerHello messages are encrypted. Figure 5.9 depicts the exchanges.

The ClientHello message contains the session identifier of the session that will carry the connection. Should this identifier be absent from the server's tables, either

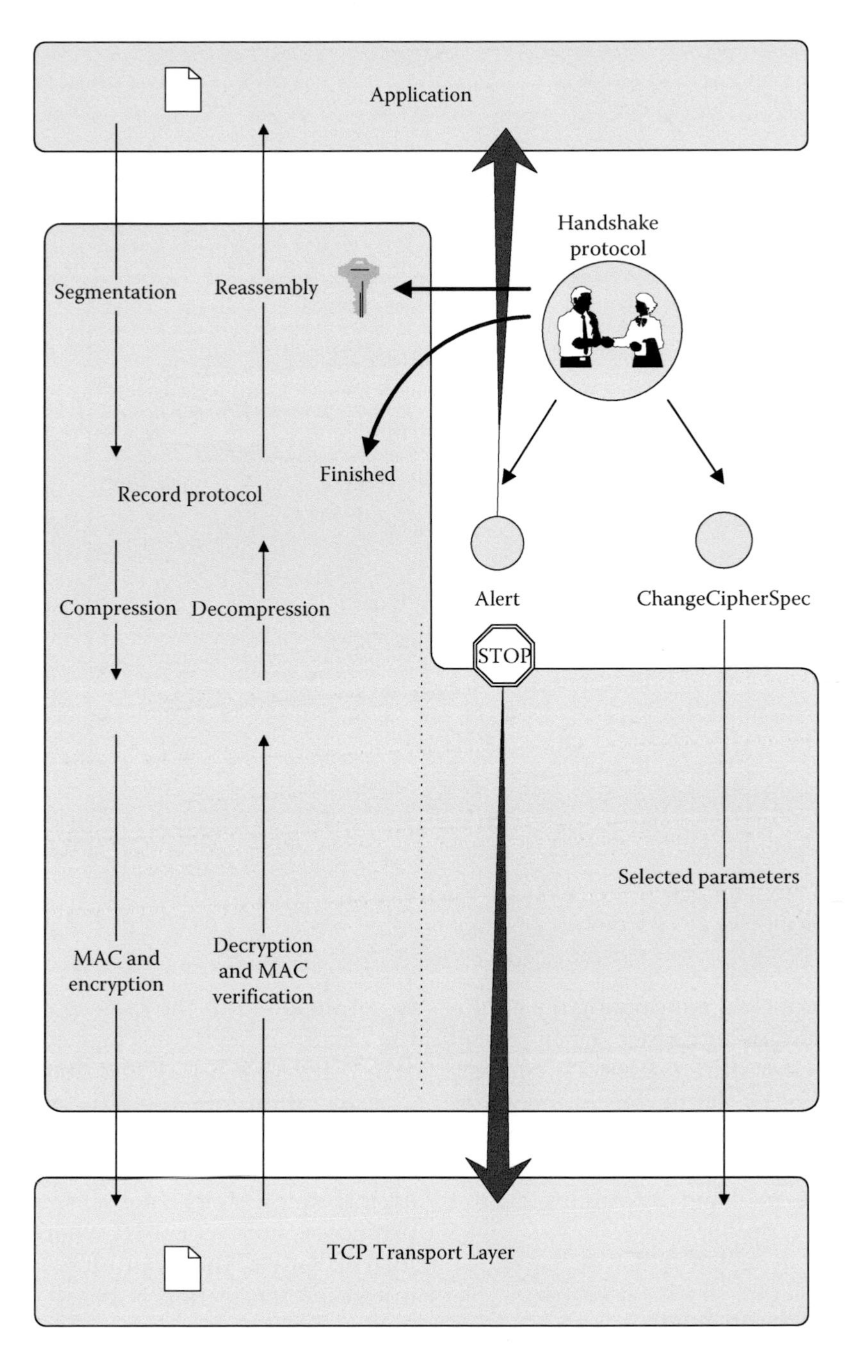

FIGURE 5.8
Formation of an SSL/TLS record.

because it is incorrect or because the session that it refers to has expired, the client is not rejected and the server starts a new complete Handshake to establish a new session.

The ClientHello message includes the identifier of the session for which the connection is being established. If this server does not recognize the session identifier, either because of an error or because the session has expired or was closed, both sides go through a full Handshake to establish a new SSL/TLS session.

If the server recognizes the session identifier, the client and the server confirm their agreement by sending the ChangeCipherSpec message from each side and end the abbreviated Handshake with the Finished message as before.

Opening a connection can thus be over an already established session or over a suspended session. When a session is resumed, the compression algorithm is retained but must be reinitialized, that is, its history erased and its state variables reset.

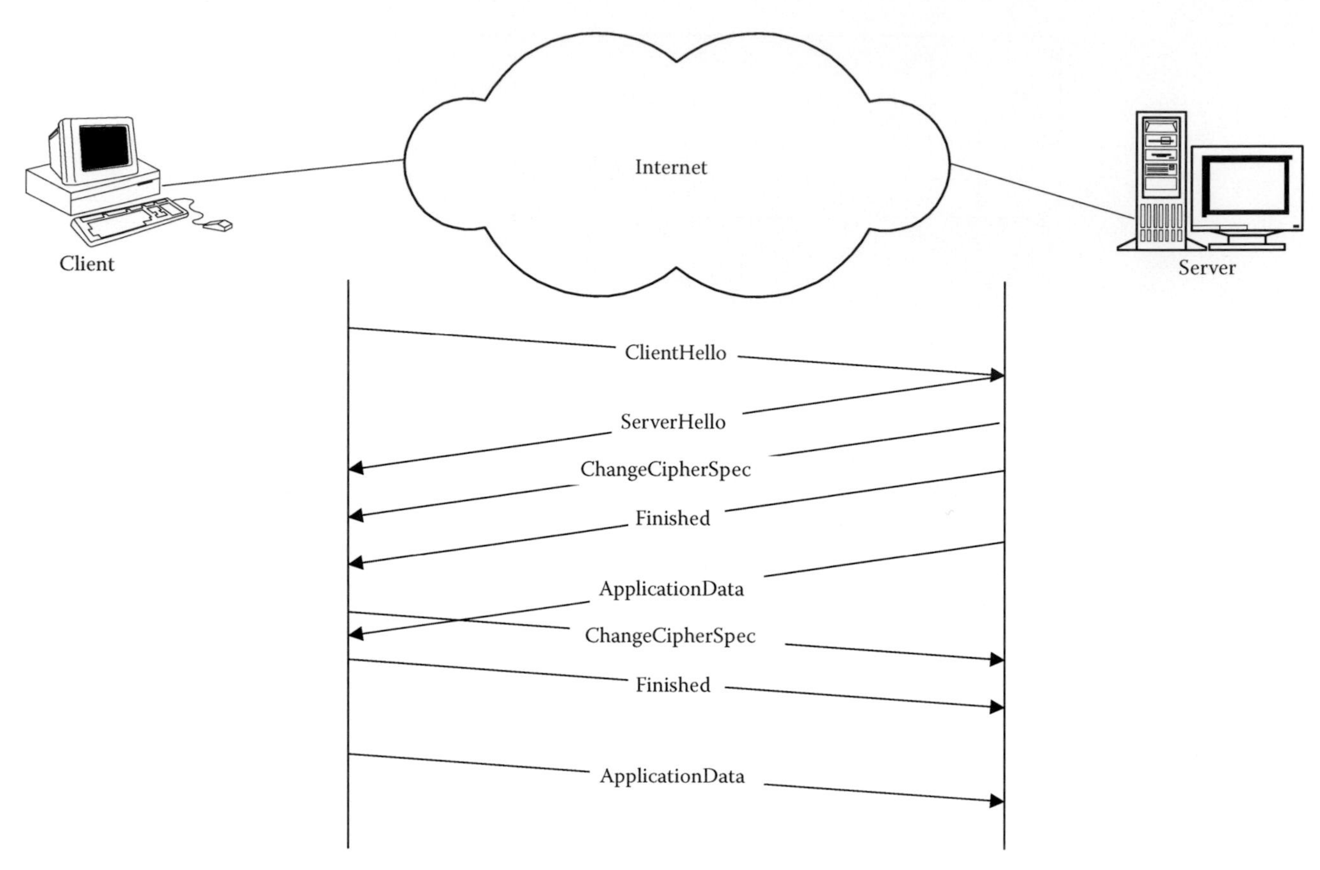

FIGURE 5.9
Functions of the Record layer in the SSL/TLS protocol.

The specifications are not clear concerning the appropriate action if the exchanges introduce a new cipher suite. Similarly, it is not clear if in a session previously established with a Version 3.0 the new connection can be of Version 2.0. In fact, downgrading the version can degrade the security due to the weaknesses of Version 2.0, and some attacks can be initiated by forcing such a downgrade (Rescorla, 2001, pp. 137, 308).

5.3.9 Renegotiation or Rehandshake

Renegotiation is an optional feature by which a server or a client can request new client credentials to supersede those already in use (Rescorla, 2001, pp. 100, 242). These new credentials may, for example, establish stronger cryptography for a specific transaction, accord more restricted access, or prevent attacks on long-lived connections by refreshing the cryptographic parameters, such as a fresh session key or new cryptographic parameters. In contrast to the initial handshake, the exchanges during the renegotiation are encrypted using the existing cryptographic parameters.

To initiate the renegotiation, the client transmits a ClientHello, while the server forces the client to restart the handshake with a HelloRequest. The HelloRequest is not included in the message hashes that are used in the CertificateVerify or Finished messages.

In 2009, Marsh Ray and Steve Dispensa discovered a vulnerability in the TLS renegotiation procedure; this was also independently discovered by Martin Rex (Salowey and Rescorla, 2009). This vulnerability gives an intruder a short time window to hijack application protocols, such as HTTP, where authentication is performed only at the beginning. In the case of HTTP, the initial authentication is based on the combination of a user name and a password, after which the server sends an authentication cookie with a unique session identifier to be stored in the client's web browser. The cookie is enclosed in subsequent HTTP requests to circumvent the need for subsequent authentication, thus reducing the computational load on the server.

The attack can take two scenarios. The first is depicted in Figure 5.10a and consists of the following steps:

1. The attacker establishes a legitimate TLS connection with the server.
2. In the case of HTTP, for example, to commandeer products or services, the attacker sends an incomplete HTTP request (i.e., without a

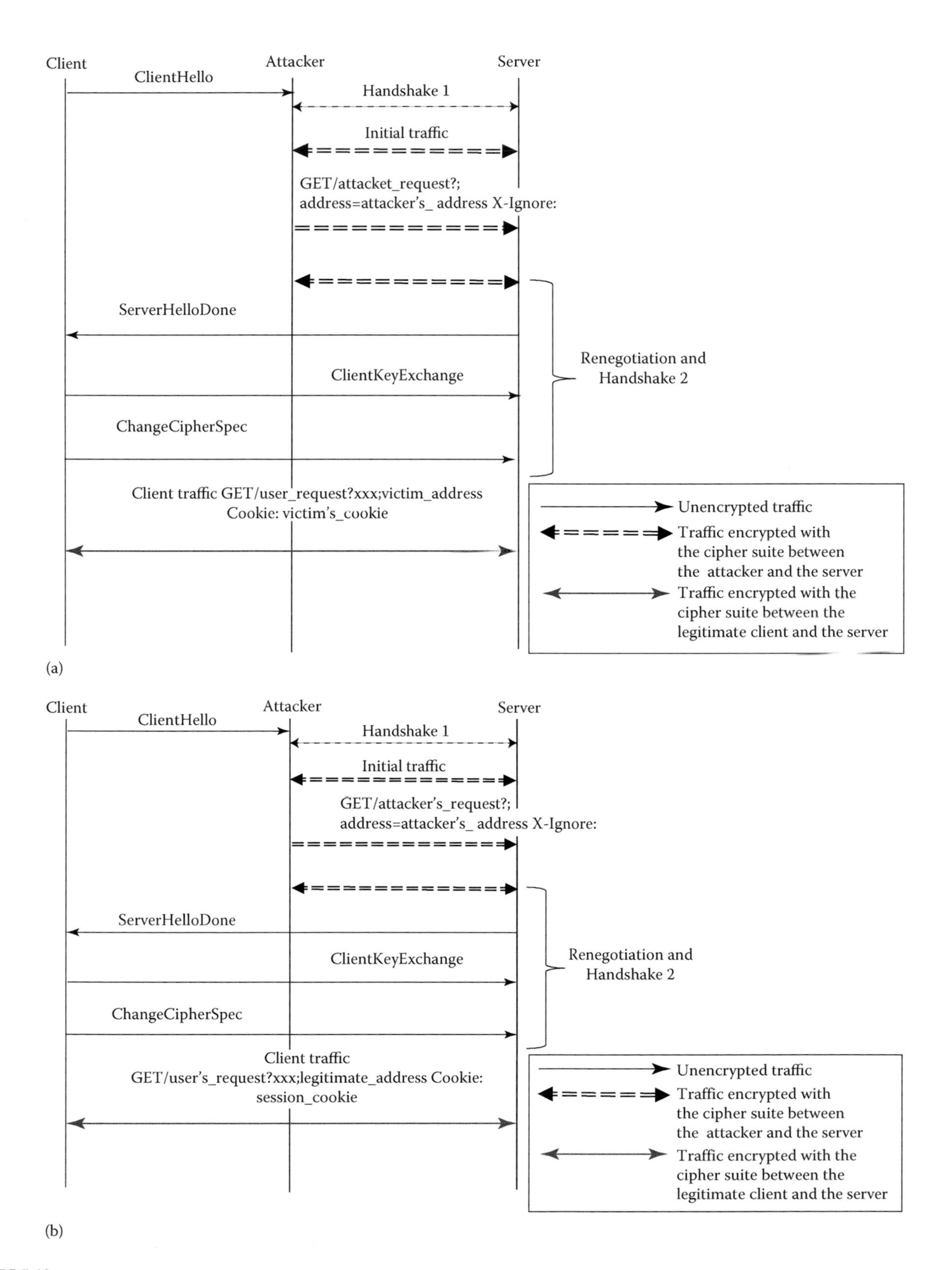

FIGURE 5.10
(a) First scenario of a renegotiation attack. (b) Alternate scenario for a renegotiation attack.

carriage return) to act as a prefix to the first client's HTTP request. The attacker request takes the following form:

```
GET /resource_requested;address=
attacker's_address HTTP/http_ver-
sion X-Ignore-This:
```

The "X-Ignore-This" prefix is an invalid HTTP header and without a new line character is concatenated with the first line of the next request.

3. When the user starts a TLS session, the attacker intercepts the ClientHello and sends the user's data to the server protected with the cryptographic parameters already negotiated between the attacker and the server.
4. The TLS handshake between the legitimate user and the server establishes a new set of cryptographic parameters. The new parameters are opaque to the attacker, and the attacker loses the capability to decrypt or modify any intercepted communication between the user and the server. The server, however, cannot readily distinguish between the legitimate and spurious traffic. As a consequence, when it receives an HTTP request from the legitimate user,

```
GET /second_item?address=user's_
address HTTP/http_version Cookie:
session_cookie
```

it will assume it is linked to the previous incomplete request from the attacker, that is,

```
GET /resource_requested:address=
attacker's_address HTTP/http_
version X-Ignore-This: GET /
second_item?address=user's_
address HTTP/http_version Cookie:
session_cookie
```

Because HTTP is a stateless protocol, it treats each request independently, the only link being achieved through the use of the cookie. Thus, the presence of the cookie will deceive the server into interpreting the linked requests as coming for the same user. Accordingly, the attacker's request will be accepted and processed using the credentials of the legitimate user. Note that the attack succeeds against one request only.

The second scenario takes place as follows (Suga, 2012):

1. The intruder intercepts the client's ClientHello to extract the user's data.
2. The intruder establishes a regular TLS channel between itself and the server.
3. The intruder sends an incomplete HTTP request using the cryptographic parameters of the TLS session that it has initiated.
4. Next, the attacker initiates a renegotiation as if it were the legitimate user using the user's data. The client does not know about that renegotiation but completes the exchanges as defined in the Handshake protocol. The renegotiation is also transparent to the application layer (i.e., HTTP).
5. Once the new TLS session is underway, the attacker loses the capability to view or modify the user's traffic. However, the server is tricked into interpreting the spurious and the legitimate HTTP requests as linked and to process them accordingly.

Figure 5.10b illustrates the steps that comprise the alternate scenario.

The defense against the renegotiation attack is either disabling the feature or implementing the mechanism described in RFC 5746 (2010). The mechanism relies on a TLS extension—the Renegotiation Information Extension (RIE)—that ties the renegotiation exchanges with the connection already established over which the renegotiation is conducted. This extension contains one of three values. If the party is not renegotiating, then it contains a fixed "empty" string. When a party would like to start renegotiation, it includes the verification data from the previous handshake. Specifically, the ClientHello that a client sends contains the value of the client_verify_data field from the Finished message of the precedent Handshake, that is,

$$\text{PRF}\,(\text{MasterSecret} \,\|\, \text{"client\_finished"} \,\|\, \text{Hash}\,(\text{handshake\_messages}))$$

Similarly, if the server is initiating the renegotiation, the ServerHello contains the concatenation of client_verify_data and the server_verify_data of the previous Handshake exchanges, where the server_verify_data field is defined as follows:

$$\text{PRF}\,(\text{MasterSecret} \,\|\, \text{"server\_finished"} \,\|\, \text{Hash}\,(\text{handshake\_messages}))$$

This will make sure that both sides have the same view of the previous handshake so that any mismatch will signal intrusions.

The specifications of SSL Version 3.0 and TLS 1.0 and 1.1 require implementations to ignore data following the ClientHello, that is, any extension, but some server implementations incorrectly fail the handshake in this case. To avoid this possibility, the ClientHello includes the code for a special cipher suite, called Signaling Cipher Suite Value (SCSV), in the list of supported cipher suites. The SCSV is not a true cipher suite, that is, it does not correspond to any valid set of algorithms and servers that do not implement the SCSV mechanism will ignore it. Servers that implement this method will know that the client is able to secure renegotiation through the RIE mechanism.

5.3.10 The Alert Protocol

The role of the Alert protocol is essentially to signal channel events to generate alarm messages in response to errors and to indicate changes in the state of a connection such as the closing of a connection.

The Record layer encrypts the Alert messages using the encryption attributes in place for all the messages coming from the upper layers. Depending on the seriousness of the threat, the alarm can be a simple warning or can cause the disconnection of the session. A warning message does not warrant a specific action. In contrast, a fatal message forces the transmit side to close the connection immediately without waiting for an acknowledgment from the other party. From its side, the receiver will close the connection as soon as the alarm message arrives. This feature, however, makes SSL/TLS vulnerable to denial-of-service attacks, should an intruder succeed in substituting the canonical messages with nonconforming messages that would provoke session disconnects. It should be noted that both SSL and TLS Alert messages are not authenticated. As a consequence, an attacker can send false alert messages using the same sequence numbers as those used by encrypted data packets, causing their removal for the data stream (Saarinen, 2000). Also, some Alert can leak side information that could be used to mount attacks as described later in this section.

The Alert protocol can be invoked in one of the following ways:

- By the application, for example, to indicate the end of a connection
- By the Handshake protocol if it encounters a problem
- By the Record protocol directly, for example, if the integrity of a message is in question

Table 5.6 lists the messages of the protocol Alert in SSL arranged in alphabetical order.

Table 5.7 lists the additional Alert messages of TLS 1.0 also in alphabetical order.

The messages no_certificate and export_restriction have been dropped in TLS 1.1.

The TLS 1.2 ClientHello message may contain the extensions to request additional functionalities from the server. The server checks that the size of data in the message matches one of the standard formats; otherwise, it returns a fatal decode_error alert. If the ServerHello contains extensions that the client does not support, the client sends the fatal message, unsupported_extension.

When a TLS 1.2 client's ClientHello requests the establishment of a connection with a server does not support TLS 1.2, the server will respond with a ServerHello indicating the version that it supports. If the client agrees to use this version, the handshake will proceed. Otherwise, the client must send a protocol_version alert message and close the connection.

Some changes to Alert messages followed the identification of some attack schemes, in particular the Bleichenbacher attack and the padding attack, which are described in the following.

TABLE 5.6

Messages of the Alert Protocol of SSL

Message	Context	Type
bad_certificate	Failure of a certificate verification.	Fatal
bad_record_mac	Reception of an incorrect MAC.	Fatal
certificate_expired	Expired certificate.	Fatal
certificate_revoked	Revoke certificate.	Fatal
certificate_unknown	Invalid certificate for other reasons than those mentioned earlier.	Fatal
close_notify	Voluntary interruption of a session.	Fatal
decompression_failure	The decompression function received improper data (data too long).	Fatal
handshake_ failure	Inability to negotiate common parameters.	Fatal
illegal_parameter	A parameter in the Handshake is out of range or is inconsistent with other parameters.	Fatal
no_certificate	Negative response to a certificate request.	Warning/fatal
unexpected_message	Inopportune arrival of a message.	Fatal
unsupported_certificate	The received certificate is not supported.	Warning/fatal

TABLE 5.7

Additional Messages of the Alert Protocol of TLS 1.0

Message	Context	Type
access_denied	Valid certificate received, but access was refused due to unsatisfactory checks.	Fatal
decode_error	Message not decoded for incorrect size or out of range parameter.	Fatal
decrypt_error	Failure of one of the cryptographic operations in the Handshake, for example, the decryption of *ClientKeyExchange*, signature verification, or verification of the *Finished* message.	Fatal or warning
decryption_failed	Failure in decryption of a block.	Fatal
export_restriction	Negotiation violating U.S. export restrictions.	Fatal
internal_error	Internal error independent of the protocol or the peer.	Fatal
insufficient_security	Cipher suite that the client proposes is less than what the server requires.	Fatal
no_renegotiation	Renegotiation refused following the initial contact.	Warning
protocol_version	The server recognizes the protocol version that the client requests but does not support it. Alternatively, the server suggests a protocol version that the client does not accept.	Fatal
record_overflow	The size of the block used exceeds the specifications.	Fatal
unknown_ca	Unknown certification authority.	Fatal
user_cancelled	Handshake is being canceled for some reason unrelated to a protocol failure.	Warning

5.3.10.1 The Bleichenbacher Attack

An SSL Version 3.0 server using RSA with the encryption coding standards PKCS #1v1.2 can be coerced into revealing whether the format of particular message, when decrypted, conforms to the format specified by PKCS #1 because the server stops processing if the message is not conforming to the format but proceeds with the signature verification if it is (Bleichenbacher, 1998). The Bleichenbacher attack takes advantage of flaws within the PKCS #1v1.2 function to gradually reveal the content of an RSA encrypted message using the timing difference in both cases. The time differential is exhibited also when different Alert messages are used to indicate if the message is not conforming or if the integrity verification fails. Following the capture of the ClientKeyExchange message with the encrypted PreMasterSecret, a man-in-the-middle adversary can use this knowledge to recover the PreMasterSecret and hence the session key by sending a large number of chosen ciphertexts (at least 2^{13} attempts) to the server. This is why this attack also became known as the *million message attack* (Rescorla, 2001, p. 168). However, that large number is still less than if a randomly chosen probe message was constructed. Figure 5.11 illustrates the SSL Version 3.0 block format when it is encrypted using PKCS #1v1.2.

The Bleichenbacher attack is a specific form of adaptive chosen-ciphertext attacks (CCA2) in which an attacker sends a number of ciphertexts to be decrypted and then uses the results of these decryptions to improve the selection of subsequent ciphertexts.

RSA PKCS #1v2.1 encryption uses one of the two methods specified in RFC 3447: the improved encryption/decryption scheme using Optimal Asymmetric Encryption Padding (OAEP) denoted as RSAES-OAEP or RSAES-PKCS1-V1_5. RSA PKCS #1v1.5 replaces the older RSA PKCS #1v1.2 encryption, which is susceptible to the Bleichenbacher attack. PKCS #1v1.5 has a padding that is sensitive to changes in the plain text so that the probing message that an attacker sends will also fail the integrity check and be rejected. A systematic analysis has shown that TLS with RCA PKCS #1v1.5 is secure (Krawczyk et al., 2013).

TLS 1.0 thwarts the threat by treating incorrectly formatted messages in the same manner as correctly formatted RSA blocks, thereby hiding decryption failures.

An extension of the Bleichenbacher attack revolves around the two left-most octets of the PreMasterSecret

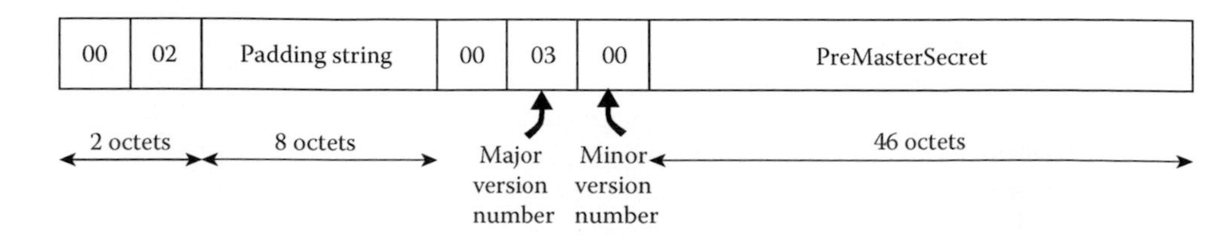

FIGURE 5.11
TLS session establishment with protection against denial-of-service attacks with puzzles.

used to store the major and minor version numbers (Klima et al., 2003). From the definition of the protocol, the attacker knows the expected value of the version numbers and that, according to the PKCS #1 specification, the preceding octet contains "0x00." This knowledge can be used to reduce the number of chosen plaintexts to be tried to recover the PreMasterSecret. Treating nonconforming message blocks and/or mismatched version numbers in a manner indistinguishable from the case of a conforming message block avoids this vulnerability, and this approach is used in TLS 1.2. It should be noted that the complexity of these attacks may render them impractical in many situations.

5.3.10.2 Padding Attacks

Padding consists in aligning the length of the final block with the size required by the encryption algorithm (zero to 255, inclusive). This is done by several possible schemes. For example, n octets are added to the plain text such that the total is a multiple of the block size B. The plain text is padded by appending the n octets, each n containing the value n, that is, 1 block of "1," 2 blocks for "2," 3 blocks of "3," and so on. Thus, the last block ends with a string of octets: 0x01, or 0x0202, or 0x030303, and so on. This is the padding method of PKCS #7. Another possibility is to add a single octet of 0x128 and then as many 0 octets as needed to make the overall length a multiple of the block size (Ferguson et al., 2010, p. 64).

If padding errors are treated different from bad MACs, an adversary may be able to launch a padding attack. The attack works in two phases (Vaudenay, 2002). As shown in Figure 3.16, the decryption operation is described by

$$P_i = C_{i-1} \oplus D_k(C_i)$$

Assume, for example, that PKCS #7 padding is used. The attacker takes the penultimate code C_{B-1} and flips its bits randomly and submit the modified code C'_{B-1} and C_B to the TLS server. If the padding is correct, then this means that the decoded plain text P_B^i ends with 0x01 with probability $1/2^8$ or 0x0202 with probability $1/2^{16}$, or 0x030303 with probability $1/2^{64}$, and so on. The correct padding will be found in at most 255 attempts, and 0x01 is the most likely last octet of the P_B. (The less probable cases can be detected with a few additional attempts.) Once the value of the final octet of P_B is known, it is possible to get the last octet of the penultimate block in the original message P_{B-1} as the final octet of $P'_{B-1} \oplus 0x01$. Other padding schemes are discussed in Black and Urtubia (2002).

TLS 1.0 defines two distinct alert messages one for bad padding (decryption_failed) and the other for MAC errors (bad_record_mac). Even though the error codes are encrypted, padding verification takes much less time than MAC verification, and this distinction may be exploited to perform some attacks on padding in CBC-based cipher suites.

An attacker can send repetitive requests to conduct a dictionary attack over several sessions to recover the secret message without knowing the encryption key under the following conditions (Canvel et al., 2003; Moeller, 2004):

- The signature is constructed using a block algorithm in the CBC mode of encryption.
- Traffic is intercepted near the receiving end.
- The underlying exchange protocol is known.
- The same plaintext is transmitted in the same position for all the sessions. This is the case, for example, of a fixed length password used for authentication.

TLS 1.1 defends against this timing attack by specifying that HMAC be computed even when the padding is not valid before rejecting the packet. However, what data to use for the calculation are not specified. Also, TLS 1.1 eliminated explicit error messages that could indicate whether the padding check or the integrity check is what causes a decryption failure. Thus, the bad_record_mac message refers to both padding errors and a bad signature. The use of decryption_failed is optional. Finally, in TLS 1.2, the decryption_failed and the no_certificate alerts are not used. However, when the padding check fails, there is no way to determine the size of the actual message and the number of padding octets to be removed, so the padding is assumed to be of zero length and the integrity check uses the whole block. Thus, when the padding is damaged, the computations can take longer, which opens the way for some other attacks (AlFardan and Paterson, 2003a,b).

5.3.11 Denial-of-Service Attacks

The computational load on a server during the establishment of a TLS/SSL session can be substantial, particularly after receiving the ClientKeyExchange message. By submerging the server with requests, an attacker can overwhelm the processing resources of an SSL Version 3.0 server. TLS 1.0 or 1.1 implementations can ignore received Record blocks if their type is unknown (§6 of RFC 2246, 1999; RFC 4346, 2006). Using this property, it is possible to thwart a denial-of-service attack during the establishment of a TLS session (Dean and Stubblefield, 2001). The defense consists in sending a *puzzle* to the client when the computation load on the server exceeds a certain threshold. As a consequence, clients are obliged to abate transmissions, allowing the server some respite for recuperation.

The puzzle is the following simple cryptographic problem. The server selects a random number x, computes its hash $h(x)$, and then forms a new number x' by zeroing the n lower order bits of x. The triplet $\{n, x', h(x)\}$ forms the puzzle that is sent following the Certificate message. The client has to discover the value of x by trying all the values for the missing n bits and then comparing hash for each combination with the received hash $h(x)$. The average number of hashes that the client computes is 2^{n-1} (see Section 3.7.1) while the server computes only two: the first is part of the puzzle and the second to verify that the hash of the value that the client discovers corresponds to the hash that was sent. These exchanges are shown in Figure 5.12.

As long as the server congestion lasts, only clients that react to the puzzle will have a chance to connect. The

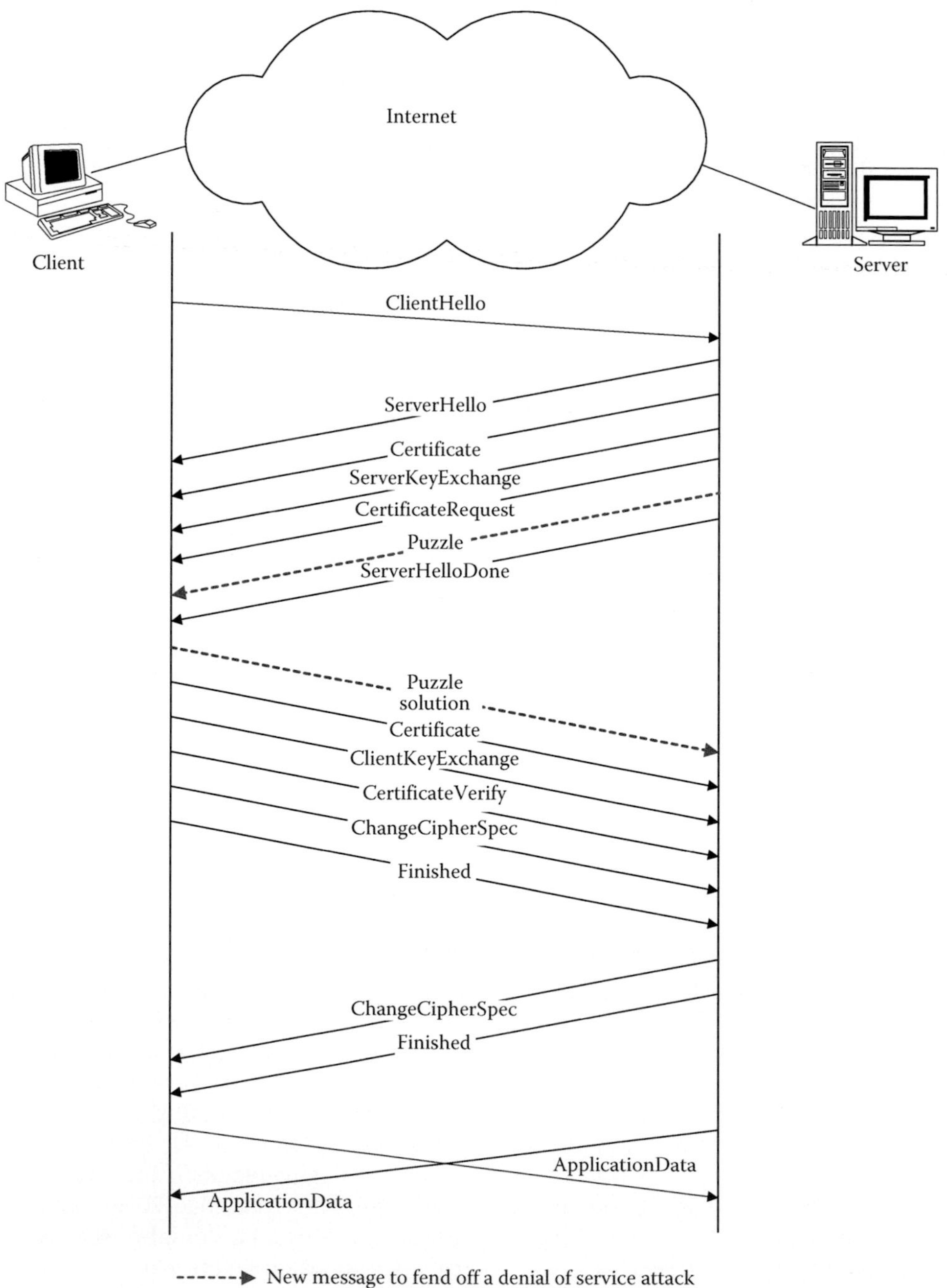

FIGURE 5.12
Format of an SSL Version 3.0 block with RSA encryption using PKCS#1v1.2.

server will timeout clients that do not process the puzzle, thereby freeing the resources allocated to these aborted sessions. Once the congestion has cleared, the server will respond connection attempts following the normal TLS procedures, that is, without imposing the puzzles on the clients.

With TLS 1.2 (RFC 5246, 2006), an unknown Record type must result in the sending of the unexpected_message alert message and the termination of the connection. So, a possible strategy under congestion is to send an unknown Record type to force the client to abort the session.

5.4 Performance of SSL/TLS

The main computational load in SSL/TLS comes from the cryptography, particularly, during session establishment. The ephemeral Diffie–Hellman algorithm for key establishment is exceptionally costly and is used only if the perfect forward secrecy is necessary; in that case, a fresh key is used for each handshake.

The performance of an e-commerce server is measured by three criteria: the number of simultaneous encrypted sessions (i.e., the number of transactions per second), the response time to requests, and the available throughput for each of these sessions. The SSL/TLS protocols use significant computing resources due to the use of public key algorithms. The total load on the server is substantial, because it will have to deal simultaneously with multiple requests for session establishment. The load increases with the number of simultaneous sessions saturating the server and leading to a substantial drop in the performance.

In general, for a given client, the load on the server can be twice or three times that of the client (but if the client has an authentication certificate, the client load will be quadrupled, while that of the server will increase by about 20%; Rescorla, 2001, pp. 187–188). More specifically, the use of the DSA signature algorithm and the exchange of keys with the ephemeral Diffie–Hellman algorithm increase the load on the server by a factor ranging from 5 to 7 (Rescorla, 2001, p. 192).

Among the measures used for load reduction are session resumption to avoid a new Handshake and the use of accelerators to off-load the cryptographic computations. These accelerators come either in the form of separate machines or as boards to be inserted in the server. Separate machines act as proxies between the server and the clients and can be organized in series or in parallel to increase the computing power by distributing the load over several machines. A cluster arrangement has an additional advantage: in case of failure, the computation can automatically be switched from the failed machine to a standby machine. To achieve this recovery however, each machine of the cluster must know the instantaneous states of all the connections established through all the machines of the same cluster (Rescorla, 2001, pp. 204–209; Rescorla et al., 2002; Qing and Yaping, 2009). This arrangement is given many names such as hot standby arrangement, server-aided RSA, or client-aided RSA.

Inter-machine communications add an extra computational surcharge that can mask the gain in speed if the number of machines per cluster exceeds three to four machines. In addition, because SSL/TLS are Point-to-Point Protocol, the proxy has to terminate the connection, that is, decrypt the messages from the client to extract the information to be distributed to the other machines (or to a mirror machine) so as to ensure recovery in case of failure, before forwarding them to the server. Clearly, it becomes imperative to protect all these machines against potential physical or logical intrusions; otherwise, the whole security edifice will be affected.

The choice of an accelerator board is tricky: expected performance gains vary with the board, the server configuration, and the traffic profile. The improvement depends on the utilization of the central processors, which affects their response to unexpected traffic peaks. Other factors include the rate with which their performance saturates and their tolerance to faults. As a consequence, even when the average response time improves, the response time may be significant in some cases. Also, the presence of additional machines between the client and the server to carry out the cryptographic operations adds new security risks. Therefore, the use of accelerator boards requires a good understanding of the service context; otherwise, they may be useless if not harmful (Bontoux, 2002).

Other techniques to improve the performance are designed to speed up the decryption (Qing and Yaping, 2009). For example, with rebalanced RSA, the algorithm parameters e, d are chosen such that, while d is large, $d \bmod (p-1)$ and $d \bmod (q-1)$ are small numbers. Another approach to speed RSA decryption is to use batching techniques.

5.5 Implementation Pitfalls

SSL was originally proposed by Netscape and evolved into TLS. In general, implementations of SSL/TLS comprise two modules: the first for the cryptographic functions and the second for the protocol library. This software architecture allows the modification of the properties of the composite SSL/TLS server according to legal restrictions or technical choices.

Many weaknesses were discovered in SSL Version 2.0 (RFC 6176):

- Message authentication uses MD5.
- Handshake messages are not protected. This permits a man-in-the-middle attack to trick the client into picking a weaker cipher suite than it would normally choose.
- Message integrity and message encryption use the same key, which is a problem if the client and server negotiate a weak encryption algorithm.
- Sessions can be easily terminated. A man in the middle can easily insert a TCP FIN to close the session, and the peer is unable to determine whether it was a legitimate end of the session.

This is why RFC 6176 (2011) prohibits the use of SSL Version 2.0 in new implementations. While SSL 3.0 is more secure, it was not approved at the U.S. federal level because it relies in part on cryptographic algorithms that are not on the approved list (such as RSA for key establishment or RC4 for confidentiality). In contrast, TLS, when properly configured, is approved for the protection of U.S. federal information (Chernick et al., 2005, p. 1, n. 1).

No formal protocol description language was used in the specification of either SSL or TLS, but a mixture of English and pseudo-code. Furthermore, there are no standard reference implementations nor conformance test suites to verify implementations before they are deployed. Inevitably, this has led to misinterpretations as well as interoperability and security problems. In fact, many differences were noticed between the Netscape version of SSL and the specifications that Netscape itself had published (Freier et al., 1996; Rescorla, 2001, pp. 50, 79). Furthermore, before TLS 1.2, it was not clear what version to be used in the messages, the one that was in the ClientHello message or the one negotiated for the session. Appendix D.4 of RFC 5246 lists some pitfalls, including:

- Treatment of Handshake messages, such as ClientHello, Certificate, and Certificate Request, that require fragmentation into multiple TLS records because their size exceeds 2^{14} octets.
- In the ClientKeyExchange message, when the PreMasterSecret is encrypted with RSA, the version used must be the same that was in the ClientHello message and not the one negotiated. This is to prevent rollback attacks.
- Not sending an empty Certificate message if the client does not have a suitable certificate in response to the server's CertificateRequest.
- Behavior when the handshake encounters an error (the handshake should continue).
- TLS 1.2 allows the client to specify to request additional functionalities from the server. The rules for handling these extensions are quite complicated and could lead to many interoperability problems. For example, some servers will refuse the connection if any TLS extensions are included in ClientHello.
- Correct checking by the client implementation of the parameters that the server has sent for the Diffie–Hellman key exchange.
- The use of unpredictable initialization vectors in CBC mode ciphers or of a strong and, most importantly, properly seeded random number generator.
- Some server implementations incorrectly close the connection when the client offers a version newer than TLS 1.0. Earlier versions of the TLS specification were not fully clear on what to do in this situation.

The OpenSSL is today the most widely used open-source cryptographic library. It is used in many open-source operating systems such as Ubuntu, Fedora, or OpenBSD, in web servers such as Apache, as well as in client software. Its genesis is a collaborative effort to develop a commercial quality implementation of SSL—and later TLS—starting from the library that Eric A. Young and Tim J. Hudson created under the name of SSLeay. The initial stimulus was to bypass the U.S. restrictions on the export of encryption software. However, the OpenSSL Foundation is now based in the United States. The code is available free of charge and can be downloaded from http://www.openssl.org. Members of the core development team are listed at http://www.openssl.org/community/team.html, last accessed January 18, 2016).

In 2014, it was discovered that Releases 1.0.1* to 1.0.1f of OpenSSL have an error in the implementation of the Heartbeat Extension mechanism; the fix was hurriedly made available in Release 1.0.1g on April 7, 2014. This bug, nicknamed "Heartbleed," was independently discovered by security engineers at Codenomicon and Neel Mehta of Google Security. Adam Langley and Bodo Moeller prepared the fix.

When a Heartbeat Request arrives containing a payload, the receiver responds with a message carrying an exact copy of the arriving payload. The payload consists of an arbitrary bit pattern. The bug is the lack of

* Release 1.0.1 was made available on March 14, 2012.

check of whether there are actual data in the incoming request or that the payload is not too large. This can be represented as

```
buffer = malloc (1 + 2 + payload + padding)
```

The message type is coded in 1 octet and 2 octets are used to represent the payload length. If there is no check on the payload length, an attacker can request the allocation of system memory in chunks of 64K octets, which are then sent back in the response. Another problem is that the memory allocated is not cleared before it is sent. The attacker repeats the request until enough secrets are revealed thereby comprising the secret keys used to encrypt the traffic, the names and passwords of the users, and the actual content. Exploitation of this security hole leaves no abnormal traces in the system logs. When this bug was discovered in 2014, the Federal Financial Institutions Examination Council urged banks that use OpenSSL to upgrade their software as soon as possible to address the vulnerability and to replace all the private keys for each service. The fix is to simply add a bounds check and drop the packet if the payload length does not fall within the required limits.

The incident revealed the extent of the use of OpenSSL; social sites, banks, network equipment manufacturers, service companies, network appliances, and even government sites all rushed to install the repaired version of the software. It should be noted that, even with this bug, PFS protects past communications from retrospective decryption. Unfortunately, PFS is not commonly used and when used, it depends on the quality of the random number generator in generating truly random numbers. Luckily, however, there had been no reports that the vulnerability was exploited.

5.6 Summary

TLS is the mostly widely used cryptographic protocol to secure communication over the Internet because it supplies a relatively simple mechanism to protect exchanges between two points over TPC. The client and server negotiate the parameters of a stateful connection using the Handshake protocol. During that handshake, the authentication of the server takes place (authentication of the client is optional), and both parties negotiate the cipher suite to establish a shared key (MasterSecret) and other cryptology parameters. Data are fragmented and secured using the Record protocol; each fragment is encrypted individually after adding an integrity check. To protect against replay attacks or reordering attack, the computation of the message authentication code includes the sequence number of the fragment to be transmitted.

TLS 1.0 was an upgrade of SSL Version 3.0. Its exchanges follow the same scheme of SSL, but the main differences concern the following items:

- The instant at which the encryption of the data starts
- The available cipher suites
- The method of computation of the MasterSecret and derivation of the session key
- The number of Alert messages
- The reaction to Record blocks of unknown types

The evolution of TLS has been largely driven by the discovery of security vulnerabilities. The first set of vulnerabilities concerned the protocol. Later, the focus was on attacks that rely on side information that is unintentionally revealed as the cryptographic operations take place; this is why they are often called *side channel attacks*. Many recent attacks require a large number of sessions to conduct a timing analysis of the decryption processes or an analysis of the statistical biases in the encryption stream to recover the plaintext, partially or fully. Table 5.8 summarizes the main vulnerabilities and how their treatment guided the evolution of SSL/TLS.

TLS 1.1 added some security measures by replacing the implicit definition of the initialization vector with an explicit specification, by changing the handling of padding errors, and by the use of AES encryption. TLS 1.2 made significant cryptographic changes: the use of EEC for encryption, the use of the SHA-256 to replace the MD5/SHA-1 combination in the generation of the pseudo-random function, and by giving the client the ability to indicate to the server the hashing function to be used in the digital signature independently from the encryption algorithm. TLS extensions are also supported. In the current state of knowledge, it seems that authenticated encryption, which was only added in TLS 1.2, remains the only possibility with no known vulnerabilities.

In applications that involve several actors simultaneously, such as a client, a merchant, and a gateway with banking networks, SSL/TLS associations must be established in parallel, which increases substantially the complexity of the operation. At each association, the data are decrypted and reencrypted. Thus, in electronic commerce applications, the merchant will have access to all the account information of the buyer.

TABLE 5.8

List of Vulnerabilities Guiding the Evolution of SSL/TLS Specifications

Vulnerability	Resolution
SSL 3.0 allows a party to send application data as soon as that party sends its Finished message. This gives an intruder the possibility to send a spurious ClientHello to renegotiate a less robust cipher suite.	TLS 1.0 requires the party to wait for the Finished message from the other side. In TLS 1.2, the version in the ClientKeyExchange must be identical to the version in the ClientHello.
SSL 3.0 uses a key derivation model with MD5, which was shown to be vulnerable.	TLS 1.0 uses a combination of MD5 and SHA-1. TLS 1.2 removed MD5 altogether and uses SHA-256.
Compression allows the discovery of secret information based on the size of the compressed HTTP request.	Disable compression.
Renegotiation gives an intruder a short window to hijack application protocols such as HTTP, where authentication is performed once at the beginning of the session.	Either disable renegotiation or tie the renegotiation to the connection already established.
A denial-of-service attack can be conducted on SSL V3, TLS 1.0 and 1.1 using unknown blocks.	In TLS 1.2, an unknown block causes the termination of the connection.
Having two alert messages one indicating padding errors and the other integrity failures can be used to recover the secret message without knowing the encryption key.	TLS 1.1 specifies that the integrity check must be carried out even when the padding is not valid before rejecting the packet. Also, the same error should be used for padding error and integrity failures.
SSL V3 using RSA with PKCS #1v1.2 can be coerced to reveal the PreMasterSecret.	Treat incorrectly formatted and correctly formatted messages the same. Use PKCS #1v1.5.

Another structural weakness is due to the fact that the encryption parameters are not necessarily updated during a session; the longer the session, the higher the chance that the keys can be broken.

Questions

1. What are the advantages and disadvantages of using SSL/TLS in electronic commerce?
2. Define perfect forward secrecy.
3. What is a cipher suite in SSL/TLS?
4. What is the main difference in key generation between SSL Version 3.0 and TLS 1.0?
5. What are the main improvements of TLS 1.0 over SSL Version 3.0?
6. How does the establishment of a TLS/SSL connection differ from the establishment of a TLS/SSL session?
7. What are the advantages and disadvantages of the ephemeral mode of operation in TLS/SSL?
8. Explain the challenges faced in the use of SSL accelerators.
9. Why are TLS/SSL implementations so difficult to secure?
10. Why did RC4 continued to be used in TLS even though the U.S. federal government prohibited its use in government installations?
11. What is the advantage of delaying data transmission until both sides have exchanged and verified their Finished messages?
12. How can a denial-of-service attack be mounted against some versions of SSL/TLS? How can the use of cryptographic puzzles offer some protection?

6

Wireless Transport Layer Security

This chapter reviews the technical aspects of Wireless Transport Layer Security (WTLS), a specification by the Wireless Application Protocol (WAP) Forum to secure transactions in mobile networks (Wireless Application Protocol Forum, 2001). The first version of WTLS was a complete revision of TLS to meet the constraints of data communications over Global System for Mobile Communication (GSM) networks using the Short Message Service (SMS) for bit rates of 9.6 kbit/s and the General Packet Radio Service (GPRS) for bit rates between 28 and 56 kbit/s. Cellular technologies of the third generation (3G) or Universal Mobile Telecommunication System (UMTS) were also considered. The protocol was designed taking into consideration the limitation of handsets in terms of memory, computational power, battery life, screen displays, and keyboards. There were several revisions, the latest was WTLS Version 6 in April 2001. However, given the operational complexities of WAP 1.x, particularly that of end-to-end security with WTLS in tandem with TLS, WAP 2.0 was published in August 2001 based on the Transmission Control Protocol/Internet Protocol (TCP/IP) stack. This necessitated some extensions to TLS to accommodate mobile communication. All these considerations will be discussed in the following.

Note: The WAP Forum was consolidated along many other organizations of the mobile industry into the Open Mobile Alliance (OMA).

6.1 Architecture

The first version of WTLS was part of the Wireless Application Protocol (WAP) stack of April 1998. This stack was designed to satisfy the following requirements:

- Maintain the operation over the User Datagram Protocol (UDP) even with packet loss.
- Operate even when the round-trip transmission delays are important.
- Take into account the reduced computation power or storage capacity of mobile terminals.

Figure 6.1 depicts the position of WTLS in the WAP 1.0 stack and its correspondence with the TCP/IP stack. At the top layer, the Wireless Markup Language (WML), which is more adapted than HyperText Markup Language (HTML) to the constraints of limited bandwidth and terminal capabilities, is used to describe access pages. Furthermore, a microbrowser (WAP browser) was needed and its communication with the WAP gateway was binary encoded for efficiency.

The use of WTLS in banking transactions requires a public key infrastructure under the control of the user's financial institution. As will be explained in Chapter 9, typical mobile devices have a Subscriber Identity Module (SIM), a tamper-resistant integrated circuit card that securely stores the International Mobile Subscriber Identity (IMSI) and all keys used to identify and authenticate subscribers in the network. A personal identification number (PIN) is needed to access the content stored in the SIM card. Terminals used for WAP services have a Wireless Identification Module (WIM) to hold the necessary keys and the user's credentials for the financial transaction. The relation between the SIM and the WIM can be in one of the following configurations:

- The SIM and WIM are in the same integrated circuit card.
- The handset is a dual-slot phone with two separate card readers, one for the SIM card and the other for the WIM card.
- The WIM reader is external that communicates with the handset using a wireless protocol, such as Bluetooth.

6.2 From TLS to WTLS

The main modifications that WTLS introduces to Secure Sockets Layer (SSL)/Transport Layer Security (TLS) relate to the following elements:

- The format of the identifiers and the certificates
- The cryptographic algorithms
- The content of some Handshake messages
- The exchange protocol during the Handshake
- The calculation of secrets
- The size of the parameters

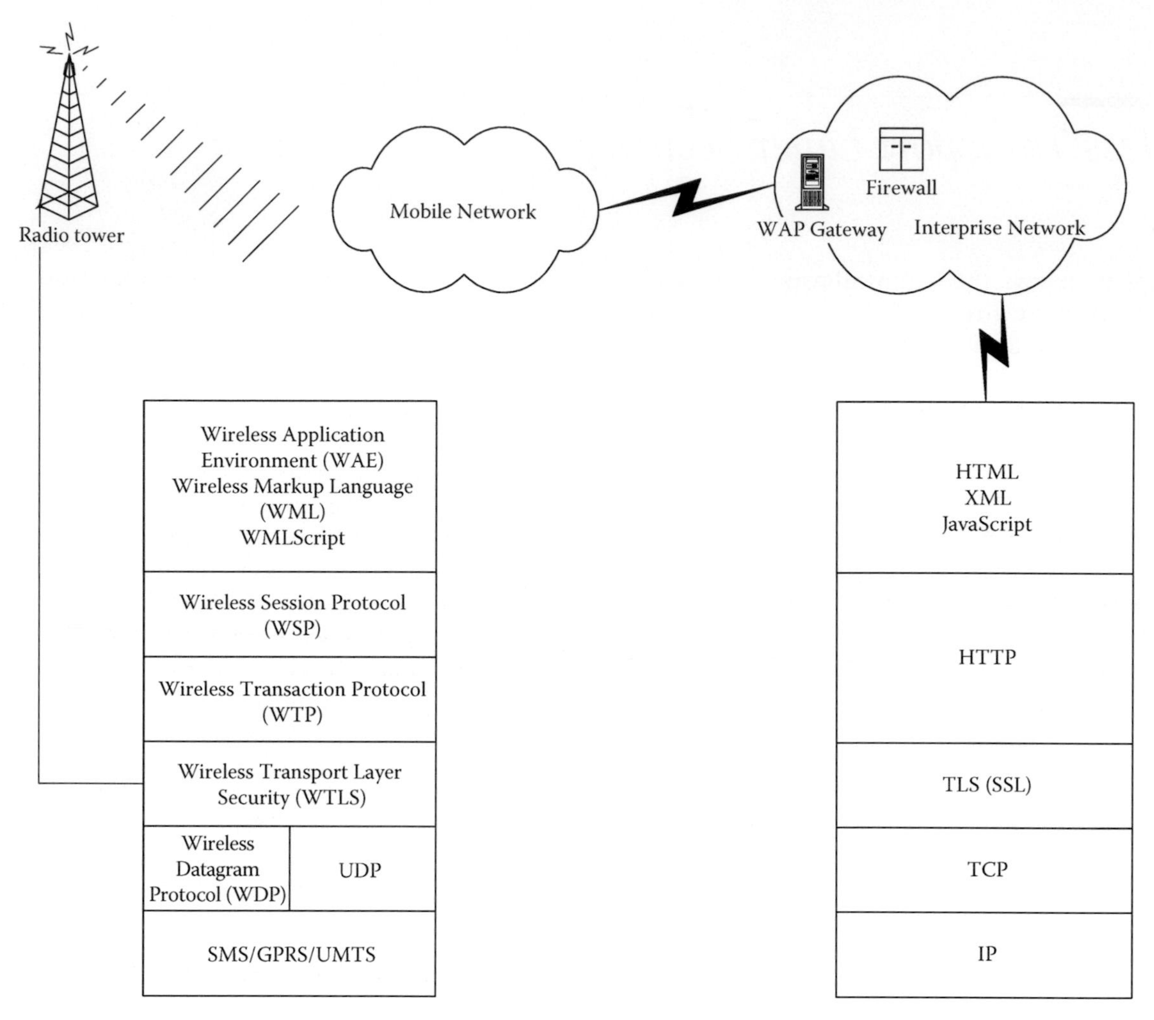

FIGURE 6.1
Position of WTLS in the protocol stack of WAP 1.0 and its correspondence with the TCP/IP stack.

- The Alert messages
- The role of the Record protocol

All these modifications make WTLS totally incompatible with TLS/SSL even though, like them, WTLS is composed of four protocols: Handshake, Record, ChangeCipherSpec (CCS), and Alert. Figure 6.2 depicts its position in the protocol stack between the WAP application called Wireless Transaction Protocol (WTP) and the transport layer, which can be either Wireless Datagram Protocol (WDP) or UDP. The combination of WSP/WTP/UDP (Wireless Session Protocol/Wireless Transaction Protocol/User Data Protocol) reduces by more than half the number of packets that HTTP/TCP/IP would have used to perform the same functions.

WDP is responsible for data fragmentation and not the Record protocol as in the case of SSL/TLS. Record adds an unencrypted header of 1–5 octets to each message arriving from the higher layers to indicate the message type and to signal the presence of optional fields. These include 2 octets for the sequence number in the case of explicit numbering and 2 octets for coding the length of the block of encapsulated data.

6.2.1 Identifiers and Certificates

There are several WTLS identifiers: X.509 distinguished names, SHA-1 hash of the public key, a secret binary key known only to the two parties, and a textual name in a character set known to the two parties. Just like SSL and TLS, the exchanges can be anonymous, but this mode of operation is subject to man-in-the-middle attacks.

WTLS recognizes three types of certificates:

1. Certificates defined in ITU-T Recommendation X.509 Version 3 (1996)
2. Certificates defined in ANSI X9.68 (2001) for certificates signed with the elliptic curve digital signature algorithm (ECDSA)
3. WTLS' own certificates, optimized to save on the wireless bandwidth

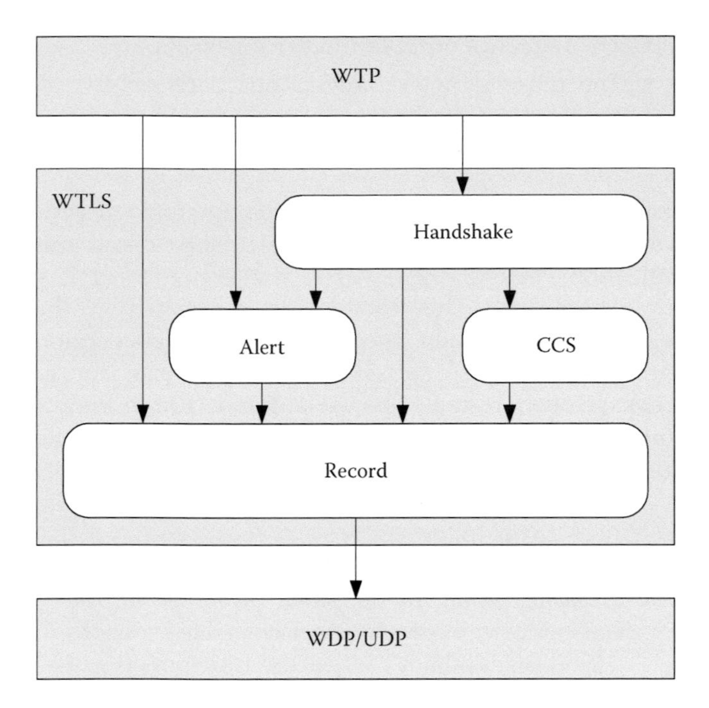

FIGURE 6.2
Protocol stack for WTLS subprotocols.

6.2.2 Cryptographic Algorithms

Table 6.1 contains the Key Exchange Algorithms used in charge by WTLS. It is seen that WTLS adds the Elliptic Curve Diffie–Hellman (ECDH) algorithm to those used in TLS. In addition, the key size for the RSA exchanges is 512, 768, or 1024 bits.

Table 6.2 lists the encryption algorithms used in WTLS. They all use blocks of 64 bits with an initialization vector of the same size. Notice the presence of several variations of the RC5 protocols as well as of IDEA. Some algorithms have been deliberately weakened to satisfy the restrictions that the U.S. authorities imposed at the time on the export of encryption algorithms.

Finally, the hash functions SHA-1 or MD5 are used to verify the integrity of the transport. However, the procedures of Table 6.3 are much weaker than those used in SSL/TLS.

As shown in Table 6.4, WTLS parameters have been shortened compared to those of TLS/SSL.

As a result, the size of the Finished message is reduced to 12 instead of 36 octets.

TABLE 6.1
Key Exchange Algorithms Negotiated by the Handshake Protocol of WTLS

Code for the Key Exchange Suite	Description	Key Size Limit in Bits
NULL	No exchange; the length of the PreMasterSecret is zero.	—
SHARED_SECRET	The secret is exchanged by other means than WTLS, for example, it may be cast in a smart card of the mobile terminal.	None
DH_anon	Diffie–Hellman key exchange without authentication; each side sends to the other its temporary Diffie–Hellman public key; each party calculates the PreMasterSecret using its private key and the public key of the counterpart.	None
DH_anon_512	Diffie–Hellman key exchange without authentication; each side sends to the other its temporary Diffie–Hellman public key; each party calculates the PreMasterSecret using its private key and the public key of the counterpart.	512
DH_anon_768	Diffie–Hellman key exchange without authentication; each side sends to the other its temporary Diffie–Hellman public key; each party calculates the PreMasterSecret using its private key and the public key of the counterpart.	768
RSA_anon	RSA without authentication; the PreMasterSecret is the secret value that the client generates encrypted with the public key of the server appended with the server's public key	None
RSA_anon_512	RSA without authentication; the PreMasterSecret is the secret value that the client generates encrypted with the public key of the server appended with the server's public key.	512
RSA_anon_768	RSA without authentication; the PreMasterSecret is the secret value that the client generates encrypted with the public key of the server appended with the server's public key.	768
RSA	RSA with authentication with certificates.	None
RSA_512	RSA with authentication with certificates.	512
RSA_768	RSA with authentication with certificates.	768
ECDH_anon	ECDH without authentication	None
ECDH_anon_113	ECDH without authentication	113
ECDH_anon_131	ECDH without authentication	131
ECDH_ECDSA	ECDH with certificates signed by ECDSA.	None

TABLE 6.2

Encryption Algorithms Used in WTLS

Representation of the Cipher	Exportable	Effective Length of the Key in Bits	Use in SSL/TLS
NULL	Yes	—	Yes
RC5_CBC_40	Yes	40	
RC5_CBC_56	Yes	56	
RC5_CBC	No	128	
DES_CBC_40	Yes	40[a]	Yes
DES_CBC	No	56	Yes
3DES_CBC_EDE	No	168	Yes
IDEA_CBC_40	Yes	40	
IDEA_CBC_56	Yes	56	
IDEA_CBC	No	128	Yes

[a] The encryption use only 35 bits of the key, and the remaining form parity bits (Saarinen, 2000).

6.2.3 Handshake Messages and Exchanges

The *ClientHello* message adds the following fields to those already present in SSL/TLS:

- A list of the key exchange methods that the client accepts in the order of decreasing preference; in SSL/TLS, the whole cipher suite is negotiated at once.
- The list of identifiers that the client accepts in the order of decreasing preference.
- The list of certificates.
- The list of compression methods.
- The sequence number mode for packets.
- The number of exchanged bits before the cryptographic parameters are refreshed (key_refresh).

The additional fields in the ServerHello message contain the following information:

- The sequence number mode for packets
- The interval in exchanged bits between two consecutive updates of the cryptographic parameters (key_refresh)

The renewal of the cryptographic parameters (encryption key, key for hashed MAC calculations, and initialization vectors) occurs after the exchange of 2 × key_refresh bits. This cadence is negotiated at the beginning of a session. Modification of the encryption parameters during a session reduces the risk due to breakage of short keys. The variables `client_random` and `server_random` remain constant throughout the duration of the associated connection.

The main modifications that WTLS introduces in the exchanges are as follows:

1. Messages going in the same direction can be consolidated in one transmission. This consolidation is mandatory when the transport is not reliable and is used for text exchanges in SMS.
2. Retransmission of messages is allowed under some conditions.
3. The ClientKeyExchange message is no longer mandatory. In fact, it is redundant when the key exchange uses ECDH and the client is certified.
4. The packet sequence number mode can be selected. When the transport layer is TCP, each side maintains the sequence numbers separately for each connection and increment them with each message transmission for that connection. This is the same for SSL/TLS, and the mode is called implicit. In contrast, when the transport layer is UDP or any other unreliable protocol, the explicit mode is mandatory. In this mode, the sequence numbers are sent in the

TABLE 6.3

Hashing Algorithms Used in WTLS

Representation of the Hash Function	Description	Key Size in Bits	Size of the Hash Bits	Use in SSL/TLS
SHA_0	No hashed MAC.	—	—	
SHA_40	Only the first 40 bits of the output are used.	160	40	
SHA_80	Only the first 80 bits of the output are used.	160	80	
SHA	All output bits are used.	160	160	Yes
SHA_XOR_40	The input data are divided into blocks of 5 octets; an *exclusive OR* operation is successfully performed on different blocks; this function is not recommended and does not protect stream ciphers (Saarinen, 2000); it is kept for terminals with little computing power.	0	40	
MD5_40	Only the first 40 bits of the output are used.	128	40	
MD5_80	Only the first 80 bits of the output are used.	128	80	
MD5	All output bits are used.	128	128	Yes

TABLE 6.4

Comparison of the Sizes of Some Variables in SSL/TLS and WTLS

	Size in Octets	
Variable	**SSL/TLS**	**WTLS**
symmetric encryption key	5–16	5–21
`client_random`	32	16
session identifier	3	2
MasterSecret	48	20
sequence number	8	2
PreMasterSecret	48	Variable (20 for RSA)
`server_random`	32	16

clear in Record messages so that they can be used to compute the hash values. A third choice is not to have any numbers, but this mode is not recommended because it is exposed to replay attacks, in particular when the higher layers do not include the necessary defenses

6.2.4 Calculation of Secrets

6.2.4.1 Computation of the PreMasterSecret

In case of a shared secret, the PreMasterSecret is stored in the terminal smart cards. Otherwise, it is exchanged using one of the algorithms of Table 6.1. The size of the PreMasterSecret varies according to the method used for the key exchange.

6.2.4.2 Computation of MasterSecret

The pseudo-random function (PRF) is constructed with one hash function only. Thus, we have

PRF (secret, label, seed) = P_hash (secret, label||seed)

where

$$\begin{aligned} \mathrm{P\_hash}(\text{secret, rand}) &= \mathrm{HMAC\_hash}\left[\text{secret, } \mathrm{A}(1) \,\|\, \text{rand}\right] \,\| \\ &\quad \mathrm{HMAC\_hash}\left[\text{secret, } \mathrm{A}(2) \,\|\, \text{rand}\right] \,\| \\ &\quad \mathrm{HMAC\_hash}\left[\text{secret, } \mathrm{A}(3) \,\|\, \text{rand}\right] \,\| \,\ldots \end{aligned}$$

where

rand = label || seed

P_hash represents P_MD5 or P_SHA-1 depending on whether MD5 or SHA-1 is used, while HMAC_hash indicates HMAC_MD5 or HMAC_SHA-1 depending on the case

The seed is a random number, and the label is a character string. In this particular case, the label is the text "master secret." Once that MasterSecret is obtained, the encryption key can be extracted for the encryption block:

KeyBlock = PRF {MasterSecret, expansion label, sequence number||`server_random`||`client_random`}

The block is recomputed at regular intervals according to the period for key refresh. On the server side, the expansion label is *server expansion*, while on the client side it is *client expansion*. The operations are repeated enough times so that the output can be divided into the required fields. From the client side, the partitions are the following:

```
client_write_MAC_secret(size of the hash key)
client_write_encryption_key(size of the
  encryption key)
client_write_IV(size of the initialization
  vector)
```

On the server side, the partitions are as follows:

```
server_write_MAC_secret(size of the hash key)
server_write_encryption_key(size of the
  encryption key)
server_write_IV(size of the initialization
  vector)
```

At the time WTLS was defined, U.S. regulations prevented the export of some cryptographic algorithms. For those encryption algorithms that were exportable, the encryption key and the initialization vector are capped according to the following formulae:

```
final_client_write_key =
  PRF(client_write_key, "client write key",
  client_random||server_random)
client_write_iv = PRF("","client write IV",
  client sequence
  number||client_random||server_random)
final_server_write_key = PRF(server_write_
  key, "server write key",
  client_random||server_random)
server_write_iv = PRF("","server write IV",
  server sequence
  number||client_random||server_random)
```

Thus, for these algorithms, the pseudo-random function used in the last step of the computation of the initial vectors does not contain a secret. As a consequence, their value depends on variables that were sent in the clear, that is, the sequence numbers of the client or the server and the random numbers `client_random` and `server_random`.

Finally, the initialization vector for the CBC (cipher block chaining) encryption mode is formed from the vector calculated earlier using an exclusive OR operation applied on a new vector S. S is constructed by concatenating the 2 octets of the sequence numbers enough times to form the number of octets needed in the initialization vector. For example, if we start with an initialization vector IV_0 of 64 bits, the initialization vector used IV will be calculated with the corresponding vector S according to the following formula:

$$IV = IV_0 \oplus (S\|S\|\|S\|S)$$

6.2.5 Alert Messages

Table 6.5 contains an alphabetical list of the Alert messages that WTLS added to those of the TLS 1.0 of RFC 2246 (1999). The message close_notify of SSL/TLS was separated into connection_close_notify and session_close_notify to distinguish between the termination of a connection and the closure of a session. WTLS introduced another type of messages (critical message) whose significance is left for the recipient to define. As for SSL/TLS, the Alert type is susceptible to truncation attacks because it is not authenticated.

Two alert messages, bad_record_mac and decryption_failed, are common to both WTLS and TLS 1.0. The first indicates that a record was received with an incorrect MAC, while the second is sent when the decryption has failed, either because the decrypted text was not a multiple of the block length or because the padding values were not correct. In TLS, the decryption_failed error is encrypted and it indicates a fatal alert forcing a session abort, while in WTLS both messages are sent as warnings and in the clear. As a consequence, WTLS is susceptible to a side channel attack using the error information in the padding verification (Vaudenay, 2002).

6.3 Operational Constraints

WAP was designed so that mobile workers can maintain contact with the information systems of their enterprises, banks, and so on. However, because TLS and WTLS are incompatible, servers will have to maintain two instances of each application for equal access on the mobile network or on the web: one in HTML or XML and the other in WML. On the network side, a gateway must intervene to ensure the interoperability between TLS and WTLS. For example, the message transmitted from the mobile terminal and secured by WTLS must be decrypted by the gateway and then encrypted again according to TLS. This conversion is not risk-free, because for a short interval, the message is in the clear so that end-to-end security cannot be guaranteed without additional efforts to protect the information in the clear during the conversion.

From the security viewpoint, WAP 1.1 did not provide the means for digital signatures using public key cryptography, while WAP 1.2 allowed the generation of digital signatures according to the PKCS #1 specifications.

6.3.1 Positioning of the WAP/Web Gateway

The conversion gateway can be located with respect to the WAP platform and the enterprise firewalls according to several possibilities:

TABLE 6.5

Additional Alert Messages in WTLS

Message	Context	Type
connection_close_notify	Voluntary termination of the connection.	Fatal
disabled_key_id	The keys that the client supplied are disabled administratively.	Critical
duplicate_finished_received	The server received a second Finished in an abbreviate Handshake.	Warning
key_exchange_disabled	Key exchange is administratively disabled to protect an anonymous key exchange in progress.	—[a]
no_connection	No secure connection with the sender (sent as a clear text).	Fatal or critical
session_close_notify	Voluntary termination of the session.	Fatal
session_not_ready	Secure session temporarily not available due to maintenance.	Critical
time_required	Sent by the server to inform the client that the Handshake requires more time.	Warning
unknown_key_id	None of the client keys are recognized.	Fatal
unknown_parameter_index	Server does not know the indicator that the client supplied for the available key exchange suite.	—[a]

[a] Left to the sender's appreciation.

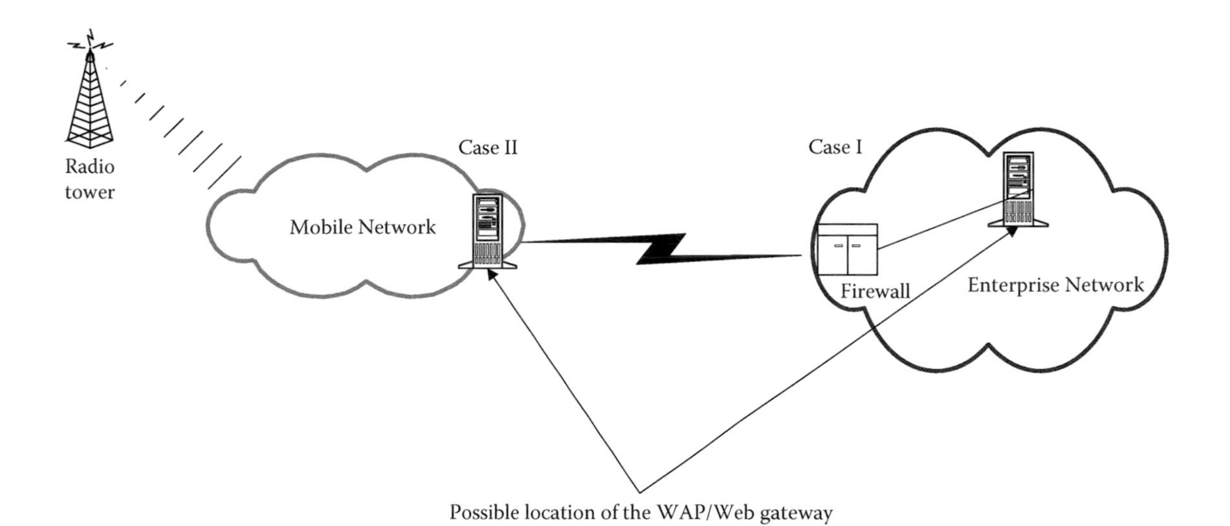

FIGURE 6.3
Possible locations of the WAP/Web gateway between the mobile network and the enterprise network.

1. The enterprise manages the access by deploying the gateway and the WAP platform behind its firewall. In this case, the security problem related to the WTLS/TLS conversion is resolved. This is shown as Case I in Figure 6.3. Mobile users connect to a telephone number at the enterprise, and the WTLS protocol is used up to the gateway. Either the radio link extends to the enterprise network or the employer houses its equipment in a secure location that the mobile operator provides. In either case, the employer takes care of all the links in the security chain. The advantage for the enterprise is that it is completely independent of the mobile network whose only role is to supply the radio channel. However, the enterprise must be able to master all the necessary skills to manage and maintain the system and evolve it with technology.
2. The mobile operator is responsible for the radio communication including the conversion from WTLS to TLS, as shown in Case II of Figure 6.3. The telephone operator must be a trusted party and capable of carrying out the responsibilities that befall it.
3. The mobile operator is only responsible for the radio link and the mobile network. The gateway WAP/web is hosted at the Internet service provider (ISP) of the enterprise before the firewall on the enterprise's premises. This corresponds to Case III in Figure 6.4, in which the ISP controls the conversion from WTLS to TLS.

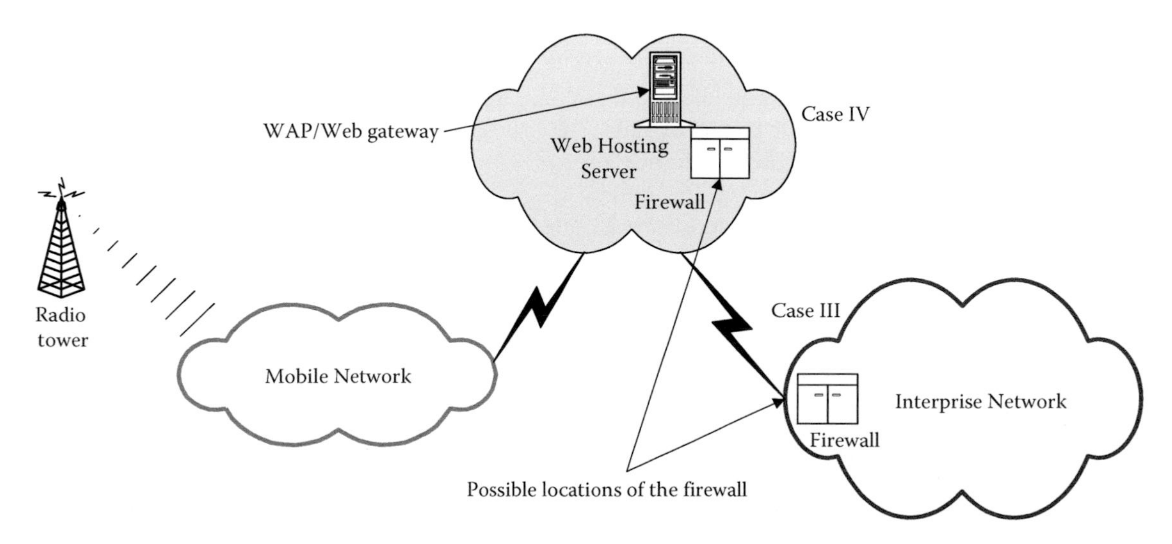

FIGURE 6.4
Possible arrangements of the firewall and the WAP/Web gateway between the enterprise and its Internet access provider.

4. Finally, as shown in Case IV of Figure 6.4, it is possible for the enterprise to rely completely on its ISP for the whole security outside the mobile network: the WAP/web conversion, application hosting, and firewall.

None of these configurations is totally satisfactory. This is why the problem has also been addressed from a protocol point of view. We present two approaches in the following. The first is called Integrated Transport Layer Security (ITLS) and the second Network Assisted End-To-End Authentication (NAETEA).

6.3.2 ITLS

The ITLS solution consists of encrypting the message twice using two different keys. The first key is a secret shared between the mobile terminal and the WAP/web gateway; the second is the key between the mobile terminal and the server (Kwon et al., 2001). This solution taxes the terminal in terms of computation power management. It requires some modifications in the WTLS protocol at the client level to exchange a secret with the gateway. More precisely, the WAP gateway must transmit to the client a new message IntCertificate following the Certificate message coming from the server, while the client responds with another new message, IntClientKeyExchange. A third message, Hash_Handshake, is used to verify the integrity of the session establishment between the client and the gateway. Finally, the gateway intervenes to modify the content of the ClientHello, Server Hello, and Finished messages as well as the Record exchanges with the user's data. These modifications are depicted in Figure 6.5.

6.3.3 NAETEA

The NAETEA solution has the advantage of avoiding any changes to WTLS and of saving the terminals from additional cryptographic computations through an intervention of the mobile network, once the connection encrypted with WTLS has been established. In this protocol, the terminal shares a secret session key with the network (Ks) (established with WTLS) and has a pair of public/private keys for digital signature, PSK_{SIGT} and SSK_{SIGT}. The web server has a pair of public/private encryption keys, PK_W and SK_W, and a pair for signature, PSK_{SIGW} and SSK_{SIGW}. The corresponding network keys are PK_N/SK_N and PSK_{SIGN}/SSK_{SIGN}, respectively. In the following, K{X} represents the encryption of the content *X* with the key *K* and || indicates concatenation:

1. The terminal produces a random number $Rand_T$ and encrypts it with the public encryption key of the web server PK_W. This operation gives $PK_W\{Rand_T\}$.
2. The terminal adds to the cryptogram its certificate and encrypts the whole with the secret key shared Ks with the network:

   m1 = Ks{$PK_W\{Rand_T\}$||signature certificate of the terminal}

3. The terminal signs the hashed message with its private signature key SSK_{SIGT} using a hash function H:

   sigT = SSK_{SIGT}\{H(m11||identifier of the web server)}

4. The message M1 sent to the network is formed as follows:

   M1 = (m1||sigT||signature certificate of the terminal||identifier of the web server)

5. After receiving the message M1, the network verifies the integrity of the signature sigT with the help of the public signature key extracted from the received certificate. Next, it decrypts m1 with the key Ks to verify that the certificate is identical to the one that was encrypted. Once this verification has been accomplished, the network forwards the data received in full confidence to the web server. It makes a new message by signing the data with its private signature key SSK_{SIGN}:

   m21 = ($PK_W\{Rand_T\}$||sigT||terminal certificate||identifier of the web server)

   sigN = SSK_{SIGN}\{H(m21, signature certificate of the network)}

6. The message M2 is constructed as follows:

   M2 = m21||sigN ||network certificate

7. Having received the message M2, the web server examines the network signature to ensure that the message M2 really comes from the network. Once this verification has been done, it can extract the random number $Rand_T$ with its private encryption key SK_W.
8. The web server generates another random number $Rand_W$ and constructs the session key $K_{WT} = H(Rand_T || Rand_W)$ and a hash $H(Rand_T || K_{WT})$. This session key will serve to encrypt the exchanges between the server and the mobile

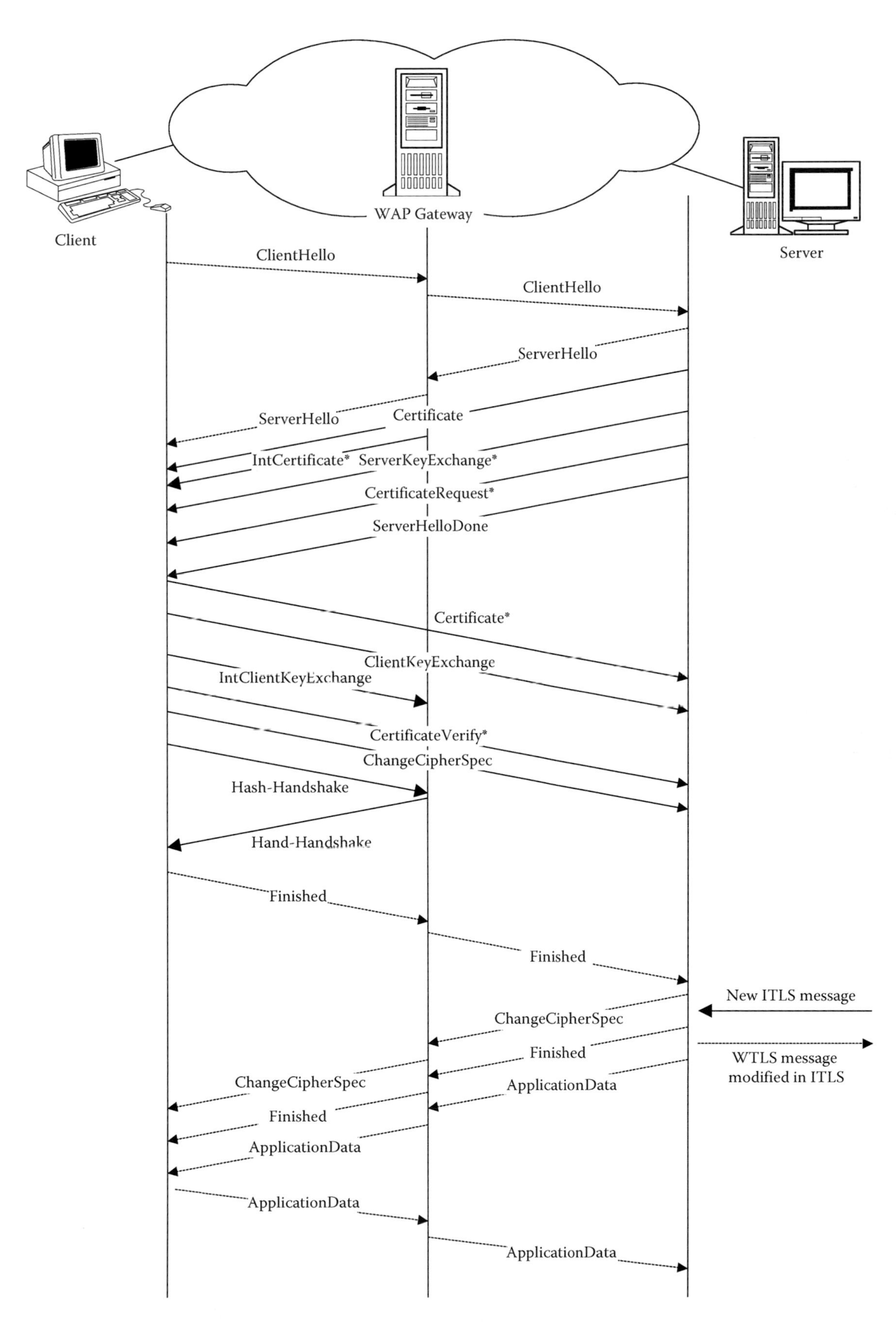

FIGURE 6.5
Additional exchanges in the ITLS protocol. * Optional message.

terminal. The concatenation of the two random numbers and the server encryption certificate gives m31:

$$m31 = Rand_W \| H(Rand_T \| K_{WT}) \| \text{signature certificate of the web server}$$

9. It computes $sigW = SSK_{SIGW}\{H(m3)\}$ with its private signature key SSK_{SIGW}. The message M3 returned to the network is formed by concatenation as follows:

$$m31 \| sigW \| \text{signature certificate of the web server}$$

and the whole is encrypted with the public encryption key of the network PK_N. Thus, we have

$$M3 = PK_N \{m31 \| sigW \| \text{signature certificate of the web server}\}$$

10. The network decrypts the message M3 with its private key SK_N, verifies the signature, and prepares accordingly the message to the terminal. This message is encrypted with the key Ks established with the terminal:

$$M4 = Ks \{Rand_W \| H(Rand_T \| K_{WT}) \| \text{encryption certificate of the network}\} \| \text{encryption certificate of the network}\}$$

11. The terminal proceeds to authenticate the message by comparing the certificate received in the clear with the one encrypted with the key Ks. Once this has been verified, the terminal extracts the random number $Rand_W$ and then calculates the encryption key K_{WT} with $H(Rand_T \| Rand_W)$. Finally, it compares the value of the hash $H(Rand_T \| K_{WT})$ with the one enclosed in the message M4. If both values are identical, the terminal can be confident that it is the server that has received the random number $Rand_T$ that it has originally sent and that, therefore, the key K_{WT} can guarantee the confidentiality of the exchanges with the server.

The exchanges associated with this protocol are depicted in Figure 6.6.

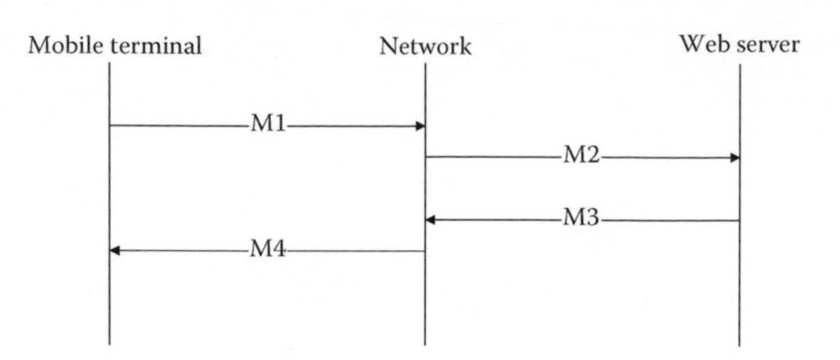

FIGURE 6.6
Authentication with the NAETEA protocol.

6.4 WAP 2.0 and TLS Extensions

WAP 2.0 introduced several changes to bring mobile services closer to the Internet Protocol stack. While it provides backward compatibility with WAP 1.0 and WAP 1.2, the specification uses XHTML Basic for describing the content of websites and the WAP cascading style sheets (CSS) to describe the layout, colors, and fonts (i.e., the presentation semantics) of the site as distinct of the content. Furthermore, the WAP gateway was replaced with a WAP proxy server and WTLS with TLS. The WAP proxy performs the same functions of typical Internet proxies such as filtering and caching.

XHTML Basic is a simplified version of the Extensible HyperText Markup Language (XHTML) that the W3C has defined on the basis of the XML markup language. Later, the OMA adapted XHTML for mobile applications as the XHTML Mobile Profile (XHTML MP) to support WAP CSS. An additional benefit of the use of XHTML is to take advantage of the vast experience with the development tools for regular websites.

The differences between WAP 1.x, WAP 2.0, and the conventional web are summarized in Table 6.6.

Because mobile operation was not considered when TLS was originally designed, RFC 4366 added some

TABLE 6.6
Main Differences between WAP 1.x, WAP 2.0, and the Web

	WAP 1.x	WAP 2.0	Conventional Web
Origin of the specifications	WAP Forum	WAP Forum/ OMA	IETF/W3C
Content	WML	XHTML Basic	HTML
Server	WAP server	WAP server or web server with a WAP proxy	Web server
Application	Wireless Application Environment (WAE)/Wireless Markup Language (WML)/WML Script	XHTML Mobile Profile (XHTML MP), Wireless Profiled HTTP (WP-HTTP),	HTML, XML
Transport layer	WDP/UDP/ WTLS	TLS/Wireless Profiled-TCP (WP-TCP)	TCP/TLS

TLS extensions to conserve bandwidth in wireless networks. Thus,

- TLS clients and servers can negotiate the maximum fragment length to reduce it from 2^{14} to 2^9, 2^{10}, 2^{11}, or 2^{12} octets.
- TLS clients and servers can negotiate the use of a truncated HMAC of 10 octets.
- TLS clients can provide the TLS server with the name of the server they are trying to contact. This is to secure connections to servers that host multiple *virtual* servers at a single network address, such as in enterprise networks.
- To conserve memory on constrained clients, TLS clients and servers can negotiate the use of client certificate URLs to be sent at the start of a session.
- TLS clients can indicate to TLS servers which certificate of authority's root keys they possess. In this way, TLS clients can store a small number of root keys of certification authorities within the constraints of their memory and avoid repeated Handshake failures.
- The server can verify the status of the client certificate during a Handshake using the Online Certificate Status Protocol (OCSP) of IETF RFC 6960. This capability avoids sending a Certificate Revocation List (CRL) and saving bandwidth.

These capabilities are negotiated with extensions to the ClientHello and ServerHello messages that indicate two new optional messages, CertificateURL and CertificateStatus:

1. The CertificateURL message contains a sequence of URLs where the client certificate resides and, optionally, its hash.
2. The CertificateStatus message contains the OCSP response.

Figure 6.7 illustrates the TLS exchanges with the extensions that RFC 4366 had defined.

The new Alert messages associated with the extensions of RFC 4366 are listed in Table 6.7.

6.5 WAP Browsers

WAP browsers use WAP 1.2 or WAP 2.0; WAP 1.2 browsers connect to a WAP gateway, which in turn uses regular HTTP to communicate with the web server. With WAP 2.0, the WAP browser connects to a WAP proxy, which usually resides at the carrier premises, to query the web server with HTTP to retrieve the content on behalf of the client. When the HTML content is retrieved, it is converted back to the format of the browser. An example of a WAP 2.0 browser is Openwave (Hernandez, 2009). For iPhones, browsers can be discovered through the iTunes Store. A list of WAP browsers can be found at http://webcab.de/wapua.htm, last accessed January 18, 2016.

An important browser is Wapzilla Internet Browser from Red Hot Bits. It supports browsing with WAP 1.x through standard WML pages and Wireless Application Protocol Bitmap (WBMP) format. It supports browsing WAP 2.0 standard XHTML pages with XHTML Mobile Profile (XHTML MP).

Multimedia messaging (MMS) also uses WAP. To send a multimedia message, the cellphone delivers the message using WAP 2.0 (or WAP 1.2) to a WAP proxy (or a WAP gateway), which then sends the MMS to the carrier's Multimedia Messaging Center (MMSC). When an MMS is sent to a cellphone recipient, the cellphone receives a text message (SMS) to notify the recipient of the arrival of a new MMS message. The recipient must use the cellphone to establish a data connection using WAP to download the actual MMS message using WAP 2.0 (or WAP 1.2) through a WAP proxy or a WAP gateway.

Most mobile browsers today, just like desktop browsers, use Web 2.0 applications to generate and share content in real time, such as JavaScript, Asynchronous JavaScript and XML (Ajax), cascading style sheets (CSS), and HTML5. JavaScript defines aspects to alter the content displayed on the client side. Ajax is a group of interrelated web development techniques to query the web server with HTTP asynchronously (i.e., in the background) without interfering with the displayed page so that it stays live. Finally, CSS is a language for describing the presentation semantics of an HTML document. Rendering engines, such as WebKit, are embedded as part of the device firmware for Safari (iPhone) and Google Android (Hernandez, 2009).

Mobile browsers with proprietary protocols or engines such as Skyfire use the technique of server transcoding but rely on TLS for transport security (Hernandez, 2009).

6.6 Summary

WAP and WAP-based services were designed to take into account the technological limitations of mobile phones and mobile networks in the 1990s. The widespread use of TLS/SSL encouraged its adaptation as WTLS WAP-based mobile communications. Because of

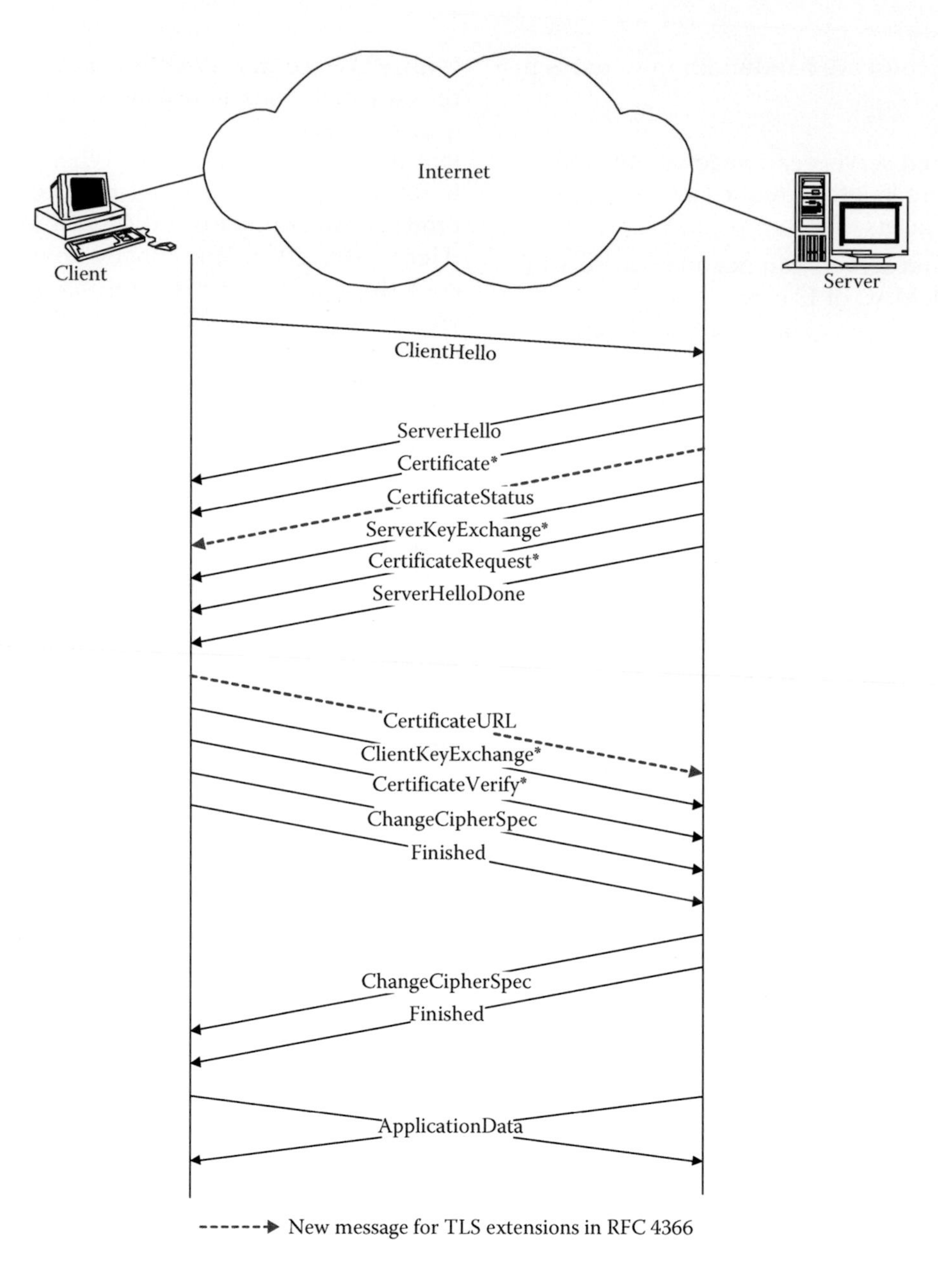

FIGURE 6.7
TLS exchanges with the extensions of RFC 4366. * Optional message.

TABLE 6.7
Additional Alert Messages in the TLS Extensions of RFC 4366

Message	Context	Significance
unsupported_extension	This alert is sent by clients that receive an extension in the ServerHello that they did not request in the corresponding ClientHello.	Fatal
unrecognized_name	This alert is sent when servers do not recognize the server name indicated in the client's extension.	May be fatal
certificate_unobtainable	This alert is sent by servers who are unable to retrieve a certificate chain from the URL that the client supplied.	May be fatal; for example, if client authentication is required
bad_certificate_status_response	This alert is sent when clients receive an invalid certificate status response.	Fatal
bad_certificate_hash_value	This alert is sent by servers when the certificate hash that the client provided does not match the certificate hash retrieved.	Fatal

these constraints on wireless communications, WTLS 1.x was from the outset, incompatible with either TLS or SSL. This restricted the deployment of WAP services because of the complexity of network planning and the administration of end-to-end security. The WAP 2.0 specifications specify that TLS adopted a profile adapted to wireless transactions. Finally, DTLS of RFC 4347 may offer another standard alternative to WTLS.

Questions

1. List some sources of incompatibilities between WTLS and TLS.
2. Why are some weak encryption algorithms used in WTLS?
3. Compare and contrast solutions to the problem of the unencrypted message in WTLS/TLS conversion.
4. In the NAETEA protocol,
   - What is the function of the shared key between the terminal and the mobile network?
   - How does the web server authenticate the terminal?
   - Why is the session key between the mobile terminal and the web server based on two random numbers?
   - How is the session key protected from the mobile network?

7

The SET Protocol

This chapter discusses the Secure Electronic Transaction (SET), a protocol designed to secure bank card transactions initiated on open networks. SET was sponsored jointly by Visa and MasterCard in collaboration with important players in business software such as IBM, GTE, Microsoft, SAIC (Science Applications International Corporation), Terisa Systems, and VeriSign (SET specification, 1997).* Secure Electronic Transaction LLC (SETCo) was responsible for maintaining the specifications, for conformance and interoperability testing, and for issuing licensing agreements. Despite this impressive list of promoters, the market opted TLS/SSL rather than SET. From a technical point of view, however, the study of SET is still worthwhile because it introduced many innovations that could be reutilized in the future.

There are three parts to this chapter. The first contains a description of the security services of SET, the cryptographic algorithms, particularly the concept of the dual signature and the protocol exchanges during various transactions. Next, efforts to promote SET are described, including the use in hybrid solutions with TLS/SSL. Finally, some ideas are offered on the possible reasons behind the failure of SET in the market place.

7.1 SET Architecture

SET secures bank card transactions over the Internet. It operates at the application layer with an exclusive focus on payment transactions, that is, other exchanges, such as those related to search or to selection of goods, are out of scope. SET also provides a secure interface to the banking infrastructure for authorization and remote payment.

Issuer institutions, typically a bank affiliated with Visa or MasterCard, give their customers payment cards conforming to the SET specifications. Cardholders are authenticated with public key cryptography using a certificate from a certification authority that is stored in the hard disk of their computer or on an outside storage medium (at the time, 31/4″ diskettes).

SET is a transaction-oriented protocol that operates in a request/response mode; that is, messages are paired. The message structure follows the DER (Distinguished Encoding Rules) of ASN.1 (Abstract Syntax Notation 1) (Steedman, 1993). The ASN.1 version that is used dates from 1995 and is described in ISO/IEC 8824-1 (1998), 8824-2 (1998), 8824-3 (1998), and 8824-4 (1998), whereas the DER rules are defined in ISO/IEC 8825-1 (1998). The messages are encapsulated with MIME as described in PKCS #7 (IETF RFC 2315). In particular, SET uses the following PKCS #7 structures:

- SignedData, for data that are signed
- EnvelopedData, for clear data that are in a digital envelope
- DigestedData, for digests
- EncryptedData, for encrypted data

The SET specifications cover the roles and responsibilities of the following six entities:

1. The cardholder
2. The merchant's server
3. The payment gateway
4. The certification authority
5. The issuer institution of the cardholder's bank card
6. The acquirer institution, which is the merchant's bank

Figure 7.1 depicts the functional architecture of SET. The cardholder, the merchant, the certification authority, and the payment gateway are all connected by the Internet. The client communicates with the payment gateway using a tunnel that goes through the merchant's server. Each participant has a certificate from a SET certification authority. These certificates are enclosed in each of the messages exchanged among the cardholder, the merchant, and the payment gateway.

The issuer and acquirer institutions are linked by a closed and secure bank network. The payment gateway bridges the open and the closed networks. It has

* At the time, SAIC also owned the Internet Network Information Center (InterNIC), which, until 1999, was the sole administrator of Internet domain names .com, .org, and .net. It was also the owner of Telcordia (formerly BellCore). A fully owned subsidiary of SAIC, Tenth Mountain Systems, Inc., was the first SET Compliance Certification Authority (SCCA).

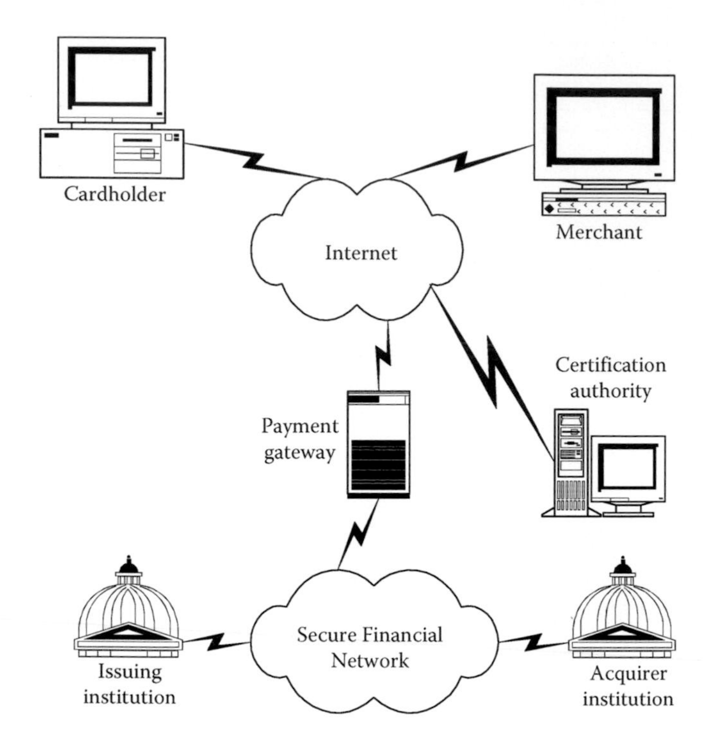

FIGURE 7.1
Actors in a SET transaction.

two interfaces, one conforms to the SET specifications, on the Internet side and the other conforms to the proprietary protocol used within the secure financial network side.

SET secures the exchanges between the client and the merchant and the exchanges between the merchant and the payment gateway. The payment gateway manages the payments on behalf of the banks, whether issuer or acquirer. Clearly, the gateway must be approved by the banking authorities, if not be under the responsibility of a financial institution to carry out these functions.

Figure 7.2 depicts the position of SET in the protocol stack TCP/IP (Sherif et al., 1997).

On the client side, the SET functionalities are implemented in a special package often denoted as a SET wallet.

7.2 Security Services of SET

SET transactions provide the following services:

- Registration of the cardholders and the merchants with the certification authority
- Delivery of certificates to cardholders and merchants
- Authentication, confidentiality, and integrity of the purchase transactions
- Payment authorization
- Payment capture to initiate the request for financial clearance on behalf of the merchant

SET employs the techniques of public key cryptography to guarantee simultaneously the following:

- Confidentiality of the exchanges, that is, that they cannot be read online by an entity external to the transaction
- Integrity of the data exchanged among the client, the merchant, and the acquirer bank
- Identification of the participants
- Authentication of the participants

A necessary but not sufficient condition for nonrepudiation of the transactions is that the cardholder be certified. Other conditions are a trusted time-stamping mechanism and an irreproachable certification authority. Finally, because it is the payment gateway that verifies the exactness of the Payment Instructions and not the merchant, the gateway will be called upon to arbitrate disputes. Note that, however, to facilitate the deployment of the SET protocol, buyer's certification is optional in Version 1.0 of the specifications.

Message confidentiality is achieved with symmetric encryption algorithms (also called secret key encryption algorithms). The secret key itself is distributed with public key cryptography algorithms. For instance, when the payment gateway wants to send confidential information to the merchant, it generates a symmetric encryption key with which it encrypts the data. This same key is encrypted with the public key of the merchant who, being the only entity with the corresponding private key, is the only party capable of retrieving the symmetric key and decrypting the data.

Message integrity aims at guaranteeing that the received data are exactly what the sender has transmitted and that they were not corrupted by mischief or by error during their transit on the network. SET uses the digital signature of the sender to ensure message integrity, that is, the digest of the message encrypted with the private key of the sender. Any entity having access to the corresponding public key is able to verify the message integrity by comparing the calculated digest of the message with the one obtained by decrypting the signature. In this way, if the public/private key pair is unique and attempts to steal the sender's identity were not successful, the digital signature simultaneously guarantees the sender's identity and the data integrity.

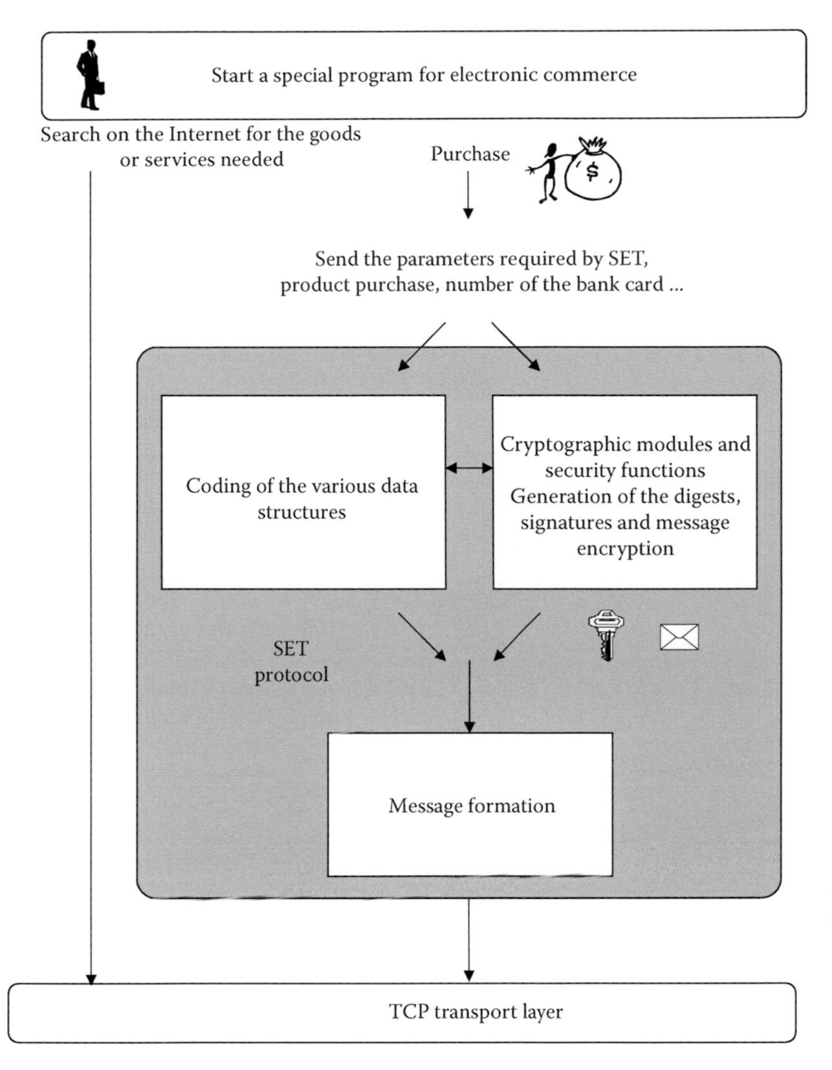

FIGURE 7.2
SET in the TCP/IP protocol stack.

The identification of the participants in a SET transaction corresponds to a preestablished relation between an encryption key and an entity. Each entity attaches to its message, whether encrypted or not, a digital signature that only that entity could have generated but that can be verified by the peer entities. As seen earlier, in SET, the sending entity constructs this signature by encryption with its private key using a public key algorithm; this signature is verified with the corresponding public key.

In SET, the certificate signed by the certification authority gives credence to the association of a public key with its owner, thus providing authentication. The authentication procedures of SET are based on Version 3 of ITU-T Recommendation X.509. Each certificate contains the identity of its owner, a public key related to the public key encryption algorithm used, and the signature of the authority that issued the certificate. For mutual authentication, two parties have to go backward along the certification path until they encounter a common authority.

7.2.1 Cryptographic Algorithms

As shown in Table 7.1, SET employs existing cryptographic algorithms. It is seen that SET uses SHA-1 for ordinary hashing and HMAC-SHA-1 keyed hashing.

TABLE 7.1
Cryptographic Algorithms Used in SET

Algorithm	Services
DES	Confidentiality
RSA	Authentication, identification, and integrity
SHA-1	Hashing
HMAC-SHA-1	Keyed hashing

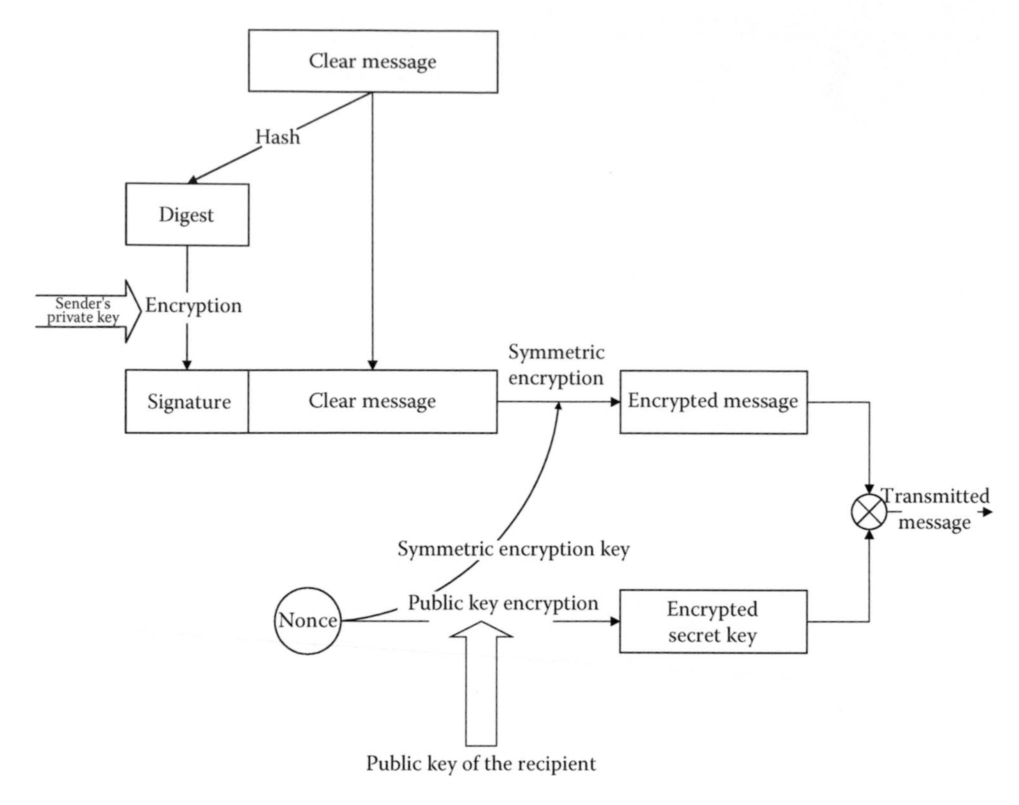

FIGURE 7.3
Processing of a typical message in SET.

In a SET transaction, some participating entities, such as the merchants, have two pairs of public encryption keys: one for signing documents and the other to exchange keys during the identification phase. The first pair is called the signature key, and the second is called key encryption key (or key for key exchange). Figure 7.3 summarizes the main steps for processing a typical message to implement the services of authentication, confidentiality, and integrity. The private key used for signature is from the pair of signature keys, whereas the public key for encrypting the symmetric key comes from the pair of encryption keys.

Once created, the digest is coded according to the DER rule and then arranged to conform to the content type *DigestedData* and to a *digest* with the various fields shown in Table 7.2. As stated earlier, this syntax is based on the PKCS #7 (IETF RFC 2315) syntax as redefined in Book 2 of the SET specifications. The digest length then jumps from 20 to 46 octets, which corresponds to an increase in the transmission overhead of more than 100%.

The ensemble formed with the digest and the clear message is encrypted with the Data Encryption Standard (DES) algorithm using as a symmetric key a random number of 8 octets (64 bits) generated anew for each message. This key is then sent to the recipient within a digital "envelope" with the format specified in PKCS #7. This envelope is in turn encrypted with the public RSA key of the message's recipient. The length of the RSA key is 1024 bits.

TABLE 7.2
Digest for the PKCS #7 Format in SET

Field	Length (in Octets)
Header of *DigestedData*	2
Version	3
Header of *DigestAlgorithm*	2
Algorithm	7
Parameters of the algorithm	2
ContentInfo	2
ContentType	6
Header of hash	2
The hash proper	20
Total	46

To give additional protection to the symmetric DES key, before encrypting the DES key with the public key of the recipient, the numerical envelope is processed with the OAEP (Optimal Asymmetric Encryption Padding) technique that Bellare and Rogaway (1995) have proposed. The OAEP processing makes all bits of the cryptogram equally resistant to attack. The cryptogram is formed of 128 octets that include the symmetric key, the buyer's primary account number

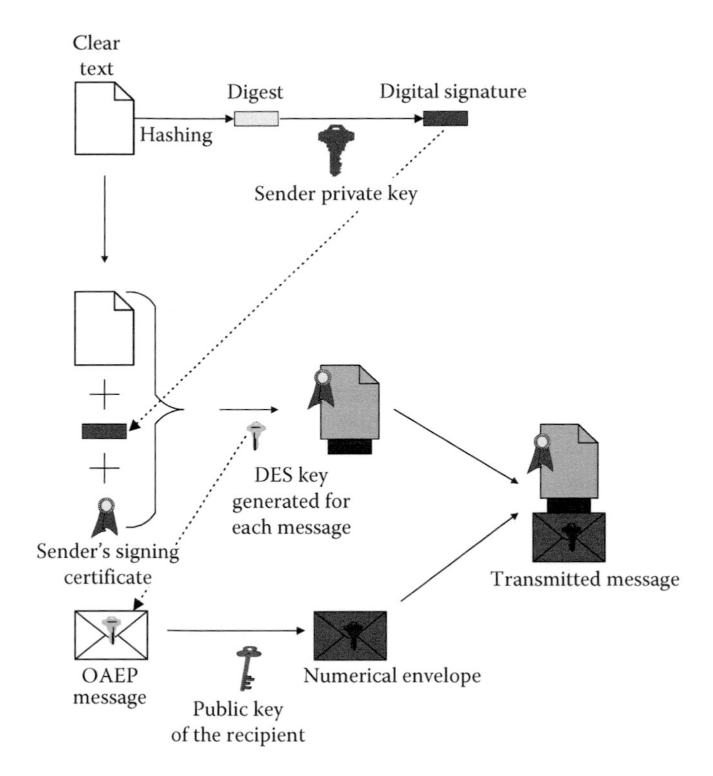

FIGURE 7.4
Cryptographic processing of SET messages sent by the buyer.

(PAN), and other variables that are combined with padding, random numbers, and the like to offer enhanced resistance to brute-force attacks (SET, 1997, Book 3, pp. 15–23). Thus, the SET protocol uses the private key of the sender for the digital signature and the public key of the recipient to encrypt the digital envelope that contains the DES key that was used to encrypt the clear text. If the cardholder is certified for signature, the certificate is attached to the message encrypted with the symmetric key, as shown in Figure 7.4. In this case, each message is authenticated.

7.2.2 Dual Signature

One innovation that SET introduced is the dual signature to link elements of two encrypted messages, each sent to a different recipient, with the same signature, thus avoiding unnecessary exchanges. With this procedure, a recipient reads the message that is addressed to it and verifies the integrity of the other message without knowing its content. In this case, a buyer sends the purchase order to the merchant and the Payment Instructions to the buyer's bank (through the payment gateway) simultaneously. The Payment Instructions are executed only after the merchant has accepted the purchase order without knowing the buyer's account number, which is revealed to the bank only. Once the bank has ascertained that the merchant has accepted the purchase, it processes the transaction without knowing the nature of the purchase.

Let m_1 be the message destined to the merchant and m_2 be the message for the payment gateway (by way of the merchant). The buyer concatenates the digests of both messages to construct a new m_3:

$$m_3 = H(m_1) \mid\mid H(m_2)$$

The buyer then applies the hashing function $H()$ to the new message m_3. Accordingly, the message sent to the merchant is composed of the following elements:

$$m_1, H(m_1), \{m_2, H(m_2)\} PK_G, \{H(m_3)\} SK_C,$$
$$H(m_1) \oplus H(m_2)$$

where
PK_G is the public key of the payment gateway
SK_C is the private key of the buyer
$\oplus$ is the exclusive OR operation.

The merchant reads the message m_1 that is addressed to him and verify its integrity by comparing the received $H(m_1)$ with the recalculated value. Following verification, the merchant decrypts $H(m_3)$ with the client's public key, PK_C, and extracts $H(m_2)$ by the exclusive OR operation $H(m_1)\oplus\{H(m_1)\oplus H(m_2)\}$. This allows it to concatenate the computed digest $H(m_1)$ with the extracted digest $H(m_2)$ to reconstitute $m_3 = H(m_1) \mid\mid H(m_2)$. If $H(m_3)$ gives indeed the result obtained by decrypting $\{H(m_3)\}\ SK_C$ with the client's public key, then the integrity of the message m_2 to the payment gateway has been established without accessing the content of m_2. If the merchant accepts the integrity of the message, it sends to the gateway

$$\{m_2, H(m_2)\}PK_G, \{H(m_3)\}SK_C, H(m_1)$$

As a consequence, the gateway can extract $H(m_2)$ and, by concatenating $H(m_1)$, it verifies that the quantity $H(m_3)$ is the same as the one obtained with the client's public key, which is an indication that the merchant has accepted the offer, because it is the merchant that relays the client's offer to the payment gateway.

Thus, the payment is not executed unless the merchant accepts the offer, whereas the sale is effective only if the bank approves the Payment Instruction. At the same time, the payment gateway cannot access the purchase order, and the Payment Instructions remain opaque to the merchant.

7.3 Certification

In this section, SET procedures for the certification of the cardholder and of the merchant are presented.

Note: The word "certification" here is used differently than in the case of the "certification" of SET components. The latter use is associated with the conformance of a SET implementation to the specifications.

7.3.1 Certificate Management

As shown in Figure 7.5, the hierarchy of certificate management consists of nine entities, with the root certification authority at the top of the certification structure and at the bottom, the participating end entities and the SET cardholders.

The root certification authority is the ultimate verifier of the participants' authenticity and controls the delivery of the electronic certificates. CertCo & Digital Trust Company (CertCo) was selected to fulfill this role and to issue the 2048-bit root key for public key cryptography that serves to generate the certificates that authenticate all other actors. The root public key was to be divided into fragments, with each fragment probably owned by one of the large card issuers. CertCo was founded in December 1996 as a subsidiary of Bankers Trust, which merged in 1998 with Deutsche Bank.

The certificates of the root authority are self-signed and are accessible to all end entities of the SET. The signature certificate of the root entity contains the digest of the master key, as well as a replacement key. Whenever the master key is renewed, new certificates are issued with the replacement key as the master key.

The brand certification authorities constitute the next level of the certification hierarchy. This is the level of the various organizations for the management of bank cards principally, Visa and MasterCard, or payment cards, such as American Express. These authorities can delegate their responsibilities in a given geographic zone to geopolitical certification authorities whose role is to accredit local authorities. The size of the keys for public

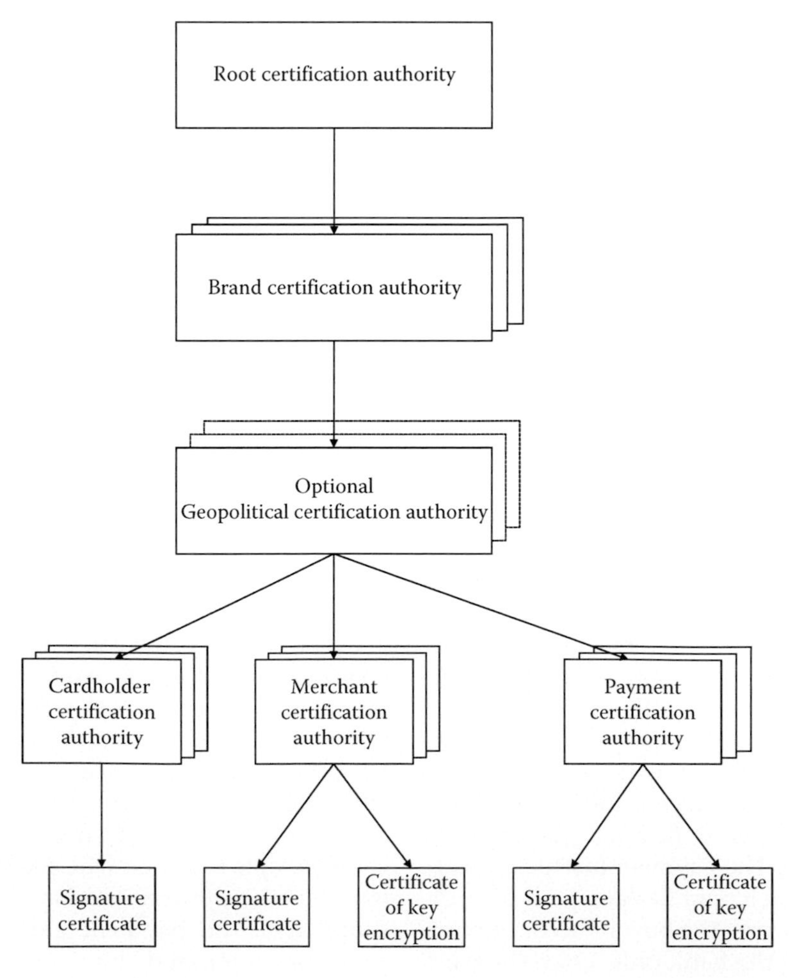

FIGURE 7.5
Hierarchy of certificate management.

key cryptography of all brand certification authorities is 1024 bits. The financial institution that issues the cardholders' cards must have a certificate from the brand certification authority that allows it to certify its clients.

Figure 7.5 also shows that end entities in the SET have two types of certificates: signature certificates and key encryption certificates. Signature certificates are used to verify the signatures, while key encryption certificates are used when the secret key of the symmetric encryption is encrypted. Thus, the key encryption certificate contains a public encryption key that is used to encrypt the DES symmetric key used for encrypting messages, such as the Payment Instructions. A pair of keys corresponds to each type of certificates so that there are signature keys and key encryption keys; for legal reasons, the signature keys have larger sizes.

The authorities of the three higher hierarchical levels also have two certificates: the first authorizes them to certify the low hierarchical levels while the second gives them the right to sign the revocation lists.

Local certification authorities own a third type of certificates to show that they are allowed to certify end entities. Any end entity can, at any time, verify the validity of the certificate of the root authority by querying its own certification authority. The latter is supposed to know the digest of the certificate of the root authority and, therefore, can verify if the two digests are identical. If the comparison shows that the root certificate is not valid, the user can request a new certificate.

SET cardholders have signature certificates. For encryption, they use the public keys that the merchant or gateway provide in their certificates. The cardholder signature certificate does not contain the name of the cardholder but an encrypted value that acts as a pseudonym. Similarly, the certificate does not contain the financial references of the payment, but their digest calculated with a one-way hash function starting with the account number, the expiration date, and the PIN:

$$\text{Hash} = H(\text{Account number} \,||\, \text{Expiration date} \,||\, \text{PIN})$$

where || represents the concatenation operation.

Because the bank knows the financial references, it is able to ascertain whether the certificate really belongs to the cardholder.

Merchants and payment gateways have signature certificates and certificates of key encryption. The number of certificates that a merchant needs depends on the number of gateways it interfaces with. To reduce the amount of processing, the merchant server may stock the certificates of all its payment gateways. If the merchant is in touch with several acquirer institutions, the savings can be significant, because reconstruction of the certification path for each one of them at every exchange that concerns them is time consuming.

The authority that certifies the payment gateway, typically the acquirer bank, has an additional certificate to sign the Certificate Revocation List. Although the protocol specifies the procedure for certification, the procedures for certificate revocation are left to the individual authorities.

Thus, an acquirer bank that plays the role of a certification authority has four certificates: a signature certificate, a key encryption certificate, a certificate for signing certificates, and a certificate to sign the Certificate Revocation List.

7.3.2 Registration of the Participants

7.3.2.1 Cardholder Registration

Registration can be either by e-mail or by clicking on a special button on the cardholder's bank. Registration of the cardholder comprises the three phases shown in Figure 7.6: initialization, request for the registration form, and presentation of the filled-in form. Listed in Table 7.3

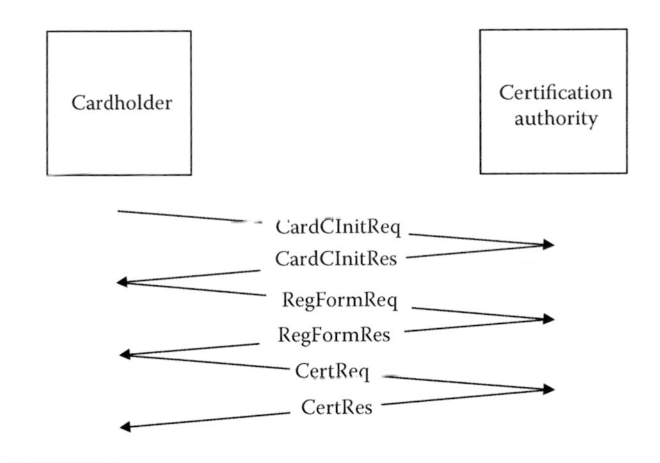

FIGURE 7.6
Messages exchanged during cardholder registration.

TABLE 7.3
Messages for Registering the Cardholder

Message	Direction of Transmission	Significance
CardCInitReq	Cardholder → Certification authority	Request to initiate a registration
CardCInitRes	Certification authority → Cardholder	Response to the request to initiate a registration transaction
RegFormReq	Cardholder → Certification authority	Request for a registration form
RegFormRes	Certification authority → Cardholder	Response to the request for a registration form
CertReq	Cardholder → Certification authority	Certificate request
CertRes	Certification authority → Cardholder	Response to a certificate request

are the SET messages exchanged between the certification authority and the cardholder during registration.

Details of the cardholder registration are presented as follows.

7.3.2.1.1 Initialization

The registration starts when the cardholder software sends a *CardCInitReq* to the certification authority to begin the certification procedure. The certification authority responds with the message *CardCInitRes,* which is signed with the authority's private signature key and which contains the certification authority's certificates for the key exchange and for signature. Figure 7.7 depicts the composition of the *CardCInitRes* message.

Upon reception of this message, the cardholder software verifies the authority's signature, either by using the public key that was obtained in a safe manner or by following the certification path to the root authority.

After verification of the certificate, the cardholder is able to protect the confidentiality of the subsequent exchanges. Each message is encrypted with a secret encryption key using the DES algorithm. This secret key itself will be encrypted with the public key of the authority so that only the latter is able to decrypt. This processing is identical to that depicted in Figure 7.2.

7.3.2.1.2 Request for the Registration Form

The cardholder must complete a registration form to which the issuer of the bank card has agreed. To obtain this form, two sets of exchanges between the cardholder software and the certification authority are needed. The first message is RegFormReq and is used to request the registration form. It contains the identity of the financial institution or the issuer bank and the cardholder's payment data, particularly the PAN. The message is encrypted with a random secret key by applying the symmetric encryption algorithm DES. Next, the secret key and the card number are encrypted with the public key of the authority. Figure 7.8 depicts the formation of the RegFormReq message. Note that the messages originating from the cardholder are not signed as yet, because the requested signature was not yet awarded. The first four numbers of the PAN form the BIN (bank identification number) that identify the issuer bank and the card brand (Visa, MasterCard, etc.).

After receiving and processing the *RegFormReq* message, the certification authority extracts the pertinent registration form from its archives, if it already has it, or retrieves it from the competent sources. It then signs it with its private signature key and encloses its response in the *RegFormRes* message.

7.3.2.1.3 Completing and Sending the Form

The cardholder's software verifies the authority's signature, following the certification path all the way back to the root authority, if necessary. With authenticity verified, the cardholder's software extracts the registration form and verifies its integrity by decrypting the signature with the certification authority's public key and comparing the decrypted digest with the digest recalculated on the basis of the received form. If everything is correct, the cardholder enters the required data in the form to request the certificate. The cardholder's software generates the pair of private and public keys and three random numbers. The first random number is the secret encryption key that will serve to encrypt the message before its transmission with the symmetric algorithm DES. The second random number is the symmetric key that the certification authority will use to encrypt the certificate. The third random number is attached to the

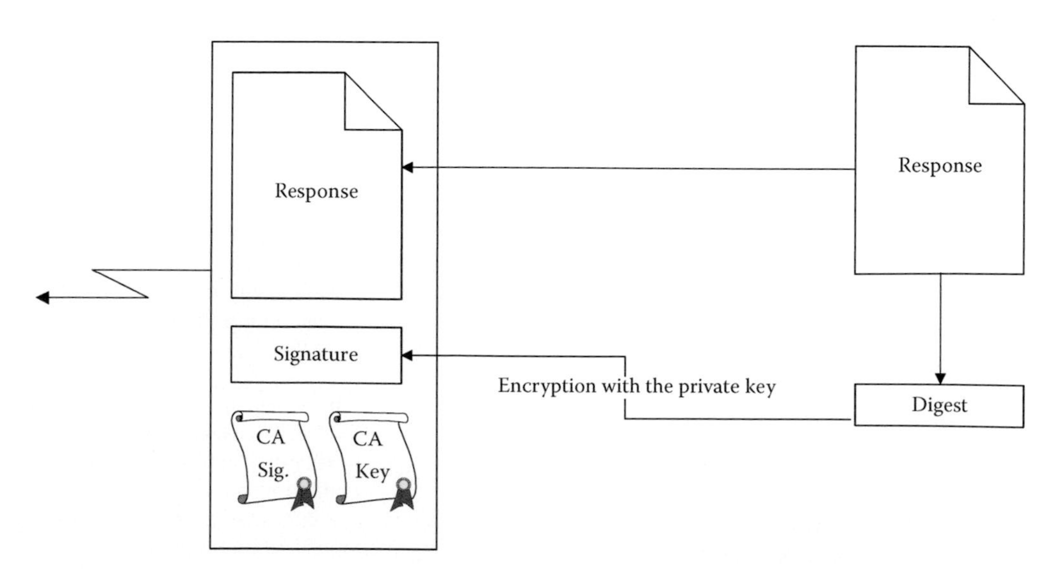

FIGURE 7.7
Formation of the CardCInitRes message.

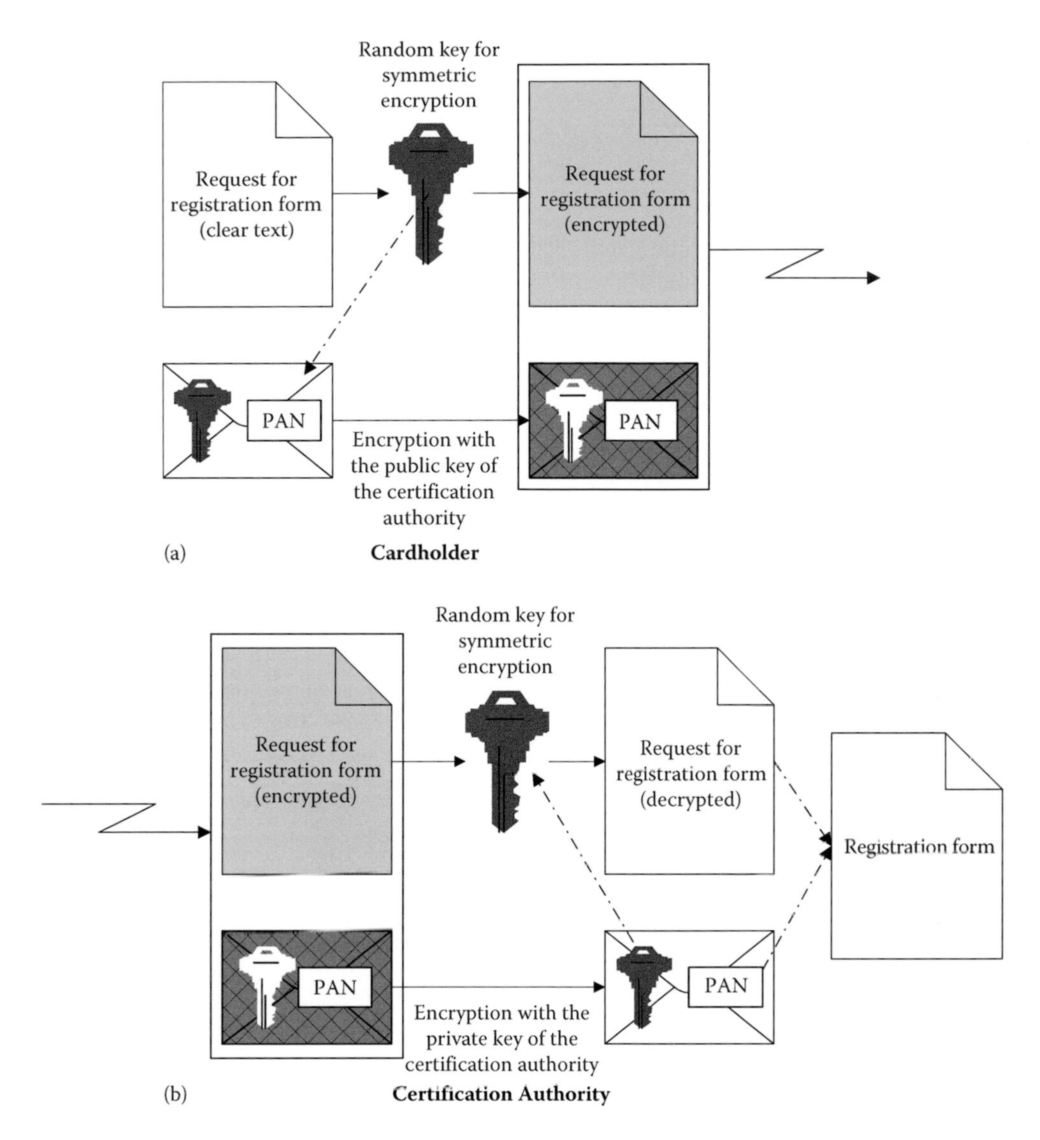

FIGURE 7.8
Processing of the RegFormReq message by (a) the cardholder's software and (b) the certification authority.

registration form and will be used by the certification authority as explained in the following.

Thus, the CertReq from the cardholder contains the registration form, the just generated cardholder's public key for the signature and the encryption key. The whole message is hashed and the digest encrypted with the cardholder's private signature key to produce the message signature. Next, the message is encrypted with the first secret key, which, in turn, is encrypted with the public key of the certification authority. Figure 7.9 summarizes the processing steps to form the CertReq message.

At this stage, the authority has all the information necessary to check the certification request according to the regulations of the card issuer institution; these rules are independent of the SET. Once the verification has been done, the authority proceeds to issue the certificate. The authority uses the third random number it received from the cardholder to construct another secret that it keeps to itself. It then computes the hash of this new secret, the PAN, and the card expiration date, that is, *H(PAN, expiration date, random number)*. This hash is included in the certificate to establish a protected link between the certificate and the user's data.

Note that the use of the random number, or *salting*, defends against dictionary attacks on the account number. In fact, the possible values for the expiration date are limited and could be guessed on the basis of the certificate expiration dates. Similarly, the number of issuing banks is limited. Without the random number (or *salt*), it would be relatively easy to construct a dictionary of all possible numbers formed of the concatenation of account numbers and expiration dates and then systematically compare the calculated hash values with the value in the certificate. When the two numbers match, the PAN would be obtained.

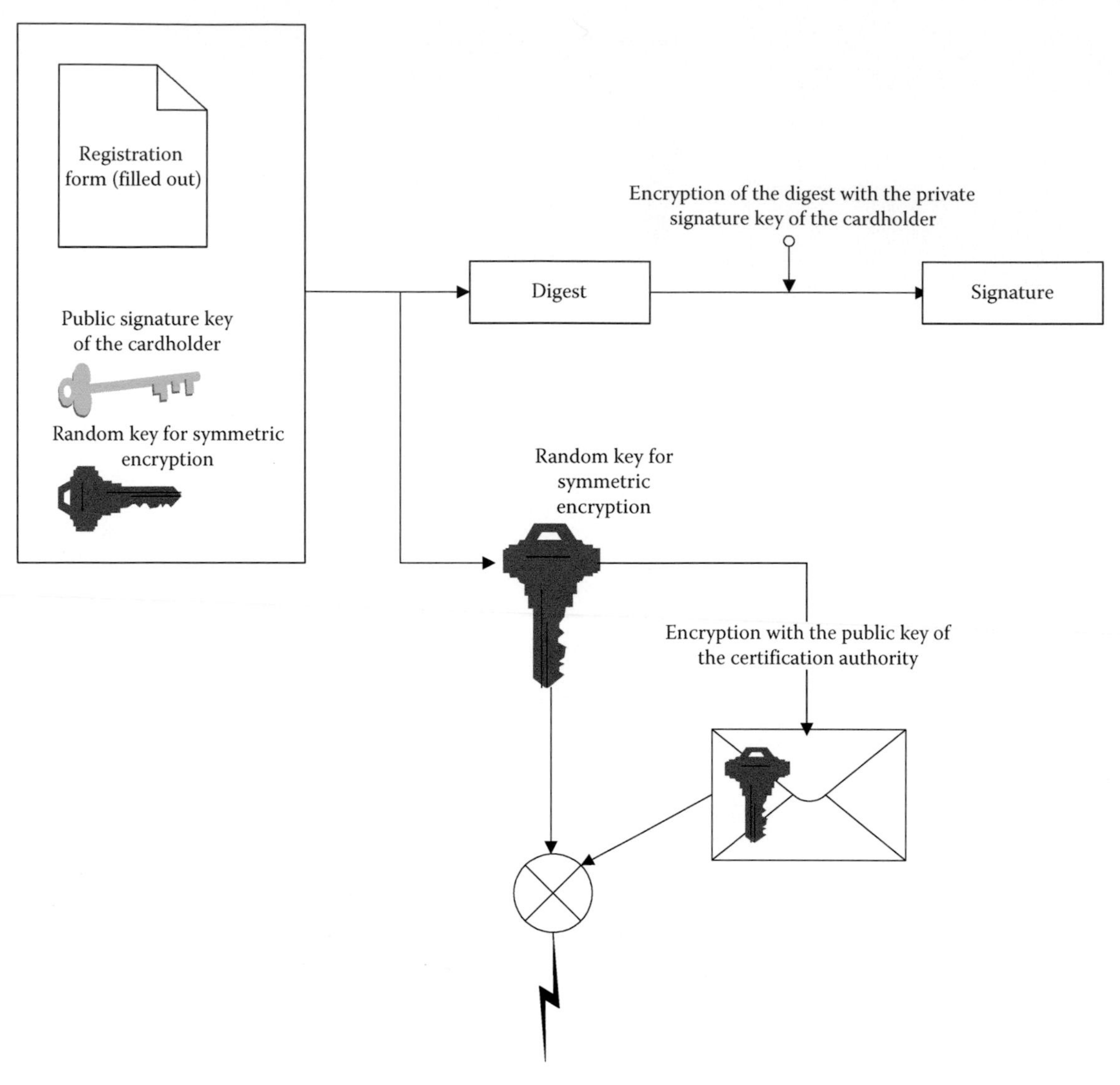

FIGURE 7.9
Formation of the CertReq message.

The authority signs the certificate with its private signature key and assigns it an expiration date, according to the operating rules. The third random that the cardholder generates will be resent to the cardholder (to protect against replay attacks). The CertRes is encrypted with the symmetric DES that the cardholder includes in the CertReq message. After receiving and processing the CertRes message, the cardholder stores the certificate in order to participate in SET transactions.

7.3.2.2 Merchant's Registration

Merchants must be registered by a certification authority before they can exchange SET messages with buyers and payment gateways. In a manner similar to the client's registration, the merchant engages in a series of exchanges of SET messages with its certification authority to obtain a signature certificate and a key encryption certificate. Table 7.4 contains the set of messages that the merchant's server exchanges with the certification authority during the merchant's registration and certification. These exchanges are depicted in Figure 7.10.

The merchant's software starts the conversation by sending the Me-AqCInitReq message. This command is simultaneously a request for registration and for the registration form. The authority responds by attaching the form to the Me-AqCInitRes message that also contains the authority's signature certificate so that the merchant verifies the signature by following the certification path all the way back to the root authority.

As explained earlier, the merchant has two certificates: one is for signatures and the other for encryption. Thus, the merchant's software transmits the filled-out

TABLE 7.4

Messages for Merchant's Registration

Message	Direction of Transmission	Significance
Me-AqCInitReq	Merchant → Certification authority	Initialization request for certification
Me-AqCInitRes	Certification authority → Merchant	Response to the initialization request
CertReq	Merchant → Certification authority	Certificate request
CertRes	Certification authority → Merchant	Answer to the certificate request

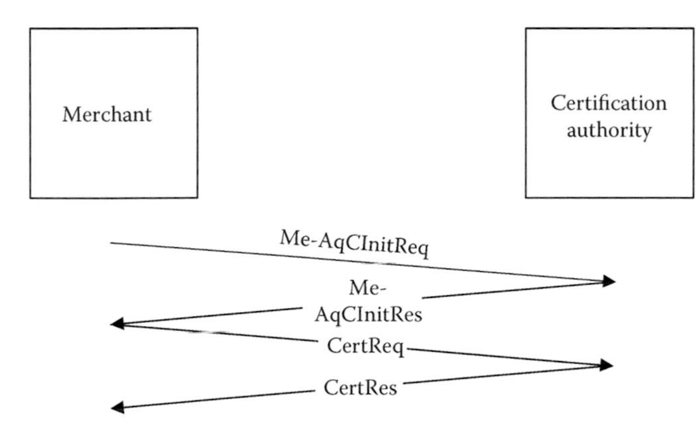

FIGURE 7.10
Merchant's registration and certification.

registration form with the two public keys attached, as well as a signature calculated over the whole transmission with the private key for signature. The whole message is encrypted with the secret key according to the DES algorithm. This secret key and the merchant's account number are put in a digital envelope and then encrypted with the public key for key encryption of the certification authority. The whole set constitutes the CertReq message as indicated in Figure 7.11.

Following the reception of the CertReq message, the authority decrypts the envelope to extract the secret key and then decipher the certification request. Once the signature has been successfully checked, the request is submitted for approval by the acquirer bank according to the rules established by the financial institution; these rules are outside the scope of the SET. Once the authorization is obtained, the authority constructs the certificates that the merchant requested, signs them with its private signature key, and then transmits them in the CertRes message, joining its signature certificate. Figure 7.12 shows how this message is prepared and then transmitted to the merchant.

Upon receipt of the CertRes, the merchant's software verifies the signature with the public key for signature of the authority, to prevent fraud. After this check, the software extracts the certificates and save them for later use.

7.4 Purchasing Transaction

In a purchasing transaction, the SET intervenes after the client chooses the desired item and selects the means of payment. The SET is responsible first for securing the transport of the payment authorization over the network. Next, the SET confirms the transaction in a protected manner. Finally, the SET is responsible for securing the reimbursement of the merchant through the financial clearance process. The SET can also secure other operations, for example, payment adjustment, but the principles are the same as for the straightforward case of purchasing and clearance. Therefore, these additional operations will not be discussed, and the reader is invited to refer to SET specifications or to books on SETs (Loeb, 1998).

7.4.1 SET Payment Messages

There are two types of payment messages: the mandatory and the optional. Mandatory messages occur in every purchasing and payment transaction, whereas the optional messages are for complementary services that are not needed in each implementation. Table 7.5 lists the obligatory payment messages, and Table 7.6 lists the optional messages.

Mandatory messages can be grouped into three categories of request/response pairs related to purchase orders, payment authorizations, and financial settlement. The basic exchange for a payment includes the following messages: PReq, AuthReq, AuthRes, PRes, CapReq, and CapRes (Figure 7.13). The PReq and PRes messages are exchanged between the merchant and the cardholder, whereas the other messages are exchanged between the merchant and the payment gateway.

The messages from the cardholder toward the payment gateway pass by the merchant's server. To preserve the confidentiality of the client's banking data from the merchant, the client has to have the gateway certificate for key encryption to establish an encrypted link with it.

7.4.2 Transaction Progress

7.4.2.1 Initialization

With the optional PInitReq message, the cardholder requests from the merchant and the payment gateway their certificates. This message is transmitted in the clear on the network. The response is the PInitRes message

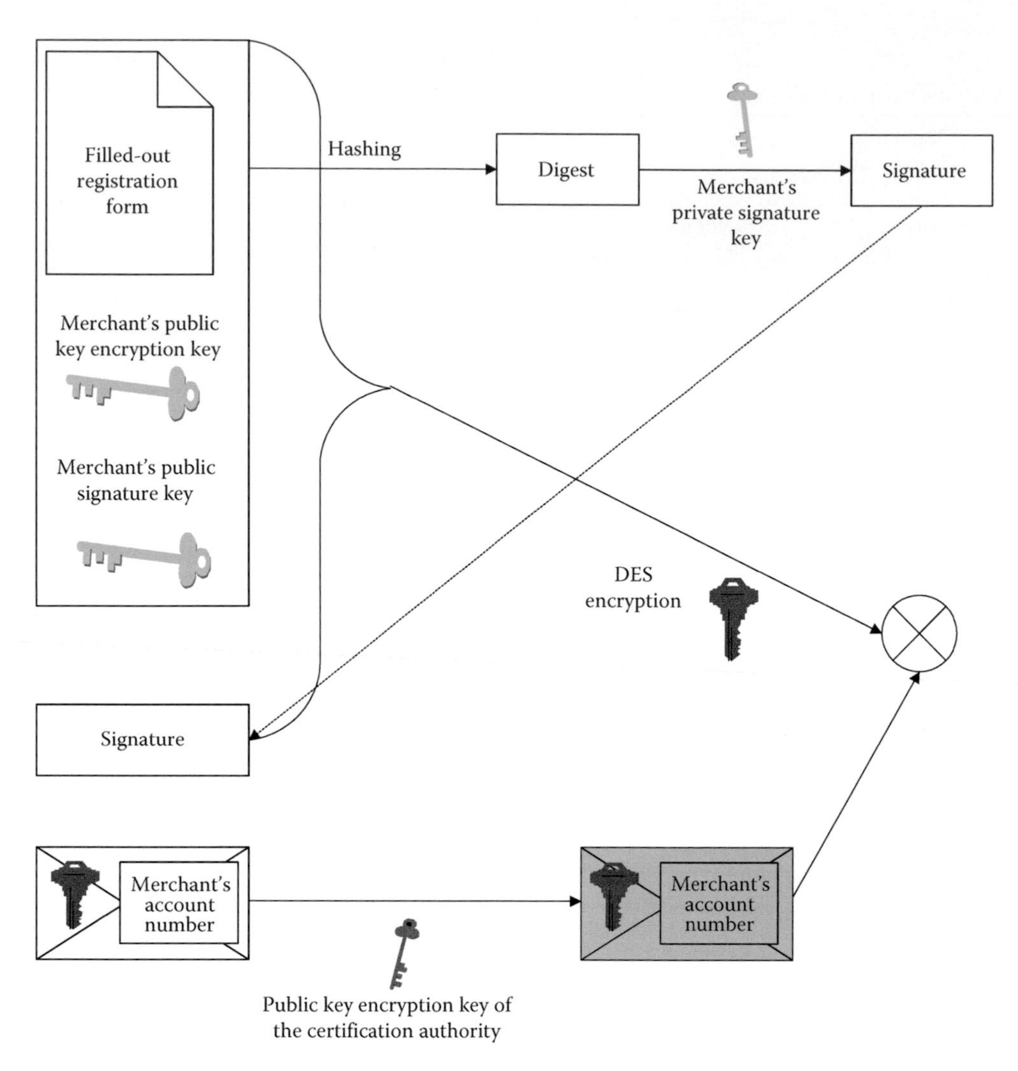

FIGURE 7.11
Processing of the CertReq message at the merchant's site.

that contains the merchant's signature certificate, the gateway's key encryption certificate, and a transaction identifier, TransID. This message is signed with the merchant's private signature key.

Upon reception of the PInitRes message, the cardholder verifies the merchant's signature and the merchant's signature certificate, as well as the key encryption certificate of the gateway by following the trust chain backward to the first trusted authority or even back to the root authority. Once authenticity is proven, the merchant may store the certificates to avoid going back on the itinerary for any new transaction.

7.4.2.2 Order Information and Payment Instructions

The Order Information (OI) and the Payment Instructions (PI) are transmitted in the PReq message. The Order Information identifies the goods to be purchased, whereas the Payment Instructions contain the PAN, the price of the items, and the transaction identifier TransID, if it has been supplied by the merchant. Otherwise, the cardholder software creates the transaction identifier.

The digests of the Order Information and the Payment Instructions are concatenated, then encrypted with the private signature key of the cardholder to construct the dual signature of the PReq message. Next, this message is signed with the private signature key of the cardholder. Illustrated in Figure 7.14 is the construction of the dual signature of the PReq message.

The Payment Instructions, together with the digest and the dual signature, are transmitted after their encryption with a secret key generated with a random number generator. This key, in turn, is encrypted with the public key encryption key of the payment gateway before transmission, which puts it beyond the

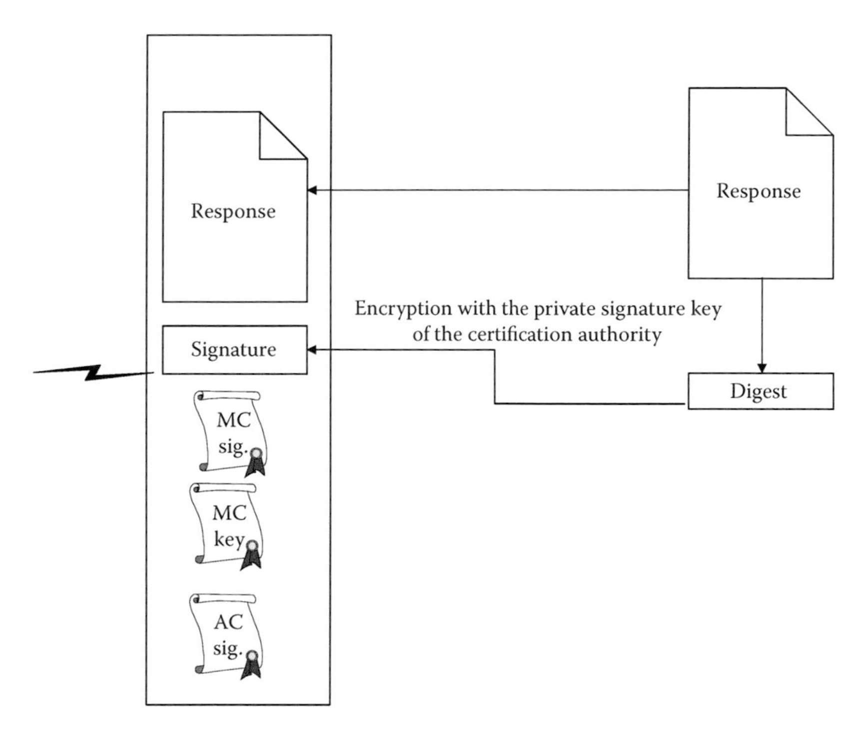

FIGURE 7.12
Composition of the CertRes message by the certification authority.

TABLE 7.5
Mandatory Payment Messages in SETs

Messages	Direction of Transmission	Meaning
PReq	Cardholder → Merchant	Purchase request
PRes	Merchant → Cardholder	Response to the purchase request
AuthReq	Merchant → Gateway	Authorization request
AuthRes	Gateway → Merchant	Response to the authorization request
CapReq	Merchant → Gateway	Capture request
CapRes	Gateway → Merchant	Response to the capture request

merchant's reach. When the cardholder has a signature certificate, this certificate is transmitted as well. Figure 7.15 depicts the components of the PReq message, and it is seen that this is one of the most complex messages of the protocol.

More details on the PReq message are given in the following to convey an idea on the complexity of the SET protocol and the redundancies that it introduces as a way to guarantee security. Keep in mind that the content of the PReq message depends on whether the cardholder has a signature certificate. If the cardholder is certified, the PReq will comprises two parts as indicated in Figure 7.16.

The symbols used in Figure 7.16 are those used in SET specifications. Thus,

- The operator $DD(A)$ designates the digest of A reorganized according to the *DigestData* content type of PKCS #7 and denoted as *digest* of A.
- The operator $L(A, B)$ is a linkage operator that represents a link or reference pointer to B attached to A, produced by concatenation of A to the digest of B

$$L(A, B) = \{A, DD(B)\}.$$

- The operator $SO(s, t)$ represents the signature of participant s on message A of type PKCS #7 SignedData.
- The operator $EX(r, t, p)$ represents the following sequence of operations: encryption of message t with a DES key, k; insertion of this key and the variable p in the PKCS#7 envelope with OAEP to form OAEP$\{(k, p)\}$; and the encryption of this envelope with r, the public RSA key of the recipient.
- Finally, fields between the square brackets [ ] are optional.

The first component of the PReq message, PIDualSigned, contains the Payment Instructions and is directed

TABLE 7.6
Optional Payment Messages in SETs

Messages	Direction of Transmission	Meaning
PInitReq	Cardholder → Merchant	Initialization message to allow the cardholder to obtain the merchant and the payment gateway certificates
PInitRes	Merchant → Cardholder	Response to the PInitReq message
AuthRevReq	Merchant → Gateway	Message used by the merchant to cancel an authorization and to reduce the amount of a transaction already authorized
AuthRevRes	Gateway → Merchant	Response to the AuthRevReq message
InqReq	Cardholder → Merchant	Inquiry on the status of the transaction in progress
InqRes	Merchant → Cardholder	Response to the inquiry in InqReq
CapRevReq	Merchant → Gateway	Message used to cancel a capture
CapRevRes	Gateway → Merchant	Response to the CapRevReq message
CredReq	Merchant → Gateway	Message used to claim credit on a transaction already captured
CredRes	Gateway → Merchant	Response to the CredReq message
CredRevReq	Merchant → Gateway	Request to cancel a CredReq message
CredRevRes	Gateway → Merchant	Response to the CredRevReq message
PCertReq	Merchant → Gateway	Request by the merchant for the payment gateway certificate
PCertRes	Gateway → Merchant	Response to the PCertReq message
BatchAdminReq	Merchant → Gateway	Message for administering capture request
BatchAdminRes	Gateway → Merchant	Response to the BatchAdminReq message
Error	All	Error message

Note: The *Error* message is sent when the received message is not recognized or if the message content is ambiguous and requires clarification.

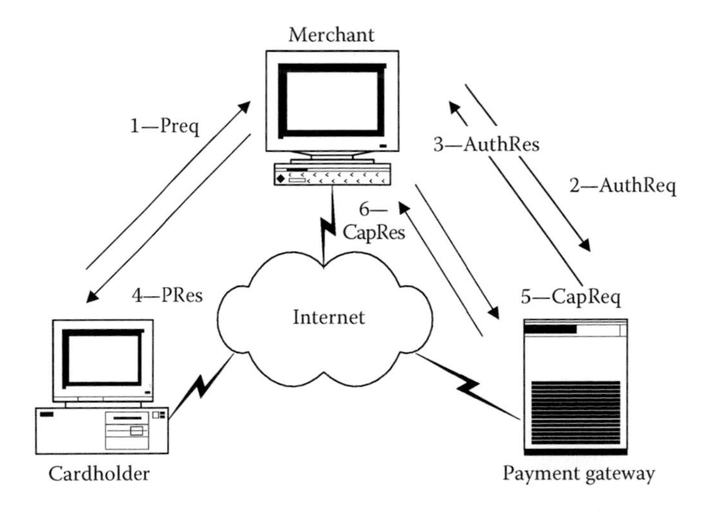

FIGURE 7.13
Mandatory messages in a SET purchase transaction.

to the acquirer bank with all the necessary indications to identify the cardholder. The second part of the message, OIDualSigned, contains the Order Information, signed and encrypted, as indicated earlier. Listed in Table 7.7 are the mandatory fields shown in Figure 7.16 and their contents. For more details, the reader is invited to refer to Book 3 of the SET specifications.

7.4.2.3 Authorization Request

The authorization request initiates a dialog between the merchant server and the payment gateway on one side and between the gateway and the issuer bank on the other side. The merchant server treats the PReq differently depending on whether it contains a signature certificate of the cardholder. If a certificate is available, the server verifies it by going back to the certification path and checking the signature on the Order Information using the public key of the cardholder that is mentioned in the certificate. Then, the server verifies the dual signature, and when the security checks are met, it sends back an acknowledgment to the cardholder with the PRes message. This message is transmitted as a clear text and is signed to verify the integrity of transmission upon reception. The message must contain the merchant's signature certificate.

The merchant's server next relays the Payment Instructions to the gateway in *the AuthReq message.* This request is signed with the private signature key of the merchant and the whole message is encrypted with a symmetric DES key. Next, this key is enveloped electronically together with the merchant's signature and encryption certificates and, if available, the signature certificate of the cardholder. The envelope then is encrypted with the public encryption key of the

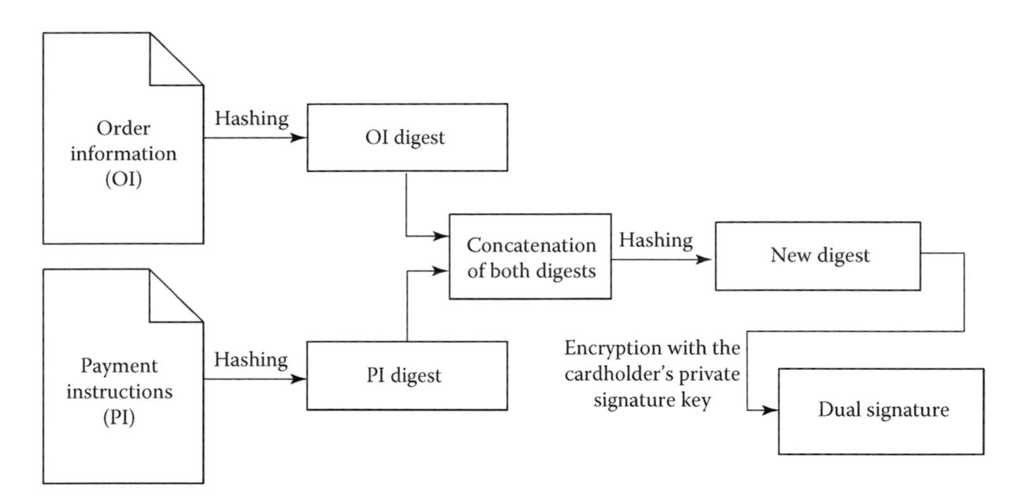

FIGURE 7.14
Dual signature of the PReq message.

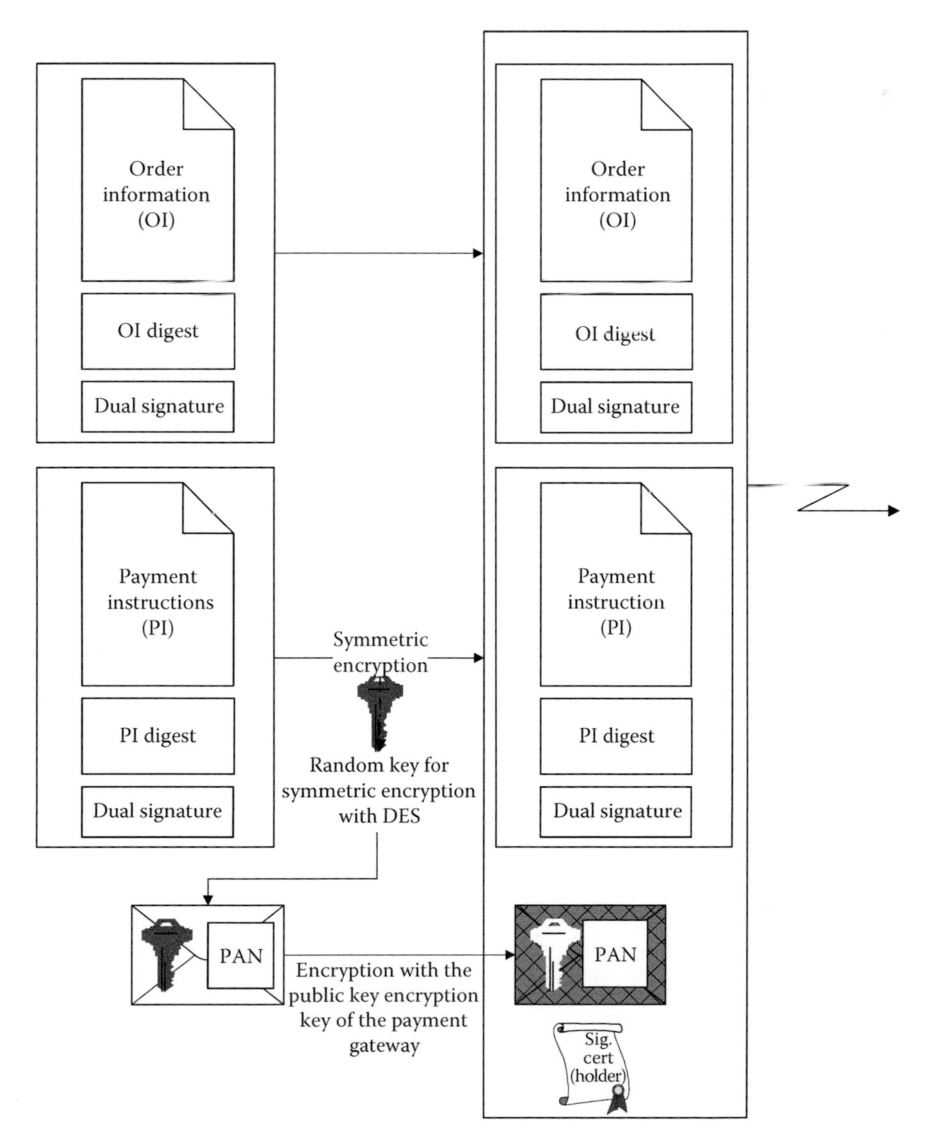

FIGURE 7.15
Construction of the PReq message.

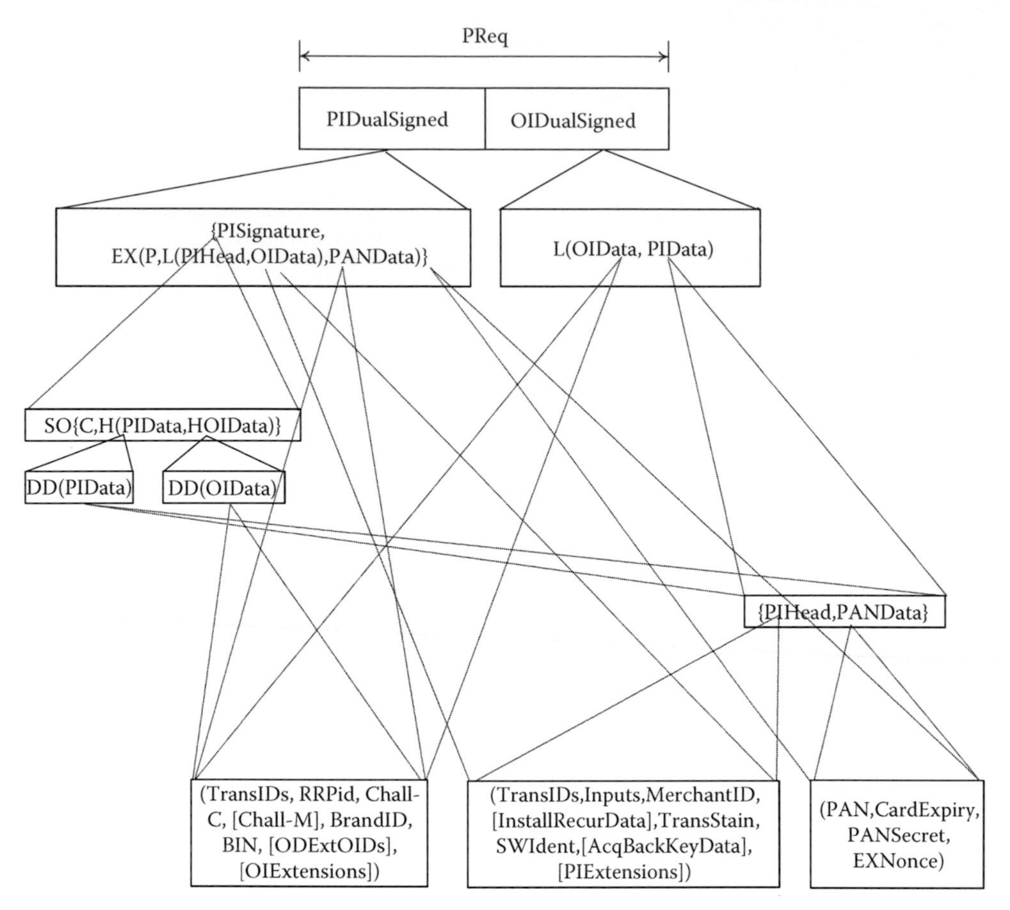

FIGURE 7.16
Composition of the PReq message.

TABLE 7.7
Content of the Mandatory Fields of PReq

Field Name	Contents
Inputs	Digest of the data included in the Order Information and the purchase amount
OIData	Digest of the Order Information, the challenge to the merchant to ensure the freshness of the cardholder signature, the card brand, and the issuer bank identifier
OIDualSigned	OIData and digest of PIData
PANData	Card number, expiration date, secret code, nonce to protect against dictionary attacks
PIDualSigned	PISignature and linkage with the Order Information
PIHead	Merchant identifier, hashed digest HMAC of the global transaction identifier XID and the cardholder secret code, identifier of the software provide, data on the algorithm of the acquirer bank indicated in the gateway certificate
PISignature	Digest of the cardholder signature
TransID	Transaction identifiers

Note: HMAC, hashed message authentication code.

payment gateway. Figure 7.17 depicts the different elements that constitute the AuthReq message.

7.4.2.4 Granting Authorization

Upon receiving the AuthReq message, the gateway checks the received certificates by going back along the certification path. It decrypts the envelope of the request for payment authorization with the help of the private key of the pair of key encryption keys to extract the symmetric encryption key. This symmetric key is then used to decrypt the request whose integrity is verified by the public signature key of the merchant.

The payment gateway performs similar verifications for the Payment Instructions, extracts the symmetric key used by the cardholder and the cardholder's PAN, and then verifies the dual signature with the public signature key of the cardholder. Furthermore, the gateway must check the consistency of the request for payment authorization with the Payment Instructions, for example, that the transaction identifier is the same in both parts.

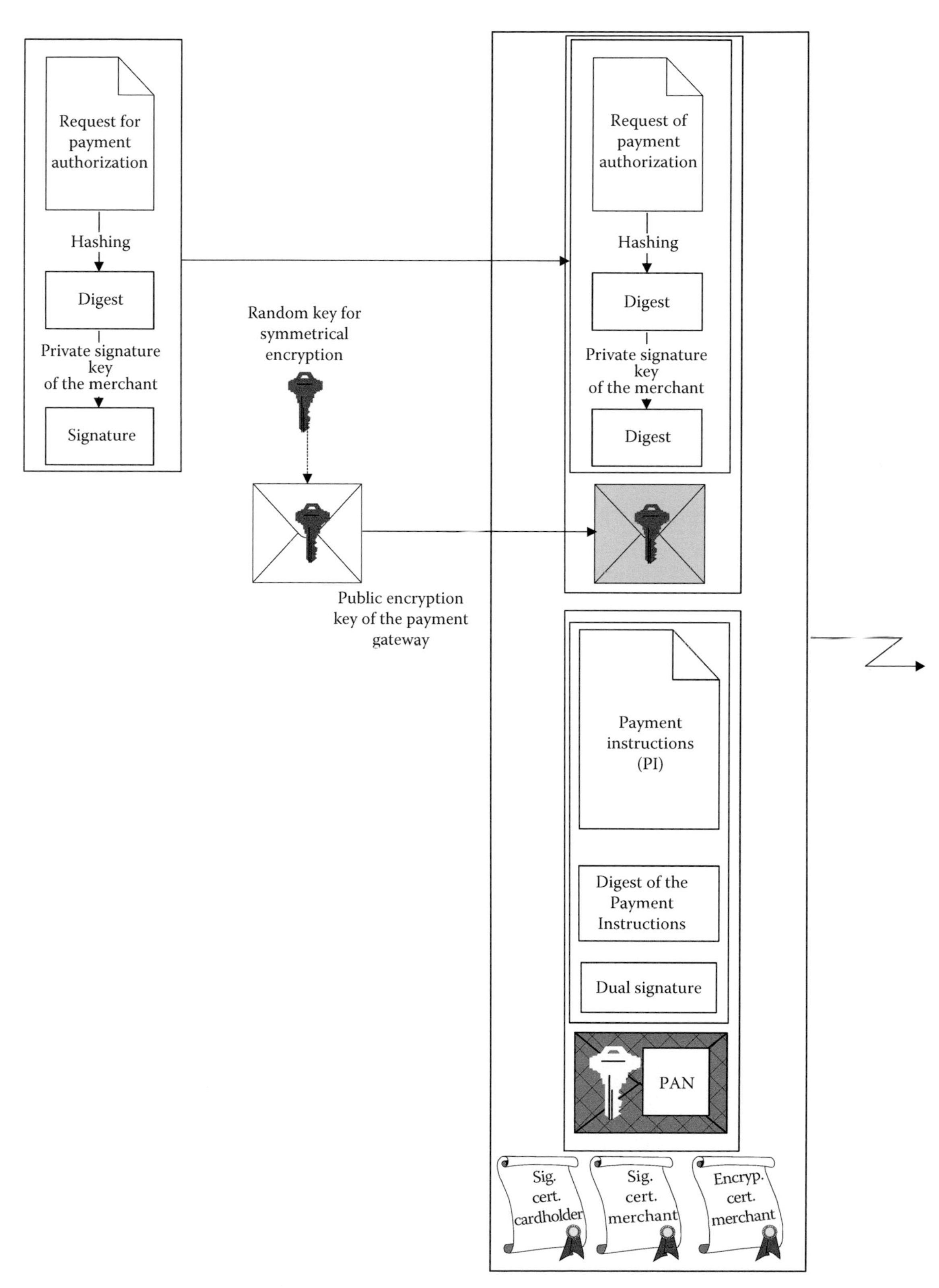

FIGURE 7.17
AuthReq message transmitted to the payment gateway by the merchant.

Once this stage is successfully terminated, the payment gateway prepares an authorization request and sends it to the issuer bank through the bank card network. The steps are not part of a SET and will not be explained here. Suffice it to say that bank cards have a spending cap, so if the expense ceiling of a cardholder is reached, a referral procedure kicks in with an attendant intervention by phone before authorizing the payment (Dragon et al., 1997, pp. 171–179). The merchant software is supposed to take care of these eventualities.

Once the authorization arrives, the gateway sends the response message AuthRes. This message is encrypted with the DES algorithm with a secret random number, which is put in an envelope that is encrypted with the public key of the key encryption pair of the merchant. The AuthRes message comprises two distinct parts:

1. A mandatory part that is the response of the issuer bank authorizing the payment.
2. An optional part that contains a capture token signed with the private signature key of the payment gateway. This token must accompany any capture request associated with this transaction to facilitate daily remote collections. The token is encrypted with a second secret key that the gateway generates. As before, the secret key will be inserted in an envelope that is encrypted with the public encryption key of the payment gateway. Thus, this envelope can be only opened by the payment gateway.

When it receives the AuthRes message, the merchant server extracts from this message the response of the issuer bank to the request for authorization as well as the encrypted token. The server decrypts, with its private encryption key, the first envelope to recover the secret key and to decipher the response of the issuer banks to the authorization request.

In contrast, the merchant server cannot recover the capture token because the DES encryption key remains in the envelope encrypted with the public encryption key of the payment gateway. The envelope containing the DES key and the encrypted token is stored as is and then used to accompany capture requests.

7.4.2.5 Capture

This is the procedure whereby the merchant claims the credit that was implicitly accorded during transaction authorization. The financial settlement is initiated by sending the CapReq message to the payment gateway to request capture or collection. This message includes, among others information, the amount of the transaction, the transaction identifier, the reference to the authorization, the encrypted and signed capture token, and the envelope containing the encryption key previously saved. Similar to all other SET messages, the capture request is encrypted with a secret key that is inserted in an envelope that is encrypted with the public encryption key of the payment gateway.

This request can take place directly after the purchase or at regular intervals by grouping several purchases for efficiency reasons. One capture message can thus include several capture tokens from several transactions. The requests are given a sequence number to distinguish among them and to associate a response with its associated request.

Upon receipt of the message, the payment gateway extracts the secret key from the envelope to decrypt the capture request. It then proceeds to check the signature as indicated previously. The gateway then extracts the envelope that contains the symmetric key used to encrypt the capture token and decrypts it with its private key encryption key. Thus, it will be the only entity capable of extracting this symmetric key to decrypt the capture token. The payment gateway sends a clearance request using all the information that it has collected to the acquirer bank.

To conclude, the gateway responds to the merchant with the CapRes message. This message contains the response of the acquirer bank with the amount finally credited to the merchant account. This message is signed with the private signature key of the payment gateway. The CapRes message is encrypted with a secret key that is included in the envelope encrypted with the public key encryption key of the merchant. Verification of the message by the merchant's server proceeds according to the procedures explained several times earlier.

7.5 Optional Procedures

A SET defines the optional requests from the merchant to initiate the following actions:

1. Modification or cancellation of a previous authorization with the help of the exchange of AuthRevReq and AuthRevRes messages. The merchant can place a request at any moment after the authorization is obtained but before the capture request.
2. Modification or cancellation of a capture with the messages CapRevReq and CapRevRes.
3. Refund of the cardholder with the messages CreditReq and CreditRes. This exchange occurs

after reconciling the accounts of the merchant with the issuer bank. The response message confirms or rejects the request for refund.

4. Cancellation of a refund. The optional message CredRevReq gives the merchant the possibility of requesting the cancellation of a refund already accepted. The message from the gateway CredRevRes contains the decision of the issuer bank regarding the cancellation.
5. Request for the gateway's key encryption certificate. The merchant request is in the PCertReq, whereas the response of the payment gateway is in the PCertRes message.
6. Grouped settlement of a batch of capture tokens. The merchant sends the AdminReq intended to the acquirer bank through the payment gateway by joining the accumulated capture tokens, usually every 24 hours. The message can indicate either the start or the end of the processing of a batch of tokens or an inquiry on the status of a batch. The response from the bank is relayed through the gateway within the AdminRes message.

7.6 Efforts to Promote SETs

This section summarizes some activities to encourage SET adoption. As explained earlier, buyer's certification was made optional in Version 1.0 of the specifications. The other efforts for SET promotion are as follows:

1. SET promoters developed tools to help software developers verify that an implementation adhered to the protocol, in particular:
   a. They devised a SET reference implementation.
   b. They made conformance test suites available to indicate the expected behavior of compliant implementations.
2. In France, SETs were adapted to the integrated circuit cards used by French banks.
3. Finally, the SET wallet was replaced with a browser incorporating TLS/SSL in a hybrid SET/SSL architecture.

7.6.1 SET Reference Implementation (SETFEF) and Conformance Tests

Visa and MasterCard funded the development of a reference implementation, SETREF (Secure Electronic Transaction Reference Implementation), over several platforms (UNIX, Windows and Windows NT). SETREF is also distributed in a CD accompanying Loeb's book (1998). S/PAY was a toolbox with a demonstration software to show how to integrate SET with payment software.

SETREF utilized the cryptographic toolkit BSAFE from RSADSI, which at the time was subject to export restrictions by the United States. To overcome this drawback, the worldwide academic community developed an interface compatible with BSAFE with the use of the CAPI of SSLeay.

SETCo also provided conformance tests to verify compliance to the protocol and managed interoperability testing among various commercial implementations. These tools were needed because the protocol was not specified with a formal language but in English so that ambiguities and misinterpretations were likely.

7.6.2 SETs and Integrated Circuit Cards

The initial focus of SETs was on security through software implementation. Adaptations of SETs to integrated circuit cards (chip or smart cards) originated from Europe. A joint project of the Groupement d'intérêt Économique (Gie) Cartes Bancaires and EuroPay aimed at establishing a new standard Chip-Secured Electronic Transaction (C-SET) (Groupement, 1997). The second project, E-Comm, was a commercial offer supported by BNP, Société générale, Crédit lyonnais, Visa, France Telecom, and Gemplus. These two efforts converged into Cyber-COMM, which is technically based on C-SET and which includes the BO' cards used in France, as well as the EuroPay, MasterCard, and Visa (EMV) cards.

SETCo published two documents to adapt SET to smart cards (SETCo, 1999a,b) with C-SET and with the EMV specifications (SETCo, 1999a,b). The specifications considered online identification of cardholders using personal identification numbers (PINs) entered from the keyboard or via secure PIN-pad readers.

These two extensions retain the fundamental characteristics of SETs:

1. The cardholder secrets, such as the bank card number, are hidden from the merchant during a transaction.
2. The merchants and, optionally, the clients are certified.

In addition, C-SET exchanges were integrated with SET exchanges so that a transaction could take place even if one of the parties conforms to the SET and the other to the C-SET. The security level is the same as when both parties conform to the C-SET.

7.6.3 Hybrid TLS/SET Architecture

Bridging SET and TLS was proposed to encourage market adoption of SET. The use of a hybrid TLS/SET architecture was motivated by the following considerations:

1. The hierarchy of SET authorities introduces a rather heavy certification structure that causes computational delay for each exchange. A joint operation of TLS and SET would reduce the cryptographic load on the client software, particularly because the certification of clients in SET Version 1.0 is optional.
2. TLS/SSL plug-ins were readily incorporated into browsers, while users had to download and integrate the SET wallet. Because of the reduction of computational requirements on the client software, the plug-in installed on the clients' posts is called a *thin wallet* (Inza, 2000).
3. A hybrid mode SET/TLS of operation, such as 3D SET, would be attractive to small businesses or those that deal with small amounts because of its low cost and ease of registration.

In this architecture, clients and merchants could continue to secure their exchanges at the transport layer using TLS/SSL through a payment intermediary. SET services would be assured on the bank side. With three independent servers, a three-party dialog has to involve the consumer, the payment intermediary, and the merchant server (Sherif et al., 1997, 1998). A TLS/SSL session is established between the client and the intermediary server so that the data input (client name, order details, name of the merchant, the transaction amount, number of the bank card to be charged, etc.) is not accessible by the merchant. Similarly, the intermediary establishes another TLS session with the merchant to transfer the payment order without the financial details or, possibly, without the purchaser identity. Once the merchant accepts the order and the intermediary authenticates and verifies the response, the payment intermediary establishes a SET communication channel with the SET payment gateway to request a payment authorization. As a proxy for buyers and merchants before SET certification authorities, the intermediary can justify investment in more powerful machines.

Two hybrid architectures have been proposed: 3D SET and SET Fácil (Easy SET).

7.6.3.1 3D SET

The 3-Domain SET or 3D SET is a hybrid TLS/SET architecture where the merchant's bank and the payment intermediary are associated. The intermediary is the technical agent of the bank to relay and secure the messages exchanges and to originate the SET exchanges.

As shown in Figure 7.18, in 3D SET, the payment intermediary acts at the intersection of three domains: that of the client, that of the merchant, and that of SET proper. Thus, the payment gateway plays the following roles:

1. TLS server with respect to the client and the merchant.
2. Web host that initiates the SET transaction for the merchant.
3. TLS certification authority for the consumers.

From the SET viewpoint, the payment intermediary has two pairs of private/public keys, one pair for the exchange of data encryption keys and the other pair for signing the transmitted data. Because the payment intermediary plays the role of a TLS certification authority for the merchant and the client, it will own a third pair of private/public keys to be used with the TLS certificate.

Because the SET credentials (certificates, keys, etc.) reside at the site of the payment intermediary, they can be protected from failures in the hardware of the individual users.

As a proxy, the intermediary could also take charge of the merchant registration by the SET authorities and store the merchant's certificates and pairs of private/public keys. In this way, the merchant would avoid downloading the SET software and going through the associated certification procedures; it would only need a browser to present the TLS certificate. This solution was thought to be of interest to small business owners; the inconvenience would be that they would not be able to use their SET certificates without passing by the payment intermediary.

Furthermore, the intermediary could act as a trusted third party to arbitrate disputes, provided that all traces of the communications are kept with time stamps. An additional advantage is to be able to change the protocol on the server side without bringing about parallel changes on the client side. In particular, the solution circumvents restrictive laws in the area of cryptography. The payment intermediary would operate where the SET encryption parameters are allowed, but TLS connections could use weaker cipher suites depending on the locations of merchants and their clients.

The main drawback of this method is that the client does not control the certificate that is issued in his or her name, because it is stored by the intermediary. Furthermore, if the merchant and the client always go through the same intermediary, in the long run, the intermediary may be able to reconstruct their marketing

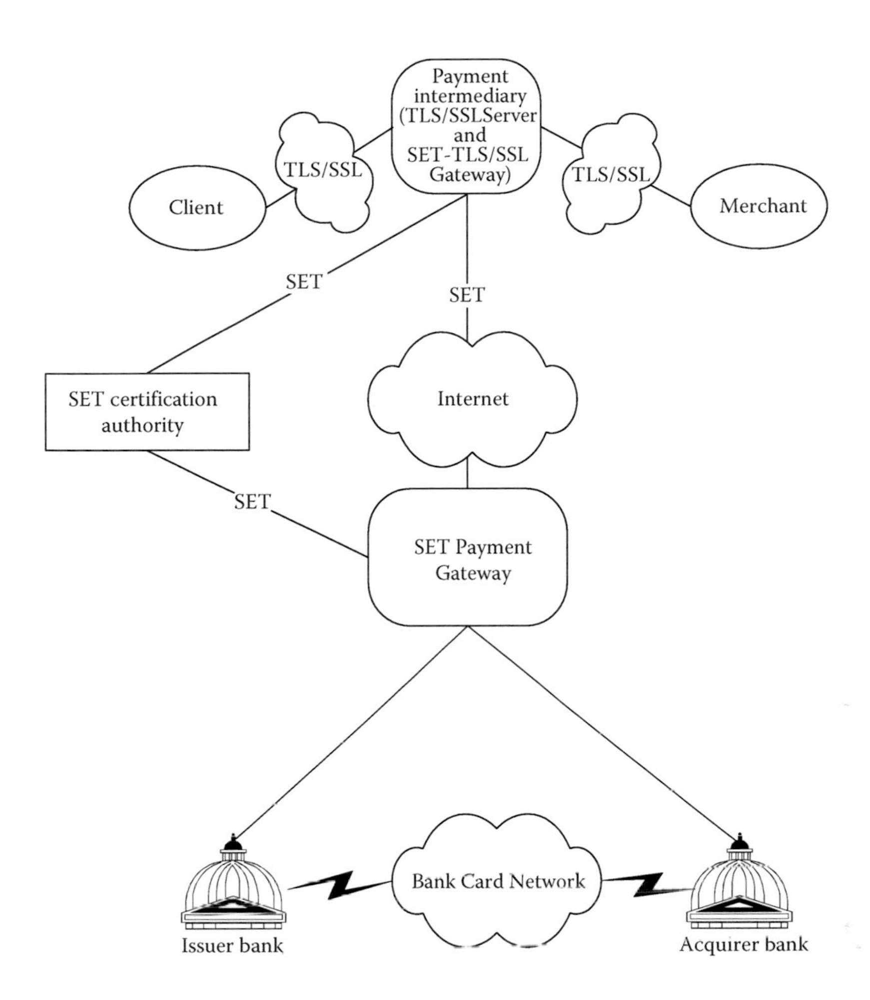

FIGURE 7.18
Architecture of 3D SET.

profiles. In the absence of legislation on the subject, this may be objectionable to individual parties.

7.6.3.2 SET *Fácil*

In SET *Fácil* (Easy SET), a solution championed by the Spanish bank Banesto (now part of the Santander group in cooperation with IBM), the bank is the payment intermediary. As a consequence, the server takes care of all operations that need intense computations, such as processing of client's certificate (Merkow, 2000).

7.7 SET versus TLS/SSL

SET was designed to be the main payment method with bank cards on the Internet between the merchants, the buyers, and their financial institutes. It failed to gain widespread use, however. There are several possible explanations for this outcome:

1. *The SET introduced two new entities*: Certification authorities to certify the actors and the payment gateway in the authorization network. In contrast, the infrastructure for TLS/SSL certification is much easier to establish, and no additional entities were needed in the payment networks.
2. The SET requires authentication of each message, which adds delays and computational costs to each transaction. In addition, SET computational overload in software-only implementations may reduce the response time to inacceptable levels (Schneier, 1998b; Inza, 2000).
3. While the use of the payment intermediary reduces the cryptographic load on both the client and the merchant by an order of magnitude, payment intermediaries did not get any benefit from implementing SET.
4. The secrets on the cardholder side are stored on the hard disk of a computer, which introduce another set of risks. Furthermore, at the time of

SET introduction, the use of integrated circuit payment cards (smart cards) was mostly outside the United States, the main market at the time of payments over the Internet.

5. For SET implementations, the client software (the e-wallet) needed to be installed in addition to the browser. In contrast, the client software for TLS/SSL solutions is integrated in all browsers.
6. The cost of operating SET solutions favored TLS/SSL-based alternatives. These costs included
   a. The cost of new terminals with specialized hardware
   b. Client-side certification
   c. Maintenance and support of purchaser and merchant software
7. TLS/SSL solutions reduced the risk of fraudulent transactions to a level that could be tolerated without going through the complication of SET implementation. Furthermore, TLS/SSL offered a blanket protection for transactions over TCP.
8. No lead segment of the target population could be identified as a lead user that would encourage the adoption of SET. On the contrary, information specialists outside the banking community were not convinced that the computations associated with SET were needed or could be justified.
9. There was no legal obligation to reduce fraud to the levels that SET achieved.

Table 7.8 summarizes the main differences in the characteristics of TLS/SLL and SET.

TABLE 7.8

Comparison between TLS/SSL and SET

TLS/SSL	SET
Simple and easy to use.	Complex and requires a certification infrastructure.
Generalist protocol.	Banking payment protocol.
Distributed with browsers.	A wallet has to be installed on the client side.
Authentication infrastructure is not mandatory.	Infrastructure for authentication is mandatory.
Authentication is at the beginning of the session.	Each exchange is authenticated.
Point-to-Point Protocol.	Several parties participate in a transaction.
The merchant receives all the details of the order and the payment.	With the dual signature, access to information is restricted to those who need it.

7.8 Summary

SET was intended to be a solution dedicated to electronic commerce that preserves the existing interfaces with bank networks. Accordingly, the protocol gives the buyer the ability to verify the authenticity of the merchant site with the help of certificates. In addition, the method of the dual signature in SET links the purchase order that the client sends to the merchant, including as the Payment Instructions directed to the issuer bank via the SET payment gateway with the same signature. Thus, the Payment Instructions are sent encrypted to the merchant, which merely forwards them to the payment gateway. Furthermore, only the merchant can read the purchase order, which remains opaque to the gateway, and only the payment gateway is allowed to extract the purchaser's bank card number to submit an authorization request to the bank network.

The principal characteristics of SET are summarized as follows:

- The merchant keeps the Order Information that is signed with the private signature key of the client. The merchant also retains the response of the gateway (in the *AuthRes* message) signed with the private signature key of the gateway. If the client is certified, the merchant has a copy of the cardholder certificate and the public key that it cites. Nevertheless, the merchant does not have the details of the client's bank card.
- The cardholder receives a response to the purchase order, which is signed by the private signature key of the merchant. The cardholder receives a copy of the merchant signature certificate but not the encryption certificate, which is used for financial settlement.
- The payment gateway knows the financial details of the transaction between the merchant and the cardholder without being aware of the subject of the transaction.
- Each of these transactions has a unique transaction number that is encrypted, which prevents replay attacks.

To ensure the security of all payments by bank cards on the Internet, SET designers attempted to overcome the fundamental weakness of IP Networks (where management and control information are mixed with users' traffic) by requiring the strong authentication of each exchange of messages. Despite these security advantages, solutions based on point-to-point security using SSL/TLS have been adopted over SET. Thus, market fragmentation was avoided, but the market converged on a suboptimal solution from the point of bank card security.

There has not been any systematic study for SET's lack of success, but some ideas have been advanced such as the following (Ferguson et al., 2010, pp. 10–12):

1. There is a disincentive for merchants to use SET because a merchant can verify the integrity to the banking information without reading it (it is destined to the bank). This prevents the shopkeeper from leveraging customer data for targeted advertising and promotions. Most existing commercial systems assume that the merchant has access to the customer's card number, and so, the use of SET would have required a massive rewrite of these systems.
2. SET does not protect the buyer from viruses that could infect the buyer's machine and force the SET wallet to sign spurious transactions.
3. SET does protect against online theft of card numbers through phishing or spy software or by massive data breaches that target the merchant's database.

SET was attempted during the early days of online shopping where the exact threats were not well understood. However, it offered some elegant technical solutions, such as the dual signature approach, which would probably be used in the future for other applications.

Questions

1. What is the purpose of the dual signature in SET?
2. Enumerate the keys of an acquirer bank if it is also a certification authority.
3. Compare the payment solutions using TLS/SLL solution to the solution that SET offers.
4. What are the main characteristics of the SET wallet (stored on a hard disk or a smart card)?
5. What is 3D SET?
6. How was SET adapted to integrated circuit cards?
7. Explain the market preference of TLS/SSL solutions over SET.

8

Payments with Magnetic Stripe Cards

Magnetic stripe cards are still used in many applications—cash withdrawal, credit/debit prepayments, storage of currency or tokens, access control, and so on—particularly in the United States. This chapter surveys the security measures to protect payments at Point-of-Sale Terminals and online. The efforts of the Payment Card Industry Security Standards Council are highlighted. Online security measures are presented next, including monitoring criteria to detect possible fraud in real time as well as some new architectures such as the 3-D Secure initiative. This chapter concludes with a description of U.S. efforts to migrate to integrated circuit payment cards, using the Europay, MasterCard, and Visa (EMV) specifications.

8.1 Point-of-Sale Transactions

Magnetic stripe cards consist of rectangular plastic support with elements for identification and authentication of the issuer on both of its sides. Some of the hallmarks on the face side are as follows:

- Logos of the card issuer, the financial institution, and the bank card schema operator
- Reliefs of the card primary account number (PAN), the cardholder name, and the expiration date produced by embossment
- Ultraviolet marks that glow when highlighted by a special light
- A hologram to increase security and make counterfeiting more difficult

The security features, however, have not prevented counterfeiting of cards. A website egregiously called fakeplastic.net was launched in 2012 to sell counterfeit credit and debit cards, as well as holographs used to make fake driver's licenses in the United States (Zambito, 2014).

The PAN consists of 10–19 digits divided into groups of four. The first digit of the first group identifies the network of the bank card schema (4 for Visa, 5 for MasterCard, etc.). Following are codes for the country, the representative of the bank card schema in that country, and the bank issuing the card. The last digit forms a verification code called "Luhn's key" and is selected to meet the following condition:

$$\sum_{i\,even} n_i + \sum_{i\,odd} (2n_i) \bmod 9 = 0 \bmod 10$$

The reverse side has a place for the cardholder's signature, the return address in case of loss and, in North America, the list of associated bank networks. In addition, the backside has a magnetic stripe with three recording tracks, each 0.28 cm wide and separated by a small distance. The personalization data are recorded on the first two tracks. Track 1, used in the airline industry and optional in most payment applications, consists of 79 octets (i.e., 79 ASCII characters) separated into field to describe the account number, expiration data, the cardholder's full name, some service code to specify how the terminal should read the card and a redundancy check to verify the integrity of the data being read by the magnetic stripe reader. Track 2 contains a shorter version of the data on Track 1; it is 40 octets long and includes the PAN, the expiration date, a service code, and a redundancy check. The use of a truncated version of the data is a legacy from the days of low-speed dial-up payment terminals and was used to expedite the communication with authorization centers. Track 3 is a read/write track and is sued for driver's licenses and the savings and loan industry (Dragon et al., 1997, pp. 150–153).

The transaction flow in a point-of-sale transaction consists of the steps depicted in Figure 8.1:

1. At the point-of-sale device or the automated cash dispenser, the data recorded in the magnetic tracks are captured.
2. For identity verification, the cardholder may be asked to enter a personal identification number (PIN).
3. The terminal compares the entered PIN with the one retrieved from the magnetic strip.

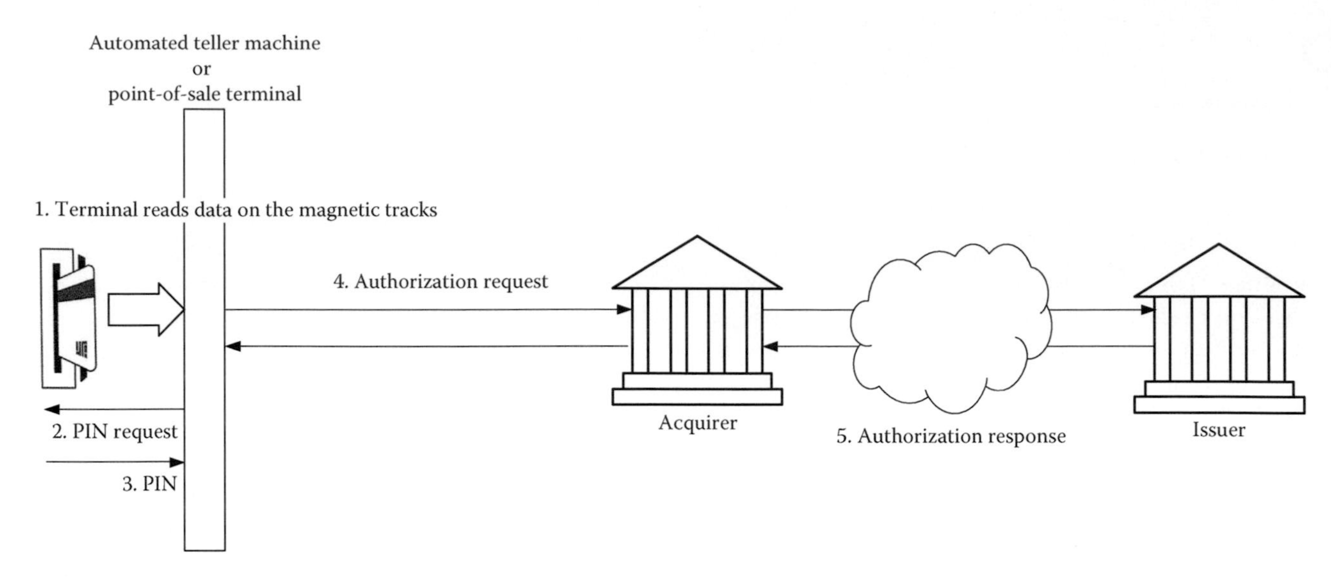

FIGURE 8.1
Transaction flow in a transaction with a magnetic stripe card.

4. If the PIN is valid, an authorization request is prepared with all the data from the card as well as the transaction amount. The request is sent to the acquirer that forwards it to the issuer.
5. The issuer makes an authorization decision based on the data in the authorization request. The response is sent to the acquirer that forwards it to the terminal.

The real-time approval of payment is specific to the United States and is a legacy of the wide availability of telephone services. When the authentication server is unavailable, card issuers cap the amount of a transaction without authorization at the so-called "floor limit." Limits on daily spending or cash withdrawals can be encoded in the magnetic stripe. Other limits concern the maximum allowed value for a single transaction or international transactions.

The main vulnerabilities of point-of-sale transactions with magnetic stripe cards are as follows:

1. The cardholder credentials are not encrypted. As a consequence, any magnetic card reader can retrieve the personalization data stored in the tracks.
2. The card lacks computational power and is a passive participant in the transaction. This is different from integrated circuit cards that can perform cryptographic computations for confidentiality, integrity, and authentication.
3. The exchanges between the card and the card reader are unencrypted.
4. Once the data from the card have been read, there are no further interactions between the magnetic stripe reader or the terminal and the card.

As a consequence, it is relatively easy to read the data on the magnetic strip and clone the card with a magnetic stripe encoder plugged into a computer. Also, when the merchant does not have access to the physical card or cannot authenticate the cardholder in person (*card not present*), such as for Internet, mail order and telephone order (IMOTO) transactions, verification of the authenticity and integrity of transactions with magnetic stripe cards is not always possible.

A PIN is needed for cash withdrawals from automated tellers. Debit card payments may require either a signature or a PIN, while credit card transactions typically require a signature. Signature debit transactions, like credit card transactions, are routed over the financial processing network, typically the Visa or MasterCard networks. PIN debit payment networks require PIN encryption in the authorization request. PIN debit transactions can be routed either on the Visa or MasterCard networks or on the cash withdrawal/PIN debit networks, such as Interlink, STAR, Maestro, or New York Currency Exchange (NYCE).

There are several ways to enter the cardholder data at a point-of-sale terminal:

1. *Manual entry*: In this method, the cashier punches on a keyboard the account number and the expiration date embossed on the front of the card. This method is often used if the magnetic stripe is damaged and cannot be read. Cardholder authentication is based on a

comparison of the customer signature with the signature on the backside of the card as well as on any official identification.

2. *Magnetic stripe reader*: A magnetic stripe reader scans the magnetic strip to extract the personalization data. Some readers have encryption capabilities that are used for point-to-point encryption.
3. *PIN pad*: A PIN pad is a device with firmware to identify the user through a PIN. The equipment exists either freestanding or integrated into cash registers. Many PIN pads have a Tamper-Resistant Security Module where the cryptographic computations are done. The use of a PIN increases the level of security because forging of handwritten signatures is easier than stealing PINs.

Figure 8.2 depicts the evolution of the volume of debit and credit card transactions in the United States, while Figure 8.3 shows the corresponding evolution of the value of the transactions (Bank for International Settlements, 2014). As seen, debit transactions compared to credit transactions are mostly used as a cash replacement for small purchases. Tables 8.1 and 8.2 present the relevant data.

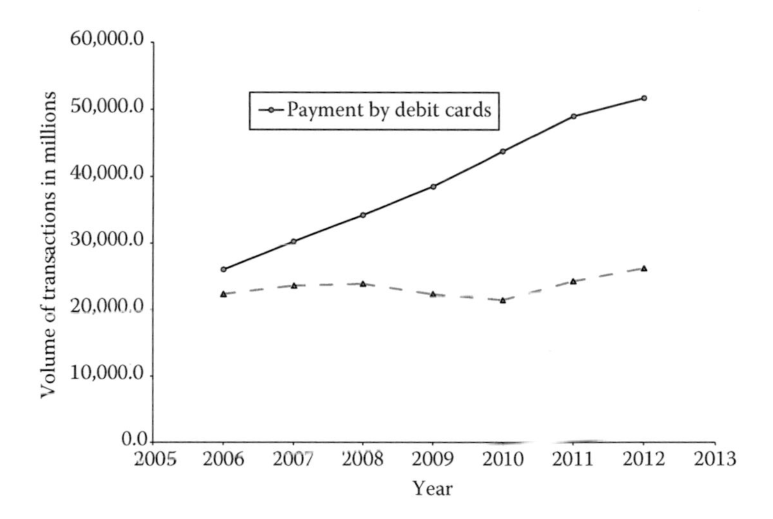

FIGURE 8.2
Evolution of the use of payment cards in the United States: volume of transactions in millions. (From Bank for International Settlements, Committee on Payment and Settlement Systems, Statistics on payment, clearing and settlement systems in the CPSS countries, Figures for 2013, Basel, Switzerland, September 2014, available at http://www.bis.org.)

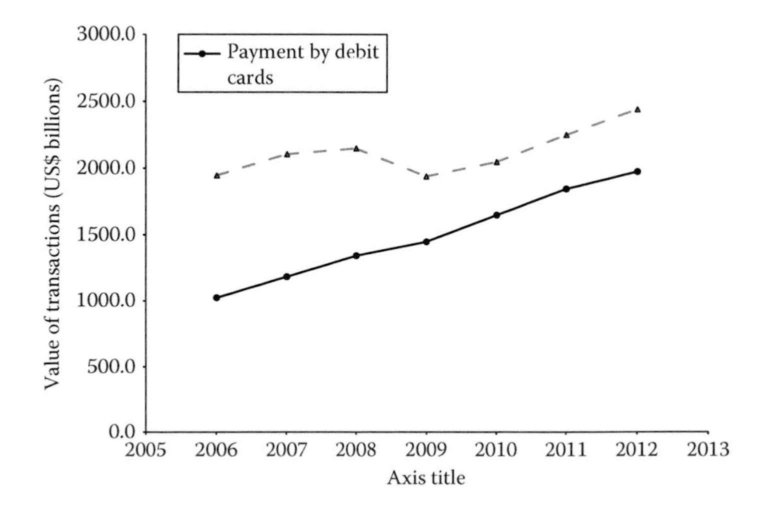

FIGURE 8.3
Evolution of the use of payment cards in the United States: value of transactions in US$ billions. (From Bank for International Settlements, Committee on Payment and Settlement Systems, Statistics on payment, clearing and settlement systems in the CPSS countries, Figures for 2013, Basel, Switzerland, September 2014, available at http://www.bis.org.)

TABLE 8.1

Evolution of the Use of Payment Cards in the United States: Volume of Transactions in Millions among Nonbanks

	2006	2007	2008	2009	2010	2011	2012
Payment by debit cards	26,037.6	30,247.8	34,215.7	38,518.9	43,780.4	49,006.1	51,717.2
Payment by credit cards	22,356.7	23,611.6	23,900.6	22,352.6	21,447.6	24,279.8	26,221.4

Source: Bank for International Settlements, Committee on Payment and Settlement Systems, Statistics on payment, clearing and settlement systems in the CPSS countries, Figures for 2013, Basel, Switzerland, September 2014, available at http://www.bis.org.

TABLE 8.2

Evolution of the Use of Payment Cards in the United States: Value of Transactions in US$ Billions among Banks

	2006	2007	2008	2009	2010	2011	2012
Payment by debit cards	1023.7	1182.9	1342.2	1447.3	1648.8	1846.8	1975.8
Payment by credit cards	1944.9	2104.9	2148.5	1938.6	2047.0	2250.3	2441.8

Source: Bank for International Settlements, Committee on Payment and Settlement Systems, Statistics on payment, clearing and settlement systems in the CPSS countries, Figures for 2013, Basel, Switzerland, September 2014, available at http://www.bis.org.

8.2 Communication Standards for Card Transactions

ISO 8583 is the standard that financial organizations use to communicate and complete card transactions whether from an automated teller machine (ATM), a point-of-sale terminal, on the Internet or on a mobile network. Its first version dates from 1987 with successive revisions in 1993, 1998, and 2003; most implementations, however, conform to the 1987 specifications. The standard also defines system-to-system messages for secure key exchanges, reconciliation of totals, and other administrative purposes.

A message consists of three components: a message type identifier, a variable length bitmap, and up to 192 data elements. The first 2 octets of the message indicate the total message length.

The message type identifier field consists of 4 octets:

1. The first octet represents the version of the standards.
2. The second octet represents the transaction type. Transaction types include purchases, withdrawals, deposits, refunds, reversals, balance inquiries, payments, and interaccount transfers.
3. The third octet indicates the message function.
4. The last represents the origin of the message.

The data elements represent the actual specifics of the transaction, each being reserved for a particular information. For example, a data element can indicate the primary account number, the amount of the transaction, the data and time of transmission, the transaction amount, the currency, the conversion rate of the currency, the merchant identity, the type of business, the capture method, the type of transaction, and so on. Should a particular element be needed, its corresponding bit pattern must be included in the bitmap. The original specification of 1987 included 128 data elements, and the number was increased to 192 in later revisions.

The length of data elements may be fixed or variable. A length indicator precedes a variable length field.

One or more bitmaps indicate which data elements are present. The presence of up to 64 data elements is indicated in a primary bit map. A secondary bitmap is needed for up to 128 data elements. Finally, a tertiary bitmap is needed for up to 192 data elements.

ISO 8583 leaves various implementation details, and this raises interoperability issues (Mian et al., 2015):

- ISO 8583 defines the message format and the communication flow but does not specify the encoding scheme which can be ASCII, hexadecimal, or any other encodings.
- Some vendors add proprietary information in the message length field.
- The placements of fields, such as the currency element, may vary with the versions used.
- Financial card networks (e.g., Visa, MasterCard) have adapted the standard by adding customized proprietary fields.

Thus, for two different ISO 8582 implementations to interoperate, some interworking function needs to be added. This is not a scalable option as the process has to be repeated for each separate implementation. As mentioned in Chapter 4, ISO 20022 is one possible approach because it provides a common platform based on the Unified Modeling Language (UML). ISO 20022 uses an XML wrapper around the messages, which expands the bandwidth requirements particularly when the number of real-time transactions is large. Also, no major card payment network is currently accepting transactions conforming to ISO 20022. Thus, global adoption of ISO 20022 for card payments is unlikely to take place in the near future.

8.3 Security of Point-of-Sale Transactions

Payment cards were first applied for face-to-face commerce. The issuers relied on merchants to conduct cardholder authentication, using an official identity card or a driver's license. Merchants were also supposed to compare the signature on the receipt with the holder's signature on the back of the card. The handwritten signature was used as another authentication measure as well as a nonrepudiation tool. Yet, discarded receipts allowed fraudsters to reuse the imprinted information imprinted to counterfeit cards and conduct spurious transactions, particularly from unattended terminals such as cash dispensers (Count Zero, 1972; Gomzin, 2014, pp. 114–121). One obvious measure was for the imprint to reproduce only the last four numbers of the PAN. Another was to add a code for fraud control, typically but not necessarily, recorded in Track 2. The code is referred to as card verification value (CVV) or card verification code (CVC) and consists of a cryptogram generated by a cryptographic algorithm using a key under the issuer control from the PAN, the expiration date, and other identifiers. Because the code was available only on the magnetic track and not embossed, the receipt imprints could not be used to extract all the information to be encoded in counterfeit cards. It should be noted that this is different from the new cryptogram added later to secure online transactions.

Optional security measures were also introduced to deter fraud using visual inspection such as inclusion of the cardholder's picture, complex graphics, and holograms. Due to their costs, these measures are not uniformly followed; moreover, they are not effective in the case of unattended terminals or IMOTO transactions. Finally, in high turnover environments, security measures are relaxed to improve the flow.

If the PIN associated with the legitimate card is compromised, fraudulent debit transactions can be conducted, including cash withdrawal from automated cash dispensers.

8.3.1 PCI Standards

In 2006, several global payment brands—American Express, Discover, Japan Credit Bureau (JCB), MasterCard, and Visa—established the Payment Card Industry Security Standards Council to develop security standards for the payment industry (Gomzin, 2014, pp. 55–84). The payment card industry (PCI) comprises organizations that store, process, and transmit cardholder data.

One of the outputs of the PCI Council is the PCI Data Security Standard (PCI DSS), now in its third version as of November 2013. PCI DSS focuses on the transaction processing environment in retail stores and payment processing centers, which have been historically the loci of data breaches.

A large number of rules and recommendations pertain to "data at rest," that is, storage of sensitive data in merchant systems, virtual terminals, and payment gateways after a transaction was authorized. Its directives aim at securing the processing of card payments, protecting cardholder's data, preventing unauthorized use of personal information, and responding to security incidents. In particular, the following data elements cannot be stored even in an encrypted form: the full magnetic stripe data, the CVV, and the PIN. Other data elements retained on a need basis only include the cardholder's name, the PAN, the expiration date, and the service code. For traceability and accountability, each individual accessing an application or a server that processes or stores payment card information must have a unique access code and password.

All parties and entities (e.g., third parties or contractors) engaged in the processing of payment cards must be certified to comply with the PCI DSS requirements and renew their certification annually. Merchants fall under four categories of PCI compliance, depending on the number of transactions they process each year, and whether those transactions are performed from a brick and mortar location or over the Internet. Each payment card brand (Visa, MasterCard, etc.) has their own requirements and definitions of PCI compliance levels. Visa's PCI compliance level definitions are shown in Table 8.3.

TABLE 8.3

PCI Compliance Levels for Visa

PCI Compliance	Number of Visa Transactions Processed Annually
Level 1	6 million (all channels)
Level 2	1–6 million (all channels)
Level 3	20,000–1 million (e-commerce transactions)
Level 4	Up to 20,000 e-commerce transactions or up to 1 million transactions on all channels

Some of the security provisions for "data in transit" are as follows:

- The PAN must be encrypted during transmission and storage using any standard encryption method that the National Institutes of Standards and Technology (NIST) has approved.
- The CVC/CVV must be hashed during entry into an application and are not stored after the authorization has been received.

Another PCI standard is the Payment Application Data Security Standard (PA-DSS), formerly referred to as the Payment Application Best Practices (PABP). It guides the development of payment applications to meet the PCI DSS requirements, including the test procedures and the user documentation. Signing the code and the configuration and data files protects the integrity of all aspects of the software applications. Signing certificates from a public certification authority will obviate the need to install proprietary root certificates on each target system.

Testing program ensures that the security requirements have been met. For example, the PIN Transaction Security (PTS) device testing and approval program focuses on the PIN pads: their security modules including Tamper-Resistant Security Modules, cryptography, and firmware. Other point-of-sale devices considered include magnetic stripe readers, keyboards, customer displays, and unattended payment terminals.

The PCI Council maintains a list of validated payment applications on its website. However, the cycle of software development is much shorter than that of testing and validation, so this list lags behind the current commercial offers.

Both PCI DSS and PA-DSS define general principles. Guidance on specific technologies, such as mobile payments, point-to-point encryption, tokenization, and so on, is left for special initiatives by the corresponding PCI Special Interest Groups.

8.3.2 Point-to-Point Encryption

The point-to-point encryption (P2PE) PCI standard provides detailed security testing requirements and test procedures for application vendors and solution providers to verify that their products protect payment data. The standard covers all aspects end to end, that is, from the data entry device to the data center of the payment gateway or payment processor via the point-of-sale terminal, including the retail environment and the transport network.

At the encryption end, the hardware module is denoted as a Tamper-Resistant Security Module (TRSM) because it is designed to detect physical intrusions and, in that case, to destroy the keys. The decryption is performed at the data center in a hardware security module (HSM), which is an appliance or an extension card installed in the payment processor dedicated to high-speed cryptographic computations.

TABLE 8.4
Variants in P2PE Implementations

P2PE Type	Encryption End	Decryption End
Hardware/hardware	Encryption and key management in hardware	Decryption and key management in hardware
Software/software	Encryption and key management in software	Decryption and key management in software
Software/hardware	Encryption and key management in software	Decryption and key management in hardware
Hardware/hybrid	Encryption and key management in hardware	Decryption in software and key management in hardware

In hybrid solutions, the hardware performs some functions while others are done in software. For example, key management is in hardware while the encryption or decryption or both are done in software. By convention, the systems are described as encryption side/decryption side as shown in Table 8.4 (Gomzin, 2014, p. 210).

It should be noted that P2P encryption does not resolve the main weakness of the magnetic stripe cards, that is, they store data in the clear.

8.3.3 Point-of-Sale Fraud

There are several ways to steal the data recorded on the magnetic tracks on payment cards, some more sophisticated than others. Prepaid gift cards with magnetic stripe are anonymous, so their cloning does not pose significant difficulties, because the identity of the holder is not available.

Methods for harvesting PINs include "shoulder surfing" at cashier's terminals or at ATM terminals to observe what the combination that a card owner has entered, compromised PIN pads, clandestine cameras, and so on. Fraudsters have also placed doctored point-of-sale devices to capture all the necessary data: the identity, account data, and security codes (CVV1 or CVC1). Many elaborate and miniaturized devices have been attached to ATMs to steal the card data and the associated PIN. For example, fake machines can be stacked directly on top of legitimate cash dispensing machines. These machines would have pinhole cameras to record unaware cardholders punching their PIN, while fake readers would access the data recorded on the magnetic stripe. In some cases, special devices would even trap the real card and not return it to the user after the transaction had been conducted (Slawsky and Zafar, 2005, pp. 104–106). The stolen data would be transmitted immediately via text messages to create thousands of phony cards and withdraw millions of dollars from various banks. Of course, the transactions would have to be made before the card is reported missing or the cardholders detect unauthorized items in their statements.

Another fraud mechanism is to intercept newly issued card stolen en route to the legitimate cardholder. To counter this fraud, card issuers have instituted card activation procedures. Upon receipt of their cards, cardholders must call and authenticate themselves to activate their new card. The call center agent typically asks a set of question to verify that the card is in the hands of the legitimate recipient. Many issuers use automated activation systems that recognize the calling telephone number and to punch on the telephone keypad basic information to be validated. Also, the PIN and the card are sent by post separately to reduce the risk of interception.

Application fraud occurs when fraudsters create fictitious identities using the personal information available in public records. Attacks on centralized databases such as those maintained by large retail chains or credit bureaus applicants contribute to the phenomenon of ID theft. The stolen information is sold to make fraudulent applications in the name of another person without their permission, which is much more difficult to detect. Typically, card issuers rely on a number of shared databases to screen applicants and to flag submissions that may be suspicious. In the United States, the Issuers' Clearing House Service (ICS) maintains a centralized, nationwide database that provides reports to Visa members concerning consumer risks. It was developed in 1989 to help reduce bank card credit and fraud losses. In the United Kingdom, the major lenders have formed a nonprofit organization called CIFAS to pool the information available concerning fraud prevention.

Over the 2006–2010 period, card counterfeiting rose to become the foremost source of signature debit fraud in the United States (41% compared to 25% for lost or stolen cards and 22% for IMOTO). For PIN transactions (cash withdrawal and debit), it is ranked second (44%) behind lost or stolen card (49%) (Sullivan, 2013).

The consequences of fraud vary according to the card type. Because forging signatures is easier than stealing PINs, the loss per dollar for signature-authorized payment is significantly higher than losses for payments authorized by PIN, as shown in Table 8.5. PIN debit payment networks require PIN encryption and as a result, fraud in face-to-face purchases with PIN debits (i.e., *card-present* transactions) is not large unless both the card data and the PIN have been compromised.

TABLE 8.5

Losses Due to Fraudulent Card Transactions in the United States (2009)

Payment Card	Loss per US$	Value of Transaction in Millions of US$	Value of Losses in Millions of US$
PIN debit card	0.0319	563,100	179.6
Signature debit card	0.1271	857,500	1089.9
General-purpose credit card[a]	0.1271	1,714,000	2178.5
Prepaid cards	0.0401	140	0.6
Total		13,134,740	3448.1

Source: Sullivan, R.J., *Econ. Rev.*, 98, 59, First Quarter 2013, Federal Reserve Bank of Kansas City, available at http://www.kc.frb.org/publicat/econrev/pdf/13q4Sullivan.pdf, last accessed May 11, 2014.

[a] The loss rate is assumed to be the same as that of signature debit cards.

It should be noted that there is no comprehensive source on the losses due to payment fraud in the United States. Statistics are available for PIN-authorized transactions (debit and cash withdrawals) and signature-authorized debit cards. Statistics on credit cards is not available, so it is assumed that their loss rate is the same as that of debit card payments with a signature, because they use similar authentication and approval protocols (Sullivan, 2013).

Migration to the so-called chip-and-PIN authentication with EMV imposes changes to Point-of-Sale Terminals, as well to the back-office systems that process transactions. Yet the highly fragmented retail industry and the banking industry could not agree on how to share the cost of the transition to EMV-compliant systems. There are even disagreements on the cost of that migration, ranging from $8 billion to around $26 to $30 billion, in addition to about $2 billion for card replacement (Carelli, 2014; Reagan, 2014).

8.4 Internet Transactions

The basic procedure for Internet transactions is for the cardholder to enter the authentication data including the card number, address, name of the cardholder, and the card security cryptogram on the back of the card. Screening of transactions exploits various databases to assess risk factors before proceeding with the transaction. Online security codes and one-time passwords are used to increase the robustness of the security. Some man-in-the-middle attacks consist in intercepting the messages between the card and the bank terminal by masquerading as a card terminal with respect to the user and as a user to the terminal. This type of attacks, called relay attacks, will be discussed in Chapters 9 and 10.

8.4.1 Screening for Risks

Risk analysis in financial institutions concentrates primarily on credit risk and, to a lesser extent, on fraud. Credit risk is outside the scope of the book; the focus here is on transaction screening to identify fraud possibilities. The screening rules aim at balancing convenience and speed with loss prevention. This is similar to what happens with point-of-sale low-value transactions, which typically do not require issuer approval. In fact, card networks have eliminated the signature requirements for many credit purchases under $25.

Various card networks use past fraud reports to establish filtering criteria to identify suspicious transactions. The parameters concern location, financial institution, Internet domain name, as well as abnormal behavior. The screening process detects suspicious patterns and assigns fraud probabilities to each pattern. Depending on the issuer policy, the transaction may be blocked or the customer may be contacted by phone to confirm the validity of the transaction. Yahoo established some telltale signs for fraudulent orders (Richmond, 2003):

1. Shipping address to countries with high incidence of fraud or unverifiable addresses.
2. Untraceable e-mail address.
3. The order contains expensive items or multiple items.
4. Express shipping is requested.
5. Shipping address and billing address are different.

Typical filtering criteria are listed in Table 8.6.

TABLE 8.6

Some Filtering Criteria for Fraud Prevention

Risk Factor	Screening Criterion	Comments
Bank or issuer	Bank identification number also known as issuer identification number	
Internet	Domain (host) impact	The customer had a risky IP or e-mail address.
	Internet inconsistency	The IP address and e-mail domain are not congruent with the billing address.
User identity	Excessive name change	The name changes several times in the last 6 months.
	Multiple identities	Different linkages of one identity element to another identity element. For example, multiple phone numbers are associated with the same account.
User address	Address inconsistency	The street address, city, state, or country of the billing and shipping addresses do not correlate.
	Suspicious address	The billing or shipping address is similar to an address previously marked as suspect.
	Excessive address change	The billing address changed several times in the last 6 months.
	Geolocation inconsistency[a]	The customer's e-mail domain, phone number, billing address, shipping address, and IP address are not consistent.
	Unverifiable address	The billing or shipping address cannot be verified.
User account	High number of PANs[b]	The customer used an excessive (e.g., more than six) number of cards recently (e.g., in the last 6 months).
	Velocity	The PAN was used several times in the past 15 minutes.
Merchandise	Type of merchandise	Online purchases of electronic equipment, liquor, and so on, or unusual purchases for that particular account holder.
Miscellaneous	Invalid value	The request contains an unexpected value.
	Negative list	The account number, street address, e-mail address, or IP address for this order appears on a negative list.
	Nonsensical input	The customer name and address fields contain meaningless words or language.
	Obscenities	The end-user input contains obscene words.
	Phone inconsistencies	
	Risky order	There are several high risky items in the transaction, user, bank, and merchant information.
	Time hedge	The customer is attempting at an unreasonable time.

[a] As explained earlier, this may cause a problem to purchases with perishable card numbers.

[b] For example, out-of-country purchases.

8.4.2 Online Security Code

Batches of stolen cards are sold online to counterfeiters who use them to forge new payment cards. To test the validity of a PAN, fraudsters can make a small donation to a nonprofit organization on its website. Many nonprofit organizations do not request any security code, and since this is a donation, shipment address is typically not required. If the transaction goes through, then the PAN is valid and can be used to create a fake card (Gomzin, 2004, p. 115).

Fraudsters also take advantage of the fact that card numbers are allotted to issuers in blocks. They methodically check each number in a block sequence with a card generator to test if the generated number falls within an issuer's allocation and attempt to conduct a transaction by successive guesses of passwords. Thus, multiple attempts for the same card number but with different parameters are considered an indicator of a fraud attempt. Valid numbers can also be harvested in restaurants, bars, hotels, and airport shops.

In 1997, MasterCard extended the idea of the security code to IMOTO purchases to verify that the user is in possession of the card before the transaction is authorized. Other issuers followed suit. In this method, a new data element is derived from the user identification data using a cryptographic algorithm (usually DES). The cryptogram is a 3-digit number for Visa or MasterCard that is printed in ink on the signature panel on the back of the card; for American Express, it is a 4-digit number on the front. The new security cryptogram appears only on the card or statements and is not recorded on the magnetic stripe. It does not show on receipts because it is not embossed.

The cryptogram has several designations as shown in Table 8.7. The confusion in terminology is a symptom of the lack of standardization.

During an online transaction, the merchant asks the buyer to enter the cryptogram to forward it to the issuer. If the issuer determines that the code is valid, this provides some assurance that the physical card was in the buyer's possession at the time of the online transaction.

By 2011, a majority of U.S. Internet merchants required that online buyers enter the card security code to proceed with a transaction (Sullivan, 2013). However, this protection has limitations. The code is generated once and for all for each card, that is, it does not change for each transaction. Should it be compromised, the only recourse is to block the card and replace it with a new issue. Merchant systems are also vulnerable points through which card details pass through, including the security code. Targeting these systems can also compromise the security mechanism.

It should be noted that the rules for allocating the liability for fraud in face-to-face contact purchases and online purchases are different. In the card-present case, that is, for a face-to-face transaction in a physical store, it is the issuer who is liable for the fraud. In fraudulent online purchases, however, the card issuer charges back the transaction to the acquirer bank (the merchant's banks). Moreover, under the Fair Credit Billing Act (FCBA) of 1974, in the United States, the maximum liability for unauthorized use of a credit card is $50. Moreover, if they report that the theft of a card is reported before it is used, they are not responsible for any unauthorized charges. If a credit card number is stolen, but not the card, the cardholder is also not liable for unauthorized use.

TABLE 8.7

Designations of the Cryptogram for Online Transactions

Designation	Abbreviation	Card Network
Card identification number	CID	America Express
Card verification code	CVC2	MasterCard
Card verification digit	CVD	Discover
Card verification number	CVN	JCB
Card verification value	CVV2	Visa

8.4.3 Perishable Card Numbers

Perishable card numbers, associated with virtual dynamic cards, is a concept introduced in 2001 to avoid transmitting the PAN over the Internet. It relies on a virtual, single-use card number, which expires within 1 or 2 months. E-Carte Bleue is an example of such a solution and is part of the Verified by Visa family of solutions.

At enrollment time, the bank's customer establishes a user name and a password. The user name can be stored as a cookie, but no other card information is stored on the computer. For security reasons, there is a limit on the number of virtual card per user.

There temporary card is generated by one of two ways. The client, downloaded from the issuer website and installed on the user's terminal, can generate the virtual card number. Alternatively, the issuer generates it at checkout time. The buyer clicks on a specific button on the merchant server to be directed to the issuer website. After proper identification, that site generates the number that would be used instead of the physical PAN in the merchant's payment system. A visual cryptogram—typically letters of a distorted image, sometimes with the addition of an obscured sequence of letters or digits—may be generated to make sure that there is a real human at the terminal. Reconciliation of the virtual and physical cards is done by the issuer.

This solution is more successful for micropayments, such as online game subscriptions. In the case of physical goods, some difficulties may arise (Béranger, 2003; Lazarony, 2014):

1. If the merchant requires that the buyer produces the real physical card at the moment of delivery of the goods or service, for example, theatre tickets, train or plane reservation, the PAN of the physical card bears no relation to the temporary card number that shows in the merchant ledger. So it is not possible to pick up theatre tickets or confirm a train, plane, or car rental reservation with a virtual credit card.
2. For out-of-stock items, the virtual card may have expired because the merchant could charge the account. Some card brands, such as Discover, match the expiration date on the virtual card with the expiration date on the physical card to extend the life of the virtual card and make it useful for recurrent charges such as monthly bills or preorder merchandises.
3. The temporary card is restricted to a single shipment, but the order may have to be split into multiple shipments.
4. Some fraud screening systems may flag the generation of several card numbers, each for specific transactions, as an indication of a fraud attempts and then block these transactions.
5. Finally, visual cryptograms may not be legible and may prevent visually impaired persons from conducting online transactions.

Tokenization avoids some of these drawbacks. The idea is to replace the PAN with a token that represents it in one-to-one correspondence to protect the original account number.

There are several methods to generate the token ranging from hash functions to encryption algorithms. However, this technique does not protect the sensitive authentication data in the magnetic stripe that are needed to process the payment. This is why tokenization is more useful in the case of online EMV transactions, as will be discussed in Chapter 9.

8.4.4 One-Time Passwords

One-time passwords (OTPs) may be used in online banking transactions with magnetic stripe cards to authenticate the cardholder.

At checkout, the cardholder enters the authentication data including the card number and the card security cryptogram on the back of the card and possibly a PIN. The authentication server generates a one-time password (OTP) that is sent to the user with a summary of the transaction either using a voice message on a landline or in a text message to the mobile phone using the Short Message Service (SMS). The password is called the mobile transaction authentication number (mTAN). The cardholder keys in the received password via the webpage of the bank server to complete the transaction.

This method verifies that the buyer possesses the card with its cryptogram and is near the landline phone or the mobile phone of the legitimate user. This is why it is called two-factor authentication and is used in many Internet services, such as Vodafone's m-pay micropayment service.

The method reduces the chance of man-in-the-middle attacks from the Internet since the attacker has to track the exchanges between the server and the user on two different networks at the same time. There is no protection, however, in case both the card and the mobile phone have been stolen, that is, when two or more attacks have taken place.

Furthermore, the default data format for SMS messages is plaintext. Encryption in the mobile networks (particularly, the so-called A5/1 or A5/2 algorithms) has proven to be vulnerable. Also, the transmission of SMS is not reliable, and messages can be dropped if the network is congested. Finally, depending on the mobile subscription package, text messaging may induce some additional charges.

A more robust method uses an application installed in the mobile phone to establish an encrypted channel with the bank server (Zelle, 2012). The user downloads a banking mobile application and establishes a password for access. Therefore, the SMS that contains the mTAN is protected against man-in-the-middle attacks. In addition, identity theft requires the theft of two pieces of information, the credentials to access the bank server from the terminal and the password to operate the application on the mobile phone. Figure 8.4 displays the sequence of messages.

The sequence is as follows:

1. The user enters the PIN to form an authentication request into the terminal.
2. The authentication request is sent to the bank server.
3. The bank server requests the user to start and activate the mobile application.
4. The user uses the password to activate the application.
5. The bank server sends the SMS that contains a message containing the mTAN, account number, the bank identity, and the amount to be paid.
6. The mobile decrypts the SMS and displays it to the user.
7. After reading the SMS and verifying the bank identity, the account number, and the amount, the user enters the mTAN into the mobile phone.

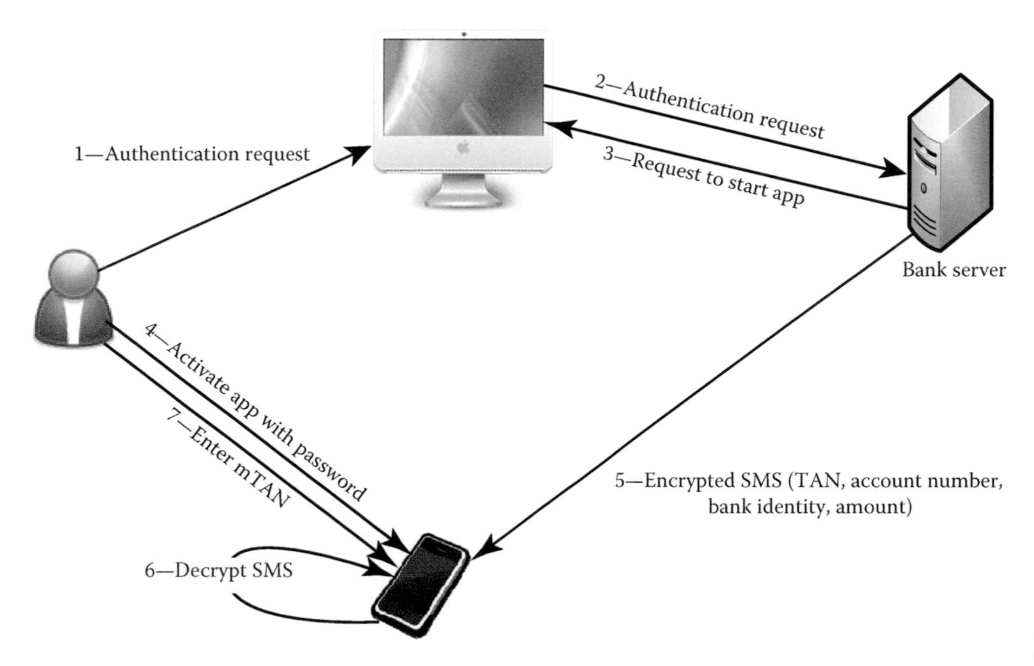

FIGURE 8.4
Communication sequence for a robust one-time password scheme.

Other possible variants would be to key in the mTAN into the web server from the terminal keyboard as well as capturing the user's fingerprints and uploading them to the server side as an additional authentication (Tsai, 2012).

8.4.5 Online Fraud in North America

Table 8.8 and Figure 8.5 show losses due to online fraud in North America, in terms of revenue loss and percentage of the total loss. These figures highlight that, in 2012, online losses constituted less than 1% of the losses due to other more consequential data breaches that expose many accounts at once.

TABLE 8.8
Online Losses Due to Fraud in North America (2001–2012)

Year	Revenue Lost (in Billions of US $)	Percentage of Revenue Lost
2001	1.70	3.20
2002	2.10	2.90
2003	1.90	1.70
2004	2.60	1.80
2005	2.80	1.60
2006	3.10	1.40
2007	3.70	1.40
2008	4.00	1.40
2009	3.30	1.20
2010	2.70	0.90
2011	3.40	1.00
2012	3.50	0.90

Source: Kiernan, J., Credit card and debit card fraud statistics, http://www.cardhub.com/edu/credit-debit-card-fraud-statistics, 2013, last accessed June 1, 2014.

Several events caused some changes in the United States. In 2012, the Federal Trade Commission (FTC) filed a lawsuit against the hotel chain Wyndham Worldwide after its databases were breached on three separate occasions between 2008 and 2009 to gain access to hundreds of thousands of debit and credit card accounts resulting in $10.6 million in fraudulent charges. The penetration of Target's computer system in the 2013 Thanksgiving season, which involved the theft of the credentials of as many as 110 million accounts followed by smaller breaches at retailers such as Neiman Marcus and Michaels, heralded a new era in cyberattacks where retail companies could be held liable for negligence. In this case, a server in the Target internal network was penetrated and programmed to poll the various Point-of-Sale Terminals to send back customer data from the magnetic stripes on an hourly basis and then send them outside the company. These accounts were then sold to the highest bidder to conduct fraudulent transactions or counterfeit payment cards.

With lawmakers calling for Congress and the Federal Trade Commission to investigate, large retailers became convinced of the need to switch to chip cards. In addition, the main payment card processors advised that, as of October 2015, they will no longer assume liability for fraudulent in-store charges at businesses that have not converted to EMV card readers.

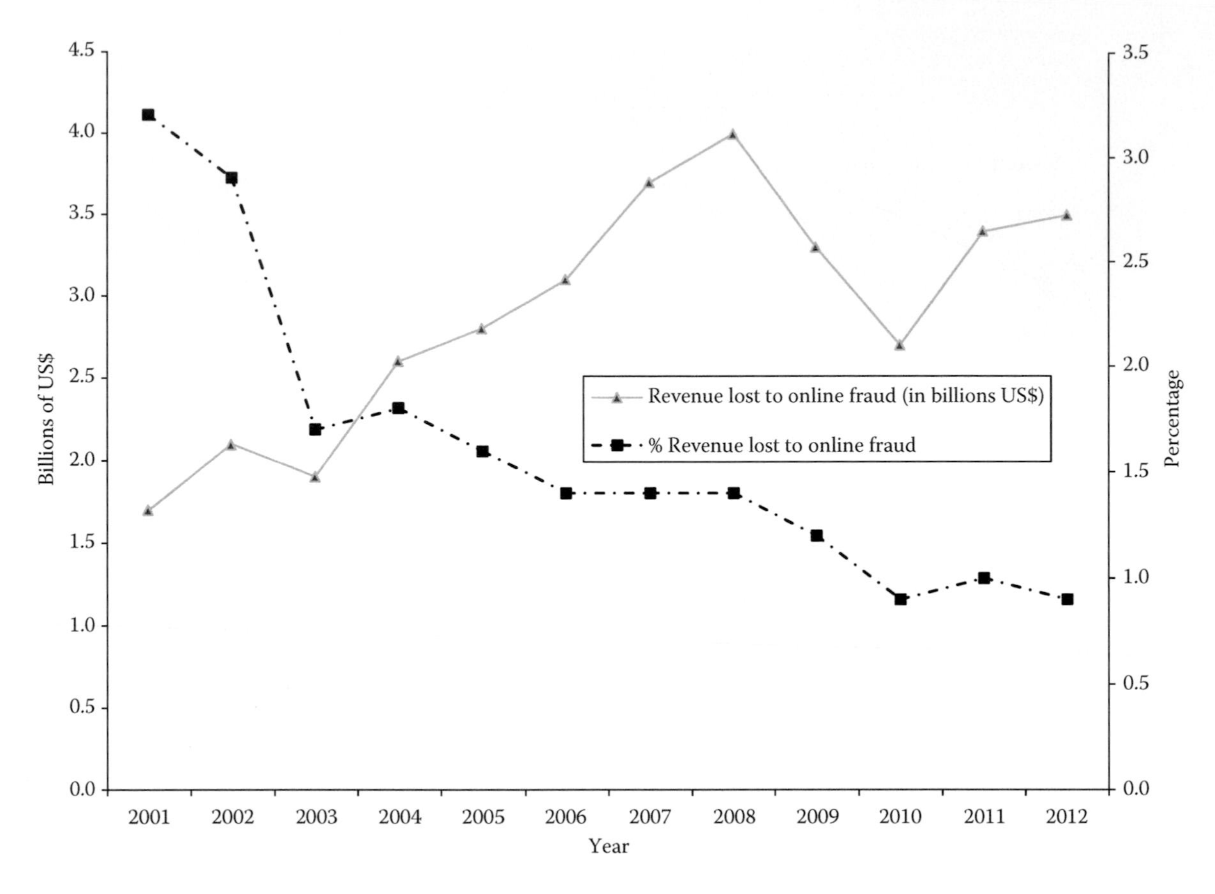

FIGURE 8.5
Online losses due to fraud in North America (2001–2012).

8.5 3-D Secure

3-D Secure is a solution that Visa introduced in 2001 to prevent three main sources of fraud: use of unauthenticated cards, theft of credit card numbers, and fraudulent claims by unethical merchants. The goal is to facilitate remote card payments irrespective of the access channel—Internet, broadband digital TV, text messages through the Short Message Service (SMS), WAP, and so on. The solution requires users to register and establish a PIN with their card issuer for authentication during an online transaction. The PIN is requested at checkout at participating online retailers. In this way, 3-D Secure allows merchants to verify remotely whether a given payment card is under the control of a specific user. 3-D Secure became operational in 2003 under the Verified by Visa brand. MasterCard established an equivalent service called SecureCode.

The architecture considers three domains: the issuer domain, the acquirer domain, and the interoperability domain through Visa payment network where the Visa intermediary for 3-D Secure verifies the parties. Within the issuer domain, each card issuer is required to maintain a special server known as the Access Control Server (ACS) to support card authentication. The Visa Directory resides in the interoperability domain and mediates the communication between the merchant and the issuer (Visa, 2002; Jarupunphol and Mitchell, 2003). Figure 8.6 shows that 3-D Secure uses four point-to-point SSL/TLS connections to connect the buyer, the merchant, and the payment gateway. The links are as follows:

- Cardholder ↔ Merchant
- Merchant ↔ Visa directory
- Cardholder ↔ Access Control Server
- Visa directory ↔ Access Control Server

This is why the solution was originally called 3-D SSL (3-Domain SSL).

The use of SSL/TLS provides the services of confidentiality, integrity, and authentication at the transport layer. This model of operation preserves existing banking channels, and the circuits for verification and for financial settlement continue to go through the card networks. The solution avoids the installation of

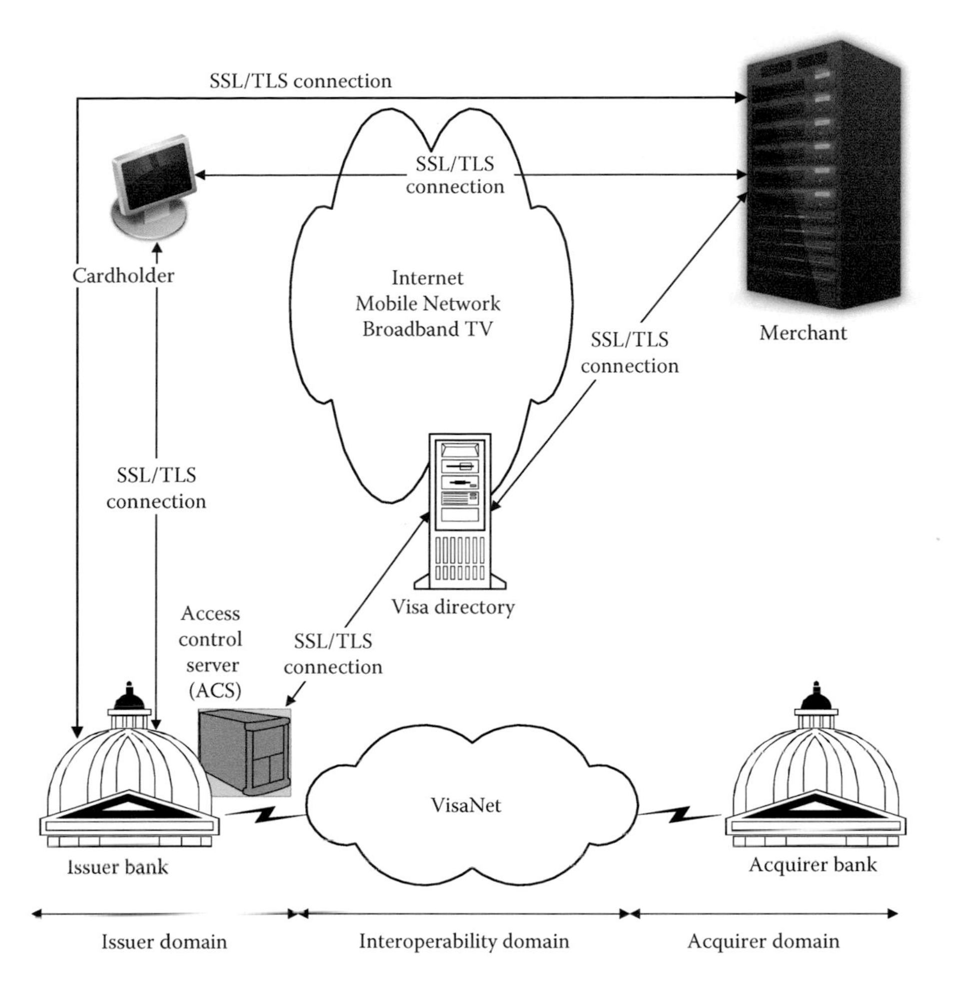

FIGURE 8.6
The domains of 3-D Secure and the SSL/TLS point-to-point connections. Visa configuration is used for illustrative purposes.

any additional software on user's terminals, while the merchants need only to add a plug-in to their payment server. Visa/MasterCard takes on the roles of the payment gateway and of the certification authority and provides the following additional services:

- A directory service that determines whether there is a range of card number that includes the primary account number (PAN) being verified
- A certification function to generate the various X.509 certificates to be used
- A repository for authentication history recording each attempted payment authentication, irrespective of whether it was successful

It should be noted that the merchant has access to all the cardholder account information.

Nonrepudiation at the application layer is achieved through the use of signatures, in particular because the cardholder enters a password or a PIN.

8.5.1 Enrollment

To participate in the 3-D Secure program, the issuer and acquirer banks provide their bank identification numbers (BINs) and the URL addresses of their respective servers.

Client enrollment is by whatever channels the issuer bank has selected. Similarly, the mechanism for cardholder authentication mechanism is left to the issuer. It can be a PIN, a smart card, an identity certificate, or a biometric measure. The default is a password that the user chooses at enrollment. If needed, the issuer bank will supply all what is necessary (equipment or software) for authentication. Visa/MasterCard will keep all the data related to this card (holder identity, banking coordinates, version, etc.) in its secure directory.

The merchant subscribes to the service through the acquirer bank, which supplies and activates the merchant server plug-in (MPI) module. The issuer assumes the risk of fraudulent transactions and chargebacks are not permitted. This means that the cardholder is ultimately responsible for payments.

Finally, the client's browser must support JavaScript to offer a conduit for the communication between the merchant and the issuer bank without the buyer's intervention.

8.5.2 Purchase and Payment Protocol

The negotiation between the buyer and the merchant is outside the scope of 3-D Secure. The message flow for payment authentication is shown in Figure 8.7. In a nutshell, if 3-D Secure has been activated for the card, the merchant site directs the buyer to a website where the card authentication takes place. The messages are coded in XML (3-D Secure, 2001).

Payment authorization proceeds as follows (Visa is used as an example):

1. 3-D Secure is started when the buyer indicates an intention to buy by clicking on the corresponding "Buy" button.
2. After receiving the purchase order with the Payment Instructions, the merchant plug-in (MPI) requests the list of participating card ranges from the Visa's 3-D Secure directory with the *Card Range Request* (*CRReq*) message. This is to verify that the primary account number (PAN) is within the range of participating cards.
3. The directory response is in the message *Card Range Response* (*CRRes*). Depending on the parameters of the initial request, CRRes can contain either the entire list of participating card ranges or the changes since the last exchange. The returned information can be used to update the MPI's internal cache. This exchange may be skipped if the MPI has the ability to store the content of the Visa directory, provided that the local cache is refreshed, every 24 hours at least.
4. Next, the MPI sends the *Verify Enrollment Request* (*VEReq*) message to the Visa directory to determine if 3-D Secure authentication is available for a particular PAN. The directory checks if the merchant, the acquirer, and the cardholder PAN can be authenticated using 3-D Secure. If the response indicated in the *Verify Enrollment*

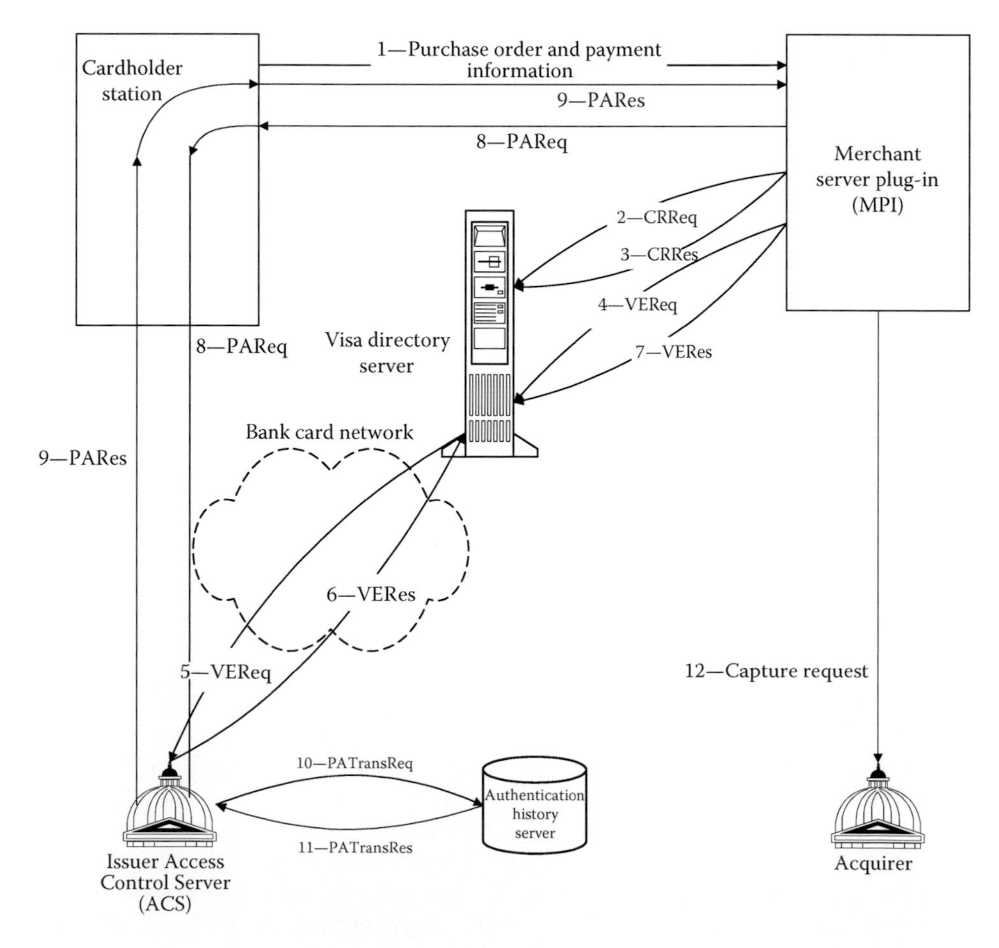

FIGURE 8.7
Flow for payments with 3-D Secure.

Response (VERes) message is negative, the transaction can proceed in the traditional way. Otherwise, the directory will query the ACS of the issuer bank for its authorization.

5. The *VERes* message from the Access Control Server of the issuer bank indicates whether the holder is registered in the 3-D Secure program. In such a case, it contains the URL address toward which the cardholder browser will post the data that the merchant plug-in module has provided. If the cardholder is not a participant in the 3-D Secure program, the transaction can continue along the typical lines.
6. Upon receiving the response from the directory, the MPI sends a *Payer Authentication Request* (*PAReq*) to the issuer Access Control Server (ACS) using the URL obtained in the preceding step. The *PAReq* contains details of the purchase that needs to be approved. This is presented in one of several ways:
   a. The information may appear to the buyer in a secondary pop-up window.
   b. The merchant may appear using an HTML technique called inline frame or *iframe*. The merchant defines the frame for the purchase authentication page and gets the URL to embed the iframe.
7. The transfer of control to the ACS goes through the user's browser using a JavaScript that resides in the authentication page. Alternatively, a more secure way would be to ask for the user explicit intervention to continue the 3-D Secure transaction. To protect the data from being inadvertently changed by the browser, they are coded with Base64 coding. Because in that coding, each 3 octets of data are expanded into 4, the data are compressed before coding. Furthermore, for mobile Internet devices, the *Condensed Payer Authentication Request* (*CPRQ*) is used to reduce bandwidth utilization.
8. The issuer ACS establishes an SSL/TLS session with the cardholder's terminal and displays the details of the transaction, as obtained from the MPI, and asks the holder for an identifier (a PIN, through biometric means, by inserting a chip card in a card reader, etc.) and for approval of the purchase. The cardholder sees a window that contains purchase details to enter the Verified by Visa password. The ACS links the *PAReq* with the *VERes* messages.
9. After the holder has been authenticated, the ACS builds a cardholder authentication verification value (CAVV) of 20 octets. The CAVV contains a cryptogram of 2 octets computed as signature to validate its integrity. The CAVV is sent to the MPI in the *Payer Authentication Response* (*PARes*), message, itself signed with the issuer's signature key. The ACS uses the CAVV to vouch to the merchant that it has authenticated the cardholder. The message uses Base64 coding, with the data compressed before encoding. For mobile devices, the *Condensed Payer Authentication Request Response* (*CPRS*) is used.
10. The response is sent to the Authentication History Server for archival within the *Payer Authentication Transaction Request* (*PATransReq*) message. The confirmation of the archival is in the *Payer Authentication Transaction Response* (*PATransRes*) message.
11. The merchant's server verifies the signature (the CAVV) before he or she submits a settlement request to the acquirer bank.

To summarize, the aforementioned exchanges comprise four phases:

1. Update of the merchant's cache with the ranges of valid accounts
2. Verification that a given card number in enrolled in the 3-D Secure and, in that case, authentication of the cardholder to authorize a specific transaction
3. Verification of the integrity of the payment information with the signature of the issuer
4. Archival of the authentication attempt

Summarized in Table 8.9 are the 3-D Secure messages and their significance.

8.5.3 Clearance and Settlement

As explained earlier, after receiving the authentication response, the merchant server extracts the necessary information to make a capture request. It sends this request to its bank with an indication that the transaction is secured according to 3-D Secure. The acquirer bank requests the approval of Visa Directory Server (or the MasterCard equivalent). The server compares the data received from the issuer and acquirer banks before its approval. Once that approval has been given, the settlement follows the usual procedures.

8.5.4 Security

3-D Secure is an attempt to strengthen remote payments with bank cards by relying on an infrastructure under

TABLE 8.9
3-D Secure Messages

Message	Direction of Transmission	Significance
CRReq	Merchant → Visa Directory	Request full or partial list of the ranges of participating cards
CRRes	Visa directory → Merchant	Return full of participating card ranges or changes from last update
VEReq	Merchant → Visa directory → Issuer	Check if 3-D Secure authentication is available for a particular PAN
VERes	Issuer → Visa directory → Merchant	Indicate if 3-D Secure authentication is available for a particular PAN
PAReq	Merchant → Issuer (via cardholder's browser)	Base64 encoded request to authenticate the payer, protected against inadvertent change by the browser
CPRQ	Merchant → Issuer (via cardholder's browser)	Same function as PAReq for mobile communications
PARes	Issuer → Merchant (via cardholder's browser)	Base64 encoded response to PAReq signed by the issuer Access Control Server
CPRS	Issuer → Merchant (via cardholder's browser)	Same function as PARes for mobile communications
PATransReq	Issuer → Authentication history server	Request archival of PARes
PARransRes	Authentication history server → Issuer	Response to archival request

the control of Visa/MasterCard and their associated banks. The various entities are given X.509 Version 3 certificates as follows:

1. Visa/MasterCard directory has a server certificate with respect to the merchant and a client certificate with respect to the issuer.
2. The MPI has a server certificate with respect to the cardholder and a client certificate with respect to the Visa directory and the issuer.
3. The ACS has a server certificate and a signature certificate to sign the *PARes* message; the ACS initiates two sessions using its server certificates: one to the cardholder and the other to the merchant.

The certification authority can be Visa/MasterCard or one of the recognized certification authorities.

The mandatory cipher suite is (TLS_RSA_WITH_3DES_EDE_CBC_SHA): SHA-1 for hashing, Triple DES for symmetric encryption, and RSA for static signature. Typically, the symmetric key used for encryption of traffic involving the cardholder is 40 bits for DES (80-bit security for Triple DES) due to the proliferation of U.S.-exportable browsers. The minimum size of the RSA key for signature is 768 bits, and the recommended value is 1024 bits. An optional cipher suite is (TLS_DH_RSA_WITH_3DES_EDE_CBC_SHA), where the Diffie–Hellman algorithm is used to exchange keys and RSA to sign the messages in the key exchange to ensure their integrity. The 128-bit SSL/TLS cipher suites are preferred whenever possible. In particular, the connections between the MPI, the cardholder terminal, and the issuer ACS use HTTPS. It should be noted that each transaction requests at least five point-to-point links secured with SSL/TLS.

Note that the iframes or pop-ups do not have an address bar, so there is no easy way to verify who is asking for the password (Murdoch and Anderson, 2010).

Finally, the issuer has all the details of the transaction, including the description of the goods or services purchased.

8.5.5 Evaluation

3-D Secure is a response to increased fraud rate in IMOTO transactions with magnetic stripe cards. Cardholder authentication is stronger in 3-D Secure transactions because of users entering an additional password to be verified by the card issuer during the authorization phase. The 3-D Secure client is easy to install and operate. Furthermore, the complexity of the security procedure is shifted from the user or merchant sides to a platform under the control of the payment intermediary.

Merchants who adopt 3-D Secure pay a fee to the card scheme operator but get the transaction treated as "cardholder present" with less risk of repudiation. The liability is thus shifted to the cardholder.

3-D Secure has several limitations. The password is static so that security is vulnerable if the password is compromised. Not all browsers, particularly on mobile devices, support JavaScript, and this capability is regularly turned off in corporate networks for security reasons. Similarly, the browser must be configured to support iframes and pop-ups (Jarupunphol and Mitchell, 2003; Petre, 2012). The iframes or pop-ups do not have an address bar, so there is no easy way to verify who is asking for the password. As a consequence, there is no protection against phishing websites that target 3-D Secure by displaying the "Verified by Visa" or "MasterCard SecureCode" inline window (Murdoch and Anderson, 2010). A password reset typically uses a method called Activation During Shopping (ADS). In this method, a weak form of authentication (such as the date of birth) is requested, which may be available from public records (Murdoch and Anderson, 2010). Finally, the issuer has all the transaction details including what was purchased. In other words, the benefits that cardholders receive are not commensurate with their increase in liability for fraud and their loss of privacy. This has limited the use of 3-D Secure, particularly in the United States, where users have no liability in the case of electronic commerce fraud.

8.6 Migration to EMV

The growing sophistication of attacks on card processing systems, merchant payment systems, or systems for intermediaries such as payment processors has created problems for the card payment industry. Finally, new possibilities for card fraud emerged in parallel with the increased use of the Internet and mobile telephony.

To face this situation, the payment card industries have pushed for the migration from magnetic stripe to microchip cards running the EMV protocol. To provide incentive to banks and merchants to upgrade their equipment to conform to EMV, the payment card brands shifted the liability of the *card not present* fraud to the non-EMV-compliant party. For example, a merchant equipped for EMV can still accept payments with a magnetic stripe card, but the cardholder's bank will be responsible in contested cases. Conversely, if the buyer has a chip card but the merchant does not have the means to accept it, the merchant's bank will be responsible for any transaction that the cardholder repudiates.

European banks moved to the EMV technology (with "Chip and PIN" authentication) in January 2005. In the United States, however, the migration has faced significant resistance, particularly from the retail industry. The experiences of countries that have adopted EMV can be a guide for the evolution of fraud with credit and debit cards in the United States. Figures from the United Kingdom published in 2014 are shown in Tables 8.10 and 8.11 and illustrated in Figures 8.8 and 8.9. In this data set, the category "card ID theft" represents accounts opened in the name of another person without permission.

TABLE 8.10

Card Fraud by Types in the United Kingdom in Millions of Pounds

	2007	2008	2009	2010	2011	2012
Telephone, Internet, and mail order	290.5	328.4	266.4	226.9	220.9	245.8
Counterfeit	144.3	169.8	80.9	47.6	36.1	42.1
Lost or stolen card	56.2	54.1	47.7	44.4	50.1	55.2
Card ID theft	34.1	47.4	38.2	38.1	22.5	32.1
Mail nonreceipt	10.2	10.2	6.9	8.4	11.3	12.8
Total	535.3	609.9	440.1	365.4	340.9	388.0

Source: http://www.theukcardsassociation.org.uk/plastic_fraud_figures/index.asp, last accessed June 1, 2014.

TABLE 8.11

Card Fraud by Source in the United Kingdom in Millions of Pounds

	2007	2008	2009	2010	2011	2012
UK fraud	327.6	379.7	317.4	271.5	261.0	286.7
Fraud outside the United Kingdom	207.6	230.1	122.6	93.9	80.8	101.3
Total	535.2	609.8	440.0	365.4	341.8	388.0

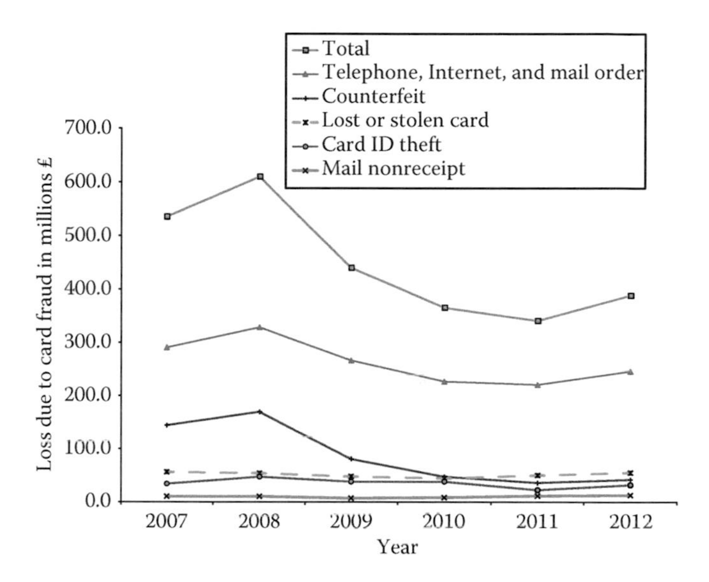

FIGURE 8.8
Evolution of credit and debit card fraud in the United Kingdom (2007–2012).

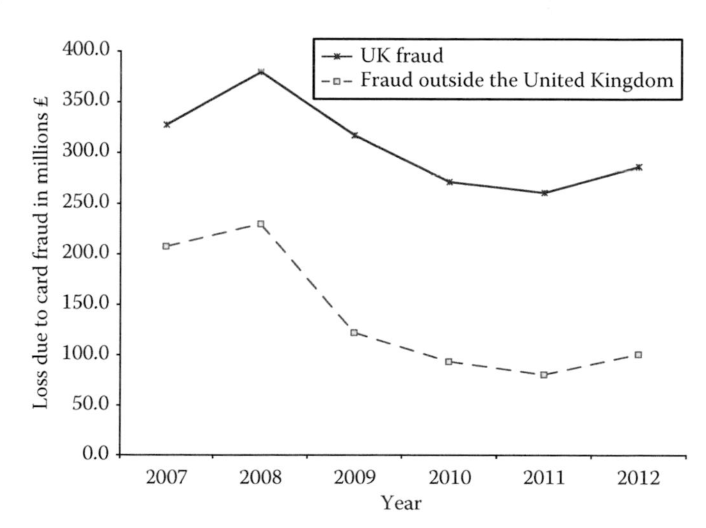

FIGURE 8.9
UK credit and debit card fraud by geography of origin (2007–2012).

"Mail nonreceipt" is the case of a newly issued card stolen en route to the legitimate cardholder and then diverted for use by fraudsters.

The time series exhibit several salient characteristics. First, the introduction of EMV reduced card counterfeiting. Second, the overall fraud increased suddenly the first year after the introduction of the EMV, after which the total fraud fell continuously until the 2011–2012 period where it rose again. The first year increase, the decrease for several years, and then the trend reversal in 2011 occurred for transactions conducted from within the United Kingdom as well from outside the United Kingdom. The rise in the number cards lost or stolen parallels fraud increases with IMOTO transactions, which leads to the belief that the main purpose of the theft is to obtain the PIN from the magnetic stripe (Sullivan, 2013, p. 71).

To explain these results, it is useful to note that EMV provides a fallback mode of operation with magnetic stripes to accommodate legacy terminals as well as cards in circulation. By concentrating on these terminals, within or outside the United Kingdom, fraudsters were able to force integrated circuit cards to use the data on their magnetic stripe. The spike in fraud just following the introduction of chip cards can be explained by attempts to fully exploit this window of vulnerability as much as possible, before issuers would block the cards and banks upgrade their terminals. The total fraud fell for a few years as all these vulnerabilities were fixed until new ways were uncovered, reflecting in rise in the fraud rate.

Similar behavior has started to show in the United States, as well, even before the introduction of EMV. Following the Home Depot data breach in 2014, some debit transactions were pushed from Brazil onto New England banks using the information retrieved from the magnetic stripes but faking the transactions as if they were with EMV cards. To do so, the fraudsters captured traffic from real EMV-based chip card transactions and then inserted the stolen card data into the transaction stream, while modifying the merchant and acquirer bank account on the fly. Clearly, they were in control of a payment terminal and had the ability to manipulate the data exchanges going through that terminal (Krebs, 2014).

As will be discussed in Chapter 9, EMV was designed for face-to-face transactions and not for online transactions. Therefore, the reduction of fraud in IMOTO transactions was not as dramatic as for other types of transactions and is attributed to other factors, such as the generalization of 3-D Secure. The same observation was made in France, where the fraud rate in IMOTO transactions increased to 61% of the total volume of all payments with French cards in 2011, while IMOTO transactions contribute only to 8.4% of the total value of all card payments. To counteract this threat, French authorities recommended the adoption of 3-D Secure. The initial rise in skimming attacks was observed in the Netherlands as well (Sullivan, 2013, pp. 70–72).

In most countries, it took 7–10 years to roll out the EMV infrastructure. The full rollout of EMV in the United States is expected to last much longer because of the scope and complexity of the deployment. In fact, there are about 15,000 financial institutions in the United States that will be going through this transition. Also, gas stations are not expected to have chip readers before 2017. This will be giving fraudsters ample time to mount their attacks. Furthermore, a comprehensive system for collecting and reporting payment fraud is lacking in the United States, which make smaller banks particularly vulnerable. To assist in this transition, MasterCard and Visa announced the creation of a cross-industry group to assist banks, retailers, point-of-sale device manufacturers, and others.

While the United States has been slot to adopt smart cards, China is phasing out magnetic stripe card using another specification called PBOC 3.0, named for the People's Bank of China. The technical documents are in Chinese but advocates say that PBOC 3.0 adds features not supported by EMV, such as contactless. This approach is consistent with China's policy to achieve technology independency through learning by doing while simultaneously avoiding royalties for technology use (Sherif, 2015). However, it is possible that Visa and MasterCard would ask the U.S. government to launch a litigation with the World Trade Organization (WTO) with the argument that this payment standards should be considered as a barrier to trade by deviating from a *de facto* standard.

8.7 Summary

Only a small portion of payment card transactions are potentially fraudulent, but the consequences of such a fraud on the industrial actors are significant in terms of monetary loss, public relations, and inconvenience to customers. When millions of records are compromised and used to carry fraudulent payments, confidence in the resilience of the card environment is always at risk.

The main sources of fraud with payment cards are lost or stolen cards, card counterfeiting, and/or online purchases. Payment cards with magnetic stripes are particularly vulnerable to card counterfeiting. Several approaches have been tried to protect transactions with magnetic stripe cards, none of them fully satisfactory. Integrated circuit cards improve the protection against counterfeiting and are finally being introduced in the United States several decades after the rest of the industrial countries. However, as long as the magnetic stripe is maintained, whether in a traditional card or in an integrated circuit card, they will continue to open areas of vulnerabilities to be exploited for fraud. As the ability to counterfeit cards dries up, fraudsters are expected to explore vulnerabilities, such as incorrect implementations of EMV or poor business controls. Based on the experience from other countries, online fraud may surge after EMV chip card rollout unless other procedures are introduced (3-D Secure, tokenization, etc.).

Questions

1. What are the main security vulnerabilities in point-of-sale transactions with magnetic stripe cards?
2. How does the card verification value (CVV)/card verification code (CVC) enhance the security of magnetic stripe cards?
3. How can the security of the CVV/CVC code be bypassed?
4. What is the difference between the CVV/CVC and the CVV2/CVC2?
5. Describe a typical fraud with PIN debit transactions.
6. How do the PCI DSS standards protect cardholder's data?
7. Why are there many names for the cryptogram used in online transaction?
8. Describe the various categories of online attacks on payment transactions.
9. Explain how 3-D Secure offers the services of confidentiality, authentication, integrity, and nonrepudiation.
10. What are the advantages and limitations of 3-D Secure?
11. What are the factors that delayed the adoption of chip payment cards in the United States?

9

Secure Payments with Integrated Circuit Cards

The use of integrated circuit cards (smart cards) in payment applications started in France for telephone cards and withdrawal cards from automated tellers. A major expansion of their use followed the success of Groupe Spécial Mobile (GSM—Global System for Mobile Communication), the European specifications for digital cellular networks. Other payment applications followed in public transit systems and the development of the EuroPay, MasterCard, and Visa (EMV) specifications for financial transactions.

This chapter covers the security of smart cards in payment applications. First, we give an overview of the different types of integrated circuit cards in terms of memory and processing power, operating systems, and their interfaces with personal computers. Next, the ISO standards of integrated circuit cards with contacts and contactless are summarized. Radio-frequency identification (RFID) and near-field communication (NFC) standards are also presented. The file structure of integrated circuit cards is illustrated with two examples: the Swedish electronic identity card and the SIM of GSM terminals. Several architectures of multiapplication cards are presented. Next, the physical and logical protection of smart cards is presented. Payment applications are reviewed, in particular, the use of RFID for electronic toll payment and the EuroPay,* MasterCard, and Visa (EMV) specifications.

9.1 Description of Integrated Circuit Cards

As the name implies, what distinguishes integrated circuit cards from all previous cards paper and plastic cards is the implantation of a microelectronic chip. This is an important advantage in payment applications because it allows the cryptographic operations necessary for security. This section presents the main items in an integrated circuit cards as well as their interfaces to personal computers.

9.1.1 Memory Types

A typical smart card uses an 8-bit processor, a read-only memory (ROM) of 16 Koctets, a random access memory (RAM) of 256–512 or even 1024 octets, and a programmable read-only memory (PROM)—also called nonvolatile memory (NVM)—of 3 up to 32 Koctets.

The ROM contains the card's operating system and a mask configured according to the application, such as telephony or banking (Guillou et al., 2001). The ROM retains its content in the absence of a power supply and has two principal forms: electrically programmable read-only memory (EPROM) and electrically erasable programmable read-only memory (EEPROM). EPROM's content is erased only after exposure to ultraviolet radiation. This type of memory is found often in disposable cards. Programmable or secure cards use EEPROM memories to store encryption keys, updates, or bug corrections. This category of memories requires two clock cycles to write an octet of data, the first to erase the existing data and the second to record the new octet. New memory technologies such as flash or Ferrite RAM (FeRAM) can reduce the write time and the power consumption.

The RAM is programmable and can contain intermediary results as long as there is a power supply, which explains its other name "volatile memory." The content of static RAM (SRAM) remains stable while the content of dynamic RAM (DRAM) must be refreshed periodically to prevent leakage.

A typical smart card used in transportation operates with an 8-bit processor, an EEPROM of 2 Koctets and a proprietary operating system, while in top-of-the-line cards, the size of the ROM can reach 128 Koctets, the RAM 4 Koctets, and the EEPROM 64 Koctets (Borst et al., 2001).

From a functional viewpoint, the memory of a secured integrated circuit card is organized in the form of a hierarchy of zones (Martres and Sabatier, 1987, pp. 91–93):

- The fabrication zone is the part of the memory that is recorded before personalization. It includes the lot number of the wafer, the name of the manufacturer, its serial number, the serial number of the card, and the identity of the card supplier.
- The secret zone locks the cardholder's confidential information, such as the PIN or the secret cryptographic keys and the personal data files.

* In 2002, Europay became MasterCard Europe.

- The transactions or working zone stores the temporary confidential information related to individual transactions, such as the amount, the balance of the stored value, and the details related to the vendor. This part is shared among the various applications in multiapplication cards.
- The access control zone records, under the control of the microprocessor, all successful access or access attempts to the secret zone or to the transactions zone (if this zone is protected). Thus, it is possible to block access to the card after three unsuccessful attempts.
- The free reading zone or the open zone is where nonconfidential information, such as the names of the holder and the issuer and the expiration date, is stored. This zone is accessible to the different applications of a multiapplication card.

Access to memory is controlled by a security logic that depends on the processing powers available on the chip.

9.1.2 Processing Capabilities

Integrated circuit cards can be divided into several categories according to their computational capability:

- *Memory cards.* These cards include a microprocessor of reduced capacity that can stock data and even offer minimal protection to the data; the processing power of this type of card is adapted to prepayments, such as for telephone cards as well as for other price-sensitive, large-scale applications.
- *Wired-logic cards.* These cards are used to offer minimum control to regulate access to encrypted television channels.
- *Smart cards* proper or *integrated circuit cards.* These are programmable machines that are capable of making complex computations. Microprocessor cards are very flexible and primarily used in security-sensitive applications such as for subscriber identification in cellular telephony or in electronic payments.

Traditional prepaid telephone cards are contact cards, monoapplication and disposable. The total value of telephone units corresponds to an equal number of memory cells. Cells are progressively erased as the card is used until all units have been depleted. Their storage capacity is of the order of 8–256 kbits but can reach hundreds of megabits. The advantage of using memory cards is that the stored data can be protected against unauthorized access. Access is supervised by the operating system according to some security logic.

Microprocessors provide additional security protections through the implementation of cryptographic algorithms. Typically however, a coprocessor is added to handle the cryptographic computations and store the secured data. In multiapplication cards, security requires compartmentalized access to the computational resources so that several applications could coexist each with its own personalized credentials.

9.1.3 Operating Systems

In the past, the operating system—which resides in ROM—used to be specific to a given manufacturer, such as the proprietary mask M4. For multiapplication cards, there are now several standard options:

- Multiapplication Operating System (MULTOS) was developed by the MULTOS Consortium that includes chip manufacturers, payment card schemes, particularly MasterCard, card management and personalization system providers, and smart card solution providers (MasterCard, Mondex, EuroPay, Gemplus, Siemens, etc.). The MULTOS technical specifications cover the operating system, the MULTOS executable language (MEL) assembler language, the API, and the chip interfaces. It is also specifically designed for secure payment applications.
- Java Card was developed by Visa International and its allies for smart cards based on the Java virtual machine (see Section 9.4.2) defined in the Java Card Forum.
- A third possibility is *Windows for Smart Card* from Microsoft, which has a library of cryptographic commands and the *Windows.Devices.Smart Cards* API to work with physical and virtual smart cards and smart card readers.

There are also several proprietary operating systems. For example, the Advantis card utilizes the operating system TIBC 3.0 (*Tarjeta inteligente y cajas de bancos*), developed by Visa Spain, for the application of an electronic purse. The German company ZeitControl provides the BasicCard with a secure proprietary system and cryptographic functions such as 4096-bit RSA encryption and 512-bit elliptic curve cryptography (Husemann, 2001).

9.1.4 Integrated Circuit Cards with Contacts

As shown in Figure 9.1, an integrated circuit card with contacts has eight contact points specified in the ISO/IEC standard 7816. The standard defines the locations

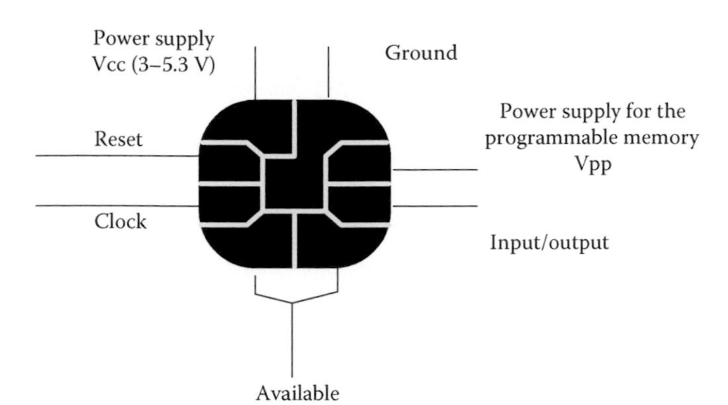

FIGURE 9.1
The eight contacts of an integrated circuit card with contacts.

of the contacts between the card and the card readers to ensure their compatibility. The contacts must resist the wear and tear due to usage and the corrosion from various factors (abrasion, corrosion, sweat, pollution, chemical substances, etc.).

In cards with contacts, the clock rate varies between 3.5 and 5 MHz. The power supply is between 3 and 5 V. Microelectronic advances have eliminated the need for an additional power source to supply the programmable memory with 5–15 V or even 21 V. This reduces the chances of external attacks through voltage fluctuations, and the freed contact was used for duplex transmission, thereby doubling the data rates.

Contact card readers are generally equipped with a mechanism to grab the card automatically as soon as it is inserted in the card reader slot. This setup enhances the reliability of the reading and strengthens the card reader's resistance to user abuse and to vandalism, although at the cost of increased complexity and operational and maintenance costs.

If the integrated circuit is programmable, cryptography may be used to secure the data exchanges. Typically however, a coprocessor is added to handle the cryptographic computations and store the secure data.

9.1.5 Contactless Integrated Circuit Cards

Contactless integrated circuit cards—or RFID-enabled cards—transmit data over short distances using radio frequencies. The card consists of an integrated circuit (called a transponder of tag in the case of RFID), an antenna, and the associated electronic circuitry for data transmission and processing, typically embedded in a plastic casing. They are available in a variety of forms such as key bobs, documents, and mobile devices in addition to plastic cards.

Contactless operations avoid the wear and tear of mechanical systems, particularly those at tollbooths that pull in tickets with magnetic stripes, thereby reducing maintenance costs. They can accommodate persons moving at small speeds close to points of sale, vending machines, parking lots, and so forth, thereby speeding the flow. In fact, a 2008 study by MasterCard Worldwide has shown that contactless payments reduce the duration of a transaction by more than 50% from 26.7 to 12.5 seconds (Lerner, 2013, p. 110).

Active contactless cards have their own power source in the form of lithium or carbon and manganese batteries or rechargeable batteries. Batteries, however, are costly, wear out, and are hard to dispose of. Passive devices are powered through inductive coupling from the reader (interrogator). When the integrated circuit comes close to the reader's antenna, the magnetic field from that reader induces a voltage in the chip's antenna to power its circuitry. This appears as an increased impedance at the reader's antenna. By varying the circuit parameters at the contactless card's antenna in proportion to the data to be transmitted, the voltage at the reader's antenna will represent the data to be transmitted back. This type of data transfer is called load modulation.

Proximity cards are used for distances under 40 cm, while vicinity cards are used for distances of 1–1.5 m. The bit rate is typically higher for proximity cards than for vicinity cards (106, 212, and 424 kbit/s versus less than 26 kbit/s), while the power consumption of vicinity cards is less. Proximity cards are used in payment applications such as automated ticketing in high-volume transportation systems and at point-of-sale terminals.

For larger distances, the communication from the transponder back to the reader uses the reflections at the integrated circuit's antenna (backscatter). A load resistor is connected in parallel with the antenna of the contactless device, and this load is switched on and off according to the data to be transmitted. Thus, the intensity of the signal reflected from the transponder can be backscatter modulated. The distances involved are of the order of 10 m, which makes the system suitable for applications such as highway toll collection system. Figure 9.2 illustrates the difference between near- and far-field methods of communication between a reader and the tag.

9.2 Integration of Smart Cards with Computer Systems

There are many independent players in open payment system architecture: card manufacturers, banks, application developers, as well as terminal providers. Thus, some specifications are needed to interface the various

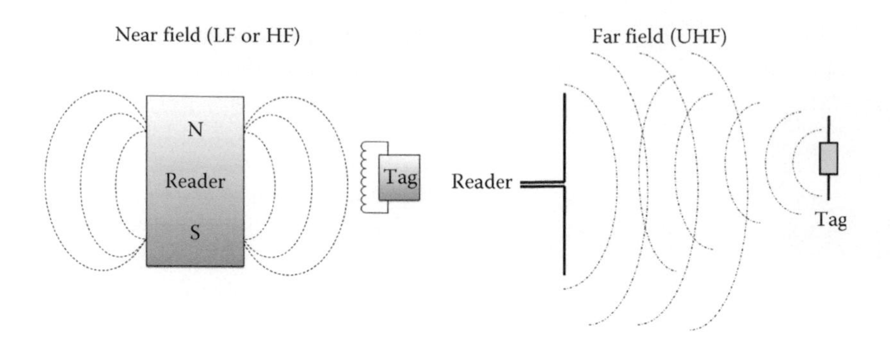

FIGURE 9.2
Near- and far-field methods of communication between reader and tag. (From Bueno-Delgado, M.-V. et al., Radio-frequency identification technology, in: Sherif, M.H., ed., *Handbook of Enterprise Integration*, Auerbach Publications, Boca Raton, FL, 2010, pp. 429–466. With permission.)

components and maintain the overall security of the payment system as well as the individual components.

Two main solutions were developed to provide a single secure interface to access integrated circuit cards with the Windows operating system, independent of the access peripherals or the applications residing in the card. These are the Java-oriented OpenCard Framework (OCF) and the PC/SC tool box for Windows. For Linux, the Movement for the Use of Smart Cards in a Linux Environment (MUSCLE) has been active to provide a standard interface.

9.2.1 OpenCard Framework

The OpenCard Framework (OCF) is the product of the OpenCard Consortium driven mainly by IBM, Sun Microsystems, and Gemplus. The objective was to create a PC interface to smart card independent of the operating system. The work ended with OCF Version 1.2 and a reference implementation by IBM.

OCF is a middleware implemented in Java with a common programming interface for all Web applications (Hermann et al., 1998). The framework allows any application stored in a smart card application to access both contact and contactless cards that use the Application Protocol Data Units (APDUs) defined by ISO/IEC 7816 in their commands. The OCF services shield the Java applications from the specifics of each smart card or access terminal (ATM terminal, computer, card reader, etc.) through abstract services: CardService for cards and CardTerminal for peripherals. This allows the grouping of complex sequences of commands to be executed with a simpler set of instructions, such as handling several simultaneous authentication requests as encountered in financial or health-care applications (Kaiserswerth, 1998). Figure 9.3 illustrates this framework.

OCF is now used primarily to send commands from Java environments to smart cards, with the commands directly coded as APDUs.

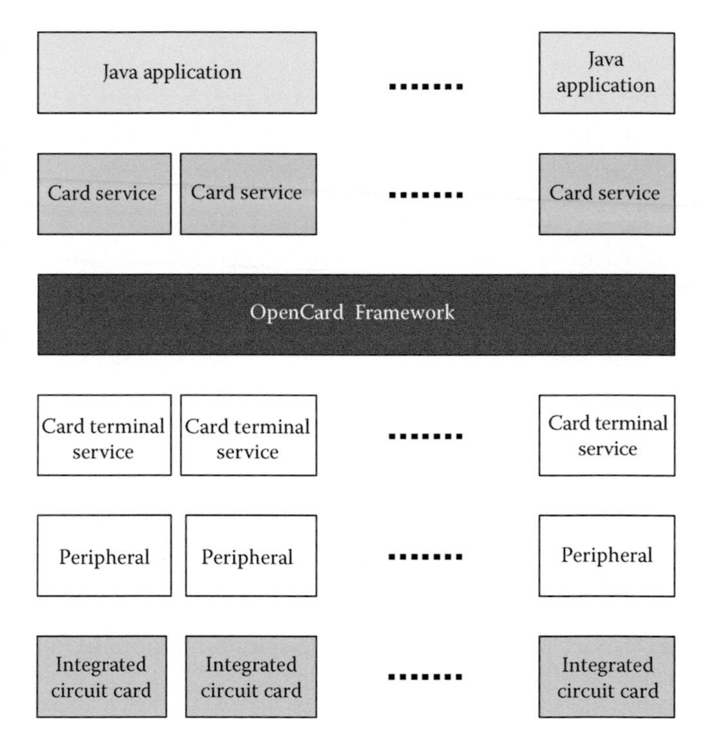

FIGURE 9.3
Framework of operation of OpenCard.

9.2.2 PC/SC

The PC/SC is a shorthand notation for the "Interoperability Specification for Integrated Circuit Cards and Personal Computer Systems." This specification was the first effort to agree on specifications for connecting smart cards to PCs. It was produced by a work group from leading integrated circuit cards and personal computer vendors such as Gemplus, Hewlett-Packard, IBM, Sun Microsystems, and Microsoft. It focuses on the Windows operating system and supports multiple languages (PC/SC, 1997). Microsoft has also published its own cryptographic library, CryptoAPI, which can be used in the Windows environment to access the

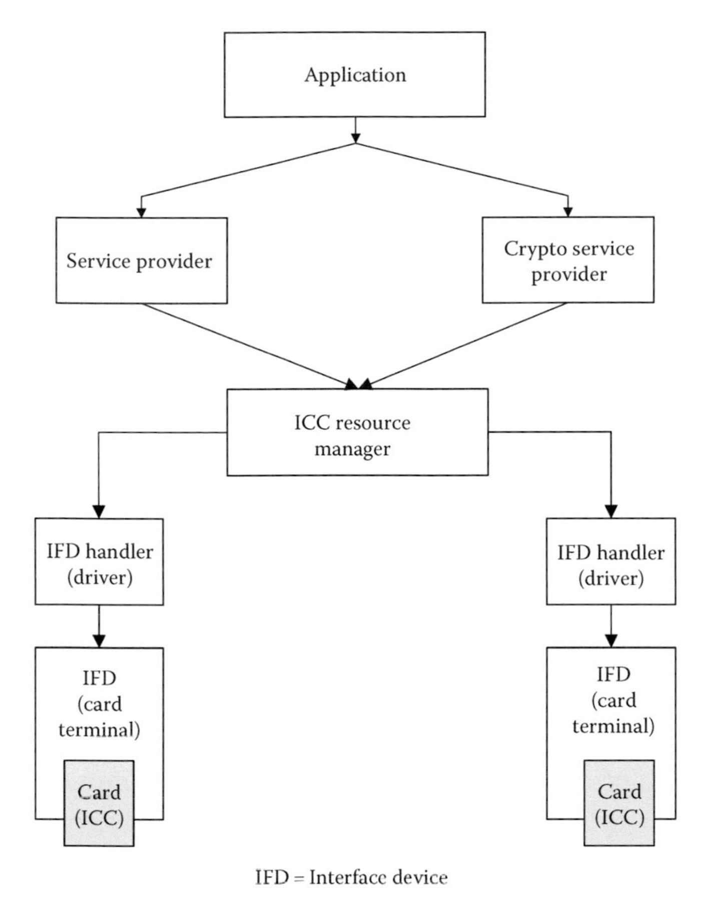

FIGURE 9.4
Framework of operation of PC/SC.

cryptographic functions and the certificates stored in a smart card. CryptoAPI also provides secure messaging and certificate management. Figure 9.4 shows the architecture of PC/SC.

The resource manager of the integrated circuit card is the hub of PC/SC exchanges; it controls the device drivers for the peripherals as well as for all the exchanges with the applications through *service providers*. These maintain the various card registers and follow the communication protocols. A specific provider, the cryptographic service provider, takes care of all the functions of authentication and security of the exchanges.

9.2.3 Movement for the Use of Smart Cards in a Linux Environment

The MUSCLE published the first version of a Linux application programming interface to smart card in 2000. The MUSCLE is based on PC/SC, but its source code is available under a General Public License (GPL) as defined by the Free Software Foundation. Thus, it can be modified and extended by other parties.

9.2.4 Financial Transactional IC Card Reader (FINREAD)

Bank terminals operate in a secure environment. The channel between the smart card reader and the terminal is assumed to be secure, provided that the terminal has not been tampered with. In contrast, a home computer may be compromised with malware to observe the exchanges as well as to record the keyboard strikes, including the PIN.

One solution is to use an offline tamper-resistant card reader equipped with a small display and its own secure PIN pad. Accordingly, after the establishment of a secure TLS channel between the user's computer and the banking server, the server sends a random n-digit challenge that a user's browser displays on the screen. The user inserts the card into the slot of the card reader and enters the PIN into the PIN pad to unlock the smart card so that it could communicate with the keypad. Next, the user manually punches the n-digit of the challenge displayed on the computer screen into the PIN pad. The card responds with an m-digit response on its display that the user enters from the computer keyboard into the special field displayed on the computer screen (Hiltgen et al., 2006).

The European program Community Research and Development Information Programme (CORDIS) has provided the second solution, naming an interoperable tamper-resistant card reader to secure online smart card transactions. The specifications consist of eight parts (CWA 14174-1 to 14174-8, 2004) and were adopted as a CEN Workshop Agreement (CWA) by the European Committee for Standardization (CEN, Comité Européen de Normalisation) in 2004.

The secure smart reader is denoted as FINREAD. It uses a Java virtual machine to provide a multiapplication environment that hosts payment applications for different banks, each with its own security credentials. External processes have to go to a special application that screens all request and blocks all those that it does not approve. Data encryption and authentication using a public key infrastructure (PKI) are also available. The user enters the PIN into the reader PIN pad and then follows the instruction on the display (CORDIS, 2013).

Authentication is done at the card level as well as at the transport and the application layers. The communication between the browser and the card reader uses a secure interface that conforms to the PKCS #11 library. The transport layer authentication takes place when the TLS channel is established with client authentication. At the application layer, a signed Java applet participates in a challenge–response authentication exchange between the bank server and the card (Hiltgen et al., 2006).

9.3 Standards for Integrated Circuit Cards

Standards for smart cards are needed for two main reasons: they increase the scale factor, which reduces the unit production costs, and they facilitate the networking of applications on a worldwide level through harmonized interfaces. Standardization has advanced on the physical aspects of the cards; however, the lack of standardization at the logical level is hampering financial applications on a large scale. In this section, the standards for cards with contacts and for contactless cards are presented.

9.3.1 ISO Standards for Integrated Circuit Cards

ISO/IEC 7816 is a multipart international standard that defines various aspects of both kinds of integrated circuit cards: with contacts and contactless. The current standard comprises 14 parts. Parts 1, 2, and 3 deal with the physical interface between integrated circuit cards with contacts and card readers to ensure their compatibilities. Parts 4, 5, 6, 8, 9, 11, 13, and 15 are relevant to cards with contacts and contactless cards. Part 7 defines a secure relational database, while part 10 concerns prepaid memory cards. There is no part 14. Because ISO/IEC 7816 is the foundation for all smart cards, the various parts are described in more detail.

ISO/IEC 7816-1:2011 specifies the physical characteristics of the card, the dimensions of the integrated circuit, the resistance to static electricity and electromagnetic radiation, the flexibility of the support, and finally, the location of the integrated circuit on the card. This includes embossing and/or a magnetic stripe and/or tactile identifier mark.

ISO/IEC 7816-2:2007 defines the dimensions and the positions of the metallic contacts on the card.

ISO/IEC 7816-3:2006 describes the electric signals (power, signal rates, shape and duration of the electric signals, voltage levels, current values, parity convention, etc.) and the transmission protocols between the card and the terminal and the card's response to a reset originating from the terminal. Four protocols are currently defined as follows:

- A character-oriented (i.e., byte-oriented) half-duplex protocol represented by the value $T=0$. This means that the smallest unit that the protocol process is 1 octet. This protocol minimizes the memory used and is simple to implement. It was initially adopted by early banking applications in France and is used in the SIM of GSM terminals.
- A block-oriented half-duplex protocol identified by the value $T=1$. This mode is also used for contactless card conforming to ISO/IEC 14443.
- A block-oriented full-duplex protocol identified by the value $T=2$; this mode, however, is rarely used.
- The value $T=14$ indicates the use of legacy proprietary protocols, such as the health-care cards in France (Sésame-Vitale) and Germany.
- The values $T=3$ to $T=13$ are reserved for future use.

The $T=0$ transmission protocol is byte oriented, while the $T=1$ is an asynchronous half-duplex protocol. Both are used in EVM. Further details are available in pp. 254–272 in *The Smart Card Handbook* (Rankl and Effing, 2010).

ISO/IEC 7816-4:2013 specifies the command–response pairs exchanged across the interface of a smart card reader and a smart card in terms of APDUs. The standard defines the means of retrieval of data elements and data objects in the card, the structures of applications and data stored, and the methods to access the files and the data in the card. The standard defines how to use an application identifier to ascertain its presence and to retrieve an application. The security architecture includes the means and mechanisms for identifying and addressing applications in the card. The specification applies to contact or contactless cards.

ISO/IEC 7816-5:2004 defines a unique universal identifier for applications residing in a smart card.

ISO/IEC 7816-6:2004 defines the data elements on integrated circuit cards both with contacts and without contacts. It gives the identifier, name, description, format, coding, and layout of each data element and defines the means of retrieval of them from the card.

ISO/IEC 7816-7:1999 defines the commands used for accessing a secure relational database in the integrated circuit card using a Structured Card Query Language (SCQL).

ISO/IEC 7816-8:2004 specifies commands that may be used for cryptographic operations in all integrated circuit cards (either with contacts or without contacts). These commands are complementary to and based on the commands listed in ISO/IEC 7816-4.

ISO/IEC 7816-9:2004 specifies commands for integrated circuit cards (both with contacts and without contacts) for card and file management, for example, creation and deletion, for the entire life cycle.

ISO/IEC 7816-10:1999 specifies the electronic signals for synchronous cards.

ISO/IEC 7816-11:2004 specifies a subset of the ISO/IEC 7816-4 commands and data objects to be used for the verification of smart cardholders through biometric methods. ISO/IEC 19785-1 complements that subset with additional data objects.

ISO/IEC 7816-12:2005 specifies the operating conditions of an integrated circuit card with a USB interface

(USB-ICC). The standard defines two protocols for control transfers. Version A supports the $T=0$ character-oriented protocol and version B supports the APDUs defined in ISO/IEC 7816-4.

ISO/IEC 7816-13:2007 specifies the commands for managing applications over their entire life cycle of the card.

ISO/IEC 7816-15:2004 specifies the storage of cryptographic data as well as the syntax, formats, and mechanisms to share this information.

9.3.2 ISO Standards for Contactless Cards

The description and operation of contactless cards are covered by a set of ISO/IEC standards (14443, 15693, 18000-3, and 18092). They define the physical, mechanical, and electrical aspects of contactless cards as well as the communication protocols.

Many contactless card applications operate in the high-frequency (HF) band at 13.56 MHz with a data rate of 106 kbit/s. In other RFID applications, the frequencies used are the low-frequency (LF) band (less than 135 kHz); the industrial, scientific, and medical (ISM) band, around 13.56 MHz and 2.4 GHz; and the super high frequency (SHF) band around 5.8 GHz.

Depending on the range of operation, contactless cards can be divided into proximity cards and vicinity cards. The distances are typically under 40 cm for close coupling cards (proximity cards) and under 1 m for remote coupling cards (vicinity cards). ISO/IEC 14443 is a four-part international standard that defines the characteristics of the cards (the *proximity cards*) used for identification and the transmission protocols. The physical characteristics and operation of vicinity cards are defined in the three-part standard ISO/IEC 15693. Vicinity cards are not used in electronic commerce applications: they are used to control the access to secure areas and as identification badges that must be worn in a visible manner.

Contactless cards can also be divided into two categories, active and passive. As explained earlier, passive cards are mute until they receive a signal from a reader; once powered up, they start transmitting. One reader at a time can energize a passive card; if more than one reader tries to wake up a passive card, a condition known as "reader collision" occurs. Also, in passive systems, the reader can communicate with multiple cards, so it must assign them an order of transmission so that they do not interfere with each other. Active contactless cards have a small battery, which allows them to start the communication and to broadcast their signals that can be received at greater distances than passive cards.

The relevant standards for proximity cards, which are more common in contactless and mobile payment applications, are as follows:

ISO/IEC 14443-1:2008, which describes the physical characteristics of proximity cards, including the environmental stresses that the card must withstand without permanent damage to its functionality.

ISO/IEC 14443-2:2010, which concerns the bidirectional communication between readers and cards. For ISO 14443A cards—the most commonly used cards—the transmission rate from the reader to the chip (the downlink) is 106 kbit/s, with 9.4 μs per bit. The width of the modulation pulses is only 2–3 μs to guarantee continuous power supply to the card.

ISO/IEC 14443-3:2011, which defines the initialization of the communication between the proximity card and the card reader and the anticollision protocols for both types A and B cards. Type A is used in contactless smart cards and electronic passports.

ISO/IEC 14443-4:2008, which specifies a half-duplex block transmission protocol (modulation and coding) to be used by higher layers and applications after the initial phase described in ISO/IEC 14443-3. The application layer communication protocol is based on the protocol $T=1$ defined in ISO 7816-3.

9.3.2.1 *Anticollision Protocols*

A collision is said to occur when two or more contactless devices located within the working range of a terminal start transmitting at the same time, thus interfering with each other. The probability of collisions is higher for passive cards/tags because they start to transmit as soon as they are powered. In contrast, active tags listen before transmitting, which reduces the chance for collisions.

Collisions can occur in a number of ways (Bueno-Delgado et al., 2010):

- Multiple card/tags are in the read range of the same reader.
- One card/tag is in the read range of multiple readers.
- Multiple readers interfere with each other.

ISO standards deal with the first case with two standard anticollision protocols, called type A and type B, to ensure that individual cards have different behaviors in time so that the terminal can identify them separately and address them individually. The anticollision protocols are half-duplex schemes operating at the media access control (MAC) layer, each with its own set of transmission parameters in terms of modulation, line coding, and transmission rate and for each direction.*

* An explanation of these parameters can be found in any communication textbooks (e.g., Proakis and Salehi, 2013).

A third protocol is associated with the FeliCa interface defined by Japanese standard JIS X 6319-4.

Situations of one tag in the read range of multiple readers and multiple readers interfering with each other are more applicable in the case of RFID contactless systems used for tracking objects than in the case of payment systems.

9.3.2.2 *Type A Anticollision Protocol*

According to type A anticollision protocol, the reader attempts to recognize the cards/tags in its coverage area in several interrogation cycles. If a collision occurs, the set of cards/tags is split into two subsets using the card/tag identification number. The reader continues the subdivision procedure until each set comprises one card only. Clearly, this variant of the binary search tree algorithm puts the computational burden on the reader.

More specifically, the reader sends a polling command (REQA) periodically at a minimum interval of 500 μs. When a card enters in the reader's field, it powers up and responds to the REQA command with an Answer to Request A (ATQA) command and transitions from the IDLE state to the READY state. To schedule the transmission of all the cards in the vicinity, the reader broadcasts an ANTICOLLISION command with, as a search criterion, a bit mask of 40 bits. Each card in the vicinity compares the bit mask with its unique identifier (UID), which in ISO 14443A can be 4, 7, or 10 octets. The various cards that fit the search criterion respond with the first 40 bits of their UIDs in the Select Acknowledge (SAK) command. If there is only one response, the process ends. If several cards respond at once, a collision occurs. Accordingly, for each bit collision of the bit mask, a branch in the binary tree is made by setting the corresponding bit in the subsequent iterations to "0." Also because the length can be 4, 7, or 10 octets while the search criteria can accommodate 5 octets (40 bits) only, a UID of 7 or 10 octets will require two or three rounds before the reader receives the full card UID. The exchanges are depicted in Figure 9.5.

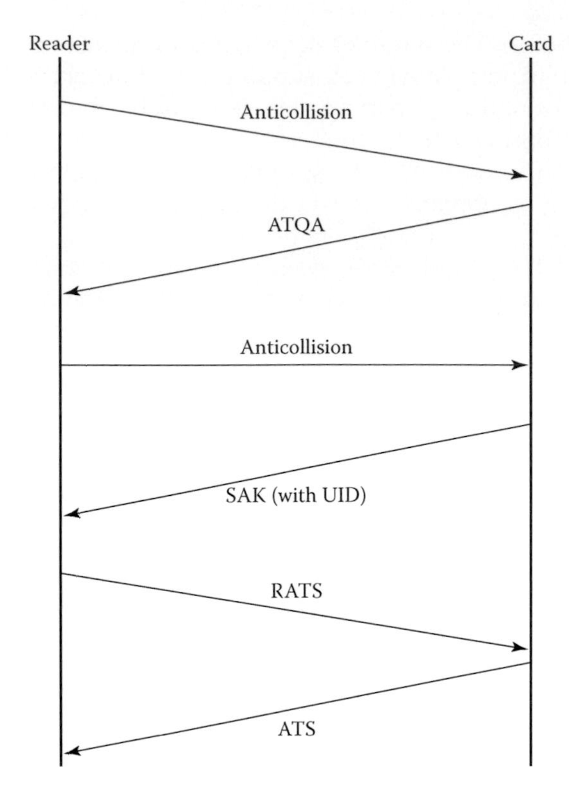

FIGURE 9.5
Initialization of ISO 14443 Type A cards.

Once the card moves into the ACTIVE state, both sides start negotiating the communication parameters on the radio link. First, the reader transmits a Request Answer to Select (RATS) command and the card returns its settings in an ATS command. If the card supports Protocol Parameter Selection (PPS), the reader can request changes as well.

To maintain bit synchronization during the anticollision process, the response time is specified as $(n \times 128 + 84)/f_c$ if the last datum that the reader sent was a "1" and is $(n \times 128 + 20)/f_c$ if the last bit was a "0." For the REQA and SELECT commands, n is 9 and $n \geq 9$ for all other commands. ISO 14443-4 specifies the Frame Waiting Time (FWT) as $(256 \times 16/f_c) \times 2^{FWI}$, where FWI is an integer from 0 to 13, with a default of 4. This corresponds to FWT of 300 μs, 5 s, and 4.8 ms, respectively.

9.3.2.3 *Type B Anticollision Protocol*

Type B cards use the variation of the Aloha anticollision protocol, called frame slotted Aloha. The reader sets a number of time slots dynamically for each card to send an ATQA command in one of these slots. For synchronization, the reader transmits a slot marker at the beginning of each slot. A short time after the transmission of a slot marker, the reader can determine whether a card has begun to transmit the ATQA command within the current slot. If not, the current slot can simply by interrupted by transmitting the next slot marker.

9.3.3 RFID Standards

RFID technology uses the same physical principles of contactless integrated circuit cards to track objects (Finkenzeller, 2010). An RFID contactless system consists of a transponder or tag, an antenna, a reader/writer, and a host computer. The transponder/tag is a compact package of a microchip connected to the antenna, which is either attached to an object or integrated directly into its fabric. The simplest tag is a memory card that stores a product's electronic data.

When the tag comes close to the RFID reader, an electrical current is induced in the antenna, powering up the chip's circuitry to perform certain computations before transmitting a response. Because the RFID tags are passive, their price can be attractive for mass applications.

RFID systems grew independently out of many industrial sectors and have been optimized for a particular field of application. As a consequence, the terminologies as well as the related standards are not completely harmonized. In particular, the rapid uptake of RFID in the retail supply chain is currently driving standardization even though contactless cards were first used in access control and payment applications. For example, RFID transponders are called commonly tags because they are applied to products before shipment for tracking and inventory purposes instead of barcodes. In toll collection applications, the transponders are called onboard units. Readers are sometimes called interrogators. They may be called validators in urban mass transit systems and roadside units in intelligent transportation systems.

The passive communication between the interrogator/reader and the transponders is established in one of two ways: interrogator talks first (ITF) or tag talks only after listening (TOTAL). In ITF systems, the tag modulates its antenna reflection coefficient with an information signal only after the interrogator allows it to do so. The system is denoted as TOTAL if the tag listens first to the interrogator modulation, determines what system it is, and, if it is ITF, transmits the information without waiting for permission.

To clarify the standards situation in payment applications, the focus will be standards from ISO and from EPCglobal.

9.3.3.1 ISO Standards

ISO/IEC 18000 defines the communication protocols and the link parameters for toll operating systems such as the operating frequency, operating channel accuracy, occupied channel bandwidth, maximum power, modulation, duty cycle, data coding, bit rate, bit rate accuracy, and bit transmission order. However, it does not define the application parameters such as data format, protocols, and encryption techniques. Patent on encryptions and various tweaks to take into account various environmental conditions make the systems incompatible.

ISO/IEC 18000 consists of six major parts, the first released in 2008. It describes several air interface communication in the following bands: below 135 kHz (part 2), at 13.56 (part 3) MHz, at 433 MHz (part 7), between 860 and 960 MHz (part 6), and at 2.45 GHz (part 4). Part 1 contains the reference architecture and defines the various parameters. Part 5 was supposed to be for the microwave frequency of 5.8 GHz but it was rejected.

Part 1 provides a framework for the common communications protocols using the internationally useable frequencies for RFID. It describes possible common system management and control and information so that it could be exchanged among systems.

ISO/IEC 18000-6 was split into various sections depending on the channel encoding used in the forward link and the anticollision/arbitration algorithms. It was republished in 2012 as follows:

- Part 61 is labeled as type A
- Part 62 is labeled as type B
- Part 63 is labeled as type C
- Part 6 is labeled as type D

Types A, B, and C are ITF, while type D is TOTAL.

ISO/IEC 18000-6 has standardized and anticollision protocol denoted as I-Code. There are, however, many other nonstandard anticollision protocols used in practice (Bueno-Delgado et al., 2010, pp. 444–450).

9.3.3.2 EPCglobal®

EPCglobal is leading the development of industry-driven standards for the Electronic Product Code™ (EPC) technology based on the use of radiofrequency identification (RFID) in the supply chain. EPCglobal is a joint venture of GS1 (formerly known as EAN International) and GS1 US (formerly known as the Uniform Code Council [UCC]). Research on global RFID is conducted at the Auto ID Labs, the successor of the MIT Auto-ID Center formed in 1999.

GS1 US publishes the various Electronic Data Interchange (EDI) standards based on ANSI X12 as well as the Uniform Communication Standard (UCS) used in the food, warehouse, and grocery industries and the Voluntary Interindustry Commerce Standards (VICS) EDI used in the general merchandise retail industry.

The Auto-ID Center and its successor, EPCglobal, have developed a different set of classification and standards for RFID tags. This can be summarized as follows.

The first group comprises passive tags with an ultra high frequency (UHF) interface incompatible with what ISO specified. These are denoted as Class 0 and Class 1. Class 0 has a one-time, field-programmable nonvolatile memory that is typically programmed at the factory. Class 1 has a programmable nonvolatile memory that can be programmed by the user. Once it is programmed, the memory is locked. Class 0 and Class 1 are also not interoperable. The simplest tags from these classes contain a single bit and are used for electronic article surveillance, mostly to prevent shoplifting. Other tags contain a product electronic code that references information stored in a database to describe the product.

In 2004, EPC global began developing a second-generation protocol (Gen 2) more closely aligned with ISO standards. A new subclass of tags, Class 1 Gen 2 represented as EPC Gen 2 (Electronic Product Code Class 1 Generation 2), conforms to these specifications to operate at the two frequency ranges: 860–868 MHz for Europe and 902–928 MHz for North America (Bueno-Delgado et al., 2010). EPC Gen 2 was adopted with minor modifications as ISO/IEC 18000-6 Type C (ISO/IEC 18000-6C). U.S. toll operators have organized as the 6C Toll Operators Committee (6CTOC) to promote the use of the ISO/IEC 18000-6C with EPC Gen 2 passive RFID tags to promote interoperability.

Class 2 tags have up to 65 Koctets of read/write memory so that they can do some security functions such as encryption.

Class 3 tags have a battery. Battery power increases the read range, that is, the distance from which a reader can communicate with the tag. The boosted transmit power is also useful in difficult radio environments. Another advantage is to allow the tag to collect data, that is, to serve as a sensor, even when the tag is not powered. The data are then transmitted back to the reader when the tag is activated. These tags are called semipassive because they do not initiate the communication with the reader. One of their uses is in wireless sensor networks.

Class 4 tags are active; they use a battery to run the chip circuitry and to power a transmitter that broadcasts to the reader.

Class 5 tags are active RFID tags that can communicate with other tags, including other Class 5 tags, that is, they function as readers/interrogators.

Tags come in a variety of shapes, types, and sizes and can be read only or rewriteable. They can be embedded into various products such as merchandise, inventories, posters (smart posters), and business cards to link them to additional contents, for example, in museums and retail stores. Table 9.1 summarizes the characteristics of the various EPCglobal classes, while Table 9.2

TABLE 9.1

Tag Classification per EPCglobal

EPC Class		Memory	Programming	ISO Compatible
Class 0 Gen 1		Read only	Programmed at factory	No
Class 1 Gen 1	Passive	Write once and read many	Programmed by the user and then locked	No
Class 1 Gen 2 (EPC Gen 2)		Read/write		ISO/IEC 18000-6C
Class 2		Read/write (65 Koctets)		
Class 3	Semipassive		Programmable	
Class 4		Read/write		
Class 5	Active			

TABLE 9.2

Key Characteristics of RFID Tags

Frequency Band	Frequency Range	Read Range	Cost	Applications	Standards
Low frequency (LF)	125–135 kHz	10 cm (passive) to 1 m (active)	Low	Access control inventory control, timing for sporting events	ISO 11784/11785 ISO/IEC 18000-2
High frequency (HF)	13.553–13.567 (13.56)[a] MHz	3 cm to 1 m	Medium	Contactless cards, near-field communication, passive transponders/tags	ISO/IEC 15963,14443A, 14443B, 18000-3, 18092 EPC Class 0/1
Ultra high frequency (UHF)	433 MHz	15–100 m		Transponders in containers	ISO/IEC 18000-7
	850–950 MHz	60 cm to 3 m (passive) >10 m (active)	High	Transponders in pallets	EPC Class 0/1 EPC UHF Gen 2 ISO/IEC 18000-6
	2.45[a] GHz	1–6 m (passive)		Container/toll collection	ISO/IEC 18000-4
Super high frequency (SHF)	5.8[a] GHz	15–100 m (active)		Toll collection	None

Sources: Bueno-Delgado, M.-V. et al., Radio-frequency identification technology, in: Sherif, M.H., ed., *Handbook of Enterprise Integration*, Auerbach Publications, Boca Raton, FL, 2010, pp. 429–466; Finkenzeller, K., *RFID Handbook*, 3rd edn., John Wiley & Sons, Chichester, UK, 2010, pp. 21–23; Smart Border Alliance, RFID Feasibility Study Final Report, Attachment D, January 21, 2005, available at https://www.dhs.gov/xlibrary/assets/foia/US-VISIT_RFIDattachD.pdf, last accessed November 16, 2014.

[a] Available world for industrial, scientific, and medical (ISM) applications.

summarizes the key characteristics of RFID tags per frequency band.

9.3.3.3 Open Specifications

The communication protocol used in the tags manufactured by Kapsch TrafficCom (formerly Mark IV Industries) is a time division multiplexing protocol called the IAG-TDM. It operates at 915 MHz with an air-interface protocol, whose specifications were released in 2013 only. They are available under the condition that any derivative or supplementary applications or code they may develop be available without restrictions. The protocol is used by the E-Z Pass system that was adopted by 25 toll agencies across the northeastern states of the United States.

9.3.3.4 Privacy Concerns

Significant privacy concerns have been raised regarding the lack of any cryptographic measures in some RFID applications, particularly Class 0/Gen 1, Class 1/Gen 1, and Class 1 Gen 2 (EPC Gen 2) tags. These tags can be read and legitimate transactions can be eavesdropped from any distance (using relay attacks, as explained in Section 9.8.4). The same can be said of the risk of tracking unprotected tags that are worldwide readable. Cards used in access control and electronic passports are also vulnerable to eavesdropping and cloning. The Wikipedia article on radio-frequency identification is a good starting point for the most recent developments in this important and controversial subject.

9.3.4 Near-Field Communication Standards

As explained earlier, near-field communication (NFC) is a subset of RFID technology, where the tag includes a small coil to extract the energy from an electric current induced by the reader. Thus, the technology marries contactless identification according to ISO/IEC 14443 and wireless communication at 13.56 MHz. As a consequence, there is an overlap between the use of proximity cards and NFC-enabled devices in payment applications. Because of this, NFC devices can read proximity cards and contactless terminals can communicate with NFC tags. Thus, NFC-enabled devices may replace with contactless cards in future applications.

As of 2014, NFC solutions are more popular in Asia and Europe than in North America. It is estimated that there are about 100 million contactless cards in Europe, which amount to a fifth of all payment cards. Visa Europe started to roll out NFC merchant terminals in 2007 and their number reached 1.5 million terminals in 2014, with 350,009 in the United Kingdom (Schäfer and Bradshaw, 2014).

NFC was coinvented by NXP Semiconductors (formerly Philips Semiconductors) and Sony.* The technology is intended for short-range (3–10 cm) contactless communications and is used by systems in the LF or HF bands. NFC readers are part of Point-of-Sale Terminals, public transit gates, industrial equipment, and so on. In contrast, far-field communication is employed in the UFH and microwave bands.

The standardization of NFC technology started in the European Computer Manufacturers Association (ECMA†) with ECMA-340 and ECMA-352, which were later adopted as ISO/IEC 18092 and ISO/IEC 21481. ISO/IEC 18092/ECMA-340 and ISO/IEC 24181/ECMA-342 define the NFC interfaces and protocols NFCIP-1 and NFCIP-2, respectively.

NFCIP-1 is a superset of ISO/IEC 14443 and FeliCa (JIS X 6319-4). Thus, it operates at 106, 212, or 424 kbit/s, with a modulation and encoding scheme for each bit rate. The NFC protocol is similar to the protocol for proximity card: half-duplex communication, collision avoidance, and so forth. One difference is that both NFC parties can assume the role of the master, while for contactless cards the terminal is always the master. Another is that the NFC protocol defines two communication modes: active and passive. In the active communication mode, each device generates its own high-frequency field at the carrier frequency to transmit data. In the passive communication mode, only the initiator generates a high-frequency field at the carrier frequency, while the target uses load modulation for data transfer, that is, by varying the load on its coil in proportion to the data signal to be transmitted (Rankl and Effing, 2010, p. 349).

The alignment of the various bit rates with the corresponding standard is summarized in Table 9.3.

NFCIP-2 of ISO/IEC 21481 (ECMA-352) supports mode switching, that is, detecting and selecting the communication mode for any of the contactless technologies operating around 13.56 MHz (reader, proximity card, vicinity card). Furthermore, it is designed to

TABLE 9.3

Relation of ISO/IEC 18092 (NFCIP-1) to Other Specifications

Bit Rate (kbit/s)	Standard
106	ISO/IEC 14443 types A and B
212, 424	JIS X 6319, ISO/IEC 14443 type B
848	ISO/IEC 14443 type B

* NXP supplies the MIFARE cards, while the FeliCa chip is from Sony.

† The European Computer Manufacturers Association (ECMA) was founded in 1961 to develop standards in information technology and communication and consumer electronics. In 1994, the name was changed to ECMA International—European association for standardizing information and communication systems.

avoid interfering with any ongoing communication at 13.56 MHz. Thus, an NFCIP-2 device will not activate its radio power when it detects another radio source nearby, which minimizes the changes for collisions.

The version of ISO/IEC 21481, published in 2012, improves the interoperability with ISO/IEC 14443. The 2013 edition of both ECMA-340 and ECMA-352 is aligned with ISO/IEC 21481:2012.

The NFC Forum has published several specifications for the technology architecture, including data formats and the four different tag types supported by all NFC devices. These will be presented in Chapter 10.

9.3.5 File System of Integrated Circuits Cards

The file system of ISO/IEC 7816-4 defines the memory structure of the standard integrated cards. It is composed to directories comprising a master file (MF) and two types of subdirectories: dedicated files (DFs) and elementary files (EFs). Each file has an identifier coded on 2 octets in hexadecimal notation. Figure 9.6 illustrates the arrangement of the various files.

The master file (MF) is at root of the tree and is always identified by the file identifier 3F 00. The file identifier of the first DF is 01 00 and the last is 3E 00. Thus, an ISO/IEC 7816 card cannot contain more than 62 dedicated files in addition to the master file. Each DF is associated with a given application and may contain one or more elementary files. Application selection may be through the SELECT FILE command with the application identifier (AID) as an argument or indirectly with the help of the special elementary DIR (directory) or ATR (Answer to Reset).

The EFs contain the data. Each EF is identified by its position in the tree, that is, by the path leading back to the master file. The identifier is also coded on 2 octets and takes the form *xx yy*, where *xx* is the identifier of the DF to which the EF belongs and *yy* is a sequential number of the EF in that particular directory. Thus, *xx* is 3F if the file depends directly on the master file and the number of elementary files in a directory cannot exceed 63.

The elementary files 2F 00 and 2F 01 under the master file have special indexing functions. The first is called DIR and the ATR. Roughly speaking, the file DIR contains elements that allow the identification of the applications while the file ATR specifies how the card can find the applications or the various objects.

The maximum number of elementary files in a card is thus $63 \times 63 = 3969$ files. Clearly, this structure is rigid and does not suit dynamic situations where files can be added or deleted corresponding to the addition or removal of applications. In fact, ISO/IEC 7816-4 does not allow the creation of new files; therefore, the various suppliers of integrated circuit cards have had to define proprietary commands for file management.

ISO/IEC 7816-4 distinguishes between two types of elementary files, internal EFs, and working EFs. The latter contain data for the exclusive use of entities external to the card. The internal EF contains data that the card uses during its operation. For example, in a monetary application, the following files can be present:

- Key files for the storage of the keys that will be used to derive a session key as specified by the payment protocol employed. Given the sensitivity of banking transactions, the applications that use purses will most probably need several keys, one for each action, such as for certification, for debit, for credit, or for electronic signature. Each key will be associated with an individual file.
- PIN files to stock the PINs that control access to the application file. The application files and the access conditions are irrevocably defined during the personalization phase.
- Purse files. For each purse, the file indicates the maximum balance, the maximum payment for each transaction, the current balance, and a

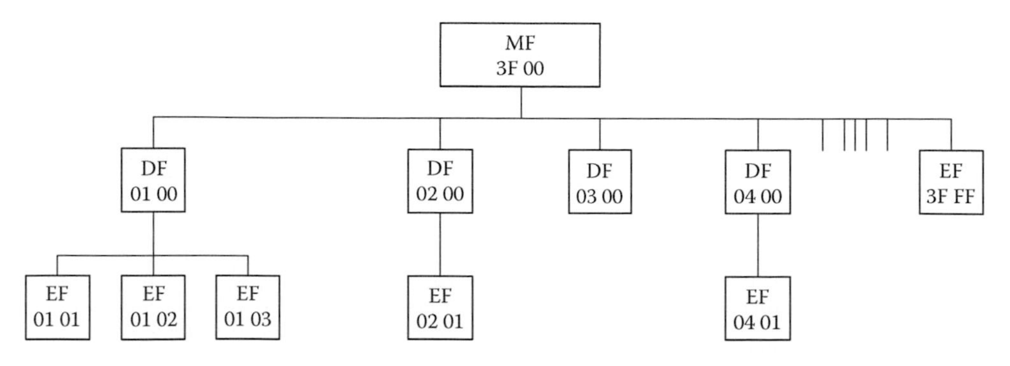

FIGURE 9.6
Tree structure of ISO 7816-4 File System.

backup balance to recover the previous value in case of a failure.

- Certificate files, in the case of public key encryption.
- Application usage files.

9.3.5.1 Swedish Electronic Identity Card

The example of the Swedish electronic identity card (EID) can illustrate the preceding presentation. This card is defined by the Swedish Standard SS 61 43 30 on the basis of the work of Secured Electronic Information in Society (SEIS). SEIS is a Swedish nonprofit organization of about 50 firms and organizations in the financial, industrial, and public administrative sectors in Sweden with the aim of taking advantage of new networking technologies.

In this application, the master file contains the following elementary files (SEIS, 1998):

- The file EF_{PAN}, which defines the account embossed on the card Private Account Number (PAN)
- The file EF_{PIN}, which defines the master PIN that all applications on the card may use
- The DIR file (2F 00), which, in conformance with the specifications of ISO/IEC 7816-4, holds the application identifier for the electronic identity card
- The dedicated file of the application with three elementary files:
  - The application usage file (AUF), EFAUF
  - The file for the private RSA key, EFPrK
  - The certificate file, EFCERT

Figure 9.7 depicts the logical organization of the files for the Swedish electronic identity card (SEIS, 1998, Annex B).

9.3.5.2 Subscriber Identity Module of GSM Terminals

The Subscriber Identity Module (SIM) cards of GSM mobile devices are fully programmable computer systems with a real-time operating system. Thanks to the Java virtual machine, they can run custom Java applications.

The GSM 11.11 specifications define the functions and parameters of a SIM (European Telecommunications Standards Institute, 1996, 1997). The SIM is a tamper-resistant chip with a unique identifier and a secure storage for the credentials that allow access to the services that a network operator provides. These parameters are the International Mobile Subscriber Identity (IMSI), of the user, the cryptographic algorithms, and a 128-bit root key K_i that it shares with the home network authentication center. The user defines a 4–8-digit personal identification number (PIN) to control access to the SIM card from the mobile phone. Knowledge of this PIN authenticates the user before some actions are defined in the specifications.

The mobile terminal drives the SIM in a *master–slave* relationship. Communication between the mobile equipment and the SIM takes place with the $T=0$ data transmission protocol. The authentication process is as follows. For each new SIM, the home network authentication center combines a 128-bit random number (RAND) with the root key K_i that is stored in the SIM to produce an expected response XRES of 32 bits and a 64-bit session key K_c. These three numbers form the so-called GSM subscriber's triplets. Now, each location area in the global land mobile network has a unique identifier, the Location Area Identity (LAI), which is broadcast periodically on the broadcast control channel. Thus, a terminal receives that LAI sometimes after it enters in that region and then stores it in the SIM. Whenever a terminal updates its location during roaming or is attempting to place a call, it sends the identity stored in the SIM to the local network. If it is not the home network, the local network inquires about the GSM triplet corresponding to the SIM of that terminal

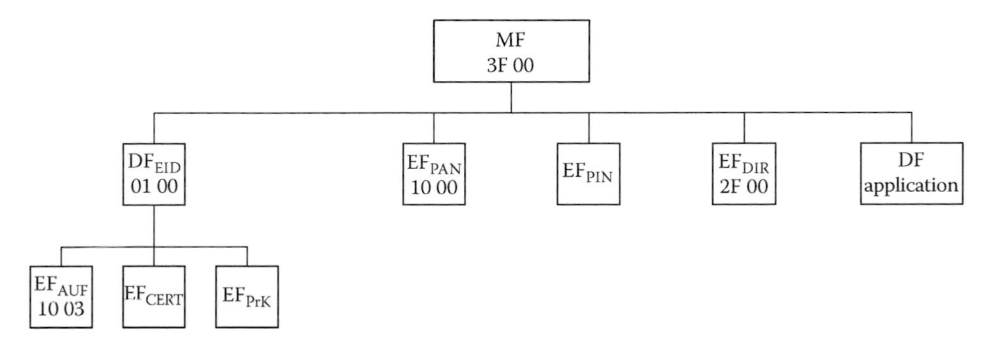

FIGURE 9.7
Logical organization of the Swedish electronic identity card.

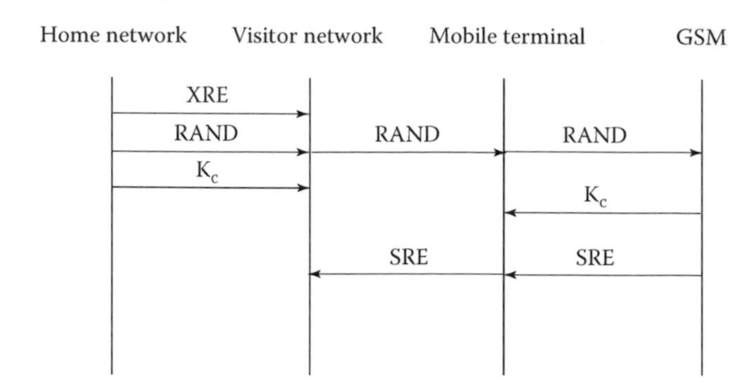

FIGURE 9.8
Protocol exchanges for the authentication of the SIM and the establishment of the symmetric session key K_c.

from the home network. Once it receives the triplet, it sends the random number (RAND) to the SIM through the mobile terminal. The SIM then uses RAND and the root key K_i to generate a 32-bit signed response SRES and the 64-bit K_c and give them to the terminal so that it forwards them to the local network. If the SRES matches the XRES, the network allows the call to proceed and establishes a secure communication channel with the SIM using the session K_c for symmetric encryption. These exchanges are shown in Figure 9.8.

Whenever the local network is not the home network, it assigns the user a Temporary Mobile Subscriber Identity (TMSI), which it sends encrypted with the session key K_c so that the SIM could store it in the appropriate EF. It should be noted, however, that the GSM protocols do not provide the means for the terminal to authenticate the network; therefore, a compromised base station in the mobile network can be used to launch a *man-in-the-middle* attack without being detected.

The SIM has a hierarchical file system, with an MF and two DFs that contain the EFs with the application data. The file identifier of the MF is 3F00 as defined in the standard. The two dedicated files have "7F" in the first octet for their file identifiers. The first, denoted as $DF_{TELECOM}$ with the file identifier "7F10," contains elementary files storing various parameters related to the service, such as abbreviated dialing numbers or stored dialing numbers. The identifier of the second dedicated file, DF_{GSM}, is "7F20." DF_{GSM} contains parameters related to the GSM operation, such as the IMSI, the session key K_c, the location information, and the TMSI. In particular, personalization uses one of two files elementary files in the EF_{SST} grouping (SST = SIM service table) in the telecom directory DF_{GSM}, EF_{GID1} "6F3E" or EF_{GID2} "6F3F" (GID = group identifier).

According to the specification, the EFs directly below the MF have the value "2F" in the first octet of their file identifier. There is only one such file, EF_{ICCID}, which has the identifier "2FE2" and stores the unique identification number for the SIM. Network operators can store in the SIM their own files for operational and administrative purposes.

User access to the SIM from the mobile phone is controlled by a 4–8-digit personal PIN under user control. However, with a special instruction and the correct PIN, PIN verification can be switched off so that it is not required before placing a voice call. In general, access conditions to the various files vary from level 0 (always accessible) to level 15 (never accessible) as shown in Table 9.4.

The size of SIM cards has been shrinking, and there are now mini, micro, and nano (also called fourth form factor or 4FF) SIM cards. (The original SIM was given retroactively the code 1FF.) The measurements of nano SIM cards are 12.3 mm × 8.8 mm × 0.67 mm, and they hold the same amount of data as earlier SIMs. GSM terminals commonly used mini (2FF) and micro SIMs (3FF), but all iPhones starting from iPhone5 contain nano SIMs.

When network operators subsidize a mobile handset, they tie it to a particular SIM for a certain length of time. This function is based on the GSM/ETSI specification TS 101 624. The technique is usually called SIM lock and is based on matching two sets of data stored in both the SIM and the mobile equipment each time the device is switched on. One method is for the mobile phone to read data in the SIM, typically in the group identifier files, and to verify that they are identical to the data stored in the mobile phone. Another is for the SIM to read unique data from the handset and compare it with the stored data. The mobile can be used if both data sets are identical.

TABLE 9.4
Levels of Access to the GSM SIM

Level	Access Condition
0	Always
1	One of the following three conditions must be true for access: A correct PIN (PIN1) used for cardholder verification was presented to the SIM during the current session. PIN verification is disabled. Unblocking of PIN verification has been successfully performed during the current session. After three consecutive false PIN presentations, not necessarily in the same card session, access is blocked until unblocking is successfully performed on PIN1.
2	One of two conditions must be true of access. A correct PIN (PIN2) was presented to the SIM during the current session. Unblocking of PIN2 has been successfully performed during the current session.
3	Reserved for future use.
4–14	These levels are under the control of the appropriate administrative authority.
15	Never.

A new specification from the GSM Association (GSMA) defines an embedded SIM, or embedded Universal Integrated Circuit Card (eUICC). Embedded SIMs are attached to the device permanently and do not need to be changed to switch between mobile operators. This new design was spurred by the growth of tablet traffic, which does not need a phone number and links to wide area network over Wi-Fi, because of cost considerations or ease of access. Most tablets, even those with SIM card slots, are not bought from a mobile operator. An embedded SIM would also facilitate machine-to-machine communication such as in utility meters, traffic lights, remote patient monitoring, and car communications. In particular, the European Commission has mandated that, starting 2015, all new cars sold in member states must be fitted with an embedded SIM to contact emergency services automatically in the event of a collision (the so-called eCall program). Thus, car manufacturers can install the embedded SIMs firmly so that they do not shake loose without binding the car buyer to a specific operator because the switch of operators does not require card replacement.

In June 2014, the Japanese mobile operator NTT DoCoMo announced the first embedded SIM. The Netherlands has allowed utilities and car companies to issue their own SIMs. Also, Apple have equipped its latest iPads with these reprogrammable SIMs. Should smartphones use these embedded SIMs, however, many regulatory and security issues will be raised. In many countries, for example, people have to show proper personal identification when signing up for mobile phone service.

An embedded SIM would give one or two possibilities. A master profile from one operator could then switch in/out the profiles to a partner or affiliate mobile network operators in other countries. Another possibility is that the device manufacturer would set up a basic bootstrap profile to enable *over-the-air* (OTA) installation of operator-specific applications. It should be noted that all these flexibilities ride on a hardware certified for its security.

Soft or virtual SIMs are a different story because in this case the operator security credentials are stored in a software emulation of a smart card running on a user device or a merchant terminal. This new architecture raises numerous security concerns because it is easier to breach software defenses.

9.4 Multiapplication Smart Cards

Multiapplication smart cards make it possible to have several applications on the same card, provided that the card resources can be shared without compromising security. The definition of standardized interfaces helps to facilitate the development of these applications independently as well as porting them from one card to another, even if they are made by different suppliers. The ISO/IEC 7816-4 standard is the preferred starting point for achieving that goal, not only in the case of open specifications such as EMV and Java Card but also for proprietary solutions. This section begins with a presentation of the file system that ISO 7816-4 has defined.

9.4.1 Management of Applications in Multiapplication Cards

There are three possible cases depending on the type of relationship among the applications that coexist on the same multiapplication card:

- A primary application may control other secondary applications.
- Several applications may be federated under the control of a central authority.
- All installed applications are independent.

9.4.1.1 Secondary Applications Controlled by the Primary Application

This situation requires perfect coordination among the providers of the various applications to share harmoniously the resources of the card. In fact, the current operating systems for smart cards are not multitask and are closer to the systems used by the computers of the 1970s.

In the distribution and organization of the data files, the secondary applications will be considered as logical subsets of the primary application, as shown in Figure 9.9. The index file (DIR or ATR) that points to the dedicated files for the secondary applications is defined, once and for all, during the personalization of the card. Information sharing is solely between each secondary application and the primary application. Security relies on the primary application authenticating the secondary applications.

9.4.1.2 Federation of Several Applications under a Central Authority

In this case, the applications share common data, for example, the personal data of the cardholder. This sharing is done under the auspices of a central authority that controls the master file, as indicated in the diagram of Figure 9.10. This authority is usually the card supplier.

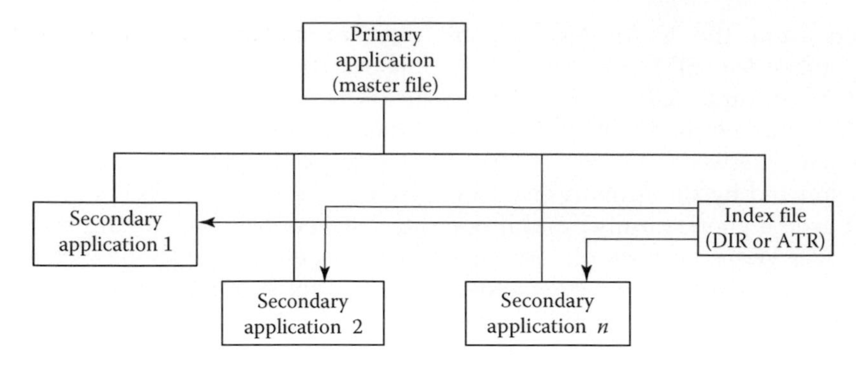

FIGURE 9.9
Logical representation of the configuration of the dedicated files for a primary application controlling secondary applications.

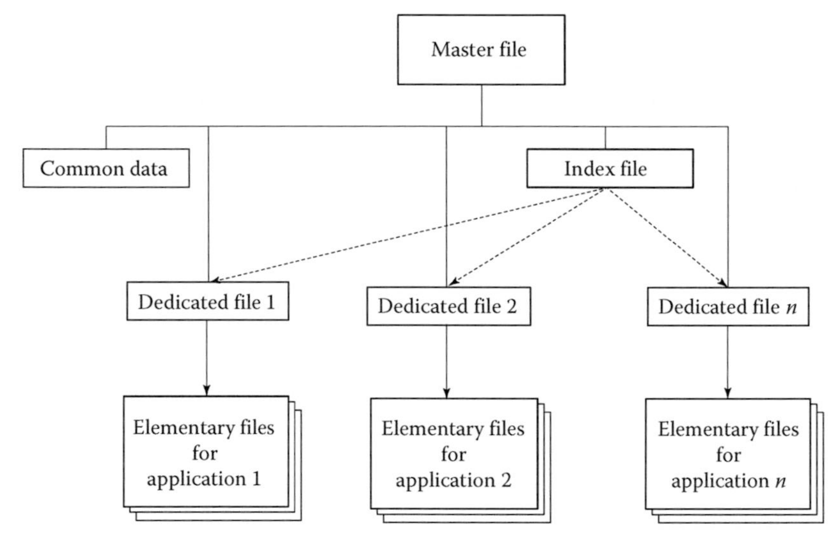

FIGURE 9.10
Several applications under the control of a central authority.

The central authority attributes a unique application identifier to each application and stores it in the index file. This file points to the dedicated file for each application. Any application has the right to read this file, but the write permissions are reserved to the central authority.

Before reactivating itself, an application has to authenticate itself to a common security module. Similarly, when an application A requests permission to access the data of another application B, the common security module has to authenticate it. This module next sends the authentication request to the application to determine whether application A owns the access privileges.

In this configuration, mechanisms have to be provided to allow multiple applications to share digital credentials such as keys and certificates effectively. Some of the important initiatives in this area is the format defined in PKCS #15, but many other aspects have to be standardized (Nyström, 1999).

9.4.1.3 Independent Multiapplications

In this case, there are no central authorities. The application identifiers are assigned in the chronological order of their addition and are maintained in an index file, whose access is protected, typically with a public key algorithm.

Each application supplier creates new applications according to the specifications of the card API and then distributes its product to the user directly or indirectly through an intermediary. One of the currently popular methods for managing the various applications is shown in Figure 9.11. It consists in superimposing a virtual machine on the card's operating system. This virtual machine is responsible for

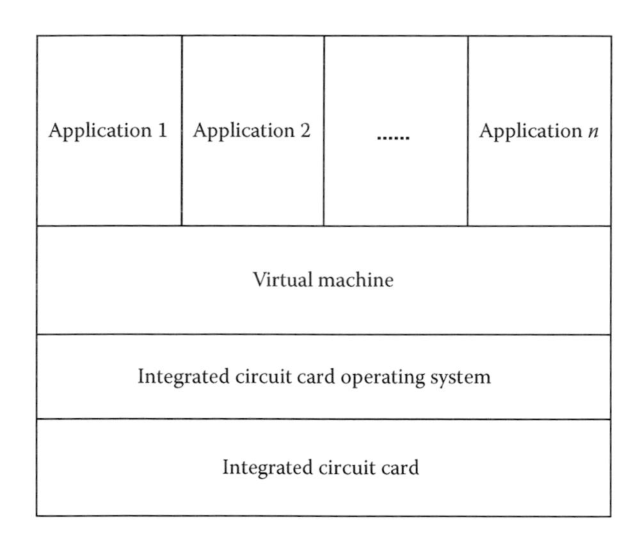

FIGURE 9.11
Management of applications with a virtual machine.

the control of the interactions among the applications as well as the communications of each one with the smart card.

9.4.2 Java Virtual Machine

The Java Card architecture specifies how to convert Java programs to run on integrated circuit cards. This facilitates the development of portable applications independent of the chip type or the underlying operating system. Exchanges with the peripheral equipment are conforming to ISO/IEC 7816.

The Java™ Platform Standard Edition, also known as Java™ 2 Platform Standard Edition or J2SE, is used to develop and deploy secure, portable applications. It has a security model known as the "sandbox" model, where untrusted code is confined to a restricted environment of *sandbox* while the trusted code is able to access all the system's resources. Security is enforced through a number of mechanisms such as automatic memory management, garbage collection, and range checking on strings and arrays. The Java Card inherits the main security features of the Java language and adds the following additional properties:

- Atomic transactions.
- A firewall mechanism can confine applets to run in their respective areas in the same card (i.e., *sandboxing*).
- Cryptography including random number generation, key and PIN management, digital signature, and user credential management.

Thus, the advantages of using a Java Card are as follows (Ghosh et al., 2005, p. 316):

- Multiple applications can coexist securely on a single card.
- New applications can be installed securely after the card has been issued.
- Programs developed for the Java Card can be run on any Java-compatible card.
- A secure object-sharing mechanism is available to support cooperative application on a single card.

To improve the smart card performance, the executable code is prepared off-card and then is executed on board in the virtual machine.

The security of a JVM depends on the separation of the various applications or *sandboxing* to separate their logic and offer individual protection.

9.5 Security of Integrated Circuit Cards

The security of the integrated circuit cards covers both the secret data stored in the card and the access rights to the service. The objectives of the security process are to prevent forgeries at all the stages of the production; to prevent the theft of the firmware for applications and security; to protect the stored information; and to detect and prevent any illegal or abusive usage. The protection must be given both during production and during the utilization.

9.5.1 Security during Production

The fabrication, personalization, and distribution of microprocessor cards take place in seven steps (Service Central de la Sécurité des Systèmes d'Information, 1997; Daaboul, 1998):

1. Design and development of the integrated circuits
2. Design and development of the card firmware
3. Fabrication of the silicon wafers
4. Insertion of the firmware, packaging of the integrated circuit, and final testing
5. Prepersonalization with addition of the programs related to the final use of the card and verification of their correct operation

TABLE 9.5

Access Conditions to the Memory of a Smart Card throughout Its Life Cycle

Phase in the Life Cycle	Fabrication	Prepersonalization	Personalization	Utilization	Invalidation
Access mode	Physical addressing		Logical addressing		
Operating system	Not accessible				
Fabrication data	Read, write, update	Read, but can be blocked			
Directories		Read, write, update			
Data		Read, write, update		Not accessible in most cards	
PIN		Read, write, update			

6. Personalization of the integrated circuit by adding the names of the issuer organization and of the holder and with addition of the application software
7. Issuance of the smart card on the plastic support with embossing, imprinting of the logos, and distribution of the card

Table 9.5 summarizes the access conditions to the memory of the smart card during the life cycle of the card.

The production of a smart card requires the intervention of the following participants:

- Designers of the integrated circuits and the security software
- Manufacturers of the integrated circuits and producers of the security software
- Certification authorities
- Developers of the application software
- Card producers (art designers, embossers, printers, etc.)
- Card issuers, which are the entities legally responsible for the card content and for the delivery of each card to its intended user

The security of a smart card has to take into account all the phases of production and transportation of the card among the various participants. A sufficient level of security must be ensured for each phase of the production cycle before passing to the next phase.

During the design phase, security relates to the protection of documents containing the requirements on the integrate circuit, the design documents, the software, the prepersonalization procedures, and the security of the production environment. The security policy has to establish an inventory of all possible threats as well as available countermeasures during fabrication and transport. Control of the production environment includes the material and tools used, the good and defective products, and the inventories. Risks include the disclosure of the requirements, the modification or theft of goods and materials, and the modification of the software including the microcode of the integrated circuit and the operating system.

Protection of the silicon wafers during their production is the work of the manufacturers of integrated circuits (ICs). During this phase, the chips are probed individually while they remain in the silicon wafers to verify the functioning of the microprocessor. The chips that are declared to be valid are blocked using symmetric encryption with a foundry key of 8–16 bits; the fabrication lot number and the number of the manufacturer are also engraved. The fabrication key is generated from a master key by using a *diversification algorithm*. Such an algorithm allows the derivation of the fabrication key from the master key using the card serial number. This method allows a central authority to authenticate all the lots without having to stock all the keys. At the end of the phase, the bad ICs are extracted from the wafer, and the verified ICs are delivered to the card manufacturer.

During prepersonalization, the card supplier cuts the wafer to separate the individual chips. These are then tested again before being molded under pressure into the milled plastic card. The logo of the application supplier is also put on the back of the card. Once it is installed, the chip is tested to verify the correct operation. The supplier unlocks the microprocessor using the foundry key, adds the operating system, etches the serial number of the card, and then blocks writing through direct memory access. From this point on, any communication with the memory will have to be through logical addressing to protect all the stored data from any modification or unauthorized access. Access to the card is then blocked again with the help of a prepersonalization key or a transport key that is associated with the issuer, before delivering the integrated circuit to the application supplier.

During personalization of the smart card, the application supplier records files related to the application on the card, in addition to the personal data of the cardholder such as the cardholder's identity and PIN as well as the unblocking keys. At the end of this phase, the cards are distributed to their holders. Depending on the commercial offer, the holder may be able to modify some of the personal parameters, such as the PIN. Details of the security during this phase will be presented later in this chapter.

Most smart cards keep a record of unsuccessful access attempts by entering an incorrect PIN. The counter is reset to 0 when a valid entry is made. When a predefined threshold is reached, access to the card or to a specific file is blocked. Some cards allow the user to choose the thresholds; in others, the threshold is predefined to a number from 3 to 7 (Ugon, 1989).

A card may be invalidated for several reasons:

- The expiration date of the card has been reached. In a monoapplication card, the expiration date of the card is the same as that of the application. However, for a multiapplication card, the card's expiration date is that of the master file. When an application expires, the operating system blocks all write and update operations, but read operations remain possible for analysis purposes.
- All available places in the memory zone reserved for the data of new transactions have been exhausted. It is still possible to read the card when the correct PIN is entered, but it is impossible to write new data. To avoid service interruption, a new card is necessary.
- The card can be invalidated by the issuing institution following fraudulent usage or a declaration of theft, which leads to a simultaneous blocking of the smart card PIN and of the unblocking key. A partial blockage, for example, blocking a specific PIN, will only affect the applications that use that PIN and can be unlocked by the owner of the application with the related unblocking key.

It should be noted that in 2015 it was revealed that the U.S. and British intelligence services were able to penetrate the internal computer network of Gemalto, the largest manufacturer of SIM cards in the world, planting malware on several computers that allowed them to steal the card encryption keys (Scahill and Begley, 2015).

9.5.2 Physical Security of the Card during Usage

Physically, the rectangular plastic support encloses elements is similar to magnetic stripe cards with the exception of the contacts for the microprocessor on the face of the card. A hologram is also available to increase security and make counterfeiting more difficult. The location of this hologram is the same in all countries.

Smart cards have a magnetic strip on the reverse side for interoperability with old readers. As for magnetic stripe cards, the elements of identification and verification of the PIN, the expiration date as well as the codes describing the user's privileges are recorded on magnetic tracks on the reverse side.

In general, smart card includes tamper-resistant circuits that inhibit output operations when a physical attack is detected. A dielectric layer offers passive protection of the integrated circuits from impurities, dust, and radiation. When this passive layer is violated, the integrated circuit may react to light, temperature, voltage, or frequency differentials.

Physical protection of the memory cells can be employed to prevent a selective erasure or to distribute the storage of sequential words in noncontiguous memory cells. Finally, there are special fuses to deactivate the test circuits that are used before their distribution of the cards.

9.5.3 Logical Security of the Card during Usage

Several measures assure the logical security during usage. In retail banking, the current standards for key management are ISO 11568 (parts 1, 2, and 4) and ISO 9564 for PIN management, including over open network.

ISO 9564 and ISO 16609:2012 specify the use of cryptographic operations within retail financial transactions for PIN encipherment and message authentication, respectively. The ISO 11568 series of standards is applicable to the management of the keys introduced by those standards. ISO 11568-1:2005 is applicable to the keys of both symmetric and asymmetric cipher systems used for protecting confidentiality, integrity, or authentication in retail financial services: Point-of-Sale Terminals, automated cash dispensing machine and automated transaction machine (ATM) transactions, and so on.

In face-to-face commerce, the identification of the card is done by the merchant using physical means (identity card of the holder, signature, etc.) or by calling an authorization server or using the holder's PIN, which is the electronic signature for cash withdrawal or for payments. In contrast, online authorization systems rely on cryptographic procedures to authenticate the participants (cardholder, card, and merchant terminal). The process consists of two phases: reciprocal authentication of the cardholder and the card and reciprocal authentication of the card and the network terminal. In case of offline verification, for example, in the South African system for paying electric consumption, the card contains the credit awarded to the cardholder, encrypted with a symmetric algorithm (Anderson and Bezuidenhoudt, 1996).

The first line of defense consists in the authentication of the card and the user. In case of online authorization, a secure logical channel is set next between the smart card and the host system through the reader. The establishment of this channel requires the reciprocal authentication of the card and the authorization server on the network. Time stamping of the transactions ensures nonrepudiation, which requires a precise clock with power supply ensured in all conditions. A second series of measures includes recording details of the transaction in an audit file and the counting of unsuccessful attempts to access the card, with blockage when the number exceeds a predetermined threshold. Finally, there is a period of validity of cards that is limited to reduce the possibility of cloning or attacks by replaying old messages.

The procedures for cryptographic authentication rely on algorithms for symmetric or asymmetric encryption as described later.

9.5.3.1 Authentication with Symmetric Encryption

The advantage of this mode of authentication is to avoid the need of a cryptographic coprocessor, and consequently, to reduce the cost of the card. The authentication exchanges take place in the following fashion (Hamman et al., 2001):

1. After insertion of the card in the slot of the machine, the card reader generates a random number and sends it to the card.
2. The card computes the message authentication code (MAC) of the concatenation of the random number and the card identification number (CID). The derivation of CID depends on the specifications of the system and depends on the chip serial number, the account number, a secret code, and the expiration date. The card sends the MAC and the CID to the authentication server.
3. Using CID, the server derives the card encryption key from the master key. It performs the inverse computations of that of the card and compares the result with the number received.
4. The result of the comparison defines the success or failure of the authentication.

9.5.3.2 Authentication with Public Key Encryption

Authentication with asymmetric encryption can be either static or dynamic. In static authentication, the data exchanged have been fixed once for all during the fabrication of the card. In contrast, the exchanges during dynamic authentication vary with each transaction that stops fraudsters from replaying past verification codes for authentication. Dynamic authentication is called offline because it involves the terminal and not the authorization center.

9.5.3.2.1 Static Authentication

The card signature is computed with the RSA public key algorithm and is stored in the chip. It is used to authenticate the card for each payment. This method is vulnerable to replay attacks because the constant cryptogram can be reused to make fraudulent payments. Also, it can be copied into blank cards allowing cloning.

9.5.3.2.2 Dynamic Authentication

In dynamic authentication, the cryptogram is generated anew for each transaction. Figures 9.3 through 9.12 show the various exchanges that take place during dynamic authentication:

1. The terminal sends a random number, RAND, to the card.
2. The card concatenates its identification number CID and the random number and calculate the digital signature of the concatenation with its private key SK_C. This signature is sent to the terminal accompanied with the card certificate signed by the corresponding certification authority (CA).
3. To ensure the card authenticity, the terminal decrypts the signature using the card's public key of the card, PK_C, as extracted from the certificate. It compares the results to the hash obtained directly with the hash function H().

This deterministic authentication method usually requires a cryptographic coprocessor to off-load the

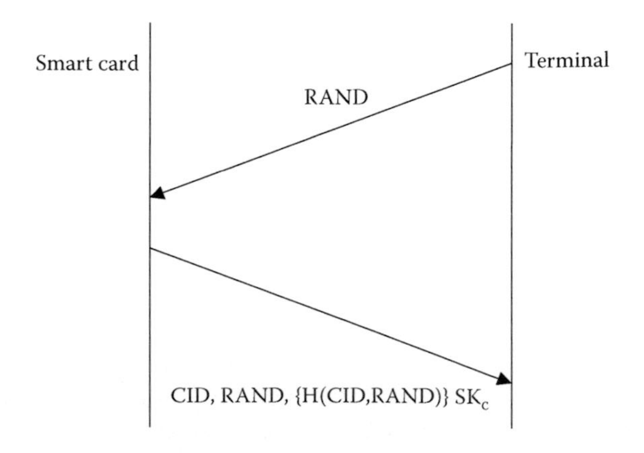

FIGURE 9.12
Message exchanges in dynamic authentication.

main microprocessor and speed up the calculations. This would increase the cost of smart cards. To overcome this constraint, Guillou and Quisquater proposed in 1988 a method for probabilistic authentication using zero-knowledge techniques. In this scheme, the authentication is interactive and consists of three exchanges: a commitment by the claimant, a challenge by the verifier, and a response by the verifier. The counterpart to the reduction in the computational load is an increase in the number of exchanges and the presence of a residual error. However, the verifier can reduce the error probability by requesting additional iterations.

The starting point is the following equation (Guillou et al., 1988):

$$G \times Q^e \equiv 1 \bmod n$$

where

G is a public number calculated from the card CID
Q is the signature of CID as computed by a banking authority with the RSA algorithm and its private key
(e, n) constitute the public key of the verifier and Q is the corresponding private key

The exchanges take place as follows:

1. The card sends to the verifier (the server or the card reader) the commitment t:

$$t = R^e \bmod n$$

with R a random number between 1 and $(n - 1)$

2. The verifier responds with a random challenge V between 0 and $(e - 1)$.
3. The card responds with T calculated as follows:

$$T = R\,Q^V \bmod n.$$

4. The verifier can now verify the authenticity of the card by reconstituting the commitment as follows:

$$\begin{aligned} &G^V T^e \bmod n \\ &= G^V R^e (Q^V)^e \bmod n \\ &= (GQ^e)^V R^e \bmod n \\ &= R^e \bmod n \end{aligned}$$

The exchanges are illustrated in Figure 9.13.

It should be noted that any legitimate claimant can finish each iteration with success without revealing the secret code Q. All what the verifier can derive is that the claimant has the necessary credential without being able to reconstitute its value. It can be shown that an impostor has only one chance in $(e - 1)$ to guess the response. In the case of $e = 2^{16} + 1$, there is one possibility of cheating in 65,536 attempts, which seems to be sufficient in banking applications (Guillou et al., 2001).

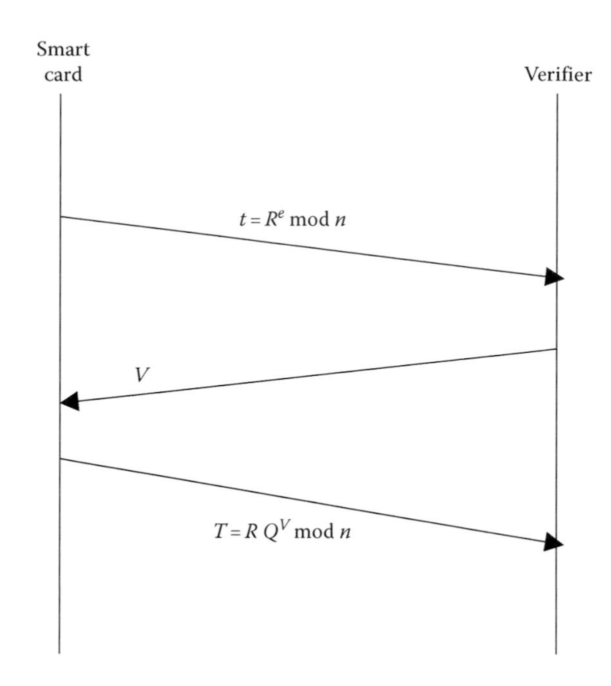

FIGURE 9.13
Interactive authentication with a zero-knowledge proof. (From Guillou, L.C. et al., *Ann. Télécommun.*, 43(9–10), 489, 1988.)

9.6 Payment Applications of Integrated Circuit Cards

This section presents some of the payment applications of integrated circuit cards starting with the historical smart card used by French banks. The EMV specifications will be presented in its own section. The integration of contactless cards with mobile devices is treated in Chapter 10. Electronic purses (e.g., FeliCa) will be discussed in Chapter 11.

9.6.1 Historical Smart Card of French Banks

French banks pioneered the use of integrated circuits in payment cards. From a security point of view, the user authentication was static authentication. The operation consists in verifying that the cubic power modulo n (where n a 320-bit public key of the banking authority) of an integer A gives another integer J called the *redundant identity of the card*. This is written as

$$A^3 \bmod n = J$$

Both A and J are 320-bit long. A is attributed during the personalization phase of the smart card, while J is derived from the card identity CID according to

$$J = (1 + 2^{160})\ \text{CID}$$

The length of CID is 160 bits, and it contains the following parameters in the clear: the serial number of the chip, the account number, the name of the cardholder, the PIN, and the expiration date (Guillou et al., 1988).

The card sends the values of CID and A to the terminal, which then verifies that A^3 and J conform to the relation indicated earlier. The protection comes from the difficulty of getting the cubic root (modulo n), which, according to number theory, requires factoring n into its prime factors. The complexity increases with the length of n.

Nevertheless, because each authentication reuses the same values, it is possible to deduce the value of the public key n as well as the correspondence between CID and card parameters by observing the exchanges between an authentic card, even when it is expired, and a payment terminal. After succeeding in factoring n and discovering the relationship between the card parameters and the value CID, one can select any arbitrary 16-digit value to be the card number as well other parameters from which the values CID and J can be computed. The cubic root of J modulo n will yield the value of A that will be sent to the payment terminal for verification. The counterfeit will fool a local verifier and will remain unnoticed as long as the authorization is carried out locally. Yet, French banks use semionline verification, that is, the authorization server is called once every 24 hours or when the transaction amount exceeds a surveillance threshold.

In 1998, Serge Humpich was able to break the static authentication system and clone a French smart card called the "Yescard." Following this affair, the French banking establishment decided to speed up their migration from the BO′ mask to a new standard called CB 5.2 based on the EMV specifications. During the transition, where BO′ and EMV coexisted, the length of the encryption key was increased to 768 bits. This transition ended in July 1, 2003, when all cards switched to the EMV environment.

9.6.2 Speedpass

ExxonMobil's Speedpass is a proprietary payment system for gasoline purchase at ExxonMobil stations introduced in the late 1990s (Meridien Research, 2001, pp. 25–26). It uses keychain fobs with an embedded radio-frequency transport linked to a credit or debit card account. The device communicates a unique identification code and a secure code to a payment terminal to initiate a payment, making it more secure than a magnetic stripe. Card numbers or other personal information are stored on the host server (i.e., in the cloud) and stored on the Speedpass device. Payment history is stored online for easy management.

9.6.3 Toll Collection Systems

RFID systems for automatic toll collection operate at any number of frequencies in the UHF range between 860 and 960 MHz (typically 900 MHz in North America) or 2.45–5.8 GHz. In many countries, 5.8 GHz is selected to achieve better transmission rates. These systems are also known as long-range systems or backscatter systems due to their physical principle (Persad et al., 2007; Finkenzeller, 2010, p. 22; Nowacki et al., 2011).

Electronic toll collection systems require highly available and reliable links, irrespective of weather and light conditions and a relatively high data transmission rate to ensure real-time identification.

Toll collection RFID systems are based on automatic vehicle identification at the tollgate using a radio-frequency transponder or tag (also called onboard unit) mounted on the windshield inside the vehicle. The onboard unit is a read-only tag that performs validation function. It has no display and cannot receive or transmit any messages. The road site unit is the interrogator/reader that starts querying the onboard unit at the indication of a sensor loop upon a vehicle's approach. The communication protocol between the onboard unit and the road site unit is denoted as a dedicated short-range communication protocol (DSRC). The back-end system consists of a large subscriber database that contains the information on all the enrolled users and their vehicles.

We now discuss the payment architecture of RFID toll collection systems.

9.6.3.1 Subscription

At subscription, the vehicle owner gives information on the vehicle to be stored in a database: make, model, year, color, license plate number, name and contacts of owner, and so on. This input will be stored in the systems database. The subscriber has to specify a payment instrument to pay for the tolls (cash, checks, debit card, credit card, or debit from a current account). If the subscription is approved based on these data, a virtual purse will be set up for the user.

9.6.3.2 Virtual Purse

In the normal case, toll collection is a micropayment system, often to replace the use of special tokens. Much larger amounts are involved in case of toll violations.

With the first violation notice, a large penalty is added to the toll amount, with further additional penalties for later notices.

To reduce the operation costs, micropayment systems often resort to service prepayment (which allows them to gain interest on the prepaid amounts) and grouping of micropayments before financial clearance. However, whether to the service is prepaid depends on the operator's policies and the payment instruments that the subscriber has chosen.

In prepayment systems, a virtual purse stores the prepaid monetary values. This virtual purse is a bank account supplied with legal tender using a credit or debit card, a direct debit or a fund transfer. The vehicle identification number is linked to that account against which the appropriate toll is charged. The back-end system deduces the toll amount from the virtual purse either on a per transaction basis or by combining several charges together. When the total monetary stored drops below a certain threshold, the purse is automatically replenished. The value of that threshold depends on the instrument used; it is usually lower for direct debits or credit transfers than for cash or checks. In postpayment systems, the vehicles have to be preregistered with the operator. In either case, penalty fees are added if an invoice remains unpaid beyond a certain limit of time.

9.6.3.3 Security

Because of the limited capabilities of the transponder, the exchange between the onboard unit (tag) and the roadside unit (reader) is not encrypted. Authorization and authentication are not supported except with certain types of RFID tags, but there are cost/benefit tradeoffs. There is no nonrepudiation mechanism because picture taking of the license plates is triggered by toll violations only. Thus, users rely on the goodwill of the operator in resolving billing disputes. It is not known, however, that RFID readers and tags do not have 100% accuracy in all instances and that environmental factors affect that accuracy.

One possible source of error in the system is the delay in updating records of the subscriber's financial instruments, such as a new credit card number in lieu of a comprised or an expired card. Defective transponders are another source of errors.

A potential extension of RFID technology would combine ticketing and toll collection, that is, individual identification and vehicular identification. This would be based on a multiapplication (universal) tag for public transit systems, for example, subways, commuter trains, and buses in addition to highways. This extension, however, would merge and compete with plastic contactless cards or NFC technology. It would also require the use of more capable onboard units to allow communication with the server and with other vehicles, to authenticate these other parties, as well as to display the received information. However, such developments would require much higher security mechanisms. A number of RFID security models have been suggested such as the RFID attributes that RSA Security, Inc., has developed, which include tag privacy, tag authenticity, reader security, and tag database security.

9.6.3.4 Interoperability

Lack of interoperability is a nuisance for any driver going through different regions and prevents the establishment of area-wide tolling programs. A driver would require a separate tag for each system as the vehicle passes through several tolling systems.

OmniAir Certification Services is a not-for-profit organization set by the toll operators, their suppliers, and systems integrators to promote the adoption of ISO/IEC 18000-6C. The organization provides testing and certification services to enable the deployment of interoperable equipment and applications.

9.7 EMV® Card

The EMV specifications started as the collaborative output of EuroPay, MasterCard, and Visa. Since February 1999, the work is overseen by EMVCo, an organization registered in the state of New York and that comprises six member organizations—American Express, Discover, JCB, MasterCard, UnionPay, and Visa.

The first EMV specifications were published in 1995 as EMV 2.0. They were upgraded to EMV 3.0 in 1996, a version also known as EMV '96; with later amendments, it became known as EMV 3.1.1 in 1998. This was further amended to version 4.0 in December 2000 (sometimes referred to as EMV2000). The current version is EMV 4.3, which became effective in November 2011. The first implementations based on EMV took place in January 2002.

Compliance to the EMV specifications is verified for terminals, software, and cards. Level 1 tests verify compliance of terminals with the electromechanical characteristics, logical interface, and transmission protocol requirements. Level 2 tests concern the software compliance with the debit/credit application requirements.

Card compliance is verified at the mechanical and functional levels as well as for the secure storage capability of the chip and the minimum security designed to withstand known attacks. The final type of approval process concerns contactless mobile payments and mobile applications.

In Europe, MasterCard and Visa completed the transition from magnetic stripe cards to EMV smart cards on January 2005. Beyond this date, banks that have made the transition are considered liable for fraud recorded on payments made with cards that they have issued. Banks have upgraded their automated teller machines to accept EMV chip cards. This migration went faster in England than in France because the rate of loss due to fraud in the former is five times the rate in the latter (Buliard, 2001).

EMV is a complex protocol with many options and several revisions (1996, 2000, 2008, 2011, etc.), documented mostly in natural language. The following is an attempt to present a coherent picture of the protocol, which is scattered through many books, but the specifications themselves are the final arbiter. Fortunately, a formal model for card and terminal is available, as well as an open-source Java Card implementation of the EMV standard at http://sourceforge.net/projects/openemv (de Ruiter and Poll, 2011).

9.7.1 EMV Cryptography

EMV uses two symmetric encryption algorithms: the Triple Data Encryption Standard (TDES) with the MAC as standardized in ISO 16609 and the Advanced Encryption Standard (AES) using a key length of 128 bits, 192 bits, or 256 bits (EMV, 2010). Both ciphers are specified in ISO/IEC 18033-3. The values for the public key exponent are 3 and $(2^{16}+1)$, which is known to speed the encryption (Menezes et al., 1997, p. 291). The moduli of the public keys of the certification authority, the issuer, the card, and the PIN encryption cannot exceed 248 octets. Hashing is done with the Secure Hash Algorithm (SHA-1) standardized in ISO/IEC 10118-3 (EMV, 2011, Book 2, Annexes B and C).

The format of the EMV public key certificates is defined in Sections 5 and 6 of Book 2 of the EMV Specification. They are more compact than X.509 certificates and are created using the ISO/IEC 9796-2 digital signature algorithm.

The security of an EMV transaction is based on the chip of the card. If the transaction is completed online, the issuer authenticates the card. If it is completed offline, it is the terminal that authenticates the card. EMV is used with contact interfaces and for contactless transactions, but the security approach is difference. The use of EMV for contactless interfaces is the subject of Chapter 10.

To ensure backward compatibility with the legacy terminals that can only read magnetic stripes, EMV specifications allow the fallback to cardholder authentication using the data on the magnetic stripe or using the cardholder's signature. To do this, bypass of PIN entry is allowed at the terminal at the discretion of either the merchant or the cardholder. Unfortunately, this feature increases the vulnerability of the EMV scheme to fraud.

Offline authentication uses one of three data authentication modes:

1. Offline static data authentication (SDA).
2. Dynamic data authentication (DDA). This is the mode that, in 2011, Visa and MasterCard mandated on all their branded EMV cards.
3. Combined dynamic data authentication (CDA) with application cryptogram generation. This mode was added in version 4.0 for both offline and online transactions.

Only one method of authentication can be performed during a given transaction.

9.7.1.1 Static Data Authentication

The terminal authenticates the card using a digital signature scheme based on public key encryption to confirm the integrity of the data stored during the card personalization. For each application, the card contains the digital signature of critical data computed with the private key of the issuer as well as the certificate of the issuer delivered by the certification authority. The terminal stores the issuer public key and authenticates the signed static application data with that public key. A terminal must be able to recognize the public keys of the six EMVCo member organizations per registered application.

Because the card does not possess cryptographic capabilities, the PIN that the user enters is sent to the card for verification in the clear, that is, unencrypted. The card records success or failure and decrements the PIN retry counter and blocks itself if it becomes negative.

SDA allows the detection of unauthorized alternations to the critical data residing on the card. The main advantage is that cards do not have to support cryptographic processing, that is, they become less costly. There are two major problems with static data authentication:

1. Static data authentication is vulnerable to replay attacks in offline transactions.
2. Also, their cryptogram can be copied to a counterfeit card (card cloning).

Attack with cloned card works as follows. A fake card inserted into an offline point-of-sale terminal will fool the terminal to authorize the transaction since the copied critical data are authentic, including the issuer certification from the brand certification authority, and all data for the payment applications are signed with the issuer private key. According to EMV, the conforming card performs PIN for offline transactions, so a stolen SDA card can be created to accept any PIN entered at the terminal. In other words, a "Yescard" compatible with EMV specifications can be created. Possible solutions are to use online authorization (perhaps through a Wi-Fi connection of the point-of-sale terminal) and/or switch to dynamic data authentication. If the amounts involved are small, however, it may be more economical (i.e., cheaper cards and terminals and non-real-time connections) to just cover the losses.

9.7.1.2 Dynamic Data Authentication

Dynamic data are unpredictable and transaction dependent. A card that supports DDA must have its own signature key pair and the capability to generate digital signatures. A card is personalized with the issuer public key certificate, its own public key certificate, and private key. The terminal (automated teller machine or point of sale) uses a stored copy of the public key of the certification authority of the card brand to verify the issuer public key certificate and to verify the issuer-signed certificate for the card public key. As in the previous case, each terminal must recognize the public key of six certification authorities per registered application.

The terminal authenticates the inserted card through a public key–based challenge and response and then verifies the integrity of the data residing on that card. To avoid replay attacks, the card receives from the terminal a random number that will be concatenated to the authentication-related data (ARD) indicated in the EMV specifications. Next, it signs the whole set with its private signature key SK_{ICC}. This signature is sent to the terminal with the card's certificate from the issuer and the issuer's certificate from the certification authority. In EMV parlance, the response is called Signed Dynamic Application Data (SDAD). Thus,

$$SDAD_{DDA} = \{ARD_1, H(ARD_1, nonce_{ICC})\}\ SK_{ICC}$$

where

ARD_1 represents some authentication-related data
$nonce_{ICC}$ is the nonce that the integrated circuit card generates
SK_{ICC} is the card's private signature key
H() represents the Secure Hash Algorithm (SHA-1)

Once the terminal verifies the validity of both certificates and the integrity of the received signature, it can confirm that the card is authentic.

In summary, the following steps are used in the DDA authentication:

1. The public key of the CA is retrieved from the terminal storage.
2. The CA public key is used to verify issuer certificate, which contains the issuer public key.
3. The card public key is extracted from its certificate from the issuer.
4. The signature is verified using the card's public signature key.

Clearly, the card must be able to perform some cryptographic computations. Therefore, it is possible to use PIN encryption during PIN verification. If that option is exercised, once the cardholder enters the PIN, the terminal encrypts that PIN with the card's public encryption key and sends it to the card. The card reports success or failure and decrements the PIN retry counter.

This mode of operation allows the verification that the card is genuine and that the data it contains have not been tampered with. Yet, the authentication is only between the card and the terminal because the issuer authorization has not been secured.

DDA allows the authentication of the card and the verification of the data exchanged, the card does not sign terminal-dependent data dynamically. Thus, after card authentication, DDA does not tie the subsequent exchanges concerning transaction authorization to that card. In other words, the terminal cannot verify that the transaction was actually carried out by the card it has authenticated. Thus, it is possible to first go through the authentication process and then respond to the terminal messages with a different card. This opens the possibility to mount an attack in which the genuine chip does the dynamic authentication while the fake chip performs the rest of the transaction. This attack is particularly easy in unattended situations.

9.7.1.3 Combined Dynamic Data Authentication

This mode was introduced with EMV 4.0 (EMV2000) as an enhancement to DDA. With CDA, the card digitally signs the data at key transaction points with its private signature key SK_{ICC} for the terminal to verify its integrity. This effectively closes one of the weaknesses of DDA and provides valid evidence on the transaction in case of a dispute. During online authorization, the issuer's authorization will be requested as well.

CDA is associated with two authentication cryptograms (AC):

- The Authorization Request Cryptogram (ARQC), which can also be sent to the issuer online to request the authorization of the transaction
- The Transaction Certificate (TC) at transaction completion

A third cryptogram, Application Authentication Cryptogram (AAC), is sent when the card refuses or aborts the transaction.

Each cryptogram consists of a MAC generated over the data it references using DES/3DES in the CBC mode. With CDA, SDAD is a signature on the card-generated nonce and other authenticated related data in addition to the terminal generated nonce. This can be represented as

$$SDAD_{CDA} = \{nonce_{ICC}, ARD_2, H(nonce_{ICC}, ARD_2, nonce_{Terminal})\} SK_{ICC}$$

where

$nonce_{ICC}$ is the nonce that the card generates

$nonce_{Terminal}$ is the nonce that the terminal generates

ARD_2 comprises data related to the authentication cryptogram, such as its type (ARQC or TC), its value, and other parameters used in the CDA calculations, as specified in the EMV specifications.

Application cryptograms are also used for online authentication.

9.7.2 EMV Operation

An EMV session comprises three phases: card authentication by the terminal, cardholder verification, and transaction authorization. Transaction authorization can be done at the terminal level (offline authentication) or at the issuer's authorization center (online authentication).

During card authentication by terminal, the terminal determines first whether the card has a chip, typically, by reading the magnetic stripe (the assumption is that many cards still do not have a chip). If it does not have a chip, then the fallback mode is either to proceed with a magnetic stripe transaction or to stop the transaction.

With integrated circuit cards, the terminal requests a list of all the applications that the card supports. The application indicates the networks that the card can use, the programs and the application it supports. If there is at least one match, then other processing restrictions are evaluated, for example, incompatibilities among the protocol versions of the card and the server, geographic restrictions, or limits that the issuer or acquirer imposes.

During cardholder authentication, the PIN that the cardholder enters is compared to the one stored in the card. The authentication method is defined in the cardholder verification method (CVM). The CVM lists in priority order the types of cardholder verification methods that the card issuer accepts. Typical verification methods are a handwritten signature, a PIN without encryption, an encrypted PIN (both offline or online), a combination of a PIN, and a signature or no verification at all. The CVM defines as well as the response if the cardholder verification fails, such as trying a lower priority method or the transaction aborted. Online PIN verification would require contacting the issuer while offline verification is at the terminal level only. From EMV 4.0 onward, offline PIN verification can be done either with plain text exchanges or with encrypted exchanges.

Notes:

1. If the CVM is not part of the signed data, it can be altered. A fraudster, who has stolen a card but does not know the PIN, can make the terminal fall back to the magnetic stripe. With the cardholder data from the magnetic stripe and after recording the PIN, a counterfeit card can be created.
2. PIN verification is not authenticated.

Table 9.6 contains an alphabetical listing of the messages used in EMV exchanges that are discussed in this chapter.

Transaction authorization concerns various techniques for risk management. According to the level of risk, certain transactions can be authorized offline, that is, by the terminal, or online, by the issuer itself. Once the transaction is completed, the chip executes commands that the issuer has sent to update its data.

9.7.2.1 Offline Authorization

Figure 9.14 shows the main exchanges of EMV during offline authorization of a transaction. The first four messages implement the dynamic data authentication of the card. The next two are associated with cardholder verification with a PIN. The final set of messages relate to transaction authorization.

9.7.2.1.1 Card Authentication

The terminal starts the transaction by sending to the card the GET_PROCESSING_OPTIONS command followed by the READ_RECORD command. In response, the card returns the identity of its certification authority CA, its public key certificate, and that of the issuer. The terminal needs both certificates to authenticate

TABLE 9.6

Alphabetical Listing of the EMV Messages Mentioned in the Text

Variable	Name	Comments
ARD	Authentication-related data	Includes date, time, card identifier, PAN, terminal identifier, etc., as specified by the EMV specifications.
ARPC	Authorization response cryptogram	An optional response to an ARPC from the issuer confirming the validity of the chip. The card can then reset its counters.
ARQC	Authorization request cryptogram	A request for the issuer's confirmation that the chip is valid (i.e., it has not be en copied or altered).
CA	Identifier of the certification authority	
CERT_C	Public key certificate of the integrated circuit card	Certification by the issuer.
CERT_I	Issuer public key certificate	Certification by the certification authority.
IAD	Issuer authentication data	Proprietary application data related to the issuer in an online transaction.
PAN	Private account number	User's account inscribed on the card.
RCP	Reference control parameter	User preferences and whether the transaction is online or offline.
SIGN_C (ARD)	Signature of the card on ARD	Uses the card's private key.
TC	Transaction certificate	Verification of the authenticity and integrity of the transaction data.
TD	Transaction data	Amount, currency, date, time, PAN, terminal identifier, *nonce2*.

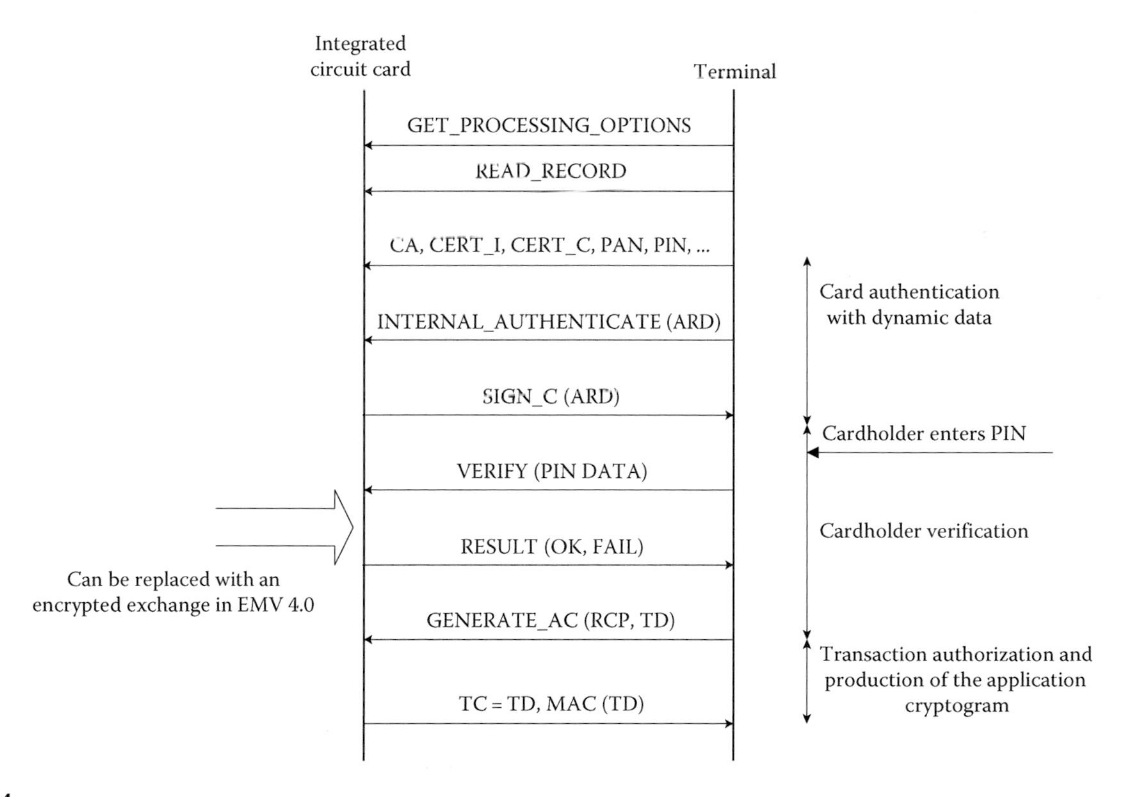

FIGURE 9.14
Exchanges in an EMV transaction with dynamic authentication and generation of the application cryptogram (offline case).

the card's public key stored in the card's public key certificate CERT_C. Other parameters concern various elements such as the primary account number (PAN) associated with the card, the expiry date, and control parameters such as the CVM. CVM indicates whether the cardholder's verification will be conducted online or offline using a PIN, a handwritten signature, or nothing at all. Both certificates are needed for the terminal to authenticate the card's public key stored in CERT_C.

The chip card is typically programmed for one more applications, each with an Application ID. Assuming that there is a match in at least one Application ID, the terminal issues an INTERNAL_AUTHENTICATE command with the authentication-related data ARD.

A dynamic data authentication data object list (DDOL) defines the ARD, which include the date and time of the transaction, the card identifier, the PAN, the terminal identifier, and a nonce that the terminal generates to avoid replay attacks. The DDOL may be supplied by the card in its initial response, otherwise a default DDOL is used.

The card responds to this challenge with its own dynamic data (i.e., the nonce that it generates) and protects the integrity of all the elements with a cryptographic signature using its private signature key. The signature covers the nonce from the card and the hash of that nonce and the ARD. All these exchanges constitute the dynamic data authentication of the card.

9.7.2.1.2 Cardholder Verification

Cardholder verification is the next step. The terminal transmits the VERIFY command to the card to compare the entered PIN with the PIN that it stores for the application and confirm whether it is correct. If the comparison fails, the response includes the number of remaining PIN verification attempts. When this number is reached, the card locks itself from further use.

If the terminal does not support the EMV specifications or it does not have a chip, the fallback is to use the information on the magnetic stripe or to allow an attendant to compare visually the signature on the back of the card with the signature on a paper receipt. This fallback mode, however, opens the way for some attacks as will be explained later.

PIN verification can be conducted offline, that is, at the terminal or online, in which case the terminal encrypts the PIN and sends it to the issuer over a payment network. In addition, EMV 4.0 (EMV2000) added two exchanges illustrated in Figure 9.15 to allow encryption of the PIN as it transits from the terminal to the card, provided that the card supports public key encryption. With the command *GET CHALLENGE*, the terminal requests that the card sends a random number *RAND1*. The terminal generates another random number *RAND2* and concatenates both numbers to the PIN that it receives. The resultant is encrypted with the card public key PK_C and sent in the *VERIFY* command. The card will then decode the cipher with its secret key SK_C to verify that the random number RAND1 is the one that it has sent and that the PIN matches the value that it has. Also, in EMV 4.0, the transaction TD must include another random number to prevent replays.

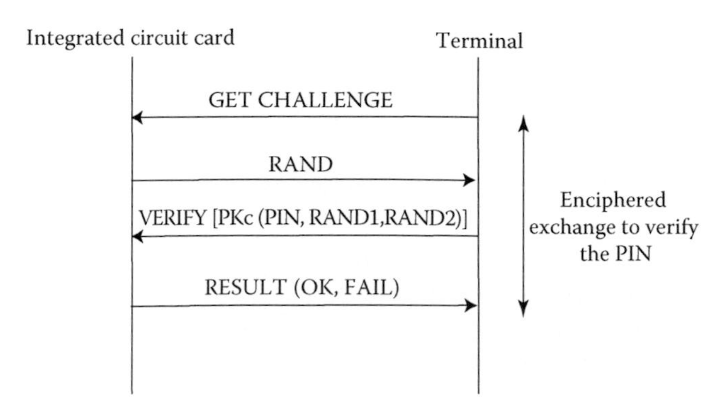

FIGURE 9.15
Exchanges to verify the PIN in EMV 4.0 (offline case).

In the case of "OK," the card response is 0X9000, while in the case of failure it is 0X63C*n*, where *n* is the number of remaining PIN verification attempts before the card locks up. Note that the result from the card is not authenticated, even though it is encrypted.

In principle, it is possible to insert a flexible circuit board that lodges between the reader and the card's contact to capture the signal and transmit it to a nearby receiver. If the PIN is unencrypted, then it can be retrieved. Furthermore, a card advertises that it supports encrypted PIN verification by placing an appropriate entry in the CVM list. If the list is not signed, however, it can be modified to prevent encryption so that the terminal will continue to send the PIN in the clear, allowing its capture (Drimer et al., 2009).

9.7.2.1.3 Transaction Authorization

The terminal uses the GENERATE_AC command to ask the card to computer one of the cryptograms. The arguments of the command tell the card which cryptogram to produce and the type of authentication method (DDA or CDA). They also include the Reference Control Parameter (RCP) and the transaction data (TD). In this case, the GENERATE_AC command indicates that the PIN verification is conducted offline. The card responds with the Transaction Certificate, which contains TD and the application cryptogram requested. TC is passed to the acquirer and used to claim payment during the clearing process. The cryptogram is computed using the card's private signature key and covers the data exchanges including the random number received from the terminal as well as the TC. The EMV specifications indicate that the MAC is computed using DES/3DES in the CBC mode, with a session key derived from a master key shared with the issuer. Often, however, the card computes the MAC with an asymmetric encryption algorithm (RSA) in the case of CDA. TC indicates that the card is allowing the transaction to proceed and will be forwarded later to the issuer (the cardholder's bank) for auditing purposes.

9.7.2.2 Online Authorization

The exchanges during an online dynamic authentication of the card with generation of the application cryptogram are depicted in Figure 9.16.

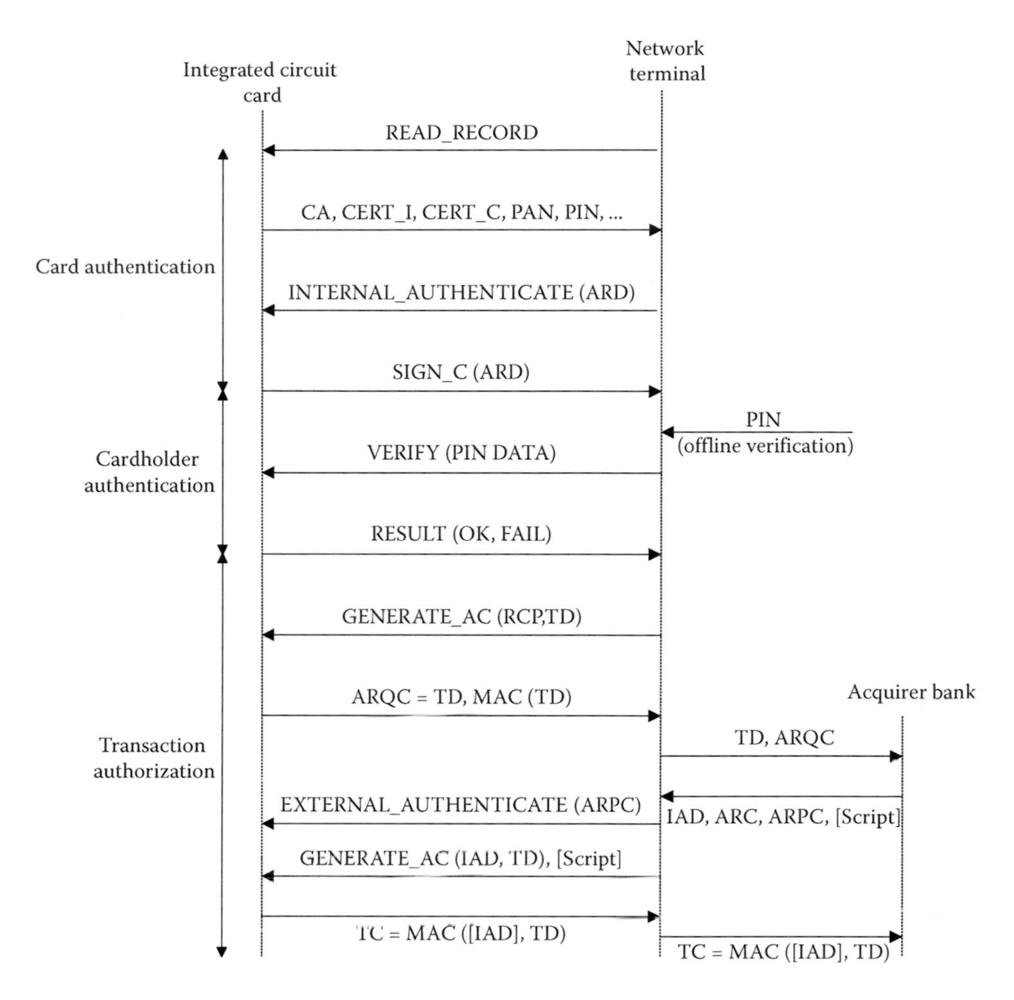

FIGURE 9.16
EMV exchanges for card authentication and cardholder verification (optional fields are enclosed in square brackets []).

The difference between online authentication and offline authentication shows in the response to the GENERATE_AC request. In the online case, the card produces an Authorization Request Cryptogram (ARQC) to be passed to the acquirer (the merchant bank) along with the transaction data (TD). The acquirer forwards the data to the issuer for validation. The ARQC includes information on the terminal, the application, the transaction (currency, amount, date), a sequence number that identifies the transaction, a nonce that the card has generated, and a MAC to protect the integrity of critical transaction data (TD) (EMV, 2011, Book 2).

In the case of SDA/DDA, the card computes the MAC with a symmetric encryption algorithm (DES/3DES in the CBC mode) with a session key derived from a master key shared with the issuing bank. The MAC is calculated with an asymmetric signature (RSA) in the case of CDA.

The Authorization Request Cryptogram (ARQC) is unique to the transaction and is the only information that the retailer has and record when the card is used. Thus, the merchant's cannot link to an account number. This is an advantage over magnetic stripe cards where the information that the merchant has includes the cardholder's full name, the primary account number, and the card's expiration date.

Note: ARQC and TC indicate whether the cardholder verification was attempted and failed but not if the verification succeeded or if it was not attempted at all. Thus, allow a man-in-the-middle attack where the cardholder PIN verification is suppressed, while the ARQC and TC are allowed between the legitimate card and the terminal. In this case, the terminal believes that PIN verification was successful while the card assumes that it was not attempted (Murdoch et al., 2010).

If the issuer accepts the transaction, it returns an authorization response message that contains the proprietary Issuer Authentication Data (IAD), a 2-octet authorization response code (ARC), and, optionally, an authorization response cryptogram (ARPC), which is typically a MAC over ARQC ⊕ ARC. The ARPC message indicates that the issuer has validated the chip so that the card could reset all its counters. Both items are forwarded from the issuer to the acquirer and then to the terminal. Optionally, the terminal can pass it to the card with the (optional) *EXTERNAL_AUTHENTICATE* command and/or a second *GENERATE_AC* command. That second *GENERATE_AC* command contains the IAD and an optional script to request a Transaction Certificate (TC) and its cryptogram.

The chip attempts to verify the ARPC. By doing so, the chip can ascertain that the response came from the issuer, that is, it was not fraudulently introduced during the transition of the response. The card responds with a message that contains the Transaction Certificate (TC) and its MAC to indicate approval of the transaction. The issuer can verify later the integrity of the TC and the authenticity of its origin.

9.7.3 EMV Limitations

EMV is currently the main protocol used for smart card payments worldwide. The specifications were developed for point-of-sale applications, where the EMV terminal is under the control of the merchant and, indirectly, the acquirer. In these situations, the transaction is face to face and the terminal and the bank communicate over a secure channel. This has the following implications:

- Because the purchase is face to face and the goods are delivered to the purchaser immediately, the protocol does not address merchant authentication.
- SDA does not protect from card cloning.
- In face-to-face situations, the physical insertion of the card into the terminal can be visually ascertained. Thus, once it authenticates the card, the terminal does not need to check the computations that the card performs. As a consequence, the protocol does not link different parts of the transaction to verify that the same card is used throughout. CDA avoids this problem that can arise particularly in the case of unattended terminals.
- Similarly, there is no provision for the card to authenticate the terminal, and the terminal does not explicitly authorize or sign any part of the transaction. CDA also avoids partially that problem.
- Because the terminal and the issuer trust each other, the terminal does not verify the data that it receives from the issuer and the issuer trusts the terminal to deliver the messages to the card.

Based on the aforementioned considerations, the EMV specifications are not suitable when the aforementioned assumptions are not fulfilled, for example, as for remote transactions over nonsecure networks such as the Internet or a mobile network (Van Herreweghen and Wille, 1999). As explained in the next section, tokenization was proposed in 2014 to resolve that gap.

9.7.4 EMV Tokenization

Payment tokens are specific *Jetons* stored in EMV chip cards or NFC devices to secure EMV transactions over the Internet, for mobile payments and, in general, for all card-not-present transactions (EMV, 2014). They are used as part of the payment chain and, when submitted in a transaction to the payment system, would cause a payment to occur. However, they are restricted to specific domains, such as a specific merchant or a specific digital wallet operator, that is, they are not used as a general payment instrument. As such, they target electronic commerce merchants and digital wallet operators.

EMV tokens replace the PAN, that is, the credit or debit card numbers in all the processing of the relevant payment transactions. Further, they can be securely mapped back to the original card account number by the provider of the payment token and authorized entities only. An additional capability is the ability to unlink the token from the original card account number in case that token is either no longer needed, or a mobile device or card has been lost or stolen.

The payment token has the same characteristics of a valid PAN, that is, it is a 13- to 19-digit numeric value that satisfies all the validity rules and checks that a PAN must satisfy, including the verification code called "Luhn's key."

However, some additional rules apply for additional security:

1. The payment token should correspond to a valid PAN.
2. It should have an expiration date.
3. The number of a payment token should be different for the number of any valid PAN.
4. When a token is requested, some identification and validation checks need to be performed. Based on these checks, a token assurance level is determined to indicate the confidence level that the PAN is indeed that of the cardholder.
5. The payment token can be limited to a specific channel, for example, for NFC only, or for a given merchant or a mobile wallet, or a combination of any of these.
6. When a token is requested, the requestor must indicate where the storage will be stored (e.g., a secured hardware storage or a consumer device).
7. A set of parameters are established during token issuance to enforce the appropriate usage of the payment token in payment transactions.

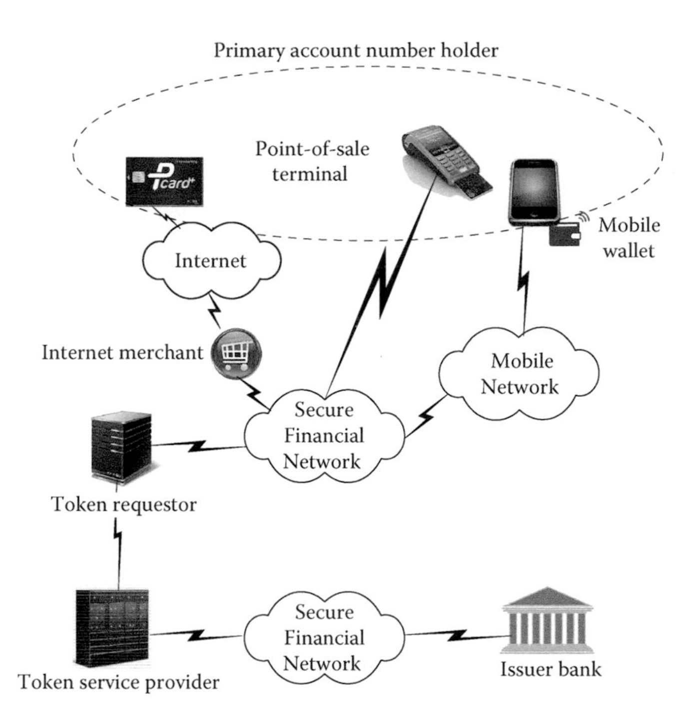

FIGURE 9.17
Flow during the request and issuance of a payment token. (From EMV, Payment tokenisation specification, Technical Framework, Version 1.0, March 2014, http://www.emvco.com/download_agreement.aspx?id=945, last accessed November 7, 2014.)

Two new intermediary roles are introduced, the token service requestor and the token service provider as shown in Figure 9.17, which displays the flow during the request and issuance of a payment token. As shown in Figure 9.17, the request may come from an online merchant, a point-of-sale terminal of the digital wallet in a smartphone. Other possible token requestors include the card issuers, the chip manufacturers for a smart card, the payment gateway on behalf of merchants, the acquirers, and so on. For example, an online merchant that has payment card data on file may seek to avoid the responsibility of protecting card data by replacing the PANs with payment tokens. In any case, the token requestor will have to register with a token service provider. According to the current specifications, each token service provider establishes its proprietary processes for the collection of identity credentials, review and approval of the payment token request. If it verifies and approves the request, the token service providers store the issued payment tokens and a secure depository together with the other relevant parameters such as the expiry date, location, domain restrictions, as well as the corresponding PANs.

During a transaction, the token service provider plays the role of the issuer in an EMV transaction from a bank terminal, that is, it approves the transaction on behalf of the issuer using the token instead of the PAN. It then maps the token to PAN and forward the information to the issuer to transfer the funds to the acquirer. If the token service provider is not the issuer, the end-to-end view that EMV had achieved is now lost.

The payment tokenization specification introduces mandatory and optional data elements in the protocol exchanges to establish payment tokens—*Token Request* and *Response to Token Request*—and to update the assurance level assigned to a payment token after issuance: *Token Assurance Level Update Request,* and *Response to Token Assurance Level Update Request.* Exchanging the payment token to obtain the original PAN and PAN expiry date is called de-tokenization and can be conducted without verification. With verification, the process involves checks on the validity of the payment token and enforcement of the domain restrictions controls associated with the payment token. De-tokenization involves two additional exchanges: *De-tokenization Query Request* and *Response to De-tokenization Query Request.* With verification, the exchanges are called *De-tokenization with Verification Request* and *Response to De-tokenization Query Request.*

However, the exchanges have not been defined to the level that is needed to ensure interoperable implementations and the language used in the specification is ambiguous (Corella and Lewison, 2014). For example, tokenization introduces another cryptogram in addition to the Authorization Request Cryptogram (ARQC) that is passed from the terminal to the issuer in an EMV transaction, as described earlier. According to the specification, the Token Requester generates this new cryptogram to validate authorized uses of the token and inserts it in the transaction message based on the type of transaction and the associated usage. However, the location of such cryptogram has not been standardized as of November 2014. Also, the protocol exchanges between the token service provider and the issuer have not been defined.

Finally, one alternative to tokenization is for issuers who can achieve the same result by provisioning a secondary account number to be used for online transactions (Corella and Lewison, 2014).

9.7.5 Other Attacks on EMV

This section describes two classes of attacks: the first exploit backward compatibility with legacy terminals and the second are relay attacks.

9.7.5.1 Attacks Due to Backward Compatibility

As mentioned earlier, the fallback position is to read the necessary data from the magnetic stripe or to compare the user's signature with the signature on the back of the card. In this type of attacks, the EMV reader is compromised to force the card to use the magnetic stripe and then capture surreptitiously the exchanges between terminals and legitimate cards during a transaction. These exchanges are then analyzed to extract the PIN and enough data to reconstruct a magnetic stripe card that could be used in a fake card.

9.7.5.2 Man-in-the-Middle Attacks

A variation of the man-in-the-middle attack is to use a programmable electronic circuit to connect a legitimate (stolen) card and a fake card inserted into the point-of-sale terminal. The fraudster does not need to know the card's PIN. During PIN verification, the circuitry intercepts the VERIFY command to trick the card into believing that the terminal does not support PIN verification and has either opted for signature verification or skipped cardholder verification altogether. Any PIN can be entered, but the dummy PIN is intercepted before it reaches the card. The terminal receives from the intermediary device the code 0X9000 as a confirmation that the PIN was verified correctly. Similarly, the issuer remains unaware because, according to the protocol, the terminal only reports failed attempts for PIN verification in the Terminal Verification Results data structure. As a result, with this contraption, any 4-digit code will do and the transactions considered normal and accepted by the bank network (Murdoch et al., 2010). The protocol exchanges in this case are shown in Figure 9.18.

The dummy PIN does not reach the card, so the PIN retry counter is unchanged. As mentioned earlier, TC—and for online approval ARQC—indicates whether the cardholder verification was attempted and failed but not if the verification succeeded or if it was not attempted at all. Thus, the terminal believes that PIN verification was successful while the card assumes that it was not attempted (Murdoch et al., 2010). For offline transactions, the issuer will not be contacted until later, after the transaction was completed.

To conduct such attack (Murdoch et al., 2010), use a fake card with thin wires embedded in the plastic substrate to contact the card's contact points to an interface chip for voltage-level shifting. The interface chip is connected to a general-purpose field-programmable gate array (FPGA) board that interfaces with a laptop, hidden perhaps below the sale counter. The laptop is connected to a smart card reader into which the genuine card is inserted. The laptop has a Python script running to intercept the VERIFY command and respond to the terminal with the OK Result. The setup used in the attack is depicted in Figure 9.19.

EMVCo's response was that the attack would be extremely difficult to conduct without some merchant collusion and that would require some bulky material to carry out successfully. It is possible, however, that new versions of the attack could be conducted with mobile devices.

9.7.5.3 Relay Attacks

Relay attacks form a special kind of man-in-the-middle attacks, particularly on contactless smart cards, called *skimming*. The purpose of the attack is to make the legitimate user think one amount is being used, while the card details are relayed to an attacker that charges a different amount. This is done by inserting wireless transmission devices to relay the exchanges between the reader and the authorization center over a radio channel so that an unauthorized reader can interact with the chip to obtain and use the data without the need of deciphering it. Thus, in a relay attack, the attackers bypass the authentication phase of EMV by simply passing along challenges and responses between legitimate participants, that is, by operating at the MAC layer instead of the application layer. The arrangement is a modification of that of Figure 9.19 as shown in Figure 9.20 (Drimer and Murdoch, 2007).

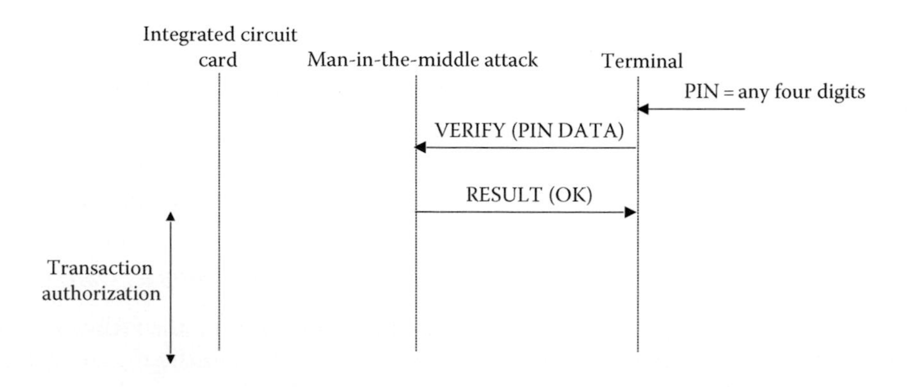

FIGURE 9.18
Protocol exchange with a man-in-the-middle attack during PIN verification. (From Murdoch, S.J. et al., Chip and PIN is broken, *31st IEEE Symposium Security and Privacy*, Oakland, CA, May 16–19, 2010, pp. 433–445.)

FIGURE 9.19
Setup to conduct a man-in-the-middle attack during PIN verification. (From Murdoch, S.J. et al., Chip and PIN is broken, *31st IEEE Symposium Security and Privacy*, Oakland, CA, May 16–19, 2010, pp. 433–445.)

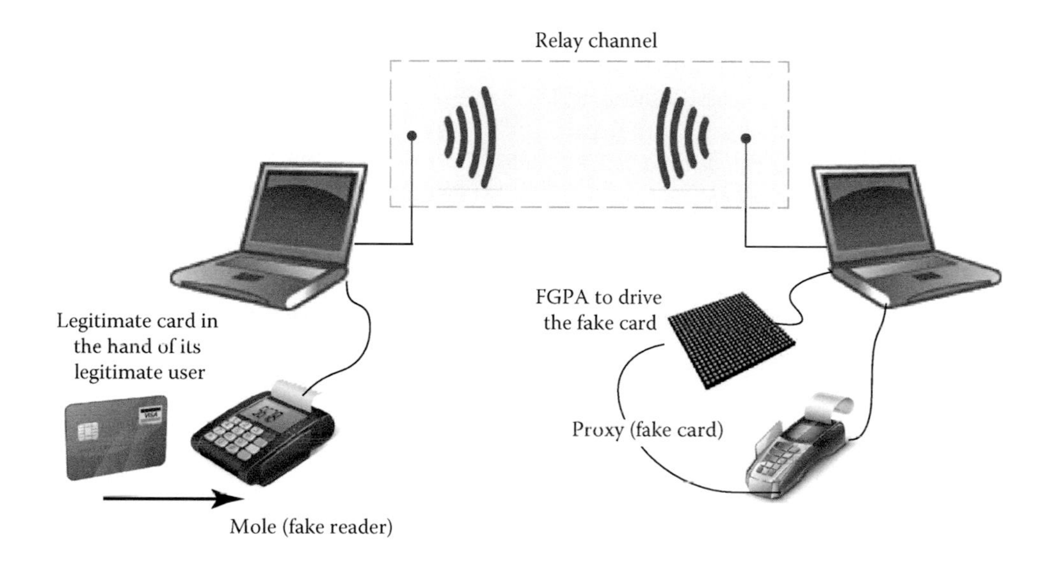

FIGURE 9.20
The equipment used to conduct a relay attack on an integrated circuit card with contacts. (From Drimer, S. and Murdoch, S.J., Keep your enemies close: Distance bounding against smart card relay attacks, *Proceedings of Security'07, 16th USENIX Security Symposium*, Boston, MA, August 6–10, 2007, pp. 87–102, available at http://static.usenix.org/events/sec07/tech/drimer/drimer.pdf, last accessed December 3, 2014.)

For example, a restaurant patron inserts the legitimate card to pay into a terminal, which looks like any other terminal except that its original circuitry has been replaced to communicate with a laptop (or a smartphone). The fake terminal is called a mole. At this instance, the waiter sends a signal to an accomplice in the same restaurant or even in a different store to insert a fake card (proxy) into a legitimate terminal to conduct a higher value transaction. A second laptop in the fraudster's backpack is connected to the fake card (proxy). The two laptops (or smartphones) communicate over a cellular network or a local area wireless network such as Ethernet or Wi-Fi. The PIN that the patron enters is captured by the first laptop and is sent over the wireless network so that the accomplice enters it when asked. This second laptop, in turn, runs the script that controls the fake card through an FPGA-based smart emulator that is connected to the counterfeit card. Thus, with both real and fake cards inserted, the user PIN is relayed to authorize the transaction, but with a larger amount than the legitimate transaction, unbeknownst to the cardholder.

For the attack to succeed, however, the actions on both sides of the radio link must be synchronized. The genuine card must be inserted in the counterfeit terminal before the counterfeit card is inserted in the genuine terminal. PIN verification at the genuine terminal should be conducted after the legitimate user has entered the PIN. The timing margins at the application layer are wide enough for the success of the attack even if there is a delay of about 3 seconds between a command and a response. Furthermore, the counterfeit terminal can impose additional delay by pretending to establish contact with the issuer for online authorization, while the counterfeit card can request an extra time by sending to the genuine terminal what the EMV specification calls a null procedure byte (0X60) before the first response.

9.8 General Consideration on the Security of Smart Cards

There are four main categories of attacks: logical (noninvasive) attacks, physical (destructive) attacks, attacks exploiting defects due to sloppy implementations, and attacks on the channels between the card and the card reader. These attacks may be made by amateurs, technical experts, and organizations specialized in reverse engineering.

In general, integrated circuit cards offer greater security than magnetic stripe cards. Although these attacks are more difficult with smart cards, a number of physical and logical attacks on the security of smart cards have been recorded and catalogued (Anderson and Kuhn, 1996; Kömmerling and Kuhn, 1999; Drimer and Murdouch, 2007; Rank and Effing, 2010, pp. 676–679).

The following is a brief presentation of these attacks.

9.8.1 Physical (Destructive) Attacks

Destructive techniques start with the extraction of the integrated circuit from the plastic support. First, a cut is made in the plastic around the chip module until the epoxy resin becomes visible. This resin is then treated with fuming nitric acid and washed in acetone until the silicon surface is fully exposed. Once the chip is uncovered, it is possible to probe the behavior of the various components and recover the cryptographic keys embedded in the cards by trial and error. The usage of laser probes or focused ion beams allows the exploration of the states of the microcontroller to extract the necessary information. However, the cost of carrying out this type of attack is relatively high, which puts these attacks outside the realm of hobbyists or typical hackers.

9.8.2 Logical (Noninvasive) Attacks

Noninvasive attacks are either active or passive. Active attacks modify the environmental conditions to perturb the functioning of the integrated circuits to compromise the hidden information. Write operations into the EEPROM memory can be affected by changing the ambient temperature, by imposing an instantaneous surcharge on the power supply, or by applying shorter clock pulses to perturb the operation of the microcontroller. Given that the encryption keys and the security software are stored in this memory, security is vulnerable to this type of attack, called glitch attacks, because they can prevent the execution of verification instructions. For example, (Anderson and Kuhn, 1996; Kömmerling and Kuhn, 1999; Drimer et al., 2009):

- The functioning of the random number generation can be perturbed to furnish a fixed number if the voltage is sufficiently decreased.
- Variations of the power supply can disable the security mechanism or even erase the content of the memory.
- Some secured processors are so sensitive to modifications in their environment that they declare many false alarms.
- The internal test circuits could be reactivated, which gives potential attackers access to the card's operational circuits from a limited number of probe points that reduces considerably the number of combinations to be examined.

Passive attacks, in contrast, concentrate on eavesdropping and observing the operation of the card to detect variations in the current supply or leakages in radiation. This is because each instruction has a specific signature, which allows the distinction, for example, of branching instructions or operations that involve a coprocessor. With contactless operations, eavesdropping on the communication between the reader and the smart card or tag can be achieved using a large loop antenna at 1–2 m (Krishenbaum and Wool, 2006).

Among the passive methods is differential power analysis (DPA), which is based on the principle that power consumption depends on the bits involved.

In the Internet era, new methods have been invented. A common swindle consists in producing card numbers using a generator program and retaining only those numbers that have been attributed to cardholders. To do so, it is sufficient to attempt a subscription of one of the online services using a given number; if the subscription is accepted, this indicates that the number is valid. The impostor can now make purchases with this card number until the real cardholder discovers the theft by reviewing the statement and contest the transactions.

9.8.3 Attacks against the Chip-Reader Communication Channel

The focus with this category of attacks is on the link between the integrated circuit and the reader. A particular case is the relay attack to be discussed in the next section. Physical protection of the reader is essential to prevent manipulations or its replacement by a doctored unit. Another possible source of fraud is the data collected from the magnetic stripe of a valid card through a doctored terminal. The corresponding PIN can be stolen by spying the user while the code is entered, either through direct observation or with a hidden camera.

Some malware may be designed to attack point-of-sale systems and collect card data briefly stored in the memory of point-of-sale machines. Even when point-of sale systems do not access the Internet, they are connected to back-office systems of the retailer. So, if an attacker is able to penetrate the corporate wide network, the malicious code can target readers and be designed to bypass common virus detection software.

9.8.4 Relay Attacks on Contactless Cards

Contactless systems have limited operational range (from a few centimeters up to 9 m, depending on the frequency card). Successful communication is established when the chip is in the proximity of the reader. Thus, the assumption is that the devices are in close proximity when they establish successful communication. However, if the exchanges between the two devices are transmitted over an extended distance without their knowledge, an attacker can prompt actions that otherwise require the physical closeness of the transponder and the reader (e.g., in access control systems, payment systems). The two legitimate participants receive valid exchanges from each other and assume that they are close.

The attack exploits the fact that when a spurious contactless card (also called *proxy*, *ghost*, or *clone*) is within the communication range of the reader, it is powered by induction by the reader. At the MAC layer, the reader cannot distinguish it from a genuine card if it conforms to the radio interface and performs the necessary modulation and demodulation procedures. Concurrently, a substitute reader (also called *mole*, *leach*, or *skimmer*) communicates by stealth with the user's card as if it were the genuine reader because at the physical and MAC layers the actions of the mole and those of a legitimate reader are indistinguishable. The mole/proxy (skimmer/clone or leach/ghost) combination can then send the reader's commands to the legitimate user's chip and relay the responses back. In this way, the intended transmission range for which the system was designed, typically 5–10 cm, is effectively extended to at least 40–50 cm, if not more. Furthermore, the mole/skimmer/leach interfaces with the card surreptitiously (i.e., without the knowledge of the legitimate owner and without touching the target) as a reader to access all the information that a genuine reader would be able to access. The delay introduced by the rogue system does not affect the operation as long as bit synchronization is maintained during the anticollision procedures within the limits that ISO/IEC 14443-3 specifies (Hancke, 2005, 2006; Krishenbaum and Wool, 2006; Hancke and Drimer, 2009).

Figure 9.21 illustrates how a relay attack can be conducted on a contactless card. A wireless relay channel is established between a fake reader and a fake card with a lightweight 40 cm diameter copper-tube antenna. At the application layer, the various EMV commands and cryptographic challenges are transferred back and forth. If the timing constraints of ISO/IEC 14443-3 are respected at the physical layer, then the presence of the relay channel will go undetected. Authentication of the fake card as a genuine card is achieved by relaying the response from the authentic card. Meanwhile, an accomplice uses a fake card close to the contactless merchant terminal. The legitimate user enters the PIN, which is relayed for verification. Once the PIN is verified, a larger amount is logged without the user's knowledge. In other words, the relay attack is conducted without knowing the cardholder PIN.

This kind of relay attacks cannot be readily prevented by cryptographic techniques at the application layer. Countermeasures are at the physical layer to ensure that the legitimate card and the legitimate reader are in close proximity. The round-trip delay of the messages can be used to estimate the distance between them, and distance-bounding protocols have been proposed. With this approach, if the round-trip between the terminal and the smart card is too long, then an alarm

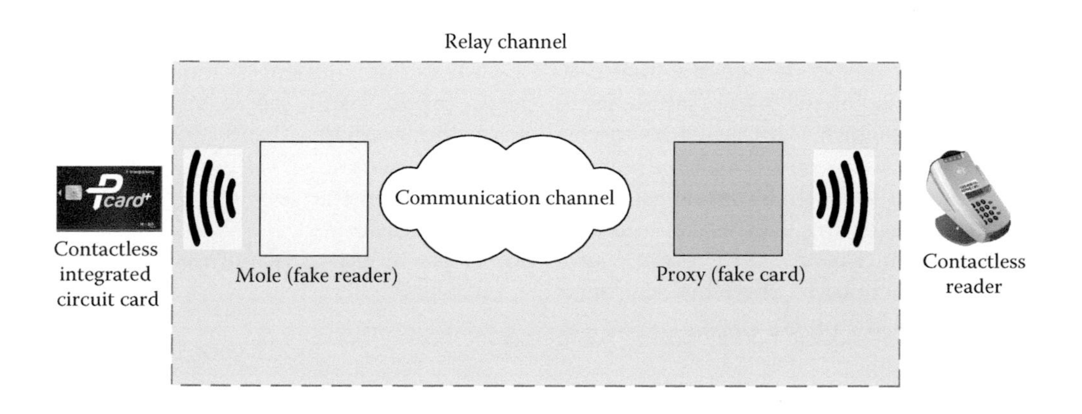

FIGURE 9.21
A relay attack on a system using a contactless integrated circuit card.

TABLE 9.7

Comparison between Replay Attacks and Relay Attacks

Characteristic	Replay Attack	Relay Attack
Attacker action	Stores previous messages and replays them.	Intercepts the communication and relays it.
Impersonation	The attacker impersonates the legitimate user.	The impersonating system is distributed and usurps the role of the card and the reader.
Locus of the attack	Application layer.	MAC layer.
Protection	Through the use of sequence numbers.	Use of distance-bounding methods in the case of contactless cards.

is raised and the transaction declined. However, this method is not standardized and is not available in commercial payment systems (Hancke and Drimer, 2009; Lima et al., 2009).

Relay attacks are distinct from replay attacks as summarized in Table 9.7.

9.9 Summary

Compared to cards with barcodes or with magnetic stripes, smart (or integrated circuit) cards have computational capacities that allow them to take complex decisions associated with payment transactions or access control. They can also support several applications in parallel. In multiapplication cards, resource sharing among the applications must be strictly controlled to maintain the level of security that is commensurate with each application. This is called sandboxing.

The security of smart cards covers the whole life cycle. Protective measures cover the data recorded in the chip as well as the exchanges in which it participates. Cryptographic means are used to authenticate the various parties (the chip itself, any reader, the user, or the authorization center) before any exchange of values can take place. Cryptography also ensures the confidentiality and integrity of the transaction. Replay attacks can be foiled by the inclusion of random numbers, time stamping, sequence numbers, and so forth. Other measures of logical security include protecting all the communication channels involving smart cards, card readers, and host systems. It is also essential to react quickly once an intrusion is detected, including aborting an ongoing transaction.

Contactless cards are exposed to an additional type of attacks, the relay attack, which targets the MAC layers. Protection against these attacks may require including distance-bounding techniques in the protocols to ensure that the legitimate card and the legitimate reader are in close proximity.

In banking applications, most systems used to be proprietary, but migration to the EMV specification is changing the picture at the application level. Also, the radio interfaces of contactless cards are gradually converging on the various ISO standards. Full interoperability, however, has not been reached in all situations as the case of electronic toll collection demonstrates.

Questions

1. Compare magnetic stripe cards with integrated circuit cards in terms of cost and security.
2. Compare the processing of payment transactions with magnetic strip and integrated circuit cards using EMV.
3. List some disadvantages for the lack of standards in electronic toll collection systems using RFID.
4. What are the main constraints on the security of smart cards?
5. Explain the differences between active and passive contactless cards.
6. What is a semipassive contactless card/tag?
7. Compare the NFC protocol with the ISO/IEC 14443 standard.
8. Compare and contrast NFC and RFID.
9. Compare the characteristics of payment with contactless cards and with near-field communication.
10. What are the main steps of dynamic data authentication (DDA)?
11. What is the main difference between static data authentication (SDA) and dynamic data authentication (DDA)?
12. Compare the characteristics of the different methods of data authentication.
13. Compare and contrast deterministic dynamic authentication and probabilistic dynamic authentication.
14. What are the advantages and disadvantages of tokenization?
15. Why cannot EMV with CDA protect against relay attacks?

10

Mobile Payments

Mobile payments are financial exchanges that involve at least one mobile terminal. They differ from payments from desktop terminals in at least four important ways. First, the number of parties increases substantially. In addition to the merchants, buyers, and their respective banks and the clearing and settlement networks, new actors include mobile network operators, handset makers, integrated circuit card manufacturers, bank card schemes, payment processors or payment service providers, trusted third parties or Trusted Service Managers (TSMs), and regulators (Nambiar and Lu, 2005; European Payments Council, 2012). Second, the rapid developments in wireless communications have led to a wide variety of access protocols and mobile devices as well as numerous technologies and standards that have to be considered. Third, wireless communications are readily exposed to environmental interferences and eavesdropping and hence are more difficult to secure. Finally, mobile commerce raises more acutely the issue of controlling the access to customer through the management of the security (Adams, 1998).

Initially, the limited capabilities of terminals and networks restricted the type of transactions to the purchase of logos and ringtones. While this minimized the cost of fraud, it reduced the incentive to compete for access to the end user, particularly because banks are less interested in small amounts. With more powerful mobile phones, cryptographic algorithms could be implemented to protect high-value transactions, and this increased the strategic value of direct access to the user. In addition, the cost of bypassing the security measures of the operating system through "rooting" or "jail breaking" to install uncertified applications could become significant.

The first section of this chapter presents a reference model for mobile payments. Next, we discuss the design alternatives for the configuration of payment security in mobile terminals. A review of the various technologies used in mobile terminals and the relevant threats follows. We then present some commercial mobile payment offers.

10.1 Reference Model for Mobile Commerce

Mobile commerce is a convergence of banking, telecommunications, and information systems. The way this convergence takes place affects the arrangements among the various parties, that is, how the converged service is offered and managed, the responsibilities of each party, and how to divide the revenue stream among the various parties.

As explained in Chapter 9, mobile terminals contain a Subscriber Identity Module (SIM), a tamper-resistant chip that stores the International Mobile Subscriber Identity (IMSI), the telephony cryptographic algorithms, and a 128-bit root key shared with the network authentication center.

With Universe Mobile Telecommunications System (UMTS) and Long-Term Evolution (LTE) networks, the authentication of the network and terminal is mutual and messages are encrypted by default. UMTS employs longer cryptographic keys and stronger cipher algorithms than GSM. Also, the SIM function is split into two parts: a software application called the Universal Subscriber Identity Module (USIM) with the associated hardware denoted as the Universal Integrated Circuit Card (UICC).

The basic security of the telecommunication applications is not sufficient to secure the payments. Therefore, payment applications require additional security credentials stored in the secure element (SE). This is another tamper-resistant chip with a separate microprocessor and a secure storage for payment applications to constitute a trusted execution environment (TEE). During manufacturing, the secure element is loaded with the critical data (keys, certificates, and personalized information) and the various cryptographic algorithms for the payment application. The SE performs the various cryptographic functions such as hashing, random number generation, key generation, signature, and encryption/decryption outside the mobile operating system. During the execution of the sensitive code, the integrity of these calculations is verified at each step so that access to the protected data is blocked, should a threat be detected.

Control of the secure element gives a commercial advantage. Accordingly, there are five possible configurations for the division of responsibilities among the stakeholders (Ondrus, 2003; GSM Association, 2007; Chaix and Torre, 2012). One party clearly dominates in the first two arrangements, while the remaining three arrangements distribute the power among the various entities:

1. The bank-centric model, in which a bank issues the chip that contains the security credentials for financial access while the mobile operator is responsible for the chip implementing the protocol for mobile communication.
2. The mobile operator–centric model, in which the same chip from the mobile operator manages the security of the communication and payment applications.
3. The third-party service provider model, in which an independent entity acts as a neutral intermediary among all parties: mobile network operators, chip manufacturers, phone makers, or other parties that may need to access the secure element.
4. The cooperative model where all parties agree on the way to establish the security mechanisms and share in the revenues from mobile commerce. For example, several mobile network operators and issuer banks can form a consortium to build a joint TSM for their respective customers.
5. Finally, the manufacturer of the mobile phone with embedded SIM provides the security and therefore is in a position to be the central player.

10.1.1 Bank-Centric Model

In this arrangement, banks control the whole payment value chain leaving the networking and transport functions to the mobile operators. This model takes advantage of a fully developed financial system, where banks have large technical and financial capabilities. This is typically translated in dual-chip phones or dual-slot mobile phones, one for the telecommunications and the other for the banking application.

Banks, however, must secure the cooperation of all mobile network operators in a given market to install and control the banking application in the secure element of handsets. Moreover, if the mobile terminal is lost or stolen, banks may depend on the mobile operator to deactivate the application.

10.1.2 Mobile Operator–Centric Model

In this configuration, mobile network operators are at the center of the operation. The same chip handles both the security of the communication and the financial aspects. The mobile operator pays the merchants and extends a credit to its customers for their purchases and charges them at the end of the billing period after adding a small transaction fee. This solution is particularly suitable for micropayments, as seen in the case i-mode®. It is also useful when the domestic financial system is underdeveloped in terms of geographic distribution or quality because it exploits the reach of mobile networks in remote areas.

Both the buyer and the seller are restricted to the same network unless the network operators have established agreements to allow the transactions to take place.

For banks, there is some danger in allowing mobile network operators' exclusive access to the end users. When there are multiple mobile network operators, banks have to manage multiple technical and business interfaces.

10.1.3 Third-Party Service Provider Model

Figure 10.1 illustrates the allocation of fees and commissions in this model (Chaix and Torre, 2012). The independent service provider acts as a trusted intermediary to manage the security credentials for each security domain. It may also offer a private currency to settle the accounts. Its role is to act as a bridge between the service providers (bank, transit company, etc.) and the various mobile network operators; it debits the buyer's account and then credits the merchant's account. This gives the user more independence from banks or mobile network operators.

The magnitude of the financial risks that the service provider assumes depends on the way merchants are compensated. The risks may be considerable if the service provider pays a merchant before the customer payment is settled and cleared. In contrast, the service provider bears no risk if the customer payments must clear before merchants are paid, but the latter's revenues will be delayed.

10.1.4 Collaborative Model

The solutions in these modes of operation are open and neutral partnerships of banks, card payment schemes, payment processors, and mobile operators in a transparent way so that each entity does not need to change its key processes.

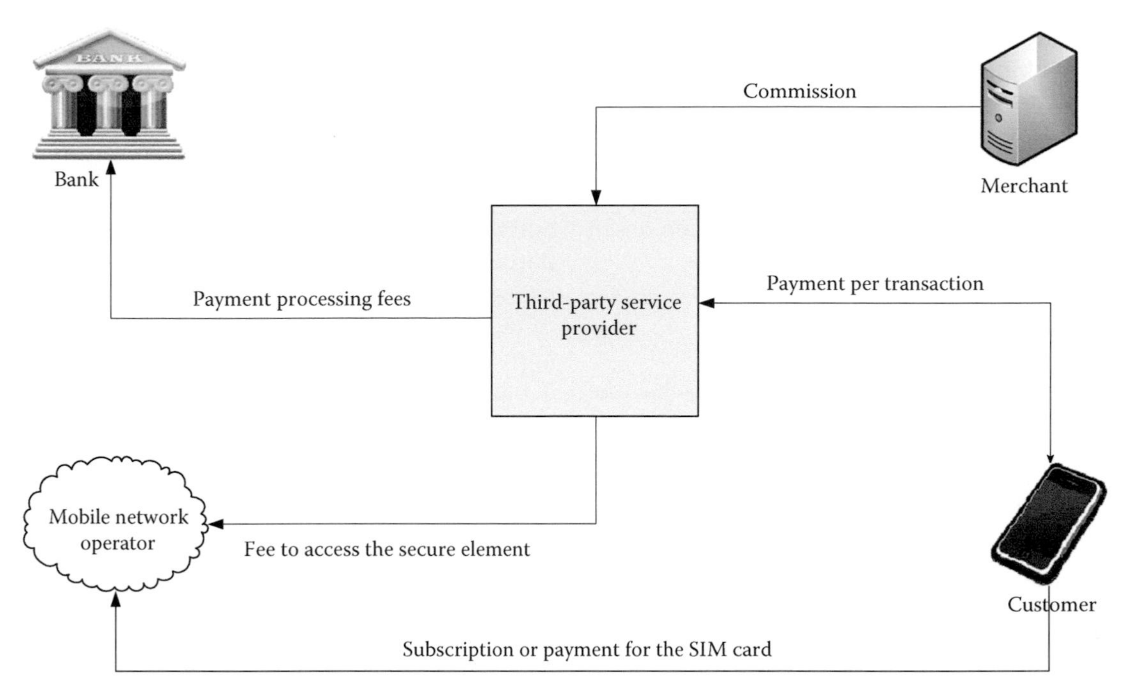

FIGURE 10.1
Distribution of financial rights in the third party service provider model.

10.1.5 Manufacturer-Centric Model

Embedded SIMs—or embedded Universal Integrated Circuit Card (eUICC)—were introduced in 2014. They make it easier for users to switch operators without having to replace their SIM card by reprovisioning network subscription through over-the-air (OTA) download and configuration.

The impetus for embedded SIM came for the development of the Internet of Things (IoT) and machine-to-machine (M2M) communication. The use of traditional SIMs for transportation, utility metering, and other applications is very cumbersome because of the large variety of mobile networks and because the sensors can be hermetically sealed or located in remote locations. Over-the-air provisioning reduces the cost of maintaining such devices.

From a manufacturer's perspective, embedded SIMs simplify the design of the mobile terminals, but they require new methods to provision access credentials remotely to ensure security. Other entities, such as utilities or car manufacturers, may issue their own SIMs, which increases the complexity of network subscription management. It is also possible that new intermediaries would emerge in the form of IoT platforms. Currently, these platforms enable car manufactures, car fleet managers, jet engine manufacturers, and vending machines suppliers to manage their inventory and service needs and to offer real-time diagnostics, safety, security, or updates on a global basis irrespective of the network providers or the geographic location. It is possible that such platforms would also be interested in offering payment services in the future.

10.2 Secure Element in Mobile Phones

Getting an application into the secure element is costly in terms of the time and effort to develop and test secure programs, to ensure adequate sandboxing of the various applications that it supports, and to obtain the necessary authentication certificates. As a result, there are several design alternatives—including software emulation—for the secure element of mobile terminals (Carat, 2000; Swchiderski-Groshe and Knospe, 2002; Chang, 2014, pp. 38–40), which are presented in the following.

10.2.1 Option 1

The terminal has a multiapplication microprocessor, a modified SIM or UICC, that encloses the secure element as well. The mobile operator has exclusive access to that multiapplication chip, including the security credentials and the transaction data for all applications. Each application provider, such as a bank, pays the mobile operator an access fee to rent a section of the multifunctional chip. The mobile operator handles the customer billing as well. Such an arrangement requires additional testing to ensure that each application is correctly sandboxed, which increases the cost of the handset. Another restriction is that the storage capacity limits the number of applications that can coexist in the same chip.

10.2.2 Option 2

In dual-chip (dual-SIM/UICC) mobile handsets, one chip is reserved for the phone service and the other for other applications such as another telephone number or for a payment application and its associated secure element (European Payments Council, 2012, pp. 41–42, 80, 94). The entity that provides the second chip determines which applications will be allowed to access the secure element in addition to the cost of that access. If that entity is a bank, it will most probably prevent its competitors from placing their application. Also, not all handsets can be designed to handle a second card slot, so a plug-in a card reader (dongle) could be used instead.

10.2.3 Option 3

In dual-slot devices, the mobile handset has a built-in smart card reader into which the buyer inserts the payment card to transmit the data and authorize the payment with their PIN. The payment transaction proceeds with the exchanges at the application level protected using the EuroPay, MasterCard, and Visa (EMV) protocol as if the handset were a banking terminal. The success of that approach depends on the manufacturers' ability to produce reasonably priced dual-slot terminals. This option has proven not to be viable even before the introduction of embedded SIMs.

10.2.4 Option 4

The payment application and its secure element are stored in a removable memory component such as a Secure Multimedia Card (Secure MMC), a Secured Digital (SD) card, or a micro SD card to be read with a separate card reader.

10.2.5 Option 5

With an embedded SIM or embedded Universal Integrated Circuit Card (eUICC), the secure element is fixed within the terminal. The applications are their associated security mechanisms that are provisioned remotely over the air. This option will be deployed widely over the next years to support global deployments of connected cars and machine-to-machine communication for consumer applications, known as the IoT.

10.2.6 Option 6

Software emulation obviates the use of a specialized hardware. The mobile operating system controls the access to the software, which can be stored in the terminal or an application server (i.e., in the Cloud). This approach is sometimes called Host Card Emulation (Chang, 2014, pp. 46–48).

10.2.7 Near-Field Communication Terminals

Similar design options are available in near-field communication (NFC)-enabled mobile phones. For example, the secure element and the NFC controller may be combined in a single multiapplication modified SIM/UICC card. Other options include putting the secure element in a secure removable digital (SD) memory card or in an embedded hardware fixed to the mobile terminal (Coskun et al., 2012, pp. 83–86). Alternatively, the secure element can be implemented in software (Guillemin, 2006; Roland et al., 2012).

10.2.8 Java™ 2 Platform Micro Edition

Java 2 Platform Micro Edition (J2ME) is specifically designed for devices with limited processing capabilities and storage capacity, such as the SIM/UICC. The communication between applications in the main memory and the secure element goes through application programming interfaces (APIs). JSR 177 (Security and Trust Services API) supports the secure communication with smart cards through Java Card–trusted applets. These applets support the identification, verification, and authentication of subscribers through cryptographic methods. Applications in the manufacturer and operator domains are granted access automatically, while applications from trusted third parties may require additional user input, such as PIN entry. Other mechanisms provide finer access control, for example, to restrict access based on certificates or access control lists (ACLs) (Java Community Process, 2007; Coskun et al., 2012, pp. 18–19, 212–215).

A mobile wallet is stored in the secure element as a J2ME application. A graphical user interface (GUI) provides access to the mobile wallet. The application is installed in one of two ways. The mobile network operator can install it in the modified SIM/UICC card and provision it over the air or users can download it and install it.

The API accesses the underlying hardware, including the secure element, through the OS. Malicious software can exploit this property if they are able to bypass the security checks that the mobile OS performs (Roland, 2012a). End users can contribute to weakening the terminal security bypassing the OS security measures to install nonqualified applications through "rooting" or "jail breaking."

10.2.9 Unauthorized Access to the Secure Element

Rooting is an operation to accord users or applications privileged control (known as "root access"), the privilege to run operations that carriers and/or hardware manufacturers have restricted normal users from executing. With root privilege, the user has administrator-level permissions and can alter or replace system applications and settings. Jail breaking allows the download of additional applications, extensions, and themes that are unavailable through official channels (e.g., through the Apple App Store). In both cases, unsigned code is able to run on a handset, thus exposing the device to potential malware or unauthorized access to personal information, particularly in case the phone is lost or stolen.

There are at least two possible attack scenarios when arbitrary applications are installed and are able to access the secure element: denial-of-service (DoS) attacks and relay attacks (Roland et al., 2012; Roland, 2013).

DoS attacks exploit the fact that the secure element permanently locks the card after several successive failed authentication attempts, a typical number is 10. The malicious code injected into an application accesses the secure element API to perform failed authentication attempts. After 10 failures, it becomes impossible to install or remove applications in the secure element, that is, card management is blocked even though installed applets continue to function.

To subject a mobile phone to the second type of attack, that is, a relay attack, one possible way is to install a malicious application on the target handset so that it would relay all communications with the secure element to an external card emulator over a wireless interface. This is called skimming, the skimmer being a hidden device that is doing the intercept. The two legitimate parties remain unaware because they continue to receive valid data and, therefore, assume that they are in close physical proximity. The data captured can be analyzed and then used to clone a fake card.

10.2.10 User Authentication

There are several possibilities depending on the arrangement among the mobile network operator, the customer's bank, and any intermediary (European Payments Council, 2012, pp. 44–52):

1. The user enters manually the details of the payment card so that the transaction proceeds as in the case of a desktop application. However, this method has a disadvantage because the keypad or screen of the typical mobile device does not meet the security requirements of banking terminals.
2. The authentication uses a one-time password (OTP). This scheme is called dynamic passcode authentication (DPA) or "strong cardholder authentication."
3. The secure element of the mobile device stores the authentication credentials, and the mobile network operator is queried to authenticate the chip in the mobile device before the transaction could proceed.
4. A special online payment application may be installed on the mobile device to handle the security aspects. The user activates the application with a password established at registration.

10.3 Barcodes

As explained in Chapter 1, 1D barcodes can be used at the point of sale to speed out the checkout process. The barcode may be on a label on the product or in an e-mail stored on the mobile phone. In that case, the user displays it on the screen of the mobile device and holds it against the scanner of the point-of-sale equipment. Some of the mobile point-of-sale systems are discussed in Section 10.11.

2D barcodes often include a text description of the product, a hot link to the merchant website, phone numbers, e-mail addresses, and other alphanumeric data. Fake tags can direct the users to infected sites. There are currently no commercial 2D barcodes that offer sufficient security against spoofing, and research is ongoing on data hiding techniques or data compression and encryption before encoding to offer some protection (Kato et al., 2010, pp. 95–96).

The availability of phones capable of reading QR codes since 2002 has led to the proliferation of QR Code generation and decoding applications. Individuals now can create their own QR Code symbols as well as decode the tags for online shopping or to check real-time information, for example, on public transport and sports events. The client software in the mobile terminal can decode the tag directly to extract the information from the local tag or it can decode the pictogram to extract the link that points to the website of the content or the service.

Payments can be made using the monetary value stored in mobile wallets or with other instruments such as a checking account, a payment card, a PayPal account, or a gift or reward card. The purchase amount and the transaction fee are then deduced from the buyer's account.

With a mobile wallet, the consumer launches the smartphone application at the point of sale so that the terminal can read the QR Code and debit the wallet for the amount of the transaction. When the balance of the account falls below a certain threshold, the consumer can authorize a reload of funds via the website or by responding with a text message.

2D barcodes can be used to authenticate transactions from an unsecure terminal connected to the Internet as illustrated in Figure 10.2 (Starnberger et al., 2009). After initiating the payment transaction, the terminal requests a nonce N from the bank server to protect against replay attacks. The terminal encrypts the concatenation of the transaction record T and the nonce N using the public key of the user's mobile phone PK_M. It displays the results as a 2D barcode that is captured with the camera of the mobile phone to be decrypted to extract the transaction data T and the nonce N.

To approve the data after verification, the user enters a password shared between the phone and the bank server (or a biometric signature such a fingerprint). The mobile phone uses this password as the cryptographic key in the calculation of Hashed message authentication code (HMAC) on the concatenation of the transaction record, the nonce, and some flags. The HMAC is used to verify both the data integrity and the authentication of a message. The first X characters are displayed as a shortened hash that inputs into the terminal.

The terminal transmits the transaction data and the shortened hash to the bank server for confirmation. The bank server computes the confirmation hash and transmits the hash to the terminal for displaying the first X characters. The user knows that the bank server has confirmed the transaction if the two shortened hashes match.

In this way, the terminal does not read any of the user's credentials, and the mobile phone is the trusted device.

10.4 Bluetooth

Bluetooth is a short-range wireless technology for low-power communication in the frequency band 2–2.4835 GHz. It does not require line-of-sight communication

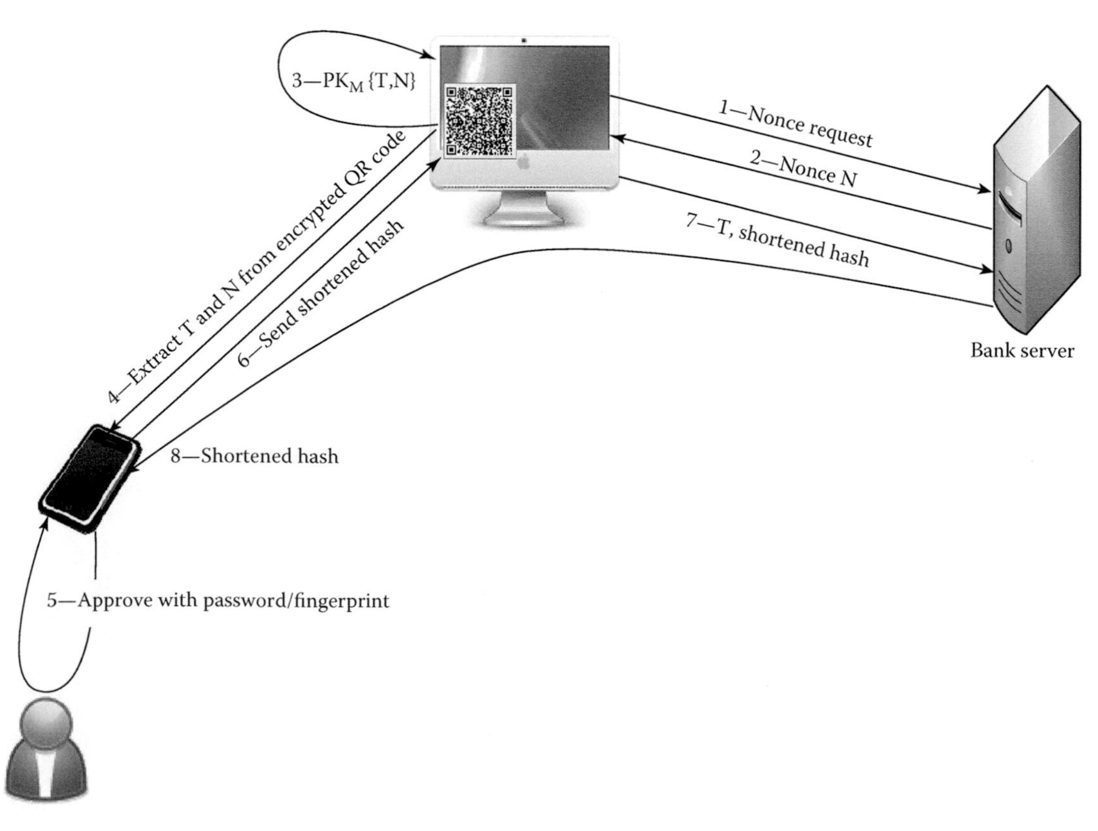

FIGURE 10.2
Transaction authentication with 2D barcodes.

between devices. Bluetooth devices can form a piconet (an *ad hoc* mobile network) without a centralized administration and security enforcement mechanism. They avoid interfering with other systems by transmitting very weak signals.

In mobile commerce, Bluetooth is used to establish a communication channel between a mobile device and a point-of-sale terminal. During the channel establishment, the Bluetooth devices pair, that is, they authenticate mutually and then bond, that is, store the resulting link key. Device discovery allows one device to detect another device. If a device is discoverable, other Bluetooth devices in range can detect it, pair with, or connect to it if connectible, that is, if other devices can initiate connections. Optionally, user authorization for incoming connection requests may be required. The settings for discoverability and connectability affect the overall security.

In a typical payment scenario, the details of the purchasing transaction are transferred from a mobile terminal to a cash register with Bluetooth device and then transmitted to a authentication server using a regular telecommunications link. Payleven, for example, is a system where card readers can pair up with iPhones, iPads, or Android devices via Bluetooth and of associated business management software. The solution accepts payments with debit or credit cards, while the software tool manages the reconciliation process at store. Payleven is available in several countries such as, in alphabetical order, Brazil, Germany, Italy, and the United Kingdom.

10.4.1 Highlights of Bluetooth History

Ericsson's vision in the early 1990s was to use Bluetooth as a wireless alternative to the RS-232 data cables operating in the unlicensed 2.4 GHz industrial–scientific–medical (ISM) band. In 1998, it collaborated with IBM, Intel, Toshiba, and other companies to form the Bluetooth Special Interest Group with the goal of writing a common specification for the technology. Several revisions of the Bluetooth specifications added multiple voice and data profiles or services (headset, cordless telephony, file transfer, printing, etc.). The first version of the specifications was released in 1998. In this version, the basic output power is 1 mW (0 dBm), but the power is raised to longer ranges to 2.5 mW (4 dBm) or 100 mW (20 dBm). The 4 dBm version is mostly used, but with 20 dBm transmission, it can reach to about 100 m under line-of-sight conditions. Due to the overhead, the net data rate is 721 kbit/s out of a raw data rate of 1 Mbit/s.

Versions 2.0, 3.0, and 4.0 followed, respectively, in 1999, 2001, and 2009. The IEEE stamped Version 1.1 as 802.15.1-2002. Version 1.2 was ratified as IEEE 802.15.1-2005 and includes a discovery mechanism to pair devices, typically by entering a preset or user-created code. Bluetooth 2.1 is called Enhanced Data Rate and has a different modulation scheme that raises the raw data rate to 2.1 or 3 Mbit/s. Version 3.0 can operate over a Wi-Fi radio link according to IEEE 802.11 with increased data rates up to 24 Mbit/s. These versions are compared in Table 10.1.

Version 1.0 uses the SAFER+ algorithm with a block size of 128 bits and with a 128-bit key length. Version 2.1 adds Elliptic Curve Diffie–Hellman (ECDH) for the exchange of public keys and for pairing. The encryption in Version 4.0 uses the Advanced Encryption Standard (AES) algorithm with 128 bits. This is summarized in Table 10.2.

Bluetooth devices are identified by a unique 48-bit address and a human readable name that users define. The devices establish a master–slave relationship with a master device in charge. The communication goes over 79 different channels separated by 1 MHz and assigned by frequency hopping using a pseudo-random sequence. The transmitters change frequencies 1600 times per second to minimize the interference among near devices. Up to seven devices can participate in a Bluetooth network or piconet under the control of a single master device for a total of eight devices per piconet. A single device can participate simultaneously in several piconets provided that it acts as a master in only one.

Devices fall into one of three classes in terms of power consumption and range for data transfer, as shown in Table 10.3. The maximum allowed power is much less than in the case of cellular telephony (3 W).

Version 4.0 (called Bluetooth Smart) is three specifications in one. The first specification concerns Bluetooth Low Energy (BLE), previously known as WiBree. This is a power-saving feature built on a protocol stack that it

TABLE 10.1

Highlights of the Successive Versions of Bluetooth

Version	Year of Issuance	Effective Data Rate (Mbit/s)
Version 1.0	1999	0.721
Version 2.0	2004	3.0
Version 3.0	2009	24.0
Version 4.0	2010	Up to 25

TABLE 10.2

Ciphers Used in the Successive Versions of Bluetooth

Version	Algorithm
Version 1.0	E_{22}, E_{21}, and E_1 algorithms based on SAFER+
Version 2.1	Elliptic Curve Diffie–Hellman
Version 4.0	AES with a key of 128 bits

TABLE 10.3

Power Consumption and Data Transfer Rates of Bluetooth

Class	Maximum Allowed Power (mW)	(dBm)	Typical Range (m)
Class 1	100	20	100
Class 2	2.5	4	10
Class 3	1	0	1

is not compatible with the other specifications. It is used in wearable devices such as heart rate or temperature monitors. The second specification is the high-speed data transfer introduced in Bluetooth 3.0 based on Wi-Fi. Finally, the so-called classic Bluetooth reproduces the technologies of Versions 1 and 2. Apple used BLE in its iBeacon, a feature in iOS 7 available with the iPhone 5 series.

Finally, in Bluetooth 4.2, the data rate is boosted to 2.5 Mbit/s, and the device can connect to the Internet.

10.4.2 Security of Bluetooth

For security considerations, Bluetooth devices must use low-power Class 2 or Class 3 transceivers without external amplifiers or high-gain antennas (Defense Information Systems Agency, 2010; National Security Agency, n.d.). Also, the U.S. Department of Defense (USDOD) prevents the use of Bluetooth 3.0 High Speed alternate MAC and PHY or the BLE technology (Defense Information Systems Agency, 2010).

Currently, Bluetooth has four different security modes, as shown in Table 10.4.

In Security Mode 3, each device initiates Bluetooth authentication immediately after the initial establishment of the Bluetooth connection. This is the safest mode for pairing with legacy Bluetooth (Defense Information Systems Agency, 2010)

In Version 1.0, the strength of the security relies primarily on a code of up to 16 characters entered during the pairing in both devices, the personal identification number (PIN) (Shaked and Wool, 2005; Becker, 2007). However, should the manufacturer ship the device with a default static code, it is usually short and simple, such as 0000 or 1234 and is identical for all devices of the same model. This method, however, provided little protection against sniffing. An attacker can eavesdrop on the entire pairing and authentication process and save the exchanged messages. The sniffer program uses brute force to guess the code by enumerating all possible values, which amount to 10,000 for a 4-digit code (Shaked and Wool, 2005). The time it takes to break a PIN is directly proportional to its length, so to crack a 16-digit PIN will take longer than for a 4-digit code.

TABLE 10.4

Four Security Modes of Bluetooth

Version	Authentication	Encryption	Notes
Mode 1	—	—	
Mode 2	Yes	Yes	Encryption is for each individual service separately.
Mode 3	Yes	Yes	Link-level security mode. Encryption is at the link layer.
Mode 4	Yes	Yes	Secure Simple Pairing.

Bluetooth Version 2.1 introduced "Secure Simple Pairing" (SSP) with Elliptic Curve Diffie–Hellman (ECDH) with an exchange of public keys between devices before pairing. After that exchange, each device uses its private key to calculate a shared 192-bit shared key (the link key) that the devices use for authentication and to derive the encryption key. Authentication is based on HMAC-SHA-256 with a 128-bit key. A 128-bit Bluetooth encryption is the minimum requirement for confidentiality (Defense Information Systems Agency, 2010).

The following are some of Bluetooth's vulnerabilities and security risks:

1. Bluetooth piconets do not have intrusion detection mechanisms.
2. Because Bluetooth is a peer-to-peer network technology, there is no centralized security enforcement and it is up to each device to implement the necessary protections.
3. Bluejacking is a technique for abusing the vCard feature on mobile phones to send unsolicited messages or business cards (vCards) to Bluetooth-enabled devices. Acceptance of vCards often requires no interaction on the receiver's end, and attackers can saturate the channels with anonymous messages without any credentials. An attack denoted as Blueper (www.hackfromacave.com/blueper.html) floods the target with file transfer requests (Dunning, 2010).
4. Bluesnarfing is a technique for hacking into a Bluetooth-enabled mobile phone and copying its entire contact book, calendar, or anything else stored in the phone.
5. A backdoor attack involves establishing a trust relationship through the pairing mechanism but ensuring that it no longer appears in the target's register of paired devices. The attacker may be free to continue to use any resource that a trusted

relationship with that device grants access to. The countermeasure is to pair devices in a secure area using long, randomly generate passkeys.

6. A virus can be designed to use Bluetooth technology for propagation to available Bluetooth devices in the vicinity.

Bluetooth introduces a number of potentially serious security vulnerabilities (Becker, 2007; Communications Security Establishment Canada, 2008; Dunning, 2010; National Security Agency [NSA], n.d.). Some of the documented attacks on systems of mobile payments are as follows:

1. *Range extension*: A high-gain antenna extends the range of the Bluetooth radio-frequency unit from meters to kilometers so that attacks can be conducted from longer distances.
2. *Obfuscation*: This attack modifies the firmware of certain Bluetooth chipsets to change the device address and disturb or hijack existing device pairings.
3. *Fuzzing*: Applying malformed packets to cause unwanted results, such as buffer overflows or increased device activity, stack crashes, etc., that could give access to secrets stored in memory.

10.5 Near-Field Communication

Mobile phones equipped with near-field communication (NFC) technology can be used for contactless payments a follows. The phone with the NFC chip touches or is held in close proximity to a point-of-sale terminal with an NFC-capable reader. The reader powers the NFC chip by inductive coupling so that transponders on both sides can establish communication and start exchanging the information. There are, however, some variations in this picture, depending on the type of tab and the operating mode as discussed in the remainder of the section.

10.5.1 Tag Types

The NFC Forum has defined four types of tags as summarized in Table 10.5 (Coskun et al., 2012, p. 102; Chang, 2013, pp. 16–17).

Type 1 and 2 tags have a small amount of memory and a low data rate. They are inexpensive and are suitable for applications with low security requirements. Type 3 corresponds to the FeliCa specifications that were intended for micropayments with some security functions and more complex applications in transit systems. Type 4 tags correspond to contactless proximity cards, that is, they are suited for multipurpose applications that require cryptographic protections.

10.5.2 Operating Modes

According to the NFC Forum, there are three modes of operation of NFC, the peer-to-peer mode, the reader/writer mode, and the card emulation mode.

In the *peer-to-peer mode* (active/active mode), each device is self-energized, that is, it must have its own source of power to generate the radio-frequency fields. Both devices can act alternatively as a contactless reader or as a virtual contactless integrated circuit card. The NFC Forum has defined the Logical Link Control Protocol (LLCP) to regulate the flow of data, the Simple NFC Data Exchange Protocol (SNEP) at the application level for sending and receiving the messages between the two NFC-enabled devices. The NFC data exchange format (NDEF) defines binary messages used to consolidate the payloads arriving from the various applications.

The peer-to-peer mode can be used to bypass the secure element, for example, to exchange virtual business cards or photos or to pair two Bluetooth devices in a single step. Bump Technologies exploited that mode to develop an application for data transfer between two smartphones by tapping them against each other. The principle is as follows: the application on each site sends readings from its phone's accelerometers, its location, IP address, and so on, to Bump's server. The server runs a matching algorithm that uses the location data and the timing of the accelerometer events to determine which

TABLE 10.5

Basic Characteristics of the NFC Forum Tag Types

Parameter	Type 1	Type 2	Type 3	Type 4
Reference specifications	ISO/IEC 14443 Type A	ISO/IEC 14443 Type A	JIS X 6319-4 (FeliCa)	ISO/IEC 14443 Type A or Type B
Maximum memory in kbit	2	2	1000	64
Data rate in kbit/s	106	106	212, 424	106, 212, 424
Size of signature in bits	16, 32	None	16, 32	Variable
Chip name	Topaz, BCM20203	MIFARE	FeliCa	DESFire, SmartMX-JCOP, Calypso B
Manufacturer	Innovision, Broadcom	NXP (Philips)	Sony	Several

devices have been in physical contact and are requesting to be linked. Bump server forwards the information to the intended recipient over the mobile network. Google acquired Bump in September 2013 and terminated the service in January 2014.

In the *reader/writer mode*, the active NFC device initiates the communication by powering the passive device through inductive coupling to read the data from the passive device residing in an NFC tag or to write data to that tag. The application program data units (APDUs) are defined by ISO/IEC 7816, and NDEF defines the format of the NFC communication over an ISO/IEC 18092 interface. The applications operating in the reader/writer mode usually do not need to access the secure area because they only read data stored in a tag and write to that tag. The discovery of the contactless RFID tag or integrated circuit card uses the standard JSR 257 API (Coskun et al., 2012, pp. 179–181).

In the *card emulation mode*, an NFC device has the software to emulate the functionalities of a contactless integrated circuit and the necessary NFC/RFID hardware to behave like a passive contactless integrated circuit card or a passive tag, Thus, it can send ISO/IEC 7816 data units over the contactless specifications of ISO/IEC 14443 or FeliCa. Software emulation avoids the use of the cost of renting space on the secure element but poses some security risks as will be shown.

The secure element communicates with the application layer in the main processor using the JSR 177 API and JSR 257. JSR 257 defines how the reader accesses, through the NFC controller, the contactless card or the virtual contactless card that the smartphone emulates, the RFID tags, or barcodes (Java Community Process, 2011). Because the standard Java APIs do not support the peer-to-peer mode, proprietary extensions of JSR 257 have to be used (Coskun et al., 2012, p. 19). Figure 10.3 illustrates the protocol stack for each of these modes.

Figure 10.4 illustrates the NFC datapaths through an NFC-enabled mobile phone as a contactless card, while Figure 10.5 shows the paths when the NFC-enabled phone acts as contactless reader. In these figures, the application processor is the main computing unit of the

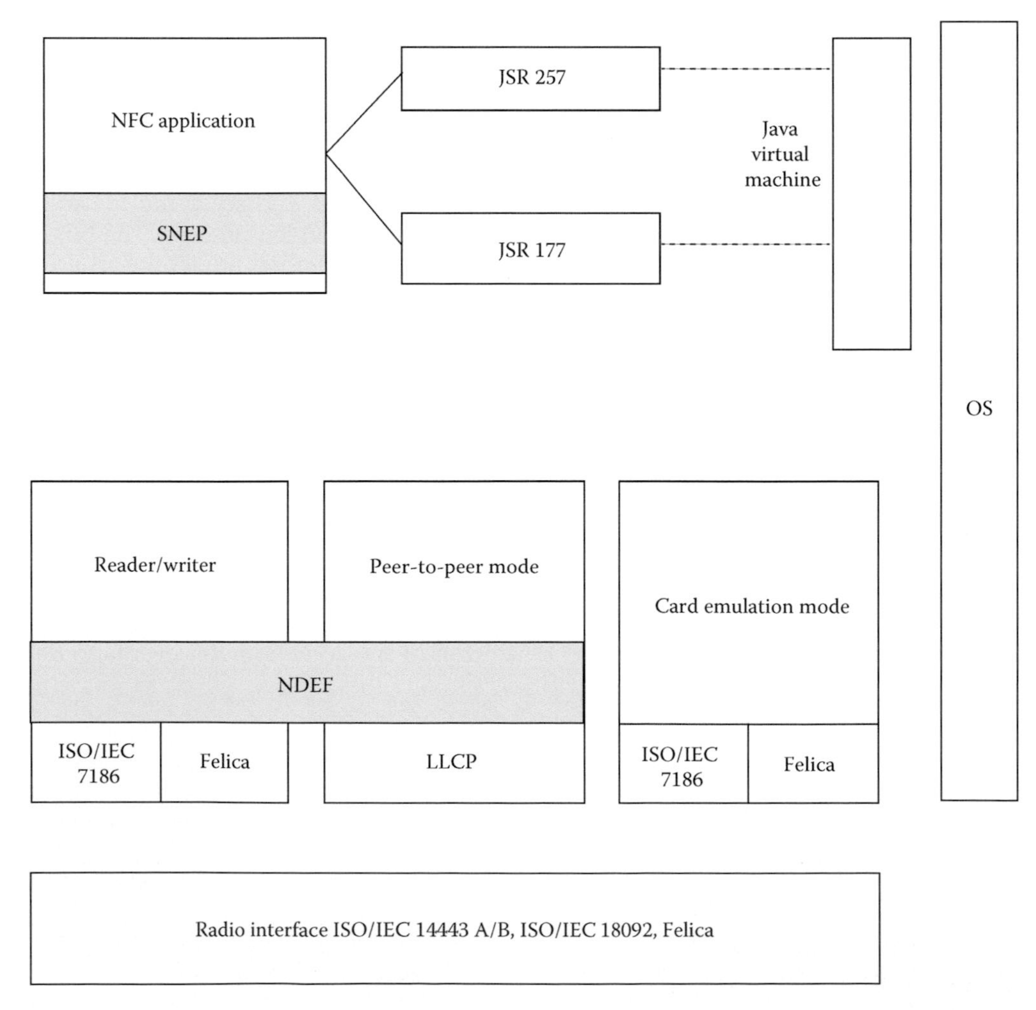

FIGURE 10.3
The protocol stacks for the various NFC modes defined by the NFC Forum.

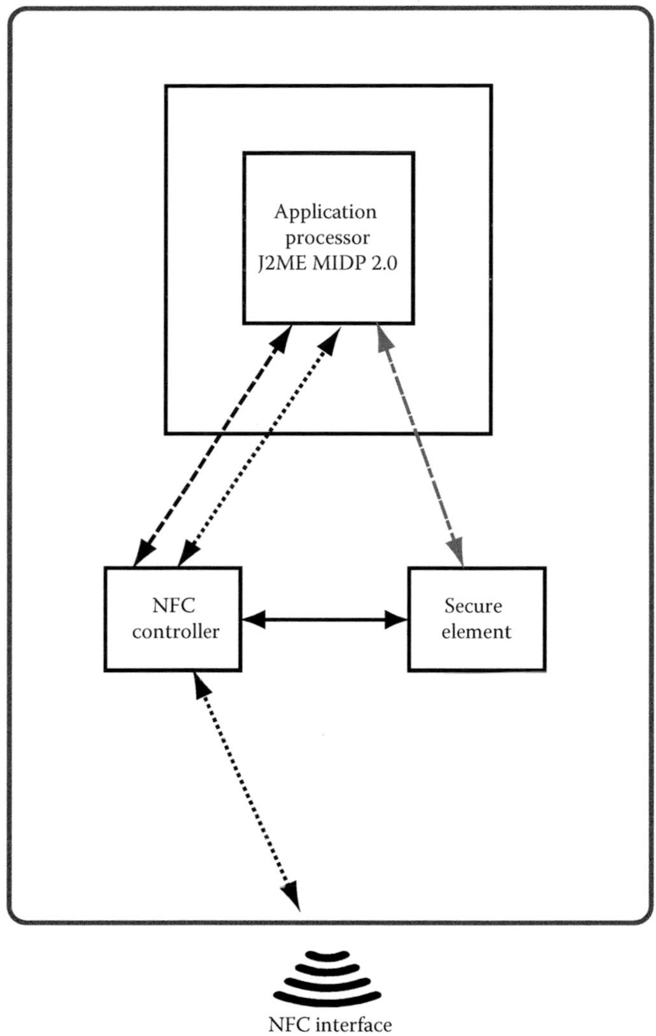

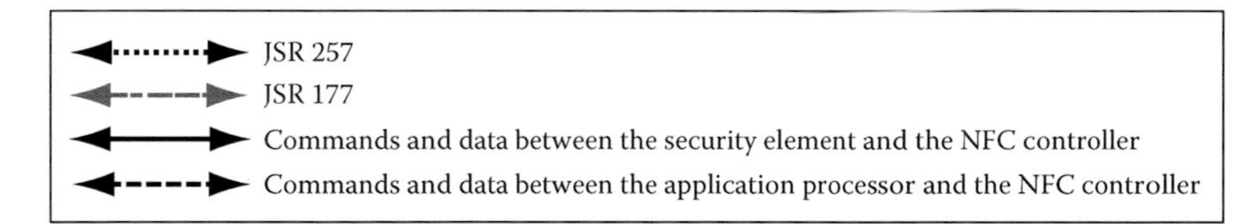

FIGURE 10.4
Datapaths through an NFC-enabled mobile phone acting as a contactless card.

mobile phone and provides data storage and processing capabilities alongside the basic mobile phone services. It constitutes the Application Execution Environment (AEE).

10.5.3 Transaction Authorization

NFC transactions can be authorized according to one of the following possibilities (European Payments Council, 2012, pp. 33–38):

1. Authentication is typically not needed for micropayments and the consumer just needs to tap the mobile terminal for the transaction to proceed. This is called "tap and go."
2. When larger amounts are involved, authentication with a PIN is required. It can be performed locally at the merchant terminal or by contacting a remote authentication server. If the transaction is approved, the amount of the transaction is withdrawn from the user's account specified

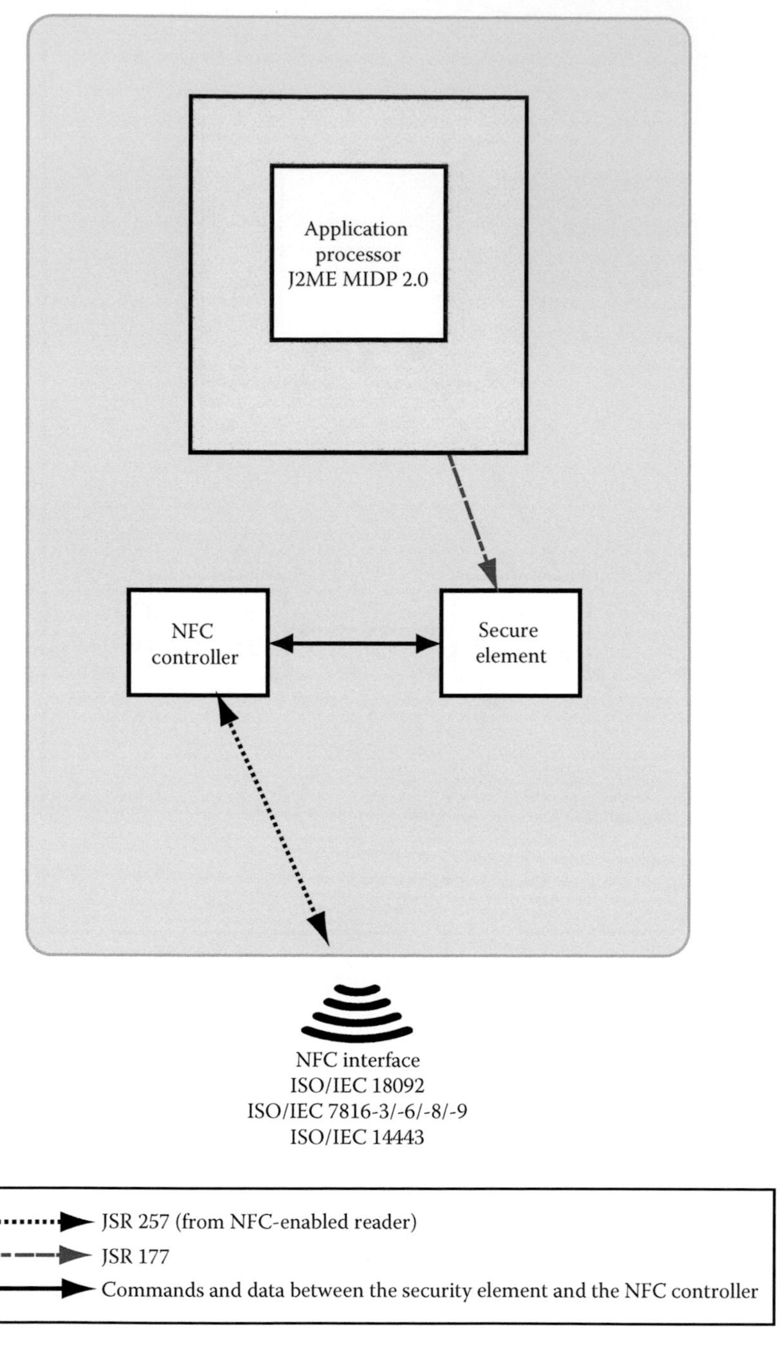

FIGURE 10.5
Datapaths through an NFC-enabled mobile phone acting as a contactless card reader.

at enrollment. However, the point-of-sale terminal must have the capability to encrypt the PIN. This is called "single tap and PIN."

3. In the "double tap with mobile code" mode of operation, the first tap initiates the transaction and the second to confirm it with a mobile code. For security reasons, the mobile code differs from the PIN used for contact-based card payments.

10.5.4 Security of NFC Communications

A complete list of possible NFC security issues is available in many publications (Coskun et al., 2012, pp. 22–24, 266–270; Mobey Forum, 2013b). Eavesdropping is not considered a severe problem in NFC because of the close proximity of the device and the reader. Data modification is also less likely when the lower bit rate of 106 kbit/s is used (Van Damme and Wouters, 2009).

However, spoofing attacks and relay attacks are of primary concern.

10.5.4.1 Spoofing of NFC Tags

The task of inducing a user to visit a fake website is much easier with posters equipped with an NFC tag, the so-called smart posters, because they already contain imbedded links that cannot be verified visually. When a phone with NFC capabilities captures the tag, its browser is directed to an advertiser's website. If the tag is tampered with or replaced by a malicious link, the browser goes to a website with the same look and feel as the authentic site that inserts some form of spying software. The spyware can snoop, for example, on payment credentials as they are keyed in. Other surreptitious consequences are modifying contacts' information, transmitting data with Bluetooth or text messages, placing a call to a preprogrammed number to record conversations, and so forth. (Ladendorf, 2013; Mobey Forum, 2013b, pp. 7, 10).

10.5.4.2 Relay Attacks

Software emulation of the secure element in NFC-enabled mobile phones opens new possibilities for relay attacks. The configuration comprises two NFC-enabled devices in the relay channel between the user NFC-enabled phone configured in the card emulation mode (or with a separate contactless card) and the NFC-enabled card reader. The devices in the relay channel are either in the peer-to-peer mode or in the reader/write mode, with one acting as a contactless card and the other as a contactless card reader (Francis et al., 2010a,b, 2011). The fake reader (also known as mole, leach, or skimmer) is placed in close proximity to the device under attack, while the fake card (also called proxy, ghost, or clone) is placed near the authentic reader. Any communication protocol can be used in the relay channel: Bluetooth with API JSR 82 (Java Community Process, 2010), Wi-Fi, the cellular communication, or the Internet. All these components can be made with off-the-shelf equipment.

Alternatively, the relay application can be inserted in the victim's mobile phone if the available security controls have been bypassed through rooting or jail breaking (Roland, 2012a,b, 2013; Roland et al., 2012, 2013). The malicious application will trick the operating system and acquire the necessary privileges to access the secure element, intercept the smart card commands from the victim's phone, and forward them to a card emulator acting as the proxy across the network. In either case, the reader at the point of sale will not be able to distinguish between the proxy (card emulator or phone enabled as a contactless card) and the legitimate integrated circuit card. Thus, every command that the proxy receives from the actual reader is forwarded to the mole, which then sends it to the phone under attack.

If the traffic is captured and analyzed, PIN harvesting is possible. Figure 10.6 is a schematic representation of the attack configuration with two NFC-enabled mobile phones in the relay channel (Francis et al., 2011). The contactless smart card is situated in an NFC-enabled phone or it can be a separate device.

Protection from relay attacks requires changes in the contactless or wireless protocol, such as inserting a time stamp to each message before sending or to put an upper limit. Countermeasures have to operate at the physical layer by limiting the round-trip delay of the messages (Coskun et al., 2012, p. 271).

In short, software emulation of the secure element is inherently risky because the software code is more vulnerable to malware than hardware. Furthermore, the mobile phone can be turned into a platform to relay attacks on other smart cards.

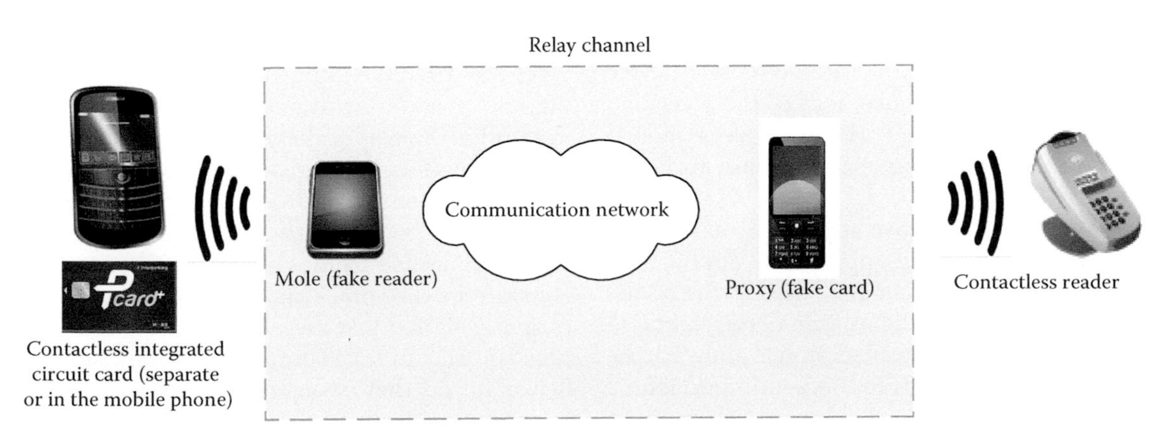

FIGURE 10.6
Setup for a relay attack with two NFC-enabled mobile phones.

10.6 Text Messages

Payment Instructions to the point-of-sale terminal or to the network server can be sent with one of several text messaging protocols: Short Message Service (SMS), SIM Application Toolkit (STK)/USIM, or Unstructured Supplementary Service Data (USSD).

10.6.1 Short Message Service

SMS is a store-and-forward messaging system that allows users to send and receive messages up to 160 characters long. It is simple to use and is compatible with all mobile phones including low-end devices, but mobile operators may surcharge text messages associated with payment transactions.

The security of an SMS transaction depends primarily on the security mechanisms of the mobile network. In particular, an authentication key is stored in the smart card of the terminal to identify it to the network. However, this security mechanism is broken if the Subscriber Identity Module (SIM) card is cloned.

The default data format for SMS messages is plaintext. Encryption can be applied during the transmission between the base transceiver station and the mobile station, but end-to-end encryption is not available and some encryption algorithms in the mobile network (e.g., the so-called A5 algorithm) are vulnerable. Furthermore, keying complicated Payment Instructions is not easy. Finally, congested networks can drop SMS messages. For these reasons, SMS is not used for large payments, particularly whenever the banking infrastructure is developed.

Typically, user authentication is based on two factors: the registered phone and a personal identification code, either a PIN or a one-time password (OTP). The OTP is received via SMS, sometimes after a successful authentication with a PIN. The SMS may also include transaction data to give the user a way to verify that the transaction has not been modified during transmission to the bank. For the method to work, the mobile device must be at hand and ready to use, that is, the reception is good and the battery charged. The OTP over a mobile network is also called the mobile transaction authentication number (mTAN).

Vending machines must have a special tag affixed for the phone to read with the appropriate application. When the buyer brings the phone near the tag, the phone application sends an SMS to the mobile network operator to request approval. If the transaction is allowed, the operator in turn sends an SMS to the vending machine to allow the purchase. After the buyer has made a selection, the vending machine informs the operator by SMS so that the amount can be recorded.

10.6.2 SIM Application Toolkit (STK/SAT/USIM)

According to the GSM specifications, terminals drive the SIM in a *master–slave* relationship. In 2G mobile networks, the SIM Application Toolkit (STK or SAT) defines commands and procedures for applications residing in the SIM to initiate data transmission and, if needed, to encrypt their messages with a private key stored in the SIM (Hillebrand, 2002, p. 403). At the network layer, data are encrypted between the terminal device and a wireless gateway. At the wireless gateway, the data are decrypted and reencrypted using another set of keys that the bank has sent to the wireless gateway. The USIM Application Toolkit (USAT) is the 3G equivalent of the STK for 2G terminals.

Ericsson and Telenor Mobile jointly developed one of the early applications with STK to pay for cinema tickets in Norway (Van Thanh, 2000). Currently, M-PESA is the most successful application of STK to secure payment messages.

10.6.3 Unstructured Supplementary Service Data

The Unstructured Supplementary Service Data (USSD) is a session-oriented protocol for real-time communications in GSM. USSD messages contain up to 182 alphanumeric characters and are sent in plain text. No session information is stored on the mobile device. USSD is often used in prepaid services to query the available balance and to refill the balance on the user's SIM card. USSD is often used to deliver OTPs or PIN codes. For example, Mobito, a mobile network operating in the Czech Republic, and WIZZIT, in South Africa, use USSD for their money transfer services.

10.6.4 Over-the-Air Application Provisioning

Over-the-air (OTA) application provisioning may provide an avenue for attackers to con users into installing undesirable applications, particularly if the attacker has the necessary certificates, perhaps obtained fraudulently. Another possibility is to send improperly signed OTA commands repetitively and to use the responses to construct a table from which the encryption key can be recovered. After cracking the key, the attacker can send properly signed text messages to download Java applets onto the SIM to take control of the handset (Nohl, 2013). It was found that two conditions make cracking of the key much easier:

1. A weak encryption algorithm, such as 56-bit DES. 3DES keys can also be cracked if their signature scheme can be downgraded to a shorter key length.
2. A faulty implementation of the Java virtual machine so that a Java applet can break out of its sandbox and access the memory space of other applications sharing the card.

It should be noted that early disclosure of these vulnerabilities to the GSM Association (GSMA) has given the network operators the chance of developing adequate countermeasures.

10.7 Bank-Centric Offers

In a bank-centric configuration, the mobile network operator plays a supporting role. It provides its subscribers with terminals that are capable of supporting bank-centric mobile payments. An additional role is to supply applications to access and control payment account details hosted on the bank server.

MasterCard PayPass is an application from MasterCard (MasterCard Worldwide, 2009). It was tested in 2002 for contactless and anonymous payments under $25 (or €25). The first commercial service was available in 2004 under the brand name TAP & GO™ in McDonald's stores.

There are two specifications: PayPass-Mag Stripe and PayPass-M/Chip. PayPass-Mag Stripe is for authorization networks that support magnetic stripe cards. Secure authentication is enhanced with the card verification code (CVC). Traffic is not encrypted.

Payments using the PayPass-M/Chip are encrypted between the mobile device and the point-of-sale terminal according to EMV. The account details are stored in the MasterCard PayPass server.

Mobile MasterCard PayPass is the mobile version of MasterCard PayPass and is also included in Google Wallet. Thus, either the bank or the mobile network operator or a third party can have direct contact with the customer, that is, verify the availability of funds and approve the transaction.

To reduce the risk of fraud, an authorization with PIN or a signature may sometimes be required, particularly following several transactions that did not require signature verification or PIN entry. The threshold is under the control of each merchant defines.

Visa payWave is the corresponding system for Visa. Users make their payments by tapping the payWave symbol on the reader at the checkout point, or wave it at a distance of about 4 cm, using a PayPass-enabled device. A green light and a beep indicate that the payment has been approved.

10.8 Mobile Operator–Centric Offers

Applications in industrialized countries have had limited success. In contrast, offers have been very successful in some countries such as Kenya as a way to provide banking to the poor. We go briefly over some offers in developed countries before focusing on the Kenyan case of M-PESA.

10.8.1 Offers in Industrialized Countries

The examples discussed are arranged in alphabetical order: Paiement CB sur Mobile (France), QuickTap (the United Kingdom), and Softcard (the United States).

10.8.1.1 Paiement CB sur Mobile

This service was among the first use of dual-slot handsets. It was launched in June 2000 by France Telecom (called now Orange) for subscribers to its Itineris mobile telephone service, as part of its prepaid Mobicarte offer.* The payments were for purchases on the web.

Users first give the merchant their mobile phone numbers via the Internet or a voice call. The merchant server responds with the details of the transaction in a test message. Should the purchase terms be accepted, buyers insert their card into the slot of their mobile phone and enter their PIN to proceed as in the case of a regular point-of-sale terminal. This triggers an authorization request through the banking system, with confirmations to follow.

10.8.1.2 QuickTap

QuickTap is the UK service from Orange for contactless payments. Users can load up to £100 at a time by linking their contactless mobile account with a payment card. Payments of up to £15 ($24) can be made at any point of sale in the United Kingdom equipped with NFC technology. The amounts are then deduced from the QuickTap balance. Gemalto is the Trusted Service Manager.

The SIM/UICC card is first registered with Orange. The QuickTap Wallet application is then downloaded

* CB means *carte bancaire*, the bank card standard of the French banking association (*Groupement des cartes bancaires*).

on a compatible Android device (e.g., Samsung Galaxy SIII). During the activation, the user selects a user name and a passcode and enters the phone and SIM/UICC number and then another security word. A five-digit activation number appears on the computer browser to be entered into the phone before a PIN is selected. This PIN is used for authorization at each transaction.

10.8.1.3 Softcard (ISIS) Mobile Wallet

The Softcard (ISIS) mobile wallet is a mobile payment joint venture with the backing of three main U.S. mobile operators, AT&T, T-Mobile, and Verizon Wireless. American Express and Chase Bank are the partners on the payment side. American Express prepaid cards, for example, can be accessed through widgets within the wallet to check and update balances. Other applications include merchant loyalty programs and coupons. For example, customers of the Subway sandwich chain can load their reward cards into the Softcard walled and then tap their phones at the point of sale to pay and receive reward points. Among the security features are remote wallet freezing, PIN protection, and dynamic security codes.

The reference model for the operation is mobile network operator centric. The secure element resides in a modified UICC that the mobile operator provides. Softcard acts as the intermediary between the service providers and the mobile operators.

The data transfer from the smartphone or the handheld device to the point-of-sale terminal uses the NFC technology.

In 2014, the venture changed its name to Softcard to avoid any confusion with the radical group ISIS (Islamic State in Iraq and Syria).

10.8.2 M-PESA

M-PESA is an operator-centric payment service in Kenya. Safaricom (the local branch of Vodafone) controls every stage of the process with banks acting as depositories of the funds.

Customers with a mobile phone register at an authorized M-PESA retail outlet (called a "certified agent") and then deposit cash into their account. The amount deposited minus a set fee is available in the form of a virtual Jeton accessible through an application in the SIM. Safaricom deposits the total value of its customers' balances in pooled accounts in two regulated banks and manages customer accounts in the virtual currency. Thanks to an agreement with Western Union, international money remittances to an M-PESA mobile phone in Kenya are also possible.

Users can transfer funds to other M-PESA users, pay bills, and purchase additional mobile airtime with their balance. Senders need to only know the mobile phone number of the recipients. The certified agents convert the virtual currency back into the legal tender and retain a transaction fee, part of which goes to Safaricom. All transactions are validated and recorded in real time using secure messages.

Safaricom charges 10% of the face value as transfer fees. Because of the lack of banking infrastructure, particularly in rural areas, there was no significant opposition even from Kenyan banks to that arrangement.

M-PESA was able to grow and reach 40% of the adult population in about 2 years; after 4 years of operation, it was used by more than two-thirds of households (Jack and Suri, 2010, 2011). It is considered to be the most widely used mobile money in the world. In 2015, the value of the transactions conducted with M-PESA was estimated to be about 40% of the Kenyan gross domestic product (*Economist*, 2015c). There are attempts to introduce M-PESA to other countries, such as India, Ghana, and Nigeria, but the success was not as spectacular (Crabtree, 2012).

M-PESA has stimulated discussions on the feasibility of replacing cash transactions with a digital equivalent as well as the formation of localized currencies tied to the local economy for micropayments (Ledgard and Clippinger, 2013).

10.9 Third-Party Service Offers

Apple Pay, Deutsche Bahn's "Touch and Travel," and Google Wallet are the main examples of this category. Paybox can be included as well (even though Deutsche Bank is the majority owner) because it is run as a separate entity. PayPal Mobile will be discussed in Chapter 12. For a banking perspective, these are wrapper services because they improve access to the existing payment architecture, for example, by linking a user's mobile phone number to an account, and do not introduce a new currency or a new core payment system (Ali et al., 2014a).

Some third-party systems use public key certificates for the authentication of mobile clients and the merchant sites. The certification authority may be the service provider, the payment server, or a Trusted Service Manager.

10.9.1 Apple Pay and Passbook

Passbook is the wallet from Apple. One of the applications is Apple Pay, a payment service announced in 2014 that uses NFC to establish a channel with the merchant terminal. Users approve the transaction by pressing with their fingerprint on a special sensor on the device. In 2015, it was announced that Apple Pay will be used for the electronic benefit transfer (EBT) card that is used for Social Security and veteran benefits. While Apple Play has made the process less arduous on the user, merchants have to upgrade their terminals to take advantage of this facility. In addition, the use of biometric identification is risky in the long term because of the consequence of a system being compromised on the status of the biometric signature.

To set up Apple Pay, users with an existing iTunes account do not need to reenter their card details, just the card security code. Other users must open first the Passbook application and take a photo of the front side of the payment card to record the number of the card they would use. The information can also be typed. However, FireEye, a cybersecurity firm, demonstrated in February 2014 how to capture every tap on the iPhone's screen or button by exploiting a security flaw in iOS 6 that Apple quickly fixed (Bradshaw and Kuchler, 2014). Capturing the image of a bank card for optical character recognition is also not considered safe (Mobey Forum, 2013a, p. 11). Furthermore, without adequate verification, the number on the card may be from a stolen card reprinted for fraudulent purposes.

The card number is not stored in the phone but is used to create another identifier or token kept in the phone's secure element. Tokens are created for each transaction using unique keys stored in the phone to replace the PANs. As explained in Chapter 9, this method is called tokenization because the token is communicated to the merchants instead of the original card number, making fraud much more difficult.

Apple has signed up many U.S. banks (Citigroup, Bank of America, Wells Fargo, Capital One, JPMorgan Chase, and Barclays) and MasterCard, Visa, and American Express to participate in the offer. However, details of the communication protocol with the card schemes are still unavailable. It is also not clear if the token is stored in software or in the secure element of the telephone. Major retail stores have also announced their support for Apple Pay. In February 2014, Apple Pay accounted for two-thirds of the mobile contactless payments in the United States (*Financial Times*, 2015a).

Soon after Apple announced its product, ShopKeep, an iPad-based retail payment system decided to include NFC in its terminals and announced that it would distribute the new equipment to its network of merchants (estimated to be around 10,000).

Card networks and banks have agreed to lower their fee for payment processing to encourage the shift away from cash and checks. The downside for them is that their revenues are reduced and their brand is no longer visible. Furthermore, issuer banks have been disintermediated in favor of token service providers that issue the tokens and then authorize online transactions based on the tokens they had issued.

A sharp rise in reports of fraudulent transactions was noted 6 months after the introduction of Apple Pay (6% compared to .1% with plastic payment cards) (Tsukayama and Halzack, 2015). This was attributed to loading stolen cards, indicating that users' authentication procedures at sign up were not strong enough.

10.9.2 Deutsche Bahn's "Touch and Travel"

Deutsche Bahn AG, Germany's main railway operator, introduced the "Touch and Travel" location-based ticketing service in 2006. The initial offer was first restricted to long trips (in excess of 50 km), but this was later relaxed in November 2013. The offer was done in partnership with mobile network operators such as Telefonica Europe, Vodafone, and Deutsche Telekom, which provided their customers with the modified chip with the necessary security profiles.

The Touch and Travel client application is first downloaded and installed on the mobile phone. Next, the user registers the payment information on the Deutsche Bahn's portal and obtains a customer number and a PIN. Thus, the payment data are not stored on the phone. The customer is billed monthly for all tickets.

Riders communicate with the Touch and Travel terminals at the start of their trip and at the destination using one of three methods: (1) NFC in phones with NFC chips, (2) 2D barcode images captured with the phone's camera, or (3) manual entry of the terminal numbers through the graphical user interface. With this information, the system calculates the price for the trip. The mobile phone stores the train ticket while the ground terminals transmit the itinerary to the billing department.

Risks are related to systemic failures. For example, how can a passenger prove the purchase of a ticket if a mobile terminal is misplaced or lost or runs out of battery? What would happen if a Touch and Travel terminal malfunctions? Finally, what happens if the network connectivity to the billing server cannot be established? (Lerner, 2013, p. 119)

10.9.3 Google Wallet

Google Wallet was introduced on September 2011 to run on Android 3.2 (Gingerbread) or higher. The initial participants were Citibank from the financial side, Sprint as a mobile network operator, MasterCard PayPass as a provider of contactless payments, First Data as a point-of-sale solution provider, and Samsung as the mobile handset manufacturer. Visa payWave and other partners were added later.

Google Wallet includes payment cards, prepaid cards, virtual gift, and reward cards from participating U.S.

retail stores with NFC-enabled Android devices. The application allows addition and management of various payment cards, finding and saving a variety of offers, and viewing the transaction history. The secure element is in software.

Google Wallet can be used to make online purchases of Google-associated services or on sites with the "Buy with Google" button. Prior to iPhone 6, the exchanges with iPhones used the Bluetooth Low Energy (BLE) of the iBeacon technology (a feature in iOS 7).

To initiate a contactless transaction with NFC, the Android phone is put within 4 cm or touches the NFC-enabled point-of-sale terminal. Users are authenticated through their PIN. The user receives only a payment receipt.

Google *Prepaid Card* was a magnetic stripe card used for online purchases only and discontinued on October 2012. The card was cobranded with MasterCard PayPass card and with Money Network/First Data as the Trusted Service Manager.

Google Wallet on the client corresponds to Google Wallet Instant Buy on the merchant side. To redeem offers received by e-mail, the merchant scans the associated barcode or enters it manually.

Google Wallet was not successful due to some security issues discussed later and the reluctance of retailers.

10.9.3.1 Account Activation

A Google account is needed for Google Wallet to be added. To activate the Google Wallet, a user signs in at wallet.google.com, enters the required personal information, chooses a 4-digit wallet PIN, and selects a method of payment.

10.9.3.2 Payment and Compensation

When a user makes a payment using Google Wallet, Google actually pays the merchant and then processes the transaction with the customer's payment instruments. This is a *split transaction* model in ACH terminology. A Customer-Initiated Entry (CIE) credit ACH transaction is performed on the user's behalf to pay the merchant, while a WEB (Internet-Initiated Entry) debit initiates the collection of the funds from the user's bank account over the Internet.

The integration of Google Wallet with Gmail allows person-to-person payments (in the United States) with transfers from a bank account or the wallet balance.

10.9.3.3 Revenue Sources

The Google Wallet service does not generate revenues directly. The first source of revenues is advertisements such as Google Offers, where subscribers as well as users of Google Wallet are informed of daily deals at their localities. More importantly, Google mines the data on the transactions because it knows the location and context of every payment and every offer redeemed through Google Wallet.

10.9.3.4 Security

Google Wallet locks after five invalid authentication attempts. If the mobile device is lost or stolen, the Google Wallet application can be disabled online at wallet.google.com or by calling customer service. When the Google Wallet application is reset, all payment credentials and transaction data are wiped out. The data can also be erased via the *Settings* menu. Finally, Google Wallet is not supported on rooted phones.

To save on battery life, the antenna is activated only when the screen is on. If the phone is not used for 30 minutes, the application logs out the user and the PIN needs to be reentered to unlock the Google Wallet and reactivate the antenna. This is also a security measure.

Sophisticated attacks exploit the fact that PIN verification in Google Wallet is done outside the phone secure element in a software hosted by Google. First, a relay function can be integrated in any application downloaded into the mobile phone. This application could then initiate a relay attack to capture all the data exchanged in a transaction and then capture the necessary credentials. With root privilege, access to sensitive data is possible, such as PIN and signing keys, credit balances, limits, expiration dates, names or cards, transaction dates, and locations that are stored as clear text in the system logs and in various databases (Hoog, 2011). Because the PIN consists of four numbers, its value could be obtained through brute force by calculating 10,000 hashes at most (Goth, 2012; Rubin, 2012). In other words, Google Wallet could be easily "skimmed" to "clone" the card.

Version 1.1-R52v7 of the Google Wallet was also susceptible to relay attacks (Roland, 2013; Roland et al., 2013), and videos of successful attacks were posted on YouTube at http://www.youtube.com/watch?v=hx5nbkDy6tc (as verified on December 28, 2013). In Version 1.6, it was also possible to unlock the Google Wallet as if the PIN had successfully been verified. Consequently, a malicious application could unlock the wallet and access the information on the financial account stored in the secure element as if the PIN verification was successful (Roland et al., 2013).

Moving PIN verification to the secure element of the phone solves these problems. However, the chip manufacturer would have had to digitally sign the code running in the secure element. Also, by moving the PIN verification into the secure element, the banks would be in the position to control the procedure and impose their security policies and extract more rent for accessing the secure element.

10.9.4 Paybox

Paybox is simultaneously a server-based technology platform and a multichannel solution that includes micropayments and mobile payments as part of a universal portfolio for electronic commerce. It is open and neutral, not tied to a particular operator or a specific bank, and has formed partnerships with banks, card payment schemes, payment processors, and mobile operators in a transparent way so that each entity does not need to change its key processes.

Paybox is an intermediary among banks, operators, merchants, and users. Its service enables customers to purchase goods and services and make bank transactions over several channels. The value of purchases is credited to the merchants' accounts and debited to the buyers' current accounts. Payments can be in multiple currencies.

Paybox was launched in Germany in May 2000 and expanded in Austria, Spain, Sweden, and the United Kingdom. Deutsche Bank owns 50% of the company and is responsible for payment clearing and settlement. Companies partnering with Paybox include Deutsche Bank (payment processing), Lufthansa Systems (central computer and data security), Oracle (software), HP (hardware), and Intershop (e-commerce systems).

Figure 10.7 illustrates the overall architecture of Paybox (Ondrus, 2003; Massoth and Bingel, 2009).

The customer's requirements for using Paybox are the possession of a mobile phone, a bank account, and enrollment in Paybox. Subscribers are charged an annual fee, and there are no transaction fees. During enrollment, each subscriber receives a unique identifier and a PIN. Thus, the PIN is independent of the mobile network operator. As a consequence, users can associate multiple accounts with their Paybox account.

Paybox can be used for payments of purchases, person-to-person transactions (including international transfers to any country where Paybox operates), bill payments, and payment to non-Paybox users.

10.9.4.1 Purchase Payments

Paybox provides several mobile payment service solutions: interactive voice response (IVR), SMS, WAP, and near-field communication (NFC) for point-of-sale applications (Massoth and Bingel, 2009). Merchants select the solution that suits them best.

A typical Paybox payment transaction with IVR authentication would proceed as follows:

1. The customer sends the mobile phone number to the merchant.
2. The merchant transmits to Paybox the buyer's phone number and the price on a channel secured with TLS/SSL using a 128-bit key and with an MD5 signature.
3. An IVR system application calls the buyer to ask for payment authorization.
4. The customer authorizes the payment by entering the PIN, which is transmitted in the form of Dual Tone Modulation Frequency (DTMF) tones.

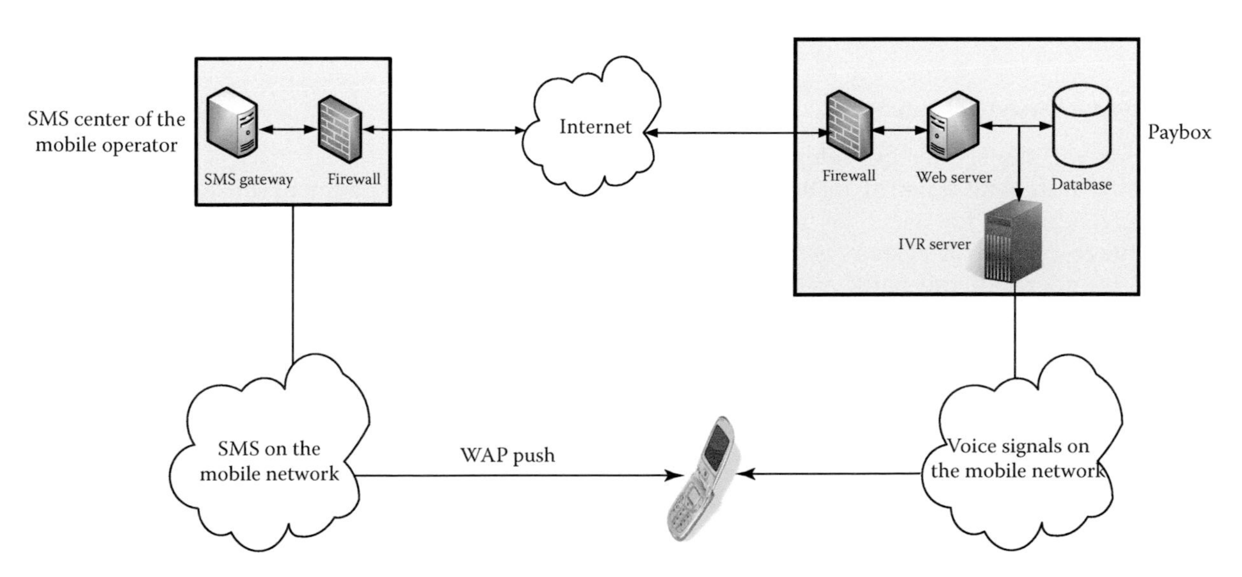

FIGURE 10.7
Overall architecture of Paybox.

5. After verification of the PIN, Paybox informs the merchant of the approval.
6. Paybox informs Deutsche Bank, the trusted third party, to settle the payment via a direct debit from the customer's account.
7. The transaction is confirmed to the customer by an automated voice message or SMS. The confirmation to the merchant server is done over the Internet.

Thus, the authentication procedure with the IVR system uses the two-factor authentication procedure.

With WAP, the consumer receives a message with a customized Universal Resource Locator (URL) that will load a web page asking for payment authorization when the buyer opens it with a WAP browser. The authentication is again with the PIN.

The one-time password application uses time synchronization between a server and an application on the phone to generate a one-time password. The buyer starts that application and enters a PIN. The application builds a hash using the PIN and other information related to the buyer for identification on the server side.

NFC is used for contactless proximity payments. First, the cashier starts a transaction with the Paybox server. Then as soon as the phone touches the tag, an application contacts a payment server using HTTPS to secure the connection. The buyer enters the PIN for authentication. The application sends an MD5 hash of the phone number, the PIN, and transaction ID. The transaction ID is used to prevent replay attacks.

Finally, the SMS application sends a short message to the consumer asking for authorization with a *Yes*. Thus, this is the only application that does not use two-factor authentication.

10.9.4.2 Person-to-Person Transactions

To initiate a transfer, the sender provides the counterparty's telephone numbers, amount to be transferred, and type of currency and, for users, its unique Paybox ID. The advantage of such a system is that only the mobile phone number, not the bank account number or credit card details, is transmitted. To transfer to an account with a financial institution, the account number must be provided.

10.9.4.3 Business Model

The current business model is to charge an annual subscription fee (€5 per annum) and merchants for each transaction. The average commission is around 3%, which is comparable to that of bank cards.

10.9.4.4 Additional Privacy Measure

Consumers can request an alias phone number from Paybox to avoid giving their mobile phone numbers to merchants. This concept is similar to the *tokenization* used in Apple Pay, whereby the primary account number (PAN) is replaced by an alternative number called payment token.

10.10 Collaborative Offers

Service European services have adopted the collaborative model of operation. We will present in alphabetical order Mobito, from the Czech Republic, Mpass from Germany, and Pay2Me from Belgium.

10.10.1 Mobito

Mobito is a collaboration of three mobile network operators (T-Mobile, Telefonica, and Vodafone) and four major Czech banks. It is active throughout the Czech Republic. Thanks to its banking connection, it was granted permission to operate as an electronic money institution on May 2012. Thus, payments can be made directly from a linked bank account, with payment cards or through the balance value stored in a Mobito account. Users confirm a transaction with a PIN. Person-to-person payments are also supported.

Mobito makes use of the USSD protocol. The TSM is from Oberthur Technologies with its MoreMagic Mobile Money platform.

Daily summaries are available in the client app with more details on the server that can be exported into Excel files.

10.10.2 Mpass

Mpass is a German payment service that started in 2008 with the support of several mobile network operators (O_2, Vodafone, and Deutsche Telekom) in partnership with MasterCard. Initially, it was used for online and point-of-sale purchases with debit cards under the label Maestro® PayPass™ as well as direct debits (Lerner, 2013, pp. 112–115). Later, contactless payments at point-of-sale terminals equipped with the NFC technology were added.

Subscribers to any participant operator can set up their Mpass account at www.mpass.de and select a login and an Mpass PIN. Payment confirmation requires the PIN and an OTP (mTAN) received by SMS. Once the transaction is consumed, users receive an SMS receipt with the transaction number and the amount paid. Operators charge for the SMS exchanges.

Other users have to go through a verification procedure to establish a prepaid account. First, they have to define an activation code and then deposit 1¢ into their bank account. Once this transaction has been verified, they can sign into the Mpass portal to select a login and a password, record their mobile phone number, and define their PIN. They then prepay their service from the validated bank account.

10.10.3 Pay2Me

Pay2Me is a collaborative offer of several banks and mobile network operators. Each participating operator supplies a modified 128 kbit SIM card to its subscribers, which includes the cryptographic certificates, keys, and algorithms that secure the exchanges.

Both vendors and buyers must participate in the scheme and use their credentials during the transaction for identification. For example, vendors respond to a purchase request with their m-banxafe reference and the amount due. Buyers confirm by entering their four-digit PIN. The receipt is then returned to the buyer by SMS. Person-to-person payments follow the same protocol.

Pay2Me limits the amounts to a minimum of €6 and a maximum of €2500.

10.11 Payments from Mobile Terminals

Mobile terminals are used to offer two kinds of configuration. Mobile Point-of-Sale Terminals accept payments from the mobile phones. Also, they can be engaged in person-to-person fund transfers. A host of start-ups as well as larger companies such as Amazon, NCR, PayPal, Verifone, or Bank of America provide similar solutions. They compete partly on the technology but mostly on the fees they charge merchants or end users. Most of these fees go to card schemes and other institutions that handle risk and fraud detection. Therefore, those who opt for lower-profit margins aim at positioning themselves on a longer-term basis, to develop advanced store management products or to acquire a deeper understanding of the underlying business, taking advantage of the customer data they collect.

In principle, all offers must conform to the Payment Card Industry Data Security Standard (PCI DSS) in terms of storage security. Yet, some do not comply such as the solution from ScanPay, which requires the user to hold the payment card briefly in front of the device's camera, to scan the information, and then to enter the identification cryptogram. After recognizing the card details (account number and expiration data), the user is prompted to enter the PIN after which the application expedites the whole information to the payment gateway (Chicheportiche, 2013). For security reasons, however, the Mobey Forum recommends against using manual card entry and analysis of the card images to extract the card number either with optical character recognition or through contactless communication using Bluetooth (Mobey Forum, 2013a, p. 13).

The following sample illustrates the diversity and the similarity of the solutions.

10.11.1 iZettle

iZettle offers a comprehensive mobile point-of-sale solution based on the EMV specifications for integrated circuit cards. Its scope covers person-to-person payments and small business transactions. The package comprises a terminal application, a chip card reader (dongle) to be plugged into USB port of the merchant terminal, and a business management software. It is available in all Scandinavian countries (Sweden, Norway, Denmark, Finland), Germany, Mexico, Spain, and the United Kingdom.

After inserting the customer card into the card reader slot, the merchant presses the *Pay* button for the transaction to proceed. For low amounts, customers authorize the payment by signing on the touch screen of the merchant's phone or terminal with their finger. For large amounts, a PIN is needed. The payment information is never stored in the mobile or the card reader.

Merchants have a virtual till in an online account that iZettle manages. The settlements are conducted daily for a flat commission.

10.11.2 Payleven

Payleven is available in several countries such as Brazil, Germany, the United Kingdom, and Italy. It uses Bluetooth to pair the merchant smartphone with a special EMV card reader that Payleven provides. The buyer's card is inserted into the reader and the buyer validates the transaction with a PIN.

In 2014, Payleven was used to place fraudulent EMV debit charge transactions on New England banks even though these banks had not issued EMV cards. The charges were put in escalating amounts—nearly doubling with each transaction—indicating the fraudsters were putting through debit charges to see how much money they could drain from the compromised accounts (Krebs, 2014).

10.11.3 Paym

Paym is a mobile person-to-person payment service offered by several UK financial institutions. Both the sender and the recipient must register and link their mobile phone numbers with their bank accounts.

The sender enters the recipient's phone number (or selects their number in their own phone's address book) and initiates the transfer. The sender's bank retrieves the sender's banking information using the associated registered phone number. The sender is shown the recipient's name for confirmation but not their account details. The sender's bank sends the money using the Faster Payment Service.

10.11.4 Square

Square was the first to popularize mobile Point-of-Sale Terminals for small and/or itinerant businesses (florists, food-truck owners, etc.) in the United States. There are about 26 million of such entities in the United States that are not registered with a bank card scheme (Dembosky, 2012).

Square offer comprises a mobile terminal application, a magnetic stripe card reader (swiper) in the form of a dongle that plugs into the mobile device, and a business management software. The system works over wireless (Wi-Fi) or a cellular network.

A mobile cash application, SquareCash, is used for person-to-person payments. Users open an account with their debit cards and can send or request money from friends and family using their phone number or e-mail address (i.e., they do not need the details of a bank account), provided that they have accounts with SquareCash. SquareCash is not a banking institution, so it deposits the sum it receives in a predefined bank account. Transactions are encrypted. Other security items include account alerts and security code locks. The social photo-sharing Snapchat has launched a product that ties with SquareCash and lets users transfer funds between debit cards by sending a chat message containing a dollar sign and the amount by dragging images of notes around the screen of the mobile terminal.

Square Register turns a merchant's tablet or mobile phone into a cash register as well as a card reader. Merchants can arrange their merchandise into virtual shelves for potential buyers to browse through. When merchants register with Square, they link their payment coordinates (bank account or bank card) to their phone numbers and define a 4-digit personal code (PIN). This information is stored with Square. Square's special card reader (or dongle) plugs into the 3.5 mm audio jack of the merchant's smartphone or terminal. When a card is swiped through, the card reader retrieves the customer's payment data and encrypts them. The primary account number (card number) can also be keyed in manually, but this method of entry is discouraged because the phone keyboards can be preyed into (see also Mobey Forum, 2013a, p. 11). Customers confirm the transaction by signing with their finger on the merchant's touch screen. The receipt can be sent as a text message or e-mailed or they can retrieve it using its unique URL.

Square Register is typically configured to automatically deposit the earnings into the merchant's current account every 24 hours. It offers inventory and invoicing tools to track sales data and to generate receipts, offer cash advances, manage refunds, and design targeted advertising. Some of the sales data are available at the merchant's client; more detailed information can be downloaded from the server. Another product, Square Capital, offers loans at a 10% interest in exchange of a percentage of future sales that is deducted automatically. Thus, Square tries to appeal to small merchants by improving their business and by tailoring discounts and loyalty programs to suit their customer base.

The client on the buyer's phone, Square Wallet, was discontinued in 2014. It operated as follows. During registration, users specified their payment cards, took their picture with their phone's camera, and stored it in their Square Wallet. Within 100 yards of a terminal equipped with Square Register, their payment card and picture are made available to the clerk. At the checkout terminal, the buyer opens his or her Square Wallet and selects the business being visited to communicate with the checkout terminal. The buyer provides his or her name to the clerk, which then verifies the identity with the picture and issues a receipt for the transactions. Small amounts did not require an explicit authorization. For large sums, a PIN confirmation is needed.

One possible reason of the lack of success of Square Wallet is that it required merchants to upgrade their checkout terminals and to change their procedures depending on whether the customer had the client application ready on their phone.

10.11.5 Starbucks Card Mobile

Starbucks Card Mobile is a mobile payment application designed to speed up the ordering process and to reduce the use of plastic gift cards. Users install an application in their mobile phones that is linked to a prepaid account with Starbucks. The Starbucks app displays a barcode that is scanned at the point of sale to debit the prepaid account with the amount indicated. The application tracks recent transactions and reward points.

10.11.6 Zoosh

Verifone uses a technology called Zoosh and an application called Way2ride for iPhone and Android terminals to pay for taxi rides in New York City.

Each taxi is assigned a unique ID that is transmitted, in the form of inaudible sound waves, to the mobile terminal that has installed Way2ride. The client forwards the code to a server for authentication and retrieval of the consumer's preferences. These are then sent to the taxi over the mobile network and then displayed on a special backseat screen for selection and confirmation. With the auto pay option, a default card is preselected, so the consumer would merely need to enter the amount desired and confirm the payment.

10.12 Summary

Mobile payments depend on a new industrial architecture that integrates many existing lines of business and technologies in mobile communication and electronic payments. Since the late 1990s, mobile operators have envisioned mobile commerce as the next logical step to increase their revenues and profits. However, early adopters of mobile commerce were mostly found in countries lacking a developed financial infrastructure.

Mobile broadband networks capable of carrying data at higher speeds can improve the customer's experience during mobile transactions. Mobile network operators would like to capitalize on these opportunities to recover their investment, but they are facing new entrants from the field of information technology. Financial institutions have likewise promoted mobile banking as a way to pay bills, check balances, and deposit checks from mobile phones.

One consequence of the use of mobile devices is to increase neighborhood commerce and offer small and medium enterprises payment capabilities that they could not have afforded.

Questions

1. What are the advantages and disadvantages of the use of mobile phones for payments over separate smart cards?
2. Why is the bank-centric operation associated with dual-chip or dual-slot phones?
3. What arrangements can be made to avoid giving a specific bank or a network operator the central player in mobile payments?
4. Define an embedded SIM.
5. Why are there many options for the inclusion of the secure element in mobile terminals?
6. What are the main options for the inclusion of the secure element in mobile terminals?
7. Evaluate the use of software emulation in the provision of the secure element in mobile terminals.
8. What is the purpose of Java2 Platform Micro Edition (J2ME)?
9. Compare and contrast rooting and jail breaking.
10. Discuss the vulnerabilities of using Bluetooth for mobile payments.
11. Describe the operating modes of NFC.
12. Compare NFC and Bluetooth.

11

Micropayments

Micropayments involve a tiny amount of money, typically between 10¢ and $10, and are uneconomical to process with typical noncash systems. However, if the cost per transaction is made commensurate with the small values processed, micropayments have the potential of replacing cash to pay for daily items such as parking fees or transportation tickets, and also virtual goods such as streaming services, ringtones, online games, or digital music. Telephone companies have long tracked usage using the number of impulses per second, and the telephone tick could have been used to track micropayments (Pedersen, 1996). In any event, micropayments were first used to replace cash to pay for calls from public telephone booths. Today, online micropayments are also imbedded in mobile phones or can use cloud-based virtual purses.

This chapter gives an overview of micropayments for both face-to-face purchases and online transactions. Electronic purses are chip cards that store prepaid monetary value in legal tender to be used at point-of-sale terminals. They are in the form of stand-alone cards or applications incorporated in a debit or credit card. We describe the standardized interfaces of electronic purses such as the Common Electronic Purse Specifications (CEPS), the GlobalPlatform, and the Electronic Commerce Modeling Language (ECML). We review some of the present commercial offers for face-to-face commerce, namely, in alphabetical order, Advantis, FeliCa, GeldKarte, and Proton. Next, we discuss the evolution of online micropayment systems and highlight some research projects from a decade ago. Based on this experience, we discuss some of the factors that may have affected market acceptance of the various solutions.

11.1 Characteristics of Micropayment Systems

The main feature of micropayment systems is the use of various techniques to reduce the cost of processing. The main approaches used are service prepayment, offline authorization, and grouping of micropayments before financial clearance. For face-to-face cards, the cost can be reduced with cheaper cards by reducing the amount of processing power needed. A method available in the United States is only to use the lower-cost Automated Clearing House (ACH) network for routing the transactions instead of the card network. Finally, there are some management techniques to reduce the load.

11.1.1 Prepayment

Prepayment allows the system operator or sponsor to invest the collected sums until the relevant service is consumed.

Prepayment for face-to-face transactions is mostly with prepaid cards, including gift cards. The value stored can be expressed in terms of a legal tender or of Jetons (telecommunication service units, transportation tickets, game currency, etc.). When legal money is stored in the electronic purse, a bank is responsible for issuing the purse and loading it with value through a secure physical device, such as an automated teller machine (ATM). When the purse issuer is not a financial institution, the loading function is still the responsibility of the credit institution, while the purse issuer (also called intermediary or broker) manages the card design and distribution. The distribution of purses included in mobile phones is often arranged with the associated mobile phone operator.

11.1.2 Offline Authorization

Offline authorization reduces the communication cost per transaction. Micropayments, in general, do not require any confirmation before authorization. Disposable prepaid cards are not typically personal identification number (PIN) protected and do not require online authentication of the card or the cardholder. In the case of PIN-protected debit cards, cardholder authentication can be performed at the terminal level using locally stored check lists that are updated daily.

Integrated circuit cards with an integrated security application module (SAM) can avoid querying the authorization server for each transaction. The SAM is a tamper-resistant device where the cryptographic secrets and/or critical information are stored. It fulfills the following functions:

- Identification of the card and the cardholder
- Authentication of the card and the cardholder

- Management of a reduced blacklist card, downloaded from the main authorization server
- Automatic generation of an authorization request based on risk factors predetermined with the bank or card scheme operator
- Generation of data to confirm the transaction
- Storage of the transaction data
- Nightly connections to the bank servers to upload the transaction data and synchronize the decision criteria, risk situation, etc.

11.1.3 Aggregation of Transactions

The idea here is to collect the amounts for several micropayments and then issue one single authorization request for all the accumulated amounts instead of requesting the authorization for each transaction separately. Settlement is done periodically, for example, at the end of the collection period. This aggregation reduces the per transaction cost of processing and communication. Grouping has also the secondary advantage of providing partial anonymity for individual transactions. This solution is common in the kiosk model originally implemented by the Minitel.

For online micropayments, the transactions may be settled periodically by the means of payment instruments determined at subscription. If the service provider has combined the micropayment services to its main offer such as mobile communication or Internet access, another billing option is to add the micropayment transactions to the main invoice.

11.1.4 Reduced Computational Intensity

In integrated circuit cards, reducing the computational intensity lowers the chip complexity and alleviates the requirements on its performance, leading to a lower unit cost per card. One way to achieve this is to use symmetric encryption algorithms instead of public key algorithms whenever possible.

In consumer applications, the absolute maximum of transaction time is 300 ms, with most transactions taking less than 150 ms. If RSA encryption is too slow, faster techniques such as the elliptic curve cryptography (ECC) may also be used.

11.1.5 Routing through the ACH Network

In the United States, routing a transaction through the Automated Clearing House (ACH) system rather than the signature credit network reduces the cost. This is the case of PIN debit cards used at Point-of-Sale Terminals. It is seen how changes in banking regulations and infrastructure as well as advances in computation and communication can reduce the niche where micropayment schemes would be desirable.

11.1.6 Management of Micropayments

Some of the approaches to reduce the cost of back-office systems is that payments cannot be revoked or disputed. Also, Jetons can have expiration date to reduce the management load and cost. This is particularly the case of Jetons used in online transactions.

11.2 Standardization Efforts

In the mid-1990s, there were over 20 European projects for electronic purses independent of each other. All of these purses offered the same functionalities but were incompatible with each other (Kirschner, 1998). This caused potential users two major drawbacks: they must have as many purses as necessary to conduct their shopping and they must mobilize parts of their holdings by paying before consumption. The lack of interoperability forced operational difficulties upon service providers and imposed additional costs on manufacturers. Service providers are locked to specific manufacturers and readers that can read multiple formats are more expensive and take a longer time to process a transaction. For purses to operate across national boundaries, they must all conform to the same international standard. Furthermore, the rise of the Internet induced a shift in the strategy of electronic purse operators to overcome the technical and commercial obstacles to international networking. The planned introduction of a common European currency in 2002 gave considerable momentum to the harmonization and standardization efforts to such a proliferation under control by common agreements at the level of the protocols, the application, and the terminals.

In this section, we review the main efforts to standardize the interfaces to electronic purses.

11.2.1 Common Electronic Purse Specifications

The first version of the Common Electronic Purse Specifications (CEPS) published in 1999 was guided by the specifications that became the four-part CEN standard, EN 1546, in 1999. This standard defines the interfaces, the security architecture, and the functionalities of electronic purses to ensure the interoperability of several applications. CEPS was optimized for simple processors while making it compatible with EMV. Using the EMV specifications to build purses ensured their interoperability and reduced the cost of their hardware and software by allowing competition among suppliers (CEPSCO, 2001).

The commonality of purses covered technical aspects such as the cryptographic operations, the commands and responses between the card and the devices that support it, the functionality for the SAM in the purse (PSAM) and in the fund loading devices (LSAM), and the data captured during transactions.

Common business procedures are related to the minimum data needed to clear and settle the transaction, the minimum responsibilities of the card issuer in detecting fraud and minimizing risks to the purse system, and the role of the certification authority (CA) and the data elements needed to request and generate a public key certificate.

The following aspects will be described in more details:

1. Authentication of the purse by the issuer
2. Loading of value
3. Point-of-sale transactions

11.2.1.1 Authentication of the Purse by the Issuer

Authentication of the purse by the issuing institution, which is not necessarily a banking institution, can be done either online or offline.

Online authentication uses a secret key shared between the issuer of the purse and the smart card of the purse itself. This key allows the derivation of a common session key that will serve to compute the MAC to protect the integrity of the data exchanges and to encrypt the authentication data. CEPS limits the key size on open networks to 8 octets (64 bits) but leaves open the choice of the encryption and hashing algorithms.

Offline authentication uses RSA-based hierarchical certificates for mutual authentication and for the exchange of a temporary session key next between the chip card and the terminal. This session key will be used to encrypt the data according to Triple DES.

11.2.1.2 Loading of Value

Loading the purse with value depends on whether the purse is linked to a bank account.

If the purse is linked to a bank account, the conversion of the monetary value to a dematerialized form stocked in the purse is under direct control of the holder's bank. Authentication of the card, verification of the holder's identity on the basis of a PIN, and authorization of the transaction can be done at once. The exchanges involve the card, the load device, and the authorization server of the issuing bank.

The purse is not linked to a bank account when the line of credit is from a totally separate account or if it entails a revolving credit. In these cases, the risk of error or fraud increases and the authentication is more complex. The communication protocol must verify the integrity of the value transfer from the client's bank (or from that of the purse issuer) to the acquirer bank in addition to the authenticity and the identity of the cardholder and of the card that the holder presents. A Loading Secure Application Module (LSAM) provides the security during the loading of value. The integrity of the communication is protected with signatures using the hashing algorithm SHA-1. Random numbers are included to protect against replay attacks.

11.2.1.3 Point-of-Sale Transactions

The protocol for point-of-sale payments defines procedure for offline reciprocal authentication of the purse and the point-of-sale terminal. This protocol conforms to the EMV specifications and covers single transactions as well as a series of periodic transactions (such as the payment for service bills).

The terminal can authenticate the card with the public certificates of the card issuer and of the issuing bank. Similarly, the card authenticates the Purchase Secure Application Module (PSAM) of the terminal with its certificate. The integrity of the exchanges is protected with MAC calculated by symmetric encryption. The terminal checks the validity of the PIN of the cardholder.

The transaction amount is displayed on a screen, and the cardholder has to indicate consent. Data relative to the transactions are exchanged between both sides of the transaction, and their integrity is ensured with the help of MACs. The amount of the value stored in the purse is updated. It should be noted that the use of screens raises some problems regarding their resistance to breaking as well as their reflectivity and energy consumption.

The specifications allow the possibility of canceling the last transaction provided that the cancellation is done from the same PSAM that has managed it in the first place and by using the same security parameters. The purpose of these restrictions is to discourage fraud.

Finally, an audit trail of the transactions is stored in the security module for the authorized operator to collect.

11.2.2 GlobalPlatform

GlobalPlatform is an association that develops technical specifications for the physical and logical security of smart cards, particularly the secure element (SE), the trusted execution environment (TEE), and the system messaging of trusted end-to-end solutions.

GlobalPlatform also manages a compliance program to confirm that products meet its functional requirements.

The Small Terminal Interoperability Platform (STIP) is an organization of terminal manufacturers and smart card manufacturers that wanted to define a Java specification for interoperable terminals and transaction-oriented devices. They later merged with GlobalPlatform. As of January 2014, GlobalPlatform was working on a new secure channel protocol, named Secure Channel Protocol 11 (SCP11) using on elliptic curve cryptography (ECC) and on AES for secure messaging.

11.2.3 Electronic Commerce Modeling Language

The Electronic Commerce Modeling Language (ECML) is described in RFC 3106 (2001) and RFC 4112 (2005). It defines a common XML format for the exchanges between applications and merchant sites so that buyer's information can be imported directly from a digital wallet. The exchanges cover billing, shipping, and payment information. This saves the customer from repeatedly filling out the same information on multiple sites.

We now go over some of the commercial offers for micropayments in point-of-sale transactions.

11.3 Electronic Purses

An electronic purse is a prepaid card that contains money in a given legal currency and can be used for any purpose (if acceptable to the seller). An electronic purse is usually protected by security mechanism to identify and authenticate the owner. Finally, an electronic purse is traceable if it points to a bank account. Van Hove (2006) presents a survey of 16 different projects and their commercialization strategies.

This section presents the main projects that are still in operation. Mondex is one of the highest profiles among these unsuccessful projects.

The first patents for Mondex were filed on April 12, 1990, in the name of T. Jones and G. Higgins, two employees of the National Westminster Bank (Natwest) in the United Kingdom. In 1991, the project started in partnership with several Japanese suppliers: Dai Nippon Printing Co. for the card, Hitachi Panasonic and Oki Electric Industry for the integrated circuits. In 1992, BT (formerly British Telecom) and the Natwest Bank declared their support. In 1993, Midland Bank jointed the project, and the two banks united to form Mondex UK with equal shares from each. Finally, Mondex International was formed in the summer of 1996 as an independent company with MasterCard as a majority owner, together with 17 major multinational corporations (Mayer, 1997).

Various aspects of the protocol were proprietary. The Mondex exchanges used a new type of MIME e-mail messages, but two Mondex cards could exchange values.

Starting in 1995, several trials were conducted at Swindon (a locality west of London) in the United Kingdom, in San Francisco and Manhattan in the United States, and at Guelph in Canada. In Hong Kong, more than 45,000 cards were distributed with the participation of more than 400 merchants in three commercial malls.

Participants did not find any advantage in replacing coins and bills with an electronic purse (*Banking Technology*, 1998; Hansell, 1998). Merchants were unhappy with the difficulty of the operation and by the time it took to conduct a transaction (less than 5 seconds) (Westland et al., 1998; Van Hove, 2001). Thus, despite the strong institutional support, Mondex failed in the market place.

11.3.1 Advantis

Advantis, or Advantis JCrypto GP card, is one of the contactless EMV microprocessor cards, which uses the GlobalPlatform specifications to support multiple applications. It offers an open platform for different forms of payments, including micropayments with legal tenders or with Jetons as well as nonmonetary functions.

The monetary functions include an application (applet) conforming to the EMV specifications, a purse conforming to the CEPS interface, as well as another purse implementing the *servicios para medios de pago* (SERMEPA, services for instruments of payment) specifications. SERMEPA is a leading Spanish technical provider for the payment industry in Spain. The EMV payment applet is integrated with the operating system in the ROM of the chip rather than being stored in the external memory. The purse application (applet) is built on the TIBC 3.0 (*Tarjetas inteligente y cajas de bancos*, Smart card and credit union) operating system from Visa Spain. The symmetric encryption algorithms are DES and 3DES, and the size of the memory can be 2, 4, 8, and 16 Koctets. The Jeton holder for public transportation uses a contactless card with the TIBC 4.0 operating and proprietary applications. The size of the memory is 8 Koctets.

The nonmonetary functions include access control to physical buildings or networks, a digital identification card, medical files, loyalty programs, and so on. The card has a cryptoprocessor dedicated to public key infrastructures as well as to authentication based on X.509 certificates.

11.3.2 FeliCa

FeliCa is a rechargeable contactless smart card from Sony built on the CXD9559 operating system (Sony, 2012). The data transmission channel uses a carrier frequency of 13.56 MHz to exchange half-duplex traffic at the data rate of 212 kbit/s. The file structure is defined in the Japanese standard JIS X 6319-4, while the radio interface is now standardized as ISO/IEC 18092 (ECMA-340). The use of FeliCa for micropayments is discussed in Chapter 11.

FeliCa is externally powered. The FeliCa card reader supplies power through inductive coupling to cards in range. When the data transfer is complete, the reader stops the supply of power.

FeliCa can accommodate multiple applications from distinct suppliers each with its own set of cryptographic keys and access rights. The exchanges are protected with the Triple DES or AES algorithms, using dynamic keys, that is, generated at each session. The UK IT Security Evaluation and Certification Scheme (2002) has confirmed that FeliCa RC-S860, ROM Version 6, OS Version 3, complies with the security requirements defined by the EAL4 Common Criteria of ISO/IEC 15408.

FeliCa was one of the first contactless integrated circuit cards used in transportation systems. It is the basis for the Octopus card in Hong Kong, the EZ-Link card in Singapore, and Suica (Super Urban Intelligent IC Card) for the East Japan Railway (JR East). With Suica, passengers pass through station gates by placing the card over a reader. In these applications, typical values of payments are less than ¥1000. The maximum value that can be stored on a card is ¥20,000. The card can be reloaded at ticket vending machines in values ranging from ¥1,000 to ¥10,000 (Bank of International Settlements, 2004b). The EZ-Link card is used as an electronic purse in addition to its role as a Jeton holder for transportation ticketing.

The Edy (euro, dollar, yen) application is an electronic purse installed on a FeliCa card. The application is managed by the bitWallet consortium formed of several Japanese firms, with NTT DoCoMo and Sony as the principal shareholders.

NTT DoCoMo integrated the FeliCa into mobile phones for its i-mode® service. Next, it introduced a "wallet phone" for the Osaifu-Keitai services in July 2004 so that multiple FeliCa systems (such as Suica and Edy) can be accessed from a single mobile phone.

Osaifu-Keitai services are based on a server-based wallet (i.e., in the cloud) that contains electronic money, identity cards, credit cards, public transport fare collection cards for railways and buses (i.e., Edy and Suica), loyalty programs, airline check-in, and others.

11.3.3 GeldKarte

The GeldKarte is the chip version of the Eurocheque card, introduced in 1968 to guarantee check payment up to a predefined ceiling (GeldKarte, 1995). In the 1980s, it evolved into a magnetic stripe card and a means for cash withdrawal from ATMs. As such, it was widely used outside Germany, particularly in Austria, the Netherlands, and Switzerland (Dragon et al., 1997, pp. 156–157).

The integrated circuit version results for an initiative by the *Zentraler Kreditausschuß* (ZKA) or Central Credit Commission, a group that the German banking industry asked to design a multifunctional chip card for face-to-face commerce. Initially, the card could be used as a debit card, as an electronic purse, or in an anonymous manner as a prepaid card unlinked to any account. More functions were added later such as online banking or online payments using a secure terminal with a display. It can also be used to print account statements or even to verify the purchaser's age when buying for a cigarette vending machine.

Access to the stored data, the structure of the messages between the card reader and the card, and the logical architecture conformed to ISO/IEC 7816-4 (Althen et al., 1996). The chip has the following capacities: 12 Koctets of ROM, 256 octets of RAM, and 8 Koctets of EEPROM (Kirschner, 1998; Rankl and Effing, 2010, pp. 783–788).

In 1996, Deutsche Telekom, together with the German train operator Deutsche Bundesbahn, and the Association of Municipal Transports (VDV), introduced the PayCard, later called the T-Card, based on the GeldKarte design (Adams, 1998). GeldKarte is also the building block for the Navigo* purse, used in the public transportation systems of the Parisian Region from 2004 instead of paper tickets. Navigo is also used for parking around Paris as well as in some shops for payments less than €30.

The following discussion is focused on the electronic purse functions.

11.3.3.1 Registration and Loading of Value

Merchants and cardholders register by signing a contract with the system operator. The merchant's terminal has a SAM in the form of a card (the merchant's card). The identification data are stored in a central file that the system operator maintains.

* Navigo combines two preceding purses—Modeus and Moneo.

The card is loaded with value at a special loading terminal using a secured line to the loading center. Loading comprises two distinct phases: a debit operation of a bank account and loading the purse proper. The parties in the loading phase are the cardholder, the loading operator (which may be distinct from the issuer bank), and the authorization center.

The debit operation consists of five main steps (Sabatier, 1997, p. 76):

1. Card authentication
2. Identification and authentication of the cardholder by the operator using the confidential code (PIN)
3. Input of the required amount
4. Request for authorization from the authorization center
5. Update and synchronization of the records of all parties

Figure 11.1 depicts the various operations during the loading of value, numbered in a chronological order (Sabatier, 1997, p. 77). In this figure, the functions of the loading operator, the authorization server, and the data store have been separated from those of the issuer bank, although the same entity may perform them all.

The value is transferred from the client's account to a clearing account associated with the card, which will be used for compensating the client.

11.3.3.2 Payment

The merchant terminal controls the exchange of value between the client's card and the merchant card with the help of the SAM that it contains. The payment transaction is considered offline because it does not depend on the intervention of an authorization server or of the issuer bank. This transaction consists of the following steps: reciprocal authentication of the client's purse and the merchant's SAM, transfer of the debit amount, and production of the encrypted electronic receipts.

Reciprocal authentication of the purse application of the cardholder and the merchant card takes the following steps:

1. Following insertion of the client's card in the terminal slot, the merchant's payment manager requests the card for its identification *CID*.
2. The SAM of the merchant terminal (the merchant card) obtains a random number *RAND* and inserts it in the command *Start debit process* that the merchant terminal sends to the client's

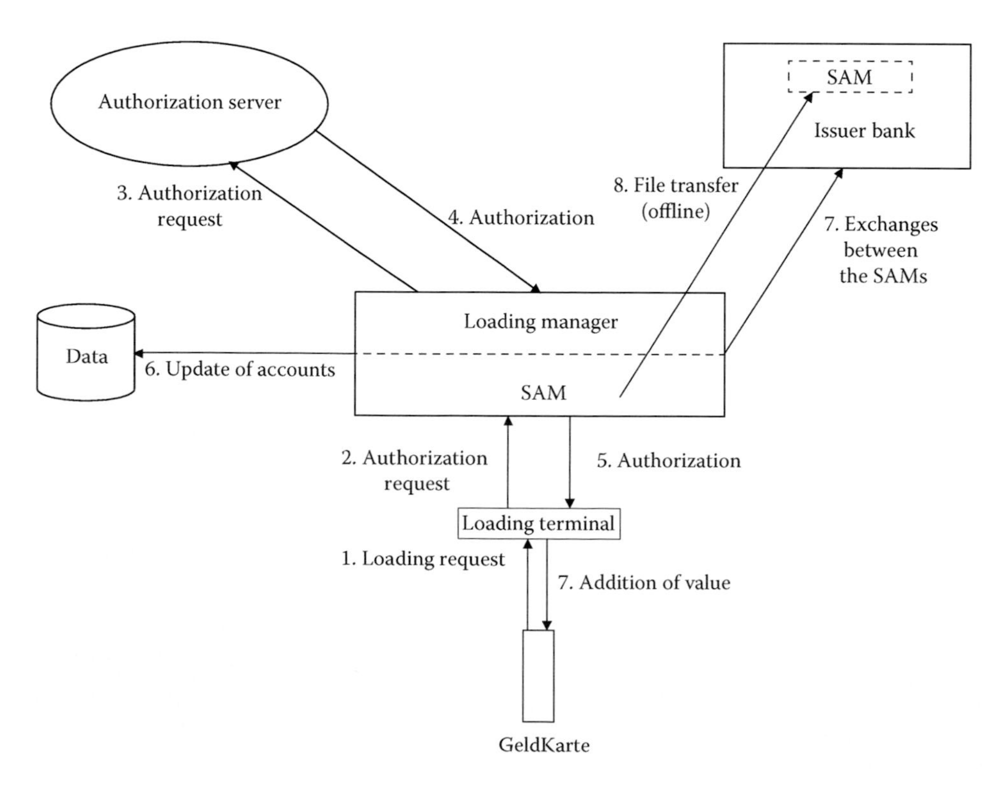

FIGURE 11.1
Charging of the GeldKarte purse.

card. The use of the random number protects against replay attacks.

3. The card responds to the command by giving its identifier *CID*, concatenated with its current sequence number *SNo*, the random number *RAND*, and the hash *H(CID, SNo, RAND)*, where *H( )* is the hash function of ISO/IEC 10118-2 (1994). The hash is encrypted using DES with the card's secret key K_C to calculate the message authentication code (MAC).
4. The merchant card derives the card key, K_c, starting with the master key K_{GM}, using a "diversification" algorithm and then decrypts the MAC. The integrity of the received message can be verified by comparing the value obtained from the decryption of the MAC with the hash value *H(CID, SNo, RAND)* from the received message.

The exchange of value takes place in the following steps:

1. The terminal sends to the merchant's SAM the command *Start payment*.
2. The merchant's SAM responds with the message *Debiting*. This message contains the merchant identifier *MID*, the transaction number *TNo* as defined by the merchant's card, the sequence number *SNo* of the client's card, and the serial number of the client's GeldKarte as retrieved from a store within the SAM. The integrity of this information is protected with a MAC using the client's key K_{RD} and then encrypted with the same key.
3. The terminal transmits the message to the card after adding the value to be transferred, *M*.
4. The client's card verifies that the MAC of the message corresponds to the received data, thereby proving the authenticity of the merchant. The purse checks that the debit is correctly attributed by verifying the identification of the card. The presence of the sequence number *SNo* enables it to detect a replay attack when the sequence number in the message is not the same as the number that it had sent to the terminal.
5. The purse verifies that the amount to be withdrawn is less than the total value stored. If this is the case, it reduces the actual amount by the value of the payment, increments *SNo* by one unit, records the transaction in a register, and computes a proof of payment. The proof is constructed by concatenating the amount *M*, the merchant identifier *MID*, the sequence numbers *SNo* and *TNo*, and the card clearing account number *ANo*. The MAC of the result is calculated with K_{RD}, and the whole set is sent to the merchant card in the message *Check payment*.
6. The merchant SAM recomputes the amount *M′* starting from the proof of payment to verify that the amount paid is the same that the amount as was requested (*M*). It will verify that the merchant *MID* is its own and that the sequence number *TNo* is the same number that it had produced.
7. If all the checks are positive, the merchant's SAM increments the transaction number *TNo* by one unit, increases the value that the merchant card stores by *M*, informs the terminal of the success of the transfer, and records the transaction data by recording it with a new key, K_{ZD}.

Figure 11.2 illustrates these exchanges.

To start the clearance process, the merchant presents, offline, to the issuer bank the encrypted proof of payment to credit its account through the financial settlement circuits. Because the proof includes the transaction number *TNo*, the receipt cannot be reused in another fraudulent clearance request.

11.3.3.3 Security

The GeldKarte protocol uses cryptography to guarantee the following:

- The cardholders and the merchants are authentic.
- The request for financial clearance is used only after the debit of an authenticated GeldKarte and only to the benefit of the card of the mentioned merchant.
- The messages are not reused for fraudulent claims.

Message integrity is verified by calculating a MAC with the symmetric encryption algorithms DES or triple DES in either the cipher block chaining (CBC) mode or the cipher feedback (CFB) mode. These computations follow ANSI X9.19, which specifies methods for message authentication for use by financial retail institutions. They use the hashing function defined in ISO/IEC 10118-2 to produce a message digest of 16 octets (128 bits). DES or Triple DES algorithms are implemented in the merchant card, while currently the client's card uses DES only. (Triple DES is planned for the future.)

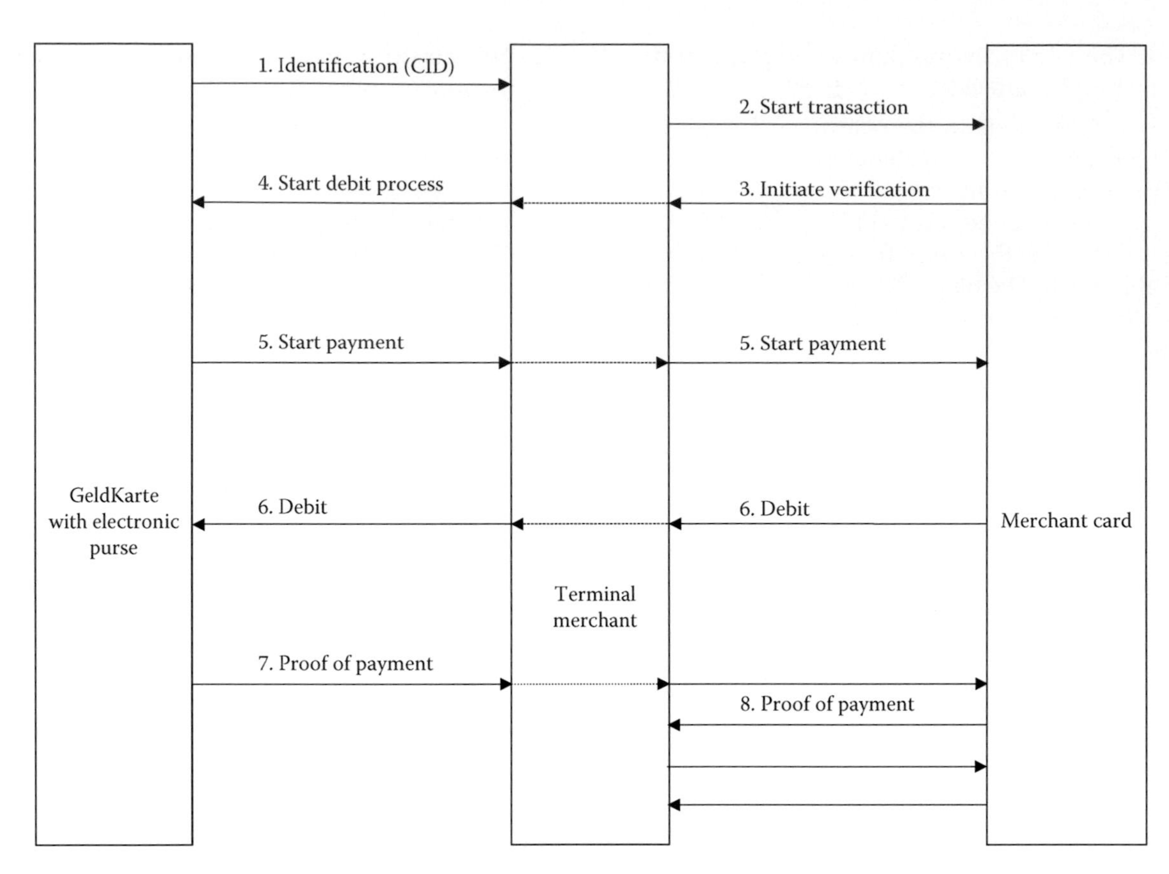

FIGURE 11.2
Message flow during a payment with GeldKarte.

The cardholder identifies himself or herself by using a secret code or PIN. The PIN is required to load the card with value but not for making a payment. In this way, payments do not have any relationship with customer's accounts. Anonymity is achieved with respect to the merchant but not with respect to the financial centers.

Each GeldKarte is personalized with an identifier and a symmetric encryption key. The parameters of the cardholder's card are the identifier *CID* and the key K_{RD}, while the corresponding merchant's parameters are *MID* and the master key K_{GM}. As mentioned earlier, with K_{GM}, the merchant card can derive the card K_{RD}, thereby avoiding the exchange of this secret key between both parties. The K_{GM} key is stored in the SAM of the merchant terminal (the merchant card).

During the personalization phase of the GeldKarte, the serial card number, the encryption keys, and the cardholder secret code are stored in the "secret zone" of the card. The same information is also stored in an encrypted file that will be stored under heavy security in the issuer bank.

Note that each party gives the transaction a unique number. This number is sent to the other party to include in its response; this mechanism builds a defense against replay attacks. The number that the merchant card defines is the transaction number *TNo* and the number that the client's card assigns is the sequence number *SNo*.

Partial anonymity of the cardholder with respect to the merchant is possible provided that the data stored in the card are not linked in any manner to the bank account of the cardholder. However, if the encrypted data are presented to the clearance center as a proof, the identity of the cardholders can be revealed.

11.3.4 Proton

In 1993, Banksys (an interbank society responsible for electronic payment in Belgium) initiated the Proton project to develop a multifunction payment card. Introduced early in 1995, it was available over the whole of Belgium in 1996. In the Netherlands, the Interpay, the Dutch equivalent of Banksys, renamed the Proton technology Chipknip. The introduction of Chipknip set a worldwide record at the time because 12 million cards were already in circulation in 1997. Several cities made the use of Chipknip compulsory for feeding parking meters.

The Chipknip card comprises a traditional debit card, loyalty programs (management of discount coupons used in marketing), and an electronic purse to pay for public transportation. The electronic purse function is for amounts less than around $20. Beyond this threshold, the function of debit card with PIN verification takes over. With offline verification, the transaction takes about 2–3 seconds (Dawirs, 1997).

The Proton electronic purse is available in Australia, Brazil, and Canada, under the name Exact brand, in Germany, New Zealand, and Sweden; in Switzerland, its name is Cash. In Hong Kong, the contactless version of Proton is used for public transportation systems (e.g., the Octopus card). However, the electronic purse was retired from Belgium on December 31, 2014. Users were requested to off-load the remaining amount back onto their bank account before the expiration date. The reason for the lack of use is because Belgian banks added the Proton application into their debit cards and some added an annual fee for the activation of the electronic purse application.

The Proton architecture is largely inspired by specifications EN 1546, but the Proton specifications document is proprietary. The chip capacity varies between 6 and 16 Koctets of ROM, from 1 to 8 Koctets of EEPROM, and 256 octets of RAM (Kirschner, 1998; Rankl and Effing, 2010).

The merchant terminals are equipped with a SAM to verify the cards and perform the mutual authentication of the card and the terminal. The security of the exchanges with the collection center is assured with the DES for confidentiality and RSA for authentication.

Proton is owned by Proton World, initially a joint venture by American Express, Banksys, ERG, Interpay, and Visa International. In 2001, Proton World became wholly owned by the Australian-based ERG, which was behind the Octopus card used in the mass transit system of Hong Kong.

11.4 Online Micropayments

Online micropayment solutions target the sale of virtual content: newspaper archives, online games, horoscopes, games, pictures, music, videos, and so forth. A trusted third party facilitates the exchanges between the buyer and the seller and guarantees a certain level of security. The intermediary gets revenues from a commission per transaction and/or fees.

Between 1995 and 2014, products for online micropayments passed through at three generations. Those of the first generation focused on technical innovations. The second generation built on the experience from telephone cards, while the following generations benefited from the growth in mobile communication.

The technical differences among the various offers concern the following points:

1. The various protocols for loading and charging the monetary value
2. The currency stored in the virtual purse or Jeton holder
3. The network access medium and protocol
4. The methods to reduce the operational cost (offline verification, grouping of transactions, etc.)
5. The security protocols

There are also business and managerial differences such as

1. Whether they are accounted or are based on the use of Jetons
2. The entity in direct contact with the end user, such as an access provider, a telecommunication company, a bank, a Trusted Service Manager, (TSM) etc.
3. The geographic coverage area of the solution
4. The conditions imposed on the merchants including commissions or per transaction fees

In accounted system, users supply the operator with their identity and financial credentials at subscription time. The system keeps track of the tiny transactions of each user and bills them at the end of the accounting period. In Jeton-based system, the value expressed in Jetons changes hands at the instant of consumption.

11.4.1 The First Generation

Systems of the first generation pioneered techniques for online micropayments. Some of these schemes evolved with the market, but others did not survive. We will discuss two examples from each category: First Virtual and KLELine that have folded and ClickandBuy and Bankpass Web that are still operational.

11.4.1.1 First Virtual

First Virtual operated between 1994 and 1998. It secured payments with credit cards without resorting to encryption by using a rudimentary version of two-factor authentication (New, 1995; Borenstein et al., 1996).

Users could subscribe by postal mail, by telephone, by facsimile, or by the Internet. First Virtual would return by e-mail an acknowledgment with a subscription number and activation instructions to be executed over the phone, by facsimile, or by postal mail. After the activation, users would send their credit card numbers. Once the credit card numbers have been verified, the First Virtual server would send to the customer a "Virtual Pin" to be used for authentication when purchasing. In this manner, the transfer of the banking coordinates occurred only once and never on the Internet.

During a transaction, the merchant would decide whether to wait for First Virtual's authorization or to supply the information immediately. In that case, First Virtual would compensate the merchant in case of fraud or lack of payments.

Service providers would tally the transactions on their site and then send the details to the First Virtual server to debit customers' accounts with the sum total of purchases exceeding $10. First USA bank collected payments in return for a commission of 2% of the transaction amount. However, because U.S. legislation allows repudiation of credit purchases within 90 days, First USA would wait that long before crediting First Virtual for the corresponding value. Next, First Virtual would compensate the merchants after deducting a flat transaction fee of 29¢ per transaction.

As the trusted service manager, First Virtual was aware of all the exchanges, of the identities of the parties and their banking references, of the content of the orders, and of the amounts charged. However, the anonymity of the purchaser with respect to the merchant as well as the anonymity of both the customer and the merchant with regard to the respective banks was assured.

The customer could also refuse the invoice, even after receiving the information, but First Virtual would cancel the users' or merchants' accounts in case of abuse, for example, when subscribers contested their bills frequently or when suppliers had many invoices challenged or had many complaints against them.

One obvious weakness of the scheme was the long delays before reimbursement of the merchants.

11.4.1.2 KLELine

KLELine was a secure payment platform, a virtual mall, and a cloud-based payment system called GlobeID. The client software was a plug-in for the browser to give access to the virtual purse. The payment system was under the control of a succession of banks (Pays and de Comarmont, 1996; Romain, 1997). KLELine covered a large range of payment instruments:

- A virtual purse for values less than 100 French francs (around $15). The virtual purse was supplied from a bank account to ensure that funds are always available.
- Bank cards for purchases exceeding 500 FF (around $5).
- Any of the two methods mentioned for purchase amounts between 100 and 500 FF.

The KLELine payment server was the intermediary between the merchant and the customer as well as the gateway between the Internet and the banking network. It was also a notary, a trusted third party, as well as the host of the virtual mall. Merchant authentication and the verification of the integrity of the messages applied RSA public key encryption with a key size of 512 bits, MD5 hashing, and an unspecified symmetric encryption algorithm. The customer was identified with a customer identifier and PIN. During the authentication phase, the integrity of the exchanges between the customer and the KLELine server was verified with public RSA key signatures and secured with a pair of RSA keys of 512 bits. During that initial exchange, a session key was established between the customer and the server for symmetric encryption of the exchanges.

KLELine accepted transactions in 183 currencies with the rates updated every 6 hours, in return of a commission ranging from 10% to 20% for micropayments (payments that use the virtual purse) and from 2% to 4% for bank card payments. The clearance and settlement procedures used in banking networks were applied to bank card purchases. For micropayments, the customer's virtual purse was debited for each transaction to the benefit of the merchant's virtual cash register. Then, 45 days after the purchase (to take into consideration delays in customer objections), KLELine emptied the virtual cash register and distributed the amount collected among the merchant, the virtual mall, and itself.

Banking support allowed KLELine to operate with legal money and various currencies. However, successive bank owners were not necessarily interested in such a small business. At the time of its liquidation, the product was available over around 500 sites, with 85 only in France.

11.4.1.3 ClickandBuy

ClickandBuy started in 2000 as FirstGate Click&Buy with a virtual purse offer and evolved in parallel with the increased use of mobile networks. It became BT Click&Buy, then a fully owned subsidiary of Deutsche Telekom in March 2010. It can be used to pay from any site that displays the ClickandBuy button, such as Britannica Online, the online version of the Encyclopedia Britannica, Apple iTunes Store, and T-Online.

The purse can be accessed from desktops and mobile phones. Purchases are grouped in a monthly bill to be debited at the end of month on a bank card or a direct debit from a back account.

To subscribe, a consumer fills an application online and provides an account to be used for charging the virtual purse with monetary value. The user also selects a PIN and a security question. The PIN is used to authorize payment transactions during a purchase.

The merchant component offers a comprehensive online administration tool. Payment data can be received online, by phone, or by e-mail.

11.4.1.4 Bankpass Web

The Bankpass Web system is an integrated solution for the management of card payments carried out on open networks. It provides the merchants with back-office applications such as reporting capabilities to track the status of the transaction at any time. It is currently a commercial offer from the SIA Group, an Italian company with services in around 40 countries including South Africa and Hungary. The service is offered as a wallet for purchasers, a management tool for merchants, and a virtual primary account number (PAN) generator for acquirers.

The customer is wallet is an add-on to the client browser. If the retailer has an agreement with Bankpass Web, identification and authentication use an identifier and a password linked to the wallet. If the retailer and/or its payment processor are not associated with the Bankpass Web system, the Bankpass Web server will be requested to assign a perishable card number for single use.

11.4.2 The Second Generation

The solutions of the second generation can be divided into the following groups: pay per click solutions, payment kiosk solutions, prepaid cards, and virtual purses.

11.4.2.1 Pay per Click

Pay per click solutions were built for Internet content providers to track transactions related to subscribed merchants and compensate them according to pre-established criteria. Internet advertisers pay a fee to website owners for traffic directed from clicking displayed links. Google AdWords is currently the most popular option because of the volume of traffic it can provide.

Click for pay schemes, however, are susceptible to fraud because there is neither identification nor authentication of the individual clicking the links. Automated script or computer program can simulate legitimate web browser clicks for the purpose of generating a charge per click without having actual interest in the subject of the advertisement. The merchant is totally dependent on the website to screen spurious traffic by detecting and removing invalid clicks, such as those coming from the same IP address or repetitive or duplicate clicks. This is why click fraud is the subject of increasing litigation.

11.4.2.2 Payment Kiosks

The second group of offers updates on the kiosk model pioneered by the Minitel. These are account systems where, at subscription time, a user populates an account profile that contains the necessary payment information. Service providers are responsible for grouping the micropayments during the collection period and for invoicing their customers per transaction or by duration. They initiate the financial compensation after receiving the payments and then reimburse the merchant and the access provider (if any) after deducting their remuneration.

One of the first successful micropayment system services was the i-mode offer from the Japanese mobile network operator NTT DoCoMo. In this service, the k operator guarantees all participants (subscribers, merchants, or intermediaries) that it identifies and authenticates. In addition, the operator plays the role of a payment intermediary by billing for the consumed services and collecting the payment on behalf of the provider for a commission. This business model proved to be judicious: in about 18 months, 12 million Japanese subscribed to i-mode, services as indicated in Figure 11.3. The decline of subscribers in the 2010s is due to migration to newer services.

Allopass, which started in 2001, originally depended on the telecommunications access provider for billing and collection and used a one-time PIN to authenticate the user. The purchaser may dial a specific vocal server to obtain the PIN or may send a text message containing a predefined keyword to a server and get in return the access code or PIN. Allopass is now HiPay Mobile, which is certified by the French banking authorities. In this way, it can collect moneys directly without going through a mobile operator.

The iTunes Store, originally the iTunes Music Store, is perhaps the most successful payment kiosk. It opened in 2003 and became the largest music vendor in the world in 2010. Payments can be made with a prepaid card (iTunes gift card) or, depending on the country, with a credit card or through PayPal.

The iTunes Store started with a uniform payment policy of 99¢ per track. The current mode of operation is built on a three-tiered pricing model—69¢, 99¢, and $1.29—with most albums priced at $9.99.

The iTunes Store is not a micropayment solution per se. Rather, it is part of a product–service–business platform bundle (Riedel, 2014). This is an integrated package of iTunes services with a series of desktop and mobile high-quality devices such as iPods, iPhones, and iPads, all of which are superbly designed. In addition, it

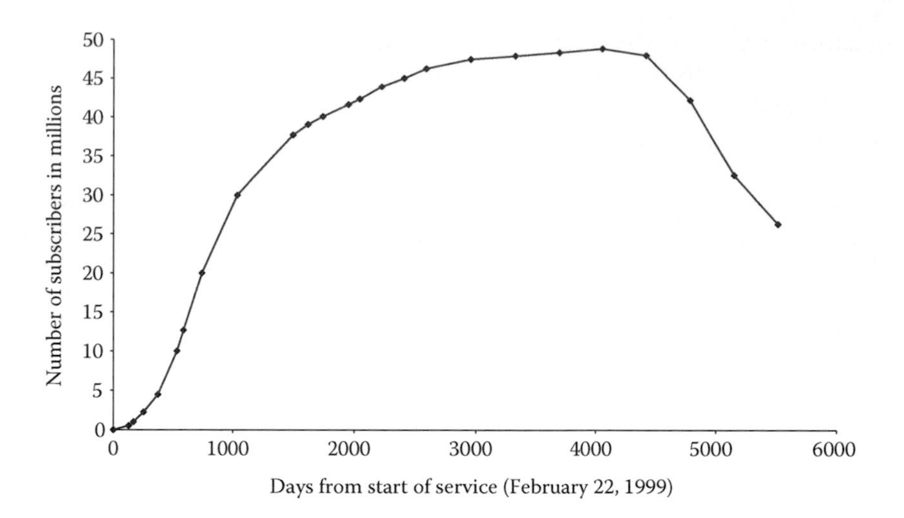

FIGURE 11.3
Evolution of Japanese subscriptions to the i-mode® service. (From NTT DoCoMo, *Financial Times*, 19, October 7, 2002; NTT DoCoMo, FY2005 (ended March 31, 2006) Quarterly Operating Data, https://www.nttdocomo.co.jp/english/corporate/ir/finance/quarter/fy2005/4q.html, last accessed, January 22, 2015; NTT DoCoMo, FY2006 (ended March 31, 2007) Quarterly Operating Data, https://www.nttdocomo.co.jp/english/corporate/ir/finance/quarter/fy2006/4q.html, last accessed, January 22, 2015; NTT DoCoMo, FY2007 (ended March 31, 2008) Quarterly Operating Data, https://www.nttdocomo.co.jp/english/corporate/ir/finance/quarter/fy2007/4q.html; NTT DoCoMo, FY2008 (ended March 31, 2009) Quarterly Operating Data, https://www.nttdocomo.co.jp/english/corporate/ir/finance/quarter/fy2008/4q.html; NTT DoCoMo, FY2009 (ended March 31, 2010) Quarterly Operating Data, https://www.nttdocomo.co.jp/english/corporate/ir/finance/quarter/fy2009/4q.html; NTT DoCoMo, Number of subscribers, November 7, 2014, http://www.ntt.co.jp/ir/fin_e/subscriber.html, last accessed January 22, 2015; Pimont, T., i-mode, ou le succès d'un service distant made in Japan, *Décision Micro & Réseaux*, 444, 20, November 13, 2000.)

provides a platform for third-party entities to distribute their products: music tracks, videos, e-books, software applications, and so on. This is consistent with Apple's overall strategy as building a family of products and services to meet various needs and uses (Mazzucato, 2013, pp. 99–101).

Users give their payment credentials once at subscription. However, phishing scams can be used to subvert this protection. For example, in March 2014, a phishing attack that targets the credentials (Apple ID user names, passwords, and credit card information) of unsuspected iTunes Store customers was uncovered. The attack used a JavaScript code that purports to validate whether an Apple ID is valid. Once the victim has entered what is considered valid credentials, it is redirected to another part of the malicious site that presents screenshots of the Apple interface to request various pieces of personal information, such as full names, dates of birth, billing addresses, and phone numbers. If the victim enters that information and clicks the "verify" button, a window pops up asking for payment card details. After the person enters the payment information, it will be redirected to the actual Apple website. However, because the site presents screenshots of the Apple interface, the links on masquerading site would not respond. Another giveaway is the lack of HTTPS when users are requested to provide sensitive information (Stanford, 2014).

PayPal, which is also an account-based system, has online micropayment solution as well. However, this is a small part of the overall business. PayPal will be discussed in Chapter 12.

11.4.2.3 Prepaid Cards

Prepaid cards access virtual purses managed by the payment operator as a trusted third party (i.e., the balance is stored in the Cloud). The main advantages for prepaid cards are as follows: (1) anonymity, (2) no subscription, (3) does not use banking or account information, (4) access to the virtual purse from landline or mobile networks, and (5) no need for a special client software, that is, the system is independent of the terminal. On the merchant side, the integrity of the communication between the merchant server and the payment intermediary must be protected with the use of electronic signatures.

The various offers can be classified according to the following criteria:

- Whether the card is rechargeable or expendable.
- The length of the card authentication code (typically from 10 to 16 characters).
- The restrictions on the purchases, such as preventing purchases from "adult" sites.

- The maximum balance that can be stored in the card.
- The possibility of combining the balance from several cards for a single payment, up to a limit set by the system.
- The possibility of exchanging the stored value with a legal tender. Such feature requires the backing of a credit institution.
- The security measures employed such as
  - Whether the use of the card is open or requires an identifier and the length of that identifier
  - The limit of authorized payment per transaction
  - The duration of card validity
  - The required conditions to transfer the balance from one card to another in case of loss or theft

Some of the limited usage cards are scratch cards. An opaque field on the reverse side of the card needs to be scratched to reveal a secret code to be used for all purchases from merchant sites affiliated with the card scheme. The payment intermediary verifies the validity of the code as well as the amount in the virtual purse before approving a transaction. The transaction data (transaction identity number, date and time, amount, merchant site, the remaining balance, etc.) are displayed on the screen to confirm the purchase.

Another option would be to include several codes on the back of the card that are revealed gradually by scratching. These codes must be used in the order of their positions on the back of the card.

Cards with microprocessor can generate a series of one-time codes to be sent to the merchant for each transaction. The one-time code would be cross-referenced with the card identification number and the amount of the transaction before validating the payment. This solution was implemented in the Smartcodes card from Sep-Tech, but this solution was not successful because of the relatively high cost of the card.

Scratch cards can be issued by independent operators. One example is the Neosurf card (www.neosurf.info) from Delta Multimédia, available in France, Switzerland, and Belgium. Each card has an identification code of 10 digits that can be used to verify the balance on the card from the operator's website as well as to transfer that balance to another card. The regular card stores values of €15, €30, €50, or €100, while the "minor" card stores €10 or €20 and is blocked from paying "adult" sites.

The rechargeable card Moneta online (www.servizi.monetaonline.it) and the expendable card CarteFacile (www.cartasi.it) are issued by a consortium of Italian banks.

Paysafecard (www.paysafecard.com) is another popular prepaid card for online transactions. It stores values at the €25, €50, and €100 denominations. The cards are purchased from local retailers who transfer the proceeds to a consolidated account held at Commerzbank AG. A special card for minors is capped at €10. With a bank sponsor, it can be loaded with legal tenders in addition to Jetons for telephony or gaming applications. Each card is identified by a 16-digit code (or PIN), concealed under a scratch foil. The 16-digit code is used to conduct purchases, to check the current balance from the Paysafecard payment server, and to query the server on the details of all transactions. A Paysafecard app is also available for Android and iOS mobile terminals.

Some of the following fraud scams that have been reported with Paysafecard are

- Imposters acting as government authorities, institutions, law firms, or the courts demanding the PIN to pay a fine or reminder fee with a Paysafecard.
- Payments for the installation of Windows update license.
- Telephone fraud (e.g., debt write-offs, prize winnings).
- YouTube videos promising to show you how to increase the balance of the Paysafecard. The real aim behind such claims is the acquisition of other people's PIN.

11.4.2.4 Virtual Purses

PayPal is perhaps the most successful system that relies on a virtual purse to conduct micropayments, as one of many payment options. PayPal will be described in Chapter 12.

Ticket Surf (www.ticket-surf.com) is an example of an online payment solution that uses a virtual purse instead of a card or a bank account. It is available in amounts of €10, €20, or €30.

First, a consumer buys a Ticket Surf voucher from the system's website or from anyone of the distributors (tobacconists, newspapers kiosks, etc.). The consumer will have a receipt with an 11- or 16-digit code (PIN). To make a micropayment online, the buyer enters the code in the corresponding field on the website. There is also the possibility of combining the contents of several virtual purses if the amount stored in one purse is not sufficient to cover the transaction.

Ticket Surf offers three security mechanisms:

- Access to the purse is protected by the PIN.
- Access to the purse is blocked for 40 minutes following an incorrect PIN entry. The new PIN is sent by e-mail to the address indicated at subscription.
- A supplementary measure is to give the user the possibility of verifying a history of the transactions that the purse has processed for verification.

11.4.3 The Third Generation

The computing capabilities of iPhones and Android smartphones as well as the growth in mobile games have accelerated the demand for online processing of micropayments. Simultaneously, mobile network operators have been looking for new ways to generate revenues after the saturation of the voice-only mobile communications. This is why the third generation of online micropayment systems considers the mobile phone as a point-of-sale terminal.

Apple Pay has been a catalyst for the broader adoption of mobile payments in the United States (e.g., as compared to the case of M-PESA in Kenya). Paydiant is building a competitor to Apple Pay and Venmo offers a way to transfer small amounts to friends (PayPal acquired both companies). Snapchat, the disappearing-photo application popular among teenagers, has added similar functionality, known as Snapcash. It is not clear which of these approaches have the potential to be embraced in the market.

Similar to the case of regular phone changes, mobile payments can be withdrawn from a prepaid balance or settled later along with the mobile phone bill and all other purchases made in the billing period. For example, with the Vodafone m-pay card system in the United Kingdom, debiting is carried out via the mobile invoice, from a prepaid account or using a payment card specified at subscription. Authentication is not required for amounts less than £15 of the payer is done using a personal identification number (PIN) when the amount exceeds.

11.5 Research Projects

In the late 1990s, several research projects investigated ways to offer micropayments for online purchases. Millicent originated from Digital Equipment Corporation (DEC), NetBill from the Carnegie Mellon University, and PayWord and MicroMint from the MIT. Each of this project introduced its own currency or Jeton to allow the immediate exchange of value. By around 2005, most of these projects were defunct.

In this section, we summarize the main features of these projects.

11.5.1 Millicent

Millicent constructs an economy of very small amounts (less than 1 cent) using a Jeton currency called the *scrip*, whose value varies from 0.1 cent to $50 (Digital Equipment Corporation, 1995; Glassman et al., 1995; Manasse, 1995). The scrip represents an account established with a particular vendor for a limited period of time; this scrip is called vendor scrip.

A Millicent transaction involves three entities: the buyer, the vendor, and a broker. Brokers are payment intermediaries that relieve customers of the obligation of managing and maintaining several individual accounts with each vendor or service provider they wish to deal with. Brokers can also offer their services to merchants by managing in their place, and for a commission, the issuance and distribution of vendor scrip. In exchange, brokers get vendor scrips at discounted price and supply them to their subscribers in the form of broker scrips. The responsibilities of a broker include issuing broker scrips, billing customers, collecting the scrips, and reimbursing vendors. They are also interfaces to the banking world on behalf of merchants, thereby alleviating their operational load by reducing the number of interfaces that they manage. Thus, long-term relationships are established between the consumers and the brokers, on the one side, and between the brokers and the merchants, on the other, instead of ephemeral relationships among vendors and buyers.

The holder of a scrip has a promise of service without a direct relationship with the banking system. Similar to the Jetons stored in telephone cards, this is a pre-consumption of service paid with the aid of traditional payment instruments. Just as for telephone cards, the Millicent scrip indicates the balance available after each purchase. One difference is that customers can close their account at any moment and ask the vendor for reimbursement in legal tender for the balance of the service that has not been consumed. Also, because vendors know the balance of the scrip that they have issued, they are able to detect double spending of a script that the vendor has issued (using the scrip twice).

The broker scrip is valid for all the transactions conducted in the web of commercial relations centered around the broker. By getting a supply of scrip from a broker, a customer avoids having distinct scrip for each vendor. Furthermore, several brokers can establish a network among themselves to allow their customers to

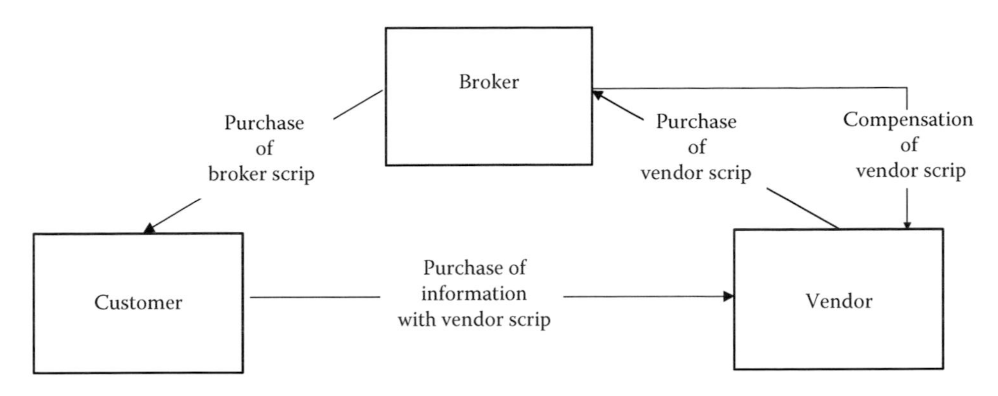

FIGURE 11.4
Cycle of the various Millicent scrips.

access vendors that are not in the same broker's network as their clients.

The function of the broker can be fulfilled by institutions known for their rigor and integrity, such as financial organizations or telecommunications operators. Some of the Millicent brokers in Europe included the Bank of Ireland, Visa, and MasterCard.

Figure 11.4 shows the cycle of scrip that underlies the Millicent model.

11.5.1.1 Secrets

The security mechanism in Millicent relies on three secrets for the production, validation, and spending of scrip. These secrets are the *customer_secret*, the *master_customer_secret*, and the *master_scrip_secret* (Glassman et al., 1995; O'Mahony et al., 1997, p. 205). The customer uses the customer_secret to prove legitimate ownership of the scrip. The vendor uses the master_customer_secret to derive the customer_secret from data in the scrip. Finally, the vendor uses the master_scrip_secret to prevent tampering or counterfeiting of scrip. The master_scrip_secret is known only to the vendor and to the broker if the latter mints the scrip on behalf of the vendor. The master_customer_secret allows calculation of the customer_secret, on the basis of the data included in the scrip. Table 11.1 summarizes the management of secrets in Millicent.

TABLE 11.1
Management of Secrets in Millicent

Secret	Producer	User	Function
master_scrip_secret	Vendor or broker	Vendor	Verification of scrip authenticity
customer_secret	Vendor or broker	Vendor and buyer	Proof of ownership of the scrip
master_customer_secret	Vendor or broker	Vendor	Calculation of customer_secret

11.5.1.2 Description of the Scrip

The vendor and the broker define the content of the scrip, but the structure remains as follows. A vendor scrip contains data fields reserved for the vendor: the expiration date, the balance, the customer, a code to verify the integrity of the scrip, as well as comments (Manasse, 1995; Millicent, 1995; O'Mahony et al., 1997, pp. 199–206). These fields are described in Table 11.2.

The field *Certificate* (also called *Scrip_stamp*) is computed using a hashing function H in the following manner:

Certificate = H(Vendor || Value || ID# || Cust_ID# || Expires || Props || master_scrip_secret)

where
|| represents the concatenation operation
H( ) is the function MD5 or SHA-1

TABLE 11.2
Fields of the Vendor Scrip

Field	Description
Vendor	Vendor identifier.
Value	Balance value of the scrip, updated after each purchase.
ID#	A unique serial number for the scrip, a part of which is used to define master_scrip_secret.
Cust_ID#	Customer identifier that serves in the computation of a secret shared between the buyer and the vendor (customer_secret); customer_id is unique to each buyer.
Expires	The date on which the scrip becomes invalid.
Props (info)	Optional details describing the profile of a buyer to the vendor (age, address, fidelity program, etc.); definition of these details is up to each broker and the associated vendors.
Certificate (Scrip-stamp)	Proof of the integrity of the scrip, that is, it has not been altered, calculated by hashing the other fields with a keyed hash function using the master_scrip_secret as a key.

This field is computed during the production of the scrip and allows the vendor or the broker to verify its authenticity when the customer presents it.

The vendor (or its broker) constructs the customer_secret with a hash using the customer identifier Cust_ID# and the master_customer_secret. This construction can be written in the following form:

$$\text{customer\_secret} = H(\text{Cust\_ID\#} \,\|\, \text{master\_customer\_secret})$$

It is the vendor (or the representing broker) that defines the correspondence between (Cust_ID# and the master_customer_secret. The vendor may also use several lists for the master_customer_secret that can be used alternately to repel falsifications and increase the level of security. In any event, knowledge of the master_customer_secret allows calculation of the customer secret on the basis of the data supplied in the scrip.

The customer receives the customer_secret when it gets the scrip originally; this can be done online if the connection is secure. The security of the links between the broker and the vendor, on the one side, and the customer, on the other, during the supply of the scrip is not strictly within the purview of Millicent. The vendor and the broker are free to choose the techniques to be used.

The convertibility of scrip from various vendors is possible at two levels. The first is for "scrip on us," that is, scrip that originates from vendors that belongs to the same broker's network. The second level of compatibility is for scrip outside the broker's network. In this case, the various brokers have to cooperate among themselves to assure this convertibility.

Note that the scrip can be made to correspond to any legal tender.

11.5.1.3 Registration and Loading of Value

At registration, the customer receives the customer's software (called *Millicent Wallet*) to be installed on the customer's computer and a customer ID (Cust_ID#). The customer chooses the broker from which to buy the broker scrip in exchange of fiduciary money. The broker defines the terms of the exchange, such as how to secure it and which algorithms to use.

The Millicent Wallet assists in the following functions:

- Purchase of broker scrip
- Exchange of broker scrip into vendor scrip and vice versa
- Payment of purchases with vendor scrip
- Acceptance and storage of vendor scrip
- Acceptance of "change" from a vendor

The Millicent Wallet can be transported from one machine to another because it keeps a record of past transactions and the balance of available scrip.

The customer obtains a vendor scrip, either directly from the vendor or by the exchange of broker scrip. Figure 11.5 depicts a simplified view of the exchanges to get a supply of scrip from the broker and to buy items from a vendor without any security mechanism.

In this example, the customer procures vendor scrip by exchanging some of the broker scrip with the broker. In exchange of broker scrip equivalent to \$5.00, the broker returns \$0.20 worth of vendor scrip and the remainder (\$4.80) in broker scrip. Thus, according to Millicent, the customer prepays for the service required using the vendor scrip, then receives the change back in the same currency.

When a purchase request is made, the vendor verifies if the scrip is its own. By comparing the number of the

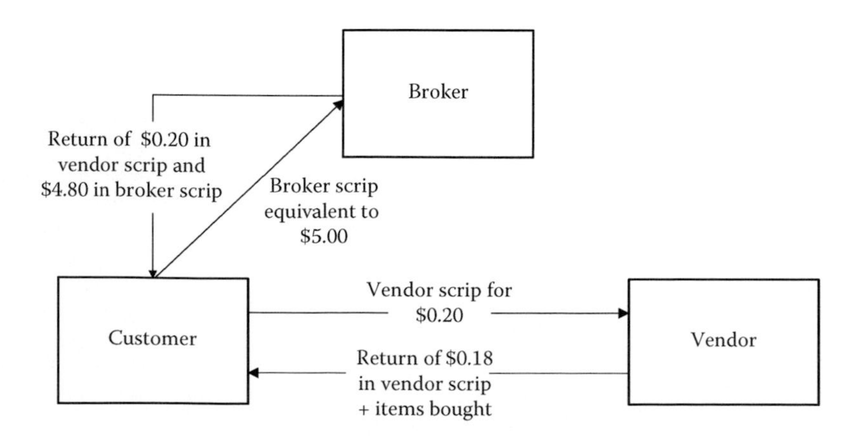

FIGURE 11.5
Purchase of vendor scrip and items without security.

scrip with the numbers in an "approved list," the vendor is able to determine if the scrip was already spent. This is a protection against double spending that does not require the vendor to query a centralized database. The cost of the verification is that of a hash computation and a local database lookup. Thus, no direct link is necessary between the vendor and the broker because all verifications are done locally.

The Millicent protocol assumes that the loss of a portion of the vendor scrip (e.g., the price of a page that was not downloaded for any reason) is negligible. In contrast, prepayment avoids the delays associated with multiple exchanges over the Internet.

This simplified view will be detailed by defining the management of security in Millicent.

11.5.1.4 Purchase

There are two ways for protecting the communication channel between the customer and the vendor during a purchase. The first method consists of encrypting the exchanges using a symmetric encryption key constructed for each session starting with the customer_secret that the customer has and that the vendor can compute. (The session key can also be identical to the customer_secret.) This key is denoted as $K_{customer\_secret}$.

Thus, the message sent to the vendor becomes

Vendor, Cust_ID#, $K_{customer\_secret}${scrip, purchase request}

The fields Vendor and Cust_ID# are sent in the clear so that the vendor can recognize that it is the message recipient and to allow reconstruction of $K_{customer\_secret}$ to retrieve the scrip and the purchase request.

The vendor's response will include a new scrip with the change, the required articles, and a copy of the initial Certificate field. This copy gives the buyer the possibility to verify the authenticity of the response. These exchanges are depicted in Figure 11.6.

This same purchase transaction can be done without assuring confidentiality but by furnishing a protection against the theft of scrip. In this case, the customer_secret and the purchase request are concatenated, and the result is hashed. The resulting digest is sent with the scrip. These exchanges are depicted in Figure 11.7.

The digest allows detection of any tampering with the scrip or with the purchase order. Even if an intruder succeeds in reading the exchange, the intruder will not be able to produce a purchase request and generate a new valid digest without the customer_secret. Accordingly, the integrity of the exchanges is protected.

11.5.1.5 Evaluation

Millicent assumes that the trust relations among the customers, the merchants, and the brokers are asymmetric. The brokers are considered to be the most trustworthy, followed by the merchants and, finally, the customers. The broker also acts as a trusted third party. The only time customers need to be trusted is when they evaluate the quality of the service. In an enterprise network, the broker may be a central directorate that manages access to various proprietary databases.

The Millicent broker intervenes each time that a subscriber wants to buy a vendor scrip from a new merchant within the broker's network. If the relations among customers and merchants are transient, the computational load on the broker's server may be excessive.

Millicent offers three levels of security. The highest level is that of *private and secure* exchanges; the intermediate level is that of *authentication without confidentiality*, that is, a cheater is prevented from benefiting from eavesdropping. Finally, the lowest level is when the scrip is sent in the clear. This can be justified when

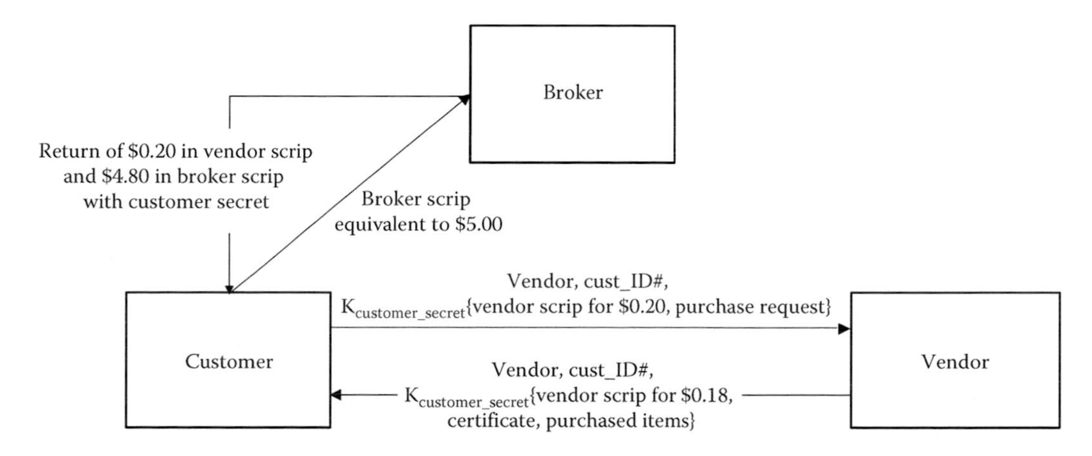

FIGURE 11.6
Purchase of vendor scrip and items (service) using an encrypted channel between the customer and the vendor.

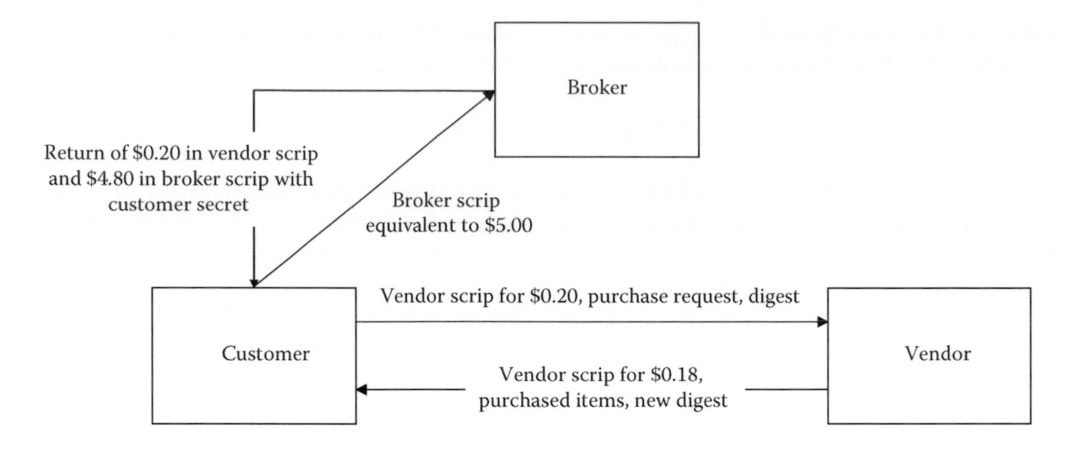

FIGURE 11.7
Purchase of vendor scrip and items (service) with protection against scrip tampering or falsification.

the amounts involved in each transaction are quite low so that many illegal uses of the vendor scrip would be needed to acquire a substantial amount of money, which increases the probability of pinpointing the source of the fraud.

Thus, the security of the scrip is partially based on the smallness of the amounts involved per transaction. In addition, a scrip is not easily reused because it can be recognized by a unique identifier and it has an expiration date. Finally, the purchase is not anonymous; the vendor is aware of the buyer's identity and profile as recorded in the scrip.

The customer is able to detect an irregularity if the ordered items are not delivered or if the balance that the vendor indicates on the scrip does not correspond to the offered service. If the abuse is from the client, the cost to the merchants is that of detection of illegal scrip and the denial of service. If the vendor is cheating, it will be the subject of many complaints so that the broker can act accordingly and refuse to honor the vendor scrip from that vendor. Finally, if the broker is cheating, vendors will observe bad scrip coming from many customers, all originating from a single broker.

Note that the secrets are not sent in the clear; however, the customer_secret is stored on the customer's computer, which increases the system vulnerability.

Millicent was able to reduce the cost per transaction with the following techniques:

- Prepayment in two ways: subscription by opening an account paid with legal money and prepayment of the service with vendor scrip before receiving the service. By assuming that users would not object to losing a micropayment, the purchase is done in one shot without traffic going back and forth on the Internet for validation.
- Reduction of the cryptographic load. Authentication avoids public key algorithms and relies on algorithms for symmetric encryption or hashing.
- The authorization for payments is done offline.
- The verification is done locally, which avoids the online interrogation of the broker for each transaction. This is possible even when the broker issues the scrip in the vendor's name because the verification lists are relatively small and can be updated daily on a local basis. Furthermore, the scrip contains the balance and a proof of its validity in the Certificate field. This saves the merchant the task of verifying the balance at each transaction and allows the client to keep track of expenses.
- Micropayments are grouped before financial settlement.
- Beyond the expiration date, the vendor scrip is not valid, which facilitates its management.
- Payments can neither be revoked nor disputed because no receipts are provided.

11.5.2 NetBill

NetBill is a set of rules, protocols, and software designed for the delivery of text, images, and software through the Internet. It was developed at the Carnegie Mellon University in partnership with Visa and the Mellon Bank to address micropayments of the order of U.S. 1 ¢ (Cox et al., 1995; Sirbu and Tygar, 1995; Sirbu, 1997). The main characteristic of NetBill is that the customer is billed only after receiving the encrypted information; however, the decryption key that allows the customer to extract the information is only delivered after the bill has been paid. This approach is particularly useful for repeated delivery of small quantities of information, such as subscription to news services or to information updates.

The NetBill server plays the role of a certification authority in addition to its function as a payment intermediary. It also manages the distribution of the pairs of RSA public/private keys as well as the session keys that are used to encrypt the exchanges between the customer and the merchant as well as their exchanges with the NetBill server.

11.5.2.1 Registration and Loading of Value

Customers subscribe by giving the data of their credit card or the coordinates of a payment instrument encrypted with the help of a downloaded security module (Money Tool). In return, the customer receives from the NetBill server an identifier and a pair of RSA public and private keys, KP_C and KS_C, respectively.

Similarly, the merchant receives after registration the program called *Product Server* as well as the pair of RSA public and private keys, KP_M and KS_M, respectively.

NetBill subscribers' accounts are prepaid from bank checking accounts, preferably the bank of NetBill. The customer's software displays the amount available on the screen.

In a future phase, NetBill will consider authorizing postpayments.

11.5.2.2 Purchase

The basic purchasing protocol uses eight HTTP messages to cover the four main phases of the commercial and financial transaction: the negotiation, the order the delivery, and the payment. These exchanges involve three parties: the customer, the merchant, and the payment intermediary (the NetBill server) or *till*. The payment intermediary plays the roles of a notary and a trusted third party and communicates with the merchant directly. Through the merchant, it communicates indirectly with the customer. The exchanges are shown in Figure 11.8.

Before the establishment of the communication channels, a phase of mutual recognition takes place to allow the partners to identify and authenticate themselves according to the public key Kerberos system. Thus, the merchant server authenticates the customer with the help of a session ticket and a certificate signed by the certification authority. The ticket itself is sealed (signed) by customer private key and then encrypted with the merchant server public key. With the session ticket, the customer and the merchant will be able to

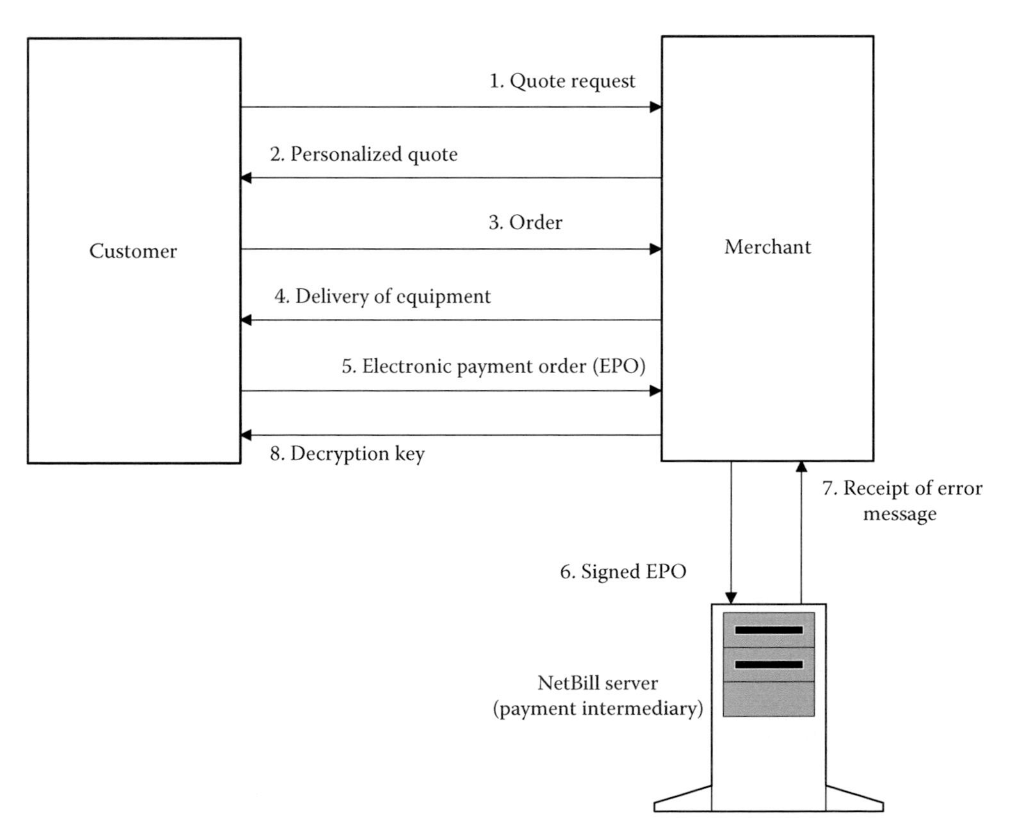

FIGURE 11.8
Messages exchanged in a NetBill purchase.

share a session key to encrypt the exchanges that follow with a symmetric encryption algorithm. The session ticket also offers a defense against any penetration of the Kerberos server. Thus, the customer constructs a symmetric key $K_{challenge}$ and sends it to the merchant in the following message:

$$K_{one\_time}\{\text{customer name, merchant name, time-stamp, } K_{challenge}\}, PK_M\{K_{one\_time}\}, Sig_C$$

where $K\{X\}$ represents the encryption of message X with the key K. The first part of the message is encrypted with the symmetric key K_{one_time}, and the second is encrypted with the public key of the merchant, PK_M, using the public key RSA algorithm. The merchant can recover the symmetric key by decrypting the second part with its private key SK_M. Sig_C represents the signature (on the full or hashed message, depending on the case) with the private key SK_C of the customer to verify the message integrity upon reception. The *time-stamp* field may also include a nonce (a random number used only one time) that will be used to fend off replay attacks using past messages.

The merchant's server verifies the signature using the public key of the customer extracted from the certificate. It recovers the symmetric key K_{one_time} to obtain the key $K_{challenge}$. Next, the merchant produces the symmetric session key K_{CM} and constructs the session ticket τ_{CM}:

τ_{CM} = merchant's name, KS_M\{customer's name, customer's address, time stamp, expiration date, K_{CM}\}

The ticket comprises two parts:

1. The merchant's name in the clear.
2. The customer's name and address, a time stamp against replay attacks, the expiration date of the ticket, and the symmetric session key, K_{CM}, all encrypted with the merchant's private key. The expiration date is reconfigurable.

The merchant's server encrypts τ_{CM} and K_{CM} with the help of the symmetric key $K_{challenge}$ and then sends them to the customer in the encrypted message:

$$K_{challenge}\{\tau_{CM}, K_{CM}\}$$

Having generated $K_{challenge}$ and sharing it exclusively with the merchant's server, the customer is sure that the message comes only from the merchant. The customer decrypts it to recover the session ticket τ_{CM} as well as the symmetric session key K_{CM}. Next, the customer verifies with the help of the merchant's public key that the session key is identical to the key that the session ticket contains.

11.5.2.2.1 Negotiation

During the negotiation phase, the customer requests the price of an item and the merchant responds with a personalized quotation. This phase begins with the identification of the customer so that the merchant can customize the offer according to the customer's profile. The exchanges take place as follows (see Figure 11.8):

1. The customer presents the ticket τ_{CM} and then encrypts the request with the symmetric session key. This request is sent in message 1:

   τ_{CM}, K_{CM}\{customer credentials, data on requested product, request qualifiers, initial price, transaction identifier TID\}

   The customer credentials are used to take advantage of special discounts offered to members of a given group. The data related to the merchandise can be composed in the request automatically by clicking on the reference to the goods. The request qualifiers are flags that indicate the customer's desires (e.g., delivery options). Other data may be added as an option, such as the initial price and the transactional identifier, and so forth. The request is then sent to the merchant server.
2. Having identified the customer, the merchant server customizes the quotation according to the customer profile (e.g., subscriber, privileged customer, volume reduction, etc.). The server then encrypts the offer as follows (message 2 in Figure 11.8):

   K_{CM}\{product description, proposed price, request qualifiers, transaction identifier TID\}

   The product description is a text that will be printed on the customer's bill. The request qualifiers are flags that indicate what the merchant accepted in response to the initial customer's request. The presence of the transaction identifier in this message is mandatory.
3. Messages 1 and 2 will be exchanged as many times as necessary for the two sides to agree on the terms of the transaction.

11.5.2.2.2 The Order

The customer indicates agreement by sending the following command (message 3 in Figure 11.8):

$$\tau_{CM}, K_{CM}\{TID\}$$

11.5.2.2.3 Delivery

After receiving the purchase order, the merchant sends the digital information encrypted with a symmetric encryption key K_{GOODS} without including this key in the message (message 4 in Figure 11.8):

$$K_{GOODS}\{GOODS\}, K_{CM}\{SHA\ (K_{GOODS}\{GOODS\})\}, EPOID$$

where {GOODS} represents the encrypted information. That file is stored on the customer's hard disk but will remain unusable without the decryption key; the client will receive this key after settling the bill. Nevertheless, the customer can verify the integrity of the received articles by applying the hashing algorithm SHA and comparing the results with the quantity SHA(K_{GOODS}{GOODS}) that is extracted from the message using the session key K_{CM}, which is a guarantee to the customer.

The merchant joins to the file an electronic payment order identifier (EPOID) to serve as an index in the NetBill registry of transactions. The EPOID is an exclusive identifier for each transaction that consists of three fields: a merchant identifier, the time stamp marking the end of goods delivery, and a unique serial number. The uniqueness of the EPOID prevents replay attacks while the time stamp is used to date transactions.

11.5.2.2.4 Payment

The objective of this phase is to close the transaction by delivering the payment to the merchant and the decryption key to the customer. During this phase, the customer's software constructs an electronic payment order (EPO) with the following fields:

1. Customer's identity.
2. Product identifier.
3. Negotiated price.
4. Merchant identity.
5. The digest of the encrypted goods SHA(K_{GOODS}{GOODS}).
6. The digest of the product request data in message 1, SHA (product request data) to forestall future disagreement over the details of the order.
7. The digest of the customer's account number with an account verification nonce, SHA (customer's account number, nonce), which is used to verify the validity of the credentials sent to the merchant, for example, the account number sent is the one that NetBill has given.
8. EPOID, the transaction index to the NetBill registry.
9. The ticket τ_{CN}.
10. K_{CN}{authorization ticket, customer's account number, account verification nonce, customer's memo field}.

Note that the authorization ticket is an access control mechanism regulated by another Kerberos server functioning in the same way as the NetBill server but is under the control of another independent authority.

The EPO consists of two fragments. The first segment is composed of the elements 1–8 of the previous list and is sent in the clear. The second segment is addressed to the NetBill server and comprises elements 9 and 10.

The exchanges take place in the following manner:

1. The customer's software constructs an EPO. The customer signs with the signature SIG_C constructed with the help of the customer's private RSA key SK_C and sends it to the merchant in message 5 of Figure 11.8. This message includes the ticket τ_{CM}, and the rest is encrypted with the symmetric session key K_{CM}:

   $$\tau_{CM}, K_{CM}\{EPO, SIG_C\}.$$

2. After verifying the integrity of the contents of the EPO, the merchant verifies the digital imprint of the file to make sure that the customer has indeed received the goods that were sent. If the merchant decides to continue the transaction, it adds the decryption key, endorses the EPO with its digital signature, and sends the new message to the NetBill server. This is message 6 of Figure 11.8, which is sent in the following form:

   $$\tau_{MN}, K_{MN}\{(EPO, SIG_C), MAcct, MMemo, K_{GOODS}, SIG_M\}$$

   where
   MAcct is the merchant account number
   MMemo is a field reserved for the merchant's comments
   The signature is constructed with the help of the RSA private key of the merchant SK_M
   K_{MN} is the secret key between the merchant and NetBill

   By endorsing this message, the merchant confirms that the transaction took place in a regular fashion.

3. The NetBill server extracts different fields for the message and checks the validity of the customer and merchant signatures, the uniqueness and the freshness of the EPOID, and determines the customer's privileges. It verifies that the customer's account balance covers the sum requested, then debits the customer's account, and credits the merchant's account. In case the payment exceeds the balance, the server returns an error message inviting the customer to supply funds to the account in order to conclude the transaction.

4. For a successful transaction, the NetBill server records the transaction, keeps a copy of the decryption key, K_{GOODS}, and digitally signs a receipt that contains the decryption key. The receipt is of the form:

   Receipt = (result code, customer identity, price, product identity, merchant identity, K_{GOODS}, EPOID) SIG_N

   To increase the computation speed, this signature uses the DSA algorithm rather than the RSA. The message sent to the merchant is message 7 of Figure 11.8, which can be described in the following manner:

   K_{MN}\{receipt\}, K_{CN}\{EPOID, CAcct, Bal, Flags\}

   where CAcct, Bal, Flags are fields that contain details on the customer's account. These details are encrypted with the symmetric key K_{CN} to prevent the merchant's server from reading them.
5. The merchant's server extracts the receipt and encrypts it with the symmetric key K_{CM} and then adds the encrypted information on the customer's account to form message 8 of Figure 11.8:

   K_{CM} \{receipt\}, K_{CN} \{EPOID, CAcct, Bal, Flags\}
6. The customer's software applies the decryption key to retrieve the information bought and updates the balance of the funds available to the customer.

11.5.2.3 Financial Settlement

The funds to be credited to the merchants are accumulated and deposited periodically through VisaNet in the bank accounts of the various merchants.

NetBill charges a commission varying between 2.5 cents for a 10-cent transaction and 7 cents for a $1 transaction (which is an overhead of between 25% and 7% of the transaction amount).

11.5.2.4 Evaluation

NetBill ensures the main security services (confidentiality and integrity of the messages, identification and authentication of the participants, as well as nonrepudiation). In addition, NetBill transactions satisfy the following requirements on monetary transactions (Camp et al., 1995):

- Atomicity, which means that the transaction must be executed as a whole for its effects to take place. The client is only billed after receiving the goods, and payment leads to access to the delivered goods. By separating the delivery of information from its decryption, customers can be assured that they have received the information before settling the bill. At the same time, NetBill guarantees to the merchant that a transaction will be refused if the balance of the virtual purse is not sufficient to cover the purchase.
- Consistency, because multiple verifications are made throughout the purchase. Examination of the digital imprints ensures that the transaction reflects exactly the terms of the agreement between the two parties. The uniqueness of the EPOID prevents the reuse of ancient orders by unscrupulous merchants. Time stamping allows purging of stale transactions.
- Isolation, given that the transactions are independent of each other.
- Durability because each party has a transaction record. In addition, the NetBill server has a record of the decryption key K_{GOODS} that the customer can claim in case of a technical problem irrespective of the origin.

The NetBill server plays the role of a trusted third party and an arbiter to resolve conflicts. Although the information relative to the customer's identity is dissociated from the goods exchanged, the EPO reveals the customer's identity to the merchant and the transactions are traceable. Because the NetBill server intervenes in each exchange, this fact can be exploited to protect the customer's anonymity with respect to the merchant. The NetBill server intercepts all the customer messages to hide all information related to their origin before reexpediting them to the merchant. The server can ensure the merchant's anonymity in the same manner.

In short, NetBill presented interesting ideas (e.g., the electronic purchase order or the distinction of information delivery from information access), but the frequent use of digital signatures, particularly public key signatures, tended to reduce the performances of commercial applications of a reasonable scale. In particular, because the intervention of the intermediary (NetBill server) is required for each transaction, the computational power of the NetBill server restricts the number of transactions that can be conducted concurrently. Scaling of the operation is a problem.

11.5.3 PayWord

PayWord is one of two variations on the theme of Millicent, the second being MicroMint. Both have

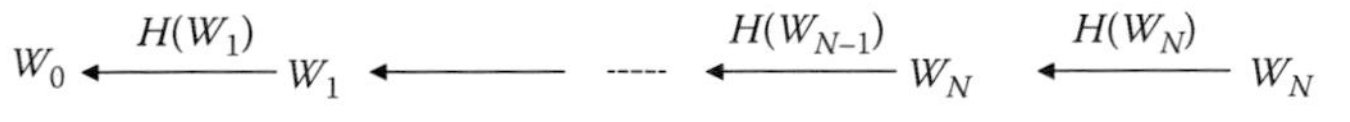

$W_0 \xleftarrow{H(W_1)} W_1 \longleftarrow \text{-----} \xleftarrow{H(W_{N-1})} W_N \xleftarrow{H(W_N)} W_N$

FIGURE 11.9
Formation of the chain of hash values in PayWord.

been proposed by Ronald L. Rivest and Adi Shamir from the Laboratory for Computer Science of MIT (Massachusetts Institute of Technology) in the United States and the Weizmann Institute of Science in Israel, respectively (O'Mahony et al., 1997, pp. 213–220; Rivest and Shamir, 1997). In both cases, the authorization is done offline.

The economy of PayWord revolves around credit in Jetons called *paywords*. Just like Millicent, the parties in a PayWord transaction are as follows: the buyer, the vendor, and the broker. Each broker authorizes users subscribing to its service to buy paywords that the broker produces to pay the vendors in the broker's network. At the same time, the broker agrees to reimburse the vendor in legal tender for the credits collected in Jetons. Thus, PayWord is a postpayment system.

A chain of paywords is formed by applying a hash function recursively starting from an initial value, W_N. Each term is a coin of unique value that represents credit at the vender affiliated to the broker's network. The series of terms is represented in Figure 11.9.

The vendor's task is to verify the signature on the initial value using a public key algorithm. With this single signature verification, the vendor authenticates the whole chain of paywords. This verification is done offline because the vendor is not required to contact the broker at each payment. The local validation of paywords requires the local storage of the revocation list (either in full or in a reduced form) and the validation parameters.

The broker receives daily the last payword consumed by each user subscribing to its service, which allows the broker to update the list of current paywords. The broker constructs a blacklist of all abusive users or of those that have declared loss or theft of their private key, SK_U.

The organization of the vendors in the broker's network and the clearance of accounts among the broker and the vendors are beyond the specifications of PayWord.

11.5.3.1 Registration and Loading of Value

The user goes to a broker to open an account by communicating on a secured channel the coordinates of the user's bank card or bank account. This channel can be postal mail or an Internet circuit secured with the help of TLS/SSL, for example. The user receives in response the subscription card C_U in the following form:

$$C_U = \{B, U, A_U, PK_U, E, I_U\}\, SK_B$$

The card is signed with the private key of the broker, KS_B. It contains the names of the broker B and the user U, the user's address A_U (which may be a postal address, an IP address, or an e-mail address), the public key of the subscriber PK_U, an expiration date E, as well as some additional information I_U (e.g., the number of the subscription card, the credit limit per vendor, the coordinates of the broker, the terms and conditions of the sale, etc.).*

This subscription card authorizes the user to create chains of paywords with the sanction of the broker. By presenting this card to the vendor, the vendor can decipher the content using the public key of the broker PK_B. This assures the vendor that the broker promises to exchange the minted Jetons with legal tender until the expiration date mentioned. As a precautionary measure, the goods sold will be delivered only to the address cited in the card.

Although the vendor knows the identity of the buyer, the system does not make any links between the sale of a particular item and the identity of the buyer, which offers certain privacy protection by reducing traceability. The subscriber has to protect, in particular, his or her private key, SK_U, although its storage on the customer's workstation is one of the weak points of the system. Nevertheless, the economy of PayWord can tolerate a certain rate of malfeasance. Given the smallness of the values that are carried by the Jetons, the cost of sophisticated mechanisms to catch cheaters may exceed the values at risk. However, systematic and large-scale counterfeiting will be detected.

11.5.3.2 Purchase

The purchase takes place in two phases: commitment to a PayWord chain and delivery of the obligation.

11.5.3.2.1 Commitment

The commitment defines an association between a vendor and a subscriber for a limited duration. This association is similar to the one defined in a bank promissory note (Fay, 1997, p. 16). This is the promise of the subscriber to pay the broker the amount in legal tender that is equivalent to all the paywords $W_1, \ldots, W_{N-1}, W_N$ that the vender presents to the broker for reimbursement before the expiration date.

* In this development, the term *subscription card* is used instead of certificate, which was used in the original description of PayWord, to avoid any confusion with the authentication certificates of X.509 type.

Each time a user wishes to contact a vendor, the user must construct a new chain of paywords $\{W_1, \ldots, W_N\}$ in the following manner:

1. Let N be the value in Jetons of the credit needed to make the purchases. Select a random number denoted as W_N that will be the nth payword and whose value will be that of one Jeton.
2. Application of the hash function H (e.g., MD5 or SHA-1) produces W_{N-1}. Thus,
$$W_{N-1} = H(W_N)$$
3. To produce W_{N-2} the hash function is applied a second time, that is,
$$W_{N-2} = H(W_{N-1}) = H\{H(W_N)\}$$
4. The chain $\{W_0, W_1, \ldots, W_{N-1}, W_N\}$ is constructed in this manner by applying the hashing function N times in sequence.

   To link the chain to a given vendor, the user signs a commitment M to this chain in the form:
$$M = \{V, C_U, W_0, D, I_M\}\ SK_U$$
   where
   V represents the vendor
   C_U the user's subscription card
   W_0 the root of the payword chain
   D the expiration date
   I_M is any additional information (e.g., the length of the chain, or the value of a payword)
   M is signed with the user's private key, SK_U, before sending it the vendor. This is represented by $\{\}\ SK_U$.

The commitment represents the largest amount of computation for the user because it implies a signature with the RSA algorithm.

By presenting the commitment as a sign of authenticity to the vendor and joining the subscription card, the vendor, with the help of the public keys of the user and the broker, respectively, PK_U and PK_B, can verify the user's signature on the commitment M and the broker signature on the user's subscription card C_B. Next, the vendor can be sure that the received values of the expiration dates D and E as well as the value of the last value of the payword chain, W_0, are correct.

11.5.3.2.2 Delivery

The user spends the paywords in an ascending order: W_1 before W_2, and so on. If the price of an item is i paywords, the payment P of the user U to the vendor V is defined by

$$P = (W_i, i)$$

Thus, the payment is not signed.

The vendor must carry (i) hashing on the root of the chain W_0 (indicated in the commitment) to verify the payment validity. However, the number of operations can be reduced by exploiting the properties resulting from the way that the chain $\{W_0, W_1, \ldots, W_{N-1}, W_N\}$ was constructed. If W_j paywords have already been spent (where $j < i$), the verification of P requires $(i - j)$ hash operations on W_j instead of (i) hash operations on W_0. Because the verification of W_j already required (j) operations on W_0, the vendor will save computations by keeping track of the index of the last value utilized for each user.

Note the following:

- Only the commitment has to be signed to guarantee the integrity of the root W_0. The payment is not signed.
- The authenticity of the user is checked using the subscription card, which is the reason it can be called a certificate.
- Verification is local and authorization is offline; that is, the vendor does not have to contact the broker for each payment.
- The payment does not mention the item to be purchased, which prevents tracing of the transactions. Although this characteristic protects the user's privacy somewhat, it does not protect from negligent or bad-intentioned vendors.
- The broker plays the role of a small-claims judge in addition to the functions of trusted third party.
- The vendor can keep a record of all paywords (even those that have been spent and redeemed) until their expiration date to protect against replay attacks.

11.5.3.3 Financial Settlement

A vendor does not have to be affiliated with a broker to be able to authenticate the user's subscription card C_U because the vendor only needs the broker's public key, KP_B. However, redemption of the paywords with legal tender requires a formal relationship between the vendor and the broker.

On a periodic basis (e.g., on a daily basis), the vendor sends the broker a reimbursement request for each subscriber of the broker's network. This message contains the commitment M of the user and the last payword received.

The broker then verifies the validity of each commitment using the public key PK_U of each user, given that the broker recognizes the subscription cards that it has issued and their expiration date. To verify the validity of the last payword per user, for example, W_K, the broker will have to do K hashing operations on W_0 of that particular user.

The broker groups small amounts before charging them to a credit card account. The interface to the bank card account is controlled by the rules and regulations of the network of bank cards from the point of view of security, financial settlement, and so on. Thus, the subscribers are responsible with respect to two authorities: the bank that has issued their credit card and their PayWord broker.

Financial settlement with legal tender is done through the banking circuits and is outside the specifications of PayWord.

11.5.3.4 Computational Load

11.5.3.4.1 Load on the Broker

For each user member of its network, the broker carries the following operations:

- Periodically (monthly)
  - Renews the subscription card using a public key signature
- Daily
  - Verifies the commitment of each user who has made a purchase, that is, the verification of a public key signature
  - Verifies the payment by computing successive hashes

These computations do not need to be performed in real time.

11.5.3.4.2 Load on the User

The user performs the following calculations:

- Periodically (monthly)
  - Verifies the subscription card and the certificate, which requires the verification of a public key signature
- For each vendor daily
  - Signs a new commitment that requires generation of a public key signature
  - Computes the hash of each payword used
  - Records the various commitments, the various payword chains, and the last payword employed

The user is the only entity that needs to sign the commitment (i.e., generate a public key signature) and calculate hashes online.

11.5.3.4.3 Load on the Vendor

The vendor must

- Verify daily the subscription cards and the commitments received from customers; these computations are online but are less intensive than those required from the customers
- Compute online the necessary hashes for the paywords received
- Keep a record of all received commitments until their expiration date as well as the last valid payword received during the day

11.5.3.5 Evaluation

As an evaluation, Table 11.3 contains a comparison between the approaches of Millicent and PayWord.

11.5.4 MicroMint

MicroMint is another scheme that Rivest and Shamir developed for micropayments (O'Mahony et al., 1997, pp. 228–236; Rivest and Shamir, 1997).

The MicroMint economy is based on Jetons called MicroMint coins. These are a sequence of bits whose validity can be easily checked but whose production is extremely expensive. Just as in the case of minting metallic coins, the per unit cost of fabrication decreases as production increases because of the economies of scale. At the same time, small-scale forgery is not economical. One difference from PayWord is that MicroMint avoids public key encryption to reduce the computational load.

TABLE 11.3

Comparison of Millicent and PayWord

Characteristics	Millicent	PayWord
Verification	Offline	Offline
Nature of the system of payment	Prepayment	Credit (postpayment)
Representation of the monetary value	Counter	Digital function (hash)
Security	Three levels to be chosen from	Prevention of double spending
Storing of value	In counters	Number of hash operations
Multiplicity of currencies	A set of counter registers per currency	A root W_0 per currency
Anonymity	No	No
Traceability	Yes	No

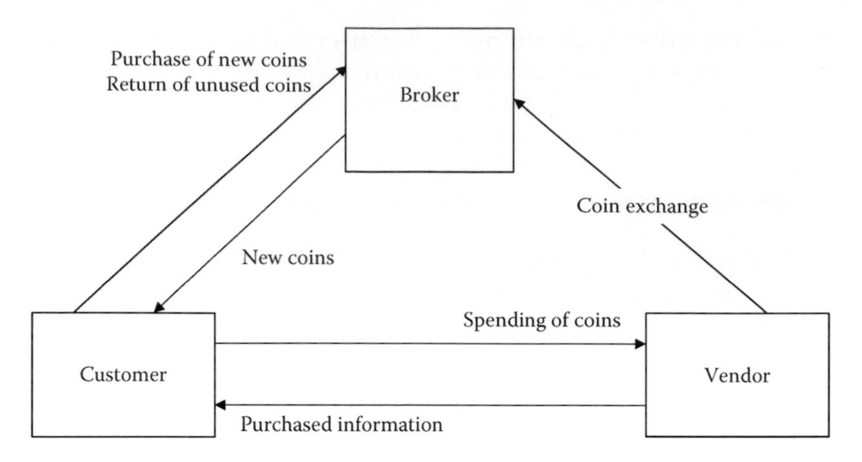

FIGURE 11.10
Cycle of coins in MicroMint.

The cycle of coins in the MicroMint economy is shown in Figure 11.10. Vendors exchange the coins collected for legal tender daily. The duration of validity of new coins is typically 1 month. Beyond that delay, unused coins are returned to the broker in exchange for legal tender or other MicroMint coins.

11.5.4.1 Registration and Loading of Value

Purchasing of coins from the broker is done by means of bank payments (credit cards, checks, etc.). The broker has to keep an inventory of all the coins purchased by each subscriber to the broker's network. MicroMint does not prescribe the relations between the broker and its clients.

If the charging of value is done through the Internet, the subscriber and the broker may resort to symmetric cryptography to fend off any attempts to steal coins during the charging. This is possible because the relations between the broker and the users that are affiliated with the broker network are generally long term.

11.5.4.2 Purchase

Purchase is done by exchanging the coins for a service or information between the customer and the vendor. The vendor has to verify the validity of the coins, but this verification does not mean that the coins are not being reused. It is up to the broker to detect later if coins were double spent and then trace the double spending to the user or to the vendor suspected to be the source of the fraud.

11.5.4.3 Financial Settlement

Every 24 hours, the vendors present the received coins to the broker. After verification of the regularity of the coins and that they have not been previously exchanged, the broker reimburses the vendor for the equivalent amount in legal tender less the broker's commission. The broker is free to accept or reject coins that have already been exchanged.

11.5.4.4 Security

It is assumed that forgers have no interest in cheating to gain a negligible amount. The security mechanisms are installed to discourage systematic fraud, such as large-scale counterfeit, the theft of coins, or sustained double spending.

In MicroMint, a coin is minted following k collisions of the hashing function $H: x \rightarrow y$, where x and y are vectors of dimension n and m, respectively, ($m < n$).

A collision is said to occur when the hash of x_1 and x_2 is the same vector y, that is,

$$H(x_1) = H(x_2) = y$$

Thus, to have k collisions, the following condition must be satisfied:

$$H(x_1) = H(x_2) = \cdots = H(x_{k-1}) = H(x_k) = y$$

where the x_i vectors are distinct.

The coin C is formed from these vectors x_i:

$$C = \{x_1, x_2, \ldots, x_{k-1}, x_k\}$$

Verification consists in assuring that the hash of all these vectors x_i is equal. For example, assume that $n = 52$ and $k = 4$. In the case of the standard hash functions such as MD5 or SHA-1, the length of the hash is 128 and 160 bits, respectively. To find the required 52 bits, the lowest-order 52 bits are retained. However, the verification of the collision does not detect whether the coin is a forgery or was previously spent.

11.5.4.4.1 Protection against Forgery

Because small-scale forgery is not attractive, the defensive measures are oriented toward industrial counterfeit. Some suggested measures are as follows:

- Monthly renewal for criteria for validity, for example,
    - Some high-order bits of the hash correspond to a certain mask.
    - The minted coins belong to a certain subset whose hash has a given structure, which allows the separation of fake pieces based on the examination of a hash.
    - The various x_i's must satisfy some conditions.
- The broker can augment the computation time by starting to mint the coins several months in advance, which will make it more difficult for counterfeiters to be ready at the required time.
- An extreme measure if the MicroMint server is compromised is to recall all the coins in circulation and replace them with a new issue.

11.5.4.4.2 Protection against Coin Theft

Encryption can protect the coins from theft during their collection by the broker. The usage of public key cryptography is possible because the relations between the vendors and the broker are relatively stable and long term. Another approach, which avoids encryption, is to personalize the coins of each subscriber to the broker's network. In return for sacrificing anonymity, the usage of the coin will require explicit permission of the owner, which makes them closer to electronic checks. On the other hand, if the coins are specific to a given vendor, the stolen coins would not have any value because the vendor would put a stop on them. Anonymity can be restored, but the universality of the coins will be sacrificed.

11.5.4.4.3 Protection against Double Spending

The broker can detect double spending because it controls an inventory of coins purchased by each subscriber in its network. The broker is capable of identifying the vendors that have returned these coins for redemption. With the cooperation of the merchants, and by analyzing the data they have collected, the broker would be able to trace the subscribers that may have used them. Because MicroMint does not use digital signatures, the identity of a forger can be repudiated. Nevertheless, the broker may refuse to supply the suspected forgers with new coins.

All this means that, in the MicroMint scheme, detection of the double spenders will not be easy. However, given the small magnitude of the amounts in play, the risks should be acceptable.

11.5.5 Evaluation of the Research Projects for Online Micropayments

Table 11.4 compares the properties of PayWord with those of MicroMin, while Table 11.5 compares the three research projects for online micropayments that have been discussed.

11.6 Market Response to Micropayment Systems

Through a series of pilot experiments, potential applications of electronic purses or Jeton holders were studied in a realistic environment. In most situations,

TABLE 11.4

Comparison between PayWord and MicroMint

Characteristic	PayWord	MicroMint
Security	Public key encryption	No encryption
	Verification against a blacklist	No easy protection against double spending
Nature of the Jeton	Specific to a vendor	General for any vendor but can be personalized for one or several users and restricted to a vendor
Financing in Jetons	Credit in paywords	Debit in coins
Nominal value of the Jeton	1 payword = 1 ¢	1 coin = 1 ¢
Verification by the payment intermediary (the broker)	Offline	Offline
Storage of value	In the number of hashes	In the number of hashes
Multiplicity of currencies or denominations	One root W_0 per currency or denomination	One condition on the x_i per currency or denomination
Anonymity	No	No
Traceability	No	Yes

TABLE 11.5

Comparison of the Research Projects for Online Micropayment Systems

Characteristic	NetBill	Millicent	PayWord	MicroMint
Services offered	Payment system	Payment system	Payment system	Payment system
Product available	Payment software	Payment software	Trusted third party, notary	—
Authorization	Online	Offline	Offline	Offline
Role of the intermediary	Trusted third party, notary	Trusted third party, notary	Notary	Notary
Security protocols	Public key Kerberos	Symmetric key cryptography, hashing	Public key cryptography; hashing	No encryption, hashing; no protection against double spending
Storage of the secrets by the customers	The payment intermediary keeps a copy of the decryption key of the items; the session keys are stored on the client machines.	The customer secret key is stored on the customer's computer.	The private key of the customer is stored on the customer's workstation.	—
Instruments for loading of value	Credit card, direct debit, fund transfer	Credit card, direct debit, fund transfer	Credit card, checks	Credit card, checks
Nature of money	Legal tender	Legal tender	Jeton	Jeton
Subscription mode	Prepayment	Prepayment	Credit	Credit
Minimum payment in legal tender	1 ¢	<0.1 ¢	0.1 ¢	—
Financing of the internal economy	N/A	Debit (prepayment) in scrip	Credit in paywords	Debit in coins
Storage of value	In a counter	In a counter	In the number of hashes	In the number of hashes
Currencies	US$	Variable	One root W_0 per currency and per denomination	One condition on the x_i per currency and per denomination
Revocability of the payment	Possible until the electronic payment order is signed	Irrevocable and without any possibility for challenge	Irrevocable and without any possibility for challenge	Irrevocable and without any possibility for challenge
Billing	Per transaction	Grouped transactions	Grouped transactions	Grouped transactions

participants did not find any advantage in replacing coins and bills because electronic purses did not offer the same level of convenience due to the absence of charging points. Merchants were unhappy with the difficulty of the operation and the time it took to conduct the transaction (Westland et al., 1998; Van Hove, 2001). Successful electronic purses or Jeton holders were supported by established entities with direct access to potential users: telecommunication operators, banks, and transport companies. The lessons learned suggest that in face-to-face situations, the chance of success of purses and Jeton holders increases when

1. For open usage, the electronic purse needs to be based on standards to take advantage of the existing infrastructure, for example, no significant changes in ATMs.
2. The audience is concentrated in geographic such as university campus, bus or train stops, etc. (Tan and Tan, 2009).
3. There is institutional support to bring customers, funding, and general public awareness either through compulsion (i.e., to ride a bus) or through inducement.
4. They are convenient to use, such as having a wide network of charge points. With the adoption of the EMV specification, it is possible to load purses from bank cash dispensers. In public transportation system, the charging point is at the entrance of the stop or station.

Important factors worked against the success of Jeton-based online micropayment systems, which delayed their acceptance (Shirky, 2000; Odlyzko, 2003; van Someren et al., 2003). Some of the factors that have slowed their acceptance are as follows:

1. The need for special hardware or software.
2. The absence of a compelling reason to use the micropayment scheme.

3. Research projects, in particular, were designed without user involvement, so there was little thought of customer after sale support, particularly when the system used encryption schemes.
4. Finally, the need to buy Jetons put some psychological barriers as explained later.

Users want predictable and simple pricing. However, in a Jeton-based economy, users must anticipate their needs and keep track of their purchases and their current balance. Because of the absence of standardization, they had to know buy more than one Jeton type to access the various websites. In such a situation, flat rate pricing is easier to use and provides protection against sudden large bills (Odlyzko, 2003). For all these reasons, accounted online micropayment systems have had a higher rate of success (Rohunen et al., 2014). Another success factor is the use of a single interface to the user, independent of the product and the supplier.

The success of iTunes shows that micropayments are piggybacked on the high quality of the offer in terms of ergonomics, user support, and reliability; they can be merged with the existing practice with ease. In this case, Apple owns and operates all the component parts of the product–service–platform bundle. It has expanded the potential audience by attracting businesses wishing to distribute their products through Apple. There is also a community of loyal owners of Apple products that are ready to provide support.

At present, micropayment solutions are not perceived to be suitable for money laundering or the financing of illegal activities, so there is little monitoring by banking and legal authorities.

Finally, it is useful to note that the failure of many early entrants belies the principle of the "first mover advantage" that many business texts promote.

11.7 Summary

Micropayment schemes address a small niche of the universe of digital payments. Many solutions were proposed in the last several decades, most of which have failed.

The technology of purses had been tried successfully in closed circuits, when Jetons were used such as for paying for telephone calls, parking, or urban transportation systems. For the general case of replacing cash *in toto*, the proposition was not attractive enough to people even when the proposed scheme had strong institutional support. To overcome some of the obstacles, standardized interfaces were developed based on EMV so that the typical ATMs could be used for charging the cards with value.

The first generation of online payment tools for small amounts established efficient and secure methods based on special Jetons and the installation of special software. The lack of customer support turned out to be an important impediment. Subsequent generations have piggybacked on existing computer and networking infrastructure and the reuse of well-mastered techniques.

Questions

1. What are the various ways of processing micropayments?
2. How can operational costs be reduced in micropayment systems?
3. Speculate on the reasons behind the lack of electronic purse projects in the United States.
4. It is sometimes said that the first generation of online micropayment systems was technical solutions in search of a problem. Do you agree?
5. What were the advantages of First Virtual? What are the possible reasons that the business folded?
6. Why is the role of the service provider important in second-generation remote micropayment systems?
7. What is the difference between the money stored in a bus card or a telephone card and the money stored in an electronic purse such as GeldKarte?

12

PayPal

PayPal is an exemplar for a new class of intermediaries in electronic commerce. It offers a free subscription to individuals looking for a payment system with simple tools, an e-mail account, and a browser, that is, no special software. Thus, it combines electronic messaging with existing payment mechanisms so that users do not have to learn new ways. With PayPal, any person connected to the Internet can transfer money to anyone with an e-mail address. PayPal also popularized some of the means for fighting online credit card fraud. However, despite the founders' dream of overhauling the world's currency markets and of liberating people from *corrupt governments* (Jackson, 2006, pp. 19–21, 173–174), PayPal remains a wrapper service, from a payment viewpoint, because it does not introduce a new currency nor a new core payment system (Ali et al., 2014a). Nevertheless, it is pursuing its goal to move from every digital device to every online merchant and becoming the "Switzerland of online payment systems" (*Economist*, 2015d).

This chapter recapitulates PayPal's evolution and contains a description of the various mechanisms offered for personal accounts and business account.

12.1 Evolution of PayPal

PayPal was originally the name of the Virtual Purse from Confinity, a company formed in 1998 with funds from Nokia Ventures, Deutsche Bank, and the venture capital fund Idealab Capital Partners. The main function of that software was encrypting financial transactions conducted with a Palm Pilot and any other personal digital assistant devices. An online service was added to allow person-to-person payments, which serendipitously enabled Confinity to reach beyond the small community of Palm owners.

In 2000, Confinity merged with X.com, an online banking company that offered e-mail-based payments and the combined company changed its name to PayPal in 2001. The next growth opportunity came from online auctions on eBay®. The idea was to benefit small sellers from the mechanisms of person-to-person transactions given that they could not qualify for a merchant's credit card processing account. This would shorten the payment cycle by avoiding paper checks or money orders. In fact, this was the motivation of eBay's acquisition of Billpoint, a credit card processor, in partnership with Wells Fargo Bank, which also provided it with efficient back-end and customer support processes (Jackson, 2006, p. 6).

Even though eBay favored Billpoint, PayPal grew to be the principal payment intermediary for the majority of eBay users. It also overtook other start-ups focusing on person-to-person payments, even those with strong banking support such as eMoneyMail (Bank One) and c2it (Citibank). Other competitors included Yahoo!'s Pay Direct, Google Checkout, and Western Union's BidPay.

Eventually, eBay acquired PayPal in 2002 and dropped Billpoint in January 2003. PayPal, in turn, acquired VeriSign's payment solution to enhance its business focus with a strong security portfolio. In 2007, PayPal partnered with MasterCard to develop the PayPal Secure Cardservice, a consumer offer of perishable card numbers. Later, these cards were replaced with a prepaid card for retail stores. In parallel, PayPal gained a banking license in Luxembourg* to conduct banking operations throughout the European Union. As a result, PayPal accounts could be maintained in 24 currencies.

In January 2008, PayPal enhanced its fraud management systems by acquiring Fraud Sciences, an Israeli start-up that develops online risk tools. In 2012, it partnered with Discover Card to allow PayPal payments at any of the stores on the Discover Card's network.

PayPal's next moves were related mobile commerce. In 2013, it bought Braintree, which processes transactions for mobile apps and which had previously acquired Venmo, a service to transfer small amounts using mobile phones. In 2015, PayPal acquired Paydiant, which provides a platform to retailers, banks, and payment processors to operate mobile wallets. Finally, it was spun out of eBay in 2015.

Table 12.1 provides some statistics on the volume of payments during the period from 2008 to 2014. The data are also depicted in Figure 12.1. The data show that mobile payments increased to constitute 20% of the volume of PayPal transactions in 2014. This, however, preceded the introduction of Apple Pay later that year.

* Luxembourg ranks second behind the State of Delaware in the United States in the list of most secretive tax havens, in terms of transparency and volume of transactions (National Geographic, 2010).

TABLE 12.1
Evolution of the Volume of PayPal Transactions (2008–2014)

Year	Total	Nonmobile	Mobile	Percentage of Mobile Transactions
2008	60.00	59.98	0.025	0.04
2009	72.00	71.86	0.141	0.20
2010	91.96	91.21	0.750	0.82
2011	118.75	114.75	4.00	3.37
2012	144.94	130.94	14.00	9.66
2013	179.66	152.66	27.00	15.03
2014	227.94	181.94	46.00	20.18

Sources: http://blog.ebay.com/wp-content/uploads/2009/01/paypal-fast-facts-q4-2008.pdf; http://www.statista.com/statistics/277841/paypals-total-payment-volume; http://www.statista.com/statistics /277819/pay pals-annual-mobile-payment-volume, last accessed January 20, 2016.

PayPal has been involved in a large number of litigations, for example, patent infringement; many of them settled out of court (Jackson, 2006, pp. 184–185). PayPal has also been involved in several class action suits over the policy of holding 30% of vendor transactions for 90 days. One of the most famous incidents involving PayPal took place in 2010, when it stopped taking donations to WikiLeaks under pressure from the U.S. federal authorities. A website Paypalsucsk.com hosts a web forum for critical postings of PayPal.

12.2 Individual Accounts

PayPal was positioned initially as a low-cost provider of peer-to-peer transactions using the interest on the float to generate revenues. Thus, registration was free with promotional bonuses. All new account holders started with a $10 deposit in their Virtual Purse (the PayPal wallet). Users would add another $10 for every new person that they referred to the service, once that person had registered (Jackson, 2006, p. 9). Users would then access their purses through PayPal by requesting a transaction through an e-mail message.*

To open an account, a user would supply an e-mail address, a bank account, and/or a credit card number. Payments could then be made from the bank account registered with PayPal, with the registered credit card or from the PayPal Virtual Purse. The Virtual Purse corresponds to a subaccount that PayPal maintains with an established bank. Strictly speaking, the funds in

* Originally, the e-mail address was the personal identifier; mobile telephone numbers were a later addition.

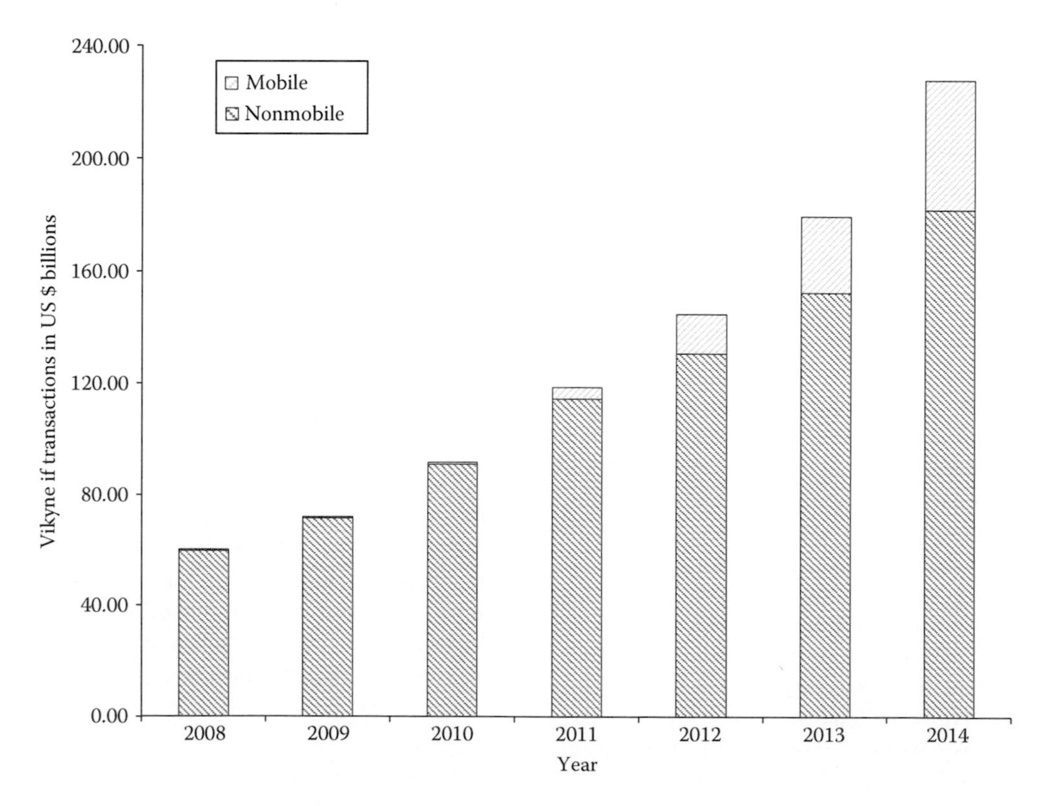

FIGURE 12.1
Evolution of the volume of PayPal transactions from 2008 to 2014.

PayPal's custody are not covered by any specific banking regulation since PayPal is not a bank. As explained later, PayPal offers some protection against fraud, but the process is under the control of PayPal and the arbitration criteria are not defined by legal authorities.

A party transfers money to another by sending a request to PayPal with the recipient's e-mail address and the amount in question on a transport session secured by TLS/SSL. The sender does not have to specify the recipient's bank routing and account number like the case of the traditional banking model. As a consequence, there is no requirement on users to learn new ways or to preload value into the account.

The recipient receives from PayPal a message by e-mail or SMS, signaling the availability of funds. To access the payment, recipients must open an account with PayPal if they are not already members. Recipients would then specify how they would like to receive the funds such as a transfer to a bank account or a payment by check from PayPal, or as a deposit in their Virtual Purse.

Cost leadership is a risky strategy because its sustainability depends on economies of scale and proprietary advantages (Porter, 1985, pp. 21, 112). Not only PayPal did not meet these conditions, but users withdrew their balances as soon as they received them, so the interest on float was not enough to sustain the enterprise. As a consequence, PayPal refocused on payments of online auctions (Jackson, 2006, p. 39).

For online auctions, PayPal intervened as a trusted third party, thereby solving two important problems. First, online buyers did not know the sellers sufficiently to trust them with their financial information. Second, sellers recorded two few sales to qualify for a merchant credit card processing account.

Figure 12.2 illustrates typical exchanges in a PayPal transaction. The payment processor handles the various authorization messages in real time and the settlement and clearing messages delivered in batch. Clearly, the transactions are not anonymous.

We discuss now some of the payment instruments in PayPal and the methods for fraud prevention.

12.2.1 Payment with Credit Cards

The procedure involves the following steps:

1. A buyer purchases items and enters a bank card number in the checkout pages of the merchant's site.
2. The transaction details are sent to PayPal for processing and forwarding to the acquirer (merchant) bank through the payment gateway.
3. The gateway encrypts the data and sends the authorization request to the acquirer bank.
4. The acquirer bank forwards the transaction information to the issuer bank (the buyer's bank) through the bank card network to authorize the transaction.
5. The issuer bank either approves or denies the transaction and sends that information back to the acquirer.
6. The payment processing network sends the transaction details and response back to the payment gateway, which encrypts them and forwards them to the merchant server.
7. If the transaction is approved, the issuer bank will deposit funds in the merchant's account at a scheduled time.

PayPal originally used the credit card networks and then switched to the Automated Clearing House (ACH) network to lower transaction cost (Jackson, 2006, pp. 93–94). Fortuitously, NACHA had just approved new types of transactions for the Internet. With ACH, the operation consists of two discrete transactions: an ACH credit to deliver funds to the merchant or recipient, for example, a customer-initiated entry (CIE) or prearranged payment and deposit (PPD) credit followed by a WEB transaction to collect funds from the buyer or sender (Benson and Loftesness, 2010, pp. 57–58).

In this procedure, PayPal advances the money before the ACH transaction ends with increases in its risk if the transaction is fraudulent or is contested. Furthermore, an ACH transaction can take several days to clear.* With phishing, account information can be stolen from account holders. To manage this risk, PayPal developed some schemes discussed in Section 12.2.5.

The recipient of a credit card payment is charged a fixed fee per transaction and a percentage on the transaction amount, for example, 3.4% of the value + $0.25 (or €0.25). Also, the fees increase with the transaction volume.

12.2.2 Payment with PayPal Account Balance

Payment with the PayPal account balance uses the PayPal Express Checkout procedure with the exchanges shown in Figure 12.3. The checkout process uses three PayPal SOAP requests/responses—SetExpressCheckout, Get ExpressCheckoutDetails, and DoExpressCheckout Payment (Abdellouai and Pasquet, 2010; Williams, 2007, pp. 151–152):

1. On the merchant site, the buyer presses the Checkout with PayPal button to start the checkout process.

* In May 2015, NACHA approved the same-day settlement of ACH transactions.

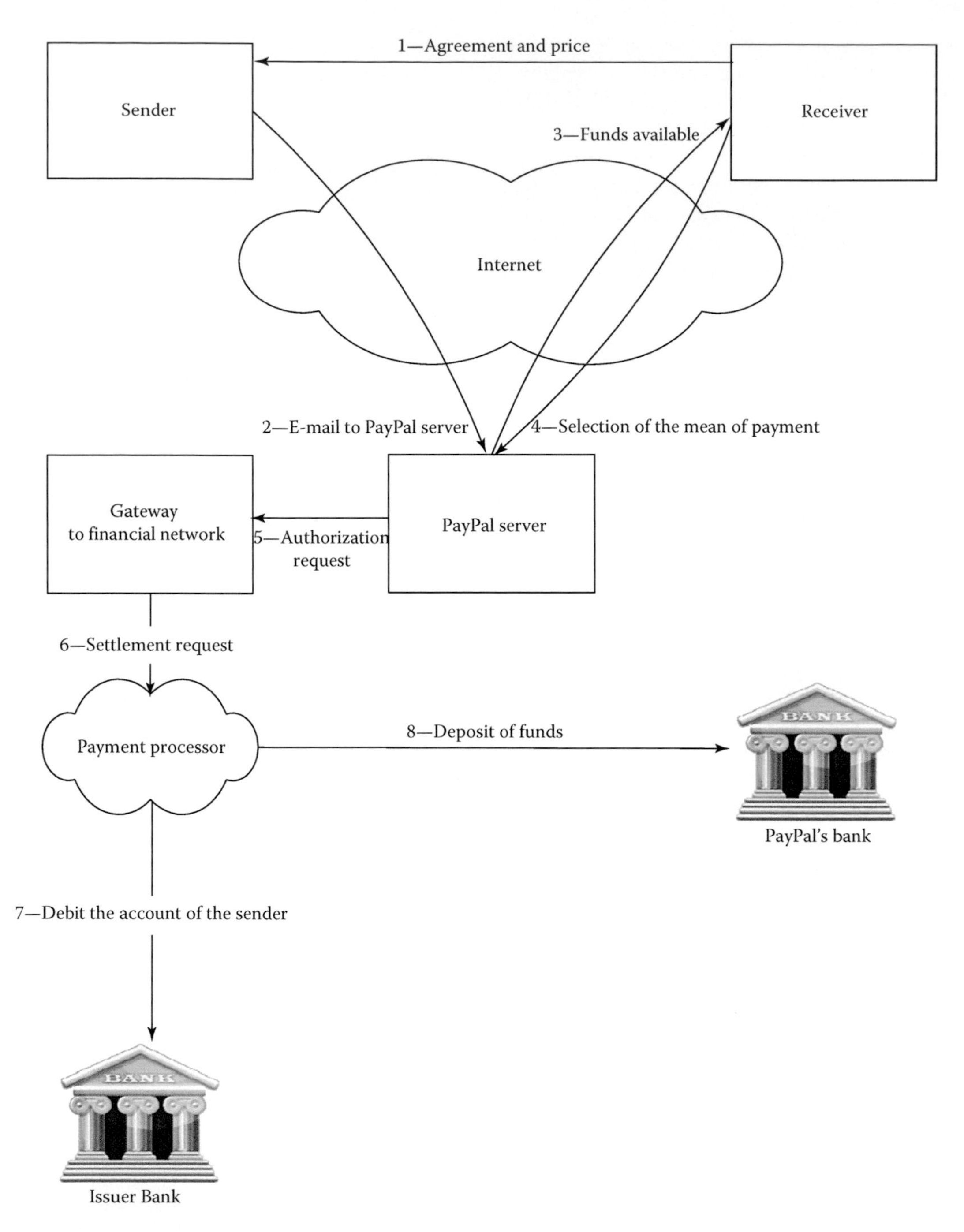

FIGURE 12.2
Typical exchanges in PayPal payments.

2. The merchant initiates the transaction with the SetExpressCheckout request, which includes three parameters to the PayPal server:
   a. The total amount of the order (exact or estimated)
   b. Return URL to specify where the buyer should be returned on the merchant site after being authenticated on the PayPal server
   c. Cancel URL on the merchant site if the buyer cancels their transaction
3. The PayPal server responds with a cryptographic token that identifies all the exchanges belonging to the same checkout flow. The token is valid for 3 hours.
4. The merchant site forwards the token to the buyer's browser and directs it to start an https session with the PayPal sever.

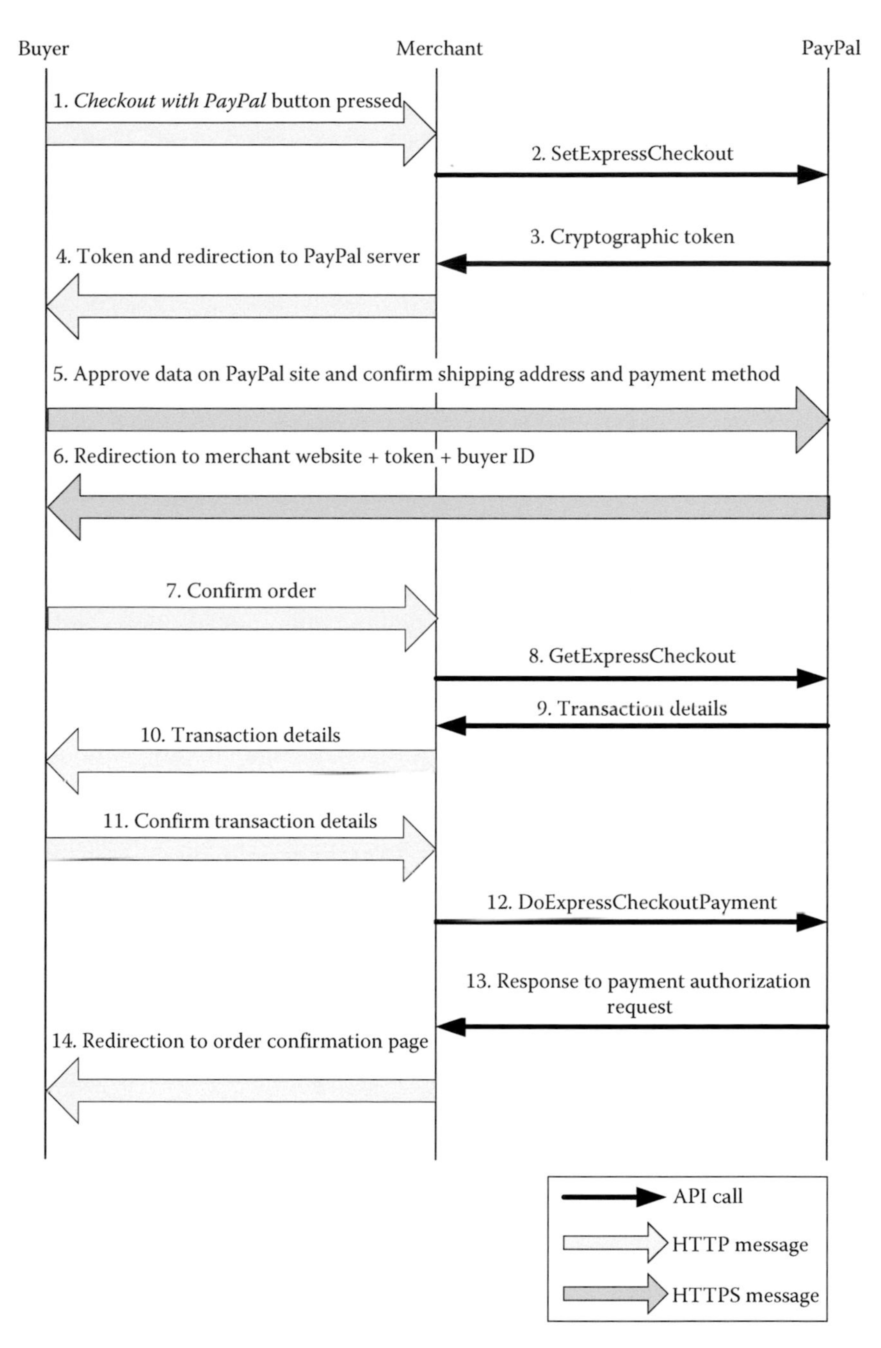

FIGURE 12.3
Exchanges for PayPal Express Checkout.

5. Using Secure HTTP, the buyer logs in for authentication and to confirm the payment method and the shipping address.
6. PayPal redirects the buyer to the merchant server using the Return URL of Step 2 and appends the Token and a Payer ID. The Payer ID is an encrypted customer account number that identifies the buyer.
7. The merchant site displays information on the order so that the buyer could confirm the order.
8. The merchant site requests from PayPal server the information that the buyer has confirmed in Step 5 with the GetExpressCheckoutDetails request.
9. The PayPal server responds includes the transaction details. In this way, the merchant sever verifies that the information with the PayPal server is consistent with the information it has.
10. The merchant site displays the transaction details alongside all the retrieved buyer data.

11. The buyer clicks to confirm the transaction.
12. The merchant requests payment authorization for the transaction with the DoExpressCheckoutPayment request. The request includes transaction details, such as the total amount, the transaction ID, and date and time of the sale.
13. The merchant receives the response to the payment authorization request.
14. The merchant site redirects the buyer's browser to the order confirmation page on its website.

12.2.3 Payment with Bank Accounts

PayPal decided to use bank account–funded payments to reduce the amount of bonus payment to new accounts by mandating additional steps in the registration process (Jackson, 2006, p. 105). In addition, this allowed PayPal to coast on the background checks of financial methods for screening and identifying individuals.

PayPal does not charge retailers for accepting checks.

12.2.4 Mobile Payments

PayPal Mobile was introduced in December 2008 to give users access to PayPal services from mobile phones. In March 2010, PayPal created an iPhone application with Bump Technologies to transfer money between two accounts by physical contact of the users' mobile phones. However, PayPal removed this capability in March 2012, probably for security reasons. In 2013, it acquired Venmo. Venmo allows groups to split bills, such as restaurant bills, easily, and then to send messages along with payment. Earlier in 2011, PayPal had acquired another mobile payment operator, Zong. By February 2015, PayPal accounted for 80% of mobile payments in the United States.

12.2.5 Fraud Prevention

Measures for fraud prevention were developed *ad hoc*, in response to the specific threats. Some of the measures apply to the sign-up process, while others are focused on the normal operating cycle, particularly to prevent phishing.

PayPal authenticates user's banking account through a micro deposit scheme. This consists in making two small, random deposits each between $0.01 and $0.99 in the given account. The account owner must confirm these deposits by logging into the PayPal account.

Similarly, to verify the ownership of a payment card, PayPal makes a small charge to that account. The charge and a 4-digit code appears on the account holder's monthly statement from the card issuer. The account holder must then report the charge and the 4-digit code to PayPal to confirm ownership.

Automatic enrollment with stolen credit card numbers is a way to take unfair advantage of the way PayPal operates (i.e., advancing the amount before recovering it) to extract money. Scripts were written to open separate account number for each number automatically. To complicate the automatic account creation used by fraud rings with stolen credit cards without hindering legitimate users, PayPal introduced in 2000 a technique coined CAPTCHA (Completely Automated Public Turing Test to Tell Computers and Humans Apart) (Jackson, 2006, p. 152). The technique uses random sequences of eight characters that are easily legible to the eye but invisible to fraud bots, as developed by Luis von Ahn, Manuel Blum, Nicholas Hopper, and John Langford from Carnegie Mellon University (Bodow, 2001). The technique proved successful in combating fraud without slowing down registrations and quickly spread to other websites susceptible to bogus account creation.

To combat the use of stolen credit cards or as a protection from nonsufficient fund (NSF) risk, PayPal places a limit on the total monthly amount a user can spend from a credit card account before verification to $250. Also, the amount of money transferred into a personal account is capped to $500 per month (Jackson, 2006, p. 129).

Currently, browsers accessing PayPal site must support authentication and encryption with https and a 128-bit key. PayPal also integrates the 3-D Secure architecture under the Verified by Visa or the SecureCode programs. As explained in Chapter 8, users register with their card issuer and establish a PIN for authentication. At the moment of consuming the transaction, they will have to provide that PIN for authentication by their card issuer before the purchase proceeds, provided that both the payment processor and the merchant bank (acquirer) support 3-D Secure. If the issuer's (buyer's) institution does not support 3-D Secure, PayPal adds a small charge to the transaction.

Many PayPal users have been targeted with fake e-mails claiming to be from PayPal *fishing* for account information from unsuspecting account holders. The e-mail may invite them to

- Visit a fake or *spoof* website to login and enter personal information. This information is then captured and replayed to access the real account.
- Call a fake customer service number to provide the information.
- Click an attachment that installs malicious or spying software (*spyware*).

Some telltale signs that an e-mail is fraudulent are as follows (Williams, 2007, pp. 25–26):

1. The mail contains a generic greeting. PayPal e-mails always use the first and last name of an individual or the business name.
2. A false sense of urgency, such as deadline for a response.
3. The invitation is to links that do not begin with https://www.paypal.com.
4. The mail contains attachments: PayPal does not include attachments in e-mail notifications.
5. Requests for reentering personal or account information such as credit or debit card numbers, bank account details, identification numbers, e-mail addresses, or passwords that PayPal already has.

Finally, PayPal has established an address (spoof@paypal.com) to enquire about a suspicious e-mail and determine whether it is authentic.

12.3 Business Accounts

Business accounts were introduced in 2000 to generate income by offering enhanced services, such as customer support and special invoicing tools for a fee (Jackson, 2006, p. 94). Later, PayPal offered payment gateway functionalities through the Payflow Payment Gateway platform that it acquired in 2005 from VeriSign. For most business products, account fees apply per transaction and some advanced products have monthly subscription fees.

At the low end, PayPal provides a virtual terminal capability so that small merchants could accept bank card payments without additional equipment (a card swipe machine).

PayPal has hosted and nonhosted solutions for merchants. In a hosted solution, PayPal offers customizable checkout pages and is responsible for the maintenance of the shopping card, the security of the payment transaction, and the reliability and availability of the service. In contrast, nonhosted solution merchants are responsible for the content of their website or mobile application, including security and protection of buyer's information. PayPal offers software development kits (SDKs) for several platforms (Java, .NET, PHP, etc.) to help merchants control and tailor their websites and integrate them with the PayPal site. PayPal also provides a list of certified suppliers that conform to its application programming interface (API).

Merchants with a PayPal account can use customized HTML "Pay with PayPal" buttons and place them on their websites. When buyers click on this payment button, they are taken to a prepopulated payment form, enter their PayPal password, and authorize the transaction (Jackson, 2006, pp. 103–104). If the merchant does not have a PayPal account but indicates that payments with MasterCard are accepted, buyers can use a "PayPal Debit Bar" facility that creates a one-time virtual MasterCard number to complete the purchase.

12.3.1 Merchant Registration

Merchants have to provide personal and commercial information including tax references. In addition, in nonhosted offers, merchants must have an Internet Merchant Account (IMA) with their bank to be able to accept payments with bank cards over the Internet. With this account, the acquiring bank underwrites the risks associated with online transactions, relieving PayPal. In both the hosed and nonhosted solutions, merchants are bound to operating rules that the relevant card association has established.

During merchant enrollment time, encryption tools are linked to the PayPal payment button on the merchant site. The encryption uses a 1024-bit RSA. The PayPal public key is included in the PayPal's public certificate that can be downloaded from the PayPal website. In general, PayPal accepts X.509 certificates from established certification authorities such as VeriSign in the Privacy Enhanced Mail (PEM) format. PayPal accepts self-signed certificates with OpenSSL or other encryption tools. The merchant public certificate is uploaded to the PayPal website in a PKCS #12 file and protected with a password using RC2-40. Generation of the PKCS12 file and its protection can be done with OpenSSL (Williams, 2007, pp. 61–64). PayPal assigns a unique Cert_ID to the merchant that will be included in all the exchanges with the merchant. The merchant's public key is extracted from its public key certificate.

The channel between the merchant and PayPal is secured using TLS/SSL with a 128-bit key. Certificates are used both on the server side and on the client side for authentication. The payment flow is as follows:

1. The merchant signs the payment information with the private key.
2. The merchant uses PayPal's public key to encrypt the payment information.
3. PayPal decrypts the payment information using its private key.
4. PayPal then verifies the merchant's signature using the merchant's public key. If the merchant identity is authenticated signature, PayPal will allow the payment to proceed.

12.3.2 Hosted Services

In the basic hosted service, PayPal payment buttons such as "Buy Now," "Donate," "Subscribe," or "Add to Cart" on the merchant site send HTML commands from the merchant site to the PayPal server. Buyers are then transferred to the PayPal server to fill in the hosted checkout pages. The merchant can write the HTML code linked to the various buttons. Alternatively, the code could be generated on the PayPal website and then copied and pasted on the merchant website.

After the transaction is completed on the PayPal site, the buyer has to click the "Continue" button on the payment page to return to the merchant site. However, the buyer may close the browser without clicking that button. Postpayment processing options give a more predictable behavior after the transaction completes, such as displaying the transaction details upon returning to the merchant website (Williams, 2007, p. 178; PayPal, 2009, pp. 13–21). The exchanges are shown in Figure 12.4 and comprise the following steps:

1. The buyer pays on the PayPal site.
2. PayPal redirects the buyer's browser to the merchant's website and appends a transaction ID for the transaction underway.
3. The merchant queries PayPal using that transaction ID and an identity token, a cryptographic string generated at enrollment.
4. PayPal returns the transaction details to the merchant.
5. The merchant displays the details to the buyer.

PayPal tracks the status of payment transactions (payment pending, received, cleared, failed, etc.). Typically, these events represent various kinds of payments; however, the events may also represent authorizations, fraud management actions, refunds, disputes, and charge backs (PayPal 2009, p. 7). The notification includes all the parameters of the transaction, such as transaction ID, date, price, and items.

To prevent man-in-the-middle attacks, a validation protocol is used based on a shared secret between the merchant server and PayPal server established at enrollment.

12.3.3 Mobile Point-of-Sale Terminals

In 2012, PayPal introduced a plug-in dongle that plugs into a merchant's mobile device (smartphone or tablet) to make it a magnetic card reader. This service offer is aimed at small merchants without card processing services. With the dongle in place and connected to the headphone jack,

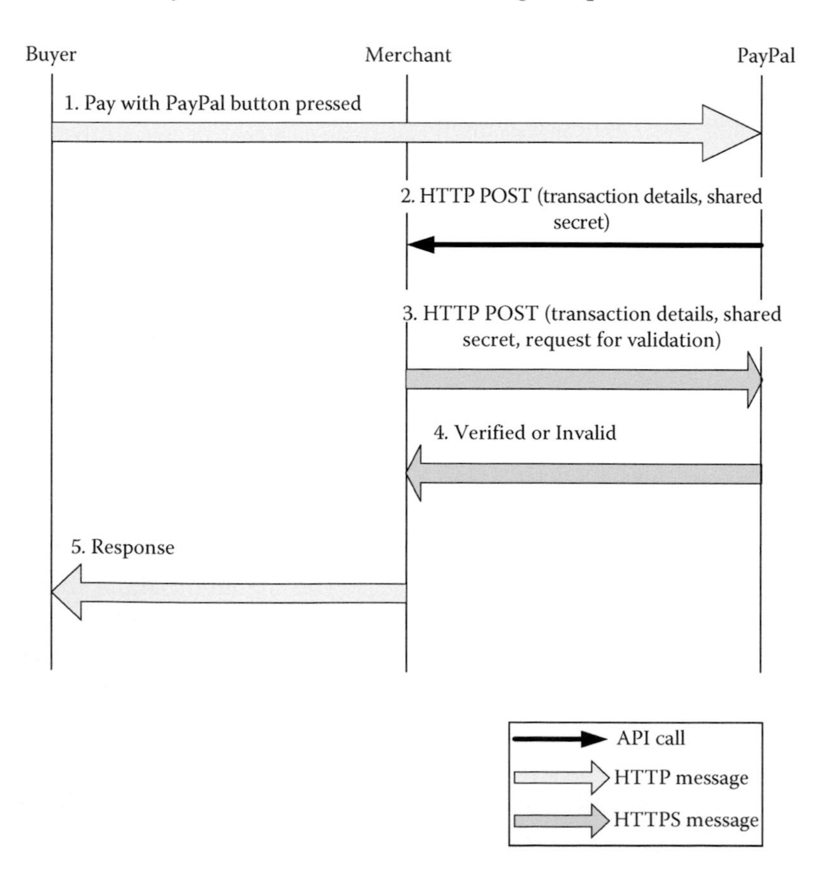

FIGURE 12.4
Exchanges associated with the basic hosted service.

the information in the debit/credit card is read and sent by an application on the merchant device called PayPal Here™. This application processes information to track sales and provide other performance measures.

During the configuration of the application, the merchant enters for each item its sale price, the sales tax, and any associated discounts. PayPal Here can invoice customers during the checkout process and can record cash payments and send receipts. A receipt can be sent to the customer's phone number or e-mail address and the application manages refunds as well.

Without the card reader, PayPal Here accepts scans of the debit/credit cards using the camera feature on the smartphone. PayPal Here accepts check payments of $1000 or less with a simple scan. PayPal Mobile Check Capture also offers the capability of clearing and processing check images through PayPal accounts.

The basic information that the merchant provides is displayed in the PayPal Local tab, which is part of PayPal's consumer application. This information will also appear on the receipts.

By default, the merchant's PayPal Here profile page shows the business as open. This means that PayPal users can use their mobile phones to find the merchant and conduct their transactions. Merchants can make their store not visible by swiping a finger across the Open sign. It will change to Closed and the store information will no longer be listed on PayPal Local.

PayPal Here tracks the sales and provides the history of transactions, and the sales figures can be broken out either with prices for individual items sold or as a total for each transaction. PayPal Here makes it easy to transfer funds from the PayPal Business account to the bank account.

With the acquisition of Braintree and Paydiant, PayPal enriched its mobile commerce offers to businesses. Braintree processes transactions for mobile apps while Paydiant provides a platform to retailers, banks, and payment processors to operate mobile wallets.

12.4 Summary

In the business, economics, or marketing literature, the concept of the first-mover advantage is often used to explain why the first entrant has become a dominant force in its field. PayPal, however, was late to the online payment field but overtook its competitors. One of the main reasons is its ease of use.

PayPal's initial strategy of reliance on auctions and person-to-person payments was found to be both risky and limiting. PayPal refocused on business intermediation by offering gateway functionalities and payment processing, competing with other online electronic commerce entities. Later, it started on an international expansion.

In summary, from a service viewpoint, PayPal's evolution to date can be subdivided into four phases as follows:

1. *Phase 1 (1998–2003)*: In this phase, PayPal offered an account-based system to allow its subscribers to send and receive payments online using a variety of instruments, including micropayments. The offer was tailored to payment transactions related to auction sites as well as to home businesses that were too small to qualify for credit card processing. In this phase, PayPal attracted traffic as a low-cost service provider. It added optional fee-bearing accounts to generate revenues through enhanced services.
2. *Phase 2 (2004–2007)*: In this period, PayPal expanding beyond eBay auctions to offer payment solutions suited to small and medium online merchants, planning to earn interest on the float from the cash accumulated in the various users' subaccounts. However, most recipients withdrew their funds immediately. Furthermore, a substantial segment of senders used their credit cards, for which the transactions could not be redirected to the ACH network and their cost could not be compressed. As a consequence, PayPal's portfolio included the functions of payment gateway and payment processor and offered solutions for securing online payments, fraud prevention, seller, protection, and financial reporting by integrating back-office systems.
3. *Phase 3 (2008–July 2015)*: In 2007, PayPal started to operate in the European Union and with other currencies. It expanded its focus to large businesses and mobile networks.
4. Phase 4 started in July 2015 after eBay divested itself from PayPal. With a series of acquisitions, it is positioning itself for transfer remittances from the United States to many developing countries as well as for cloud-based mobile commerce.

PayPal earns its money mostly from the fees it charges, typically a percentage of the transaction amount but also monthly fees for some services. A secondary stream arises from the interests accrued on funds within the PayPal system.

Questions

1. Describe the phases of PayPal's growth.
2. Evaluate the risks of PayPal in the consumer space.
3. What are the reasons that PayPal outwitted its competitors?
4. What are the strengths and weaknesses of PayPal?

13

Digital Money

Digital money is the most ambitious solution for online payment. This is an instrument that matches the speed and the ubiquity of computer networks while attempting to mimic classical fiduciary money in terms of anonymity, untraceability, and difficulty of counterfeit. The most important distinction of digital money from classical money is that its support is virtual; the value is stored in an algorithmic form in the user's computer or on a network server. It constitutes "a medium of exchange that operates like a currency in some environments, but does not have all the attributes of real currency. In particular, virtual currency does not have legal tender status in any jurisdiction" (Financial Crimes Enforcement Network, 2013). Eventually, however, the dematerialization of money could lead to the formation of a new monetary system, a disruptive potential that the success of Bitcoin and other cryptocurrencies*—also known as distributed currencies—underscores. *Crypto* indicates reliance on advanced cryptographic techniques while *distributed* shows that the transactions are processed and recorded across a network of computers through the use of a distributed and public transaction ledger (Peck, 2012).

Several systems of digital money were proposed in the early 1990s as cryptography moved into the main stream (Brands, 1993; Eng and Okamoto, 1994; Okamoto, 1995; Chan et al., 1996; Wayner, 1997). NetCash was an advanced prototype (Mevinsky and Neuman, 1993; Neuman and Mevinksy, 1995; O'Mahony et al., 1997, pp. 168–181), while DigiCash was commercialized. The 2010s saw the birth of new type of digital money, where both the production of money and the verification of the accounts are distributed.

Bitcoin is considered the foremost exemplar of this new form of digital money. It is the brainchild of Satoshi Nakamoto (2008), presumably a pseudonym of one or several cryptologists. It was designed to eliminate the role of a trusted third party, such as a central bank, in the creation of currency and in the maintenance of the system's integrity.

* More accurately, these should be called Crypto-Jetons because, at least for now, bitcoins are not used in the economy at large. However, we will use the mainstream terminology.

This chapter presents the general characteristics of digital money with a focus on DigiCash. Chapter 14 is devoted exclusively to Bitcoin and other cryptocurrencies.

13.1 Privacy with Cash and Digital Money

Money in the form of paper notes or metallic coins has no link to the nominal identity of the holder (buyer or seller) and their banking coordinates. This attribute was reproduced in some face-to-face systems as discussed in Chapter 11. Digital money aims to imitate cash in the sense that it is anonymous and untraceable at the cost of some computational complexities (Chaum, 1989, n.d.). Anonymity protects the identity of the actor (buyer or seller) by dissociating it from the completion of the transaction. Untraceability means that two payments from the same individual cannot be linked to each other by any means (Sabatier, 1997, p. 99).

In remote transactions, anonymity depends on two factors: the ability to communicate in an anonymous manner and the capability to make an anonymous payment. Clearly, an anonymous communication is a *sine qua non* condition for anonymous payments; once the source of a communication has been identified, the most sophisticated scrambling strategies will not be able to mask the identity of the intervening party or the station of origin (Simon, 1996).

Today, anonymous services use communication mixes that combine the traffic of a large number of transactions into larger ones and then splitting them again to make tracing them more difficult. Communication mixes can also be chained to substantially reduce risks, and more elaborate and secure laundries can be constructed using secure multiparty computation. Another popular anonymous communication network is Tor, which routes traffic through a worldwide, volunteer network consisting of more than 6000 relays, thereby concealing the sender's location from anyone conducting network surveillance or traffic analysis.

Tor encrypts the original data, including the destination IP address, multiple times and sends them through a virtual circuit comprising successive, randomly selected Tor relays. Each relay decrypts a layer of encryption to reveal only the next relay in the circuit to forward the remaining encrypted data. The final relay decrypts the innermost layer of encryption and sends the original data to its destination without revealing, or knowing, the source IP address.

In the first generation of digital money, verification was centralized and required that a system operator (the bank) update the account balance of each actor, even without knowing the actor's identity or linking the various transactions that actor may have performed with others. In cryptocurrencies, the account balances are stored in a distributed ledger than can be verified by all nodes.

13.2 DigiCash (eCash)

The DigiCash payment system originated with the European informatics project ESPRIT on the basis of the work of David Chaum and his team. DigiCash uses a digital currency called eCash to replace coins and notes, to retain anonymity and untraceability, and to allow secure communication over open networks.

In 1990, David Chaum formed DigiCash in the Netherlands and in the United States to commercialize the system. The Mark Twain Bank of St. Louis, Missouri, agreed to issue this digital currency in U.S. dollars. It managed a special account for DigiCash to collect the various amounts circulating in the electronic currency before depositing them in the creditor's bank accounts. About 5000 clients registered to use the currency with 300 merchants. For a while, Deutsche Bank was commercializing its own digital money in Germany. DigiCash was considered by the Conditional Access for Europe (CAFÉ) program of the European research project European Strategic Program on Research in Information Technology (ESPRIT). The objective of the CAFE program was to propose new electronic payments, including a multiapplication integrated circuit card that would serve as the electronic equivalent of a traditional wallet. DigiCash was also considered as a replacement of cash in toll road (Vigna and Casey, 2015, p. 55). In 1999, however, Mercantile Bank, which bought the Mark Twain Bank, judged the results to be insufficient and withdrew its support. The company eCash Technologies purchased all DigiCash patents and was itself acquired by other Internet companies.

The system works as follows. The user mints the eCash coins and then sends them to the bank for signature. The coins that the bank returns are then stored on the user's hard disk. They can be used to pay for purchases made at merchants that subscribe to the DigiCash system. The bank can add the role of a trusted third party to its role of a payment intermediary and a gateway to banking networks. Figure 13.1 depicts the relations that DigiCash establishes among the various parties of a transaction. In the general case, there is no guarantee to the client that the merchandise will be delivered.

These exchanges can also be used to transfer value between two peer entities, such as from one user to another.

13.2.1 Registration

The user downloads a plug-in to the browser called Cyberwallet (which DigiCash issues) to store the client's coins,

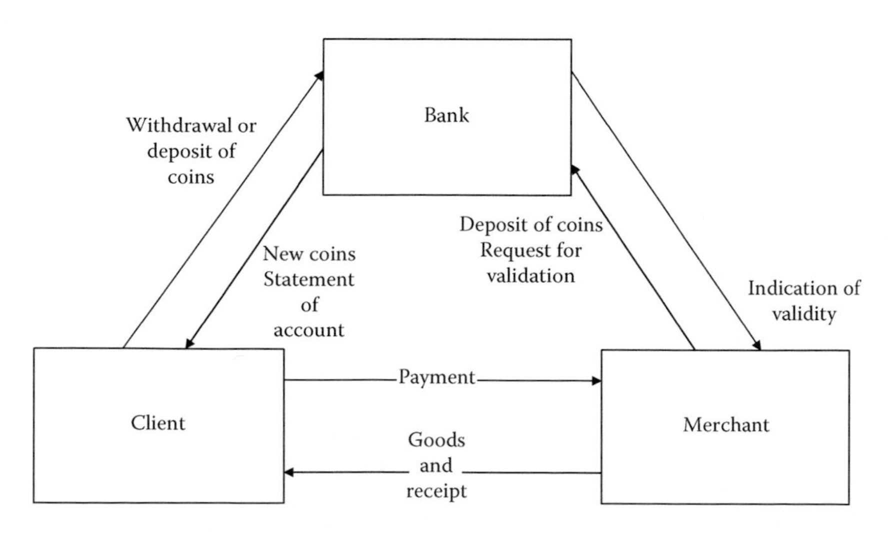

FIGURE 13.1
Relations among the various entities according to DigiCash.

track the expenditures, and verify the expiration dates of the coins. Contrary to bank notes, DigiCash coins have an expiration date as a protection against fraud including spending the same coin more than once (double spending). At registration time, the client receives all the public keys that will be used for minting and verifying the various denominations. Similarly, the merchant registers and downloads the merchant's software to manage customers, payments, and reimbursements from the bank. Both the merchant and the clients must have their accounts in the same bank because interbanking relations are not yet supported.

13.2.2 Loading of Value

To charge the digital purse with monetary value, the client software generates new coins without value, each with a unique serial number. The serial numbers are chosen at random and are large enough (100 digits) to minimize collisions with the choices of other clients. After blinding the coins, the software sends them to the bank to be signed with a blind signature as described in Chapter 3. The bank then debits the client's account for the corresponding amount (and the relevant commission) and puts its signature without uncovering the serial number of the minted coin. When they are returned to the users, the coins will be charged with the requested values and signed by the bank.

The note that the bank signs with its private key *SK* can be represented as

$$\{\text{Serial number}\}SK$$

It is seen that the note is anonymous because it depends only on the serial number and the bank's private key. The exchanges are also anonymous because they depend only on the client's identity, and only the client knows the blinding factor. The only transaction that is perhaps not anonymous is the debit of the client's account. However, the client's anonymity is only protected if the communication channel between the client and the bank does not allow a third party to guess the client's identity, for example, by revealing the terminal address or location.

The bank associates a pair of private and public keys for the same modulo N to each coin denomination; this pair of keys is called the coin denomination key. The coin is of the following form:

$$Key\_type, \{\text{Serial number}\}SK_{coin\_denomination\_key}$$

where *Key_type* indicates the coin denomination key and other related information (e.g., the currency and the expiration date).

To ensure the integrity of the exchanges, the client uses the client's private key to sign the loading value request. To ensure confidentiality, the client uses a secret session key that will itself be encrypted with the bank's public key. The bank's response, which includes the coins that have been signed blindly, is signed with the bank's private key and then encrypted with the session key.

To protect customers against failures in the operation (terminal failures or network failures), the bank can store the trail for the last n loading transactions (e.g., $n = 16$) of each customer. This facilitates error recovery and the reconstitution of the value stored in the customer's software.

13.2.3 Purchase

The merchant's software sends a payment request to the client's software in the clear. It is of the following form:

$$\text{Payment request} = \{\text{currency, amount, time stamp, merchant's account, description}\}$$

Because the message is sent in the clear although it contains details about the order amount, the currency to be used, and some description of the order, an eavesdropper may be able to see what is ordered and for how much.

Note that if several banks are allowed to sign the electronic money blindly, another field would be added to identify the emitter bank.

The payment message contains details of the transaction and the monetary notes encrypted with the bank's public key PK_{Bank}:

$$\text{Payment} = PK_{Bank}(\text{payment\_info}; \{\text{notes}, \text{H(payment\_info)}\})$$

The merchant has to join this message with the financial settlement request so that the bank can verify that the merchant and the client have accepted the same sales conditions. The *payment_info* field can be expressed as follows:

{Bank identification, amount, currency, number of notes, time stamp, merchant identifier, H(description), H(client_code)}

H(client_code) is the hash of a secret code that the customer's software selects. It could be used to show a proof of payment to the bank if the customer decides to reveal its identity.

13.2.4 Financial Settlement

To ask for reimbursement, the merchant signs the payment with its private key SK_M and then encrypts it with the bank's public key:

$$\text{Settlement request} = PK_{Bank}\{(\text{payment})SK_M\}$$

In this way, the bank can verify that the client and the merchant agree on the terms of the transaction. The bank must also verify that the notes used for purchase have not been previously used. Once all the verifications are made, the bank credits the merchant's account with the amount of the transaction and informs the merchant of the result with the message that it signs with its private key:

$$\text{Receipt} = \{\text{result}, \text{amount}\}SK_{Bank}$$

Note that the customer will behave in a similar manner if the customer desires to return the notes that have not been spent either for reimbursement with legal tender or for exchange against new notes at the expiration of the old notes.

Protection against double spending of digital money is the bank's function; the bank has to verify the serial number of each digital coin or note to be deposited and that it has not reached its expiration date. After each settlement, it records in a large database the serial numbers of all coins that have been spent. Nevertheless, a cheater can use the same coins without detection in between settlement requests.

13.2.5 Delivery

The merchant is obligated to deliver the purchased items to the client. In case this does not happen, the client can ask the bank to intervene by revealing the secret code *client_code* whose hash it has already sent, thereby giving up anonymity to trace the flow of money.

13.3 Anonymity and Untraceability in DigiCash

Anonymity and untraceability in DigiCash are based on the use of blind signatures, the principle of which was explained in Chapter 3. This mechanism is an extension of the authentication of a message with the public key encryption RSA algorithm. It allows the payer itself to mint the digital coin and the bank to seal it without the bank having access to the coin's serial number. Of course, the message exchanges are secured with the typical mechanisms of confidentiality and authentication.

13.3.1 Case of the Debtor (Buyer) Untraceability

Consider a debtor that would like the bank to sign a digital note blindly. The bank has a public verification key e, a private encryption key d, and an RSA modulus N ($N = pq$, where p and q are different prime numbers) of length 512 or 1024 bits.

13.3.1.1 Loading of Value

To request the bank's authorization without disclosing the serial number of the means of payment, s, the debtor chooses a random number r uniformly distributed between 1 and $(N - 1)$:

$$1 \leq r \leq (N-1)$$

The sizes of the numbers s and r are 200 bits. Only the debtor has this number r, which is called the blinding factor. Its role is to hide the serial number s of the payment note. This operation consists of encrypting the blinding factor with the bank public key e and then sending it to the bank for digital signing (see Figure 13.2):

$$sr^{e} \bmod N$$

The bank then seals the note by signing it with its private key d:

$$(sr^{e})^{d} \bmod N = s^{d} r \bmod N$$

and returns it to the debtor. The bank debits the client's bank account with the amount that corresponds to the digital note.

In general, the message sent to the bank contains several fields including the serial number s. These

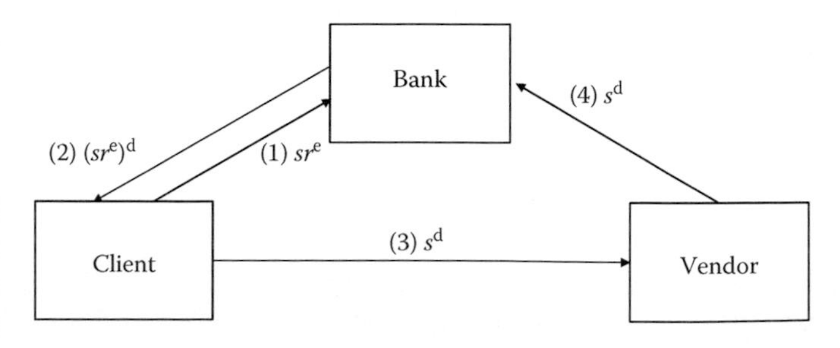

FIGURE 13.2
Simple system ensuring the buyer's untraceability.

fields indicate, among other items, the type of banking operation and the debtor's account number, the whole signed by the debtor's private key. This is represented by

$$(\text{"withdrawal"}, \text{"debtor's account number"}, sr^{e})^{c}$$

where c is the debtor's (client) private key. The corresponding public key is included in a certificate signed by the competent certification authority.

The note is anonymous because it includes only the serial number encrypted with the bank secret key. The bank knows only the account number, and because the serial number of the note was blinded, the bank will not be able to reconcile the merchant's request for financial settlement with the note that it has blindly signed.

13.3.1.2 Purchase

The debtor obtains the signed note by dividing what arrives from the bank by r to get

$$s^{d} \bmod N$$

Using the bank's public key, the debtor verifies that the received note is what was sent because

$$(s^{d})^{e} \bmod N \equiv s \bmod N$$

The buyer then sends the signed payment note to the creditor (the merchant).

13.3.1.3 Deposit and Settlement

The merchant deposits the note in the merchant's bank account by sending the following message to the bank:

$$\text{"deposit"}, \text{"merchant bank account"}, s^{d}$$

After verification that the serial number of the note has not been previously deposited, the bank credits the merchant account with the equivalent sum and withdraws the note from circulation. The bank may also send a receipt to the merchant in the following form:

$$f(\text{"deposit"}, \text{"merchant bank account"}, s^{d})^{d}.$$

This protection against double spending obligates the bank to keep a list of all the serial numbers for the notes that were deposited.

13.3.1.4 Improvement of Protection

In some circumstances, clients may be interested in forfeiting untraceability to protect their rights (such as in cases of litigation with the merchant or the theft of notes). The construction of the random number r is done by applying a one-way function (e.g., a hashing function) to a precursor number r' that the client keeps hidden. This allows the debtor to prove, if needed, that the debtor is the originator of the note, which the bank has signed blindly, and this is done by presenting the precursor r' to an arbitrator. In contrast, a merchant that refuses to fulfill its obligations or a thief that has stolen the note would not be able to do so.

Similarly, the serial number s can be formed by applying a special algorithm that allows the client to prove to an arbitrator that it is the originator of the note. If s' is a random number, s can be constituted as

$$s = \{(s' \oplus H(s')) \| H(s')\} \equiv g(s')$$

where

$H(\,)$ is a hash function
$\oplus$ is the *exclusive OR*
$\|$ is the concatenation operation
$g(\,)$ represents the total operation on the precursor s'

The debtor must then supply the precursor s' to the arbiter to prove the authenticity of the note that the merchant has deposited.

The use of the precursors allows the client to confirm to the bank that the client has indeed spent the note that the merchant is depositing. These exchanges are shown in Figure 13.3.

In this configuration, the bank verifies that the serial number of the payment note was not previously used and provides the merchant with a statement of acceptability. The merchant forwards the statement to the buyer. This statement includes an expiration date and can be signed with the bank's private key so that the merchant and the client can ascertain its integrity and its origin. If the client is convinced of the statement's authenticity, it sends the precursor s' to the bank through the vendor. The debtor verifies the precursor and forwards it to the bank, which also confirms that it is the entity that produced the serial number of the note. At this point, the bank gives its definitive approval, records the note as deposited, retains a copy of the precursor, and credits the merchant's account with the corresponding amount.

13.3.2 Case of the Creditor (Merchant) Untraceability

To achieve the creditor untraceability, the merchant issues the note and then sends it to client who, in turns,

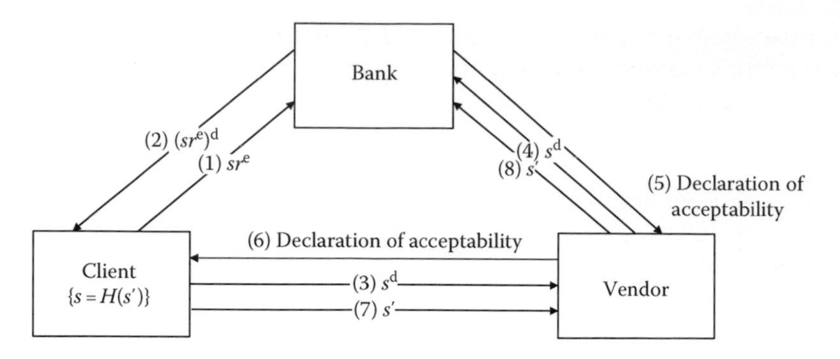

FIGURE 13.3
Protection of the buyer through the use of precursors.

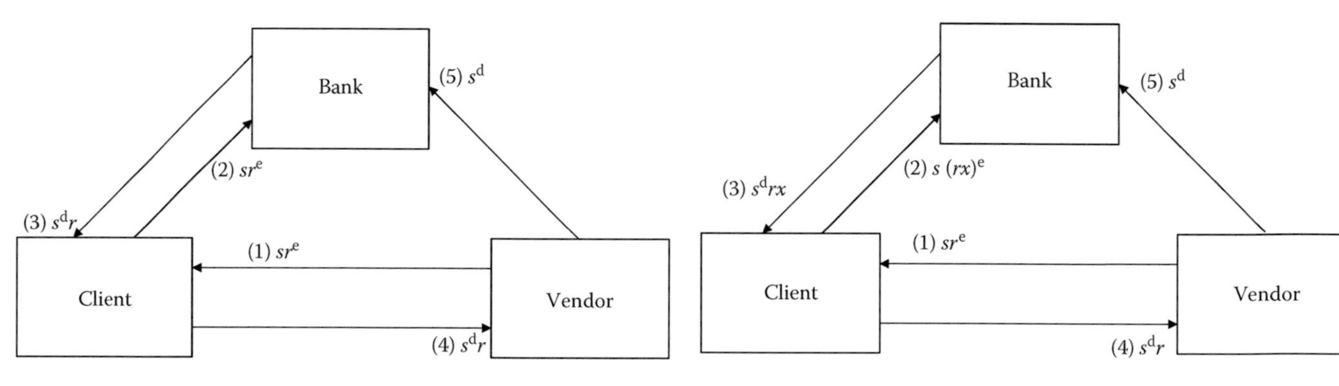

FIGURE 13.4
Merchant untraceability.

FIGURE 13.5
Mutual untraceability of the debtor and the creditor.

sends it to the bank for a blind signature and for withdrawal of the corresponding amount to the client's account. When the creditor receives the signed note, it will use the bank's public key to verify that the note has been derived from the note that was sent to the bank, as follows:

$$(s^d\, r)^e \bmod N = sr^e \bmod N$$

The client sends the note to the merchant who, in turn, sends it to the bank after removing the blinding factor as shown in Figure 13.4. The bank then credits the merchant's account with the corresponding amount. Compared with the case of buyer's untraceability shown in Figure 13.2, the exchange consists of five messages instead of four.

Notice that the message that the bank receives is identical to the one in the case of buyer (debtor) untraceability of Figure 13.2. Thus, the bank will not be able to tell which of the two parties is untraceable, the client or the vendor.

13.3.3 Mutual Untraceability

Mutual untraceability is constructed through the fusion of the cases of debtor untraceability and creditor untraceability. The debtor, after receiving the note that the creditor has issued and blinded, adds another blinding factor by choosing a random number x, uniformly distributed between 1 and $(N - 1)$. The note is sent to the bank for a blind signature and becomes $s(rx)^e$.

This technique is based on double-blinding and corresponds to the exchanges shown in Figure 13.5.

Another way to achieve mutual untraceability is to combine the two simple cases of creditor untraceability and debtor untraceability and to ask a trusted third party to be the intermediary among the parties. The client pays the third party using the debtor untraceability protocol. From its side, the trusted third party retransmits the payment to the merchant according to the rule of creditor untraceability. The bank can be this trusted third party because even in that role, it cannot trace either the payer or the payee.

13.4 Splitting of Value

The representation of several denominations of digital money utilizes the binary representation of the value of a note; a denomination is assigned to each bit that is set to 1. In this way, j denominations of digital money can represent all the values between 1 and (2^j – 1) monetary values. Each of these denominations corresponds to one of the prime numbers in the bank's public exponent of an RSA system, as shown in Figure 13.6.

Thus, to represent the value of 1, the public exponent of the bank will be 3. Similarly, the value of 2 decimal (or 10 in binary) will correspond to the public exponent of 5, the value of 4 decimal (or 100 in binary) to 7, the value of 8 decimal (or 1000 in binary) to the public exponent of 11, and so on. Thus, with the public exponent of $11 \times 7 \times 5 \times 3$, a note with 4 denominations will be capable of representing all decimal values between 1 and 15 monetary units (in binary between 0001 and 1111). The following example shows how payments made with the denominations constructed in the manner indicated solve the problem of change (Chaum, n.d.).

Consider a digital money with four denominations. Assume that a customer would like to buy two digital notes, each representing 15 monetary units. Each note has its own serial number, say, s_1 and s_2, respectively. The corresponding blinding factors are r_1 and r_2, respectively.

The bank public exponent that is used to represent the full value note is therefore

$$h = 3 \times 5 \times 7 \times 11$$

To load the value, the client mints the two notes, hides them with the corresponding blinding factors, and sends them to the bank for its blind signature. The message sent to the bank will be the concatenation of two requests:

$$s_1 \times r_1^h \,||\, s_2 \times r_2^h$$

The bank debits the client's account with the stated amount, signs the notes without knowing the serial number of the notes, and then returns the composite message:

$$s_1^{1/h} \times r_1 \,||\, s_2^{1/h} \times r_2$$

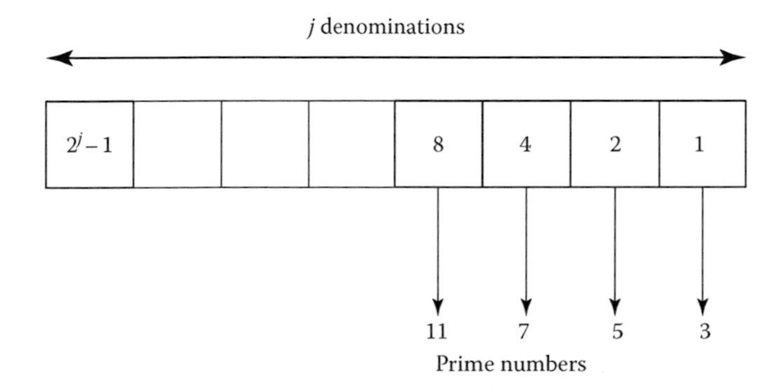

FIGURE 13.6
Representation of the various denominations with prime numbers.

The inverse $1/h$ is the bank private exponent and is calculated with the formula

$$h \times \left(\frac{1}{h}\right) = 1 \bmod ((p-1)(q-1))$$

This exponent also indicates the total value of the note.

To pay with the first note for an item valued at 10 units (whose binary representation is 1010), the client utilizes the exponent 11×5 in its purchase request. Therefore, the client composes a message to be sent to the merchant in the following form:

$$s_1^{(1/h)\,(5 \times 11)} = s_1^{1/(3 \times 7)}$$

To obtain the residue change back, the client becomes an untraceable creditor of the bank. Following the rules of the protocol for creditor untraceability, the client forms a *cookie jar* j in which to hold the residues and hides it with a blinding factor r_a. This second note takes the following form:

$$j\, r_a^{(5 \times 11)}$$

The message sent to the merchant to buy items with a total value of 10 units with a note of 15 units is composed of the concatenation of the two messages:

$$s_1^{1/(3 \times 7)} \,||\, j\, r_a^{(5 \times 11)}$$

The bank returns via the merchant the following message to indicate the change:

$$j^{1/(5 \times 11)} r_a$$

The residue change corresponds to the following exponents:

$$\frac{3 \times 5 \times 7 \times 11}{5 \times 11} = 3 \times 7$$

that is, the binary value 0101 or 5 monetary units. These exchanges associated with the first payment are shown in Figure 13.7.

Assume that the second payment is for a value of 12 units (binary representation of 1100); the public exponent is 7×11. Therefore, the composite message (using a new blinding factor r_b for the residual in the cookie jar) is

$$s_2^{1/h\,(7 \times 11)} \,||\, j^{1/(5 \times 11)} r_b^{(7 \times 11)}$$
$$s_2^{1/(3 \times 5)} \,||\, j^{1/(5 \times 11)}\, r_b^{(7 \times 11)}$$

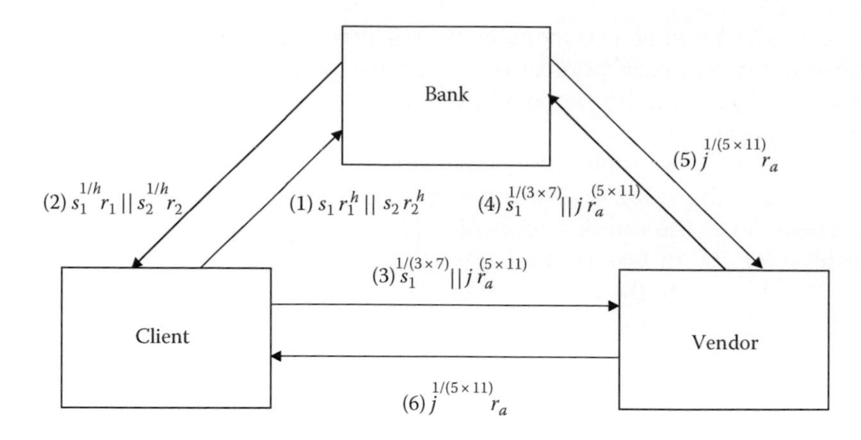

FIGURE 13.7
Exchanges with multiple denominations and returns of change.

In this case, the bank returns a message through the merchant that has the total residual change in the cookie jar:

$$j^{1/\{(5\times 11)(7\times 11)\}} r_b$$

The change for the second payment corresponds to the exponents:

$$\frac{3\times 5\times 7\times 11}{7\times 11} = 3\times 5$$

that is, the binary value 0011 or 3 monetary units. Thus, the total residual change is 5 + 3 = 8 units. This corresponds to the content of the cookie jar.

This approach depends on the withdrawal of notes of fixed value with the unspent parts later credited to the payer during a refund transaction. Other schemes have been devised to allow change to be spent without an intervening withdrawal transaction (Chaum, n.d.).

In the general case, the message sent to the bank for blind signature can take the form:

$$f(s)\times r^h (\bmod n)$$

where $f(\,)$ is a one-way function.

13.5 Detection of Counterfeit (Multiple Spending)

The fabrication of false money is the principal source of fraud for paper bills. To prevent counterfeiting, the support is perfected (paper quality, insertion of a magnetic wire, presence of watermarks, etc.) to make the creation of equivalent notes more difficult. In contrast, the duplication of digital money is much easier; it is sufficient to reproduce the file that contains the monetary data. This introduces the danger of reusing the same note, which is called "double spending." Even if the bank can distinguish between good notes and reused notes on the basis of the serial number, the identity of the counterfeiter remains hidden.

The following algorithm, due to David Chaum and his team, allows the detection of cheaters without divulging the identity of honest debtors (Chaum et al., 1990)

13.5.1 Loading of Value

Consider a debtor that owns a bank account u to which the bank associates a counter v:

1. The debtor forms k blinded messages of the following format:

   $$B_i = s_i \times r_i^e \quad 1 \le i \le k$$

   In this case, s_i is formed by $f(x_i, y_i)$ with

   $$x_i = g(a_i, c_i) \quad \text{and} \quad y_i = g\{a_i \oplus (u \,||\, (v+i)), w_i\}$$

   The variables a_i, c_i, and w_i are independent and uniformly distributed over the interval (1, N − 1), where N is the modulus for computations. The k messages are sent to the bank as candidates to make the digital notes. The functions $f(\,)$ and $g(\,)$ are two functions with two arguments that are without collisions so that each result corresponds to a single combination of the two inputs.
2. The bank chooses at random $k/2$ candidates for verification and asks the debtor to produce the corresponding variables a_i, c_i, r_i, and w_i. Let R be the set of these $k/2$ candidates.

3. Once the bank has finished all the checks and if it did not discover any irregularity, the bank sends the product of the $k/2$ candidates signed with its private key d

$$\prod_{i \notin R} B_i^d = \prod_{1 \le i \le k/2} B_i^d$$

4. The bank withdraws the corresponding amounts from the debtor's account.

As in the preceding cases, the debtor removes the blinding factors r_i, from the $k/2$ candidates ($1 \le i \le k/2$), to form the digital note as the product of all the serial numbers raised to the power of the private key of the bank:

$$\prod_{1 \le i \le k/2} s_i^d$$

The probability for detecting fraud clearly depends on the number k and the frequency of cheating. For a given value ε, if the proportion of irregular candidates among those that the bank checked exceeds ε, the probability of catching the cheat is $(1-e^{-c\varepsilon k})$ for a constant c.

13.5.2 Purchasing

When the debtor sends the merchant the digital note, the merchant starts a check procedure to catch cheats. The procedure is as follows:

1. The merchant selects a random binary string $\{z_1, z_2, \ldots, z_{k/2}\}$ and stores it in the vector $\mathbf{Z}$ of dimension $k/2 \times 1$.
2. For all the elements z_i, $1 \le i \le k/2$:
   - If $z_i = 1$, the merchant asks the client to send a_i, c_i, and

     $$y_i = g\{a_i \oplus (u \,\|\, (v+i)), w_i\};$$

   - If $z_i = 0$, the merchant asks the client to send

     $$x_i = g(a_i, c_i), a_i \oplus (u \,\|\, (v+i)), \quad \text{and} \quad w_i.$$

3. The merchant verifies that the digital note is of the correct form and then sends the note, the vector $\mathbf{Z}$, and the client responses to the bank with a request for financial settlement.

13.5.3 Financial Settlement and Verification

The bank stores the note of digital money, the vectors of the tests, and the responses. If the note is being reused, the probability of receiving two complementary values for two different merchants and for the same bit z_i is high. In other words, it is highly probable that the bank would have received both a_i and $a_i \oplus (u \,\|\, (v + i))$. With these two values, the bank can extract the debtor's account number u and then identify the debtor.

13.5.4 Proof of Double Spending

One of the difficulties of the mechanism just presented is that the discovery of the cheater's identity depends on the bank's accuracy or honesty. To protect the client from any foul play, it is necessary that a signature be used.

Several approaches are possible; it is not the intention here to review all the proposals but to highlight their ideas. For example, s_i can be formed with $f(x_i, y_i)$ and then inscribed in each note using the following equations:

$$x_i = g(a_i, c_i) \quad \text{and} \quad y_i = g\{a_i \oplus ((u \,\|\, z_i' \,\|\, z_i'') \,\|\, (v+i)), w_i\}$$

where (z_i', z_i'') are random variables.

At the time of loading the value, the debtor will have to supply the signature on the factors $g(z_i' \,\|\, z_i'')$ for each of the $k/2$ notes that the bank chooses for checking. During the purchase, the client will also have to supply the signature of the factor $g(z_i' \,\|\, z_i'')$ in the used note. This way, the bank can prove the identity of the cheat without any doubt if it can break at least $1 + k/2$ of the elements (z_i', z_i'').

13.6 Evaluation of DigiCash

DigiCash used blind stamping to verify the validity of digital coins and notes without knowing their details. Furthermore, there was no relation between the serial number of the digital money and the identity of the holder. Once a coin was deposited with the bank, it was withdrawn from circulation. The radical changes in the banking system were probably responsible for the resistance to this means of payment.

From the point of view of transactional properties, it is observed that

- The transactions are not atomic because the client is not necessarily aware of the current status of the payment. In case of a network failure, the status of the money sent is undetermined (Camp et al., 1995).
- Double spending can occur undetected.
- The bank can be a bottleneck to the whole system because it has to verify every coin to be deposited.

- The larger the size of the database that contains the serial numbers of the deposited coins, the longer the response time of the bank.

At the time, DigiCash was radically different from other instruments and raised many legal and financial issues that the financial industry, as well as regulators, were not ready to address. At the time, advanced cryptography was not legally allowed in civilian applications, especially in systems of interest to law enforcement institutions. Also, some reports attribute the lack of success to David Chaum's suspicious personality and managerial decisions, including refusal to agree with Microsoft on integrating the system with Windows 95 (Anonymous, 1999; Vigna and Casey, 2015, p. 56).

Questions

1. Name some of the legal challenges facing digital money systems.
2. How can untraceability and inheritance be made compatible within the scope of digital money?

14

Bitcoin and Cryptocurrencies

Bitcoin is the lightning rod for a new species of digital money. In this system, the operations of money creation and system supervision are distributed; that is, there is no functional equivalent to a central bank. The total number of bitcoins is fixed and the rate of issuance is programmed to decrease over time, so there is no possibility to intervene in the bitcoin economy. All value transfers are recorded in a distributed public ledger with cryptographic protection. This means that the size of this ledger increases continuously as new transactions are added. Ledger verification is distributed among the nodes of the Bitcoin network. Nodes that participate in the verification gain a reward in bitcoins if they also resolve a cryptographic puzzle.

Note: By convention, "Bitcoin" is used to represent the protocol, while "bitcoin" (BTC) represents the payment unit.

The novel techniques in bitcoin production, distribution, and supervision leave many questions unanswered regarding the scalability and security of the system. These issues need to be resolved before bitcoins could be used as viable alternatives to the current global payment system, particularly in business-to-business transactions.

This chapter presents the Bitcoin protocol and highlights operational issues that practical implementations have encountered. In particular, the risks facing bitcoin users are underlined. There are four appendices to this chapter. Appendix 14A contains the *Crypto Anarchist Manifesto,* the founding document of the Cypherpunks group where the proposal for Bitcoin was published (May, 1992). Appendix 14B analyzes Bitcoin from a socio-economic perspective. Appendix 14C presents some of the most significant cryptocurrencies selected from the constellation of proposals that followed the success of Bitcoin. Finally, Appendix 14D lists some of the service offers built around Bitcoin.

14.1 Background

Bitcoin is one form of digital money that uses the computational power of networked machines to create a new currency and to record all transactions in this currency in a public ledger, all without central coordination. It was first outlined by Satoshi Nakamoto in 2008 in the Cypherpunks mailing list, a group started in 1992 by Timothy (Tim) May and Eric Hughes to promote the defense of privacy in the digital world.

One of Bitcoin's predecessors is the B-money scheme from Wei Dai, also published on the Cypherpunks mailing list in 1998. A person creates new B-money units in proportion to the effort spent in solving a previously unsolved computational problem. The problem satisfies the following conditions:

> it must be easy to determine how much computing effort it took to solve the problem and the solution must otherwise have no value, either practical or intellectual. The number of monetary units created is equal to the cost of the computing effort in terms of a standard basket of commodities. For example, if a problem takes 100 hours to solve on the computer that solves it most economically, and it takes 3 standard baskets to purchase 100 hours of computing time on that computer on the open market, then upon the broadcast of the solution to that problem everyone credits the broadcaster's account by 3 units
>
> **Dai (1998)**

While this is the same idea behind the creation of bitcoins, the e-mail exchange between Nakamoto and Dai in August 2008 seems to indicate that the Bitcoin was designed independently of the B-money scheme.*

In September 2012, the Bitcoin Foundation was established as a U.S. nonprofit organization to promote the technology and to steward its development. Previous leaders of that foundation have been controversial: Mark Karpeles was previously of Mt. Gox, and Charlie Shrem, cofounder of the defunct company BitInstant, was charged and then sentenced in December 2014 for laundering digital currency through the Silk Road marketplace.

Bitcoin caught the attention of the mainstream media with the wild gyrations in its market price as illustrated in Figure 14.1. Early in 2013, its value increased in parallel with the shipment of the first application-specific integrated circuits (ASIC) designed to increase the efficiency and speed of the Bitcoin computations. In November to December 2013, the price of one bitcoin exceeded $1200 and then dropped significantly following the collapse of

* http://www.gwern.net/docs/2008-nakamot, last accessed April 29, 2015.

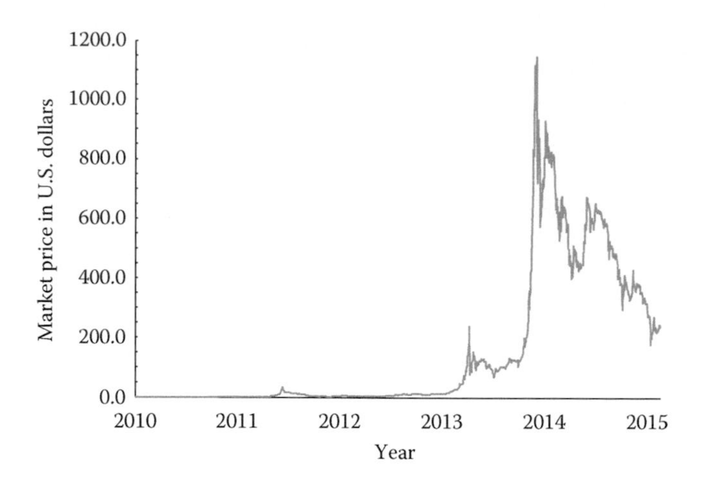

FIGURE 14.1
Market price of bitcoin in U.S. dollars as of February 19, 2015. (From https://blockchain.info/charts/market-price.)

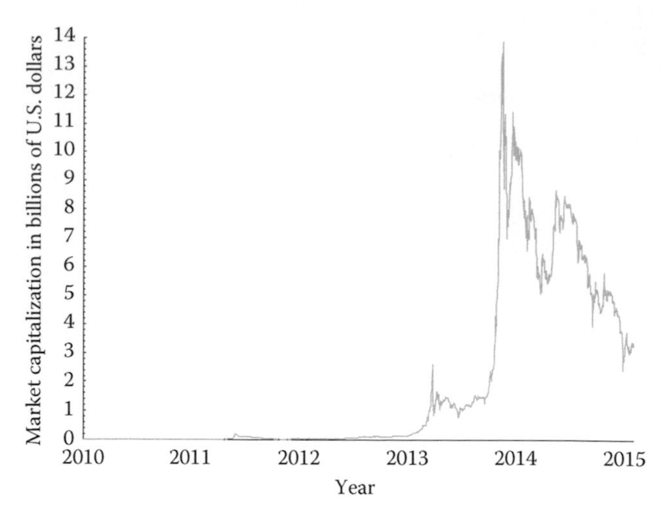

FIGURE 14.3
Evolution of the market capitalization of Bitcoin. (From https://blockchain.info/charts/market-cap.)

Mt. Gox, then the largest exchange of bitcoins, and the warnings from China's central bank to Chinese financial institutions about the risks of engaging in business with Bitcoin-related companies.

The first recorded transaction took place on January 3, 2009, and the growth in the circulation of bitcoins is depicted in Figure 14.2.

As of March 7, 2015, the total number of bitcoin transactions was 61 million, increasing from about 31 million in February 2014. This number should be compared to the yearly data for the United States in 2013: 9,026.5 million credit transfers, 13,574.6 million debit transfers through the ACH network, and 16,319.7 million check payments (Bank for International Settlements, 2014, p. 425). In the United Kingdom, the value of bitcoins in circulation was estimated to be less than 0.1% of the value of the sterling notes, while the number of transactions was about 300 (Ali et al., 2014b, p. 8). Another way of looking at it is that in the 3 months from December 2014 to February 2015, the average daily transaction was about 57 million (with a standard deviation of 0.27 million) compared to approximately 295 million per day in the Euro countries alone for conventional transactions (European Banking Authority, 2014, p. 7, §§ 6).

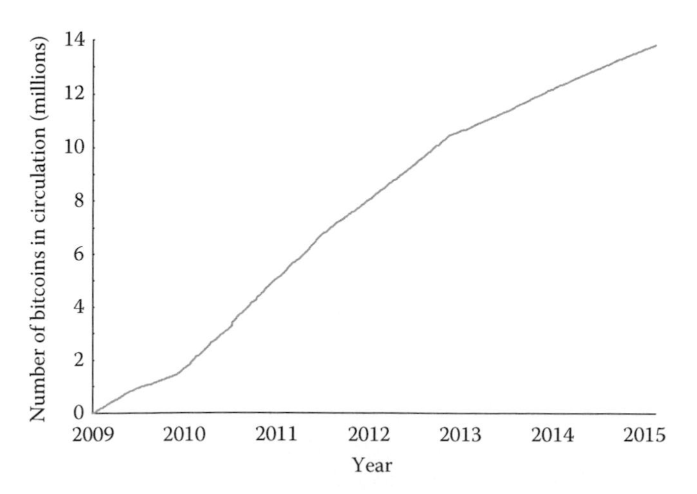

FIGURE 14.2
Evolution of the circulation in bitcoins as of February 19, 2015. (From https://blockchain.info/charts/total-bitcoins.)

Figure 14.3 depicts the evolution of market capitalization of bitcoin in U.S. dollars (number of bitcoins in circulation × value in U.S. dollars).

Table 14.1 shows the trading pattern of bitcoins against other currencies, respectively, for 24 hours and 30 days before February 19, 2015. Accordingly, about 60% of the trading is against the Chinese renminbi, 30% is against the U.S. dollars, 2.5% against the euro, and 1% against the sterling. This tendency has been consistent at least since 2014. Also in other surveys, a little less than half of respondents indicated that they live in the United States (Bohr and Bashir, 2014).

The Chinese contribution to the development of bitcoin can also be seen in that the company Canaan

TABLE 14.1
Patterns of Bitcoin Currency Trading up to February 19, 2015

Currency Traded	24-Hour Window (%)	30-Day Window (%)
Chinese Yuan Renminbi (CNY)	60.88	65.00
United States Dollar (USD)	29.66	29.31
Euro (EUR)	2.95	2.27
British Pound (GBP)	1.00	0.50
All other currencies[a]	5.51	2.92

Source: http://bitcoincharts.com/markets.

[a] Including trading against the Ripple and Litecoin cryptocurrencies.

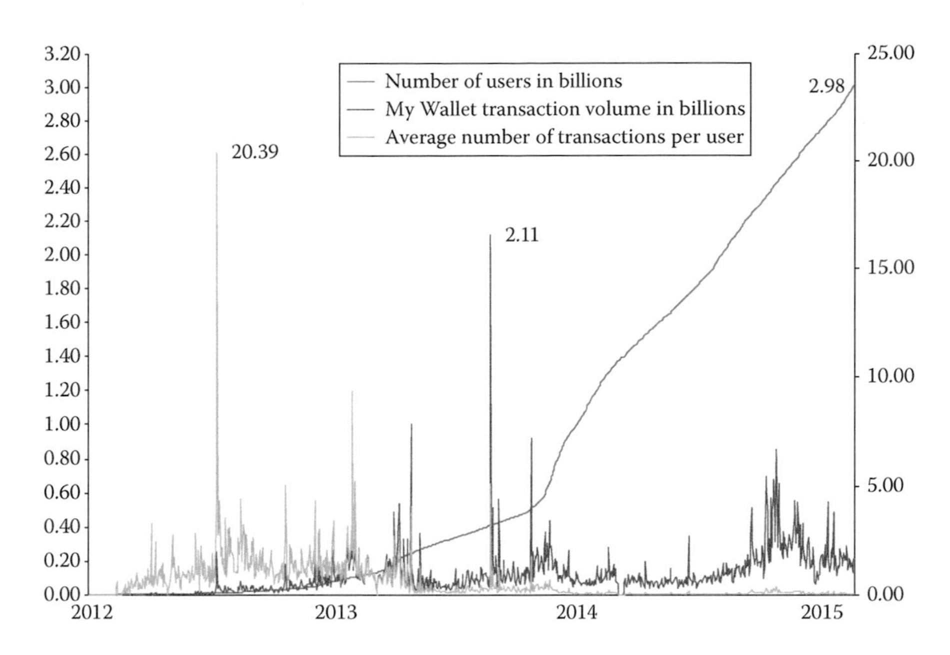

FIGURE 14.4
Usage statistics of My Wallet on February 24, 2015. (From https://www.quandl.com.)

Creative was the first to ship ASICs in January 2013 (Vigna and Casey, 2015, p. 115).

The usage statistics of My Wallet, a free online Bitcoin wallet, are shown in Figure 14.4. While the number of its users has increased steadily, as of February 24, 2015 the average transaction per user was about 0.07. Several studies have revealed that a significant amount of users have tiny amounts in their wallets and that around half of the bitcoins have not been circulated in the last 2–3 years (Hurst, 2014). For example, in August 2014, 44% of all bitcoins were assigned to less than 0.01% addresses (1538 out of 40.7 million addresses), each of which had more than a thousand bitcoins (Vigna and Casey, 2015, p. 154).

14.2 Bitcoin Protocol

The Bitcoin protocol packages features resulting from years of research in various disciplines, particularly distributed systems to resolve the problem of building a consensus among decentralized entities while exchanging information over an unreliable and insecure network (a problem known as the Byzantine General's problem). This invention is an architectural innovation, in the sense of Abernathy and Clark (1985), that is ushering follow-up innovations in payment systems, financial services, and distributed governance.

Some of the technological blocks in the Bitcoin protocol are as follows:

- Peer-to-peer networking
- Public key cryptography, particularly elliptic curve cryptography
- Cryptographic hash functions, such as SHA-256
- Hash tree techniques
- The Hashcash proof-of-work system invented by Adam Back (1997); this is the basis for reaching consensus without a central trusted authority

As mentioned earlier, the Bitcoin peer-to-peer network has a distributed open financial ledger, the blockchain, that tracks the currency units as they move among users. This ledger file is constantly growing as it stores new transactions. It is also public and protected from tampering with cryptographic means. This provides the ability of preventing respending the same bitcoin by an analysis of all past transactions.

The reference implementation written by Nakamoto is called Bitcoin core or Satoshi client and remains the most used Bitcoin client. It implements all aspects of the bitcoin system, including wallets, a transaction verification engine with a full copy of the entire transaction ledger (blockchain), and a full network node in the peer-to-peer bitcoin network. It is available from bitcoin.org.

The currency BTC generated with the Bitcoin protocol has the following characteristics:

1. The bitcoins are produced at an algorithmically defined rate until a maximum of about 21 million coins, a limit that cannot be exceeded. This artificial scarcity mimics metals such as

gold, even though the currency is not backed by any physical asset.

2. There is no central clearing system to maintain account balances and no single administrator. Payment transactions are verified in a distributed manner by the specific nodes themselves.
3. Typical transactions have one or more inputs and outputs. An output is a monetary amount and a public key that identifies the destination account. One of the outputs indicates the change returned to the payer. An input is a reference to the outputs of a previous transaction. To prevent double spending, an output can be referenced once only in the whole ledger.
4. A user's bitcoin account balance is calculated by aggregating all unspent currency locked to that user's private keys, spread over all transactions and all blocks of the blockchain.
5. New blocks are added to the blockchain in chronological order and, once written, are never changed or removed unless a longer chain that started before the block with a transaction is discovered. In that case, transactions that are not in the longer chain will have to be reverified.
6. In principle, any Bitcoin transaction can be reversed, if a longer chain is discovered. In general, however, five or more successive blocks following the block that includes the transaction are taken as an indication that the transaction will not be reversed.

14.2.1 Bitcoin Nodes

Bitcoin operation is based on peer-to-peer networking. At the networking layer, a newly joining node queries to the Domain Name System (DNS) servers to identify the bootstrap nodes. These are the nodes that supply the initial configuration. Once connected, the new node will continue to build its knowledge of the network topology from the responses of its neighbors and from the spontaneous advertisements that other nodes may send. Conversely, the addresses of nodes that leave the network persist for a while until they are purged from the routing tables. In financial applications, this may have serious consequences as payments are made to a node that is no longer in the network.

At the Bitcoin application layer, Bitcoin nodes fall into one of the following categories: full nodes, lightweight nodes, mining nodes, and web clients.

A *full node*, also called Bitcoin core or reference node or full client, stores the entire history of bitcoin transactions, that is, every block and every transaction, manages wallets, and can initiate transactions directly. As the node receives the blocks from the network, it validates them and then links them to the existing blockchain. Thus, the full node updates its copy of the blockchain continuously and then extends the blockchain. A full node can also track the movement of funds incrementally, transaction by transaction, determining unspent value.

Lightweight nodes verify transactions using a simplified payment verification procedure. They rely on full nodes to access a full copy of all transactions and therefore depend on third parties for full transaction validation. They also follow the consensus of the majority of mining power. Because of this dependency, lightweight nodes may be tricked to accepting transactions or blocks that are not actually valid. Typical users run lightweight clients or wallets using the simplified payment verification procedure to track their unspent currency but rely on the full nodes of the network to check for double spending.

Mining nodes compete to create new blocks by solving the proof-of-work algorithm and earn new money when they confirm a new block before the other miners. They collect pending transactions into a new block and then engage in the competition to confirm these blocks. The reception of a new block from the network signals the end of one round of the competition and the beginning of the next. It means that someone else won the previous round.

Web clients have even less capabilities than lightweight clients and rely completely on the third-party servers, including storing users' wallet. Dependence on service providers, however, introduces additional security risks.

Initially, all nodes were full nodes, but as the blockchain increased in size, nodes were configured differently depending on the functionalities they support.

All nodes in the Bitcoin networks perform four functions:

1. A routing function to discover peers and route traffic to them.
2. A verification function to
   a. Verify new transactions and propagate unconfirmed transactions through the Bitcoin network
   b. Verify new blocks and assemble those that meet the criteria to the blockchain
3. A selection function among contending chains to choose the one with the most cumulative difficulty as demonstrated through the proof of work (the longest chain). This is called the consensus.
4. A wallet function to create private keys and addresses.

The main network services that full nodes offer are as follows:

1. Transmitting new transactions, after verification, from users to miners
2. Broadcasting new blocks from miners, after verification, to the rest of the network
3. Filtering transactions and blocks on behalf of lightweight nodes
4. Answering requests for assistance from new full nodes to construct their own copies of the complete blockchain
5. Supplying missing blocks to any node that goes offline for any period of time

Such a node takes up a lot of disk space (currently about 30 of gigaoctets of disk space and at least 2 gigaoctets of random access memory [RAM]). Bitcoind (http://bitcoind.com) is the first full node client of network's history. It is available under the MIT license in 32-bit and 64-bit versions and is now bundled with Bitcoin-QT.

Table 14.2 lists alphabetically some Bitcoin mining software. If the mining software implements a full client, it stores the entire history of bitcoin transactions and can initiate transaction directly. If the software implements are lightweight client, this means that the miner participates in a mining pool and relies on the pool's server to maintain the full blockchain. If the software supports the Open Computing Language (OpenCL) framework, it is able to run across heterogeneous processors.

14.2.2 Bitcoin Wallets

Bitcoin wallets are programs installed on a user terminal to hold all the Bitcoin addresses of a given user as well the private keys that correspond to these addresses. The wallet stores the user's private keys, either as structured files or in a database. The private keys can be randomly generated or derived from a master key through successive applications for a one-way hash function, typically HMAC with SHA-512.

Access to wallets should be protected by a password or a passphrase. Should users lose access to their wallets, because of a forgotten password, a hardware malfunction or a malware, their bitcoins are wasted for all and there is no way to redress the situation.

Typically, users download and install the Bitcoin wallet software as a client on a computer or a smartphone. Once installed, the Bitcoin wallet generates a Bitcoin address to be disclosed to those that would like to transfer money to that user. The wallet, whether on the user's device or on a server, aggregates all unspent transaction outputs to calculate the user's balance of bitcoins.

Table 14.3 lists some of most common bitcoin wallets. In particular, Bitcoin-QT contains the original software written by Satoshi Nakamoto and is included in a full node implementation. MultiBit is a lightweight client that aims at ease of use, while Electrum focuses on speed. Blockchain is another popular wallet available from Apple Store and Google Store.

Paper wallets are Bitcoin private keys printed on paper. For example, Figure 14.5 shows the Bitcoin address created by the paper wallet at bitaddress.org. The address is presented as a long string of alphanumeric characters and uses the Quick Response (QR) 2D code. Paper wallets create backups or offline storage for bitcoins, also known as "cold storage," that is, they are printed on paper and never stored online, which is one of the most effective techniques to secure the keys.

Cold storage mechanisms protect against the loss of private keys due to accidents or attacks. The disadvantage of simple paper wallets is that the paper on which the keys are printed can be lost or stolen, so this storage has to be complemented with other offline digital media such as CDs or USB memory sticks.

TABLE 14.2

Examples of Bitcoin Mining Software

Software	Description	Locator
BFGMiner	Written in C for mining with CPU, GPU, FPGA, and ASIC. Supports the OpenCL framework	http://bfgminer.org
CGMiner	CGMiner is an ASIC/FPGA miner written in C. Runs on Linux, Windows, and OS X platforms and supports pool mining	https://github.com/ckolivas/cgminer
DiabloMiner	Written in Java for GPUs. Conforms to the OpenCL framework and work Nvidia drivers and ATI graphic cards	https://github.com/Diablo-D3/DiabloMiner
EasyMiner	Mining software from Butterfly Labs	http://www.butterflylabs.com/drivers
Phoenix2	Tuned for use with ATI GPUs	https://github.com/phoenix2/phoenix
Poclbm (Python OpenCL Bitcoin Miner)	Programmed in Python and is recommended for ATI or AMD graphics cards	https://github.com/m0mchil/poclbm
Ufasoft Coin Miner	Supported multiple currencies: Bitcoin, Dogecoin, Litecoin, Namecoin, and so on	http://ufasoft.com/coin

TABLE 14.3
Some Bitcoin Wallets

Operating System	Wallet	Location
Windows, Mac, and Linux	Armory	https://bitcoinarmory.com
	Blockchain	https://blockchain.info/wallet
	Bitcoin-QT	https://bitcoin.org/en/download
	Electrum	https://electrum.org
	Hive	https://www.hivewallet.com
	MultiBit	https://multibit.org/
Android	Bitcoin Wallet MultiBit	Google Store, for near field communication (NFC)-enabled Android phones
iOS	Blockchain Coinbase	Apple Store

Bitcoin Address

Private Key (Wallet Import Format)

SHARE

SECRET

15pXw9NZeRAmVw9fPqZXSHFv7iXPNowt5C

5JBkkCnD4UBH4upRZNaLuLxVBNggJSACZUmcbipgS8CqSxRT5iT

A Bitcoin wallet is as simple as a single pairing of a Bitcoin address with its corresponding Bitcoin private key. Such a wallet has been generated for you in your web browser and is displayed above.

To safeguard this wallet you must print or otherwise record the Bitcoin address and private key. It is important to make a backup copy of the private key and store it in a safe location. This site does not have knowledge of your private key. If you are familiar with PGP you can download this all-in-one HTML page and check that you have an authentic version from the author of this site by matching the SHA256 hash of this HTML with the SHA256 hash available in the signed version history document linked on the footer of this site. If you leave/refresh the site or press the "Generate New Address" button then a new private key will be generated and the previously displayed private key will not be retrievable. Your Bitcoin private key should be kept a secret. Whomever you share the private key with has access to spend all the bitcoins associated with that address. If you print your wallet then store it in a zip lock bag to keep it safe from water. Treat a paper wallet like cash.

Add funds to this wallet by instructing others to send bitcoins to your Bitcoin address.

Check your balance by going to blockchain.info or blockexplorer.com and entering your Bitcoin address.

Spend your bitcoins by going to blockchain.info and sweep the full balance of your private key into your account at their website. You can also spend your funds by downloading one of the popular bitcoin p2p clients and importing your private key to the p2p client wallet. Keep in mind when you import your single key to a bitcoin p2p client and spend funds your key will be bundled with other private keys in the p2p client wallet. When you perform a transaction your change will be sent to another bitcoin address within the p2p client wallet. You must then backup the p2p client wallet and keep it safe as your remaining bitcoins will be stored there. Satoshi advised that one should never delete a wallet.

FIGURE 14.5
Address generated by the virtual wallet at bitaddress.org.

Hardware wallets (e.g., www.hardwarewallets.com, last accessed December 31, 2015) are another cold storage mechanism using offline tamper-resistant equipment. A hardware wallet is dedicated to the storage of bitcoins and does not have general-purpose software or external applications.

14.2.3 Blockchain

The blockchain is the public distributed ledger of all Bitcoin transactions starting from January 3, 2009 and is continuously growing. Each block contains more than 500 transactions of at least 250 octets. The ledger is stored in all full nodes of the network in a decentralized fashion, and the protocol specifies how to update the ledger and how to synchronize all ledgers in the network.

The block structure of Bitcoin is shown in Table 14.4, and the block header is shown in Table 14.5.

The block hash identifies a block uniquely and unambiguously. It is not included in the block transmitted on the network as part of the blockchain but is computed each time a block is received from the network and is stored in a database for future use.

Each block refers to the hash of its antecedent in the blockchain, thus establishing a recursive link to all the blocks of the blockchain. Because transactions are not encrypted, it is possible to view all previous transactions and thus track every bitcoin or fraction of bitcoin spent or received. A blockchain browser (also called block explorer) is a program or a website that lets users search and navigate the blockchain.

TABLE 14.4

Block Structure of Bitcoin

Field	Description	Size in Octets
Block size	Number of the following octets until the end of the block	4
Block header	Six fields as shown in Table 14.5	80
Transaction counter	Number of transactions in the block	1–9
Transactions	The list of transactions in sequential order	Variable

TABLE 14.5

Structure of the Bitcoin Block Header

Description of the Block	Size in Octets
Software version	4
Hash of the preceding block	32
Merkle root of the transactions in the block	32
Time stamp	4
Difficulty target T	4

The Merkle root summarizes all transactions in a block and is calculated as follows. Transactions are arranged in a binary tree and then pruned to form a Merkle Tree. For an odd number of transactions, the hash of the last transaction is duplicated. At each level of the tree, the recursive procedure consists in concatenating the transaction hashes in pairs, until there is only one hash, the root or the Merkle root, that summarizes all the transactions. The hashing consists of the application of the SHA-256 algorithm twice yielding a 32-octet string. With this procedure, to check if any given transaction is in a tree of N transactions, at most $\log_2(N)$ calculations are needed instead of N calculations.

Lightweight clients use Merkle trees to verify whether a transaction is in the blockchain without downloading all the transactions. They download the block headers and retrieve the transactions of interest from the network using the Merkle path to determine which hashes are needed to complete the verification. This verification goes as follows: the transaction of interest is hashed and concatenated iteratively with the other hashes along the Merkle path until the top hash. If the top hash matches the Merkle hash, the lightweight client has a proof that the transaction is included in the blockchain.

While this procedure reduces the storage requirements and hastens the verification, it cannot detect double spending, because this detection needs data from the full blockchain. The number of confirmed blocks that follow the block of the transaction is denoted as the depth of the transaction. Usually, when six blocks follow the block that contains the transaction, lightweight clients assume that the network's full nodes have confirmed the transaction.

The first block created on January 3, 2009, is called the genesis block; it was given the height of zero. The block height refers to the number of blocks following that genesis block. On May 17, 2015, the block height was around 356,929 meaning that about 357,000 blocks were confirmed following the genesis block.

The first transaction in any block is the generation or coinbase transaction. It has no bitcoin inputs, and its output represents the reward destined to a new address that the miner's wallet has created. The coinbase transaction includes two types of rewards to the miner: new bitcoins created with each new block and transaction fees from all the transactions included in the newly confirmed block. Thus, a coinbase transaction increases the money supply with newly created bitcoins, payable to the miner that has succeeded in confirming the associated block. The monetary reward for a new block is an incentive to dedicate computation resources to keep the ledger up to date, which is also a way to increase the total supply of bitcoins in circulation.

On January 3, 2009, the value of the reward was set to 50 BTC per verified block. According to the protocol, the reward value is halved every 210,000 blocks, so the value was reduced to 25 BTC in November 2012. Assuming that, on average, the reduction takes place every 4 years, the value of the reward will decrease exponentially (less than 0.1 BTC around 2050) until approximately the year 2140 when around 21 million bitcoins* will have been issued. Mining nodes will have then to recover their cost through transaction fees. Transaction fees, which are currently low, will need to rise as usage grows and as the ceiling of the money supply is approached or reached, particularly if Bitcoin is regulated and miners incur compliance costs.

Note: Nakamoto is believed to be the owner of about 1 million bitcoins (Vigna and Casey, 2015, p. 73).

The time stamp is encoded as a Unix time, which is defined as the number of seconds that have elapsed since the midnight of January 1, 1970, in coordinated universal time (UTC) but not counting leap seconds. Due to its handling of leap seconds, some conversation is needed to get a linear representation of time.

The value encoded in the target difficulty T tells miners how small the hash should be to meet the proof-of-work criteria and succeed in confirming the block. It is represented by 4 octets as follows: the first octet represents the total length of the hash while the next 3 octets encode the prefix that precedes (length−3) octets of zeros. The value of the target difficulty is obtained from the encoding as follows (Antonopoulos, 2015, p. 194):

$$T = \text{prefix} \times 2^{8(\text{length}-3)}$$

This is illustrated in Figure 14.6.

The target difficulty field in the genesis block is 0x1D00FFFF and is the maximum value of target difficulty encoding and corresponds to the minimum difficulty D of 1. Thus, the upper bound of a valid SHA-256 hash consists of 32 octets as follows: 3 octets of zeroes followed by the prefix of 0x00FFFF followed by 26 octets of zeroes, that is,

(0x00000000FFFF0000000000000000000000000000000000000000000000000000).

Any hash smaller than the target satisfies the criteria. Therefore, the difficulty of solving the puzzle increases with the number of zeroes preceding the prefix.

Thus, the maximum target difficulty is given by

$$T_{max} = (2^{16} - 1)2^{208} \approx 2^{224}$$

Any other block has a target difficulty $T < T_{max}$. The ratio of the maximum target difficulty to the current target is defined as the difficulty D, that is,

$$D = \frac{T_{max}}{T}.$$

The role of the difficulty in the mining process will be described in the following section.

14.2.4 Mining

Mining is the process to update the common ledger in a decentralized fashion and prevent fraudulent transactions or double spending of the same bitcoin. In addition, this is the process to increase the money supply until year 2140 when all bitcoins would have been issued.

Mining is intentionally slow and energy consuming so that miners proposing changes to the ledger have to demonstrate that they have invested sufficient computational resources. After collecting pending transactions to form new blocks, miners engage in a competition to confirm these blocks. The competition consists in repeatedly processing the block header with the SHA-256 hashing algorithm until a hash that satisfies a predetermined condition is found. This is called the "proof of work." The first miner to find the solution wins the

* The exact number is 20.99999998 million.

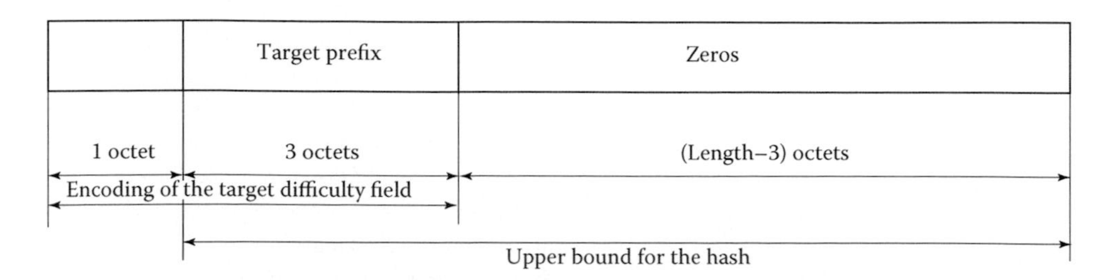

FIGURE 14.6
Relation between the encoding of the target difficulty and the upper bound of a valid hash in Bitcoin.

round and is allowed to add the block of transactions to the blockchain, with a coinbase transaction at its head. Because there may be more than one blockchain propagating in the network, however, the award is conditional on having worked on the longest blockchain. Therefore, the reward cannot be spent before the confirmation of 99 additional blocks.

Each miner starts the process as soon as a new block is received, which is a signal that the previous round of competition has ended. Miners express their acceptance of the block by working on creating the next block in the chain, using the hash of the accepted block as the previous hash. The global parameters of the protocol are adjusted adaptively so that a new block is mined every 10 minutes. This rate corresponds to 6 blocks per hour or 144 blocks per day.

After verification, the mining node announces the availability of the block to its neighbors. They, in turn, will respond by indicating the transactions that they do not have and avoid unnecessary transmission of transactions that they already have.

As the newly solved block propagates in the network, each full node and mining node of the Bitcoin network performs a series of tests before sending it to its peers. These nodes accept the block only if all transactions in it are valid and are not already spent. The independent validation of blocks constitutes the decentralized consensus, as the collective outcome of all nodes in the network, even though the bulk of the computation falls on mining nodes and that the whole blockchain is stored in full nodes only. The majority consensus is represented by the longest blockchain.

The proof-of-work algorithm is explained in the next section.

14.2.5 Proof-of-Work Algorithm

The proof-of-work principle consists in asking the sender of a message to execute a computation with a parameterizable difficulty before transmitting the message while the intended recipient (or recipients) could verify the computation easily. The idea emerged first to deter spamming by adding a micro-cost to any electronic mail message (Dwork and Naor, 1993). Adam Back (1997) adopted it in the Hashcash algorithm, which was adopted in Bitcoin.

For Bitcoin, the proof of work consists in finding a nonce that satisfies the inequality:

$$\text{SHA-256}\,[\text{SHA-256 (Previous Block header)} \,\|\, \text{Nonce} \,\|\, \text{Transaction } 1 \,\|\, \ldots \,\|\, \text{Transaction } n] < \text{Difficulty Target } T$$

T is the desired target for the difficulty and is set based on the computational power of all mining nodes in the network.

As long as the hash value exceeds T, the miner must try different nonces, until the condition is met. The average work increases exponentially with the number of leading zero bits required. If a nonce is found, the resulting block is added to the miner's Bitcoin blockchain as a proposed block and then sent for approval to participants in the Bitcoin network.

Linking the current hash to the hash of preceding block makes changes in previous transactions difficult to make. If any block is modified, all the following blocks will have to be recomputed by finding nonces that would make all previous transactions valid. The difficulty increases with time as transaction blocks accumulate.

The nonce field is 32 bits long, which gives about 4 billion values. As the difficulty increases, miners could go through all these values without finding a block. As a consequence, they have to try new headers to find a valid nonce. The block time stamp does not have to be exact anyway, so moving the time stamp back or forward, say, 10 seconds before starting to mine, is one possibility, but moving it too far into the future may cause the block to be invalid. They can also add extra nonce space in the coinbase transaction, which can store between 2 and 100 octets of data or vary the transaction set. Any of these changes would allow the miners to have another iteration of around 4 billion attempts to obtain new results.

The proof-of-work approach solves two problems. First, it provides an algorithm for an emerging consensus by allowing full nodes in the network to agree collectively on the state of the Bitcoin ledger. Second, it solves the problem of deciding how to develop this consensus, while simultaneously preventing attacks. It does this by avoiding explicit elections and registrations and with an economic barrier—by making the weight of the contribution from any node to the consensus process directly proportional to the computing power that the node brings to the network.

The resources required to confirm blocks are increasing unremittingly as computational power is added to the network by new miners or through improvements to the tools of existing miners and as blocks include more transactions. As a consequence, the central processing units (CPU) of network nodes became quickly too slow for the competitive task of mining. Switching the computations to graphics processing units (GPU) was first explored to perform many Bitcoin computations in parallel. For a short period, field programmable gate arrays (FPGAs) seemed to offer a solution, but they were replaced early 2013 by dedicated hardware in the form of ASICs customized for the Bitcoin operations (available from Avalon, Butterfly Labs* and others).† Yet as new miners join the network and existing miners upgrade their machines, the total computation power in the Bitcoin network continues to increase and mining hardware becomes obsolete in a few months.

As the block propagates across the network, full nodes and other mining nodes of the network can easily verify the new block. Once a block is verified, each of the nodes adds it to its own copies of the blockchain. As these nodes receive and validate the block, they abandon their efforts to find a block for the same block length (block height) and immediately start computing the next block of the chain. Once the majority of full nodes and miners have confirmed the block, it is added permanently to the blockchain.

The independent validation of a new block ensures that only valid blocks are propagated and that only miners that act honestly have their blocks incorporated into the blockchain. In particular, the hash in the block header must be less than the target difficulty. Also, because nodes do not synchronize their clocks, one condition for validity is that the block time stamp is less than 2 hours in the future. Other tests concern the block size, data structure, and all the transactions within the block. The independent validation of each new block by every node (full node, miner, and some lightweight) ensures that miners follow the agreed rules before claiming their award.

14.2.6 Adjustment of the Difficulty

This section describes the concept of difficulty and its adjustment to maintain an average rate of block production of 6 blocks per hour.

The hash for Bitcoin is approximately uniformly distributed values over the interval $[0, 2^{256})$, that is, any of the values has a probability $p = 1/256$ (O'Dwyer and Malone, 2014). Therefore, the probability that a hash value is $<T$:

$$P\{H(S) < T\} = \sum_{i=0}^{T-1} \frac{1}{256} = \frac{T}{256}$$

The nonce values are selected independently, and the number of trials until a nonce is found is a geometric distribution with a probability

$$p = \frac{T}{2^{256}}$$

Therefore, the expected number of hashes N until a nonce is selected that meets the criterion is

$$E(N) = \frac{1}{p} = \frac{2^{256}}{T}$$

Using $D = T_{\max}/T \approx 2^{224}/T$, we have

$$E(N) = 2^{32} \times D$$

In a system calculating hashes at the rate of R hashes/s, the average expected time τ to meet the criterion is

$$\tau = \frac{E(N)}{R} = \frac{2^{256}}{TR} = 2^{32} \times \frac{D}{R}$$

Clearly, the average expected time increases with the difficulty and is inversely proportional to the hash rate.

On April 7, 2015, the difficulty was 49, 446, 390, 688. The corresponding threshold and average time were

$$T = \frac{2^{224}}{49,446,390,688}$$

and

$$\tau = \frac{2^{32}}{R}(49,446,390,688)$$

With $R = 1$ Mhashes/s, $\tau = 2.12371 \times 10^{14}$ seconds.

For the value of the difficulty on April 7, 2015, the network hash rate R_{net} was

$$R_{net} = \frac{2^{32} \times 49,446,390,688}{600} = 3.54 \times 10^{17} \text{ hashes/s}$$

Thus, a miner with a hash rate of 1 Ghashes/s will need 56×106^2 hours, that is, about 13,031 years to discover a block.

From the point of view of miners, Bitcoin mining remains profitable as long as the cost of generating coins is less than their value, that is, if, excluding the cost of hardware and transaction fees:

$$\tau \times P \times J < \text{Reward}$$

where

τ is the average time for block discovery
P is the amount of power consumed in block discovery
J is the corresponding cost per Joule of energy

* In September 2014, the Federal Trade Commission (FTC) filed a civil lawsuit against Butterfly Labs alleging that the company has engaged in fraudulent and deceptive practices (Farivar, 2014b).

† A comparison of various ASICs in terms of hashing speed (Mhashes/s), energy efficiency (Mhash/J), and maximum power consumption in watts (W) is available at https://en.bitcoin.it/wiki/Mining_hardware_comparison.

To maintain the block generation, the difficulty of mining is adjusted to take into account these changes. The network difficulty D is adjusted automatically by the network, upward or downward, so that, on average, 6 blocks are solved per hour, that is, τ is roughly 10 minutes or 600 seconds. Every 2016 blocks (about 2 weeks), all Bitcoin clients compare the actual length of time it took to discover these blocks to estimate the hash rate of the entire Bitcoin network. If it took more than 2 weeks to find the previous 2016 blocks, the difficulty is reduced. Otherwise, it is increased to bring the time back to 2 weeks. The change is capped, however; that is, the difficulty should not be increased by or decreased by more than a factor of 4. This is summarized as follows:

$$
\begin{aligned}
&|\text{New difficulty}| \\
&= |\text{Old difficulty}| \frac{\text{Mining time for the last } 2{,}016 \text{ blocks in minutes}}{20{,}160}, \quad && |\text{New difficulty}| \leq |\text{Old difficulty} * 4| \\
&= |\text{Old difficulty} * 4|, && \text{otherwise}
\end{aligned}
$$

It should be noted that the difficulty is not directly proportional to the number of transactions or the value of the transactions.

14.2.7 Hashing Race

Miners have a strong incentive to invest in more powerful hardware to increase their chances to get the reward for discovering blocks. By doing so, they increase the total computational power of the network. As a consequence, they contribute to increasing the difficulty. In fact, this is what happened monotonically as illustrated in Figure 14.7.

The competition between miners has resulted in an exponential increase in the total hashes/s across the network (hashing power) and, as a consequence, the difficulty. So to remain competitive, miners have to invest continuously to increase their computation power. Miners have also formed mining pools to combine their computation power and share the rewards. In this arrangement, the average user is not a full-fledged Bitcoin node. Alternatively, a user can rent Bitcoin mining capabilities from cloud service providers.

An unethical option is to commandeer machines connected to the Internet to perform the calculations. In 2011, malware (e.g., ZeroAccess and Miner Bot) was detected, which would take advantage of the computation power of compromised machines to conduct mining surreptitiously (Dev, 2014). Another stealth mining program is *cpuminer 2.3.1*. In a survey of the Bitcoin users, about 10% of the sample admitted to stealing someone else's bitcoins (Bohr and Bashir, 2014).

Average miners, unable to compete with the powerful mining rings and the steadily rising difficult of Bitcoin mining, have switched to other less demanding cryptocurrencies such as Litecoin.

Mining pools are described in the next section.

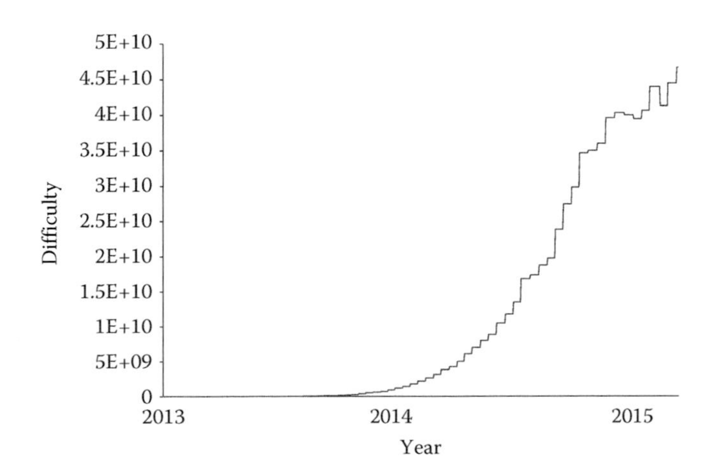

FIGURE 14.7
Evolution of the difficulty of the Bitcoin algorithm. (From https://blockchain.info/charts/difficulty.)

14.2.8 Mining Pools

Mining pools are a way to split the work of searching for a solution to a candidate block among the participant miners. Mining pools coordinate many hundreds or thousands of miners using specialized protocols, such as the Stratum, Getwork, or Getblocktemplate protocols.* The aim of these protocols is to distribute the work among the miners in an efficient manner to avoid duplication of work or wasting of efforts on a block that has already been mined. The pool server runs specialized software and the pool mining protocol and has direct access to a full copy of the blockchain so that each participant miner is relieved from running and maintaining a full node.

* An older protocol Getwork does not support mining rates above 4 Ghash/s easily.

The pool operates as follows. The pool server sends out the block to be mined, setting a lower difficulty target, typically more than 1000 times less than the Bitcoin network difficulty, and then receives updates from miners whenever they find a partial solution. Each partial solution proves that the miner is working on the problem and gives that miner a share in the final reward, in case one of the pool members succeeds in finding the solution. For instance, if Bitcoin mining requires a hash starting with 15 zeroes, the mining pool can ask for hashes starting with 10 zeroes. Eventually, one hash will start with 15 zeroes, thus winning the reward for the pool. The reward is then split based on each miner's count of shares as a fraction of the total, and the pool operator takes a small percentage for overhead.

If someone outside the pool mines a block, the pool operator sends out new data and the miners start mining the new block. In return for operating the pool server, the pool operator charges the miners a percentage of the earnings. The reward of an individual miner is calculated as follows:

$$\text{Miner's reward} = \text{Block value} \times (1 - \text{pool fee}) \frac{\text{Miner's hash rate}}{\text{Pool hash rate}}$$

where the Block value is currently 12.5 bitcoins + fees from bitcoin transactions contained in the block.

Since 2011, several protocols, P2Pool, BitPenny, Eligius, and so on, were introduced to run mining pools in a decentralized fashion. P2Pool, for example, implements a parallel blockchain called a share chain. The share chain runs at a lower difficulty than the Bitcoin blockchain and at a faster rate: 1 block every 30 seconds. Each of the blocks on the share chain records the contribution of each pool miner in doing the work. The share chain is public and accessed by all pool miners instead of having a pool server tracking the contribution and rewards of each participant miner. When one of the blocks of the share chair reaches the difficult target of the blockchain, that block is added to the Bitcoin blockchain (Antonopoulos, 2015, p. 210).

BitPenny (http://www.bitpenny.com) uses an open-source client to protect against potential abuses and security vulnerabilities resulting from centralized mining power. Automatic failover to single node mining allows continued operation if the pool server is unavailable. The Eligius pool does not charge a fee and runs on donations.

Mining pools use machines in warehouses or data centers (or data farms) where hundreds or thousands of computers combine their capacities. These warehouses are often installed in areas of cold climate to mitigate air conditioning costs or in places where relatively cheap energy is available such as geothermal in Iceland, hydropower in Washington State, and coal energy in Utah. For example, KnC Miner is a pool that operates a giant mining facility with machines placed in Boden in the north of Sweden near the Arctic Circle and next to a hydroelectric generation station to reduce the cost of cooling and energy. GHash.io is another mining pool that shot to fame when its computing capacity exceeded the 51% threshold of the total network computing power on January 9, 2014, and July 14, 2014. GHash.io, in fact, comprises Cex.io, a miner with significant computational power that rents its hardware, and a pool of other miners. This allowed it to attain a large percentage of the total network hash rate. In response to the strong reaction to its rate, it announced that it will limit its hash rate to 39.99% by asking miners to join other pools if it exceeds that threshold (Farivar, 2014a).

14.3 Operation

The highlights of the payment are summarized in the following steps (Nakamoto, 2008):

1. New transactions are broadcast to all nodes.
2. Each mining node collects new transactions into a block.
3. Mining nodes work on finding the proof-of-work for their current block.
4. Once a miner has found its proof of work, it broadcasts it all to other nodes of the network.
5. Nodes accept the block only if all transactions in the block are valid and not already spent.
6. If a block is accepted, other mining nodes work on creating the next block in the chain using the hash of the accepted block as the previous hash in the computation.

Payment information is disseminated throughout the network using two distinct types of information: transactions and blocks. A transaction relates to a specific payment, while a block relates to a group of transactions (but not necessarily all) occurring around the same time. Originally, the block size was 500 Koctets and then it was increased to 1 Moctets. Bitcoin XT is a new implementation of a full node that has raised the block length to 20 Moctets but, as of January 2016, has not been accepted by all stakeholders.

14.3.1 Getting Bitcoins

Users acquire bitcoins in three possible legitimate ways: (1) by purchasing them from other users in exchange for central bank currencies or other cryptocurrencies, (2) by obtaining them as payments for goods and services, and (3) as a reward for verifying earlier transactions (i.e., mining).

In New York City, for example, Bitcoin enthusiasts were organizing in 2015 meetings on Monday and Tuesday evenings at the Bitcoin Center* to buy and sell their Jetons (Paglieri, 2014, pp. 74–79).

Although the original design of Bitcoin aimed at eliminating third parties, exchanges are intermediaries matching buyers of bitcoins with sellers, converting bitcoins to central bank currencies or acting as depositories of their subscribers' funds in bitcoins. Some of these exchanges are (in alphabetical order) as follows: Bitstamp, Circle, Kraken, the now defunct Mt. Gox (a contraction of Magic the Gathering Online Exchange) and Virtex. Asian exchanges include Bitfnex (Hong Kong), BTC China, and OK Coin. Some (BTER, CoinCorner, and Cryptsy) exchange bitcoins into other cryptocurrencies as well. The exchanges differ in their registration requirements in terms of proofs of identity and whether they supervise the trade as well. Should an exchange face liquidity problems, it can take an inordinate amount of time to receive the funds.

Bitstamp was founded in 2011 in Slovenia and then moved its operations to the United Kingdom in 2013. It is run by two young people in their 20s and funded by the U.S. hedge fund Pantera Capital, an arm of Fortress Investments, that focuses exclusively on cryptocurrencies. Bitstamp is not regulated by the UK's Financial Conduct Authority because Bitcoin is not classified as a currency. In February 2014, the exchange suspended withdrawals for several days in the face of a distributed denial-of-service attack and in January 2015, it halted operations and reported the loss of 19,000 bitcoins.

Mt. Gox started as an exchange for card games before becoming one of the biggest exchanges of bitcoins until it collapsed on February 28, 2014, with liabilities of some 6.5¥ billion ($63.6 million). It had suffered a series of distributed denial-of-service attacks, and the U.S. authorities seized some of its funds. It was unable to account for some 850,000 bitcoins, lost to mismanagement, cyberattacks, and possibly theft by hackers who exploited a bug in the bitcoin code to double spend the same bitcoins. The size of the supposed heist shrank after some 200,000 coins turned up in an old-format Bitcoin wallet, bringing the tally to 650,000. As of March 2015, there was still no clear explanation of what happened (Paglieri, 2014, pp. 156–178; Hornyak, 2015; Vigna and Casey, 2015, pp. 116–117).

Silk Road, another exchange, operated a global marketplace to distribute drugs, weapons, and stolen identities. After closing down Silk Road, the U.S. Federal government owned more than 170,000 bitcoins that were sold over several auctions.

Among the websites that act as intermediaries without storing the bitcoins are LocalBitcoins in the United States and BitBargain and Bittylicious in the United Kingdom.

Coinbase is a platform that offers virtual wallets as a service. The service allows the payment of bitcoins via SMS from a mobile phone or from a web browser. More importantly, it charges 1% to convert bitcoins into local currency (the first $1 million are free). With this facility, businesses can minimize the effect of Bitcoin volatility by cashing them daily. This is the approach that the *Chicago Sun-Times* took as it allowed holders of Coinbase accounts to pay their yearly subscription with bitcoins (Paglieri, 2014, pp. 109–110).

In some locations, customers with a Robocoin account can buy bitcoins with central bank currencies through Robocoin terminals acting as surrogate automated teller machines. During the registration, customers provide a mobile phone number to receive a verification text message (similar to the mobile transaction authentication number [mTAN]) of mobile transactions). After punching that verification code into the machine, customers select their PIN and provide three pieces of authentication information: a government-issued identification document, a scan of their palm vein pattern and, finally, their photographic picture as captured by the machine (Ferrara, 2014). Genesis Coin is another company that is setting up machines in North America (Paglieri, 2014, pp. 71–72).

Unethical ways of acquiring bitcoins include theft by changing the payee's address from account pools (Reid and Harrigan, 2011). Another technique consists in adapting botnets used in distributed denial-of-service attacks to create a stealth network of mining nodes on the Internet. For example, in January 2014, adware was discovered in Yahoo's home page that infected the visitors' stations with code to commandeer them and create surreptitiously a huge network of Bitcoin mining nodes (Wakefield, 2014).

In November 2013, a group of MIT students led by Jeremy Rubin participated with their Tidbit program in the Node Knockout Hackathon. The program is designed to help websites generate revenue by using their visitors' computers to mine for Bitcoin as an alternative to advertisement. Tidbit won the award for innovation, but, in December 2013, the New Jersey Division of Consumer Affairs issued a subpoena to Rubin, requesting the source code of the program, other documents and agreements with any third parties, the names and identities of all Bitcoin wallet addresses associated with Tidbit, a list of all websites running Tidbit's code, and the name of anybody whose computer was mined for Bitcoins through the use of Tidbit. The Electronic Frontier Foundation agreed to represent Rubin and in May 2015, the state of

* 40 Broad Street, NYC, between Exchange Place and Beaver Street, in the southern tip of Manhattan near the financial center.

New Jersey reached a settlement provided that the program stops accessing computers without their owner's consent (Fakhoury, 2014; *Star-Ledger*, May 27, 2015).

14.3.2 Bitcoin Address

A Bitcoin address is a string of 26–34 alphanumeric characters of letters or numbers. Each address has its own balance of bitcoins. Bitcoin users are encouraged to create a new address for each transaction so that a user could receive money on one address and pay on a different address.

The address is formed by the application of the RIPEMD-160 and the SHA-256 hashing on the user's public key using the Elliptic Curve Digital Signature Algorithm (ECDSA) as represented in Figure 14.8. Most Bitcoin implementations use the cryptographic library from OpenSSL to perform the required computations. The library is described at http://wiki.openssl.org/index.php/EVP.

The private key SK can be any 256-bit (32-octet) number between 1 and $(n-1)$ where $n = 1.158 \times 10^{77}$ is the order of the elliptic curve used in Bitcoin. The public key PK is derived by the multiplication of the private key with a constant G $SK \times G$ to obtain another point on the elliptic curve. The public key is a 512-bit (64-octet) number constituted of the two 256-bit coordinates (x, y) of the new point.

Bitcoin uses an elliptic curve called secp256k1 defined by the following equation:

$$y^2 = (x^3 + 7)(\text{mod } p)$$

where p is the prime number $(2^{256}-2^{32}-2^{9}-2^{8}-2^{7}-2^{6}-2^{4}-1)$ (Antonopoulos, 2015, p. 66). NIST, however, recommends the use of P 256 (also known as secp256r1), with the prime number $p = 2^{256}-2^{224}+2^{192}-2^{96}-1$ for federal applications (National Institute of Standards and Technology, 1999, p. 8). It is not clear why secp256k1 was chosen, perhaps because of mistrust that the recommendation of secp256r1 could be a trap (honeypot). Base58 encoding is used in Bitcoin to improve human readability, whenever there is a need for a user to read and correctly transcribe long alphanumeric strings, such as a Bitcoin address, a private key, an encrypted key, or a hash. It omits characters that are can be confused for one another such as number zero (0) and capital o (O) or lower L (l) and capital i (I). Figure 14.9 show the steps in the conversion of a public key to produce the 25 octets of the Base58 Check Bitcoin address.

The sequence of steps to derive the address is summarized as follows:

1. Compute the double hash $DH_1 = RIPEMD160[SHA256(PK)]$ to produce a 20-octet (160-bit) alphanumeric string. *RIPEMD160* and *SHA256* represent, respectively, the application of the RIPEMD-160 and SHA-256 hashing.
2. Concatenate the 1-octet prefix 0x00 with DH_1, that is, $0x00 \| DH_1$. The prefix is called the version prefix and serves to identify that the data encoded relates to a Bitcoin address. The prefix for a private key is 0x80.
3. Extract the first 4 octets from the 32 octets of the double hash $DH_2 = SHA256[SHA256(0x00 \| DH_1)]$ as the checksum. The checksum is appended to the end to form the 25 octets $0x00 \| DH_1 \| Checksum$.
4. Convert $(0x00 \| DH_1 \| Checksum)$ into Base58 encoding to remove similar looking characters such a "l" (lower case L) and "I" (upper case I) as well as "0" (number zero) and "O" (upper case O). The check sum is used to prevent the wallet from accepting mistyped Bitcoin addresses, an error which leads to the loss of funds. The format with the check is called Base58 Check format (Antonopoulos, 2015, p. 73).

One technique introduced in 2012 is the ability to encode complex verification scripts using several public keys as Bitcoin addresses. An address prefix, either a "1" or a "3," indicates whether the address is special and corresponds to a verification script instead of a public key. Addresses starting with a "1," for example, 1C5LqLaDJRFf2jRpMKh9ErZgpjDB2K7Z1, are associated with a single pair of public and private keys. Addresses that begin with the prefix "3," for example, 3J98t1WpEZ73CNmQviecrnyiWrnqRhWNLy, are

FIGURE 14.8
Relation between the private and public keys and the Bitcoin address.

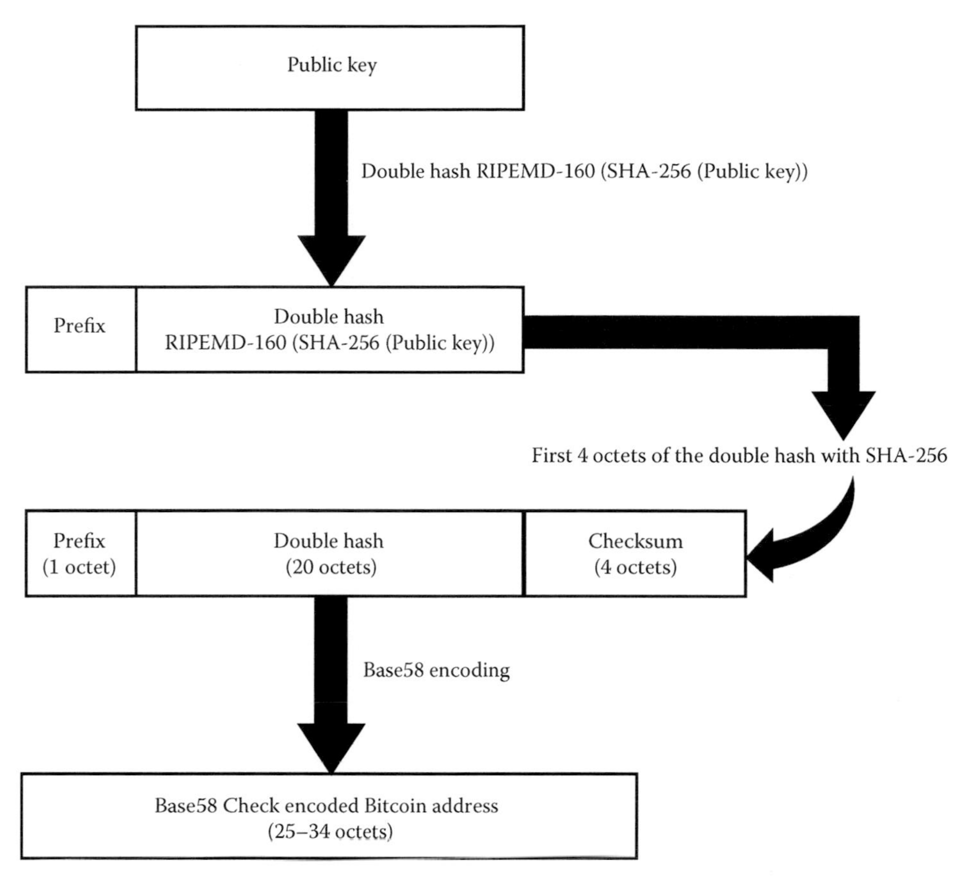

FIGURE 14.9
Conversion of a public key to a Base58 Check encoded Bitcoin address.

associated with a script that requires multiple pairs of public and private keys for accounts receivable. In this case, locked payments are released only after a minimum number of signatures is produced. This minimum number is set when the transaction is generated and cannot be changed.

A person or entity in the Bitcoin network can have many addresses, each with its own balance. In fact, a good practice for Bitcoin users is to create a new address for every transaction to increase privacy. As of July 9, 2014, there were almost 41 million addresses listed on the Bitcoin blockchain, but only 1.6 million contained a balance of more than 0.001 bitcoins (Ali et al., 2014b). This figure still overstates the number of users. As a result, there are many uncertainties regarding the usage statistics of Bitcoin and all other cryptocurrencies in general.

Some Bitcoin addresses are called vanity addresses because they contain human-readable messages. They require generating and testing an extremely large number of candidate private keys, until one has the pattern that would produce the desired pattern in the Bitcoin address. This is typically relegated to repurposed Bitcoin mining hardware "rigs" that are no longer powerful enough to compete in the mining process. Pools of vanity miners allow those with the necessary hardware to earn bitcoins by availing their hashing power to those searching for vanity addresses. However, allowing outsiders to access private keys is a risky proposition, particularly as there are no legal safeguards. For example, the website https://vanitypool.appspot.com was posting a warning (as of May 14, 2015) that its software was still in development and it could not guarantee that data and coin loss would not occur. Another site, https://bitcoinvanity.appspot.com, was implicated in key theft and was taken offline.*

14.3.3 Key Formats

The 64-octet private and public keys can be represented in Base58 Check encoding—denoted as the Wallet Import Format (WIF)—or in compressed Base58 Check encoding. Base58 Check encoding is illustrated in Figure 14.10. The value of the prefix depends on the

* From Bitcoin Wiki, https://en.bitcoin.it/wiki/Bitcoin_Vanity_Generation_Website, last accessed May 14, 2015.

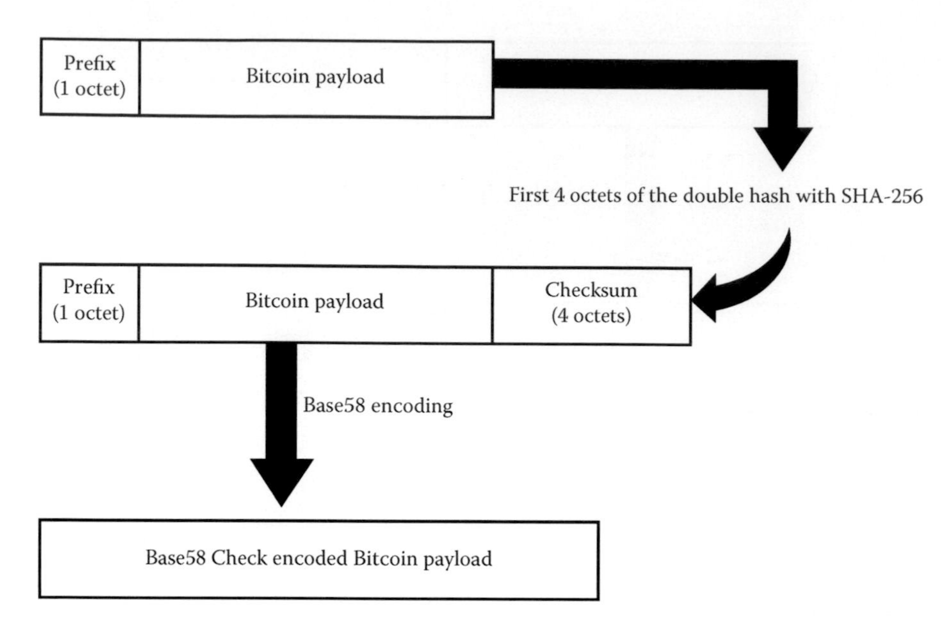

FIGURE 14.10
Base58 Check encoding of Bitcoin keys.

type of the payload to be encoded as explained in the following.

The uncompressed public key is represented as a 65-octet digit with the prefix 0x04 before the 64-octet digit constituted by concatenation of two 256-bit numbers, one for the x coordinate of the point and the other for the y coordinate. This uncompressed public key is represented with 130 digits (65 × 2) in hexadecimal format. Compressed encoding was introduced to reduce the size of the transaction and to conserve disk space on nodes that store the blockchain. A near 50% reduction in the size of the data is achieved by storing the x coordinate of the point only and just the sign of the y coordinate (odd or even bit), which will then be calculated whenever the public key is needed. Thus, the prefix of the compressed public keys will be 0x02 if y is even or 0x03 if it is odd. This gives a compressed public key that corresponds to the same public key but using 33 octets (264 bits or 66 hexadecimal digits) instead of the original 65 octets (520 bits or 130 hexadecimal digits). The scheme is summarized in Figure 14.11 (Antonopoulos, 2015, p. 79).

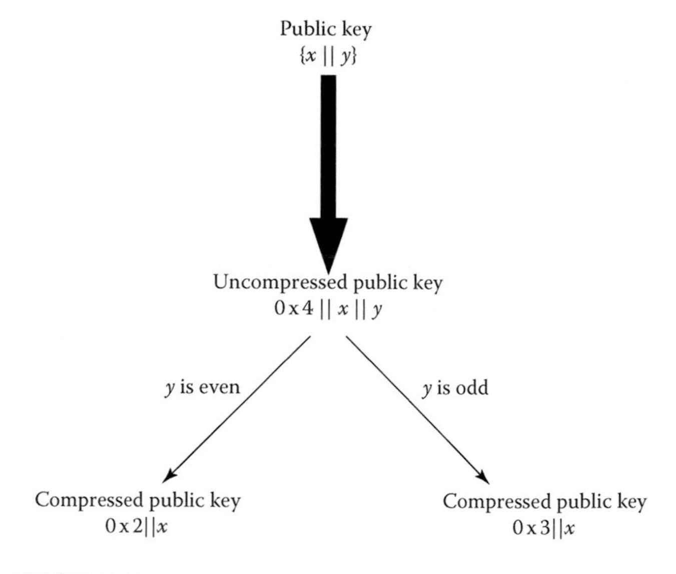

FIGURE 14.11
Compression of public keys in Bitcoin.

The representation of the private key in Base58 Check format uses the prefix 0x80 and a 32-bit checksum. The representation of the private key in compressed format is one octet longer because a suffix 0x04 is added before the Base58 encoding. The output of the encoding starts with a "K" to indicate that it is used to produce compressed public keys.

Public keys are represented in a variety of ways. If the public key of the payee is included in a transaction, it can be hashed to verify that it belongs to the intended recipient. With mobile wallets, the QR 2D code of the recipient address can be scanned with a smartphone camera to avoid typing the long string of characters.

14.3.4 Bitcoin Transaction

A Bitcoin transaction is the operation that allows the transfer of value (i.e., payment) from one owner to another. The new owner, in turn, can spend the amount received by creating another transaction that authorizes its transfer to a third owner.

Transactions are data structures that record the transfer of value from one participant to another. Their size is around 400 octets. Each transaction is an entry in the blockchain. A transaction can have multiple inputs and outputs. The inputs to a transaction reference previous deposits to that account, that is, debits against other accounts claimed from previous transactions, and proof of ownership. Table 14.6 represents the structure of a Bitcoin transaction.

To transfer bitcoins to another entity, the payer must have that entity's address and then build the transaction with the recipient's address. If needed, the transaction may also include payment instructions. To ensure the integrity of the withdrawal, the wallet signs the payment transaction with the sender's private key using ECDSA (Elliptic Curve Digital Signature Algorithm) of ANSI X9.62 and a key of 256 bits. This acts as a proof of ownership that the payee can verify with the sender's public key.

Each output consists of two parts. The first is the value of the payment in bitcoins and the second is a script that defines the conditions for the value transfer to take place. A typical condition is that the payee produces a signature with the private key associated with the public key included in the transaction. It is also possible to require multiple signatures, that is, with M private keys corresponding to M out of the N public keys included in the transaction. Once the conditions are met, the value can be transferred to the new owner's receive address.

There is no special field in the data structure of Table 14.6 to represent a transaction fee. This fee is calculated as the difference between the sum of all inputs to a transaction and the total of its outputs. Thus, if none of the transaction outputs assigns the change back to the sender, the miner can claim all the difference as a transaction fee.

The lock time field in the transaction data structure defines the earliest time that a transaction can be added to the blockchain. It is typically set to zero to indicate immediate execution.

TABLE 14.6

Transaction Data Structure in Bitcoin

Field	Description	Size in Octets
Version	Version of the software used	4
Input counter	Number of inputs	1–9
Inputs	Transaction inputs	Variable
Output counter	Number of outputs	1–9
Outputs	Transaction outputs	Variable
Lock time	Unix time stamp or block number	4

As mentioned earlier, a coinbase transaction (also called a generation transaction) is the first transaction in a block. It creates bitcoins out of nothing from a special input, called coinbase. The output of that transaction is sent to the miner's address as a reward.

In scheduling the transactions for processing, mining nodes may give precedence to those with higher fees. Most wallets use the size of the transaction in kilo octets and not the value transferred to calculate the transaction fees. Currently, the average fee is estimated to be about 1% of the total transaction, compared to 2%–4% for traditional online payment systems and 8%–9% for remittances via money transmitters (European Banking Authority, 2014, §§ 46, 47). In the current version of the protocol, the minimum transaction fee is fixed at 0.0001 bitcoin (Antonopoulos, 2015, p. 119). The low commission value will probably continue as long as miners believe that the future price of Bitcoin will rise. In the long term, however, the reward will decrease over time, so miners will have to increase the transaction fees to cover their cost. When the supply of Bitcoin is exhausted, a miner's revenues will consist of their fees only like in a centralized system.

To ensure that a malformed transaction will not go beyond one hop, thus squelching spamming or denial-of-service attacks, every Bitcoin node, either by itself or through another part, has to check a transaction it receives before forwarding it to its neighbors. Some of the criteria used for checking are as follows:

1. The coinbase transaction must be valid.
2. The inputs or outputs of regular transactions are within the allowed range of values, and the referenced inputs must be in previous transactions.
3. The size of the transaction must be within the defined bounds.
4. The signature for spent bitcoins must be valid.
5. To prevent double spending, an output must be in one transaction only and not in any other, whether pending or already in the blockchain.

If the transaction passes all the screening tests, it becomes a pending (or *unconfirmed*) transaction and is forwarded to three or four neighboring nodes. Each neighbor responds with a success or a rejection message to the sender, according to whether it accepts or rejects the transaction.

A transaction that propagates throughout the network is not included immediately in the blockchain. Each mining node schedules the unconfirmed transactions to work on them when a new round of competition starts (i.e., when a new block arrives signaling that another miner has won the current round). When a new round starts, each mining node forms a new candidate block from the unconfirmed transactions that are in its pending pool, removing those already included in previous blocks, until the maximum block size is reached. The remaining transaction in the pool will wait to the next round of competition. The selected transactions are arranged based on some kind of priority, such as the value of the inputs or of the fees.

The logical state of a block changes to *confirmed* when a miner finds a solution to the proof-of-work algorithm. The newly confirmed block is then included in the blockchain, and no other transaction can reuse the same inputs again. In principle, the new owner of the coin can also spend the transferred value to make a new payment. Typically, however, six confirmations are needed before all nodes are ready to accept the block as valid. Blocks that are considered invalid are not propagated.

It should be noted that, while the average confirmation time is about 10 minutes, some transactions may have to wait much longer to be processed and included in a block. Because transmission in the Bitcoin network is on a best effort basis, valid transactions may be lost and may have to be retransmitted. Also, when a node is restarted, all transactions not in the blockchain may be lost and may have to be retransmitted. Finally, some transactions may be left to remain in a node's pending pool beyond the 10 minute interval.

14.3.5 Orphaned Blocks

Updates to the public ledger rely on the broadcasting of transactions and new blocks. However, due to the distributed nature of peer-to-peer networks and to the Bitcoin protocol itself, inconsistencies may arise in the local copies of the public ledger. If two or more miners solve the proof-of-work algorithm within a short period of time, each miner would then broadcast its block. Nodes that receive any of these blocks will incorporate it into their blockchains. As a consequence, several distinct blockchains may be progressing in different parts of the network at the same time, a situation called *blockchain fork*.

There are several causes for a blockchain fork:

1. As the network expands, broadcasts do not necessarily reach all nodes at the same time. At each hop in the network, messages incur some delays consisting at the network layer (transmission delay) and at the application layer (for the local verification). These delays increase with the size of the block. Simulation results indicate that the median delay is about 6.5 seconds, the mean is 12.5 seconds, and the 95th percentile is around 40 seconds (Decker and Wattenhofer, 2013).
2. Messages, transactions, and blocks propagate from one node to another across the network on a best effort basis (i.e., no guarantee of delivery and the order of transmission is not preserved). As a consequence, transactions may be received out of order. For example, the node could see a reference to an unknown parent transaction in a child transaction. Also, some block broadcasts may be dropped. A node will discover this when it receives a subsequent block and realizes that it is missing one or more blocks. All such child transactions are grouped in an orphan transaction pool until their parent transactions arrive. Note that the size of the orphan pool is limited to prevent denial-of-service attacks.
3. A block may be validated in some part of the network while another contradicting block is still being propagated. If a node detects a block that conflicts what it believes to be the most recent block, it will not propagate it. Different parts of the network will have different blockchains.
4. Two or more transactions may claim the same output (i.e., spend the same bitcoin), either because of an error or by an intentional fraud. The node that receives the transaction first will accept it and commit it to its ledger and then rejects the second transaction.

It is claimed that a block fork may occur every week (Antonopoulos, 2015, p. 204), but supporting evidence is lacking. In a simulation analysis, the probability of a block fork was found to be around 1.7% or once about every 60 blocks, which is equivalent to one every 10 hours on average (Decker and Wattenhofer, 2013).

The protocol specifies that splits be resolved by switching to the branch with the largest cumulative difficulty as recorded in each block in the chain (also known as the longest branch). Nodes that were working on the other branches will switch to the longest. All other chains are discarded, or at least kept as a secondary chain for a while. Because mining nodes determine which chain to extend, the new block represents their vote.

Orphaned blocks are those that do not have a parent in the longest chain so that they are outside the consensus. These blocks are saved in the orphaned block pool until their parent block arrives. Once the parent is received and linked to an existing chain, the orphaned block is pulled and linked to the parent, making it part of a chain. If the parent does not arrive, the transactions listed in an orphaned block will have to be revalidated and all bitcoins awarded through orphaned blocks will be voided. This is a significant incentive for miners to switch to the longer branch, because the majority of the miners will not accept their contribution to the shorter branch. This has several implications.

A centrally located miner can communicate to most of the network faster than other nodes and earn share of total payments by successfully adding block to the chain. Increasing the connectivity of the network speeds up the propagation of messages, blocks, and transactions and reduces the probability of a blockchain fork by about 50% but does not eliminate this possibility (Decker and Wattenhofer, 2013).

More importantly, a sufficiently large minority of colluding miners can pool their computation power to have more hash power than the rest of the network to mount what is called a consensus attack. They can develop an alternative blockchain that they reveal at an appropriate time to force the previously public chain to be "orphaned" and to extract more bitcoins than what their computation power would have allowed (Eyal and Gün Sirer, 2013).

The attack strategy is summarized as follows:

- The colluding miners work to increase their lead as long as their private blockchain is longer than the public blockchain.
- If both the pool's blockchain and the public blockchain have the same length, as soon as the colluding pool discovers a new block, it publishes all the blocks of the private chain. In this way, it annuls the other blockchain, because it is the longest.
- If the new block is found by one miner outside the pool, the response depends on the length of the lead of the private blockchain over the public blockchain.
    - If the lead was >2, the pool reveals only the first unpublished block in its private chain, making the longest blockchain of the network.
    - If the lead was =2, the pool reveals all the unpublished blocks of its private chain. Its branch becomes the longest blockchain.
    - If the lead was =1, the pool publishes its branch of blockchain fork so that two branches of equal length can propagate in the network differing in their last block only. The miners of the colluding pool act on their own blockchain, while other miners will select whichever branch reaches them first, depending on the propagation characteristic. Either branch can end up as orphaned.

The threshold after which the colluding mining pool can disrupt the reward distribution of the Bitcoin system varies as a function of the message propagation speed in the network. The threshold is close to zero if the mining pool has good connectivity; in other words, all mining pools must be honest so that their reward remains proportional to the computing resources that they contribute. In particular, a pool that commands more than 33% of the total computation power of the network would be able to mount such an attack (Eyal and Gün Sirer, 2013). A palliative to the problem is to limit pool size and to modify the protocol to propagate all competing branches of the same length instead of propagating only one branch as is in the current protocol. In that case, the branch to be mined is selected at random (Eyal and Gün Sirer, 2013).

14.3.6 Anonymity

Bitcoin provides some measures for privacy such as the following:

- There are no individual accounts and no central user directory.
- Transacting parties are known by their addresses (their pseudonyms), which are derived from their public keys and are not linked to individual names.
- A user can generate as many public keys as needed, even one public–private key pair per transaction.
- The mapping of a user to a public key is stored on the user's node only. A payee may even generate as many public–private key pair for every transaction.

Bitcoin, however, is not anonymous because payment transactions are recorded in the clear in a public decentralized ledger, from which information on the users can be deduced. In some cases, users may voluntarily disclose their public keys, for example, to receive donations or subscriptions. Furthermore, as noted by Nakamoto, multi-input transactions indicate that all the inputs have the same owner (2008). Also, when bitcoins are purchased with a payment card, the transaction identifies the buyer, which could be traced and linked to a payment currency.

In the simplified example of Figure 14.12, transaction T_1 represents a $\{t_1\}$ payment to address Pk_1 and transaction T_2 indicates a $\{t_2\}$ payment to address Pk_2. Because transaction T_3 involves the payment $\{t_3\}$, which is the sum of $\{t_1\}$ and $\{t_2\}$, we can surmise that a single user owns both addresses, Pk_1 and Pk_2. Thus, an examination of the blockchain with the appropriate tools to trace payments could help identify users and patterns,

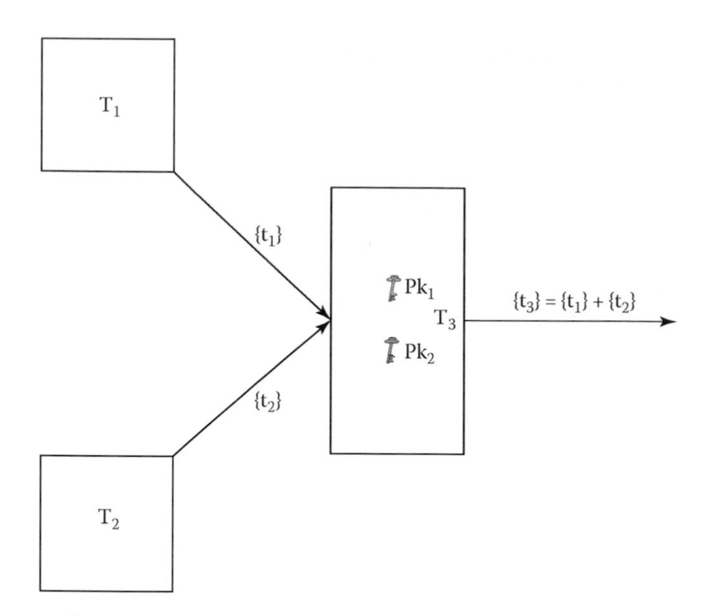

FIGURE 14.12
Identification of public key ownership through the flow of bitcoins among public keys.

years after the fact (Reid and Harrigan, 2011; Möser et al., 2013):

1. The flow of bitcoins between the various addresses allows the identification of the addresses that belong to the same user.
2. The transaction history and the flow of value among users or a group of users over time can be analyzed with network visualization techniques to assess the intensity of linkages among these users.
3. Since Bitcoin outputs can include comments and conditions, it is possible to insert information that can be used to track the flow more effectively.
4. By integrating the identifiers available in the centralized databases of exchanges, virtual wallets and other online service providers, Bitcoin addresses or public keys can be associated with other identifiers such as IP addresses, e-mail addresses, postal addresses, bank account details, and so on. This information can be subpoenaed by law enforcement agencies.

In other words, with Bitcoin, the perfect knowledge of all past transactions may be used to compensate for the imperfect knowledge of identities (Möser et al., 2013). Companies can look at the data in the ledger and pull business information of competitors (since businesses publish their addresses). The U.S. Federal Bureau of Investigation (FBI) was able to trace payments with bitcoin in the successful prosecution of the Silk Road exchange.

A common solution to this problem is to use Bitcoin *laundries* such as Bitlaundry, Bitmix, or Bitcoinlaundry. These services combine together the bitcoins of many users, for example, bundling a large number of small transactions into larger ones and then splitting them again to make tracing the bitcoins backward more difficult (Möser et al., 2013). Still, because bitcoins move in and out of the mix under the control of the service provider and because all transactions are recorded in the ledger, it is still possible to use contextual information to link addresses to users. The second disadvantage is that there is additional delay in reclaiming coins to the mixing process. Finally, laundries must be trusted not to lose or steal coins; moreover, a compromised or malicious laundry offers no anonymity.

Zerocoin is an extension of Bitcoin that provides anonymity to the payer but not to the payee and does not mask the payment amount (Miers et al., 2013). The anonymity is achieved by separating the mint and spend transactions. To mint a zerocoin, a person generates a random serial number S and hides this value with a cryptographic commitment scheme so that it can only be revealed with the knowledge of another random number r, called a trapdoor. The commitment C is the coin that is stored in a bulletin board accessible to all miners and that contains all minted zerocoins. At the same time, an amount of bitcoins equal in value to the denomination of the zerocoin in the commitment is added to an escrow pool. To redeem the coin C and use it for payment, the coin owner needs to prove two things as shown in the following by way of a noninteractive zero-knowledge proof π.

A zero-knowledge proof is a method by which one party proves to another that a given statement is true, without conveying any additional information apart from the fact that the statement is indeed true. This typically takes the form of an interactive dialogue to exchange a challenge or a set of challenges such that the responses from the prover will convince the verifier that the prover has the claimed knowledge.

Zerocoin uses the following noninteractive variant as follows. First, owners must show that they know a coin C that belongs to the set of all other minted zerocoins ($C_1, C_2, \ldots, C_n$), without revealing which coin it is. Second, they must show that they know a number r that allows the extraction of the serial number S from the commitment C. All these operations are based on SHA256 operations.

The proof π and serial number S are posted as a zerocoin spend transaction so that the miners can verify the proof π and ascertain that the serial number S has not been spent previously. After verification, the transaction is accepted to the blockchain and the amount of bitcoin equal to the zerocoin denomination is transferred from the zerocoin escrow pool. Anonymity in the transaction is assured because the link between minted coin C and the serial number S is not known and miners can verify π without knowing the payer's identity.

The Zerocoin scheme, however, has scalability problems due to performance limitations in terms of the size and duration of the transaction. It was estimated to take about 50 times more computational power than Bitcoin (Greenberg, 2013). In addition, the coins had to be of the same value, that is, they could not be divisible. Finally, the system requires a trusted party at setup.

Zerocash improves on Zerocoin by using a method called zero-knowledge succinct noninteractive arguments of knowledge (zk-SNARKs). This method is non-interactive and provides shorter proofs (less than 300 octets) that are easy to verify. Compared to Zerocoin, the protocol improves the performance compressing the size of the spend transactions by about 97% and reducing its verification time by around 99%. In addition, Zerocash hides the identities of both the payer and the payee as well the transaction amounts (Ben-Sasson et al., 2014). The drawback is that Zerocash is incompatible with the current Bitcoin protocol.

Zerocash transactions are broadcast and appended to a decentralized ledger. Zerocash provides anonymity to entries in the transaction ledger and provides two types of coins: anonymous coins and nominal (i.e., not anonymous) coins. Zerocash introduces two new types of transactions to be used with the anonymous currency: *mint* and *pour*. The nominal currency (e.g., bitcoin) is referred to as basecoin, while zerocoin is the anonymous currency. Users can convert from basecoins to zerocoins, send zerocoins to other users, and split or merge zerocoins they own in any way that preserves the total value.

In a mint transaction, users convert a specified number of nominal bitcoins (from some Bitcoin address) into the same number of zerocoins belonging to a specified Zerocash address. The conversion of basecoins to zerocash uses a cryptographic commitment based on the SHA-256 hash function and hides both the coin's value and owner address. The Coin C is formed as follows:

$$C = (a_{p_k}, v, r\ s, cm)$$

where

$$a_{p_k} = PRF_1(a_{s_k}, 0)$$
$$S = PRF_2(\rho)$$
$$k = COMM_r(a_{p_k} \parallel \rho)$$
$$cm = COMM_s(v \parallel k)$$

a_{p_k} is the user public address

a_{s_k} is a random seed the knowledge of which is needed to spend the coin

PRF_1 and PRF_2 are pseudo-random functions derived from the same function with PRF_2 collision resistant

S is the coin's serial number

ρ is a secret value

$COMM_r$ and $COMM_s$ are the commitment operations with the two random numbers (trapdoors) r and s, respectively

v is the value of the coin and $||$ is the concatenation operator

The generation of a Zerocash coin and its serial number is illustrated in Figure 14.13.

By construction, the coin commitment cm for the value of v does not reveal the owner's address a_{p_k} or the serial number S.

The mint transaction Tx_{Mint} is

$$Tx_{Mint} = (v, k, S, cm)$$

This transaction is accepted in the ledger if the user deposits (v) bitcoins.

Once the coin is put on the commitment list, the issuer can retrieve the coin after indicating knowledge related to the commitment tuple, that is, by providing a zk-SNARK proof π.

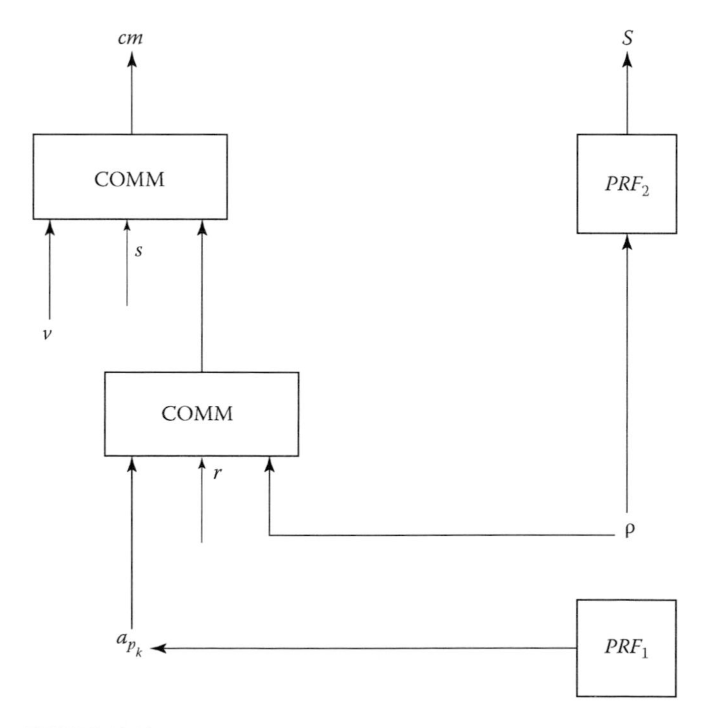

FIGURE 14.13
Representation of the generation of a Zerocash coin and its serial number.

For efficiency, mint commitments are fed into a hash function and then stored in a Merkle Tree that the Zerocash maintains to reduce the storage and retrieval effort. Zerocash uses a tree of depth 64 to support up to 2^{64} coins.

Spending coins involves a pour operation to cast their value into a set of fresh output coins with new commitments, addresses, serial numbers, and so on, as discussed earlier, such that the total value remains the same. This way the payments are not tracked because different serial numbers are used.

The pour transaction consumes the input coins by revealing their serial numbers but does not reveal any other information such as the values of the input or output coins, or the addresses of their owners. Optionally, the pour transaction can also output some nominal bitcoins. This last feature can be used to convert zerocoins back into bitcoins or to pay transaction fees.

To summarize, Zerocash transactions use a series of cryptographic operations to hide any public information about the payment's origin, destination, or amount. It accepts variable amounts and hides transaction amounts and the value of the coins that each user held. On the negative site, the verification of Zerocash transactions takes longer than for Bitcoin and, as a consequence, their propagation over the network is much slower. Large-scale network simulations have confirmed, for example, that spending (pour) transactions require up to about 2 minutes (Ben-Sasson et al., 2014). This delay makes the problem of orphaned blocks discussed earlier more severe.

Another approach to achieve anonymity has been to develop dark wallets to ensure that bitcoin transactions remain secure, anonymous, and difficult to trace.

14.3.7 Point-of-Sale Applications

Confirmation of a transaction can take at least 10 minutes, but Bitcoin never commits a transaction definitively until a majority of nodes in the network accepts the transaction. Because of the possibility of orphaned blocks, a transaction can be invalidated if a longer chain was started before the block that includes the transaction. The idea, therefore, is to wait for sufficient confirmations to reduce the likelihood of a cancelled transaction. Typically, parties accept a transaction after six confirmations, that is, after about 60 minutes on average. For small value transactions, however, retailers can decide to accept unconfirmed transactions or can subscribe to a service that would insure them against that risk. Clearly, Bitcoin is less attractive as a replacement of existing payment methods whenever low response times are important.

14.3.8 Double Spending

The Bitcoin protocol broadcasts every transaction and allows the examination of all past transactions. A double spending attack can take place either before a transaction is confirmed by six or more miners or by provoking a blockchain fork affecting the most recent blocks (around six blocks). Causing a fork at lower depth is probably not practical because the hashes of many other blocks would have to be recomputed.

To spend the same inputs twice through a blockchain fork, the fraudster replaces the payee's address associated with the transaction inputs with another address, recalculates the block of that specific transaction and all the following blocks, and then exceeds the work of all other nodes in the network to create a new longest blockchain, that is, a new consensus. Furthermore, the attacker would need to control more than half of the network's computing capacity to favor the new chain. This attack is often called the 51% attack because success is guaranteed when the fraudster controls that percentage of the total network hashing power. When a mining pool attains that level of control, the pool operator can even make the changes without the knowledge of the pool participants. As explained earlier, a consensus attack can be attempted with at least 33% of the total computation power (Eyal and Gün Sirer, 2013).

Temporary control of a majority of the computer power has occurred on a number of occasions without leading to that state. For example, on January 9 and July 14, 2014, the pool GHash.io exceeded the 51% threshold. It released a voluntary statement to assure that it will limit its processing power if its hashing rate reaches 39.9%.

To summarize, as long as no miner or a pool of miners attains a sustained majority of computing powers the voting cannot be skewed. The distribution of the hash rate among miners or mining pools is a useful indicator for the distribution of the hashing power among participants for a significant time.

As long as there is the possibility of generating coins because the upper limit has not been reached, it is probably more beneficial to spend the computation power and electricity to derive new coins instead of spending the same inputs twice. However, a consensus attack can cause other damage such as to disrupt specific services.

14.3.9 The Protocol Evolution

The Bitcoin protocol has evolved over time to remove security vulnerabilities and to introduce new capabilities for more efficiency. Some of the new capabilities are as follows (Antonopoulos, 2015, pp. 119, 132):

1. Varying the transaction fees based on network capacity and transaction volume instead of a fixed and constant value across the network.

2. Reducing the transaction size with compressed public keys.
3. Simplifying multisignature transactions with the double hash $RIPEMD160[SHA256(PK_1 || PK_2 || \cdots || PK_M)]$ to produce a 20-octet alphanumeric string instead of all the public keys.
4. Changing the way private keys are generated. In the first client, private keys were generated randomly. Next, in seeded wallets, the private keys were all derived from a common random number combined with other data and passed through a one-way hash function. Some deterministic wallets use a sequence of 12–24 words that can be used to create backups and recover all the keys. Others organize the keys in a key structure so that children keys can be derived from parent keys (Antonopoulos, 2015, pp. 85–90).

14.4 Risk Evaluation

Centralized and hierarchical systems suffer from several financial risks: credit, liquidity, and operational risks. Credit risks arise when a bank is unable to meet its obligations and pay money to other members of the payment network. A liquidity risk results from timing effects when a member bank is unable to settle a payment at a particular moment in time, even though it is solvent. Finally, operational risks relate to the situation where one bank involved in a payment transaction ceases to function, either temporarily, because of some event, such as an earthquake, a fire, or an IT failure (Ali et al., 2014a). Decentralization removes the credit and liquidity risks because payments are made directly from one entity or another. Operational risks are also reduced because the general ledger is distributed among all parties and there are many redundant copies of the general ledger. However, new risks have been introduced with decentralized operations.

In 2014, following an extensive analysis, the European Banking Authority (EBA) identified 70 risks across four categories in the current setup of cryptocurrencies. The four categories consist of 33 types of risk affecting users directly, 11 impacting other market participants, 18 threatening the integrity of the financial system, and 8 of concern to regulators. In the following, we elaborate on the following issues:

1. Limited supply
2. Loss, theft, and irreversibility of transactions
3. Volatility
4. Opacity
5. Lack of independent review
6. Unknown software risks
7. Energy consumption
8. Regulations

So far, only a limited number of businesses accept bitcoins for payments, even if their number is increasing. Overstock.com was one of the first publicly traded company to use bitcoins for micropayments to its suppliers. In September 2014, PayPal announced that it partnered with BitPay, Coinbase, and GoCoin to accept bitcoin payments for digital goods. In 2015, Genesis Trading, a New York broker dealer announced that it was fully licensed in bitcoins trading. Other mainstream institutions include Time Inc., Microsoft, Dell Computers, the Latvian airline Air Baltic, Expedia, and Dish Network. Some Silicon Valley investors have started putting money in new exchanges and digital wallet providers. Google Ventures, for example, invested in Ripple, the trading platform and the cryptocurrency. It also invested in Buttercoin, a short-lived Bitcoin exchange that closed in April 2015. The BBVA bank has invested in Coinbase, which lets people store and send payments in bitcoins. In Argentina, there are some anecdotal reports that currency exchanges use bitcoins to avoid large change fees (Vigna and Casey, 2015, p. 209).

We now discuss the new risk factors identified for Bitcoin operation.

14.4.1 Limited Supply

There are continuous debates over the fact that the Bitcoin protocol has capped the total number of bitcoins and that the rate of bitcoin issuance is reducing over time. For many centuries, money and metals were linked; a fundamental characteristic of metals, however, is that their supply is limited. In the past, breaking the link between money creation and a metallic anchor led to a huge monetary expansion and a credit boom and to a steady decline in the purchasing power of the currency.

Bitcoin reproduces many of the characteristics of the metallic anchor. First, the supply is limited. Second, the image of gold miners extracting metal to add it to circulation is reproduced with the so-called miners equipped with "rigs" (their computers). The hard limit on the total currency in the Bitcoin's design prevents any intervention to stimulate the economy if needed. If bitcoins were to evolve to play a significant role in the economy, the possible way to circumvent this limitation is to introduce bitcoin-denominated securities to compensate for the drawback of the fixed supply of the currency (Varoufakis, 2013).

Cryptographer Wei Dai, who first proposed the idea of creating money by broadcasting the solution to a previously unsolved computational problem (Dai, 1998), also criticized the fixed supply idea:

> I would consider Bitcoin to have failed with regard to its monetary policy (because the policy causes high price volatility which imposes a heavy cost on its users, who have to either take undesirable risks or engage in costly hedging in order to use the currency). (This may have been partially my fault because when Satoshi wrote to me asking for comments on his draft paper, I never got back to him. Otherwise perhaps I could have dissuaded him (or them) from the "fixed supply of money" idea.) I don't know if it's too late at this point to change the monetary policy that is built into the Bitcoin protocol or for an alternative cryptocurrency to overtake Bitcoin, but if it is, then Bitcoin is similar to self-improving AI in that it may be critical to get the first one right and it offers evidence on how hard it is for an individual or small group working outside the mainstream to do that.
>
> **Dai (2013)**

Bitcoin advocates, however, argue that deflation can be combated by discounting (Antonopoulos, 2015, p. 176).

14.4.2 Loss, Theft, and Irreversibility

To take trusted third parties out of the picture and eliminate the cost of mediation, Nakamoto's decentralized design creates nonreversible transactions. Users lose access to their wallets without their private keys, and there is no way to redress the situation. The individual shoulders the entire responsibility of keeping the money safe. Lost money is removed out of circulation, and the unclaimed bitcoins remain recorded in the blockchain. There is no mechanism to replace lost coins because it is impossible to distinguish between a lost coin and one that has not been used.

Bitcoin loss occurs for many reasons such as a corrupted hard drive and the loss of a private key. This can take place in an individual wallet or in an exchange. Even experts report bitcoin losses such as when in 2010, a group of experts implemented a complex series of encrypted backups and then misplaced the encryption keys, thus making the backup worthless and losing almost 7000 bitcoins (Antonopoulos, 2015, p. 235). While this may be acceptable for micropayments, it is hard to imagine a world where payments with bitcoins in large value transactions, particularly business-to-business transactions, cannot be refunded, even when software errors occur or cause misappropriations.

This issue is compounded in the case of multisignature Bitcoin addresses. Multisignatures are customary in governmental institutions, companies, and nonprofit organizations. The signing keys are stored in different locations and under the control of different persons. With the rapid turnover that organizations experience today, the probability of losing the keys and forfeiting the associated funds irremediably cannot be dismissed outright.

Wallets stored on a user's hard drive rather than on a server are also exposed to malware. For example, CryptoLocker is a malware that encrypts the files on the infected computer and then pops a notice asking for a ransom to decrypt the file. It comes as an attachment to a hoax e-mail pretending to be a FedEx or a UPS tracking notice. The payment is typically in instruments that are not easily tracked such as bitcoins, MoneyPak/Green Dot, Paysafe, or iTunes cards. Payment of the ransom, however, does not guarantee decryption or that the victim will not be asked repeatedly for additional payments (Bort, 2013; Kuchler, 2013b).

There have been thefts from account pools by changing the payee's address (Reid and Harrigan, 2011). A white-hat hacker was able to take 255 BTC from Blockchain, a popular wallet available from Apple Store and Google Store. This was done by exploiting a flaw in the way random numbers are generated (Southurst, 2014).

In January 2014, Elliptic Enterprises (https://www.elliptic.co) began offering insurance against the loss and theft of bitcoins and other cryptocurrencies, by storing the private encryption keys on offline servers.

One issue that has not been considered yet is that of inheritance, bequeathals, or incapacitation of the key holder. Bitcoin users are told to select complex passwords and keep their keys in a secure place and not share them with anyone. As a consequence, the user's family will not be able to unlock the funds without the user.

Sharing the details with a trusted person or a lawyer is not a valid option in most cases, since it is also recommended to use a different address (i.e., a key) for each transaction. Keeping that person aware of all the addresses instantaneously requires the establishment of a secure channel and a secure storage as well as a contingency plan in case that trusted party dies or is incapacitated, and so on *ad infinitum*.

14.4.3 Volatility

The value of bitcoin exhibits significant volatility as shown in Figure 14.14, which compares the daily variations of the exchange rates for bitcoins and the Euro versus the U.S. dollar. The wild swings reflect the absence of regulation and the fact that owners of bitcoins may be moving around the world to take advantage of arbitrage opportunities. A financial system where bitcoins form a significant contribution would be exposed to instability, particularly if the bitcoins holders have borrowed money from others to buy them or create them, or if important financial institutions have invested in bitcoins without sufficient reserves to cover the price fluctuations.

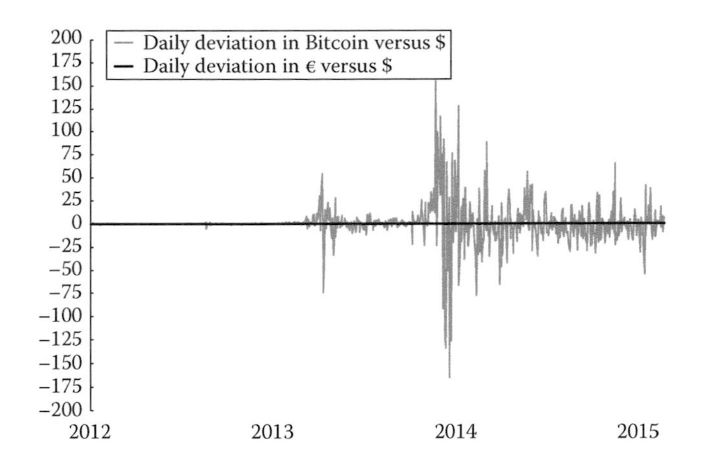

FIGURE 14.14
Comparison of the daily deviations in the exchange rates of bitcoin and the Euro versus the U.S. dollar. (From https://www.ecb.europa.eu/stats/exchange/eurofxref/html/eurofxref-graph-usd.en.html; https://blockchain.info/charts/market-price.)

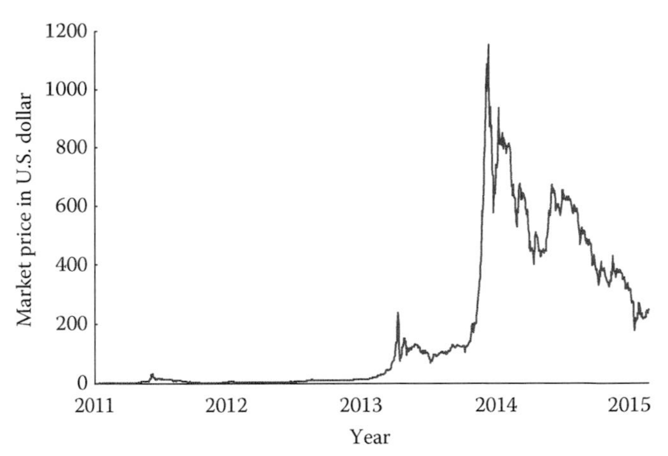

FIGURE 14.15
Daily deviations in the exchange rate of bitcoin versus the U.S. dollar. (From https://blockchain.info/charts/market-price.)

Facing these rapid and unpredictable swings, retailers quoting prices in bitcoins would have to update these prices at a high frequency to maintain a relative stable price when expressed in U.S. dollars. Some market participants could thrive on this volatility. For example, some Wall Street banks enjoy their best quarters due to diverging monetary policies around the world. Over its several years, bitcoin has performed extremely well as an investment as shown in Figure 14.3, even though derivative contracts denoted in bitcoins have constituted a source of instability (Ali et al., 2014b, p. 8). Stated differently, speculative demands for bitcoins may have outstripped transaction demands for a store of value (Varoufakis, 2013).

Because of this volatility, regulators have not allowed bitcoins for gambling in Las Vegas casinos (Ferrara, 2014). Businesses mostly use bitcoins as a means of exchange to bypass the transaction fees associated with international payments. Thus, they convert the bitcoins quickly into legal currencies—just after the transaction or at the end of the day.

In any event, the Bank of England does not consider that bitcoin and similar currencies pose at present a significant risk to monetary or financial stability, but that cryptocurrencies may evolve to compete with traditional currencies in some specific sectors (Ali et al., 2014b).

Figure 14.15 focuses on the exchange rate of bitcoins versus the U.S. dollar, illustrating the collapse of the prices from a peak of $1200 in late 2013 to around $250 in February 2015.

14.4.4 Opacity

Any individual or group of individuals can set up a bitcoin scheme anonymously. The average user has no easy way to identify the participants, including exchanges and pool operators, and no legal recourse in case of disputes. Price formation on the various exchanges is not under any control and the exchange rates differ significantly among the various exchanges without apparent explanation.

Managed pools offer additional possibilities of errors or cheating. They create centralization in a decentralized network, thus becoming points of failure of the system. Failures of the pool server including denial-of-service attacks affect all participant miners.

Bitcoin users are thus required to understand how the system works including its cryptography and how to maneuver and unregulated and unprotected world.

14.4.5 Lack of Independent Review

Despite the enthusiasm of Bitcoin proponents, the fact remains that the Bitcoin system is extremely complex, controlled by a tiny group and without any accountability. As it currently stands, neither the exchanges nor the mining pools are audited or subject to governance and probity standards. As a consequence, potential misappropriations of funds due to software errors, system failures, or outright fraud cannot be easily detected. Given the globally interconnected nature of the Internet, there are also no legally enforced grievance mechanisms.

In essence, Bitcoin eliminates both state protection and shifts control from one elite to another. In addition to those who understand the code, this tiny elite includes that own significant computation power and have access to a reliable and cheap source of electricity. As stated by Nakamoto, "the proof-of-work is essentially one-CPU-on-vote." This requires advanced computation power and cheap electricity, which are not available to all.

Bitcoin has been embraced by technologically minded early adopters that constitute the "Bitcoin community."

Changes to the protocol start within this group as a Bitcoin Improvement Proposal (BIP) and then recorded in a repository on GitHub available at https://github.com/bitcoin/bips (Antonopoulos, 2015, p. 243). In theory, participation in the discussion is open. In reality, the decision is left to the self-selected Bitcoin community of experts.

A smaller circle intervenes whenever a problem arises in the Bitcoin system. They work under the guidance of the Bitcoin Foundation's chief scientist, Gavin Andresen, who is said to have been anointed by the mysterious Satoshi Nakamoto as his successor (Vigna and Casey, 2015, p. 149). Jeff Garzik, one of these experts, described the recruitment process as follows: "It is a meritocracy. New entrants prove their worth by offering changes, accepting technical feedback, and revising their changes until they are trusted by existing developers" (Foley, 2014).

14.4.6 Unknown Software Risks

Another level of uncertainty is raised by the various protocols that govern the operation of mining pools.

The quality and reliability of the various Bitcoin systems, in terms of hardware or software, are totally at the discretion of their developers and manufacturers of the hosting service providers, the miners or the pools they belong to. There are no standard certification procedures for the implementation of the transaction ledger, the protocol, and the encryption to verify that they work correctly under all conditions. There are no certification authorities for the wallet software in terms of performance and their impact on the security of the host system (computer, laptop, or smartphone); wallets are black boxes that can be infiltrated with malware or spyware. Moreover, should changes be approved, the responsibility of ensuring that these changes are implemented correctly or that they do not affect some users adversely is undefined.

Part of the open-source philosophy is that no one owns the project and there are no expectations for prompt response to resolving problems that may arise. In fact, there is no clear structure for reporting bugs, because the members of the core development team volunteer their time and expertise and cannot be expected to devote full attention to code maintenance and repair. The Heartbleed bug with OpenSSL shows that open-source projects are underfunded and understaffed and that there is neither liability nor responsibility for the final product.

A more precise risk arises from the way a mining node encodes the 32-bit time stamp, using the Unix epoch time starting from January 1, 1970. It is expected that this clock will overflow on January 19, 2038, at 3:14:07 UTC/GMT, but mining is estimated to continue until around year 2140. The impact of this reset on the operation of Bitcoin has not been explicitly considered. Yet, the effort needed to fix this Unix and C library problem would be commensurate to the effort spend to avoid the year 2000 (Y2K) issue (Jones, 1998).

14.4.7 Energy Consumption

Given the uncertainties concerning the Bitcoin network, precise estimates of energy consumption are hard to get. One study evaluated the total power used for Bitcoin mining with the hardware available early 2014 to range from 0.1 to 10 GW. The higher range is significant, taking into account that the average electrical energy demand and production in Ireland is 3 GW (O'Dwyer and Malone, 2014). Another 2014 estimate provided the figure of about 1.46 terawatt hours per year, which is equivalent to the consumption of 135,000 average U.S. homes (*Economist*, 2015a).

To address the energy consumption issue, King and Nadal (2012) introduced a new currency called PPCoin or Peercoin. This currency is derived from Bitcoin, but it combines proof of work with a proof of stake based on the coin age, that is, amount of the currency times the holding period. According to the proof-of-stake approach, a user must hold coins in their wallets for certain duration to have the right to process transaction and gain additional coins.*

The privilege of generating a block and getting a coin can be acquired by consuming a coin age. Further to reduce energy consumption, the following changes have been made to the way peercoins are generated as compared to bitcoins:

1. The hashing operation is done over a restricted search space with one hash/s per unspent wallet per second.
2. The hash target is defined at the local level, in terms of the coin age to be consumed.
3. The targets for the proof of work and proof of stake are adjusted continuously to avoid sudden jumps in the difficulty.

The protocol includes additional changes as follows:

1. The coin age supply is expanded by 1% for each coin-year consumed to give rise to a low inflation rate.
2. There are two blockchains: one for the proof of work and the other for the proof of stake. The main blockchain for the proof-of-stake criterion is the one with the highest total consumed coin age.
3. The protocol reintroduces some centralized broadcast checkpoints.

* Some proof-of-stake schemes, such as Nautiluscoin, plan to dispense with miners altogether (Kelly, 2015, p. 110).

Jackson Palmer, a codeveloper of the alternative cryptocurrency Dogecoin, criticized the whole proof-of-work concept based on the "the amount of energy it wastes, and damage it does to the environment—without giving anything back apart from making the fiat rich richer" (Cawrey, 2014). This explains the shift away from proof of work in other decentralized digital money systems as in Appendix 14B.

14.4.8 Regulations

Despite the global nature of cryptocurrencies, there is no unified regulatory response to the Bitcoin phenomena. In December 2013, the People's Bank of China restricted Chinese banks and payment processors from transacting in bitcoins but allowed individuals to buy and sell them at their own risk. In contrast, the EBA recommended that financial institutions wishing to invest in a cryptocurrency scheme establish separate subsidiaries at arm's length from the units dealing with fiat currencies. This is part of proposal for an international regulatory regime for cryptocurrencies should their usage become generalized. The main pillars of which are as follows (European Banking Authority, 2014, pp. 38–43):

- A governance authority that oversees the scheme and guarantees the integrity of the core functional components of the scheme such as the ledger and the protocol. In addition, the authority will be responsible to maintain the reliability and integrity of the hardware and the software in the core system, the exchanges, the hosting sites, and the client software (wallets).
- The governance authority must establish payment guarantees and refund policies.
- Exchanges must be incorporated and subject to reporting requirements. They must be accountable for their actions.
- Hosting sites for virtual wallets must meet capital requirements and maintain separate accounts for their clients. They must be subject to independent audits and be responsible for their actions.
- In addition to discouraging financial institutions from engaging in the trade of cryptocurrencies, the EBA recommends that exchanges involved in cryptocurrencies be subject to the anti-money laundering and counterterrorist financing requirements.

Many of these proposals can be found to be activated in many countries. In the United States, Guidance FIN-2013-G001 of March 2013 from the U.S. Financial Crimes Enforcement Network (FinCEN) allowed the individual production of units of convertible virtual currency and their use to purchase real or virtual goods and services. By contrast, selling those units to another person for real currency or its equivalent is considered a money services business (MSB). In October 2014, FinCEN issued two additional rulings. FIN-2014-R011 specified that matching buyers and sellers of convertible virtual currency for real currency anonymously is also a money service. FIN-2014-R012 ruled that money services regulations apply to payment systems using bitcoins. Tax authorities in the United States are treating bitcoin as property and owners are required to report capital gains or losses. This means that people dealing with bitcoin need to keep track of the value of a bitcoin when they received it and when they spent it and record the gain or loss on any bitcoin separately. In other words, from the perspective of U.S. tax authorities, bitcoin is not fungible as a currency should because each unit is treated differently. Furthermore, by classifying bitcoins as property, the Internal Revenue Service (IRS) has paved the way for states or local governments to impose taxes on them as well (Paglieri, 2014, pp. 195–196).

Most U.S. banking commissions, which operate at the state level, do not consider digital currencies as money. New York State has proposed a licensing scheme for companies involved in bitcoin trading, which would require them keeping records of the identities and whereabouts of their customers and having a reserve amount of bitcoins that is larger than or equal to the sum owed to the customers. In April 2015, the New Jersey Treasury Department issued a memorandum to explain how sales tax should be applied to bitcoin transactions, treating them like barter.

Britain is an interesting exception. In March 2014, the British tax authorities removed the value-added tax on bitcoin trading to strengthen the United Kingdom as a center for global finance, including trade in cryptocurrencies. In fact, the UK chancellor, George Osborne, purchased bitcoins as a sign of support for financial innovation.

Currently, there is very little supervision over the institutions handling bitcoins such as exchanges or hosts of virtual wallets. They are not incorporated in any jurisdiction, and their business activities do not have to meet specific financial requirements nor to keep adequate records for independent audits or regulatory scrutiny. For example, they are not obligated to have sufficient funds to compensate creditors in the case of bankruptcy.

It should also be noted that, should Bitcoin mining be regulated, miners will have to incur additional expenses to satisfy regulatory compliance, including prudential and/or security requirements.

TABLE 14.7

Comparison of DigiCash and Bitcoin

Feature	DigiCash	Bitcoin
Cryptographic protection	Blind signature	Proof-of-work algorithm
Anonymity	Transactions are anonymous	Transactions are based on pseudonyms
Traceability	Transactions are untraceable	Transactions can be traced
Currency	Corresponds to central bank currencies	Created out of nothing (*ex nihilo*)
Protection against double spending	The currency has an expiration date	A transaction output must be in one transaction only
Software	Proprietary	Open-source reference implementation
Clearance and settlement	Centralized	Decentralized
Payment anonymity	Yes	No; uses a pseudonym
Payment untraceability	Optional	No

14.5 Summary and Conclusions

Until the 1990s, cryptography was mostly used to protect state secrets; today, many advanced electronic gadgets routinely use advanced encryption software. Bitcoin and similar cryptocurrencies have adopted some of the advanced cryptographic techniques to build a system for transferring digital representations of wealth securely over a decentralized computer network. They have unleashed a wave of innovations in information technology and finance engineering, which could have a long-term salutary effect on financial payment systems.

Digital money is so radically different from other instruments that it raises many legal and financial issues that the financial industry and regulators have not yet been able to address. This was one of the main roadblocks to previous schemes, such as DigiCash.

Bitcoin is an attempt to reproduce cash in the online world using a new currency but without a central control of the money supply. The money creation is controlled by an algorithm and is distributed among owners of computer power and the money transfers are recorded in a distributed ledger. This may contribute to raising the standards of performance and transparency, thus allowing the decentralized verification of the public ledger as well as to reducing the costs of international remittances.

Bitcoin introduced a new form of decentralized organization that use consensus for decision-making. This has also led to major innovations in the form of cryptocurrencies based on the same blockchain model as Bitcoin's. Other adaptations are not used primarily to generate currency but for the allocation of resources, contracts or for bartering all kinds of goods and services, thus negating the need for a currency. The main differences between Bitcoin and DigiCash are summarized in Table 14.7.

The enthusiasm that has greeted Bitcoin and other cryptocurrencies in some circles may be indicative of a general malaise and distrust of the current banking system and its political underpinnings, particularly among libertarians and information specialists. Independent currencies, however, are not yet trusted by the general population and lack a consistent system of governance. They pose new problems regarding tax collection and the funding of state expenditures. As cryptocurrencies evolve, it is possible that new business models would develop in parallel with existing institutions. The ultimate success of any currency, however, is how much faith people are willing to put into: A unit of exchange that the majority do not trust and do not use will remain confined to its core supporters.

14A Appendix: The Crypto Anarchist Manifesto

This is what is considered by many as the founding document of the Cypherpunks is the *Crypto Anarchist Manifesto*, which was modeled on the first paragraphs of the Communist Manifesto (May, 1992*).

```
From: tcmay@netcom.com (Timothy C. May)
Subject: The Crypto Anarchist Manifesto
Date: Sun, 22 Nov 92 12:11:24 PST
```

* http://activism.net/cypherpunk/crypto-anarchy.html, last accessed April 5, 2015.

Cypherpunks of the World,

Several of you at the "physical Cypherpunks" gathering yesterday in Silicon Valley requested that more of the material passed out in meetings be available electronically to the entire readership of the Cypherpunks list, spooks, eavesdroppers, and all.

Here's the "Crypto Anarchist Manifesto" I read at the September 1992 founding meeting. It dates back to mid-1988 and was distributed to some like-minded techno-anarchists at the "Crypto '88" conference and then again at the "Hackers Conference" that year. I later gave talks at Hackers on this in 1989 and 1990.

There are a few things I'd change, but for historical reasons I'll just leave it as is. Some of the terms may be unfamiliar to you...I hope the Crypto Glossary I just distributed will help.

(This should explain all those cryptic terms in my signature!)

—Tim May

The Crypto Anarchist Manifesto

Timothy C. May <tcmay@netcom.com>

A specter is haunting the modern world, the specter of crypto anarchy.

Computer technology is on the verge of providing the ability for individuals and groups to communicate and interact with each other in a totally anonymous manner. Two persons may exchange messages, conduct business, and negotiate electronic contracts without ever knowing the True Name, or legal identity, of the other. Interactions over networks will be untraceable via extensive rerouting of encrypted packets and tamper-proof boxes that implement cryptographic protocols with nearly perfect assurance against any tampering. Reputations will be of central importance, far more important in dealings than even the credit ratings of today. These developments will alter completely the nature of government regulation, the ability to tax and control economic interactions, the ability to keep information secret, and will even alter the nature of trust and reputation.

The technology for this revolution—and it surely will be both a social and economic revolution—has existed in theory for the past decade. The methods are based upon public key encryption, zero-knowledge interactive proof systems, and various software protocols for interaction, authentication, and verification. The focus has until now been on academic conferences in Europe and the United States, conferences monitored closely by the National Security Agency. But only recently have computer networks and personal computers attained sufficient speed to make the ideas practically realizable. And the next 10 years will bring enough additional speed to make the ideas economically feasible and essentially unstoppable. High-speed networks, ISDN, tamper-proof boxes, smart cards, satellites, Ku-band transmitters, multi-MIPS personal computers, and encryption chips now under development will be some of the enabling technologies.

The State will of course try to slow or halt the spread of this technology, citing national security concerns, use of the technology by drug dealers and tax evaders, and fears of societal disintegration. Many of these concerns will be valid; crypto anarchy will allow national secrets to be trade freely and will allow illicit and stolen materials to be traded. An anonymous computerized market will even make possible abhorrent markets for assassinations and extortion. Various criminal and foreign elements will be active users of CryptoNet. But this will not halt the spread of crypto anarchy.

Just as the technology of printing altered and reduced the power of medieval guilds and the social power structure, so too will cryptologic methods fundamentally alter the nature of corporations and of government interference in economic transactions. Combined with emerging information markets, crypto anarchy will create a liquid market for any and all material that can be put into words and pictures. And just as a seemingly miner invention like barbed wire made possible the fencing off of vast ranches and farms, thus altering forever the concepts of land and property rights in the frontier West, so too will the seemingly miner discovery out of an arcane branch of mathematics come to be the wire clippers that dismantle the barbed wire around intellectual property.

Arise, you have nothing to lose but your barbed wire fences!

______________________ Timothy C. May | Crypto Anarchy: encryption, digital money, tcmay@netcom.com | anonymous networks, digital pseudonyms, zero 408-688-5409 | knowledge, reputations, information markets, W.A.S.T.E.: Aptos, CA | black markets, collapse of governments. Higher Power: 2^756839 | PGP Public Key: by arrangement.

14B Appendix: Bitcoin as a Social Phenomenon

Bitcoin is an expression of the anarcho-libertarian attitude of the 2010s. It also represents a cultural and political phenomenon with its literature, poetry, artwork photography, songs, and so on. Its world comprises libertarians, anarchists, computer scientists, venture capitalists, as well as many operators in the dark side of web business. For example, from January 8 to 20, 2015, the Bitcoin Center in New York City featured an art show of physical representations of crypto-coins created as art (Coca-Cola Coin ICOKE, Disney Coin IDISNEY, etc.). The Bitcoin movement uses religious symbols and languages to express a devotion to technology.

First, we show how Bitcoin can be seen as part of a movement to privative money production and prevent any political intervention to stabilize prices or to stimulate the economy. Next, we discuss how the Bitcoin movement has taken the dimension of a crusade expressed with terms derived from Christian millennialism (Feuer, 2013).

14B.1 Anarcho-Libertarian Response to the Social and Political Environment

Libertarianism, or anarcho-capitalism, is broadly the political hue of the Bitcoin Foundation. This is a political school that favors individual initiative, private property and open markets and is wary of governments, nation-states or large institutions such as the current banking system. In this view, low-cost alternatives to financial services, particularly for international transfers, will bring about the liberation of "the poor (in developing countries) from the incompetence and corruption of bureaucrats and judges," empowering the marginalized and helping the unbanked gain access to the global economy (Vigna and Casey, 2015, pp. 6, 155, 217).

A core belief of Libertarianism is that the individual is the agent of social change through competition with other individuals. Government policies to redistribute wealth are seen as deliberate attempts to rob the hardworking people. Manipulation of the money supply is viewed with suspicion as a tool to expropriate the wealthy and to destroy wealth "by feckless governments" (Grant, 2015). Jonathan Mohan, one of the promoters of Bitcoin, stated that "Government is coercion and force. You don't fight coercion with coercion, you just ignore it. Bitcoin allows you to ignore it" (Foley and Wild, 2013). In fact, the FBI's complaint against the Silk Road marketplace discussed its philosophical underpinnings as the Austrian economic theory and the views of economists associated with the Mises Institute, including Ludwig von Mises and Murray Rothbard (Aglionby, 2013).

Nakamoto's proposal was published in the Cypherpunks mailing list, which was set up in the 1990s by a group of "crypto anarchists" gathered around Timothy May. Their aim was to fight against the government monopoly on violence through cryptographic means (Dai, 1998). Some of the members declared working to "disrupt all sorts of legal and contractual arrangements" (Vigna and Casey, 2015, p. 217). Jim Bell, an ex-Intel engineer and a member of the Cypherpunks mailing list where Nakamoto's proposal was posted, even experimented with an assignation market where anonymous donations would be collected for the purpose of eliminating unwanted politicians (Peck, 2012). In 1997, Adam Back posted the Hashcash proof-of-work algorithm on this mailing list, and this algorithm was incorporated in the Bitcoin mining function.

The enthusiastic response to Bitcoin followed a decade during which various legislations allowed increased incursions by states in the daily lives of individuals. For example, the U.S. Patriot Act, with most other jurisdictions following suit, requested the delivery of reports on some banking activities. At the same time, groups with dissimilar if not opposing political persuasions (antiwar coalitions, religious groups, etc.) concurred in their distrust of central authorities. Furthermore, the inability of governments and large financial institutions to manage the consequences of two decades of financial deregulation and globalization, particularly the sequels of the economic slowdown after 2008, has tainted their reputation and given more ammunition to the libertarian and anarchist thesis that central banks cannot be trusted.

In addition, large advanced economies shifted, starting from the 1980s, from an industrial logic to a service perspective, particularly finance. So in an environment of extremely low or negative interest rates, software and hardware projects for Bitcoin development seemed a good opportunity for designers and computer programmers as the foundation of new sources of income. Some hardware manufacturers took advantage of the opportunity that bitcoin mining offered to expand sales beyond their traditional markets, targeting investors willing to gamble that cryptocurrencies will break into the main stream.

It should be noted, however, that in Medieval Italy, coinage was under the control of the merchant class. Today, the central banks of the major economies are run independently of elected governments. In fact, since the 1980s, governments have funded their borrowing through bond issues and not through money creation, while central banks have been injecting money at record rates to save banks from bankruptcy (the Savings & Loan crisis in the United States, the fraud of the dot-com and Enron collapse, and the credit crisis of 2008). Also, in the United States, the money supply is under the control of the Federal Reserve System, which is a private institution whose chairperson is nominated by the president with the approval of Congress. Finally, crypto anarchism does not explain how to replace the other roles that a central bank plays, such as the lender of last resort in an emergency.

14B.2 Bitcoin Religion

Bitcoins do not correspond to any physical entity, and therefore, bitcoins are created out of nothing (*ex nihilo*).

TABLE 14.8

Market Capitalization of the Top 10 Cryptocurrencies on March 28, 2015

Rank	Cryptocurrency	Market Capitalization in U.S. Dollars	Price in U.S. Dollars	Type
1	Bitcoin	3,527,280,456	252.09	Altcoin
2	Ripple	284,111,510	0.01	Altcoin
3	Litecoin	64,029,318	1.70	Altcoin
4	Dash	25,762,351	4.92	Altcoin
5	BitShares	16,973,636	0.01	Metacoin
6	Dogecoin	12,892,232	0.00	Altcoin
7	Stellar	12,610,006	0.00	Altcoin
8	Nxt	10,788,669	0.01	Bitcoin 2.0
9	MaidSafeCoin	9,805,498	0.02	Bitcoin 2.0
10	PayCoin	9,429,133	0.63	Altcoin

Note: The various categories under the Type heading are explained as follows.

In theological argumentation, *creatio ex nihilo* (creation out of nothing) contrasts with *creatio ex materia* (creation out of some preexistent, eternal matter). The Bitcoin's public ledger has also a divine attribute, omniscience, by recording every single transaction and keeping it forever. As mentioned earlier, the root block is called the "genesis block," the community leaders are called "evangelists," and the followers "faithful." The names of the early converts are recorded in the history of Bitcoin and an early investor, Roger Ver, is even called Bitcoin Jesus (Feuer, 2013; Foley and Wild, 2013; Paglieri, 2014, pp. 21–22).

The mystery surrounding Nakamoto's appearance in 2008 and his occultation in April 2011 have also increased the mythological dimension of Bitcoin. Nakamoto appeared out of nowhere, was never seen, and vanished without revealing his identity, leaving traces such as e-mail exchanges* with cryptographer Wei Dai published in the *Sunday Times* (Smith, 2014). Attempts to locate him have generated significant press coverage and, at times, frenzy: Joshua Davis in the *New Yorker*, Adam Penenberg in *Fast Company* and, particularly, the article in the relaunched print edition of *Newsweek* by Leah McGrath Goodman, which claimed to have found Satoshi Nakamoto living in Temple City, outside Los Angeles under the name of Dorian Nakamoto (Goodman, 2014; Paglieri, 2014, pp. 24–27; Vigna and Casey, 2015, pp. 74–75).

Bitcoin also fits with a long tradition in the "religion of technology" (Noble, 1999, p. 207) to replace humans with intelligent machines, as expressed by the mathematician Rudy Rucker†: "the manifest destiny of mankind is to pass the torch of life and intelligence on to the computer" (Noble, 1999, p. 171). Bitcoin creation is controlled by an inflexible rule that does not take into account changes in societal needs or in the economic cycle and bypasses human control. The Bitcoin system of trust is based on cryptographic computations with the purpose of replacing human decisions on money creation and validation with the "infallibility of the blockchain" (Vigna and Casey, 2015, p. 217, l. 25).

* The emails are posted at http://www.gwern.net/docs/2008-nakamoto#section-1, last accessed April 28, 2014.

† Rudolf von Bitter Rucker (born March 22, 1946) is a U.S. mathematician computer scientist, science fiction author, and one of the founders of the cyberpunk literary movement.

14C Appendix: Other Significant Cryptocurrencies

The number of cryptocurrencies in use around the world is estimated to be around 600 for a total market capitalization of $4,704,436,547 on March 28, 2015.‡ These currencies are built around social networks and, as such, there could be multiple currencies operating in the same social network simultaneously. The top 10 currencies by market capitalization on March 28, 2015, are shown in Table 14.8. It is seen that the capitalization in bitcoins is almost eight times that of the next nine combined.

Some currencies not shown in the list, such as Nautiluscoin, are experimental devices to test various economic theories. Some of the studies concern the effect of monetary policies on the stable and sound growth of an economy, the equilibrium point of competing private currencies, as well as mechanisms to replace gold and fiat currencies with cryptocurrencies (Kelly, 2015). Devcoin is another currency not shown in Table 14.8. It was started in 2011 as a side chain of Bitcoin to provide a currency to pay open-source programmers and writers. On March 29, 2015, it was ranked 100 by

‡ http://coinmarketcap.com/all/views/all.

market capitalization on the website coinmarketcap.com for a total of $109,550 and priced at $0.000012.

The development of the various cryptocurrencies has been stimulated by a general awareness of the potentials and limitations of the Bitcoin protocol. Each of these currencies is associated with an infant economy that has yet to face the typical issues of maturing societies, such as specialization or the redistribution of wealth through inheritance or expropriation.

Another factor that stimulated the growth of these alternatives was that the source code for Bitcoin was open and can be modified to create new currencies. In fact, at one time, www.coingen.io (now defunct), automated the creation of custom currencies, with their own names and logos.

Offshoots of bitcoin, called alternative coins or *altcoins*, are based on the same design as Bitcoin. Some use a blockchain derived from that of Bitcoin and are pegged to bitcoin. Others have adopted different consensus rules. Several have not seen widespread acceptance and a few are *pump and dump* schemes, which is a form of fraud by artificially inflating an item through false and misleading statements. Finally, a different group has its own understanding of the blockchain in terms of the transaction semantics for contracts or name registrations as their primary purpose and may be called *alt chains* (Antonopoulos, 2013, 2015, pp. 215–216). These alternative cryptocurrencies differ from Bitcoins in three main aspects (Antonopoulos, 2015, p. 219):

1. The monetary policy
2. The consensus mechanism through a different proof-of-work algorithm
3. Special features such as anonymity

It is now accepted that a distributed open ledger could have wider applications by making financial transactions more transparent in banking and asset management (Ali et al., 2014b). Alternative uses of the blockchain, outside of purchasing and money transfer, are sometimes known as Bitcoin 2.0. In these applications, products and services use a distributed ledger to run autonomous and decentralized operations under software control, such as for the verification of certificates of ownership—for example, deeds or titles. This notion, called a Decentralized Autonomous Organization (DAO), is similar in the structural sense of the *Internet of Things (IoT)*, whereby physical objects equipped with electronic labels, microprocessors, and antennae remain in constant communication.

Ethereum is a representative of the Bitcoin 2.0 activities, which are not aimed at the creation of a currency, even though it has a built-in currency or *ether*. Its design, while inspired by the Bitcoin protocol, is a totally different entity. The goal is to build an environment for decentralized applications based on a blockchain and an associated programming language to build smart contracts. Ethereum's blockchain records contracts. These contracts can store data, send and receive payments, store ether units, and execute computable actions. Some of these potential applications include the transfer of securities, such as stocks and bonds, and notarized messaging. It is run as a Swiss nonprofit organization.

14C.1 Ripple

Ripple XRP was established in 2005 by Chris Larsen as a trading platform managed by Ripple Labs (Andrews, 2013).* The Ripple network today comprises traders, denoted as gateways, that exchange cryptocurrencies and convert these cryptocurrencies into central bank currencies. Gateways can be banks, remittance operators, money transmitters, and so on. Gateways create the Ripple coin, backed by gold or contracts denominated in central bank currencies. They also build a shared ledger to tally the amount that one gateway owes to another as well as information about offers to buy or sell assets and they validate this ledger by consensus every 2–5 seconds.

Customers request that one of the gateways sends money or any other digital asset to their correspondent, who receives the payment in the chosen currency from the closest gateway in the network. This arrangement reproduces the *hawala* system in a cryptographic architecture, thus achieving lower global exchange rates than the current system (Vigna and Casey, 2015, pp. 234–235).

Ripple was designed to consume less electricity than Bitcoin and to perform transactions much faster. The source code for the Ripple's peer-to-peer node software is available as open source.

As a side note, one of Ripple's founders was Jim McCaleb, who also started the Mt. Gox exchange and later founded the Stellar cryptocurrency (Vigna and Casey, 2015, pp. 82, 234, 238, 301).

14C.2 Litecoin

Litecoin (LTC) was introduced in October 2011 as an alternative to Bitcoin, which makes it one of the oldest alt coins. The main differences between Litecoin and Bitcoin are as follows:

1. The use of an alternative proof-of-work algorithm called *scrypt*.
2. Blocks are generated every 2.5 minutes instead of 10 minutes.
3. The ceiling for the number of Litecoins produced will be increased to 84 million by 2140 compared to 21 million bitcoins.

* It was founded as OpenCoin in 2012, but the name was changed to Ripple Labs in 2013.

Faster block generation allows for speedier confirmations and less opportunities for double spending attacks. It also increases the size of the blockchain and the amount of storage as well as the likelihood of orphaned blocks.

Scrypt is a password-based key derivation function designed to make brute-force attacks computationally prohibitive. The algorithm uses a large amount of memory because it constructs a large vector of pseudo-random bit strings, whose elements are accessed and combined in a pseudo-random order to produce the derived key. The generation of each element is computationally intensive, and each element is accessed many times throughout the execution of the function. Thus, there is a significant trade-off between speed and memory usage in the implementation of the algorithm.

An implementation that does not require a large amount of memory can be massively parallelized with limited expense but would run slowly. Faster implementations would require a large storage capacity to run in a parallel fashion. The cost of running a parallel operation limits large-scale attacks using hundreds of implementations in hardware, each searching in different subsets of the key space (Percival, 2009; Percival and Josefsson, 2015).

In money applications, the scrypt function reduces the advantage of mining with ASICs and, as a consequence, Litecoin is geared to software implementations. Litecoin mining has attracted many miners unable or unwilling to compete with the powerful rigs engaged in Bitcoin mining.

The most common wallet is Litecoin-Qt.

Litecoin addresses are strings of 33 alphanumeric characters that begin with the letter "L."

14C.3 Dash (Darkcoin)

Dash is the new name of Darkcoin (DRK) developed by Evan Duffield. It was launched in January 2014 and changed its name in March 25, 2015, to dissociate itself from the "dark web"* (Cuthbertson, 2015).

Dash/Darkcoin expands on the design of Dark Wallet, the brainchild of Cody Wilson and Amir Taaki, to break transactions into smaller pieces and process each piece separately to create an indecipherable array of data (Greenberg, 2014; Vigna and Casey, 2015, p. 271). It uses 11 rounds of different hashing functions and is claimed to consume 30% less electricity than scrypt. To ensure anonymity, each user's payment is split into smaller denominations and pooled with the split-up payments of other users. All information on the origin of the coin, such as IP addresses and wallet addresses, is removed. This makes it very difficult to determine their origin, hence the wallet they are attached to. Furthermore, the steps are conducted in a decentralized fashion and the user specifies the desired degree of anonymity by choosing the number of mixing rounds (2–11). At each round, random Masternodes are assigned the mixing task so that no single node has full knowledge of both inputs and outputs in the transaction process. As a result, anyone viewing the blockchain will see payments being made but they would not be able to see who paid whom. This system is called DarkSend or the X11 mining algorithm. Another innovation is InstantX, which allows a transaction to be verified in around 20–30 seconds (Greenburg, 2014).

The block generation interval is 2.5 minutes and the total currency units are limited to 22 million. Masternodes earn 20% of the mining rewards and 1000 darkcoins are required to own and operate one. This is thought to be a deterrent to infiltrators that want to de-anonymize transactions.

Dash wallet is available as open source at https://github.com/darkcoin/darkcoin/blob/master/src/darksend.cpp.

14C.4 BitShares

BitShares is a private currency as well as a platform for a decentralized marketplace. Different applications can be built upon this shared foundation of the BitShares Toolkit. BitShares PTS uses the core implementation of the BitShares Toolkit. A person building an application on BitShares software is encouraged to *give back* a percentage of the value they create to current holders of BitShares PTS.

Originally, the private currency was called BitShares (BTS), which, like bitcoin, has high volatility. This volatility was exploited to define contracts with payouts depending on the accuracy of predicting the evolution of an asset in BTS, even though in terms of central bank currencies, the values are more stable. The idea was to allow BTS holders to use their holdings to place buy orders, automatically conducted when a buyer and a short seller are matched at an agreed price. The payment received from the asset buyer and any contribution from the short seller is a collateral to be returned later to the seller when assets are purchased back from the market.† However, such a contract could not be enforced legally.

* The dark web is the section of the World Wide Web, also called the Tor Network, accessible through specialized tools, such as a Tor browser. The Tor Network is designed to protect privacy, but it is claimed to be associated now with illicit activities. The dark web is distinct from the deep web (around 95% of the World Wide Web) that Google does not index.

† In short selling, the seller does not own the asset under consideration but takes on the obligation buying the same quantity of assets at a predefined date from the market.

As of March 2015, the new currency of BitShares is called bitAssets and it trades near par value for the U.S. dollar, the Euro, the Yuan, and gold. In addition, bitAssets pay a daily interest rate to depositors because it charges short sellers interest and they post collaterals equal to three times the value of the loan and then compete to offer the highest interest rates to other short sellers (!).

The BitShares software program records the transactions on the shared ledger with the cryptographic protections of the blockchain. BitShares provides a platform for what is called a Decentralized Autonomous Corporation (DAC). DACs are entities with multiple shareholders tied together with *smart contracts*. These allow the automation of routine financial decisions according to a formula based on data derived from the market. Changes in strategy require modifications to the software, which are approved by the blockchain mechanisms in place (Vigna and Casey, 2015, p. 230). DACs, however, does have any legal status.

14C.5 Dogecoin

Billy Markus and Jackson Palmer started Dogecoin as a joke in 2013 as a fork of Litecoin with an emphasis on philanthropy, such as funding the Jamaican bobsled team to the Winter Olympics of 2014 or digging water wells in Kenya. Its mining algorithm uses scrypt to generate the keys used in hashing but does not require miners to compete (Vigna and Casey, 2015, pp. 90–91).

Note: Dogecoin is pronounced dojecoin as indicated by one of the founders, Jackson Palmer on January 15, 2014 (https://www.youtube.com/watch?v=kVDcOI0-gdQ).

Dogecoin has a production schedule of 1 minute per block, and there is no upper limit on the number of coins that can be produced. The reward to a miner follows a given schedule; it was a random value between 1 and 1,000,000 coins for the first 100,000 blocks.

The difficulty algorithm was changed in 2014 to discourage pool formation. Currently, the difficulty changes once per block. Dogecoins are used as Internet tips to users providing interesting or noteworthy content in social media.

14C.6 Stellar

Stellar is an open-source protocol for the exchange of value developed in early 2014, based on the Ripple payment protocol. It includes modifications to the consensus algorithm to overcome some of the problems that were discovered with Ripple's algorithm.

14C.7 Nxt

Nxt is part of the so-called Bitcoin 2.0 projects that provide an infrastructure in addition to the cryptocurrency. Nxt is related neither to Bitcoin nor to any other cryptocurrency. The design includes a name registry, a decentralized asset exchange, secure messaging, and a method for stake delegation to delegate the process of reaching a consensus.

Nxt's protocol defines an unlimited supply of coins. It uses the proof-of-stake approach to reach a consensus instead of the poof of work to allocate currency. The block generation interval is 1 minute.

The implementation itself was built using Java to allow other applications to use it as a platform for the exchange of assets, to create a marketplace, or to build a new cryptocurrency.

14C.8 MaidSafeCoin

MaidSafe (Massive Array of Internet Disks–Secure Access For Everyone) is an open-source program that organizes the sharing of computing resources securely over Maidesafe.net, a secure network. Maidesafe.net matches requests for computing resources or storage capacity to what is available at any given instant. Users providing resources to the Maidsafe.net network earn safecoins in proportion to their contribution.

Maidesafe.net is a self-healing distributed data network developed by David Irvine to allow people to access digital resources they own or share securely without centralized management. It handles the allocation of disk space, ensures redundancy of data, and performs switchovers upon failure. The network management mechanisms prevent virus infections, denial-of-service attacks, and spam messaging. Users vote anonymously and securely to define the direction of development of the network (Irvine, 2010).

In April 2014, David Irvine organized a crowdfunding campaign to raise funds for the development of MadeSafe by selling 400 million safecoins (reputedly 10% of all safecoins that can ever be produced) labeled as MaidSafeCoin. The campaign was conducted over Mastercoin, a platform that supports decentralized applications using a protocol layer on top of Bitcoin. Mastercoin uses bitcoins and its own internal currency, mastercoin (MST), as a sidechain of bitcoin. Using the exchange rates at the time, the campaign was able to raise the equivalent of $7 million in mastercoins and bitcoins. However, the subsequent decline in the value of mastercoin left it with $5.5 million, short of the target $8 million (Hill, 2014; Vigna and Casey, 2015, pp. 239–240).

14C.9 Paycoin

Paycoin (PYC) uses the scrypt hashing algorithm. Its blockchain is updated every 24 hours.

14D Appendix: Service Offers Based on Bitcoin

Some of the ideas available in the Bitcoin protocol have been used to provide security to distributed services. Some of these services follow arranged in alphabetical order.

14D.1 Bitmessage

Bitmessage is a peer-to-peer system using a Bitcoin alt chain that implements a decentralized secure message service to protect the privacy of users' electronic messages and their subscriptions to broadcast channels (Warren, 2012).

User's addresses are the hash of their public key. Users send a short message of around 36 characters, which constitutes their address (a hash of their public key) and some other data that include a version number of the protocol, a stream number, and a checksum. The stream number is dynamically constructed to arrange the network nodes in a hierarchical tree based on the mail traffic. The binary address is displayed as a text using Base58 encoding, prepended with recognizable characters, and made into a QR Code. The sender sends the message encrypted with the receiver public key. All messages include the sender's public key and a signature of the whole message so that the sender cannot be spoofed.

A proof of work must be completed in the form of a partial hash collision before sending a message to make spamming uneconomical. The difficulty of the proof of work is made proportional to the size of the message and is set such that an average computer must expend an average of 4 minutes of work in order to send a typical message. Each message is time-stamped to protect the network against denial-of-service attacks by flooding old messages. Messages that have not been delivered for more than 2 days are discarded.

All users would receive all messages, but they can only decode those that are destined to them. Thus, they are required to attempt the decoding of each message with their private keys to see whether it is bound for them. The protocol prevents sender spoofing by authentication.

Version 1.0 of the protocol had several cryptographic vulnerabilities such as, no authenticated encryption. Use of the ECB mode of encryption, with each block independent of the other blocks, made it vulnerable to block reordering or mixing blocks from different messages without detection. Acknowledgement of decrypted message exposes the system to the Bleichenbacher attack. Use of the same key for signature and encryption and of 2048-bit RSA for encryption provide less security than the 256-bit ECC used in Bitcoin, and so on (Buterin, 2012; Lerner, 2012). These problems were fixed with Version 2.0 that uses ECC encryption implemented in the OpenSSL library.

14D.2 Bitnotar

Bitnotar is a notary service to confirm the existence of a certain document at a certain point in time. The use of the blockchain allows the creation of a digital register of deeds that are fully administered without a central administration or a trusted third party.

The idea is to use the blockchain of Bitcoin as a distributed time stamp protocol with a new block created about every 10 minutes. Each block references its predecessor and contains references to other transactions occurring at the same time. So, if the Bitcoin address associated with a given document is N blocks deep in the blockchain, this means that it was in existence $N \times 10$ minutes in the past. To verify that the address is valid, a small amount of bitcoins is sent to the public Bitcoin address derived from the hash of the document itself.

14D.3 Blocktrace

Blocktrace, now called Everledger, is a chain of certified digital records to identify and accurately describe each diamond for the benefit of the insurance industry. These records are combined in a blockchain associated with insurance claims and crime records.

14D.4 ChronoBit

ChronoBit uses the Bitcoin blockchain to secure time stamps.

14D.5 CoinSpark

CoinSpark allows the attachment of private messages to transactions so that they can be notarized on the blockchain.

14D.6 Namecoin

Namecoin is a distributed naming and information registration system based on the Bitcoin protocol. It was established in 2011 by Daniel Kraft from the University of Graz, Austria's second oldest university. As such, it is considered the first fork of the bitcoin code, with which it shares the same characteristics: block generation time is 10 minutes, total currency is 21 million, and consensus is reached through the proof-of-work SHA-256 algorithm.

Namecoin is intended to manage a decentralized Domain Name System (DNS) (the.bit domain) to prevent censorship on the Internet. It is used to secure and transfer identity information such as Bitcoin addresses, encryption keys, TLS/SSL certificates, file signatures, stock certificates, and electronic votes. Possible applications could be electronic voting, escrow notary services, exchange of bonds, stocks and shares, and so on. In addition, it has its own currency, represented as N or NMC, whose exchange rate was $0.407867 on March 29, 2015.

Namecoin is currently used as an alternative DNS for the root-level domain.bit and can also store human readable Tor.onion names/domains,

On March 29, 2015, Namecoin was ranked 15 by market capitalization for a total of $4,573,617.

Questions

1. What are the four types of nodes in a Bitcoin network?
2. What are the functions of the full nodes in a Bitcoin network?
3. What are the common functions of all nodes in a Bitcoin network?
4. Show that the maximum number of currency units in Bitcoin is 21×10^{14}.
5. Define mining and explain its role in the Bitcoin process.
6. Explain the various ways a person could participate in the mining of bitcoins.
7. Show that if the rate of block discovery is 6 blocks per hour, then 2016 blocks are discovered in 2 weeks.
8. Show that on average the miner reward is halved every 4 years approximately.
9. Show that the expected value of a geometric distribution with a probability of an event p is 1/p.
10. Explain the advantages of Base58 encoding.
11. What is a blockchain browser?
12. Calculate the difficulty D corresponding to an encoded target difficulty of 0x1B3CC3668.
13. Calculate the target hash with an encoded target difficulty of 0x1903A30C.
14. How is a user's Bitcoin balance calculated?
15. What are the limitations of mixes in obfuscating the Bitcoin transaction history?
16. Compare fiat/government currencies and cryptocurrencies.

15

Dematerialized Checks

Check dematerialization is a comprehensive overhaul of the payment cycle from check processing to the calculation of cash flow and float values. This is why it took several decades of sustained effort to replace paper checks with their electronic images. To understand the scale of the change, the classical treatment of paper-based checks is summarized in the first section of this chapter.

The data presented in Chapter 2, based on the Bank for International Settlements (BIS), have demonstrated that check usage varies widely on an international basis even though the contribution of checks to the total volume of scriptural transactions by nonbanks has been decreasing uniformly across all countries. Today, the highest usage among the industrialized countries remains in the United States with 15.6% of the volume in 2012 (compared to 58.3% in 2000), immediately followed by France, with 13.6% of the volume in 2013 (compared to 43.6% in 2000). In many other countries, the volume of checks is insignificant, while checks are no longer used in the Netherlands. These data explain why the United States and France have been the main countries interested in the dematerialization of checks so as to increase operational efficiencies and decrease cost. These efforts are described in the rest of the chapter.

15.1 Processing of Paper Checks

The classical processing of paper checks comprises three phases: the send phase, the clearance phase, and the return phase (Dragon et al., 1997, pp. 112–124). Although the precise details vary from one country to the other, a general view helps in appreciating the efforts exerted to automate the treatment. Figure 15.1 illustrates the case of a beneficiary that receives checks from users U_1 and U_2 drawn on bank A and users U_3, U_4, and U_5 drawn on bank B.

The send phase begins when the payer issues a check by filling in the amount and the date and then signs it, thereby transforming the paper-based check into a payment instrument. The issuer sends the check to the beneficiary, which then endorses the check and deposits it in the beneficiary's bank. By endorsing each check, the payees are mandating their bank to cash the amounts indicated from the various payers' banks.

The payee's bank prepares for check clearance by sorting the checks and classifying them according to the paying banks as follows. Checks destined to the banks that belong to the same clearinghouse as the beneficiary's bank are put in one group, while the other checks are grouped according to their clearinghouses and then routed to their destinations.

The physical exchange of checks takes place during the manual clearance phase. A limited number of banks send their representatives to represent them and other corresponding banks to a daily meeting at the clearinghouse where they bring the checks to be exchanged. The checks are transported by cars, trains, trucks, or planes. Each bank of deposit prepares a cash letter to describe the checks and to summarize the total amount for each receiving institutions.

Finally, in the return phase, the paying bank verifies the signature and archives the checks, for example, by microfilming the check's front and back sides and then debits the drawer. During exception conditions, such as when checks are to be tracked or are written for large amounts, additional verifications are performed. Rejected checks are treated separately; these checks are drawn, for example, on accounts with insufficient funds, on closed accounts, on invalid accounts, on blocked accounts (accounts for businesses under bankruptcy or under financial reorganization), or on accounts whose overdraft protection has been exceeded.

15.2 Dematerialized Processing of Checks

Techniques for the dematerialized processing of checks depend on the degree to which the end-to-end treatment has eliminated paper. For electronic check presentment, also called *check truncation*, the bank of first deposit, typically the beneficiary's bank, sends the drawer's bank the information that allows faster processing of the check. In the case of check imaging, the beneficiary's bank sends a scanned image of both sides of the check and a file containing other relevant data. Remote deposit has become within consumer's reach with the spread of mobile phones equipped with cameras.

In the United States, ANSI has accredited the Accredited Standards Committee X9 (ASC X9) to support

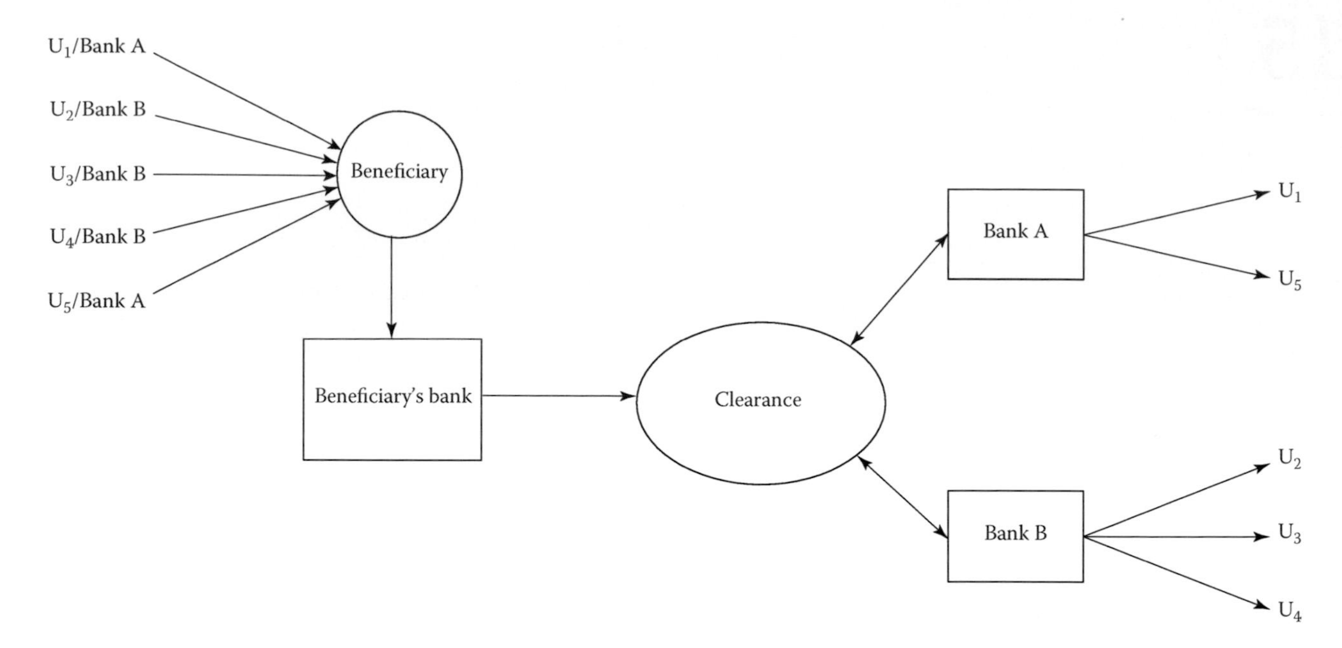

FIGURE 15.1
Classical processing of paper checks.

technical standards for financial institutions. The evolution of the standards, formats, and supporting technology resulted in incompatibility issues. As a result, financial organizations established the Electronic Check Clearing House Organization (ECCHO) in 1990 to coordinate the implementation of these standards to ensure interoperability.

15.2.1 Electronic Check Presentment (Check Truncation)

Electronic check presentment is the electronic representation of the data contained in the paper-based check. The presenting bank, which is usually the beneficiary's, has received the check, converts the payment data in the check into an electronic format to be transmitted into the clearing network.

In France, the banking information on the check is called an "image of the check," an unfortunate terminology because of potential confusion with the digital picture of the check. The information is a line of alphanumeric magnetic characters denoted as CMC7 (Caractères Magnétiques Codés à 7 Bâtonnets—Magnetic Characters Coded with 7 Sticks) and coded according to the NF Z63-001 standard from AFNOR (Association Française de Normalisation—French Association for Standardization) (it is also ISO 1004-2:2013). The coded data indicate the check number, the amount drawn, and the various banking references (the issuer bank, branch, and account information).

The exchange of these "images" was operational since 1990 through the Creic (Centres Régionaux d'Échange d'Images-Chèques—Regional Chamber of the Exchange of Check Images). On June 30, 2002, all clearing houses have closed, and French banks are now obliged to send and return the images of scanned checks via the electronic exchange SIT (Système Interbancaire de Télécompensation—Interbanking Clearance and Settlement System), either directly or through one of the 17 direct participants. The payee's bank is responsible for archiving the checks instead of the payer's bank in the manual system. The elimination of physical exchange eliminated check loss, damage, or routing errors, and one settlement date is now uniformly applied: $T + 1$. As a consequence, all the checks returned before 6 p.m. of a working day are cleared and settled the next day. Thus, the delays for "out of area" checks have been divided by a factor of 2–3.

The U.S. version is called electronic check presentment (ECP) or check truncation. It was introduced in 1996 by the New York Clearing House Association (NYCH). Here, the check information is on a line printed with magnetic ink and is processed according to the MICR (magnetic ink character recognition) technique. Associated with that data is a file of payment data defined in ANSI X9.37-2003.* The verification of the MICR line before the physical exchange of checks gives the payer's bank the chance to detect irregularities within the legal delay for clearing a check. Check truncation was expanded with the Check 21 Act to include a

* The original specification is from 1994. The financial industry designated X9.37-2003 as a Draft Standard for Trial Use (DTSU) until the final standards X9.100-180 and X9.190-187 were approved.

digitized picture of the original check as explained later in this chapter.

The objective of ECP was not so much cost reduction as it is the prevention of fraudsters taking advantage of the time delay between the physical exchanges of checks and the availability of funds to depositors. This is because of a dramatic increase in check fraud followed the Expedited Funds Availability Act of September 1990. This legislation obliged banks operating in the United States to make the funds of deposited checks available to the depositors within a certain time interval defined by law, even though the checks may not have been physically exchanged.

The electronic equivalent of the cash letter is a file called an image cash letter (ICL) and transmitted as part of the X9.37 exchanges. ICLs encapsulate the captured MICR line as an image data. In addition, a Universal Companion Document (UCD) is attached to clarify how to interpret the payment data included in the exchange. In the past, each service provider defined its own companion document detailing the rules on how to format the ICL file. This has resulted in many different variations. The UCD was developed by the collaboration of ECCHO, the Federal Reserve System, the Small Value Payments Co. (SVPCo), and other financial industry organizations to provide a clear and consistent approach for using X9.37.

15.2.2 Check Imaging

Check imaging consists in replacing paper checks with their digitized pictures in the clearance and settlement networks. This digital picture is linked to a presentment file with the relevant banking information and both are transmitted over the banking exchange network instead of the physical deposited check. Both the picture and the companion file are created at the bank of first deposit, which is also responsible for archiving the original paper checks. At the destination, the image can be displayed on screen or printed on a printer. The success of such a scheme requires the use of compatible image formats, in addition to securing the exchanges. In this way, check imaging improves the efficiency of clearance and settlement and reduces the cost of check processing by eliminating the physical transport of paper checks.

The standard images can be either in black and white (bilevel) or in color using one of several compression formats such as the algorithms defined in the ITU-T Recommendations T.4 or T.6, which are used for the transmission of Group III and Group IV facsimile or the Joint Photographic Expert Group (JPEG) algorithm. The presentation format is the Tagged Image File Format (TIFF). However, many other image file formats are currently in use, such as PDF, BMP, JPEG, JBIG, DCX, and so on.

During the 1990s, many contributors participated in the development of check imaging. In 1992, the FSTC (Financial Services Technology Consortium) started a series of projects to study the feasibility of establishing a national system for the exchange of checks. For more than a decade, the legal status of the dematerialization of checks remained in a state of flux (Harrell, 2001). The FSTC was absorbed in 2009 into the policy division of the Financial Services Roundtable (FSR) and participates currently in the IT Acquisition Advisory Council. Finally, the Check Clearing for the 21st Century Act, known as "Check 21," was passed in October 2003 and implemented in October 2004. Check 21 allows U.S. banks to clear checks based on images of the original items through a new legal or negotiable instrument, the substitute check. The substitute check, also known as an Image Replacement Document (IRD), is created from images of the front and back of the original check or from the ICL files received.

More specifically, a substitute check is a paper reproduction of the original check with the following characteristics:

1. It contains an image of the front and back of the original check.
2. It has an MICR line with all the information of the original MICR line needed to process the substitute check.
3. It conforms to the generally applicable industry standards in terms of paper stock, dimension, and other physical characteristics.
4. It is suitable for automated processing in the same manner as the original check.

Substitute checks satisfy the needs of banks and bank customers that need to have a paper document because they are unable or unwilling to process electronic images of check. IRD print locations can be distributed while the control continues from centralized processing centers. Only foreign checks are not eligible to be replaced with substitute checks.

In parallel, the Federal Reserve replaced the physical delivery of return checks with electronic delivery using one of two forms, either, the ICL files containing the return items or the image files containing images of the printed IRDs.

Because paper checks can be converted to electronic data (i.e., truncated) and reprinted as an Image Replacement Document, there is the possibility of duplicate transactions. This can be accidental in nature, but there is also the risk of fraud. For example, the item can be presented twice by the original paper check or by its substitute using the IRD or an ICL.

When Check 21 was adopted, the financial industry did not have all the necessary standards, so as an interim the financial industry adopted X9.37-2003 as the format for exchanging electronic checks as "a draft standard for trial use" until the desired specifications could be developed. X9.37-2003 defines the file sequences, the record types, the field formats and data representation and the compression algorithms for data and images, and the encryption techniques. In parallel to the check image transmission, data extracted from the MICR line are collected and processed by electronic means.

Despite its complexity, X9.37 did not meet all possible applications, and some proprietary extensions of X9.37 were introduced. For example, there are many ways depositing institutions could return endorsed checks depending on which entity has truncated the check (the bank of first deposit, intermediary process, corresponding bank, etc.) and how banks respond to incoming ICLs. Large institutions such as the Federal Reserve System or the Canadian Payments Association issued companion documents to interpret and amend X9.37. Additional standards were introduced to bring some order. In particular, ANSI X9.100-180-2006 (R2013) was designed to enhance the electronic exchange of checks with non-U.S. banks. ANSI X9.100-181-2007 restricted the formats of check images to the T.6 Group IV facsimile bilevel image (black/white) compression and to a specific subset of TIFF 6.0. Finally, ANSI X9.100-187-2013, originally issued in 2008, restricted some of the options available in X9.100-180 while remaining compatible of previous standards to encourage the migration from X9.37.

15.2.3 ICL File Structure

The ICL is an X9 file (X9.37 or X9.100-180 or X8.100-187) with the following structure:

```
File Header
   Cash Letter #1
      Header of Cash Letter #1
      Structure of Bundle #1
             Header of Bundle #1
             Item #1
                Detail Record of Check #1
                Addendum Records of Check #1
                Image View Records of Check #1
                Image Quality Analysis (IQA)
                  Records of Check #1
             Item #2
             ......
             Item # n
             Credit Item Record (X9.100-180
             only)
             Control Records of Bundle #1
      Structure of Bundle # 2
      ............
      Structure of Bundle # n
      Control Record of Cash Letter #1
   Cash Letter # 2
   .........
   Cash Letter # n
File Control Record.
```

The cash control record for each cash letter contains the total number of bundles, items, and images it contains as well as the total value.

For each item, the Detail Record and the Addendum Records contain various banking references. The Image View Records include the front and rear images of the paper check. The Image Quality Analysis (IQA) records contain the results of an analysis of the image quality of the items' images. Automated image quality analysis is not perfect, and there is a possibility of rejecting valid transactions if they do not correspond to high qualities.

The optional Credit Item Record represents the electronic version of a deposit slip and includes various account details indicating where the remote deposit should be posted. Support of this feature resulted in different implementations of X9.37.

To prevent duplicate cash letters or retransmission of the same cash letter, some validation checks are performed using the various header and control records.

15.2.4 Remote Deposit Capture

Remote deposit capture is a service that allows users to scan checks and transmit the scanned images to a bank for posting and clearing. The source document is converted to an Automated Clearing House (ACH) transaction. This is different from check truncation and electronic check presentment that are subject to U.S. law.

Remote deposit capture has different names depending upon how the service is applied within a particular environment such as corporate capture or point-of-sale check approval.

In corporate environments, ICL files are built from the image and MICR data captured from the paper checks and then transmitted in an encrypted form either through exchange network or directly to the issuer bank (the user's bank). Corporations give particular attention to the practical steps for capturing high-quality images.

In a point-of-sale check approval, a reader terminal captures the MICR data while the cashier enters the purchase amount manually. The data are transmitted to the issuer bank (the customer's bank) to verify the availability of funds. The approved check is canceled and returned to the customer.

For consumers, check deposits can be done from a mobile terminal if the user's bank provides an application with that capability to be downloaded to a mobile terminal. Users open the application and enter the amount of the check and select the account where the funds should be deposited to conduct the transfer. With camera-based applications, the user snaps photos of the front and the back of an endorsed check in a well-lit area. The application usually provides a square frame to guide the user. Alternatively, a smartphone equipped with an MICR reader extracts the bank routing number and the account number. The mobile application combines this information with the amount of deposit that the user enters and then transmits the payment information to the acquirer bank through the ACH (Automated Clearing House) infrastructure. In the United States, if the paper check is returned to the issuer the transaction is coded as a Point-of-Purchase (POP) transaction. If the check is not returned, the ACH transaction is coded as BOC (Back-Office Conversion). Alternatively, the application can initiate an ACH WEB debit to the check holder's account (NACHA, 2011). This mobile transaction includes two steps: check truncation by using MICR, which is facilitated by the Check 21 Act, followed by check conversion. The first step remains subject to check law, while the second is subject to the rules of the National Automated Clearing House Association (NACHA).

Remote deposit captures leverages the existing infrastructure for debit transactions to facilitate the use of checks of purchases. The difficulty arises from the wide variety of MICR or image formats. Furthermore, in consumer applications, the use of inexpensive check scanners may produce lower-quality pictures compared to those captured in bank operational centers or in corporate environment. While banks or service providers have no control over the capture environment, nevertheless, they are responsible of the data integrity before transmitting the data for clearance. Also, in case of malfunctions, the undefined division of responsibilities between the users, the application developers or providers, and the banks could also make it difficult to identify the root cause.

Another easy mistake in remote deposit applications is to capture the same item twice. This may require upgrading the data processing infrastructure of banks and/or service providers.

15.3 Virtual Checks

Virtual checks are used for transactions initiated by telephone or over the Internet. In the latter case, they are output from a special software that the user's bank provides for installation on the user's computer. They are signed by the payer and endorsed by the payee with digital signatures instead of handwritten or machine-stamped signatures. It should be noted that from a legal viewpoint, a virtual check is not a substitute check according to the Check 21 Act. Parties, such as banks and customers, cannot refuse to accept a substitute check that meets the Check 21 requirements, but they can refuse virtual checks. We now review some of the main forms of virtual checks.

NetCheque was an experimental system for virtual checks from the Information Science Institute (ISI) of the University of Southern California (USC) (Neuman, 1993). It was based on the methods of distributed management of the access to computation resources in a university computation center using Kerberos for authentication and for the production and distribution of session keys. A prototype is available at http://www.netcheque.org.

Many virtual checks for Internet-initiated payments are ACH transactions, as is the case of PayPal, which was the subject of Chapter 12. Another example is OFX (Open Financial Exchange), a unified specification for the electronic exchange of financial data between financial institutions, businesses, and consumers via the Internet. It was jointly developed by Microsoft, Intuit, and CheckFree in 1997 to support a wide range of financial activities including consumer and small business banking. CheckFree started in 1981 to replace paper checks from consumers with electronic payments and is now serving many of the world's largest financial institutions.

The FSTC sponsored several projects for the development of virtual checks. One of them, eCheck, was cosponsored by the U.S. Treasury (Jaffe and Landry, 1997). The eCheck technology was licensed to Clareon, which incorporated it into its PayMode secure electronic payment engine. When the pilot project ended in 2000, a private corporation was formed to build Paymode-X, a web-based electronic payment and remittance platform. The product found a temporary home at Bank of America, from which Bottomline Technologies acquired it in the fall of 2009.

The rest of the section focuses on the results of the eCheck project.

15.3.1 Representation of eChecks

The representation of the eCheck uses the Financial Services Markup Language (FSML™) (Kravitz, 1999). FTML is focused on financial applications using microprocessor cards where storage memory and transmission bandwidth are restricted. It is defined within the framework of the Signed Document Markup Language (SDML) that specifies how to sign digital documents (Kravitz, 1998). Although SDML, in turn, is inspired by the SGML of ISO 8879, it is not compatible with XML (Liu, 1998).

FSML describes the checks using a sequence of blocks to represent the data related to the check. These blocks can be nested; they are signed with public key cryptography by encrypting the message digest with the private key of the sender. Signatures are done with RSA after hashing with MD5 or with a SHA-1 hash and the DSA algorithm. The X.509 version 1 certificate signed by the U.S. Treasury Department is utilized for authentication of banks; banks in turn certify their clients. As usual, e-mail messages conform to the MIME protocol.

FSML represents the alphanumeric characters that are part of the check data using 7-bit ASCII, which limits the alphabets (e.g., no accents are used).

There are two categories of blocks in an eCheck. The first comprises the action, the check data, the debtor's account number, any attached documentation, and the invoice. It is signed with the private key of the sender, whose corresponding public key is included in the user's certificate. The second category comprises the account number and the debtor's certificate; it is signed with the private key of the bank. Figure 15.2 depicts the various blocks in the check.

Endorsement of an eCheck means adding an endorsement block, with the identity of the endorser, its banking coordinates, its certificate, and the certificate of its bank. The endorser and its bank sign the block indicated in Figure 15.3.

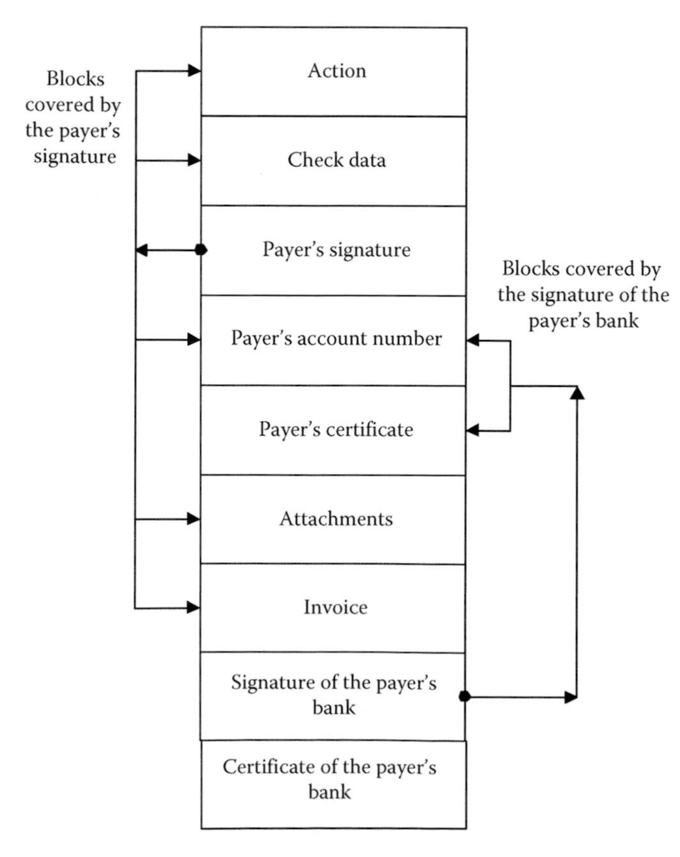

FIGURE 15.2
Representation of virtual checks in eCheck.

Typical sizes of the eCheck files are shown in Table 15.1.

15.3.2 Payment and Settlement with eChecks

An institution or an individual can initiate a fund transfer using a standard message sent by e-mail (FSTC, no date). From the exchanges in Figure 15.4, it is seen that the circuit of paper-based checks is reproduced with a dematerialized base both for the check itself and for the control message. Clearance and interbank settlement of the virtual checks uses ANSI X9.46 (1997) and X9.37 (1994). This scenario is called "deposit and clear."

The flow of exchanges for the deposit-and-clear scenario is similar to the case of paper checks. The payer sends a signed eCheck to the payee. The payee then deposits it in his bank. Clearance and settlement go through the traditional financial network. Because banking clearance takes the usual electronic path, payment to the creditor does not occur before the debtor's bank completes all the necessary verifications.

A variation of the preceding circuit can be useful if the creditor's bank is not equipped to process electronic checks, while the debtor's bank is so equipped. In this case, called "cash and transfer," the debtor sends the check to the creditor who endorses it and then sends it to the debtor's bank. After making the prerequisite verifications, the debtor sends the necessary funds through the banking network as shown in Figure 15.5.

Certified eChecks can be treated in one of two ways. Debtors send their checks to their banks to verify the availability of funds and then the banks block them before the debtors sign their checks. The banks can then return the checks to the debtors (option 1) or send them directly to the creditors (option 2). The corresponding flows are represented in Figure 15.6, where the network between the creditor, the debtor, and the debtor's bank is not shown to avoid unnecessarily cluttering the diagram.

Direct fund transfers are initiated when the payer send the eCheck to his or her bank. The actual fund transfers take place through the usual interbank circuits.

A lockbox is a service offered by commercial banks and payment processors to organizations that receive a large number of checks. In the case of paper checks, customers mail their payments to a Post Office Box that the bank/payment processor accesses. The bank/payment processor collects and processes these payments and deposits them to the company's account. In the case of eChecks, payers send their eChecks to a special account under the control of the organization's bank. The bank

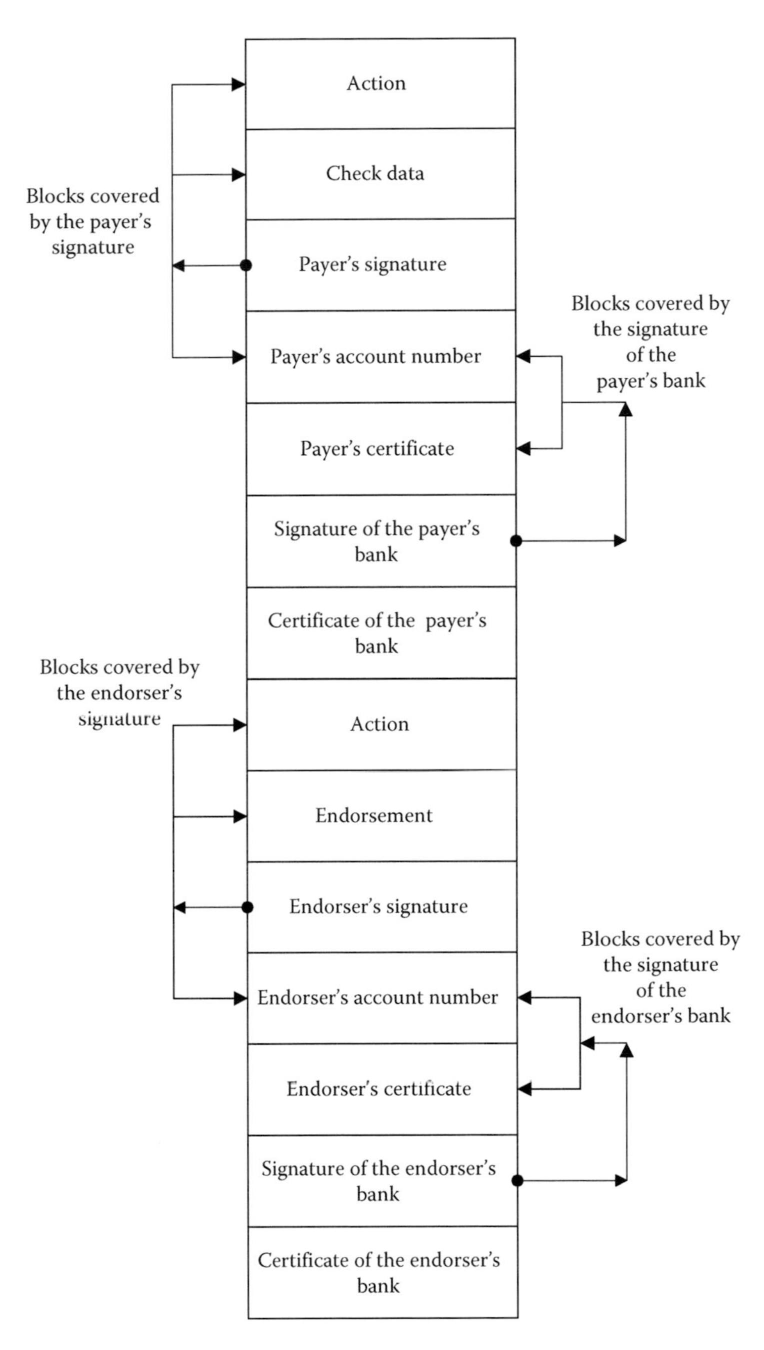

FIGURE 15.3
Representation of an endorsed virtual check in eCheck.

TABLE 15.1
Size of eCheck Files

Description	Total Size (in Octets)	Compressed Size (in Octets)
Single check	4762	2259
Single check with remittance	7052	3135
Single check with remittance and endorsement	1443	4839

will then initiate the clearance and settlement through financial networks for all the checks that have been proven valid. Figure 15.7 illustrates the corresponding exchanges.

One of the most critical aspects of payment with virtual checks is the reconciliation of the data that the customer's computer or microprocessor card records with the data that the customer financial application records and with the bank statement.

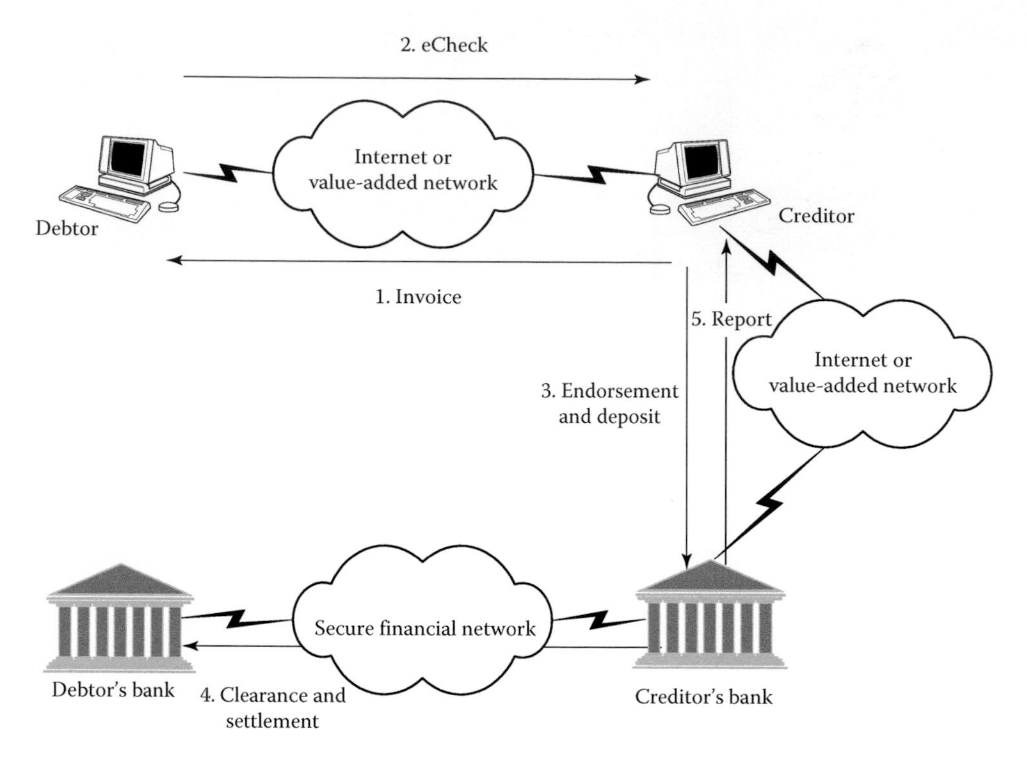

FIGURE 15.4
Exchanges in a payment by eChecks (deposit and clear).

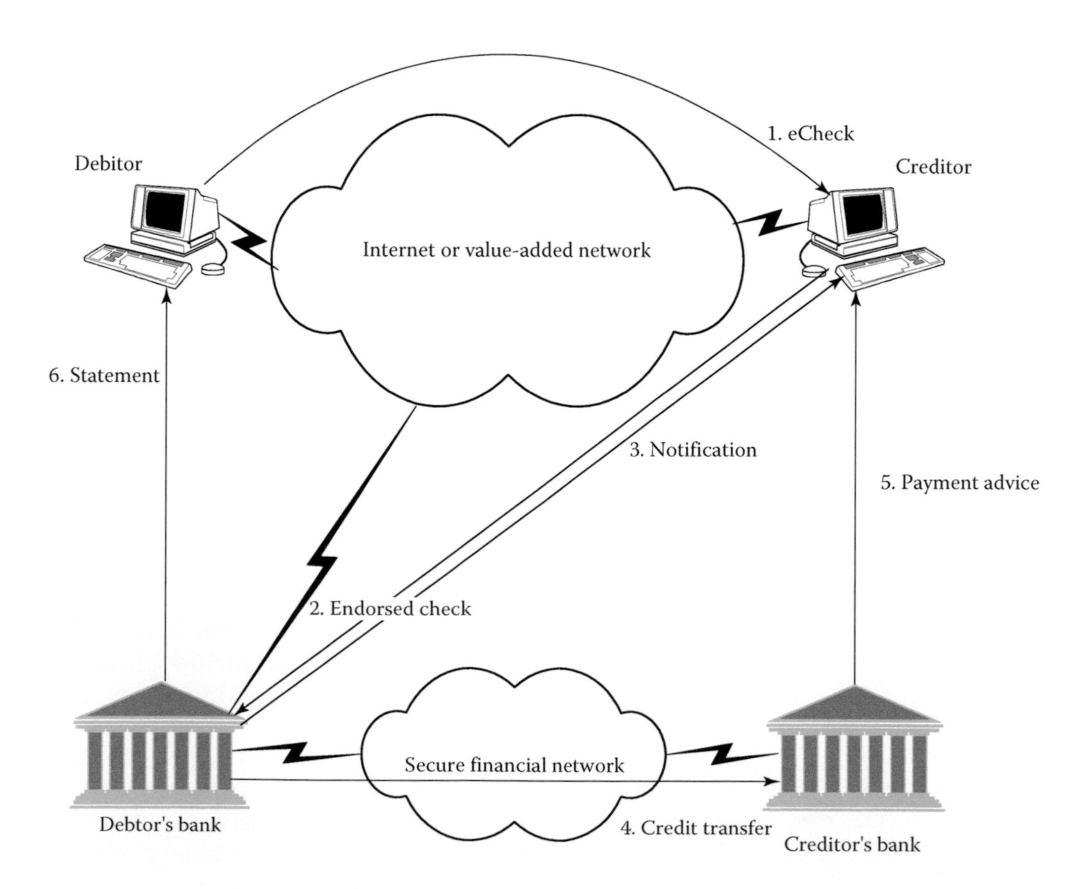

FIGURE 15.5
Exchanges when the creditor's bank cannot process eChecks (cash and transfer).

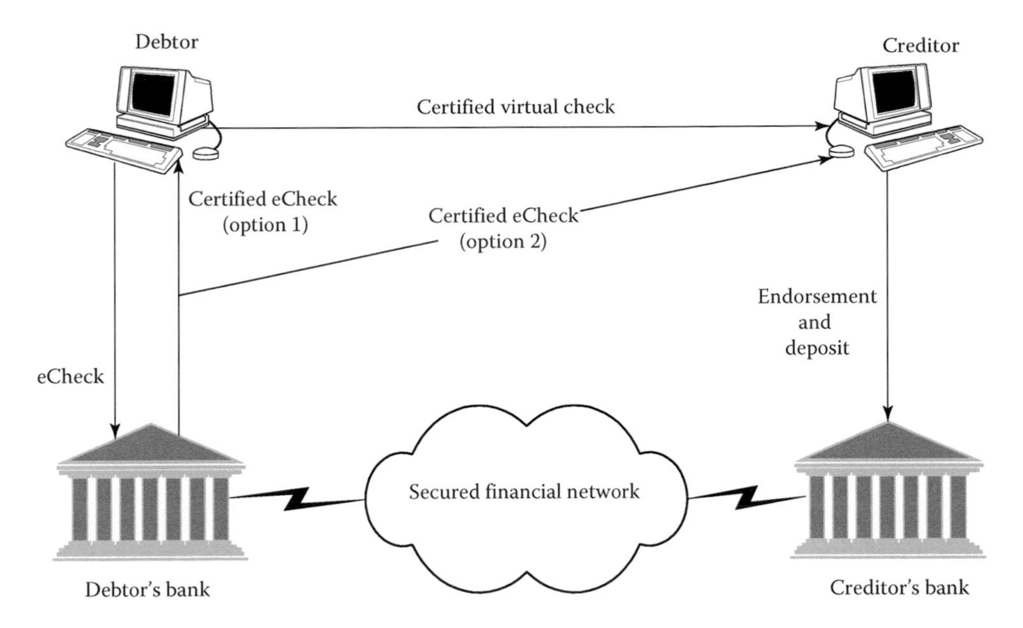

FIGURE 15.6
Exchanges for certified eChecks.

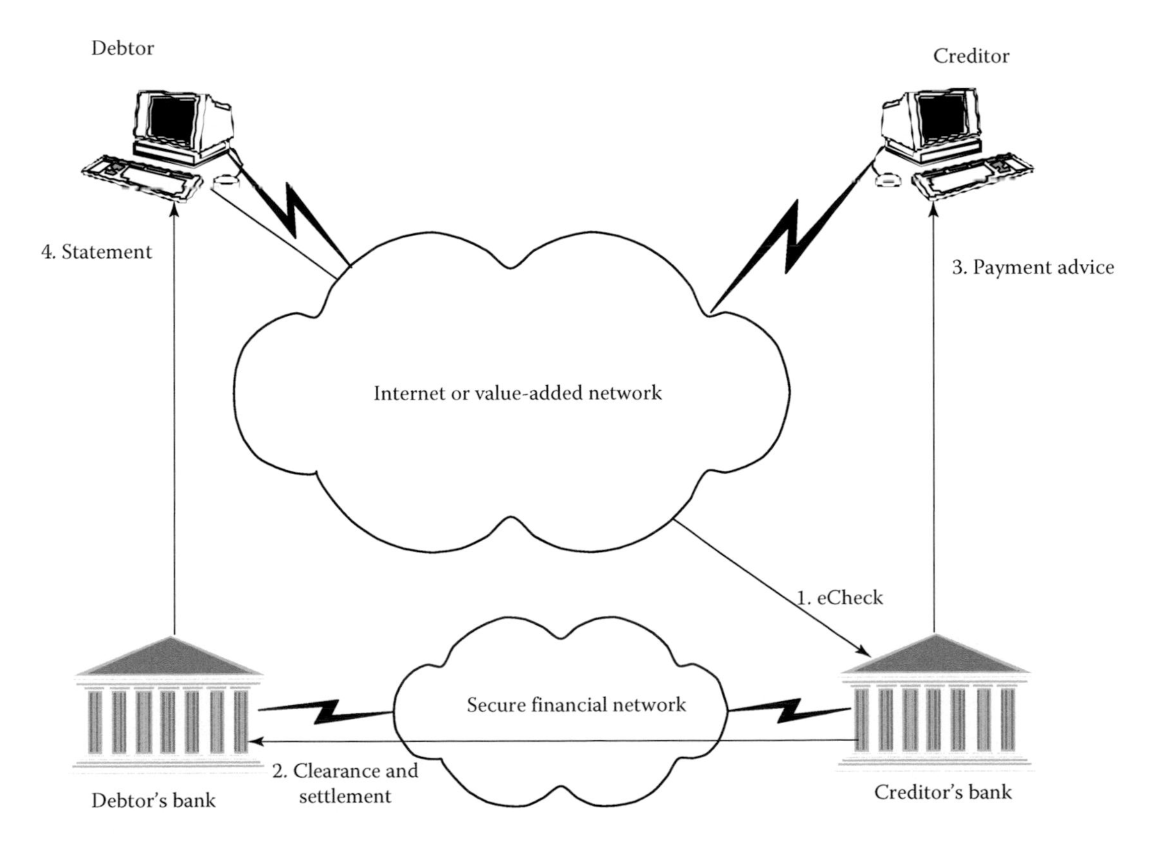

FIGURE 15.7
Exchanges for the lockbox scenario with eCheck.

15.4 Summary

In countries where checks are still widely used, banks, merchants, and governments are promoting new payment processes to bridge paper checks into the world of electronic banking. Several approaches are used to dematerialize checks. This has increased the importance of the reliability and accuracy of the software applications that support the capture of data and the processing of the companion files. The quality of this software has some legal implications when, such as specified by the Check 21 Act, some forms of dematerialized checks are now equivalent to paper-based checks.

Questions

1. Name three components for the cost of paper checks.
2. What are the advantages of dematerialized checks?
3. What is check truncation? What are the difficulties in implementing this technique?
4. What is point-of-sale check approval?
5. What are the differences between electronic check presentment and check imaging?
6. Discuss the factors favoring or opposing the use of check imaging.
7. What is check conversion?
8. What specific fraud risk is specific to check imaging and/or check truncation?
9. Why is a remote deposit capture not a legal replacement of a paper check?
10. What is the main difference between a substitute check and a virtual check?
11. Fill in the following table appropriately from a user's viewpoint:

Feature	Electronic Bill Presentment	Bank Card	Cash
Cost			
Security			
User action			

16

Electronic Commerce in Society

The experience acquired during the last several decades shows that electronic commerce (e-commerce) transforms the organization of tasks, modifies the balances within administrations or enterprises, and brings about a reevaluation of the roles played by economic agents. Success does depend not only on technical mastery but also on good management of sociological and cultural factors, particularly when integration of the processes used in several independent entities is at stake.

In the 1980s, the Minitel was the first success in the consumer arena but was confined to France. Today, the wireless infrastructure and the Internet have blanketed the whole planet with monetized applications as well as an economy based on freeware, with programs, applications, and advice available at no cost and free of commercial constraints, whether explicit or implicit. The virtual economy of a worldwide deregulated market of information satisfies the cultural and ideological necessity of growth without depletion of natural resources or pollution risks and meets the ambition of a social class whose wealth arises the power to manipulate symbols (Lash, 1994; Haesler, 1995, pp. 315–316, 325).

In the following, we will consider certain aspects regarding the wide-scale use of e-commerce; in particular,

1. The harmonization of the communication architectures and interfaces
2. The governance of electronic money
3. The protection of intellectual property
4. Electronic surveillance and privacy
5. Content filtering and censorship
6. Taxation of electronic commerce
7. Trust promotion
8. Dematerialization of records

16.1 Harmonization of Communication Interfaces

Electronic commerce relies on a high-quality telecommunication infrastructure with worldwide coverage. Conversely, without ubiquitous and reliable transmission pipes, e-commerce will be hampered except in applications prepaid cards could be used.

The telecommunication infrastructure includes several physical and virtual networks, overlayed or associated and managed by distinct administrative authorities. A series of participants must coordinate their interventions from one end of the connection to the other to ensure successful exchanges. As shown in Figure 16.1, they include the following:

- One or several bandwidth providers whose role is to establish the physical pipes between the two communication endpoints.
- One or more network operators responsible of the operation and maintenance of the routing/switching and transport network; this network, which is overlayed on the physical network of links and trunks, usually includes equipment from several manufacturers implementing equivalent techniques that should be interoperable.
- Service providers (e.g., trusted third party, certification authority, payment intermediary, etc.) responsible for the installation and management of the message exchanges; security may be incumbent on several certified providers to manage access using certificates provided by recognized certification authorities; other providers may host security applications, merchant sites, give guarantee labels, etc.
- Suppliers for the content sold online (databases, information, games, etc.).
- Retailers or brokers between content suppliers or payment services and the users.

In such a fractured architecture, trouble localization and repair is a delicate operation. There is a wide variety of network element management systems, of trouble tickets, and of diagnostic procedures. There is no single entity responsible for trouble reports, particularly when the services are provided free of charge. The problem is even more acute with cryptocurrencies, where the individual responsibility is diluted because of the lack of central organization.

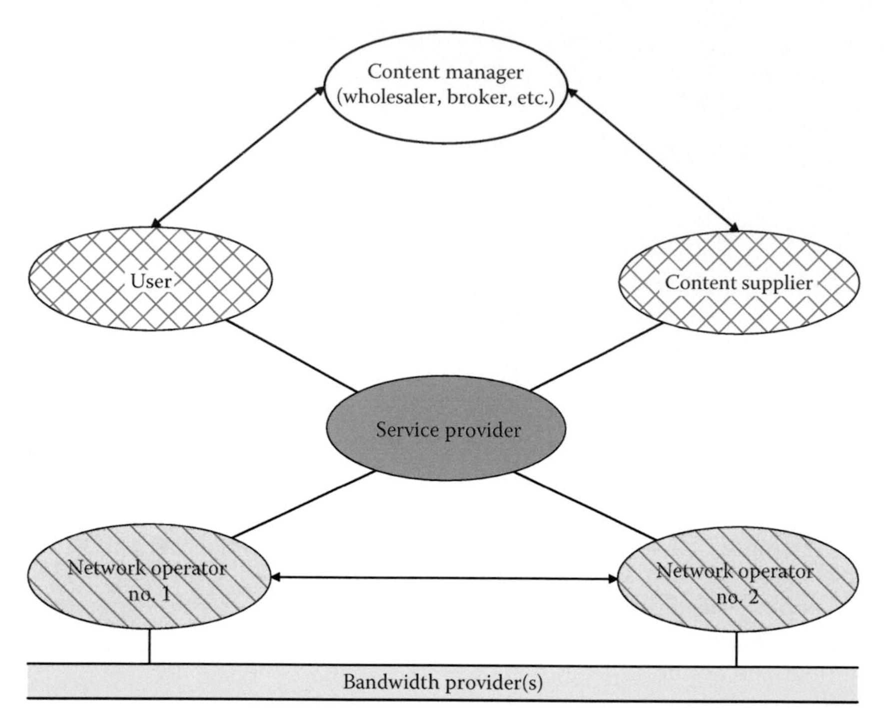

FIGURE 16.1
Contemporary organization of telecommunication services.

Beyond the confines of telecommunications and information technology, interoperability includes aspects such as

- Interfaces among the same category of applications, such as payment systems or cryptocurrencies
- Operating systems of terminals (mobile telephones, computer, personal digital assistants, smart cards, etc.)
- Protocols for charging with value electronic or virtual purse or Jeton holder
- User interfaces to applications
- Certification and certificate revocation procedures

Harmonization may be difficult because of the proliferation of interested parties. The overabundance of standards, and sometimes their competition, may constitute a barrier to the spread of e-commerce.

16.2 Governance of Electronic Money

The modern monetary system was freed of the material support by replacing it with a generalized system of trust based on continuous growth. In the past, monetary creation was absolutely limited by the scarcity of materials to be exchanged or of precious metals for minting coins. On the contrary, money liberated from a material basis, in particular electronic money, gives the impression of a creation *ex nihilo* of money that can multiply autonomously without any physical work.

The right to issue money is a privilege that benefits the state because the cost of manufacture of currency is much less than the face value of the money. The difference represents a loan without interest that the holder of the currency gives to the issuer. *Seignorage* (from the French *seigneuriage*) is the government revenue due to this difference.

Designers of electronic commerce systems will have to take into consideration these national and geographic preferences. Acceptance of means of payment cannot be decreed but depends on the confluence of economic, sociological, political, and certainly human factors. Cultural factors affect political choices, thus affecting the future of each offer of electronic commerce. In Chapter 2, it was shown how the preference for individual instruments of payment varied according to the country. Similarly, all surveys have confirmed that the Internet has been better received in the countries that Hall calls low-context countries (the Anglo-Saxon and the Nordic countries) than in high-context countries (France, Italy, and Spain).

Many private monies are used alongside legal monies in daily life. These are often promises of service in the form of airline miles, cryptocurrencies, coupons, phone Jetons, fidelity points, preferential stocks, and so on. Sometimes, people have to resort to alternate currencies when cash is rare because of a severe banking crisis (Argentina, Greece, etc.). The withdrawal of global banking services under the pressure of U.S. laws is opening the field to alternative fund transfer systems, sometimes using new currencies, such as WebMoney. Finally, a new type of money has taken the form of personal information that is becoming the chief way to pay for services that are "free," that is, not paid with fiduciary money.

In principle, the area of action of these private monies could be expanded provided that their issuers mutually recognize their respective instruments. This would have two important consequences: Banks would be excluded from the circuits of financial exchanges and private monies would benefit private entities at the expense of the public treasury, thereby increasing the deficit of state budgets.

16.3 Protection of Intellectual Property

With globalization and digitization, protection of intellectual property requires rules that are followed worldwide regarding the reproduction and the broadcast of virtual works, such as music recordings. The widespread use of digital optical disks and the spread of the Internet facilitate the storage, copying, editing, and distribution of all kinds of digital contents. Hypertext links on websites can lead the users to contents that are protected by copyrights.

Signed in 1996 under the auspices of the World Intellectual Property Organization (WIPO) and coming into force in 2002, two treaties—called the Internet treaties—update copyrights and related rights (rights of performers, publishers, and broadcasters) to fit the era of e-commerce. These are the WIPO Copyright Treaty (WCT) and the WIPO Performance and Phonogram Treaty (WPTT) that allow authors, artists, and performers to control the broadcast of their works on digital networks such as the Internet. The most controversial aspects of these two treaties are the limits they put on the ability to copy digital productions and the prohibition of any action to intentionally circumvent the technical means to protect copyrights and related rights, in particular, the manufacture, distribution, importation with the intent to distribute, and the use of any equipment for circumvention.

Google Books (previously known as Google Print) is a service from Google to scan books and magazines in public and academic libraries and store them in a digital database. The books then are printed on demand. In 2012, the scope of the project was changed to accommodate copyright holders in response to years of litigation. For French publishers, digitization was narrowed to out of print books, after securing the publisher's permission. In the United States, the settlement gave publishers a choice whether Google digitizes their books and journals. The agreement was considered a step in defining copyright in the Internet age and an evidence of the shift from printed books to e-books. However, Google's digital database is not corrected for the errors introduced into the scanned texts, including missing or badly scanned pages.

The Digital Millennium Copyright Act of 1998 integrated the stipulations of these treaties into U.S. law. However, the excessive zeal in the application of this law has led to grotesque situations, such as the case of Professor Edward Felten of Princeton University. A consortium of musical publishers started the Secure Digital Music Initiative (SDMI) to prevent piracy by developing a file format that includes coding and watermark techniques to protect the content by controlling the playback, storage, and broadcast of digital music. To prevent access without prior authorization, the rights to be exercised are recorded in an indelible manner in the music file. This allows playback control through *a posteriori* verification that the owner is legitimate; were it not the case, playback will be prevented. In September 2001, SDMI announced a public challenge to encourage attempts to break certain algorithms proposed for content protection. A team of researchers from Princeton and Rice Universities led by Professor Felten concentrated on the task and was able to reveal the mechanisms of certain algorithms. In a panic mode, SDMI tried to block publication of the results on the basis of the previously mentioned law and did not change its mind until the intervention of the Electronic Frontier Foundation (EFF) (Craver et al., 2001).

E-commerce had an indirect effect on journalism and college-level humanities. The massive copyright violation on the web has led to the "commodification of writing itself" (MacArthur, 2013) because publishers gave it away free to search engines in the quest for more advertising. This result has been the decline of editorial standards and the collapse of major publications.

From a technical viewpoint, the methods considered to protect the digital content are based on encryption or on watermark algorithms. Broadcast encryption allows the secure distribution of content, but the system design depends on several factors (Elkamchouchi and Abouelseoud, 2007):

1. The application (pay TV or online subscription)
2. The processing power and the memory available at the receiver

3. The key assignment and revocation algorithms, in particular, whether the keys used are stored in software or in tamper-resistant hardware (i.e., smart cards)
4. The ability to trace and identify users that constructed pirate decoders

Online copyright protection is another aspect of the clash with privacy, particularly if the names and personal information of targeted people are subpoenaed from Internet service providers.

16.4 Electronic Surveillance and Privacy

In the past, modern societies had prided on shielding personal privacy from unwarranted intrusions. Today, however, there is not enough political and social will to protect it in the digital age. Even though identity theft emerged in the general discourse in the 1990s (*Harper's Magazine*, 2000), privacy protection has been conspicuously absent from the legislative agenda and the commercial sector was left to self-regulation.

The integration of data from banking, mobile phone usage, payments for tolls or parking, access to buildings or offices, and usage of digital content (books, music, films, etc.) generates a detailed dossier of an individual web life and interests. The collected information is used for targeted advertisement based on the information users reveal about themselves online. The data gathered can be sold to data brokers who, in turn, can sell them to interested parties, including health insurers.

Public and private sector organizations collect all these data without public accountability nor effective controls to ensure that the available information is correct or rectified (Rossigneux and Simonnot, 2006). The only criterion so far is how much profit they can return to their shareholders. Yet this information is shared and synthesized and then used for algorithmic decisions on morality, credit worthiness, or even targeted assassination. Yet the response from a legislative view has been very timid, and there is a prevailing assumption that the solution can only come from technology not law.

We now review some of the sources for the erosion of personal privacy in particular:

1. The willingness of people to share their information online
2. Data breaches
3. Pressures to monetize data collected online
4. Government mass spying program
5. The exclusive focus on technological solution

In the absence of legislative and legal means, surveillance will continue to be routinely applied by governments, businesses, hackers, and identity thieves.

16.4.1 Disclosure of Personal Information Online

Today, people are more willing than ever to reveal a surprising amount of information about themselves online, particularly when there are no explicit fees or they perceive benefits they value even though "if the service is free, then you're the product." People voluntarily supply information to personalize the interfaces to online services (the "Cloud") or to receive location-related content. Thanks to social networks, blog sites, and photograph and video-sharing communities, many broadcast personal information on themselves, much of it intimate. Online digital profiles are accessible to a wide audience. "Sexting" is another phenomenon by which teens share nude photos of themselves and others. Some sites allow teenagers access to nude photos of others provided they post one of themselves. While surveys point that a majority opinions express concern that existing laws and organizational practices do not provide adequate protection to personal information, consumers do not necessarily accordingly (Spiekermann and Cranor, 2009).

Information gathered from social media, digital data brokers, and online trails offer clues about a person's identity or socioeconomic status (single versus in a relationship, race, sexual orientation, religion, etc.) and can be used to determine the credit worthiness of individuals (Kosinski et al., 2013; Alloway, 2015; Chalandon, 2015). Yet this is not the electronic equivalent of a traditional village, where everyone knew everyone else, because privacy has not been breached symmetrically. The information available is exploited by those who have the resources to collect the data and process them. U.S. federal laws do not protect U.S. citizens from being spied upon in the workplace (Herbert, 1998), and the U.S. courts have even upheld the right of employers to monitor all exchanges on their networks (Waldmeir, 2001). As a consequence, employers routinely monitor the online activities of their employees and read their postings or those of potential recruits in social media. There are mobile applications to spy a person's social media activity, address book contacts, Skype calls, and text messages remotely. Parents also use software to monitor their children remotely including sending parents any photo their children take. Furthermore, the information gathering and storage is international in scope and unaccountable to legislations or customary laws.

Advances in automotive technology mean that cars are collecting as much data as smartphones. The stake of data is set to expand with "connected vehicles" equipped with communication devices and self-driving functions. The ownership of these data and access to them are left unanswered. In 2014, the U.S. Federal Trade Commission (FTC) issued a report asking Congress to consider enacting legislation on these issues.

16.4.2 Data Breaches

Despite the explosion in the amount of personal information stored in digital form and the changes to online business transactions, efforts to protect the data have not been sufficient. Data security breaches are becoming common, whether due to accidental disclosure, misuse, or targeted attacks. Incident ranges from stealing personal medical data from databases or captured with health insurance smart card such as Sésame-Vitale to pilfering data from U.S. federal tax returns stored by the Internal Revenue Service (IRS). The knowledge of health records, in addition to payment information, Social Security Numbers, and medical histories, can make false identities more believable.

The Internet of Things (IoT) is built around billions of sensors in equipment to track and monitor daily activities around the clock, and it is not clear that the security implications of this project have been completely understood. IoT will compound the problem on the protection of private digital information because people will be exposed to hacking constantly in their daily routines. It has been demonstrated that cars' electronic entertainment systems can be hacked to disable its dashboard functions, steering, brakes transmission, and other systems critical to the vehicle safety from a remote laptop. It is even possible to remotely control a car's brakes, accelerator, locks, and so forth (Halpern, 2014).

In 2007, the Revenues and Customs Agency of the United Kingdom revealed that it had lost the personal data of 25 million people while the Transport ministry announced the loss of the records of more than 3 million learner drivers (Peel and Allison, 2007). In September 2013, an investigation revealed that the records of about 4 million U.S. citizens were hacked from LexisNexis, Dun & Bradstreet, and Kroll Background America. LexisNexis maintains the world's largest electronic database for legal and public records. Dun & Bradstreet aggregates information on businesses and corporations for use in credit decisions, business-to-business marketing, and supply chain management. Kroll Background provides employment background, drug and health screening. An identity theft service (ssndob.ms) sold Social Security Numbers, birth records, credit, and background checks for prices ranging from 50 cents to $2.50 per record, and from $5 to $15 for credit and background checks, paid with largely unregulated and anonymous virtual currencies, such as Bitcoin and WebMoney (Krebs, 2013a).

Table 16.1 lists selected data breaches between 2005 and 2015 in the United States (CNN, 2005; Waters, 2011; Jopson and Taylor, 2013; Kuchler, 2013a; Identity Theft Resource Center, 2015; Kitten, 2015; Kuchler, 2015; Lederman and Gillum, 2015; Popa and Zeldovich, 2015; Star-Ledger, 2015).

It is clear from this partial list that personal privacy is under constant attack and that this requires a reevaluation of the access control protocol, particularly on the Internet. In fact, it is refreshing to hear from Melissa Hathaway, who was then the acting senior director for cyberspace on the U.S. National Security Council, that the Internet was not designed with security in mind (Menn, 2009). However, the ease and frequency at which data breaks occur suggest that there is less interest in preventing fraud than managing the status quo and its consequences.

16.4.3 Monetizing Personal Data

Advertising is the principal commercial industry on the web, and the desire to leverage customer data for targeted advertising has breached all ramparts of privacy and all codes of good behavior. The movement called "Big data" aims at using all this information, combined with data gleaned from public records such as home valuation, vehicle ownership, customer surveys, and so on, to find patterns that have predictive values. Accordingly, the focus of controlling payment systems has shifted to harvesting user's data. Every action online is avidly recorded, catalogued, repackaged, used, and also sold and exchanged in specialized markets, both legal and illegal (*Economist*, 2014a).

Web companies track the preferences of users and build their profiles using the large amount of information that they collect. For example, Alexa Internet, an Amazon subsidiary, provides commercial web traffic data using its toolbar to collect data on browsing behavior that are then transmitted to its website, where they are stored and analyzed. Internet service providers have access to users web browsing histories, including sites visited and terms entered into search engines. Search engines are programmed to record the queries from users as well as the connection parameters (date, time, IP address, etc.) and then to build dossiers on users' interests and health status. When Google acquired DoubleClick in 2008, the largest search engineer and the largest supplier of advertisement to web users became intertwined. Card companies use customer data for targeted advertising, selling to retailers the ability to send messages to consumers based on their past transactions.

Smartphones add to the capabilities that the web offers the potential of tracking users as they move around the world with an accuracy of just 10 m. Some

TABLE 16.1

Selected Thefts of Personal Records in the United States and the United Kingdom

Year	Company Attacked	Millions of Records	Compromised Data
2005	LexisNexis	0.032	Social Security Number, past addresses, date of birth, voter registration information
2007	T.J.Maxx	94	Credit card records
2009	Heartland	130	Payment processing records: card numbers and cardholders' names
2013	Adobe	38	Encrypted passwords, source code, and credit card details of 2.9 m customers
	Evernote	50	Names, e-mail addresses, and encrypted passwords
	Target	40	Customer names, credit or debit card numbers, expiry dates, CVV security codes
2014	Home Depot	56	Credit and debit card information
	JPMorgan Chase	76	E-mail, online personal and credit card information
	Sony Pictures	10	E-mail, online and personal information of customer, Sony employee Social Security Numbers, and contracts with celebrities and some forthcoming movies
	Sony PlayStation Network	77	Personal information of gamers
	Target	70	E-mail, online and personal information
2015	Anthem	80	Customers' names, dates of birth, Social Security Numbers, member ID numbers, addresses, phone numbers, e-mail addresses, and employment information. Some of the customer data may also include details on their income
	Ashley Madison	37	E-mail, online, personal credentials, and credit card information
	Carphone Warehouse (UK)	2.4	Customer names, addresses, dates of birth, and bank account details
	eBay	233	E-mail addresses, birthdates, and other identity information
	IRS	0.1	Personal tax information
	Medical Informatics Engineering	3.9	Patient information: name, home address, e-mail address, date of birth, and for some patients a Social Security Number
	Office of Personnel Management	25.7	Personal information of U.S. government employees and security clearances (initial estimate was 4.5 M)
	University of California, Los Angeles Health Center	4.5	Names, addresses, Social Security Numbers, and medical data—such as condition, medications, procedures, and test results

companies develop smartphone apps and distribute them free just to gather the user's personal data and sell them. Social media sites evaluate the effectiveness of advertisement by tracking whether people buy goods after viewing them. Platforms for the "sharing economy," such as Uber and Airbnb, monitor participants' rate on the basis of accumulated behavioral data. Uber's employees, in particular, can look up the ride history of individual users or even watch them during trips in a live "God view" mode. Personal data contained in X.509 certificates can be decrypted by the encryption algorithm (Renfro, 2002). "Wearable devices" such as smart watches or wristband collect intimate data from around the body related to real-time physical activities, heart rates, and sleep patterns and then relate them to health records, individual moods, and online purchase history. Intimate or personal space is becoming an advertising platform such as the connectivity of everyday devices such as refrigerators, spectacles, umbrellas, or cars, generating a continuous stream of data for those who are willing to pay for it. Fitness apps sell information and data brokers focus on health conditions such as pregnancy, diabetes, and high cholesterol. Some services monitor students; follow them on campus; monitor their online reading habits and how long they spend working, socializing, exercising, or sleeping; and report their studiousness to university administrators (Warrell, 2015).

The current flood started with a trickle: pieces of information that browsers would reveal to visited websites. Next, servers started to plant "cookies" in the computers to record and remember information about the users to personalize the experience based on the user's references and past choices and to avoid repeated requests for identifiers (login and password) at each visit. Around 2005, browser developers added a "private browsing" mode to give users the option of preventing sites from adding long-term cookies and to block advertisements. Also, many users stated to delete their browser cookies on a regular basis. Flash cookies specific to Adobe's Flash plug-in were used as a sneaky way to regenerate deleted cookies until browsers included the capabilities to delete all flavors of cookies.

More recent techniques are under the general category of Browser fingerprinting: collecting identifying information about the individual computer used for accessing a website, including screen size, time zone, browser plug-ins, installed fonts, and so on. Some fingerprinting plug-ins are installed surreptitiously after a user downloads and installs a program, such as online gambling applications. Other code exploits Adobe Flash to determine whether people are communicating through proxies. Coinbase, a Bitcoin exchange site, offers a fingerprint product, in a sense reducing the "pseudonymity" aspects of the cryptocurrency. Ad brokers use this information to keep track of users' online activities anytime that user visits a site that displays one of its advertisements (Nikiforakis and Acar, 2014).

What is interesting today is that there is a merger of spying by large enterprises and by governments. In the current deregulated environment, there is more concern with government surveillance than with marketers or data brokers collecting personal information. This ignores two facts. Authorities can make request access to personal data from web and mobile telephone companies. Also, governments can purchase data from these companies. In 2011, TomTom, the Dutch satellite navigation company, was forced to issue an apology to its customers the press revealed that it had sold driving records it collected to help police set speed traps to motorists (Palmer, 2011). Also, the U.S. State Department purchased records on millions of Latin American citizens, from ChoicePoint, an aggregator of personal and business data (Marwick, 2014). It is also known that a large percentage of the intelligence budget in the United States goes to private industry (Ramonet, 2014).

Before focusing on government spying, it should be noted that these private companies may also sell the records they have to crooked operators. In 2005, ChoicePoint was investigated for selling 145,000 personal records to an identity theft ring (Gross, 2005). In 2013, a Vietnamese identity theft ring successfully tricked Experian, one of the three major credit bureaus, into selling them the personal records of hundreds of thousands, if not millions, U.S. citizens. The attackers then resold these data to underground sites, such as Superget.info* (Krebs, 2013b; Schwartz, 2013).

16.4.4 Government Spying

The 1970s saw intensified legislative activities concerning the protection of privacy with respect to the collection and use of personal data. For example, following the revelations on the spying of dissidents and political opposition, Federal Privacy Act of 1974 was voted to prevent government agencies from collecting personal information on individuals without their knowledge. The Electronic Communications Privacy Act and the Stored Wire Electronic Communications Act, commonly referred together as the Electronic Communications Privacy Act of 1986, updated the Federal Wiretap Act of 1968, concerning the interception of conversations using telephone lines, but did not apply to interception of computer and other digital and electronic communications. The Consumer Internet Privacy Protection Act of 1997 prohibited Internet service providers from disclosing to a third party any personally identifiable information that a subscriber had provided without that subscriber's informed written consent. It also gave individuals the ability to revoke such a consent at any time and requires the service to cease disclosing such information. Finally, the Children Online Privacy Protection Act (COPPA) of 1998 gave parents control over what information websites can collect from their children.

In the European Union, a 1995 data protection directive put strict limits on how Internet groups can use the information they collect about their users. Furthermore, in November 2001, the European Parliament voted to forbid the usage of a computer network to collect information on a client station. This prohibition covers spying software ("spyware") whose purpose is to monitor user activity secretly as well as identification or tracking mechanisms hidden in the client station. The protection of data collected by mobile telephony operators came into force in all the member states on January 16, 2002. The only data that can be gathered are those necessary to localize the user. Later use of the data is forbidden without the explicit approval of subscribers. In addition, subscribers must be able to delete these data or request that they can be processed anonymously. Finally, the operators must install the mechanisms to ensure data confidentiality.

Many of the safeguards on government spying were discarded starting 2001 by invoking moral justification (e.g., battle against pedophilia or the war on drugs) or political considerations (fight against hooligans or organized crime, etc.). The Patriot Act gave the U.S. federal authorities the right to monitor electronic transactions and to mobilize numerous service providers: banks, travel agencies, car rental agencies, casinos, and so forth. Libraries and bookstores are ordered to indicate to the authorities who has borrowed or bought what with the obligation of hiding the inquiry from the targeted person. The U.S. authorities pressured SWIFT to release details of financial international transactions to U.S. terrorism investigators. Equivalent legislations were written in many countries as well. In general, cooperation between businesses and U.S. federal law agencies is not advertised and customers are seldom aware of it (Block, 2005). NSA has developed a tool called XKeyscore to search for and analyze information on the global Internet. The program was shared with intelligence agencies of the Anglosphere (the United Kingdom, Canada, Australia, and New Zealand), as well as the German *Bundesnachrichtendienst*.

* Superget.info was a site that sold the personal information of U.S. citizens, that is, Social Security Numbers, birthdays, driver's license numbers and banking and financial records. Registration was free, and the service was funded via WebMoney and other virtual currencies.

In 1997, the Federal Bureau of Investigation (FBI) established the National Cyber-Forensics and Training Alliance (NCFTA) in Pittsburgh as a nonprofit corporation to identify, mitigate, and neutralize cybercrime threats. Through this organization, field agents cooperate with subject matter experts from banks and other companies as well academic sectors, such as Carnegie Mellon University. The organization was involved in many high-profile operations such as the take down of the DarkMarket site in 2008, the hacker forum Darkode, and the GameOver Zeus peer-to-peer botnet, both in 2015. DarkMarket was an online forum for carding, that is, buying and selling stolen identities and credit card information online. GameOver Zeus was stealing money from businesses across the United States.

In March 2002, the European Commission agreed, without Parliament authorization, to share with the United States the data that European airlines would have on their passengers flying to the United States including bank card numbers and dietary restrictions. In the United Kingdom, for example, video surveillance of public areas is widespread. Navigo, the electronic purse used in the public transportation systems of the Parisian Region and to pay for parking around Paris, gives transport authorities an exact knowledge on the trajectory of an individual. This is also the case of the E-Z Pass system that was adopted by 25 toll agencies across the northeastern states of the United States.

16.4.5 Technologies for Privacy Protection

In 1980, the Organisation for Economic Co-operation and Development (OECD) published guidelines on the protection of privacy, which were updated in 2013 (OECD, 2013). These principles emphasize the need to minimize the collection and use of personal data, to inform individuals about data collection, to provide adequate maintenance and protection for the data collected, and to define the conditions under which the data can be released to third parties, for example, for data storage. In the absence of enforced legal restrictions on the use of personal data, the approach relies on business voluntary compliance and self-regulation.

System activities that affect user privacy can be divided into the following areas: data transfer, data processing and data storage (Spiekermann and Cranor, 2009). Edward Snowden had documented extensively how the massive surveillance program by the U.S. National Security Agency (NSA) affected all these areas. The main reaction has been technological and not legislative.

One of the first solutions was the Platform for Privacy Preference (P3P) specification from the W3C to give users some control over the collection and the commercial use of the personal data that commercial websites gather. This is done by an automatic analysis of the policies of the websites about the use and the protection of these data (Cranor, 2002). Accordingly, each site describes the potential uses of the information they collect from users and the measures taken for its protection, while users indicate their preferences. The site policy is coded with XML. The automatic comparison of the site policies with the user's preferences reveals any potential discrepancy so that users make their decision knowingly; in other cases, the browser's response is already programmed.

P3P prescribes a client–server architecture and an exchange protocol based on HTTP. It also defines a vocabulary and special data elements to describe a website's policy about personal information. However, P3P was not widely accepted.

In 2014, Apple and Google announced that they will offer encryption of the data on their smartphones. Google announced that it will develop a tool on the basis of the PGP protocol to encrypt all the messages using its Gmail service (but did not give any indication on the protection afford to the messages stored on its servers). The mobile instant-message application WhatsApp, acquired by Facebook, adopted TextSecure from Open Whisper Systems, also recommended by Edward Snowden himself, to encrypt its services on Android-base devices. However, data stored in the cloud servers of Dropbox, Google Drive, and others are encrypted with keys that the providers control, not their users. In principle, these keys can be subpoenaed by the jurisdictions where the servers are located.

Facebook has launched as an experiment to improve access its services over the Tor Network, the section of the web, also called the dark web, that requires special tools such as a Tor browser to ensure anonymity.

In November 2014, the Electronic Frontier Foundation (EFF) led the formation of a new consortium of technology companies to automate the installation and management of digital certificates on web servers. The initiative, called Let's Encrypt, will automate the renewal of the digital certificates upon expiration without any administrative intervention. Other aspects of the initiative are to make the encryption by default of browser and e-mail traffic. Apple and Microsoft have been offering full-disk encryption of their desktop operating systems and have recently improved the encryption speed with the use of faster processors and custom-made chips.

Many other applications claim to offer privacy protection, but cryptography experts have expressed doubts concerning their claims of protection, particularly when the code is kept proprietary and cannot be independently evaluated (Untersinger, 2014).

On the legislative front, the European Commission is planning to introduce data protection regulations by 2016 to update the existing legal framework, which dates back to 1995. The proposal is to consider as "personal information" the mobile device identification numbers, the IP address, and most other data used for targeted advertising. This means that explicit consent is needed before companies can collect these data. It is assumed that this protection will be sufficient, even though it had not been very effective on the web.

Also, following a 4-year legal battle by a Spanish citizen, Mario Costeja González, the European Court of Justice in May 2014 instituted a "right to be forgotten," to force Google to remove links to potentially embarrassing Internet content. The irony here is that the American passion for the frontier and the reinvention of the self in a new land, which many of the Internet pioneers thought they had found in cyberspace, was being preserved by a ruling from the Old World.

However, no laws nor tools will be able to protect people who are ready to disclose their data not guarding it. In 1787, the philosopher Jeremy Bentham proposed a Panopticon, a circular building with cells along the outer walls and at the central an inspector's lodge from which people in cells could be observed at any time, without realizing that they were observed. Bentham thought that this constant surveillance in prisons, hospitals, and mental asylums would improve morals and diffuse instruction. Electronic commerce, social media, data breaches, and government spying are inexorably leading to the realization of Bentham's vision on a global scale. Perhaps we will know in a few decades whether this is a perfection or perversion of openness and whether uncontrolled this loss of privacy to total strangers (and not only within the "in-group") has been beneficial.

16.5 Content Filtering and Censorship

Content or traffic filtering raises an interesting problem because it questions a founding myth of the Internet concerning the free circulation of information. It imposes significant technical hardship and legal responsibilities on computer professionals (network administrators and access providers).

In the case of business-to-business commerce, professional services and value-added services providers have several roles: consulting, training, installation, or maintenance. These intermediaries can also offer security services: certification of parties, notarization of exchanges, time stamping of transactions, archival or information to satisfy legal requirements.

In the case of consumer commerce, many countries impose filtering of exchanges to prevent the broadcast of contents deemed illegal. Following a lawsuit started in 1999 by the Jewish Student Union of France (UEJF, *Union des Étudiants Juifs de France*) and the International League against Racism and anti-Semitism (LICRA, *Ligue Internationale Contre le Racisme et l'Antisémitisme*), a French court ordered Yahoo! in May 2000 to block access to its virtual auction site for Nazi memorabilia from France and its territories, even though the incriminated site was destined to U.S. audiences only. In 2006, the federal appeals court in San Francisco decided not to attempt to override this judgment. These regional variations of what is prohibited and what is not are clearly a hardship for managing online e-commerce on a worldwide basis.

Another subject of concern is the spam that comes for publicity campaigns destined for the largest number of Internet users without their prior consent and without considering any embarrassment that the content may cause them. For telemarketers, spam is attractive because messages are easy to reproduce and quick to spread, thereby reducing the cost of the campaign. In contrast, the resulting traffic may congest the network and overload the mail box of the recipients. This electronic pollution ends up by costing the enterprises in terms of bandwidth, storage, time to read and to delete.

"Spammers" construct their lists by systematic collections of e-mail addresses from mailing lists, web pages, newsgroups, files purchased from other online merchants, and so on.

16.6 Taxation of Electronic Commerce

The fiscal consequences of e-commerce have dwarfed sales by mail to order because of the ubiquitous and permanent nature of the cyberspace. This is why the growth of e-commerce exchanges poses the issue of taxing the associated transactions. On the one side, tax exemption of e-commerce favors online enterprises at the expense of traditional stores and reduces the income of all the governments involved. On the other side, taxation may put a break on the expansion of electronic markets. This is the context in which are expressed the divergent viewpoints of the European Union and the United States on that issue.

A directive from the European Union requires a value-added tax (VAT) on digital products delivered online (sales of software, remote maintenance, etc.) even when the merchants are outside Europe. Web hosting and other online services—including selling advertising space on a website—also require compliance with the new guidelines. U.S. vendors of online articles or digital services are obliged to establish a physical presence in one of the countries of the European Union and apply the VAT according to the rules of the countries where the buyer resides. This directive was initiated in the early 2000s and went into effect on January 1, 2015, after long negotiations with the United States.

In the United States, the 1998 Internet Tax Freedom Act, which was extended four times the latest in 2014, prevents federal, state, and local governments from taxing Internet access and from imposing discriminatory Internet-only taxes such as bit taxes, bandwidth taxes, and e-mail taxes. Nevertheless, the large budget deficits of the early 2000s convinced California and other states that taxing Internet sales might not be a bad idea after all. In May 2013, the Senate passed the Marketplace Fairness Act to allow states to collect sales taxes for purchases made online; however, it has not progressed in the House, as of August 2015.

Taxation is related to the general issue of self-regulation of the Internet, that is, no state intervention. This has allowed many practices such as manipulating the ranking by search engines, which in any other industry would have caused prosecution.

16.7 Trust Promotion

Fraud in electronic commerce has multiple sources: fraudulent description of articles sold online, rigged auctions, nondelivery of purchased items, false payments, propagation of malicious rumors, and so forth. In addition to the technical means that were presented in this book, the repression of fraud usually takes two approaches, the legislative approach and self-regulation. Thus, legislation defines the overall responsibility of the participants so that the party with grievances can ask the authorities to intervene to redress torts. Industry self-regulation relies on private institutions to accord "approval seals" to sites that promise to follow a specific line of conduct with respect to fraud prevention and the protection of data collected from the clients. In Table 16.2, a partial list of such labeling organizations for online commerce is presented.

Authorities have closed several exchanges for money laundering. Liberty Reserve was based in Costa Rica and had more than a million users worldwide, including 200,000 in the United States. Between 2006 when it was established and 2013 when the U.S. authorities shut it down, it handled 55 million transactions, many of which were illegal.

16.8 Archives Dematerialization

In a business environment, the management of electronic documents includes data archival in addition to day-to-day management. This is a critical issue whenever the archives must survive and remain accessible over long periods (30–50) years. In case of a tax audit, for example, companies are obligated to present all records that contributed, whether directly or indirectly, to the consolidated statement, including accounting, payroll, inventory, billing, and so on. Should the file be a proprietary format, the application programs may have to be available in the version of the time. What is needed, therefore, is a guarantee for the readability of the archives by guaranteeing their integrity, their longevity, and the possibility of restoring them at will. Some support media, however, are precarious. Magnetic supports are not recommended because they can be reused, and hence, the data that they contain can be falsified. In contrast, traditional storage media, such as paper or punched cards, are only used once. Throwaway integrated circuit cards offer the required

TABLE 16.2

Partial List of Labeling Organizations for Online Commerce

Label	Description	URL
AECE	Site of the Spanish Association of Electronic Commerce (*Asociación Española de Comercio Electrónico*). It awards the Confianza Online Trust Mark to websites that comply with a specific ethics code. It also provides a conflict resolution procedures to those accredited sites.	http://www.aece.org
TRUSTe	Founded in 1997 as nonprofit industry association by the Electronic Frontier Foundation and CommerceNet, in 2008, it changed its structure to a for-profit organization with a focus on privacy protection and confidentiality. It also verified conformance to COPPA for the protection of minors under 13 years of age.	http://www.truste.com
WebTrust	Site of the organization formed through the collaboration of the American Institute of Certified Public Accountants and the Canadian Institute of Chartered Accountants.	http://www.webtrust.org

guarantees because there is a progressive destruction of the memory cells as data are recorded. In contrast, the rechargeable microprocessor card calls for a public key infrastructure to ensure nonrepudiation.

Nevertheless, the readability of the archives does not depend solely on the physical condition of the support media but also on the availability of peripherals and software for reading. As a consequence, it is important to ensure the presence of adequate reading terminals and, if the format of the files is proprietary, to preserve a copy of the application software that was used to produce them.

On the consumer side, digital assets include online accounts, collections on iTunes, balances in PayPal or eBay, airline points and increasingly, cryptocurrencies. The vast majority of tech companies do not have an explicit policy concerning the ownership of data after death in their terms of service. Furthermore, it is not clear what happens when a company folds.

The necessity to deal with digital afterlife has spawned some businesses to manage the details of an individual digital life for possible succession or as virtual executors to carry out the instructions from departed individuals. In 2013, Google has unveiled an Inactive Account Manager to allow individuals to inform Google what to do with their digital assets when they die or no longer are able to access their accounts because of a mental or a physical incapacity. People can select a deactivation option after 6, 9, or 12 months of inactivity, or they can appoint a digital executor and determine what they would like to do with their e-mail accounts and the photos and videos available.

Cirrus Legacy (http://www.cirruslegacy.com) offers a service to keep track of digital assets such as e-mail accounts, online banking, PayPal, eBay, and so on, by storing passwords and their corresponding access control lists. SecureSafe (http://www.securesafe.com), a file storage and sharing site, employs a wide range of security measures including encryption methods, strong user authentication, and redundant data storage.

The legalization of electronic currencies raises the problem of escheat, that is, the reversion of personal tangible abandoned property to the state after a period of inactivity. The situation becomes more complicated in the case of anonymous cards, electronic or virtual purses, or cryptocurrencies. Different cases have to be considered: residual values, full values in cards purchased as collectibles, anonymous cards, lost cards with access control, and so on (Lorenz, 1997).

In the United States, many have had to go to court to access the digital information of a deceased relative. The legal rights are divided between states' property law, federal privacy laws, and corporate policies of the companies housing the data, as defined in their terms-of-use agreements. The Uniform Law Commission* has proposed that a personal representative of the deceased be allowed access (but no control, i.e., cannot delete files or send e-mails) to a person's legacy digital files so long as the deceased did not prohibit this access in a will.

Some five U.S. states have laws to address the treatment of digital inheritance. Rhode Island and Connecticut were the first to consider the problem, but their statutes are restricted to e-mail accounts. A 2007 statute from Indiana includes electronic document, while a 2010 statue from Oklahoma covers the broader notion of digital assets. In 2011, Idaho passed a bill similar to that of Oklahoma.

16.9 Summary

E-commerce forces a complete revision of the whole value chain: relations between production and demand, the power relations within the enterprise, the collaboration among partners, modalities for the dissemination of the information, and so forth. These changes go beyond the technical domain and modify financial or legal aspects of the conduct of business, thereby questioning the nature of society. Human behavior can now be tracked, processed, and sold on a massive scale.

Today, personal digital data constitute a valuable resource to be exploited without scruples. Big data and analytics allow intrusive spying by governments and commercial companies. Identities are being stolen and sold using unregulated monies. As the use of behavioral data collected from mobile devices and the Internet becomes the norm, merchandizing and the surveillance state will prosper hand in hand. Unregulated electronic commerce has led us to an age of surveillance.

* The Uniform Law Commission (ULC), also called the National Conference of Commissioners on Uniform State Laws, is a nonprofit, unincorporated association with headquarters in Chicago, Illinois. It was established in 1892 to provide states in the Union with legislation that promotes uniformity in state laws. It consists of commissioners appointed by each state, the District of Columbia, the Commonwealth of Puerto Rico, and the U.S. Virgin Islands. All of its members are lawyers, who may also serve as legislators, judges, or legal scholars.

Questions

1. What is the role of standards in e-commerce? Discuss their advantages and disadvantages.
2. What is escheat? How is it affected by the various types of e-commerce?
3. Can copyright protection clash with the free flow of information for e-commerce? Explain with an example.
4. What are the similarities between the United States and the European position regarding the protection afforded to data collected from users in e-commerce?
5. What are the difficulties regarding taxation of e-commerce?

References

3-D Secure, 3-D Secure™ - System Overview V. 1.0.3, December 3, 2001. Available at http://www.visanet.com.pe/verified/demovisanet-web/resources/3DS_70015-01_System_Overview_external_v1.0.3.pdf, last accessed January 21, 2016.

Abdellaoui, R. and Pasquet, M., Secure communication for Internet payment in heterogeneous networks, in *Proceedings of the 24th IEEE International Conference on Advanced Information Networking and Applications*, Perth, Western Australia, Australia, April 20–23, 2010, pp. 1085–1092.

Abernathy, W. J. and Clark, V. B., Marking the winds of creative destruction, *Res. Policy*, 14(1), 2–22, 1985.

Adams, J., Friends or foes? *Eur. Card Rev.*, 5(2), 15, 1998.

AFNOR NF Z63-001, Caractères magnétiques imprimés CMC 7—Spécifications—Jeux de chiffres, de symboles et de lettres, France, 1964, available at http://www.boutique.afnor.org/norme/nf-z63-001/information-processing-cmc-7-printed-magnetic-characters-specifications-digit-symbol-and-letter-sets/article/672984/fa009725, last accessed January 1, 2016.

Aglionby, J., From libertarian economics to alleged assassination request, *Financial Times*, October 4, 2013, p. 22.

Ahmed, A. A. and Traore, I., Biometric recognition based on free-text keystroke dynamics, *IEEE Trans. Cybern.*, 44(4), 458–472, 2014.

Al-Masri, E. and Mahmoud, Q. H., Interoperability among service registry standards, *IEEE Internet Comput.*, 11(3), 74–77, 2007.

AlFardan, N. J., Bernstein, D. J., Paterson, K. G., Poettering, B., and Schuldt, J. C. N., On the security of RC4 in TLS and WPA, July 8, 2013c, available at www.isg.rhul.ac.uk/tls/RC4biases.pdf, last accessed August 23, 2014.

AlFardan, N. J. and Paterson, K. G., Lucky thirteen: Breaking the TLS and DTLS record protocols, February 27, 2013a, available at http://www.isg.rhul.ac.uk/tls/TLStiming.pdf, last accessed August 30, 2014.

AlFardan, N. J. and Paterson, K. G., Lucky thirteen: Breaking the TLS and DTLS record protocols, in *Proceedings of the 2013 IEEE Symposium on Security and Privacy*, San Francisco, CA, May 19–22, 2013b, pp. 526–540.

Ali, R., Barrdear, J., Clews, R., and Southgate, J., Innovations in payment technologies and the emergence of digital currencies, *Bank Engl. Q. Bull.*, 2014a, available at http://www.bankofengland.co.uk/publications/Documents/quarterlybulletin/2014/qb14q3digitalcurrenciesbitcoin1.pdf, last accessed March 8, 2015.

Ali, R., Barrdear, J., Clews, R., and Southgate, J., The economics of digital currencies, *Bank Engl. Q. Bull.*, 2014b, available at http://www.bankofengland.co.uk/publications/Documents/quarterlybulletin/2014/qb14q3digitalcurrenciesbitcoin2.pdf, last accessed March 8, 2015.

Alloway, T., Big credit: Credit where credit's due, *Financial Times*, February 5, 2015, p. 7.

Althen, B., Enste, G., and Nebelung, B., Innovative secure payments on the Internet using the German electronic purse, in *Proceedings of the 12th Annual Computer Security Applications Conference*, San Diego, CA, December 9–13, 1996, pp. 88–93.

Anderson, R. and Kuhn, M., Tamper resistance—A cautionary note, in *Proceedings of the Second USENIX Workshop on Electronic Commerce*, Oakland, CA, 1996, pp. 1–11.

Anderson, R. J. and Bezuidenhoudt, S. J., On the reliability of electronic payment systems, *IEEE Trans. Softw. Eng.*, 22(5), 294–301, 1996.

Andrews, E. L., Chris Larsen: Money without borders, Insights by Stanford Business, Operations, Information & Technology, Stanford Graduate School of Business, September 24, 2013, available at https://www.gsb.stanford.edu/insights/chris-larsen-money-without-borders, last accessed April 24, 2015.

Anonymous, How DigiCash blew everything, *Next!*, January 1999, translated from the Dutch, available at http://cryptome.org/jya/digicrash.htm, last accessed April 29, 2015.

ANSI INCITS 358-2002[R2012] Information technology—BioAPI specification (Version 1.1), Information Technology Industry Council, Washington, DC, 2002 and revised in 2012 ANSI, New York, NY, available from webstore.ansi.org.

ANSI INCITS 398-2008[R2013] Information technology—Common Biometric Exchange Formats Framework (CBEFF), Information Technology Industry Council, Washington, DC, 2008 revised in 2013 ANSI, New York, NY, available from webstore.ansi.org.

ANSI X3.92:1981, Data Encryption Algorithm (DEA), American National Standards Institute, Washington, DC, 1981.

ANSI X3.105:1983, Data link encryption, American National Standards Institute, Washington, DC, 1983.

ANSI X3.106:1983, Data encryption algorithm, modes of operations, American National Standards Institute, Washington, DC, 1983.

ANSI X9.9:1986 (revised), Financial institution message authentication (wholesale), American Bankers Association, Washington, DC (replaces X9.9-1982), 1986.

ANSI X9.19:1986, Financial institution retail message authentication, American Bankers Association, Washington, DC, 1986.

ANSI X9.30:1-1997, Public key cryptography for the financial services industry: Part 1: The Digital Signature Algorithm (DSA), (Revision of X9.30:1-1995), American Bankers Association, Washington, DC, 1997.

ANSI X9.37:1994, Specification for electronic check exchange, American National Standards Institute, Washington, DC, 1994.

ANSI X9.46:1997, Financial image interchange: Architecture, overview, and system design specification, American National Standards Institute, Washington, DC, 1997 (Revoked in 2004).

ANSI X9.52:1998, Triple data encryption algorithm modes of operation, American National Standards Institute, Washington, DC, 1998.

ANSI X9.62:2005, American National Standard for Financial Services—Public key cryptography for the financial services industry: The Elliptic Curve Digital Signature Algorithm (ECDSA), American Bankers Association, Washington, DC, 2005.

ANSI X9.68:2-2001, Digital certificates for mobile/wireless and high transaction volume financial systems: Part 2: Domain certificate syntax, American National Standards Institute, Washington, DC, 2001.

ANSI X9.84-2010, Biometric information management and security, American National Standards Institute, Washington, DC, 2010.

ANSI X9.100-180-2006 (R2013), Specifications for electronic exchange of check and image data (non-domestic), American National Standards Institute, Washington, DC, 2006.

ANSI X9.100-181-2007, Specifications for TIFF image format for image exchange, American National Standards Institute, Washington, DC, 2007.

ANSI X9.100-187-2013, Electronic exchange of check and image data, American National Standards Institute, Washington, DC, 2013.

Antonopoulos, A., The crypto-currency ecosystem: A taxonomy of alt-coins, meta-coins and blockchain-riders, June 27, 2013, available at http://radar.oreilly.com/2013/06/the-crypto-currency-ecosystem.html, last accessed March 5, 2015.

Antonopoulos, A., *Mastering Bitcoin: Unlocking Digital Cryptocurrencies*, O. Reilly, Sebastopol, CA, 2015.

Arnold, M. and Ahmed, M., Upstarts target bank's lunch, *Financial Times, FT Special Report, The Connected Business*, September 24, 2004, pp. 1, 3.

Back, A., A partial hash collision based postage scheme, March 28, 1997, available at http://www.hashcash.org/papers/announce.txt, last accessed August 17, 2015.

Baldwin, R. W. and Chang, C. V., Locking the e-safe, *IEEE Spectr.*, 34(2), 40–46, 1997.

Bank for International Settlements, Committee on Payment and Settlement Systems, Statistics on payment systems in the Group of Ten countries, Figures for 1995, BIS, Basel, Switzerland, December 1996, available at http://www.bis.org, last accessed December 31, 2015.

Bank for International Settlements, Committee on Payment and Settlement Systems, Statistics on payment systems in the Group of Ten countries, Figures for 1996, BIS, Basel, Switzerland, December 1997, available at http://www.bis.org, last accessed December 31, 2015.

Bank for International Settlements, Committee on Payment and Settlement Systems, Statistics on payment systems in the Group of Ten countries, Figures for 1998, BIS, Basel, Switzerland, February 2000, available at http://www.bis.org, last accessed December 31, 2015.

Bank for International Settlements, Committee on Payment and Settlement Systems, Statistics on payment systems in the Group of Ten countries, Figures for 1999, BIS, Basel, Switzerland, March 2001, available at http://www.bis.org, last accessed December 31, 2015.

Bank for International Settlements, Committee on Payment and Settlement Systems, Statistics on payment and settlement systems in selected countries, Figures for 2000, BIS, Basel, Switzerland, July 2002, available at http://www.bis.org, last accessed December 31, 2015.

Bank for International Settlements, Committee on Payment and Settlement Systems, Statistics on payment and settlement systems in selected countries, Figures for 2001, BIS, Basel, Switzerland, April 2003, available at http://www.bis.org, last accessed December 31, 2015.

Bank for International Settlements, Committee on Payment and Settlement Systems, Statistics on payment and settlement systems in selected countries, Figures for 2002, BIS, Basel, Switzerland, March 2004a, available at http://www.bis.org, last accessed December 31, 2015.

Bank for International Settlements, Committee on Payment and Settlement Systems, Survey of developments in electronic money and internet and mobile payments, BIS, Basel, Switzerland, March 2004b, available at http://www.bis.org, last accessed December 31, 2015.

Bank for International Settlements, Committee on Payment and Settlement Systems, Statistics on payment and settlement systems in selected countries, Figures for 2003, BIS, Basel, Switzerland, March 2005, available at http://www.bis.org, last accessed December 31, 2015.

Bank for International Settlements, Committee on Payment and Settlement Systems, Statistics on payment and settlement systems in selected countries, Figures for 2004, BIS, Basel, Switzerland, March 2006, available at http://www.bis.org, last accessed December 31, 2015.

Bank for International Settlements, Committee on Payment and Settlement Systems, Statistics on payment and settlement systems in selected countries, Figures for 2005, BIS, Basel, Switzerland, March 2007, available at http://www.bis.org, last accessed December 31, 2015.

Bank for International Settlements, Committee on Payment and Settlement Systems, Statistics on payment and settlement systems in selected countries, Figures for 2007, BIS, Basel, Switzerland, March 2009a, available at http://www.bis.org, last accessed December 31, 2015.

Bank for International Settlements, Committee on Payment and Settlement Systems, Statistics on payment and settlement systems in selected countries, Figures for 2008, BIS, Basel, Switzerland, December 2009b, available at http://www.bis.org, last accessed December 31, 2015.

Bank for International Settlements, Committee on Payment and Settlement Systems, Statistics on payment and settlement systems in the CPSS countries, Figures for 2009, BIS, Basel, Switzerland, March 2011, available at http://www.bis.org, last accessed December 31, 2015.

Bank for International Settlements, Committee on Payment and Settlement Systems, Statistics on payment, clearing and settlement systems in the CPSS countries, Figures for 2010, BIS, Basel, Switzerland, January 2012a, available at http://www.bis.org, last accessed December 31, 2015.

Bank for International Settlements, Committee on Payment and Settlement Systems, Innovations in retail payments, Report of the working group on innovations in retail payments, BIS, Basel, Switzerland, May 2012b, available at http://www.bis.org, last accessed December 31, 2015.

Bank for International Settlements, Committee on Payment and Settlement Systems, Statistics on payment, clearing and settlement systems in the CPSS countries, Figures for 2011, BIS, Basel, Switzerland, September 2012c, available at http://www.bis.org, last accessed December 31, 2015.

Bank for International Settlements, Committee on Payment and Settlement Systems, Statistics on payment, clearing and settlement systems in the CPSS countries, Figures for 2012, BIS, Basel, Switzerland, December 2013, available at http://www.bis.org, last accessed December 31, 2015.

Bank for International Settlements, Committee on Payment and Settlement Systems, Statistics on payment, clearing and settlement systems in the CPSS countries, Figures for 2013, BIS, Basel, Switzerland, September 2014, available at http://www.bis.org, last accessed December 31, 2015.

Banking Technology, UK business slow on e-commerce, *Bank. Technol.*, 15(4), 14, 1998.

Banking Technology, A step in the unknown, *Bank. Technol.*, 18(6), 44–48, 2001.

Bard, G. V., The vulnerability of SSL to chosen-plaintext attack, Cryptology ePrint Archive: Report 2004/111, May 11, 2004, available at https://eprint.iacr.org/2004/111.pdf, last accessed August 31, 2014.

Barthélemy, P., Rolland, R., and Véron, P., *Cryptographie, Principe et Mises en Œuvre*, Lavoisier, Paris, France, 2005.

Becker, A., Bluetooth Security & Hacks, Seminararbeit, Ruhr-Universität Bochum, August 16, 2007, available at http://gsyc.es/~anto/ubicuos2/bluetooth_security_and_hacks.pdf, last accessed December 17, 2013.

Bellare, M., Canetti, R., and Krawczyk, H., Keying hash functions for message authentication, in *Advance Cryptology—CRYPTO'96*, Lecture Notes in Computer Science, Vol. 1109, Kobliz, N., Ed., Springer-Verlag, Berlin, Germany, 1996, pp. 1–15.

Bellare, M. and Rogaway, P., Optimal asymmetric encryption—How to encrypt with RSA, in *Advance Cryptology—Eurocrypt'94, Workshop on the Theory and Application of Cryptographic Techniques*, Perugia, Italy, 1994, Lecture Notes in Computer Science, Vol. 950, DeSantis, A., Ed., Springer-Verlag, Heidelberg, Germany, 1995, pp. 92–111.

Ben-Sasson, E., Chiesa, A., Garman, C., Green, M., Miers, I., Tromer, E., and Virza, M., Zerocash: Decentralized anonymous payments from Bitcoin, in *Proceedings of the 35th IEEE Symposium on Security and Privacy (SP)*, San Jose, CA, May 18–20, 2014, pp. 459–474, available at http://zerocash-project.org/paper, last accessed March 29, 2015.

Benson, C. and Loftesness, S., *Payment Systems in the U.S.*, Glenbook Press, Menlo Park, CA, 2010.

Béranger, A.-L., Paiement en ligne: Les fausses notes de l'e-Carte Bleue, *Journal du Net*, January 21, 2003, available at http://www.journaldunet.com/0301/030121ecartebleue.shtml, last accessed December 12, 2014.

Berber, M. Le Minitel résiste, *RFI*, October 6, 2003, available at http://www.rfi.fr/actufr/articles/046/article_25145.asp, last accessed August 16, 2015.

Berget, P. and Icard, A., *La Monnaie et ses Mécanismes, Que sais-je?* 12th ed., Vol. 1217, Presses Universitaires de France, Paris, France, 1997.

Birch, D. G. W., Let a thousand currencies bloom, *IEEE Spectr.*, 49(6), 30–34, 2012.

Bird, J., Phones help level playing field for the "unbanked," *Financial Times, FT Special Report, The Connected Business*, October 17, 2012, p. 5.

Black, J. and Urtubia, H., Side-channel attacks on symmetric encryption schemes: The case for authenticated encryption, in *Proceedings of the 11th USENIX Security Symposium*, San Francisco, CA, August 5–9, 2002, pp. 327–328, available at http://www.cs.colorado.edu/~jrblack/papers/padding.pdf, last accessed August 9, 2015.

Blanco-Gonzalo, R., Sanchez-Reillo, R., Miguel-Hurtado, O., and Liu-Jimenez, J., Performance evaluation of handwritten signature recognition in mobile environments, *IET Biometrics*, 3(3), 139–146, 2013.

Bland, B., Mobile banking hits the road in Indonesia, *Financial Times*, August 24, 2014, p. 13.

Bleichenbacher D., Chosen ciphertext attacks against protocols based on RSA encryption standard PKCS #1, in *Advance Cryptology—CRYPTO'98*, Lecture Notes in Computer Science, Vol. 1462, Krawczyk, H., Ed., Springer-Verlag, Berlin, Germany, 1998, pp. 1–12.

Block, R., U.S. finds ally in terrorism fight: FedEx, *Wall Street Journal*, May 30, 2005, p. A6.

Bodow, S., The money shot, *Wired*, 9(9), 86–97, 2001, available at http://archive.wired.com/wired/archive/9.09/paypal.html, last accessed February 13, 2015.

Bohr, J. and Bashir, M., Who uses Bitcoins? An exploration of the Bitcoin community, in *Proceedings of the 12th Annual Conference on Privacy, Security and Trust (PST)*, Toronto, Ontario, Canada, July 23–24, 2014, pp. 94–101.

Bouillant, O., Messageries électroniques, Eyrolles, Paris, 1998.

Boneh, D. and Shacham, H., Fast variants of RSA, *CryptoBytes*, 5(1), 1–9, 2002, available at http://www.hovav.net/dist/survey.pdf, last accessed August 16, 2015.

Bonnet, P., Detavernier, J.-M., and Vauquier, D., *Le Système D'information Durable. La Refonte Progressive du SI avec SOA*, Hermès/Lavoisier, Paris, France, 2008, pp. 82–87.

Bontoux, S., Accélerez le chiffrement SSL de votre serveur, *Internet Prof.*, 64, 60–63, 2002.

Borchers, A. and Demski, M., The value of Coin networks: The case of Automotive Network Exchange®, in *Organizational Achievement and Failure in Information Technology Management*, Khosrowpour, M., Ed., IDEA Group Publishing, Hershey, PA, 2000, pp. 109–123.

Bordo, M. D., Monetary standards, in *The Oxford Encyclopedia of Economic History*, Vol. 3, Mokyr, J., Ed., Oxford University Press, New York, 2003, pp. 531–535, available at econ-web.rutgers.edu/bordo/encyclopedia.doc, last accessed January 21, 2015.

Borenstein, N. S. et al., Perils and pitfalls of practical cybercommerce, *Commun. ACM*, 39(6), 37–44, 1996.

Borst, J., Preneel, B., and Rijmen, V., Cryptography on smart cards, *Comput. Netw.*, 36, 423–435, 2001.

Bort, J., Hackers are attacking millions of computers and demanding ransom in bitcoins, *Business Insider*, December 2, 2013, available at http://www.businessinsider.com/cryptolocker-ransom-in-bitcoins-2013-12, last accessed April 22, 2015.

Bosak, J., UBL: A standards-based approach to ecommerce, in *The Standards Edge: Future Generation*, Bolin, S., Ed., Bolin Communications, Felton, CA, 2005, pp. 397–409, www.thebolingroup.com.

Box, D., Ehnebuske, D., Kakivaya, G., Layman, A., Mendelsohn, N., Nielsen, H. F., Thatte, S., and Winer, D., Simple Object Access Protocol (SOAP) 1.1, W3C Note 08, May 8, 2000, available at http://www.w3.org/TR/2000/NOTE-SOAP-20000508, last accessed August 16, 2015.

Bradshaw, T. and Kuchler, H., iPhone software security flaws exposed, ft.com>reports>Cyber Security, February 28, 2014, available at http://www.ft.com/intl/cms/s/0/2b0b2158-9e45-11e3-95fe-00144feab7de.html?siteedition=intl#axzz3w7S0OzvB, last accessed January 2, 2016.

Brands, S., Untraceable off-line electronic cash in wallets with observers, in *Advance Cryptology—CRYPTO'93*, Springer-Verlag, Berlin, Germany, 1993, pp. 302–318.

Bresse, P., Beaure d'Augères, G., and Thuillier, S., *Paiement Numérique sur Internet*, Thomson Publishing France, Paris, France, 1997.

Breton, T., *Les Téléservices en France: Quels Marchés pour les Autoroutes de l'Information*? La documentation française, Paris, France, 1994.

Bridges, V., Amazon gives small sellers a big boost, *News and Observer* (Raleigh, NC), reproduced in *The Star-Ledger*, How to fiddle on Amazon? Small businesses need to offer something special, August 4, 2013, Section Three, p. 3.

Bryan, M., Ed., Guidelines for using XML for Electronic Data Interchange, Version 0.05, January 25, 1998, available at http://www.eccnet.com/xmledi/guidelines.xml, last accessed July 5, 2014.

Bueno-Delgado, M.-V., Egea-López, E., Vales-Alonso, J., and García-Haro, J., Radio-frequency identification technology, in *Handbook of Enterprise Integration*, Sherif, M. H., Ed., Auerbach Publications, Boca Raton, FL, 2010, pp. 429–466.

Buliard, F., French to migrate cards, *Bank. Technol.*, 18(3), 8, April 2001.

Burg, F. M., Personalized text-to-speech services, U.S. Patent 8,918,322, December 23, 2014.

Burr, W., Dodson, D., Nazaria N., and Polk, W. T., Minimum Interoperability Specifications for PKI Components (MISPC), Version 1, NIST Special Publication 800-15, National Institute of Standards and Technology, Gaithersburg, MD, September 3, 1997.

Buterin, V., BitMessage: A model for a new Web 2.0? *Bitcoin Magazine*, December 1, 2012, https://bitcoinmagazine.com/2855/bitmessage-a-model-for-a-new-web-2-0, last accessed May 30, 2015.

Cafiero, W. G., International standards for EDI/EFT, *EDI Forum*, 4(2), 74–77, 1991.

Camp, L. J., Sirbu, M., and Tygar, J. D., Token and notational money in electronic commerce, in *Proceedings of the First USENIX Workshop on Electronic Commerce*, New York, 1995, pp. 1–12.

Canvel, B., Hiltgen, A., Vaudenay, S., and Vuagnoux, M., Password interception in a SSL/TLS channel, in *CRYPTO 2003*, Lecture Notes in Computer Science, Vol. 2729, Boneh, D., Ed., Springer-Verlag, Berlin, Germany, 2003, pp. 583–599.

Carat, G., Mobile payments: Alternative platforms and players, *The IPTS Report*, 49, 24–31, November 2000, Institute for Prospective Technological Studies (IPTS), Seville Spain, available at http://aei.pitt.edu/57238/1/ipts.49.pdf, last accessed January 23, 2016.

Carelli, J., Will retailers be ready for EMV by Oct 2015? October 16, 2014, available at http://www.paymentsleader.com/will-retailers-be-ready-for-emv-by-oct-2015, last accessed December 13, 2014.

Cawrey, D., How consensus algorithms solve issues with Bitcoin's proof of work, *CoinDesk*, September 11, 2014, available at http://www.coindesk.com/stellar-ripple-hyperledger-rivals-bitcoin-proof-work, last accessed May 28, 2015.

CEN EN 1546-1:1999, Identification card systems—Inter-sector electronic purse—Part 1: Definitions, concepts and structures, European Committee for Standardization, Brussels, Belgium, 1999.

CEN EN 1546-2:1999, Identification card systems—Inter-sector electronic purse—Part 2: Security architecture, European Committee for Standardization, Brussels, Belgium, 1999.

CEN EN 1546-3:1999, Identification card systems—Inter-sector electronic purse—Part 3: Data elements and interchanges, European Committee for Standardization, Brussels, Belgium, 1999.

CEN EN 1546-4:1999, Identification card systems—Inter-sector electronic purse—Part 4: Data objects, European Committee for Standardization, Brussels, Belgium, 1999.

CEPSCO, Common Electronic purse specifications—Technical specification, Version 2.3, May 2001, available at http://www.it.iitb.ac.in/~tijo/seminar/cepstechspecv2.3.pdf, last accessed January 2, 2015.

CERT/CC CA-2001-19, CERT® Advisory CA-2001-19—"Code Red" worm exploiting buffer overflow in IIS indexing service DLL, last revised January 17, 2002, available at http://www.cert.org/advisories/CA-2001-19.html, last accessed October 11, 2014.

Chaix, L. and Torre, D., Which economic model for mobile payments? in *23rd European Regional Conference of the International Telecommunication Society*, Vienna, Austria, July 1–4, 2012, available at http://hdl.handle.net/10419/603, last accessed October 11, 2014.

Chalandon, S., Réseaux soucieux, *Le Canard enchaîné*, March 18, 2015, p. 7.

Chan, A., Frankel, Y., MacKenzie, P., and Tsiounis, Y., Misrepresentation of identities in E-cash schemes and how to prevent it, in *Advance Cryptology—ASIACRYPT'96*, Springer-Verlag, Heidelberg, Germany, 1996, pp. 276–285.

Chang, H.-H., *Everyday NFC, Near Field Communication Explained*, 2nd ed., Self-published, Lexington, KY, 2014.

Chang, R. K. C., Defending against flooding-based distributed denial-of-service attack: A tutorial, *IEEE Commun. Mag.*, 40(10), 42–51, 2002.

Chaos Computer Club, Chaos Computer Club breaks Apple TouchID, September 21, 2013, available at http://www.ccc.de/en/updates/2013/ccc-breaks-apple-touchid, last accessed March 18, 2015.

Charmot, C., EDI, échange de données informatisé: Définition et enjeux, *Télécom Interview*, 33, 8–10, 1997a.

Charmot, C., *L'Échange de Données Informatisé (EDI), Que Sais-je?* Vol. 3321, Presses Universitaires de France, Paris, France, 1997b.

Chaum, D., Blind signatures for untraceable payments, in *CRYPTO 82*, Plenum Press, New York, 1983, pp. 199–203.

Chaum, D., Privacy protected payments: Unconditional payer and/or payee untraceability, in *Smart Card 2000*, Chaum, D. and Schaumüller-Bichl, I., Eds., Elsevier/North Holland, Amsterdam, the Netherlands, 1989, pp. 69–93.

Chaum, D., On line cash checks, n.d., available at http://chaum.com/publications/Online_Cash_Checks.html, last accessed August 16, 2015.

Chaum, D., Fiat, A., and Naor, M., Untraceable electronic cash, in *Advance Cryptology—CRYPTO'88*, Springer-Verlag, Heidelberg, Germany, 1990, pp. 319–327.

Cheng, P.-C., An architecture for the Internet key exchange protocol, *IBM Syst. J.*, 40(3), 721–746, 2001.

Chernick, C. M., Edington, C., III, Fanto, M. J., and Rosenthal, R., Guidelines for the selection and use of Transport Layer Security (TLS) implementations, NIST special publication 800-52, National Institute of Standards and Technology, Gaithersburg, MD, June 2005, available at http://www.hhs.gov/ocr/privacy/hipaa/administrative/securityrule/nist80052.pdf, last accessed August 30, 2014.

Chicheportiche, O., M-commerce: Le décollage semble bien avoir eu lieu, *ZDNet.fr*, January 23, 2013a, available at http://www.zdnet.fr/actualites/m-commerce-le-decollage-semble-bien-avoir-eu-lieu-39786524.htm, last accessed December 26, 2013.

Chicheportiche, O., M-commerce: Scanpay invente le payment par scan de carte, *ZDNet.fr*, March 19, 2013b, available at http://www.zdnet.fr/actualites/m-commerce-scanpay-invente-le-paiement-par-scan-de-carte-39788361.htm, last accessed December 26, 2013.

Christensen, E., Curbera, F., Merideth, G., and Weerawarana, S., Web services description language (WSDL) 1.1, W3C Note, March 15, 2001, available at http://www.w3.org/TR/wsdl, last accessed June 2, 2015.

Cisco, IOS Internet key exchange vulnerability, Cisco Security Advisory, Advisory ID: Cisco-sa-20120328-ike, Version 1.0, March 28, 2012, available at http://tools.cisco.com/security/center/content/CiscoSecurityAdvisory/cisco-sa-20120328-ike, last accessed June 2, 2015.

CNN, LexisNexis customer IDs stolen, March 9, 2005, available at http://money.cnn.com/2005/03/09/news/midcaps/lexisnexis/index.htm, last accessed August 14, 2015.

Comer, D. E., *Interworking with TCP/IP Vol. I: Principles, Protocols, and Architecture*, 3rd ed., Prentice Hall, Englewood Cliffs, NJ, 1995.

CommerceNet, Lessons learned from implementing MIME-based secure EDI over Internet, CommerceNet, Palo Alto, CA, August 31, 1997.

Communications Security Establishment Canada, Technical publication: Bluetooth vulnerability assessment—ITSPSR-17A, June 2008, available at http://www.cse-cst.gc.ca/its-sti/publications/itspsr-rpssti/itspsr17a-eng.html, last accessed December 14, 2013.

Community Research and Development Information Service (CORDIS), Embedded FINREAD report summary, April 10, 2013, available at http://cordis.europa.eu/result/rcn/29542_en.html, last accessed February 1, 2015.

Corella, F. and Lewison, K, Interpreting the EMV tokenisation specification, October 19, 2014, available at http://pomcor.com/whitepapers/EMVTok.pdf, last accessed November 9, 2014.

Coskun, V., Ok, K., and Ozdenizci, B., *Near Field Communication: From Theory to Practice*, John Wiley & Sons, Chichester, UK, 2012.

Count Zero, Card-O-Rama: Magnetic stripe technology and beyond or "A day in the life of a flux reversal," November 22, 1992, available at http://phrack.org/issues/37/6.html#article, last accessed May 25, 2014.

Cox, B., Tygar, J. D., and Sirbu, M., NetBill security and transaction protocol, in *Proceedings of the First USENIX Workshop on Electronic Commerce*, New York, 1995, pp. 77–78.

Crabtree, J., M-Pesa's cautious start in India, *Financial Times*, December 27, 2012, p. 8.

Cranor, L. F., *Web Privacy with P3P*, O'Reilly and Associates, Sebastopol, CA, 2002.

Craver, S. A., Wu, M., Liu, B., Stubblefield, A., Swartzlander, B., Wallach, D. S., Dean, D., and Felten, E. W., Reading between the lines: Lessons from the SMDI challenge, in *Proceedings of the 10th USENIX Security Symposium*, Washington, DC, August 13–17, 2001.

Cuthbertson, A., Darkcoin to rebrand as DASH to disassociate itself from dark web, *International Business Times*, March 17, 2015, available at http://www.ibtimes.co.uk/dark-coin-rebrand-dash-better-represent-cryptocurrencys-digital-cash-platform-1492299, last accessed March 28, 2015.

CWA 14174-1:2004, Financial transactional IC card reader (FINREAD), Business requirements, January 1, 2004.

CWA 14174-2:2004 (Edition 2), Financial transactional IC card reader (FINREAD), Functional requirements, October 1, 2004.

CWA 14174-3:2004 (Edition 2), Financial transactional IC card reader (FINREAD), Security requirements, January 1, 2004.

CWA 14174-4:2004, Financial transactional IC card reader (FINREAD), Architectural OVERVIEW, January 1, 2004.

CWA 14174-5:2004 (Edition 2), Financial transactional IC card reader (FINREAD), October 1, 2004.

CWA 14174-6:2004, Financial transactional IC card reader (FINREAD), Definition of the virtual machine, October 1, 2004.

CWA 14174-7:2004 (Edition 2), Financial transactional IC card reader (FINREAD), Card reader application programming interfaces (APIs), October 1, 2004.

CWA 14174-8:2004 (Edition 2), Financial transactional IC card reader (FINREAD), Client application programming interfaces (APIs), October 1, 2004.

Daaboul, T., Spécification d'une Plate-Forme de Personnalisation de Carte à Puce, rapport de thèse professionnelle, ENST, Paris, France, 1998.

Dai, W., B-money, Cypherpunks mailing list, November 1998, available at http://www.weidai.com/bmoney.txt, last accessed April 3, 2015.

Dai, W., Comment posted on April 20, 2013 at 07:56:08 a.m., available at http://lesswrong.com/r/discussion/lw/h8z/bitcoins_are_not_digital_greenbacks/8tga, last accessed April 28, 2015.

Daugman, J. G., Recognizing persons by their Iris patterns, U.S. Patent 5,291,560, 1994.

Daugman, J. G., Biometric personal identification system based on iris analysis, in *Biometrics: Personal Identification in Networked Society*, Jain, A., Bolle, R., and Pankati, S., Eds., Kluwer, Dordrecht, the Netherlands, 1999, pp. 104–121.

Davida, G. I., Chosen signature cryptanalysis of the RSA (MIT) public key cryptosystem, Technical Report TR-CS-82-2, Department of Electrical Engineering and Computer Science, University of Wisconsin, Milwaukee, WI, 1982.

Dawirs, M., Porte-monnaie électronique: Le procotole normalisé par le CEN et l'implantation Proton, *in 6ème Colloque Francophone sur l'Ingénieurie des Protocoles*, Liège, Belgium, 1997.

de Galzain, P., Electronic Data Interchange (EDI) in the automotive industry, in *EuroComm 88, Proceedings of the International Congress on Business, Public and Home Communication*, Amsterdam, the Netherlands, Schuringa, T. M., Ed., Elsevier/North Holland, Amsterdam, the Netherlands, 1989, pp. 259–264.

de Oleivera, A. E. and Mota, G. H. M. B., A security API for multimodal multi-biometric continuous authentication, in *Proceedings of the 2011 Seventh International Conference on Computational Intelligence and Security* (*CIS*), Sanya, Hainan, China, December 3–4, 2011, pp. 988–992.

de Ruiter, J. and Poll, E., Formal analysis of the EMV protocol suite, in *Theory of Security and Applications, Joint Workshop, TOSCA 2011*, Saarbrücken, Germany, March 31–April 1, 2011, Lecture Notes in Computer Science, Vol. 6993, Moedersheim, S. and Palamidessi, C., Eds., Springer, Berlin, Germany, 2012, pp. 113–129.

Dean, D. and Stubblefield, A., Using client puzzles to protect TLS, in *Proceedings of the 10th USENIX Security Symposium*, Washington, D.C., August 13–17, 2001.

Decker, C. and Wattenhofer, R., Information propagation in the Bitcoin network, in *Proceedings of the 13th IEEE International Conference on Peer-to-Peer Computing* (*P2P*), Trento, Italy, September 9–11, 2013, pp. 1–10.

Defense Information Systems Agency, Department of Defense, DoD defense peripheral device security requirements, July 16, 2010, available at http://iase.disa.mil/stigs/downloads/pdf/dod_bluetooth_requirements_spec_20100716.pdf, last accessed December 17, 2013.

del Pilar Barea Martinez, M., EDI: Des organismes fédérateurs pour le commerce international, *Télécom Interview*, 33, 12–14, 1997.

Dembosky, A., Square faces big rush to spread card payments, *Financial Times*, September 19, 2012, p. 16.

Demirgurc-Kunt, A. and Klapper, L., Measuring financial inclusion: The global findex database, Development Research Group, Finance and Private Sector Development Team, Public Research Working Paper 6025, The World Bank, Washington, DC, April 2012, available at http://www-wds.worldbank.org/external/default/WDSContentServer/IW3P/IB/2012/04/19/000158349_20120419083611/Rendered/PDF/WPS6025.pdf, last accessed August 15, 2015.

Department of Defense (DoD), Public Key Enabling (PKE) and Public Key Infrastructure (PKI) Program Management Office (PMO), DoD root certificate chaining problem, March 15, 2010, available at http://iase.disa.mil/pki-pke/getting_started/downloads/unclass-faq_dod_root_cert_chaining_issue.pdf, last accessed July 5, 2015.

Dev, J. A., Bitcoin mining acceleration and performance quantification, in *Proceedings of the 27th Canadian Conference on Electronic Computer Engineering* (*CCECE 2014*), Toronto, Ontario, Canada, May 4–7, 2014, pp. 1–6.

Dey, P. K., Clegg, B., and Bennett, D., Managing implementation issues in Enterprise Resource Planning projects, in *Handbook of Enterprise Integration*, Sherif, M. H., Ed., CRC Press/Auerbach Publications, Boca Raton, FL, 2010, pp. 563–578.

Dickie, M., Crackdown as Beijing fears virtual money's real influence, *Financial Times*, March 7, 2007, p. 5.

Diffie, W. and Hellman, M. E., New directions in cryptography, *IEEE Trans. Inf. Theory*, 22(6), 644–654, 1976.

Digital Equipment Corporation, The Millicent Microcommerce System: Defining a New Internet Business Model, 1995.

Dobbertin, H., Bosselaers, A., and Preneel, B., RIPEMD-160: A strengthened version of RIPEMD, in *Proceedings of the Fast Software Encryption Workshop*, Cambridge, UK, February 21–23, Lecture Notes in Computer Science, Vol. 1039, Gollmann, D., Ed., Springer-Verlag, Berlin, 1996, pp. 71–82.

Doraswmy, N. and Harkins, D., *IPSec: The New Security Standard for the Internet, Intranets, and Virtual Private Networks*, Prentice Hall PTR, Upper Saddle River, NJ, 1999.

Dorn, J. et al., A survey of B2B methodologies and technologies: From business models towards development artifacts, in *Proceedings of the 40th Hawaii International Conference Systems Science* (*HICSS'07*), Waikoloa, HI, January 3–6, 2007.

Dowland, P. S., Furnell, S. M., and Papadaki, M., Keystroke analysis as a method of advanced user authentication and response, in *Security in the Information Society: Visions and Perspectives*, Ghonaimy, M. A., El-Hadidi, M. T., and Aslan, H. K., Eds., Kluwer, Dordrecht, the Netherlands, 2002, pp. 215–226.

Dragon, C., Geiben, D., Kaplan, D., and Nallard, G., *Les Moyens de Paiement: Des Espèces à la Monnaie Électronique*, Banque Editeur, Paris, France, 1997.

Drake, C., Oliver, J., and Koontz, E., Anatomy of a phishing email, in *Proceedings of the First Conference on Email and Anti-Spam* (*CEAS 2004*), Mountain View, CA, July 30–31, 2004, available at http://ceas.cc/2004/114.pdf, last accessed June 2, 2015.

Dreifus, H. and Monk, J. T., *Smart Cards*, John Wiley & Sons, New York, 1998.

Drimer, S., Murdoch, S., and Anderson, R., Failures of tamper-proofing in PIN entry devices, *IEEE Secur. & Priv.*, 7(6), 29–45, 2009, available at http://www.cl.cam.ac.uk/~sjm217/papers/ieeesp09tamper.pdf, last accessed December 3, 2014.

Drimer, S. and Murdoch, S. J., Keep your enemies close: Distance bounding against smartcard relay attacks, in *Proceedings of the Security'07, 16th USENIX Security Symposium*, Boston, MA, August 6–10, 2007, pp. 87–102, available at http://static.usenix.org/events/sec07/tech/drimer/drimer.pdf, last accessed December 3, 2014.

Dunning, J. P., Taming the blue beast. A survey of Bluetooth-based threats, *IEEE Secur. & Priv.*, 8(2), 20–27, 2010.

Duong, T. and Rizzo, J. Here come the ⊕ Ninjas, May 13, 2011, available at https://bug665814.bugzilla.mozilla.org/attachment.cgi?id=540839, last accessed August 30, 2014.

Dupoirier, G., *Technologie de la GED: Technique et Management des Documents Électroniques*, 2nd ed., Hermès, Paris, France, 1995.

Dwork, C. and Naor, M., Pricing via processing or combatting junk mail, preliminary version in *Advance Cryptology—CRYPTO'92*, Lecture Notes in Computer Science, Vol. 740, Brickell, E. F., Ed., Springer-Verlag, Berlin, Germany, 1993, pp. 139–147.

Eaglen, C., The relationship between EDI and electronic funds transfer, in *The EDI Handbook*, Gikins, M. and Hitchcock, D., Eds., Blenheim Online, London, UK, 1988, pp. 85–103.

ECMA-340:2013, Near Field Communication Interface and Protocol (NFCIP-1), 3rd ed., June 2013, available at http://www.ecma-international.org/publications/standards/Ecma-340.htm, last accessed November 25, 2013.

ECMA-352:2013, Near Field Communication Interface and Protocol-2 (NFCIP-2), 3rd ed., June 2013, available at http://www.ecma-international.org/publications/standards/Ecma-352.htm, last accessed August 15, 2015.

Economist (The), Little brother, Special report on advertising and technology, 412(8904), September 13, 2014a.

Economist (The), Leaving dead presidents in peace. Abolishing notes and coins would bring huge economic benefits, 412(8905), 72, September 20, 2014b, available at http://www.economist.com/news/finance-and-economics/21618886-abolishing-notes-and-coins-would-bring-huge-economic-benefits-leaving-dead, last accessed March 8, 2016.

Economist (The), Bitcoin, the magic of mining, 413(8920), 58–60, January 10, 2015a.

Economist (The), The Kaspersky equation, 414,(8926), 63, February 21, 2015b.

Economist, A new East Africa campaign, *The Economist*, 416(8946), 57–58, July 11, 2015c.

Economist (The), EBay and PayPal. Better off alone, *The Economist*, 416(8947), 56–57, July 18, 2015d.

Edgecliffe-Johnson, A., Amazon reports first quarterly profit, *Financial Times*, January 23, 2002, p. 21.

ElGamal, T., A public-key cryptosystem and a signature scheme based on discrete logarithms, in *Advance Cryptology—CRYPTO'84*, Springer-Verlag, Heidelberg, Germany, 1985, pp. 10–18.

Elkamchouchi, H. and Abouelseoud, Y., Digital rights management system design and implementation issues, in *Proceedings of the International Conference on Computer Engineering and Systems* (ICCES'07), Cairo, Egypt, November 27–29, 2007, pp. 120–125.

Emmelhainz, M. A., *EDI: A Total Management Guide*, 2nd ed., Van Nostrand Reinhold, New York, 1993.

EMV'96, Integrated circuit card specification for payment systems, Version 3.1.1, May 1998, available at http://www.ttfn.net/techno/smartcards/termspec.pdf, last accessed August 16, 2015.

EMV2000, Integrated circuit card specification for payment systems, Version 4.0, December 2000, Book 1, available at http://www.mathdesc.fr/documents/normes/emv_book1.pdf, Book 2, available at http://www.di.unisa.it/~ads/corso-security/www/CORSO-9900/smartcard/_pages/Iccbook2.pdf, Book 3, available at http://read.pudn.com/downloads90/doc/comm/343405/Book3.pdf, last accessed August 16, 2015.

EMV, AES option in EMV, Specification Update Bulletin No. 74, 2nd ed., July 2010, last updated April 4, 2011, available at http://www.emvco.com/specifications.aspx?id=23, last accessed November 28, 2014.

EMV, Integrated circuit card specifications for payment systems, Version 4.3, Book 1, Book 1—Application Independent ICC to Terminal Interface Requirements, Book 2, Security and Key Management, November 2011, available at https://www.emvco.com/specifications.aspx?id=223, last accessed August 16, 2015.

EMV, Payment Tokenisation Specification, Technical Framework, Version 1.0, March 2014, available at http://www.emvco.com/download_agreement.aspx?id=945, last accessed November 7, 2014.

Eng, T. and Okamoto, T., Single-term divisible electronic coins, in *Advance Cryptology—Eurocrypt'94*, Springer-Verlag, Heidelberg, Germany, 1994, pp. 306–319.

Enoki, K., Concept of i-mode service—New communication infrastructure for the 21st century, *NTT DoCoMo Technol. J.*, 1(1), 4–9, 1999.

Erl, T., *Service-Oriented Architecture, A Field Guide to Integrating XML and Web Services*, Prentice Hall, PTR, Upper Saddle River, NJ, 2004.

European Banking Authority, EPA opinion on "virtual currencies," EBA/Op/2014/8, July 4, 2014, available at http://www.eba.europa.eu/documents/10180/657547/EBA-Op-2014-08+Opinion+on+Virtual+Currencies.pdf, last accessed March 8, 2015.

European Commission, Recommendation 94/820 on 19th October, *Official Journal* L338, December 28, 1994, pp. 98–117, http://eur-lex.europa.eu/LexUriServ/LexUriServ.do?uri=CELEX:31994H0820:en:HTML, last accessed December 31, 2015.

European Payments Council, Mobile Payments, White Paper, Document RPC492-09, Version 4.0, October 18, 2012, available at http://www.europeanpaymentscouncil.eu/knowledge_bank_detail.cfm?documents_id=564, last accessed December 22, 2013.

European Telecommunications Standards Institute, ETSI TC-SMG, Digital cellular telecommunications system (Phase 2+)—Specification of the Subscriber Identity Module—Mobile Equipment (SIM-ME) interface, (GSM 11.11) Version 5.30, TS/SMG-09111QR1, July 1996, available at http://www.etsi.org/deliver/etsi_gts/11/1111/05.03.00_60/gsmts_1111v050300p.pdf, last accessed February 9, 2014.

European Telecommunications Standards Institute, Digital cellular telecommunications system (Phase 2+)—Personalisation of GSM mobile equipment (ME)—Mobile functionality specification, GSM 02.22, version 6.0.0, Release 1997, available at http://www.etsi.org/deliver/etsi_ts/101600_101699/101624/06.00.00_60/ts_101624v060000p.pdf, last accessed December 1, 2014.

Eurosmart, Press Release, May 6, 2015, available at http://www.eurosmart.com/images/doc/EurosmartPR/Eurosmart%20-%20Press%20Release%20General%20Assembly%206%20May%202015.pdf, last accessed January 14, 2015.

Eyal, I. and Sirer, G., Majority is not enough: Bicoin mining is vulnerable, Department of Computer Science, Cornell University, November 15, 2013, available at http://www.cs.cornell.edu/~ie53/publications/btcProcArXiv.pdf, last accessed September 11, 2015.

Faddis, K. N., Matey, J. R., Maxey, J. R., and Stracener, J. T., Performance assessments of iris recognition in tactical biometric devices, *IET Biometrics*, 2(1), 1–10, 2013.

Fakhoury, H., EFF challenges New Jersey subpoena issued to MIT student Bitcoin developers, *Electronic Frontier Foundation*, February 5, 2014, available at https://www.eff.org/deeplinks/2014/02/eff-challenges-new-jersey-subpoena-issued-mit-student-bitcoin-developers, last accessed May 30, 2015.

Fallon, T. and Welch, B., BACS—Practical control issues, in *Electronic Banking and Security*, Welch, B., Ed., Blackwell Scientific, Oxford, UK, 1994, pp. 31–42.

Farivar, C., Bitcoin pool GHash.io commits to 40% hash rate limit after its 51% breach, *Ars Technica*, July 16, 2014a, available at http://arstechnica.com/business/2014/07/bitcoin-pool-ghash-io-commits-to-40-hashrate-limit-after-its-51-breach, last accessed March 12, 2015.

Farivar, C., Feds say Bitcoin miner maker Butterfly Labs ran "systematic deception," *Ars Technica*, September 23, 2014b, available at http://arstechnica.com/tech-policy/2014/09/feds-label-bitcoin-miner-maker-butterfly-labs-as-systematic-deception, last accessed April 3, 2015.

Faundez-Zanuy, N., Biometric security technology, *IEEE A&E Syst. Mag.*, 21(6), 15–26, 2006.

Fay, A., *Dico Banque*, La Maison du Dictionnaire, Paris, France, 1997.

Federal Reserve System, The 2010 Federal Reserve Payments Study, Noncash Payment Trends in the United States: 2006–2009, April 5, 2011, available at http://www.frbservices.org/files/communications/pdf/press/2010_payments_study.pdf. *Official Journal* L 338, December 28, 1994, pp. 98–117, http://eur-lex.europa.eu/LexUriServ/LexUriServ.do?uri=CELEX:31994H0820:en:HTML, last accessed December 31, 2015.

Federal Reserve System, The 2013 Federal Reserve Payments Study, Recent and Long-Term Trends in the United States: 2003–2012, Summary Report and Initial Data Release, December 19, 2013, available at https://www.frbservices.org/files/communications/pdf/research/2013_payments_study_summary.pdf, last accessed August 16, 2015.

Federal Reserve System, The 2013 Federal Reserve Payments Study, Recent and Long-Term Trends in the United States: 2000–2012, Detailed Report and Updated Data Release, July 2014, available at https://www.frbservices.org/files/communications/pdf/general/2013_fed_res_paymt_study_detailed_rpt.pdf, last accessed August 16, 2015.

Ferguson, N., Schneier, B., and Kohno, T., *Cryptographic Engineering*, Wiley Publishing, Indianapolis, IN, 2010.

Ferrara, D., Robocoin, first Bitcoin ATM in a casino, launches at The D, *Las Vegas Review-Journal*, May 22, 2014, available at http://www.reviewjournal.com/business/casinos-gaming/robocoin-first-bitcoin-atm-casino-launches-d, last accessed April 26, 2015.

Feuer, A., The Bitcoin ideology, *The New York Times, Sunday Review*, December 15, 2013, p. 12.

Financial Crimes Enforcement Network, FIN-2013-G001, Application of FinCEN's regulations to persons administering, exchanging, or using virtual currencies, United States Department of Treasury, March 18, 2013, available at http://fincen.gov/statutes_regs/guidance/pdf/FIN-2013-G001.pdf, last accessed March 7, 2015.

Financial Crimes Enforcement Network, FIN-2014-R011, Request for administrative ruling on the application of FinCEN's regulations to a virtual currency trading platform, United States Department of Treasury, October 27, 2014a, available at http://www.fincen.gov/news_room/rp/rulings/pdf/FIN-2014-R011.pdf, last accessed March 8, 2015.

Financial Crimes Enforcement Network, FIN-2014-R012, Request for administrative ruling on the application of FinCEN's to a virtual currency payment system, United States Department of Treasury, October 27, 2014b, available at http://www.fincen.gov/news_room/rp/rulings/html/FIN-2014-R012.html, last accessed March 8, 2015.

Financial Times, Mobile payments: Tap, tap, tap, *Financial Times*, February 25, 2015a, p. 10.

Financial Times, The case for retiring another "barbarous relic," Editorial, *Financial Times*, August 24, 2015b, p. 10.

Finkenzeller, K., *RFID Handbook*, 3rd ed., John Wiley & Sons, Chichester, UK, 2010.

Flom, L. and Safir, A., Iris recognition system, U.S. Patent 4,641,349, 1987.

Foley, S., Bitcoin hack exposes virtual currency's weak spot, *Financial Times*, February 15/16, 2014, p. 14.

Foley, S. and Wild, J., Evangelism is common currency of Bitcoin faithful, *Financial Times*, December 9, 2013, p. 3.

Ford, W. and Baum, M. S., *Secure Electronic Commerce: Building the Infrastructure*, Prentice Hall, Upper Saddle River, NJ, 1997.

Forrester, S. E., Palmer, M. J., McGlaughlin, D. C., and Robinson, M. J., Security in data networks, *BT Technol. J.*, 16(1), 52–75, 1998.

France Télécom, La télématique française: bilan de l'année 1994, *La Lettre de Télétel & Audiotel*, Hors série 13, 6–8 June 1995.

Francis, L., Hancke, G. P., Mayes, K. E., and Markantonakis, K., Practical NFC peer-to-peer relay attack using mobile phones, in *Radio Frequency Identification: Security and Privacy Issues, Sixth International Workshop RFIDSec2010*, Istanbul, Turkey, 2010a, Lecture Notes in Computer Science, Vol. 6370, Berna, S. and Yalcin, O., Eds., Springer-Verlag, Berlin, Germany, 2010, pp. 35–49, available at http://eprint.iacr.org/2010, last accessed August 16, 2015.

Francis, L., Hancke, G. P., Mayes, K. E., and Markantonakis, K., On the security issues of NFC enabled mobile phones, *Int. J. Internet Technol. Secured Trans.*, 2, 3/4, 336–356, 2010b.

Francis, L., Hancke, G. P., Mayes, K. E., and Markantonakis, K., Practical relay attack on contactless transactions by using NFC mobile phones, *Cryptology ePrint Archive*, Report 2011/618, received November 17, 2011, revised February 24, 2012, available at http://eprint.iacr.org/2011/618, last accessed August 16, 2015.

Freier, A. O., Karlton, P., and Kocher, P. C., The SSL protocol Version 3.0, 1996, available at http://tools.ietf.org/html/draft-ietf-tls-ssl-version3-00, last accessed August 9, 2014.

FSTC (Financial Services Technology Consortium), Echeck, available at http://www.echeck.org/echeck-guide, last accessed December 31, 2015.

Fu, K., Sit, E., Smith, K., and Feamster, N., Dos and don'ts of client authentication on the Web, in *Proceedings of the 10th USENIX Security Symposium*, Washington, DC, August 2001.

Fumer, W. and Landrock, P, Principles of key management, *IEEE J. Sel. Areas Commun.*, 11(5), 785–793, 1993.

Galbally, J., Cappelli, R., Lumini, A., Gonzalez-de-Rivera, G., Maltoni, D., Fierrez, J., Ortega-Garcia, J., and Maio, D., An evaluation of direct attacks using fake fingers generated from ISO templates, *Pattern Recognit. Lett.*, 31(8), 725–732, 2010.

Galbally, J., Fierrez, J., Alonso-Fernandez, F., and Martinez-Diaz, M., Evaluation of direct attacks to fingerprint verification systems, *Telecommun. Syst.*, 37(3–4), 243–254, 2011.

Galbally, J., Marcel S., and Fierrez, J., Image quality: Assessment for fake biometric detection: Application to iris, fingerprint and face recognition, *IEEE Trans. Image Process.*, 23(2), 710–724, 2014.

Galbally, J., Ross, A., Gomez-Barrero, M., Fierrez, J., and Ortega-Garcia, J., From the iriscode to the iris: A new vulnerability of iris recognition systems, in *Proceedings of the Black Hat USA*, Las Vegas, NV, July 6, 2012, available at https://www.beat-eu.org/publications/index.php/attachments/single/24, last accessed July 8, 2015.

Gao, J., Kulkarni, V., Ranavat, H., Chang, L., and Mei, H., A 2D barcode-based mobile payment system, in *Proceedings of the Third International Conference on Multimedia and Ubiquitous Engineering*, Qingdao, China, 2009, pp. 320–239.

Garfinkel, S. L., *PGP: Pretty Good Privacy*, O'Reilly and Associates, Sebastopol, CA, 1995.

GeldKarte, Schnittstellenspezifikation für die GeldKarte mit Chip, debis Systemhaus GEI, August 9, 1995.

Ghosh, R. K., Arora, A., and Barua, G., Smart card based protocol for secure and controlled access of mobile host in IPv6 compatible foreign network, in *Advances in Security and Payment Methods for Mobile Commerce*, Hu, W.-C., Lee, C.-W., and Kou, W., Eds., Idea Group Publishing, Hershey, PA, 2005, pp. 312–337.

Giot, R., Hemery, B., and Rosenberger, C., Low cost and usable multimodal biometric system based on keystroke dynamics and 2D face recognition, in *Proceedings of the 20th International Conference Pattern Recognition* (*ICPR 2010*), Istanbul, Turkey, August 23–26, 2010, pp. 1128–1131.

Glassman, S., Manasse, M., Abadi, M., Gauthier, P., and Sobalvarro, P., The Millicent protocol for inexpensive electronic commerce, in *Proceedings of the Fourth International World Wide Web Conference*, Boston, MA, 1995.

Gomzin, S., *Hacking Point of Sale: Payment Application Secrets, Threats, and Solutions*, John Wiley & Sons, Indianapolis, IN, 2014.

Goodin, D., Serious OS X and iOS flaws let hackers steal keychain, 1Password contents, *Ars Technica*, June 17, 2015, available at http://arstechnica.com/security/2015/06/serious-os-x-and-ios-flaws-let-hackers-steal-keychain-1password-contents, last accessed July 15, 2015.

Goodman, L. M., The face behind Bitcoin, *Newsweek*, 162(10), 7–24, March 14, 2014.

Goth, G., Mobile security issues come to the Forefront, *IEEE Internet Comput.*, 16(3), 7–9, 2012.

Granet, J., Procédure de transfert de données fiscales et comptables, *Télécom Interview*, 33, 15–18, 1997.

Grant, J. Monetary activism is a virus that infects politics and destroys wealth, *Financial Times*, January 6, 2015, p. 7.

Gray, B., Privacy, *Login*, 24(6), 74–76, 1999.

Greenberg, A., "Zerocoin" add-on for bitcoins could make it truly anonymous and untraceable, *Forbes*, April 12, 2013, available at http://www.forbes.com/sites/andygreenberg/2013/04/12/zerocoin-add-on-for-bitcoin-could-make-it-truly-anonymous-and-untraceable, last accessed April 17, 2015.

Greenberg, A., "Dark Wallet" is about to make Bitcoin money laundering easier than ever, *Wired*, April 29, 2014a, available at http://www.wired.com/2014/04/dark-wallet, last accessed March 29, 2015.

Greenburg, A., Bitcoin's nefarious cousin Darkcoin is booming, Wired.com, May 22, 2014b, available at http://www.wired.co.uk/news/archive/2014-05/22/darkcoin-is-booming, last accessed March 28, 2015.

Gross, G., ChoicePoint's error sparks talk of ID theft law, *PCWorld*, February 23, 2005, available at http://www.pcworld.com/article/119790/article.html, last accessed August 15, 2015.

Grother, P. J. and Ngan, M., Face recognition vendor test (FRVT), Performance of face identification algorithms, NIST interagency report 8009, National Institute of Standards and Technology (NIST), May 26, 2014, available at http://www.nist.gov/customcf/get_pdf.cfm?pub_id=915761, last accessed January 10, 2015.

Grother, P. J., Quinn, G. W., and Phillips, P. J., Report on the evaluation of 2D still-image face recognition algorithms, NIST interagency report 7709, National Institute of Standards and Technology (NIST), June 17, 2010, available at http://www.nist.gov/customcf/get_pdf.cfm?pub_id=905968, last accessed January 10, 2015.

Groupement des Cartes Bancaires, Chip-Secured Electronic Transaction (C-SET) security architecture, Version 1.0, Paris, France, January 29, 1997.

Grover, A., Braeckel, P., Lindgren, K., Berghel, H., and Cobb, D., Parameters effecting 2D barcode scanning reliability, in *Advances in Computers*, Vol. 80, Zelkowitz, M., Ed., Academic Press, Burlington, MA, 2010, pp. 209–235.

GSM Association, Pay-buy-mobile business opportunity analysis, Version 1.0, November 2007, available at http://www.mobile-ecosystem.org/wp-content/files/gsma_pbm_wp.pdf, last accessed October 11, 2014.

Guillemin, C., Le Crédit Mutuel et NRJ testent le paiement par mobile, *ZDNet*, October 23, 2006, available at http://www.zdnet.fr/actualites/le-credit-mutuel-et-nrj-testent-le-paiement-par-mobile-39364170.htm, last accessed August 16, 2015.

Guillou, L. C., Davio, M., and Quisquater, J.-J., L'état de l'art en matière de techniques à clé publique: Hasard et redondance, *Ann. Télécommun.*, 43, 9–10, 489–505,1988.

Guillou, L. C., Ugon, M., and Quisquater, J.-J., Cryptographic authentication protocols for smart cards, *Comput. Netw.*, 36, 437–451, 2001.

Haesler, A. J., *Sociologie de l'Argent et Postmodernité*, Droz, Genève-Paris, Switzerland, 1995.

Hall, E. T. and Hall, M. R., *Understanding Cultural Differences*, Intercultural Press, Inc., Yarmouth, ME, 1990.

Halpern, S., The creepy new wave of the Internet, *N Y Rev. Books*, 61(18), 22–24, November 20, 2014.

Hamman, E.-M., Henn, H., Schäck, T., and Seliger, F., Securing e-business applications using smart cards, *IBM Syst. J.*, 40(3), 635–647, 2001.

Hancke, G., A practical relay attack on ISO 14443 proximity cards, January 2005, available at www.rfidblog.org.uk/hancke-rfidrelay.pdf, last accessed November 12, 2013.

Hancke, G., Practical attacks on proximity identification systems, in *Proceedings of the 2006 IEEE Symposium on Security and Privacy SP'06*, Oakland, CA, 2006, pp. 328–333.

Hancke, G. P. and Drimer, S., Secure proximity identification for RFID, in *Security in RFID and Sensor Networks*, Zhang, Y. and Kitsos, P., Eds., Auerbach Publications/Taylor & Francis Group, Boca Raton, FL, 2009, pp. 171–194.

Handa, R., Maheswari, K., and Saraf, M., *Google Wallet—A Glimpse into the Future of Mobile Payments*, GRIN, Norderstedt, Germany, 2011.

Hansell, S., Got a dime? Citibank and Chase end test of electronic cash, *New York Times*, November 4, 1998, p. C1.

Harbert, T., New king of security algorithms crowned, *IEEE Spectr.*, 49(12), 12–13, 2012.

Harper's Magazine, The searchable soul. Privacy in the age of information technology, Forum, *Harper's Mag.*, 300(1796), 57–68, January 2000.

Harrell, A. C., Electronic checks, 55 consumer finance law quarterly report, January 2001, pp. 283–289, available at http://works.bepress.com/alvin_harrell/151, last accessed July 31, 2015.

Hasselbring, W., The role of standards for interoperating information system, in *Information Technology Standards and Standardization: A Global Perspective*, Jakobs, K., Ed., Idea Group Publishing, Hershey, PA, 2000, pp. 116–130.

Hawke, J. D., Jr., Testimony of the treasury under secretary for domestic finance, House Government Reform and Oversight, Subcommittee on Government Management, Information and Technology, RR-1768, June 18, 1997.

Hendry, M., *Implementing EDI*, Artech House, Boston, MA, 1993.

Herbert, B., What privacy rights? *The New York Times*, Week in Review, Section 4, September 27, 1998, p. 15, available at http://www.nytimes.com/1998/09/27/opinion/in-america-what-privacy-rights.html, last accessed August 14, 2015.

Hermann, R., Husemann, D., and Trommler, P., OpenCard Framework 1.1, October 1998.

Hernandez, E. A., War of the mobile browsers, *Pervas. Comput.*, 8(1), 82–85, 2009.

Hill, K., The first "Bitcoin 2.0" crowd sale was a wildly successful $7 million disaster, *Forbes*, June 3, 2014, available at http://www.forbes.com/sites/kashmirhill/2014/06/03/mastercoin-maidsafe-crowdsale, last accessed March 29, 2015.

Hill, R., Minitel: An example of electronic commerce compared to the World Wide Web, *EDI Forum*, 9(2), 89–96, 1996.

Hill, R., Retina identification, in *Biometrics: Personal Identification in Networked Society*, Jain, A., Bolle, R., and Pankati, S., Eds., Kluwer, Dordrecht, the Netherlands, 1999, pp. 123–124.

Hillebrand, F., Ed., *GSM and UMTS: The Creation of Global Mobile Communication*, John Wiley & Sons, Chichester, UK, 2002.

Hiltgen, A., Kramp, T., and Weigold, T., Secure Internet banking authentication, *IEEE Secur. & Priv.*, 4(2), 21–29, 2006.

Hoog, A, Forensic security analysis of Google Wallet, December 12, 2011, via Forensics, Mobile Security Blog, available at https://viaforensics.com/mobile-security/forensics-security-analysis-google-wallet.html, last accessed October 17, 2013.

Horiuchi, T. and Hada, T., A complementary study for the evaluation of face recognition technology, in *Proceedings of the 47th International Carnahan Conference on Security Technologies* (*ICCST*), Medellin, Columbia, October 8–11, 2013, pp. 1–5.

Hornyak, T., One year later, we're no closer to finding Mt. Gox's missing millions worth of bitcoins, *PCWorld*, March 4, 2015, available at http://www.pcworld.com/article/2892892/one-year-later-were-no-closer-to-finding-mtgoxs-missing-millions.html, last accessed March 6, 2015.

Huemer, C., Liegl, P., Schuster, R., Zapletal, M., and Hofreiter, B., Service-oriented enterprise modeling and analysis, in *Handbook of Enterprise Integration*, Sherif, M. H., Ed., CRC Press/Auerbach Publications, Boca Raton, FL, 2010, pp. 307–321.

Hurst, B., How many people really own bitcoins—And why does it matter? *Bitscan*, February 10, 2014, available at https://bitscan.com/articles/how-many-people-really-own-bitcoins-and-why-does-it-matter, last accessed March 7, 2015.

Husemann, D., Standards in the smart card world, *Comput. Netw.*, 36, 473–487, 2001.

Huynh, S., Forrester research mobile commerce forecast, 2012 to 2017 (U.S.), August 8, 2012, http://www.forrester.com/Forrester+Research+Mobile+Commerce+Forecast+2012+To+2017+US/fulltext/-/E-RES80861, last accessed November 15, 2013.

IDABC, Solution profile—Global trust authority (GTA), European Federated Validation Service Study, European eGovernment Services, July 2009, available at http://ec.europa.eu/idabc/servlets/Doc2162.pdf?id=32169, last accessed July 5, 2015.

Identity Theft Resource Center (ITRC), Data breach reports, August 11, 2015, available at http://www.idtheftcenter.org/images/breach/DataBreachReports_2015.pdf, last accessed August 14, 2015.

Impedovo, D. and Pirlo, G., Automatic signature verification: The state of the art, *IEEE Trans. Syst. Man Cybern. Part C*, 38(5), 609–635.

International Civil Aviation Organization, Deployment and specification of globally interoperable biometric standards for machine assisted identity confirmation using machine readable travel document, Technical report, Version 2.0, Technical Advisory Group (TAG), Machine Readable Travel Document (MRTD)/New Technology Working Group (NTWG), May 21, 2004, available at http://www.policylaundering.org/archives/ICAO/Biometrics_Deployment_Version_2.0.pdf, last accessed August 16, 2015.

International Civil Aviation Organization, Machine reading options for td1 size MRtds, ICAO/NTWG sub-working group for new specifications td1 card, Technical report, Version 1.0, April 7, 2011, available at http://www.icao.int/Security/mrtd/Downloads/TR-New%20Specifications%20td1/TR-New%20specifications%20td1%20Card%20V%201%200%20%282%29.pdf, last accessed July 16, 2015.

International Labour Organization, Seafarers' identity documents convention (revised), 2003 (No. 185), 2006.

International Telecommunication Union, *The Mobile Money Revolution Part 2: Financial Inclusion Enabler*, ITU-T Technology Watch Report, Geneva, May 2013, available at http://www.itu.int/en/ITU-T/techwatch/Pages/mobile-money-standards.aspx, last accessed January 15, 2016.

International Telecommunication Union—Recommendation X.209, Specifications of basic encoding rules for abstract syntax notation one (ASN.1), ITU, Geneva, Switzerland, 1988.

International Telecommunication Union—Recommendation X.800, Security Architecture for Open Systems Interconnection for CCITT Applications, 1991.

International Telecommunication Union—Recommendation X.811, Information Technology—Open System Interconnection—Security Frameworks for Open Systems: Authentication Framework, November, 1995.

International Telecommunication Union, Recommendation X.812, Information technology—Open systems interconnection security frameworks for open systems: Access control framework, ITU, Geneva, Switzerland, November 1995.

International Telecommunication Union, Recommendation X.813, Information technology—Open systems interconnection security frameworks for open systems: Non-repudiation framework, ITU, Geneva, Switzerland, October 1996.

International Telecommunication Union, Recommendation X.1083, Information technology—Biometrics—BioAPI interworking protocol, ITU, Geneva, Switzerland, November 2007.

International Telecommunication Union, Recommendation X.1084, Telecommunication security—Telebiometrics system mechanism—Part 1: General biometric authentication protocol and system model profiles for telecommunications systems, ITU, Geneva, Switzerland, May 2008.

International Telecommunication Union, Recommendation Y.2740, Global information infrastructure, Internet protocol aspects and next generation networks—Security requirements for mobile remote financial transactions in next generation networks, ITU, Geneva, Switzerland, January 2011a.

International Telecommunication Union, Recommendation Y.2740, Global information infrastructure, Internet protocol aspects and next generation networks—Architecture of secure mobile financial transactions in next generation networks, ITU, Geneva, Switzerland, January 2011b.

International Telecommunication Union— Telecommunication Development Sector, Percentage of individuals using the Internet (excel), July 8, 2015, available at http://www.itu.int/en/ITU-D/Statistics/Pages/stat/default.aspx, last accessed January 28, 2016.

Inza, J., Banesto Easy SET project, July 6, 2000. Available at http://docslide.us/download/link/banesto-easy-set-project-julian-inza-jinzabanestoes-technological-strategy, last accessed January 21, 2016.

Irvine, D., Maidsafe.net, U.S. Patent 20,100,064,354 A1, 2010.

ISO Contribution N1737, Mechanisms for Bridging EDI with SGML, ISO/IEC JTC1/SC18/WG8, ISO, Geneva, Switzerland, August 29, 1994.

ISO 1004-2:2013, Traitement de l'information—Reconnaissance des caractères à encre magnétique—Partie 2: spécifications d'impression CMC7, ISO, Geneva, Switzerland, June 2013.

ISO 7372, Trade Data Interchange—Trade data elements directory, ISO, Geneva, Switzerland, 2005.

ISO 7498-2:1989, Information technology—Open systems interconnection—Basic reference model—Part 2: Security architecture, ISO, Geneva, Switzerland, 1989.

ISO 8372:1987, Information processing—Modes of operation for a 64-bit block cipher algorithm, ISO, Geneva, Switzerland, 1987.

ISO 8583-1:2003, Financial transaction card originated messages—Interchange message specifications—Part 1: Messages, data elements and code values, ISO, Geneva, Switzerland, 2003.

ISO 8583-2:1998, Financial transaction card originated messages—Interchange message specifications—Part 2: Application and registration procedures for Institution Identification Codes (IIC), ISO, Geneva, Switzerland, 1998.

ISO 8583-3:2003, Financial transaction card originated messages—Interchange message specifications—Part 3: Maintenance procedures for messages, data elements and code values, ISO, Geneva, Switzerland, 2003.

ISO 8731-1:1987, Banking—Approved algorithms for message authentication—Part 1: DEA, ISO, Geneva, Switzerland, 1987.

ISO 8879:1986, Information processing—Text and office systems—Standard Generalized Markup Language (SGML), ISO, Geneva, Switzerland, 1986.

ISO 9564-1:2011, Financial Services—Personal Identification Number (PIN) management and security—Part 1: Basic principles and requirements for PINs in card-based systems, ISO, Geneva, Switzerland, 2011.

ISO 9564-2:2005, Banking—Personal identification number management and security—Part 2: Approved algorithms for PIN encipherment, ISO, Geneva, Switzerland, 2005.

ISO 9735:1998, Electronic Data Interchange for Administration, Commerce and Transport (EDIFACT)—Application level syntax rules, ISO, Geneva, Switzerland, 1998.

ISO 9735-1:2002, Electronic Data Interchange for Administration, Commerce and Transport (EDIFACT)—Application level syntax rules (syntax version number: 4, syntax release number: 1)—Part 1: Syntax rules common to all parts, ISO, Geneva, Switzerland, 2002.

ISO 9735-2:2002, Electronic Data Interchange for Administration, Commerce and Transport (EDIFACT)—Application level syntax rules (syntax version number: 4, syntax release number: 1)—Part 2: Syntax rules specific to batch EDI, ISO, Geneva, Switzerland, 2002.

ISO 9735-3:2002, Electronic Data Interchange for Administration, Commerce and Transport (EDIFACT)—Application level syntax rules (syntax version number: 4, syntax release number: 1)—Part 3: Syntax rules specific to interactive EDI, ISO, Geneva, Switzerland, 2002.

ISO 9735-4:2002, Electronic Data Interchange for Administration, Commerce and Transport (EDIFACT)—Application level syntax rules (syntax version number: 4, syntax release number: 1)—Part 4: Syntax and service report message for batch EDI (message type—CONTRL), ISO, Geneva, Switzerland, 2002.

ISO 9735-5:2002, Electronic Data Interchange for Administration, Commerce and Transport (EDIFACT)—Application level syntax rules (syntax version number: 4, syntax release number: 1)—Part 5: Security rules for batch EDI (authenticity, integrity and non-repudiation of origin), ISO, Geneva, Switzerland, 2002.

ISO 9735-6:2002, Electronic Data Interchange for Administration, Commerce and Transport (EDIFACT)—Application level syntax rules (syntax version number: 4, syntax release number: 1)—Part 6: Secure authentication and acknowledgment message (message type—AUTACK), ISO, Geneva, Switzerland, 2002.

ISO 9735-7:2002, Application level syntax rules (syntax version number: 4, syntax release number: 1)—Part 7: Security rules for batch EDI (confidentiality), 2002.

ISO 9735-8:2002, Electronic Data Interchange for Administration, Commerce and Transport (EDIFACT)—Application level syntax rules (syntax version number: 4, syntax release number: 1)—Part 8: Associated data in EDI, ISO, Geneva, Switzerland, 2002.

ISO 9735-9:2002, Electronic Data Interchange for Administration, Commerce and Transport (EDIFACT)—Application level syntax rules (syntax version number: 4, syntax release number: 1)—Part 9: Security key and certificate management message (message type—KEYMAN), ISO, Geneva, Switzerland, 2002.

ISO 9735-10:2002, Electronic Data Interchange for Administration, Commerce and Transport (EDIFACT)—Application level syntax rules (syntax version number: 4, syntax release number: 1)—Part 10: Syntax service directories, ISO, Geneva, Switzerland, 2002.

ISO 11568-1:2005, Banking—Key management (retail)—Part 1: Principles, 2005.

ISO 11568-2:2012, Financial services—Key management (retail)—Part 2: Symmetric ciphers, their key management and life cycle, ISO, Geneva, Switzerland, 2012.

ISO 11568-4:2007, Banking—Key management (retail)—Part 4: Asymmetric cryptosystems—Key management and life cycle, ISO, Geneva, Switzerland, 2007.

ISO 11784:1996, Radio frequency identification of animals—Code structure, revised in 2010, ISO, Geneva, Switzerland, 1996.

ISO 11785:1996, Radio frequency identification of animals—Technical concept, ISO, Geneva, Switzerland, 1996.

ISO 15000-5:2014, Electronic Business Extensible Markup Language (ebXML)—Part 5: Core Components Specification (CCS), ISO, Geneva, Switzerland, 2014.

ISO 20022-1:2004, Financial services—Universal financial industry message scheme—Part 1: Overall methodology and format specifications for inputs to and outputs from the ISO 20022 repository, ISO, Geneva, Switzerland, 2004.

ISO 20022-2:2007, Financial services—Universal financial industry message scheme—Part 2: Roles and responsibilities of the registration bodies, ISO, Geneva, Switzerland, 2007.

ISO/IEC 7498-1:1994, Information technology—Open systems interconnection—Basic reference model: The basic model, ISO, Geneva, Switzerland, 1994.

ISO/IEC 7501-1:2008—Identification cards—Machine readable travel documents—Part 1: Machine readable passport, ISO, Geneva, Switzerland, 2008.

ISO/IEC 7816-1:2011, Identification cards—Integrated circuit cards—Part 1: Cards with contacts—Physical characteristics, ISO, Geneva, Switzerland, 2011.

ISO/IEC 7816-2:2007, Identification cards—Integrated circuit cards—Part 2: Cards with contacts—Dimensions and location of the contacts, ISO, Geneva, Switzerland, 2007.

ISO/IEC 7816-3:2006, Identification cards—Integrated circuit cards—Part 3: Cards with contacts–Electrical interface and transmission protocols, ISO, Geneva, Switzerland, 2006.

ISO/IEC 7816-4:2013, Identification cards—Integrated circuit cards—Part 4: Organization, security and commands for interchange, ISO, Geneva, Switzerland, 2013.

ISO/IEC 7816-5:2004, Identification cards—Integrated circuit cards—Part 5: Registration of application providers, ISO, Geneva, Switzerland, 2004.

ISO/IEC 7816-6:2004, Identification cards—Integrated circuit cards—Part 6: Interindustry data elements for interchange, ISO, Geneva, Switzerland, 2004.

ISO/IEC 7816-7:1999, Identification cards—Integrated circuit(s) cards with contacts—Part 7: Interindustry commands for Structured Card Query Language (SCQL), ISO, Geneva, Switzerland, 1999.

ISO/IEC 7816-8:2004, Identification cards—Integrated circuit cards—Part 8: Commands for security operations, ISO, Geneva, Switzerland, 2004.

ISO/IEC 7816-9:2004, Identification cards—Integrated circuit cards—Part 9: Commands for card management, ISO, Geneva, Switzerland, 2004.

ISO/IEC 7816-10:1999, Identification cards—Integrated circuit(s) cards with contacts—Part 10: Electronic signals and answer to reset for synchronous cards, ISO, Geneva, Switzerland, 1999.

ISO/IEC 7816-11:2004, Identification cards—Integrated circuit cards—Part 11: Personal verification through biometric methods, ISO, Geneva, Switzerland, 2004.

ISO/IEC 7816-12:2005, Identification cards—Integrated circuit cards—Part 12: Cards with contacts—USB electrical interface and operating procedures, ISO, Geneva, Switzerland, 2005.

ISO/IEC 7816-13:2007, Identification cards—Integrated circuit cards—Part 13: Commands for application management in a multi-application environment, ISO, Geneva, Switzerland, 2007.

ISO/IEC 7816-15:2004, Identification cards—Integrated circuit cards—Part 15: Cryptographic information application, ISO, Geneva, Switzerland, 2004.

ISO/IEC 8824-1:1998, Information technology—Abstract Syntax Notation One (ASN.1): Specification of basic notation, ISO, Geneva, Switzerland, 1998, updated as ISO/IEC 8824-1:2008.

ISO/IEC 8824-2:1998, Information technology—Abstract Syntax Notation One (ASN.1): Information object specification, ISO, Geneva, Switzerland, 1999, updated as ISO/IEC 8824-2:2008.

ISO/IEC 8824-3:1998, Information technology—Abstract Syntax Notation One (ASN.1): Constraint specification, ISO, Geneva, Switzerland, 1998, update as ISO/IEC 8824-3:2008.

ISO/IEC 8824-4:1998, Information technology—Abstract Syntax Notation One (ASN.1): Parameterization of ASN.1 specifications, ISO, Geneva, Switzerland, 1998, updated as ISO/IEC 8824-4:2008.

ISO/IEC 8825-1:1998, Information technology—ASN.1 encoding rules: Specification of Basic Encoding Rules (BER), Canonical Encoding Rules (CER) and Distinguished Encoding Rules (DER), ISO, Geneva, Switzerland, 1998, updated as ISO/IEC 8824-1:2008.

ISO/IEC 8859-1:1998, Information technology—8-bit single-byte coded graphic characters sets—Part 1: Latin alphabet no. 1, ISO, Geneva, Switzerland, 1998.

ISO/IEC 8859-2:1999, Information technology—8-bit single-byte coded graphic characters sets—Part 2: Latin alphabet no. 2, ISO, Geneva, Switzerland, 1999.

ISO/IEC 8859-3:1999, Information technology—8-bit single-byte coded graphic characters sets—Part 3: Latin alphabet no. 3, ISO, Geneva, Switzerland, 1999.

ISO/IEC 8859-4:1998, Information technology—8-bit single-byte coded graphic characters sets—Part 4: Latin alphabet no. 4, ISO, Geneva, Switzerland, 1998.

ISO/IEC 8859-5:1999, Information technology—8-bit single-byte coded graphic characters sets—Part 5: Latin/Cyrillic alphabet, ISO, Geneva, Switzerland, 1999.

ISO/IEC 8859-6:1999, Information technology—8-bit single-byte coded graphic characters sets—Part 6: Latin/Arabic alphabet, ISO, Geneva, Switzerland, 1999.

ISO/IEC 8859-7:2003, Information technology—8-bit single-byte coded graphic characters sets—Part 7: Latin/Greek alphabet, ISO, Geneva, Switzerland, 2003.

ISO/IEC 8859-8:1999, Information technology—8-bit single-byte coded graphic characters sets—Part 8: Latin/Hebrew alphabet, ISO, Geneva, Switzerland, 1999.

ISO/IEC 8859-9:1999, Information technology—8-bit single-byte coded graphic characters sets—Part 9: Latin alphabet no. 5, ISO, Geneva, Switzerland, 1999.

ISO/IEC 8859-10:1998, Information technology—8-bit single-byte coded graphic characters sets—Part 10: Latin alphabet no. 6, ISO, Geneva, Switzerland, 1998.

ISO/IEC 8859-11:2001, Information technology—8-bit single-byte coded graphic character sets—Part 11: Latin/Thai alphabet, ISO, Geneva, Switzerland, 2001.

ISO/IEC 8859-13:1998, Information technology—8-bit single-byte coded graphic characters sets—Part 13: Latin Alphabet no. 7, ISO, Geneva, Switzerland, 1998.

ISO/IEC 8859-14:1998, Information technology—8-bit single-byte coded graphic characters sets—Part 14: Latin alphabet no. 8 (Celtic), ISO, Geneva, Switzerland, 1998.

ISO/IEC 8859-15:1999, Information technology—8-bit single-byte coded graphic characters sets—Part 15: Latin alphabet no. 9, ISO, Geneva, Switzerland, 1999.

ISO/IEC 8859-16:2001, Information technology—8-bit single-byte coded graphic characters sets—Part 16: Latin alphabet no. 10, ISO, Geneva, Switzerland, 2001.

ISO/IEC 9594-1, International Telecommunication Union—Recommendation X.500, Information technology—Open systems interconnection—The directory: Overview of concepts, models and services, ISO, Geneva, Switzerland, 2012.

ISO/IEC 9594-2, International Telecommunication Union—Recommendation X.501, Information technology—Open systems interconnection—The directory: Models, ISO, Geneva, Switzerland, 2012.

ISO/IEC 9594-8, International Telecommunication Union—Recommendation X.509, Information technology—Open systems interconnection—The directory: Public-key and attribute certificate frameworks, ISO, Geneva, Switzerland, 2012.

ISO/IEC 9594-3, International Telecommunication Union—Recommendation X.511, Information technology—Open systems interconnection—The directory: Overview of concepts, models and services, ISO, Geneva, Switzerland, 2012.

ISO/IEC 9594-4, International Telecommunication Union—Recommendation X.518, Information technology—Open systems interconnection—The directory: Procedures for distributed operation, ISO, Geneva, Switzerland, 2012.

ISO/IEC 9594-4, International Telecommunication Union—Recommendation X.519, Information technology—Open systems interconnection—The directory: Protocol specifications, ISO, Geneva, Switzerland, 2012.

ISO/IEC 9594-6, International Telecommunication Union—Recommendation X.520, Information technology—Open systems interconnection—The directory: Selected attribute types, ISO, Geneva, Switzerland, 2012.

ISO/IEC 9594-7, International Telecommunication Union—Recommendation X.521, Information technology—Open systems interconnection—The directory: Selected object classes, ISO, Geneva, Switzerland, 2012.

ISO/IEC 9594-9, International Telecommunication Union—Recommendation X.525, Information technology—Open systems interconnection—The directory: Replication, ISO, Geneva, Switzerland, 2012.

ISO/IEC 9796-2:2010, Information technology—Security techniques—Digital signature schemes giving message recovery—Part 2: Integer factorization based mechanisms, ISO, Geneva, Switzerland, 2010.

ISO/IEC 9797-1:1999, Information technology—Security techniques—Message authentication codes (MAC), Part 1: Mechanism using a block cipher, ISO, Geneva, Switzerland, 1999.

ISO/IEC 9797-2:2002, Information technology—Security techniques—Message authentication codes (MAC) Part 2: Mechanism using a hash-function, ISO, Geneva, Switzerland, 2002.

ISO/IEC 10116:1997, Information technology—Security techniques—Mode of operation for an n-bit block cipher algorithm, ISO, Geneva, Switzerland, 1997.

ISO/IEC 10118-1:1994, Information technology—Security techniques—Hash-functions—Part 1: General, ISO, Geneva, Switzerland, 1994.

ISO/IEC 10118-2:1994, Information technology—Security techniques—Hash-functions—Part 2: Hash-functions using an n-bit block cipher algorithm, ISO, Geneva, Switzerland, 1994.

ISO/IEC 10118-3:1998, Information technology—Security techniques—Hash-functions—Part 3: Dedicated hash-functions, ISO, Geneva, Switzerland, 1998.

ISO/IEC 10118-4:1998, Information technology—Security techniques—Hash-functions—Part 4: Hash-functions using modular arithmetic, ISO, Geneva, Switzerland, 1998.

ISO/IEC 14662:2010, Information technology—Open-EDI reference model, 2010.

ISO/IEC 14443-1:2008, Identification cards—Contactless integrated circuit cards—Proximity cards—Part 1: Physical characteristics, ISO, Geneva, Switzerland, 2008.

ISO/IEC 14443-2:2010, Identification cards—Contactless integrated circuit cards—Proximity cards—Part 2: Radio frequency power and signal interface, ISO, Geneva, Switzerland, 2010.

ISO/IEC 14443-3:2011, Identification cards—Contactless integrated circuit cards—Proximity cards—Part 3: Initialization and anticollision, ISO, Geneva, Switzerland, 2011.

ISO/IEC 14443-4:2008, Identification cards—Contactless integrated circuit cards—Proximity cards—Part 4: Transmission protocol, ISO, Geneva, Switzerland, 2008.

ISO/IEC 15408-1:2009, Information technology—Security techniques—Evaluation criteria for IT security—Part 1: Introduction and general model, ISO, Geneva, Switzerland, 2009.

ISO/IEC 15408-2:2008, Information technology—Security techniques—Evaluation criteria for IT security—Part 2: Security functional components, ISO, Geneva, Switzerland, 2008.

ISO/IEC 15408-3:2008, Information technology—Security techniques—Evaluation criteria for IT security—Part 3: Security assurance components, ISO, Geneva, Switzerland, 2008.

ISO/IEC 15438:2006, Information technology—Automatic identification and data capture techniques—PDF417 bar code symbology specification, ISO, Geneva, Switzerland, 2006.

ISO/IEC 15693-1:2010, Identification cards—Contactless integrated circuit cards—Vicinity cards—Part 1: Physical characteristics, ISO, Geneva, Switzerland, 2010.

ISO/IEC 15693-2:2006, Identification cards—Contactless integrated circuit cards—Vicinity cards—Part 2: Air interface and initialization, ISO, Geneva, Switzerland, 2006.

ISO/IEC 15693-3:2009, Identification cards—Contactless integrated circuit cards—Vicinity cards—Part 3: Anticollision and transmission protocol, ISO, Geneva, Switzerland, 2009.

ISO/IEC 16022:2006, Information technology—Automatic identification and data capture techniques—Data Matrix bar code symbology specification, ISO, Geneva, Switzerland, 2006.

ISO/IEC 16023:2000, Information technology—International symbology specification—MaxiCode, ISO, Geneva, Switzerland, 2000.

ISO/IEC 18000-1:2008, Information technology—Radio frequency identification for item management—Part 1: Reference architecture and definition of parameters to be standardized, ISO, Geneva, Switzerland, 2008.

ISO/IEC 18000-2:2009, Information technology—Radio frequency identification for item management—Part 2: Parameters for air interface communications below 135 kHz, ISO, Geneva, Switzerland, 2009.

ISO/IEC 18000-3:2010, Information technology—Radio frequency identification for item management—Part 3: Parameters for air interface communications at 13.56 MHz, ISO, Geneva, Switzerland, 2010.

ISO/IEC 18000-4:2008, Information technology—Radio frequency identification for item management—Part 4: Parameters for air interface communications at 2.45 GHz, ISO, Geneva, Switzerland, 2008.

ISO/IEC 18000-6:2013, Information technology—Radio frequency identification for item management—Part 6: Parameters for air interface communications at 860 MHz to 960 MHz general, ISO, Geneva, Switzerland, 2013.

ISO/IEC 18000-61:2012, Information technology—Radio frequency identification for item management—Part 61: Parameters for air interface communications at 860 MHz to 960 MHz type A, ISO, Geneva, Switzerland, 2012.

ISO/IEC 18000-62:2012, Information technology—Radio frequency identification for item management—Part 62: Parameters for air interface communications at 860 MHz to 960 MHz type B, ISO, Geneva, Switzerland, 2012.

ISO/IEC 18000-63:2013, Information technology—Radio frequency identification for item management—Part 63: Parameters for air interface communications at 860 MHz to 960 MHz type C, ISO, Geneva, Switzerland, 2013.

ISO/IEC 18000-64:2012, Information technology—Radio frequency identification for item management—Part 61: Parameters for air interface communications at 860 MHz to 960 MHz type D, ISO, Geneva, Switzerland, 2012.

ISO/IEC 18000-7:2009, Information technology—Radio frequency identification for item management—Part 7: Parameters for active air interface communications at 433 MHz, ISO, Geneva, Switzerland, 2009.

ISO/IEC 18004:2006, Information technology—Automatic identification and data capture techniques—QR Code 2005 bar code symbology specification, ISO, Geneva, Switzerland, 2006.

ISO/IEC 18033-3:2005, Information technology—Security techniques—Encryption algorithms—Part 3: Block ciphers, ISO, Geneva, Switzerland, 2005.

ISO/IEC 18092:2013, Information technology—Telecommunications and information exchange between systems—Near Field Communication Interface and Protocol (NFCIP-1), ISO, Geneva, Switzerland, 2013.

ISO/IEC 19784-1:2006, Information technology—Biometric application programming interface—Part 1: BioAPI specification, ISO, Geneva, Switzerland, 2006.

ISO/IEC 19794-2:2011, Information technology—Biometric data interchange formats—Part 2: Finger minutiae data, ISO, Geneva, Switzerland, 2011.

ISO/IEC 19794-5:2011, Information technology—Biometric data interchange formats—Part 5: Face image data, ISO, Geneva, Switzerland, 2011.

ISO/IEC 19794-6:2011, Information technology—Biometric data interchange formats—Part 6: Iris image data, ISO, Geneva, Switzerland, 2011.

ISO/IEC 19794-7:2007, Information technology—Biometric data interchange formats—Part 7: Signature/sign time series data, ISO, Geneva, Switzerland, 2007.

ISO/IEC 21481:2012, Information technology—Telecommunications and information exchange between systems—Near Field Communication Interface and Protocol -2 (NFCIP-2), ISO, Geneva, Switzerland, 2012.

ISO/IEC 24728:2006, Information technology—Automatic identification and data capture techniques—MicroPDF417 bar code symbology specification, ISO, Geneva, Switzerland, 2006.

ISO/TR 9564-4:2004, Banking—Personal Identification Number (PIN) management and security—Part 4: Guidelines for PIN handling in open networks, ISO, Geneva, Switzerland, 2004.

ISO/TS 20022-3:2004, Financial services—Universal financial industry message scheme—Part 3: ISO 20022 modelling guidelines, ISO, Geneva, Switzerland, 2004.

ISO/TS 20022-4:2004, Financial services—Universal financial industry message scheme—Part 4: ISO 20022 XML design rules, ISO, Geneva, Switzerland, 2004.

ISO/TS 20022-5:2004, Financial services—Universal financial industry message scheme—Part 5: ISO 20022 reverse engineering, ISO, Geneva, Switzerland, 2004.

Jack, W. and Suri, T., The economics of M-PESA: An update, October 2010, available at http://www.mit.edu/~tavneet/M-PESA_Update.pdf, last accessed December 26, 2013.

Jack, W. and Suri, T., Mobile money: The economics of M-PESA, National Bureau of Economic Research, Working Paper 16721, January 2011, available at http://www.nber.org/papers/w16721, last accessed December 26, 2013.

Jackson, D., Preparing the organization for EDI, in *The EDI Handbook*, Gikins, M. and Hitchcock, D., Eds., Blenheim Online, London, UK, 1988, pp. 149–155.

Jackson, E. M., *The PayPal Wars*, 2nd ed., World Ahead Publishing, Los Angeles, CA, 2006.

Jaffe, F. and Landry, S., Electronic checks: The best of both worlds, *Electron. Commerce World*, July 1997, available at http://www.echeck.org/library/wp/bestofboth.html, last accessed July 30, 2015.

Jarupunphol, P. and Mitchell, C. J., Measuring 3-D Secure and 3-D SET against ecommerce end-user requirements, in *Proceedings of the Eighth Collaborative Electronic Commerce Technology and Research Conference*, National University of Ireland, Galway, Ireland, 2003, pp. 51–64, available at www.chrismitchell.net/m3sa3s.pdf, last accessed May 31, 2014.

Java Community Process, JSR 177: Security and trust services API for J2ME™, August 20, 2007, available at http://jcp.org/en/jsr/detail?id=177, last accessed December 31, 2015.

Java Community Process, JSR 82: JavaTM APIs for Bluetooth, draft review 4, April 12, 2010, available at https://jcp.org/en/jsr/detail?id=82, last accessed December 31, 2015.

Java Community Process, JSR 257: Contactless communication API, April 14, 2011, available at http://jcp.org/en/jsr/detail?id=257, last accessed December 31, 2015.

JIS (Japanese Industrial Standard) X 6319-4:2010, Specification of implementation for integrated circuit(s) cards—Part 4: High speed proximity cards, available at http://www.proxmark.org/files/Documents/13.56%20MHz%20-%20Felica/JIS.X.6319-4.Sony.Felica.pdf, last accessed November 25, 2013.

Jones, C., Bad days for software, *IEEE Spectr.*, 35(9), 47–52, 1998.

Jopson, B. and Taylor, J., Credit card data stolen from up to 40m shoppers at retailer Target, *Financial Times*, December 2, 2013, p. 1.

Juels, A, and Brainard, J., A cryptographic countermeasure against connection depletion attacks, in *Proceedings of the Internet Social Symposium Network Distributed System Security*, Kent, S., Ed., IEEE Computer Society Press, Washington, DC, 1999, pp. 151–165, available at http://www.emc.com/emc-plus/rsa-labs/staff-associates/client-puzzles.htm, last accessed August 19, 2015.

Kaiserswerth, M., The OpenCard Framework and PC/SC, IBMReport RZ 2999 (#93045), March 8, 1998, available at http://domino.watson.ibm.com/library/cyberdig.nsf/papers/2D63CAE9875AFB8A852565E9006BAB99/$File/RZ2999.pdf, last accessed February 5, 2014.

Kaplan, S. and Sawhney, M., E-Hubs: The new B2B marketplaces, *Harv. Bus. Rev.*, 78(3), 97–103, May–June 2000.

Kato, H. and Tan, K. T., Pervasive 2D barcodes for camera phone applications, *IEEE Pervas. Comput. Mobile* 6(4), 76–85, 2007, available at http://ro.ecu.edu.au/do/search/?q=Pervasive%202D%20barcodes%20for%20camera%20phone%20applications&start=0&context=302996, last accessed January 26, 2014.

Kato, H., Tan, K. T., and Chai, D., *Barcodes for Mobile Devices*, Cambridge University Press, Cambridge, UK, 2010.

Keizer, G., DigiNotar dies from certificate hack caper, *Computerworld*, September 21, 2011, http://www.computerworld.com/s/article/9220175/DigiNotar_dies_from_certificate_hack_caper, last accessed August 24, 2014.

Kelly, B., *The Bitcoin Big Bang: How Alternative Currencies Are about to Change the World*, John Wiley & Sons, Hoboken, NJ, 2015.

Kelly, E. W., The future of electronic money: A regulator's perspective, *IEEE Spectr.*, 34(2), 21–22, 1997.

Kiernan, J., Credit card and debit card fraud statistics, 2013, available at http://www.cardhub.com/edu/credit-debit-card-fraud-statistics, last accessed June 1, 2014.

Kimberley, P., *Electronic Data Interchange*, McGraw-Hill, New York, 1991.

King, S. and Nadal, S., PPCoin: Peer-to-peer crypto-currency with proof-of-stack, August 19, 2012, available at http://peercoin.net/assets/paper/peercoin-paper.pdf, last accessed April 16, 2014.

Kirschner, L., The battle of the electronic purses: Who will win? *Card Forum Int.*, 2(3), 33–41, 1998.

Kitten, T. Carphone Warehouse hack exposes data of 2.4 million customers, *Data Breach Today*, August 10, 2015, available at http://www.databreachtoday.com/carphone-warehouse-hack-exposes-data-24-million-customers-a-8463, last accessed August 15, 2015.

Klíma, V., Pokorný, O., and Rosa, T., Attacking RSA-based sessions in SSL/TLS, in *Proceedings of the Fifth International Workshop on Cryptographic Hardware and Embedded Systems CHES 2003*, Cologne, Germany, September 8–10, 2003, Lecture Notes in Computer Science, Vol. 2779, Walter, C. D., Koç, Ç. K., and Paar, C., Eds., Springer-Verlag, Berlin, Germany, 2003, pp. 426–440.

Kömmerling, O. and Kuhn, M. G., Design principles for tamper-resistant smart card processors, in *Proceedings of the USENIX Workshop on Smartcard Technology*, Chicago, IL, 1999, pp. 9–20.

Kosinski, M., Stillwell, D., and Graeple, T., Private traits and attributes are predictable from digital records of human behavior, *Proc. Natl. Acad. Sci. USA*, 110(15), 5802–5805, 2013, available at http://www.pnas.org/content/110/15/5802.full.pdf, last accessed August 14, 2015.

Kravitz, J., SDML-Signed Document Markup Language, Version 2.0, April 1998, available at http://www.w3.org/TR/NOTE-SDML, last accessed July 29, 2012.

Kravitz, J., Ed., FSML™—Financial Services Markup Language, Version 1.50, July 14, 1999, available at http://www.echeck.org/library/ref/fsml-v1500a.pdf, last accessed July 29, 2015.

Krawczyk, H., SKEME: A versatile secure key exchange mechanism for Internet, in *Proceedings of the Internet Social Symposium Network and Distributed System Security*, IEEE Computer Society Press, Washington, DC, 1996, pp. 114–127.

Krawczyk, H., Paterson, K. G., and, Wee, H., On the security of the TLS protocol, in *Proceedings of CRYPTO 2013*, Part 1, Lecture Notes in Computer Science, Vol. 8042, Canetti, R. and Garay, J. A., Eds., Springer-Verlag, Berlin, Germany, 2013, pp. 429–448.

Krebs, B., Data broker giants hacked by ID theft service, *KrebsonSecurity*, September 25, 2013a, http://krebsonsecurity.com/2013/09/data-broker-giants-hacked-by-id-theft-service, last accessed August 14, 2015.

Krebs, B., Experian sold consumer data to ID theft service, *KrebsonSecurity*, October 20, 2013b, http://krebsonsecurity.com/2013/10/experian-sold-consumer-data-to-id-theft-service, last accessed August 15, 2015.

Krebs, B., "Replay" attacks spoof chip card charges, *KrebsonSecurity*, October 27, 2014, https://krebsonsecurity.com/2014/10/replay-attacks-spoof-chip-card-charges, last accessed December 14, 2014.

Krishenbaum, I. and Wool, A., How to build a low-cost, extended-range RFID skimmer, in *Proceedings of Security'06: 15th USENIX Security Symposium*, Vancouver, British Columbia, Canada, 2006, pp. 43–57, available at http://static.usenix.org/events/sec06/tech/full_papers/kirschenbaum/kirschenbaum_html, last accessed December 31, 2015.

Kuchler, H., Security breach at Adobe hit 38m users, *Financial Times*, October 31, 2013a, p. 14.

Kuchler, H., Rise of Bitcoin fuels wave of cyberattack for ransom, *Financial Times*, December 9, 2013b, p. 3.

Kuchler, H., Hacking back, *Financial Times*, July 28, 2015, p. 7.

Kwon, E.-K., Cho, Y.-G., and Chae, K.-J., Integrated transport layer security: End-to-end security model between WTLS and TLS, in *Proceedings of the 15th International Conference on Information Networking*, Beppu, Japan, 2001, pp. 65–71.

Kwon, T., Yoon, M., Kang, J., Song, J., and, Kang, C.-G., A modeling of security management system for electronic data interchange, in *Proceedings of the Second IEEE Symposium on Computers and Communications (ISCC)*, IEEE Computer Society Press, Washington, DC, 1997, pp. 518–522.

Ladendorf, K., Austin looks to get out in front of mobile payments wave, *Austin American Statesman*, April 6, 2013, reproduced in Phoning it in: Mobile payments while becoming more popular, not without risks, *The Star-Ledger*, April 11, 2013, pp. 34, 32.

Lai, X. and Massey, J., Markov ciphers and differential cryptanalysis, in *Proceedings of Eurocrypt'91*, Lecture Notes in Computer Science, Vol. 547, Springer-Verlag, Heidelberg, Germany, 1991a.

Lai, X. and Massey, J., A proposal for a new block encryption standard, in *Proceedings of Eurocrypt'90*, Lecture Notes in Computer Science, Vol. 473, Springer-Verlag, Heidelberg, Germany, 1991b.

Landais, Y., Les produits et services EDI offerts par un opérateur, Transpac, *Télécom Interview*, 33, 60–68, 1997.

Langlois, M., Faverio, D., and Lesourd, M., *XML dans les échanges électroniques: Le Framework ebXML*, Hermes/Lavoisier, Paris, France, 2005.

Lash, C., The revolt of the elites: Have they canceled their allegiance to America? *Harper's Mag.*, 289, 1734, 39–77, November 1994.

Lazarony, L., Perishable credit card numbers take the fear out of Web shopping. Bankrate.com, December 1, 2014, available at http://www.bankrate.com/brm/news/emoney/technoguide2004/virtual-cc1.asp, last accessed December 12, 2014.

Ledgard, J. and Clippinger, J., How a digital currency could transform Africa, *Financial Times*, October 30, 2013, p. 9.

Lederman, J. and Gillum, J., U.S. personnel chief resigns in wake of massive data breach, *AOL*, July 10, 2015, http://www.aol.com/article/2015/07/10/us-personnel-chief-resigns-in-wake-of-massive-data-breach/21207891, last accessed August 23, 2015.

Lerner, S. D., Bitmessage v1.0: Completely broken crypto, *Bitslog*, November 30, 2012, available at https://bitslog.wordpress.com/2012/11/30/bitmessage-completely-broken-crypto, last accessed May 31, 2015.

Lerner, T., *Mobile Payment: Technologien, Strategien, Trends und Fallstudien*, Springer Vieweg, Weisbaden, Germany, 2013.

Levin, R., Richardson, J., Warner, G., and Kerley, K., Explaining cybercrime through the lens of differential association theory: Haddidi44-2.php PayPal case study, in *Proceedings of the eCrime Researchers Summit* (*eCrime*), Las Croabas, Puerto Rico, October 23–24, 2012.

Lima, A., Mitri, A., and Nevins, M., RFID relay attacks: System analysis, modeling, and implementation, in *Security in RFID and Sensor Networks*, Zhang, Y. and Kitsos, P., Eds., Auerbach Publications/Taylor & Francis Group, Boca Raton, FL, 2009, pp. 49–75.

Lindley, R., *Smart Card Innovation*, Saim Pty, Ltd., Australia, 1997.

Liu, M., SDML and XML, April 1998, available at http://www.fstc.org/projects/sdml/sdml_comp.html.

Loeb, L., *Secure Electronic Transactions: Introduction and Technical Reference*, Artech House, Norwood, MA, 1998.

Lorentz, F., Commerce électronique: Une nouvelle donne pour les consommateurs, les entreprises, les citoyens et les pouvoirs publics, January 1998, available at http://www.telecom.gouv.fr/francais.

Lorenz, G. W., Electronic stored value payment systems, market position, and regulatory issues, *Am. Univ. Law Rev.*, 46, 1177–1206, 1997.

MacArthur, J. R., Publisher's letter, *Harper's Mag.*, 327, 1965, 7–9, 2013.

Maltoni, D., Maio, D., Jain, A. K., and Prabhakar, S., *Handbook of Fingerprint Recognition*, Springer Science+Business Media, New York, 2003.

Manasse, M. S., The Millicent protocols for electronic commerce, in *Proceedings of the First USENIX Workshop on Electronic Commerce*, New York, 1995, pp. 117–123.

Manouvier, B. and Ménard, L., *Intégration Applicative, EAI, B2B, BPM et SOA*, Hermès/Lavoisier, Paris, France, 2007, p. 152.

Mariotto, C. and Verdier, M., Innovation and competition in the retail banking industry: An industrial organization perspective, October 5, 2014, available at Social Science Research Network, available at http://ssrn.com/abstract=2505682 or http://dx.doi.org/10.2139/ssrn.2505682, last accessed November 1, 2014.

Martres, D. and Sabatier, G., *La Monnaie Électronique, Que sais-je?* Vol. 2370, Presses Universitaires de France, Paris, France, 1987.

Marwick, A. E., How your data are being deeply mined, *N Y Rev. Books*, 61(1), 22–24, 2014.

Massoth, M. and Bingel, T., Performance of different mobile payment service concepts compared with a NFC-based solution, in *Proceedings of the Fourth International Conference on Internet and Web Applications and Services*, Venice/Mestre, Italy, May 24–28, 2009, pp. 205–210.

MasterCard Worldwide, Mobile MasterCard, PayPass product guide, January 2009, available at http://www.mastercard-mobilepartner.com/pdf/PPG_Manual.pdf, last accessed November 17, 2013.

Matsumoto, Y., Matsumoto, H., Yamada, K., and Hoshino, S., Impact of artificial "gummy" fingers on fingerprint systems, in *Proceedings of the SPIE 4677, Optical Security and Counterfeit Deterrence Techniques IV*, San Jose, CA, April 18, 2002, pp. 275–289.

Matsunaga, M., i-mode media concept, *NTT DoCoMo Technol. J.*, 1(1), 10–19, 1999.

Mayer, M., *The Bankers: The Next Generation*, Truman Talley Books/Dutton, New York, 1997.

Mazzucato, M., *The Entrepreneurial State. Debunking Public vs. Private Sector Myths*, Anthem Press, London, 2013.

McCrindle, J., *Smart Cards*, Springer-Verlag, Heidelberg, Germany, 1990.

Menezes, A., *Elliptic Curve Public Key Cryptosystems*, Kluwer, Dordrecht, the Netherlands, 1993.

Menezes, A. J., van Oorschot, P. C., and Vanstone, S. A., *Handbook of Applied Cryptography*, CRC Press LLC, Boca Raton, FL, 1997.

Menn, J., Obama aide hints at incentives to raise cyber security, *Financial Times*, April 23, 2009, p. 4.

Meridien Research, Wireless payments at the point of sale: Expanding the possibilities, *ePayments*, 4, 4, June 7, 2001.

Merkow, M., EasySET, September 7, 2000, available at https://inza.wordpress.com/2010/09/07/easyset-el-rediseno-de-banesto-del-protocolo-set, last accessed August 16, 2015.

Messerges, T.S., Dabbish, E.A., and Sloan, R.H., Investigations of power analysis attacks on smartcard, in *Proc. USENIX Workshop on Smartcard Technol.*, Chicago, IL, 1999, pp. 151–161.

Mevinsky, G. and Neuman, B. C., NetCash: A design for practical electronic currency on the Internet, in *Proceedings of the First ACM Conference on Computers and Communications Security*, 1993, available at http://nii.isi.edu/info/netcheque/documentation.html, last accessed August 15, 2015.

Mian, A. N., Hameed, A., Khayyam, M. U., Ahmed, F., and Beraldi, R., Enhancing communication adaptability between payment card processing networks, *IEEE Commun. Mag.*, Commun. Stands. Suppl., 53, 58–64, March 2015.

Michard, A., *XML Langage et Applications*, Eyrolles, Paris, France, 1999.

Miers, I., Garman, C., Green, M., and Rubin, A. D., Zerocoin: Anonymous distributed e-cash from Bitcoin, in *Proceedings of the 2013 IEEE Symposium on Security and Privacy* (*SP*), Berkeley, CA, pp. 397–411.

Millicent, Millicent-specific elements for an HTTP payment protocol, V.1.3, 1995.

Mishkin, S. and Ahmed, M., Amazon joins battle for card payments, *Financial Times*, August 14, 2014, p. 11.

Mishkin, S. and Fontanella-Khan, J., Apple Pay set to trigger rush of deals among rivals, *Financial Times*, January 30, 2015, p. 14.

Mitchell, J. C., Shmatikov, V., and Stern, U., Finite-state analysis of SSL 3.0, in *Proceedings of the Seventh USENIX Security Symposium*, San Antonio, TX, 1998, pp. 26–29.

Mobey Forum, The MPOS breakthrough: How the power of mobile has disrupted payments, MPOS task forum, May 2013a, available at http://www.mobeyforum.org/whitepaper/the-mpos-breakthrough-how-the-power-of-mobile-has-disrupted-payments, last accessed August 16, 2015.

Mobey Forum, A security analysis of NFC implementation in the mobile proximity payments environment, Security Task Force (STF), Version 2.0, June 2013b, available at http://www.mobeyforum.org/w/wp-content/uploads/NFCSecurityAnalysis_Part1_FINAL1.pdf, last accessed August 16, 2015.

Moeller, B., Security of CBC ciphersuites in SSL/TLS: Problems and countermeasures, available at http://www.openssl.org/~bodo/tls-cbc.txt, last updated: May 20, 2004, last accessed July 27, 2014.

Moore, D., Voelker, G. M., and Savage, S., Inferring Internet denial-of-service activity, in *Proceedings of the 10th USENIX Security Symposium*, Washington, DC, August 13–17, 2001.

Möser, M., Böhme, R., and Breuker, D., An inquiry into money laundering tools in the Bitcoin ecosystem, in *Proceedings of the eCrime Researchers Summit (eCRS)*, San Francisco, CA, Septemeber 17–18, 2013, pp. 1–14.

Murdoch, S. J. and Anderson, R., Verified by Visa and MasterCard SecureCode: or, How not to design authentication, in *14th International Conference on Financial Cryptography and Data Security'10*, Tenerife, Canary Islands, January 25–28, 2010, available at http://www.cl.cam.ac.uk/~rja14/Papers/fc10vbvsecurecode.pdf, last accessed June 1, 2014.

Murdoch, S. J., Drimer, S., Anderson, R., and Bond, M., Chip and PIN is broken, *31st IEEE Symposium on Security and Privacy*, Oakland, CA, May 16–19, 2010, pp. 433–445.

Musil, S., Hackers claim to have defeated Apple's Touch ID print sensor, *CNET*, September 22, 2013, available at http://www.cnet.com/news/hackers-claim-to-have-defeated-apples-touch-id-print-sensor, last accessed March 18, 2015.

NACHA, The Electronic Payments Association, Understanding the ACH network: An ACH primer, 2003.

NACHA, The Electronic Payments Association, Mobile and person-to-person ACH payment, Request for information, executive summary and scenario description, June 27, 2011.

Nakamoto, S., Bitcoin: A peer-to-peer electronic cash system, 2008, available at https://bitcoin.org/bitcoin.pdf, last accessed March 7, 2015.

Nambiar, S. and Lu, C.-T., M-payment solutions and m-commerce fraud management, in *Advances in Security and Payment Methods for Mobile Commerce*, Hu, W.-C., Lee, C.-W., and Kou, W., Eds., Idea Group Publishing, Hershey, PA, 2005, pp. 192–211.

National Geographic, Geography: Guarded treasure, May 2010, p. 24.

National Institute of Standards and Technology (NIST), FIPS publication 180, Secure Hash Standard (SHS), NIST, Gaithersburg, MD, May 1993.

National Institute of Standards and Technology (NIST), FIPS publication 140-1, Security requirements for cryptographic modules, NIST, Gaithersburg, MD, January 1994.

National Institute of Standards and Technology (NIST), FIPS publication 180-1, Secure Hash Standard (SHS), NIST, Gaithersburg, MD, April 1995.

National Institute of Standards and Technology (NIST), Recommended elliptic curves for federal government use, July 1999, available at http://csrc.nist.gov/groups/ST/toolkit/documents/dss/NISTReCur.pdf, last accessed June 20, 2015.

National Institute of Standards and Technology (NIST), Federal Information Processing Standards (FIPS) publication 197, Advanced Encryption Standard (AES), NIST, Gaithersburg, MD, November 26, 2001a.

National Institute of Standards and Technology (NIST), Recommendation for block cipher modes of operation: The CCM mode for authentication and confidentiality, Dworkin, M., NIST special publication SP 800-38C, NIST, Gaithersburg, MD, December 2001b.

National Institute of Standards and Technology (NIST), Recommendation for block cipher modes of operation: Galois/Counter Mode (GCM) and GMAC, Dworkin, M., NIST special publication SP 800-38D, NIST, Gaithersburg, MD, 2007.

National Institute of Standards and Technology (NIST), Federal Information Processing Standards (FIPS) publication 198-1, The Keyed-Hash Message Authentication Code (HMAC), NIST, Gaithersburg, MD, July 2008.

National Institute of Standards and Technology (NIST), Recommendation for the Triple Data Encryption Algorithm (TDEA) block cipher, NIST special publication SP 800-67, Revision 1, Barker, W. C., and Barker, E., NIST, Gaithersburg, MD, January 2012a.

National Institute of Standards and Technology (NIST), Federal Information Processing Standards (FIPS) publication 180-4, Secure Hash Standard (SHS), NIST, Gaithersburg, MD, March 2012b.

National Institute of Standards and Technology (NIST), Recommendation for key management—Part 1: General, (Revision 3), NIST special publication SP 800-57, Barker, E., Barker, W., Burr, W., Polk, W., and Smid, M., NIST, Gaithersburg, MD, July 2012c.

National Institute of Standards and Technology (NIST), Federal Information Processing Standards (FIPS) publication 186-4, Digital Signature Standard (DSS), NIST, Gaithersburg, MD, July 2013.

National Security Agency (NSA), Bluetooth security, systems and network analysis center, Information assurance directorate, n.d., http://www.nsa.gov/ia/_files/factsheets/i732-016r-07.pdf, last accessed June 20, 2015.

National Security Agency (NSA), The case for elliptic curve cryptography, Central security service, January 15, 2009, https://www.nsa.gov/business/programs/elliptic_curve.shtml, last accessed June 20, 2015, no longer accessible.

Nechvatal, J., Baker, E., Bassham, L., Burr, W., Dworkin, M., Foti, J., and Roback, E., Report on the development of the Advanced Encryption Standard (AES), National Institute of Standards and Technology, Gaithersburg, MD, October 2, 2000.

Neuman, B. and Ts'o, T., Kerberos: An authentication service for computer networks, *IEEE Commun. Mag.*, 32(9), 33–38, 1994.

Neuman, B. C., Proxy-based authorization and accounting for distributed systems, in *Proceedings of the 13th International Conference on Distributed Computing Systems*, Pittsburgh, PA, May 25–28, 1993, pp. 283–291.

Neuman, B. C. and Mevinsky, G., Requirements for network payment: The NetCheque™ perspective, in *COMPCON'95*, San Francisco, CA, March 5–9, 1995, pp. 32–36.

New, D., Internet information commerce: The first virtual (TM) approach, in *Proceedings of the First USENIX Workshop on Electronic Commerce*, New York, 1995, pp. 33–68.

Nightingale, J., DigiNotar removal follow-up, September 2, 2011, https://blog.mozilla.org/security/2011/09/02/diginotar-removal-follow-up, last accessed August 24, 2014.

Nikiforakis, N. and Acar, G., Browse at your own risk, *IEEE Spectr.*, 51(8), 30–35, 2014.

Noble, D. F., *The Religion of Technology. The Divinity of Man and the Spirit of Invention*, Penguin Books, Harmondsworth, UK, 1999, first published in the U.S. by Knopf in 1997.

Nohl, K., Rooting SIM cards, Security Research Labs, July 22, 2013, available at https://media.blackhat.com/us-13/us-13-Nohl-Rooting-SIM-cards-Slides.pdf, Black Hat 2013 presentation, August 3, 2013, available at http://www.youtube.com/watch?v=scArc93XXWw, both last accessed November 7, 2014.

Nowacki, G., Mitraszewska, I., Kamiński, T., Niedzicka, A., Smoczyńska, E., Ucińska, M., Kallweit, T., and Rozesłaniec, R., National automatic toll collection system—Pilot project (Part 1), *Arch. Transp.*, 23(3), 335–352, 2011.

NTT DoCoMo, NTT DoCoMo is positioned to became the global leader in multimedia communications, *Financial Times*, October 7, 2002, p. 19.

NTT DoCoMo, FY2005 (ended March 31, 2006), Quarterly operating data, https://www.nttdocomo.co.jp/english/corporate/ir/finance/quarter/fy2005/index.htlm, last accessed December 31, 2015.

NTT DoCoMo, FY2006 (ended March 31, 2007), Quarterly operating data, https://www.nttdocomo.co.jp/english/corporate/ir/finance/quarter/fy2006/index.html, last accessed January 22, 2015.

NTT DoCoMo, FY2007 (ended March 31, 2008), Quarterly operating data, https://www.nttdocomo.co.jp/english/corporate/ir/finance/quarter/fy2007/index.html, last accessed December 31, 2015.

NTT DoCoMo, FY2008 (ended March 31, 2009), Quarterly operating data, https://www.nttdocomo.co.jp/english/corporate/ir/finance/quarter/fy2009/index.html, last accessed December 31, 2015.

NTT DoCoMo, FY2009 (ended March 31, 2010), Quarterly operating data, https://www.nttdocomo.co.jp/english/corporate/ir/finance/quarter/fy2009/index.html, last accessed December 31, 2015.

NTT DoCoMo, Number of subscribers, November 7, 2014, http://www.ntt.co.jp/ir/fin_e/subscriber.html, last accessed January 22, 2015.

Nyström, M., PKCS #15: A cryptographic token information format standard, in *Proceedings of the USENIX Workshop on Smartcard Technology*, Chicago, IL, 1999, pp. 37–42.

OASIS, eXtensible access control markup language (XACML), Version 3.0, 22 January 2013, http://docs.oasis-open.org/xacml/3.0/xacml-3.0-core-spec-os-en.pdf, last accessed January 1, 2016.

OASIS, Security Assertion Markup Language (SAML) V2.0, OASIS Standard, March 15, 2005, https://docs.oasis-open.org/security/saml/v2.0/saml-core-2.0-os.pdf, last accessed January 1, 2016.

OASIS, WS-security, Web services security: SOAP message security 1.1, OASIS standard specification, February 1, 2006, http://docs.oasis-open.org/wss/2004/01, last accessed December 31, 2015.

Obaidat, M. S. and Sadoun, N., Keystroke dynamics based authentication, in *Biometrics: Personal Identification in Networked Society*, Jain, A., Bolle, R., and Pankati, S., Eds., Kluwer, Dordrecht, the Netherlands, 1999, pp. 213–225.

O'Callaghan, R. and Turner, J. A., Electronic data interchange—Concept and issues, in *EDI in Europe: How it Works in Practice*, Krcmar, H., Bjørn-Andersen, N., and O'Callaghan, R., Eds., John Wiley & Sons, New York, 1995, pp. 1–19.

Odlyzko, A., The case against micropayments, in *Proceedings of the Seventh International Conference on Financial Cryptography* (*FC2003*), Guadeloupe, French West Indies, January 27–30, 2003, Lecture Notes in Computer Science, Vol. 2742, Wright, R. N., Ed., Springer-Verlag, Berlin, Germany, 2003, pp. 77–83.

O'Dwyer, K. J. and Malone, D., Bitcoin mining and its energy foot print, in *Proceedings of the 25th IET Irish Signals and Systems Conference 2014 and 2014 China-Ireland International Conference on Information and Communications Technologies* (*ISSC 2014/CIICT 2014*), Limerick, Ireland, June 26–27, 2014, pp. 280–285.

OECD (Organisation for Economic Co-operation and Development), The OECD privacy framework, 2013, available at http://oe.cd/primary, last accessed August 13, 2015.

Okamoto, T., An efficient divisible electronic cash scheme, in *Advances Cryptology CRYPTO'95*, Springer-Verlag, Heidelberg, Germany, 1995, pp. 438–451.

O'Mahony, D., Pierce, M., and Tewari, H., *Electronic Payment Systems*, Artech House, Norwood, MA, 1997 (2nd ed., 2001).

Ondrus, J., Mobile payments: A tool kit for a better understanding of the market, École des Hautes Études Commerciales (HEC), University of Lausanne, Lausanne, Switzerland, July 2003, available at http://www.janondrus.com/wp-content/uploads/2008/06/ondrus-licence-mpayment.pdf, last accessed January 27, 2015.

Open Mobile Alliance, White paper on the M-commerce landscape, December 21, 2005, available at http://technical.openmobilealliance.org/document/OMA-WP-McommerceLandscape-20051221-A.pdf, last accessed December 15, 2013.

Paar, C. and Pelzl, J., *Understanding Cryptography*, Springer-Verlag, Berlin, Germany, 2010.

Paglieri, J., *Bitcoin and the Future of Money*, Triumph Books, Chicago, IL, 2014.

Palme, J., *Electronic Mail*, Artech House, Norwood, MA, 1995.

Palmer, M., Face recognition software gaining a broader canvas, *Financial Times*, May 22–23, 2010, p. 9.

Palmer, M., TomTom apologises to customers after selling driving data to police, *Financial Times*, April 29, 2011, p. 13.

Pasini, P. and Chaloux, J., EDI and activity based management for business—The case of Whilrpool in Italy, in *EDI in Europe: How it Works in Practice*, Krcmar, H., Bjørn-Andersen, N., and O'Callaghan, R., Eds., John Wiley & Sons, New York, 1995, pp. 85–112.

PayPal, Pay order management integrated guide, February 2009, https://www.paypalobjects.com/webstatic/en_US/developer/docs/pdf/archive/PP_OrderMgmt_IntegrationGuide.pdf, last accessed August 26, 2015.

Pays, P.-A. and de Comarmont, F., An intermediation and payment system technology, *Comput. Netw. ISDN Syst.*, 28(7–11), 1197–1206, 1996.

PC/SC Workgroup, Interoperability specification for ICCs and personal computer systems, December 1997, Version 2.01.14 released in June 2013, available at http://www.pcscworkgroup.com, last accessed May 1, 2015.

Peck, M. E., The cryptoanarchists' answer to cash, *IEEE Spectr.*, 49(6), 50–56, 2012.

Pedersen, T., Electronic payments of small amounts, in *Proceedings of the International Workshop on Security Protocols*, Cambridge, UK, April 1996, Lecture Notes in Computer Science, Vol. 1189, Lomas, M., Ed., Springer-Verlag, Heidelberg, Germany, 1997, pp. 56–68.

Peel, M. and Allison, K., Devil in the details: Why personal data are more open to loss and abuse, *Financial Times*, December 24–25, 2007, p. 5.

Pellegrini, A., Bertaccio, A. and Austin, T., Fault-based attack of RSA authentication, in *Proceedings of the Conference on Design, Automation and Test in Europe (DATE'10)*, Leuven, Belgium, 2010, pp. 855–860.

Percival, C., Stronger key derivation via sequential memory-hard functions, in *Proceedings of BSDCan 2009*, Ottawa, Canada, May 8–9, 2009, available at http://www.tarsnap.com/scrypt/scrypt.pdf, last accessed March 29, 2015.

Percival, C. and Josefsson, S., The scrypt password-based key derivation function, Internet draft, November 11, 2015, available at https://datatracker.ietf.org/doc/draft-josefsson-scrypt-kdf, last accessed December 31, 2015.

Persad, K., Walton, C. M., and Hussain, S., Toll collection technology and best practices, Technical Report, Project 0-5217, Center for Transportation Research, The University of Texas at Austin, Austin, TX, 2007, available at http://www.utexas.edu/research/ctr/pdf_reports/0_5217_P1.pdf, last accessed November 14, 2014.

Petre, V., The 3-D secure protocol, May 29, 2912, available at www.slideshare.net/vladpetre88/the-3d-secure-protocol, last accessed May 31, 2014.

Phillips, C. and Meeker, M., *Collaborative Commerce*, Morgan Stanley Dean Witter, Equity Research North America, New York, April 2000.

Pierson, J., *La biométrie, l'identification par le corps*, Lavoisier, Hermes-Science, Paris, France, 2007.

Podio, F. L. et al., Common Biometric Exchange Formats Framework (CBEFF), National Institute of Standards and Technology Internal/Interagency Report (NISTIR) 6529-A, Gaithersburg, MD, April, 5, 2014.

Popa, R. A. and Zeldovich, N., Web applications could increase security by keeping encrypted data even duration computations, *IEEE Spectr.*, 52(8), 2015, 42–47.

Popovic, E. C., Stancu, L. A., Guta, O. G., Arseni, S. C., and Fratu, O., Combined use of pattern recognition algorithms for keystroke-based continuous authentication system, in *Proceedings of the 10th International Conference on Communications (COMM)*, Bucharest, Romania, May 29–31, 2014, pp. 1–4.

Porter, M. E., *Competitive Advantage*, Free Press, New York, 1985.

Potgieser, P. G. L., Standards for international trade and enterprise interoperability, in *Handbook of Enterprise Integration*, Sherif, M. H., Ed., CRC Press, Boca Raton, FL, 2010, pp. 511–531.

Presidential Executive Memorandum, *Streamlining Procurement through Electronic Commerce*, Washington, DC, October 26, 1993, available at https://www.gpo.gov/fdsys/pkg/WCPD-1993-11-01/pdf/WCPD-1993-11-01-Pg2174.pdf, last accessed December 31, 2015.

Proakis, J. G. and Massoud, S., *Fundamentals of Communication Systems*, 2nd ed., Prentice Hall, Upper Saddle River, NJ, 2013.

Qing, L. and Yaping, L., Analysis and comparison of several algorithms in SSL/TLS Handshake protocol, in *Proceedings of the 2009 International Conference on Information Technology and Computer Science (ITCS 2009)*, Kiev, Ukraine, July 25–26, 2009, pp. 613–617.

Rabin, M. O., Digital signatures and public-key functions as intractable as factorization, Technical report MIT/LCS/TR-212, MIT Laboratory for Computer Science, Cambridge, MA, 1979.

Rambure, D. and Nacamuli, A., *Payment Systems: From the Salt Mines to the Board Room*, Palgrave Macmillan, Basingstoke, Hampshire, UK, 2008.

Ramonet, I., Google nous espionne et en informe les États-Unis, entretien exclusive avec Julian Assange, *Mémoires des luttes*, December 4, 2014, available at http://www.medelu.org/Google-nous-espionne-et-en-informe, last accessed August 14, 2015.

Rankl, W. and Effing, W., *The Smart Card Handbook*, 4th ed., John Wiley & Sons, Chichester, UK, 2010.

Ratha, N. K., Connell, J. H., and Bolle, R. M., Enhancing security and privacy in biometrics-based authentication systems, *IBM Syst. J.*, 40(30), 614–634, 2001.

Reagan, C., Shift to chip and PIN doesn't solve all for retail, *CNBC*, April 9, 2014, available at http://www.cnbc.com/id/101569008, last accessed December 13, 2014.

Redish, A., Money and coinage in *The Oxford Encyclopedia of Economic History*, Mokyr, J., Ed. in Chief, Oxford University Press, New York, 2003, Vol. 3, pp. 535–537.

Reid, F. and Harrigan, M., An analysis of anonymity in the Bitcoin system, in *Proceedings of the Third IEEE International Conference on Privacy, Security, Risk, and Trust (PASSAT) and Third International Conference on Social Computing (SocialCom)*, Boston, MA, October 9–11, 2011, pp. 1318–1326.

Remacle, F., Swift, actif stratégique du monde bancaire, *Banque Stratégie*, 131, 18–21, October 1996.

Remery, P., A system of payment using coin purse Cards, in *Smart Card 2000, Proceedings of the IFIP WG 11.6 International Conference on Smart Card 2000: The Future of IC Cards*, Laxenburg, Austria, October 1987, Chaum, D. and Schaumüller-Bichl, I., Eds., Elsevier/North Holland, Amsterdam, the Netherlands, 1989, pp. 49–55.

Renfro, S. G., VeriSign CZAG: Privacy leak in X.509 certificates, in *Proceedings of the 11th USENIX Security Symposium*, San Francisco, CA, 2002.

Rescorla, E., *SSL and TLS: Designing and Building Secure Systems*, Addison-Wesley, Reading, MA, 2001.

Rescorla, E., Cain, A., and Korver, B., SSLACC: A clustered SSL accelerator, in *Proceedings of the 11th USENIX Security Symposium*, San Francisco, CA, August 5–9, 2002.

RFC 821, Simple mail transfer protocol, Postel, J., Ed., August 1982.

RFC 822, Standard for the format of internet text messages, Crocker, D., Ed., 1982.

RFC 1035, Domain names—Implementation and specification, Mockapetris, P. V., Ed., November 1987.

RFC 1320, The MD4 message digest algorithm, Rivest, R., Ed., April 1992.

RFC 1321, The MD5 message digest algorithm, Rivest, R., Ed., April 1992.

RFC 1492, An access control protocol, sometimes called TACACS, Finseth, C., Ed., July 1993.

RFC 1661, The Point-to-Point Protocol (PPP), Simpson, W., Ed., July 1994.

RFC 1767, MIME encapsulation of EDI objects, Crocker, D., Ed., March 1995.

RFC 1951, DEFLATE compressed data format specification version 1.3, Deutsch, P., Ed., May 1996.

RFC 1967, PPP LZS-DCP compression protocol (LZS-DCP), Schneider, K. and Friend, R., Eds., August 1996.

RFC 1991, PGP Message Exchange Formats, Atkins, D., Stallings, W., and Zimmerman, P., Eds., August 1996.

RFC 2015, MIME security with Pretty Good Privacy (PGP), Elkins, M., October 1996.

RFC 2045, MIME (multipurpose internet mail extensions) Part one: Mechanisms for specifying and describing the format of internet message bodies, Freed, N. and Borenstein, N., Eds., November 1996.

RFC 2046, Multipurpose Internet Mail Extensions (MIME) Part two: Media types, Freed, N. and Borenstein, N., Eds., November 1996.

RFC 2047, Multipurpose Internet Mail Extensions (MIME) Part three: Message header extensions for non-ASCII text, Moore, K., Ed., November 1996.

RFC 2048, Multipurpose Internet Mail Extensions (MIME) Part four: Registration procedures, Freed, N., Klensin, J., and Postel, J., Eds., November 1996.

RFC 2049, Multipurpose Internet Mail Extensions (MIME) Part five: Conformance criteria and examples, Freed, N. and Borenstein, N., Eds., November 1996.

RFC 2104, HMAC: Keyed-hashing for message authentication, Krawczyk, H., Bellare, M., and Canetti, R., Eds., February 1997.

RFC 2109, HTTP state management mechanism, Kristol, D. and Montulli, L., Eds., February 1997.

RFC 2246, The TLS protocol version 1.0, Dierks, T. and Allen, C., Eds., January 1999.

RFC 2311, S/MIME version 2 message specification, Dusse, S., Hoffman, P., Ramsdell, B., Lundblad, L., and, Repka, L., Eds., March 1998.

RFC 2315, PKCS #7: Cryptographic message syntax version 1.5, Kaliski, B., Ed., March 1998.

RFC 2401, Security Architecture for the Internet Protocol, Kent, S. and Atkinson, R., November, 1998.

RFC 2437, PKCS #1: RSA Cryptography Specifications Version 2.0, Kaliski, B. and Staddon, J., Eds., October, 1998.

RFC 2440, OpenPGP Message Format, Callas, J., Donnerhacke, L., Finney, H. and Thayer, R., Eds., November, 1998.

RFC 2585, Internet X.509 public key infrastructure–Operational protocols: FTP and HTTP, Housley, R. and Hoffman, P., Eds., May 1999.

RFC 2632, S/MIME Version 3 certificate handling, Ramsdell, B., Ed., June 1999.

RFC 2633, S/MIME Version 3 message specification, Ramsdell, B., Ed., June 1999.

RFC 2634, Enhanced security services for S/MIME, Hoffman, P., Ed., June 1999.

RFC 2660, The secure hypertext transfer protocol, Rescorla, E. and Schiffman, A., Eds., August 1999.

RFC 2661, Layer two tunneling protocol L2TP, Townsley, W., Valencia, A., Rubens, A., Pall, G., Zorn, G., and Palter, B., Eds., August 1999.

RFC 2712, Addition of Kerberos cipher suites to Transport Layer Security (TLS), Medvinsky, A. and Hur, M., Eds., October 1999.

RFC 2773, Encryption using KEA and SKIPJACK, Housley, R., Yee, P. and Nace, W., Eds., February, 2000.

RFC 2986, PKCS #10: Certification request syntax specification, version 1.7, Nystrom, M. and Kaliski, B., November 2000.

RFC 3106, ECML v1.1: Field specifications for E-commerce, Eastlake, D. and Goldstein, T., Eds., April 2001.

RFC 3193, Securing L2TP using IPSec, Patel, B., Adoba, B., Dixon, W., Zorn, G., and Booth, S., Eds., November 2001.

RFC 3335, MIME-based secure peer-to-peer business data interchange over the Internet, Harding, T., Drummond, R., and Shih, C., Eds., September 2002.

RFC 3447, Public-Key Cryptography Standards (PKCS) #1: Specifications version 2.1, Jonsson, J. and Kaliski, B., Eds., February 2003.

RFC 3610, Counter with CBC-MAC (CCM), Whiting, D., Housley, R., and Ferguson, N., Eds., September 2003.

RFC 3647, Internet X.509 public key infrastructure certificate policy and certification practices framework, S. Chokhani, S., Ford, W., Sabett, R., Merrill, C., and Wu, S., Eds., November 2003.

RFC 3749, Transport layer security protocol compression methods, Hollenbeck, S., Ed., May 2004.

RFC 3943, Transport Layer Security (TLS) protocol compression using Lempel-Ziv-Stac (LZS), Friend, R., Ed., November 2004.

RFC 3962, Advanced Encryption Standard (AES) encryption for Kerberos 5, Raeburn, K., Ed., February 2005.

RFC 4112, Electronic Commerce Modeling Language (ECML), Version 2, Eastlake, D., Ed., 3rd ed., June 2005.

RFC 4120, The Kerberos network authentication service (V5), Neuman, C., Yu, T., Hartman, S., and Raeburn, K., Eds., July 2005.

RFC 4132, Addition of camellia cipher suites to Transport Layer Security (TLS), Moriai, S., Kato, A., and Kanda, M., Eds., July 2005.

RFC 4210, Internet X.509 public key infrastructure Certificate Management Protocol (CMP), Proposed standard, Adams, C., Farrell, S., Kause, T., and Mononen, T., Eds., September 2005.

RFC 4301, Security architecture for the internet protocol, Proposed standard, Kent, S. and Seo, K., Eds., December 2005.

RFC 4302, IP authentication header, Proposed standard, Kent, S., Ed., December 2005.

RFC 4303, IP Encapsulating Security Payload (ESP), Proposed standard, Kent, S., Ed., December 2005.

RFC 4306, Internet Key Exchange (IKEv2) protocol, Proposed Standard, Kaufman, G., Ed., December 2005.

RFC 4346, The Transport Layer Security (TLS) protocol version 1.1, Dierks, T. and Rescorla, E., April 2006.

RFC 4347, Datagram transport layer security, Rescorla, E. and Modadugu, N., Eds., April 2006.

RFC 4366, Transport Layer Security (TLS) extensions, Blake-Wilson, S., Nystrom, M., Hopwood, D., Mikkelsen, J., and Wright, T., Eds., April 2006.

RFC 4422, Simple Authentication and Security Layer (SASL), Proposed standard, Melnikov, A. and Zeilenga, K., Eds., June 2006.

RFC 4511, Lightweight Directory Access Protocol (LDAP): The protocol, Proposed standard, Sermersheim, J., Ed., June 2006.

RFC 4512, Lightweight Directory Access Protocol (LDAP): Directory information models, Proposed standard, K. Zeilenga, Ed., June 2006.

RFC 4513, Lightweight Directory Access Protocol (LDAP): Authentication methods and security mechanisms, Proposed standard, Harrison, R., Ed., June 2006.

RFC 4871, DomainKeys Identified Mail (DKIM) signatures, Allman, E., Callas, J., Delany, M., Libbey, M., Fenton J., and Thomas, M., Eds., May 2007.

RFC 4880, OpenPGP message format, Proposed standard, Callas, J., Donnerhacke, L., Finney, H., Shaw, D., and Thayer, R., Eds., November 2007.

RFC 5035, Enhanced Security Services (ESS) update: Adding CertID algorithm agility, Schaad, J., August 2007.

RFC 5116, An interface and algorithms for authenticated encryption, McGrew, D., Ed., January 2008.

RFC 5246, The Transport Layer Security (TLS) protocol version 1.2, Dierks, T. and Rescorla, E., Eds., August 2008.

RFC 5746, Transport Layer Security (TLS) renegotiation indication extension, Resorla, E., Ray, M., Dispensa, S., and Oksvo, N., Eds., February 2010.

RFC 6071, IP security (IPSec) and Internet Key Exchange (IKE) document roadmap, Frankel, S. and Krishnan, S., Eds., February 2011.

RFC 6151, Updated security considerations for the MD5 message-digest and the HMAC-MD5 algorithms, Turner, S. and Chen, L., Eds., March 2011.

RFC 6176, Prohibiting Secure Sockets Layer (SSL) version 2.0, Turner, S. and Polk, T., Eds., March 2011.

RFC 6347, Datagram transport layer security version 1.2, Rescorla, E. and Modadugu, N., Eds., January 2012.

RFC 6520, Transport Layer Security (TLS) and Datagram Transport Layer Security (DTLS) heartbeat extension, Seggelmann, R., Tuexen, M., and Williams, M., Eds., February 2012.

RFC 6637, Elliptic Curve Cryptography (ECC) in OpenPGP, Proposed standard, Jivsov, A., Ed., June 2012.

RFC 6649, Deprecate DES, RC4-HMAC-EXP, and other weak cryptographic algorithms in Kerberos, Hornquist Astrand, L. and Yu, T., Eds., July 2012.

RFC 6929, Remote Authentication Dial in User Service (RADIUS) protocol extension, Proposed standard, De Kok, A. and Lior, A., Eds., April 2013.

RFC 6960, X.509 internet public key infrastructure online certificate status protocol—OCSP, Proposed standard, Santesson, S., Myers, M., Ankney, R., Malpani, A., Galperin, S., and Adams, C., Eds., June 2013.

RFC 7208, Sender Policy Framework (SPF) for authorizing use of domains in email, Version 1, Kitterman, S., Ed., April 2014.

RFC 7296, Internet Key Exchange Protocol Version 2 (IKEv2), Kaufman, C., Hoffman, P., Nir, Y., Eronen, P., and Kivinen, T., Eds., October 2014.

RFC 7321, Cryptographic algorithm implementation requirements and usage guidance for Encapsulating Security Payload (ESP) and authentication header, Proposed standard, McGrew, D. and Hoffman, P., Eds., August 2014.

RFC 7427, Signature authentication in the Internet Key Exchange Version 2 (IKEv2), Proposed standard, Kivinen, T. and Snyder, J., Eds., January 2015.

Richmond, R., Scammed! Web merchants use new tools to keep buyers from ripping them off, *The Wall Street Journal*, January 27, 2003, p. R6.

Riedel, J. C. K. H., Towards an understanding of Apple's success—Conceptualising product-service-business bundles, in *Proceedings of the 2014 International Engineering, Technology and Innovation (ICE) Conference*, Bergamo, Italy, June 23–25, 2014, pp. 1–6.

Rivest, R. L., The RC5 encryption algorithm, *CryptoBytes*, 1(1), 9–11, 1995.

Rivest, R. L. and Shamir, A., PayWord and MicroMint: Two simple micropayment schemes, in *Security Protocols, Proceedings of the International Workshop*, Cambridge, UK,

April 1996, Lecture Notes in Computer Science, Vol. 1189, Lomas, M., Ed., Springer-Verlag, Heidelberg, Germany, 1997, pp. 69–87, available at http://people.csail.mit.edu/rivest/RivestShamir-mpay.pdf, last accessed January 18, 2015.

Rivest, R. L., Shamir, A., and Adleman, L. M., A method for obtaining digital signatures and public key cryptosystems, *Commun. ACM*, 21(2), 120–126, 1978.

Roberts, N., Shopper test Euro alternatives, *The Wall Street Journal*, August 14, 2015, p. A9.

Rohunen, A., Eteläperä, M., Liukkunen, K., Tullpo, T., and Chan, K. W., Implementing and evaluating a smart-M3 platform-based multi-vendor micropayment system pilot in the context of small business, *J. Digit. Inform. Manage.*, 12(1), 44–51, 2014.

Roland, M., Practical attack scenarios on secure element-enabled mobile devices, in *Proceedings of the Fourth International Workshop on Near Field Communication*, Helsinki, Finland, March 13, 2012a, presentation available at http://www.mroland.at/fileadmin/mroland/papers/201203_SecureElementAttackScenarios_slides.pdf, last accessed August 16, 2015.

Roland, M., Software card emulation in NFC-enable mobile phones: Great advantage or security nightmare? in *Proceedings of the Fourth International Workshop on Security and Privacy in Spontaneous Interaction and Mobile Phone Use (IWSSI/SPMU)*, Newcastle, UK, June 18, 2012b, available at http://www.medien.ifi.lmu.de/iwssi2012/papers/iwssi-spmu2012-roland.pdf, last accessed August 16, 2015. Presentation available at http://www.mroland.at/fileadmin/mroland/papers/201206_SoftwareCardEmulation_slides.pdf, last accessed August 16, 2015.

Roland, M., Applying recent secure element relay attack scenarios to the real world: Google wallet relay attack, Technical Report, University of Applied Sciences Upper Austria, March 25, 2013, available at http://arxiv.org/pdf/1209.0875.pdf, last accessed December 28, 2013.

Roland, M., Langer, J., and Scharinger, J., Practical attack scenarios on secure element-enabled mobile devices, in *Proceedings of the Fourth International Workshop on Near Field Communication (NFC)*, Helsinki, Finland, March 13, 2012, pp. 19–24.

Roland, M., Langer, J., and Scharinger, J., Applying attacks to Google Wallet, in *Proceedings of the Fifth International Workshop on Near Field Communication (NFC)*, Zurich, Switzerland, February 5, 2013, pp. 1–6.

Romain, H., Kléline: Un système français de paiement électronique on-line, in *l'Internet et la vente*, Aimetti, J. P., Ed., Les Éditions d'Organisation, Paris, France, 1997, pp. 96–99.

Rossigneux, B. and Simonnot, D., Des fichiers de police sans foi ni loi, *Le Canard enchaîné*, December 20, 2006, p. 4.

Rubin, J., Google Wallet security: Pin exposure vulnerability, zveloBlog™, February 8, 2012, available at https://zvelo.com/blog/entry/google-wallet-security-pin-exposure-vulnerability, last accessed October 17, 2013.

Saarinen, M.-J., Attacks against the WAP WTLS protocol, 2000, available at http://www.mjos.fi/doc/saarinen_wtls.pdf, last accessed July 26, 2014.

Sabatier, G., *Le porte-monnaie électronique et le porte-monnaie virtuel, Que sais-je?* Vol. 3261, Presses Universitaires de France, Paris, France, 1997.

Salowey, J. and Rescorla, E., TLS renegotiation vulnerability, presentation at IEFT-76, November 8–13, 2009, available at https://www.ietf.org/proceedings/76/slides/tls-7.pdf, last accessed September 7, 2014.

Sandoval, V., *Technologie de l'EDI*, Hermès, Paris, France, 1990.

Sarkar, P. G. and Fitzgerald, S., Attacks on SSL. A comprehensive study of BEAST, CRIME, BREACH, Lucky 13 and RC4 biases, August 15, 2012, available at https://www.isecpartners.com/media/106031/ssl_attacks_survey.pdf, last accessed September 12, 2014.

Scahill, J. and Begley, J., The great SIM heist. How spies stole the keys to the encryption castle, *The Intercept*, February 19, 2015, available at https://firstlook.org/theintercept/2015/02/19/great-sim-heist, last accessed August 15, 2015.

Schäfer, D. and Bradshaw, T., Apple in talks to bring pay technology to Europe, *Financial Times*, September 12, 2014, p. 18.

Schneier, B., *Applied Cryptography*, 2nd ed., John Wiley & Sons, New York, 1996.

Schneier, B., Why cryptography is harder than it looks, 1997, available at https://www.schneier.com/essays/archives/1997/01/why_cryptography_is.html, last accessed September 12, 2014.

Schneier, B., Security pitfalls in cryptology, 1998a, available at https://www.schneier.com/essays/archives/1998/01/security_pitfalls_in.html, last accessed September 12, 2014.

Schneier, B., Cryptography for the Internet, in *Third USENIX Workshop Electronic Commerce*, Boston, MA, Tutorial, 1998b.

Schneier, B., Customers, passwords and web sites, *IEEE Secur. & Priv.*, 2(4), 88, 2004.

Schoen, S., New research suggests that governments make fake SSL certificates, *Electronic Frontier Foundation*, March 24, 2010, https://www.eff.org/deeplinks/2010/03/researchers-reveal-likelihood-governments-fake-ssl, last accessed July 1, 2015.

Schuba, C. L., Krsul, I. V., Kuhn, M. G., Spafford, E. H., Sundaram, A., and Zamboni, D., Analysis of a denial of service attack on TCP, in *IEEE Symposium on Security and Privacy*, 1997, Oakland, CA, May 4–7, pp. 208–223.

Schwartz, M. J., Experian sold data to Vietnamese ID theft ring, 2013, available at http://www.darkreading.com/attacks-and-breaches/experian-sold-data-to-vietnamese-id-theft-ring/d/d-id/1112016, last accessed August 15, 2015.

Segev, A., Porra, J., and Roldan, M., Financial EDI over the Internet, case study II: The Bank of America and Lawrence Livermore National Laboratory pilot, in *Proceedings of the Second USENIX Workshop Electronic Commerce*, Oakland, CA, 1996, pp. 173–190.

SEIS (Secured Electronic Information in Society), *Specification SEIS S1, SEIS Cards, Electronic ID Application*, February 19, 1998.

Selignan, M., France's precursor to the Internet lives on; '80s-vintage Minitel network upgraded to "complement" the Web, *The Washington Post*, September 25, 2003, p. E02.

Selzer, L., DKIM: Useless or just disappointing? *Zero Day*, August 14, 2013, available at http://www.zdnet.com/article/dkim-useless-or-just-disappointing, last accessed July 2, 2015.

Service Central de la Sécurité des Systèmes d'Information, Common criteria for IT security evaluation protection profile—Smartcard integrated circuit protection profile, October 1997 (registered at the French Certification Body under the number PP/9704).

SET (Secure Electronic Transaction) specification, 1997, *Book 1: Business Description, Book 2: Programmer's Guide, Book 3: Format Protocol Definition*, Version 1.0, May 31, 1997. Retrievable from http://www.exelana.com/set, last accessed January 1, 2016.

SETCo, Online PIN extensions to SET Secure Electronic Transaction™ version 1.0, May 25, 1999a.

SETCo, Common Chip Extension SET™ 1.0, September 29, 1999b.

Shah, G., Molina, A., and Blaze, M., Keyboards and covert channels, in *Proceedings of the 15th USENIX Security Symposium Security'06*, Vancouver, British of Columbia, Canada, July 31–August 4, 2006, pp. 59–75, available at http://www.crypto.com/papers/jbug-Usenix06-final.pdf, last accessed June 2, 2015.

Shaked, Y. and Wool, A., Cracking the Bluetooth PIN, May 2, 2005, available at http://www.eng.tau.ac.il/~yash/shaked-wool-mobisys05, last accessed December 17, 2013.

Sherif, M. H., L'Internet, la Société de Spectacles et la Contestation, Institut de Recherches Internationales, Groupe de Recherche Économique et Sociale, Université Paris-Dauphine, Paris, France, February 1997.

Sherif, M. H., Standards for business-to-business electronic commerce, in *Handbook of Enterprise Integration*, Sherif, M. H., Ed., CRC Press, Boca Raton, FL, 2010, pp. 534–560.

Sherif, M. H., ICT standardisation strategies and interactive learning spaces, *Int. J. Technol. Market.*, 10(2), 113–136, 2015.

Sherif, M. H., Sehrouchni, A., Gaid, A. Y., and Farazmandnia, F., SET et SSL: Echanges sécurisés sur l'Internet, *Document Numérique*, 1(4), 421–440, 1997.

Sherif, M. H., Sehrouchni, A., Gaid, A. Y., and Farazmandnia, F., SET and SSL: Electronic payments on the Internet, in *Proceedings of the Third IEEE Symposium on Computers and Communication (ISCC'98)*, Athens, Greece, IEEE Computer Society Press, Washington, DC, 1998, pp. 353–358.

Shirky, C., The case against micropayments, December 19, 2000, available at http://www.openp2p.com/pub/a/p2p/2000/12/19/micropayments.html, last accessed January 17, 2015.

Simon, D., Anonymous communication and anonymous cash, in *Advance Cryptology—CRYPTO'96, Proceedings of the 16th Annual International Cryptology Conference*, Santa Barbara, CA, Springer-Verlag, New York, 1996, pp. 61–73.

Simpson, W. A., IKE/ISAMP considered harmful, *Login*, 24(6), 48–58, 1999.

Single, R., Law enforcement appliance subverts SSL, *Wired*, March 24, 2010, available at http://www.wired.com/2010/03/packet-forensics, last accessed August 9, 2014.

Sirbu, M. A., Credits and debits on the Internet, *IEEE Spectr.*, 34(2), 23–29, 1997.

Sirbu, M. A. and Chuang, J. C.-I., Distributed authentication in Kerberos using public key cryptography, in *Proceedings of the Symposium on Network and Distributed System Security*, San Diego, CA, February 10–11, 1997, pp. 134–141.

Sirbu, M. A. and Tygar, J. D., NetBill: An Internet commerce system optimized for network delivered services, *IEEE Pers. Commun.*, 2(4), 6–11, 1995.

Sivori, J. R., Evaluate receipts and settlements at Bell Atlantic, *Commun. ACM*, 39(6), 24–28, 1996.

Slawsky, J. and Zafar, S., *Developing and Managing a Successful Payment Cards Business*, Gower Publishing Ltd, Surrey, UK, 2005.

Smart Border Alliance, RFID feasibility study final report, Attachment D, January 21, 2005, available at https://www.dhs.gov/xlibrary/assets/foia/US-VISIT_RFIDattachD.pdf, last accessed November 16, 2014.

Smith, A., Desperately seeking Satoshi, *The Sunday Times (London)*, March 2, 2014, available at http://www.thesundaytimes.co.uk/sto/Magazine/article1379779.ece, last accessed April 28, 2014.

Soghoian, C. and Stamm, S., Certified lies: Detecting and defeating government intercept attacks against SSL, March 24, 2010, available at http://files.cloudprivacy.net/ssl-mitm.pdf, last accessed July 1, 2015.

Southurst, J., Hacker returns 225 BTC taken from Blockchain wallets, *CoinDesk*, December 10, 2014, http://www.coindesk.com/hacker-returns-225-btc-taken-blockchain-wallets, last accessed March 6, 2015.

Spiekermann, S. and Cranor, L. F., Engineering privacy, *IEEE Trans. Softw. Eng.*, 35(1), 67–82, 2009.

Stanford, A., Apple Credential Phishing via appleidconfirm.net, March 27, 2014, available at https://isc.sans.edu/diary/Apple+Credential+Phishing+via+appleidconfirm.net/17869, last accessed January 17, 2015.

Star-Ledger (The), State settles with developer of bitcoin-mining software, *The Star-Ledger*, May 27, 2015, p. 15.

Star-Ledger (The), Hackers of IRS tax data seen as based in Russia, *The Star-Ledger*, May 28, 2015, p. 9.

Starnberger, G., Froihofer, L., and Goeschka, K. M., QR-TAN: Secure mobile transaction authentication, in *Proceedings of the 2009 International Conference* on *Availability, Reliability and Security* (*ARES'09*), Fukuoka, Japan, March 16–19, 2009, pp. 578–583.

Steedman, D., *Abstract Syntax Notation One ASN.1: The Tutorial and Reference*, Technology Appraisals, Twickenham, UK, 1993.

Stevens, M., Sotirov, A., Appelbaum, J., Lenstra, A., Molnar, D., Osvik, D. A., and de Weger, B., Short chosen-prefix collisions for MD5 and the create of a rogue CA certificate, in *CRYPTO 2009*, Lecture Notes in Computer Science, Vol. 5677, Halevi, S., Ed., Springer-Verlag, Berlin, Germany, 2009, pp. 55–69.

Suga, Y., Countermeasures and tactics for transitioning against the SSL/TLS renegotiation vulnerability, in *Proceedings of the Sixth International Conference on Innovative Mobile and Internet Services in Ubiquitous Computing*, Palermo, Italy, July 4–6, 2012, pp. 656–659.

Sullivan, R. J., The U.S. adoption of computer-chip payment cards: Implications for payment fraud, *Econ. Rev.*, 98, 59–87, First Quarter 2013, Federal Reserve Bank of Kansas City, available at http://www.kc.frb.org/publicat/econrev/pdf/13q4Sullivan.pdf, last accessed May 11, 2014.

Sunday Star-Ledger (The), System glitch halts use of food stamps debit cards, *The Sunday Star-Ledger*, October 13, 2013, Section One, p. 5.

Swchiderski-Groshe, S. and Knospe, H., Secure mobile commerce, *IEE Electron. Commun. Eng. J.*, 14(5), 228–238, 2002, available at http://www.it.iitb.ac.in/~annanda/papers-280905/Secure%20m-commerce%20ECEJ.pdf, last accessed October 9, 2014.

Symantec Corporation, Symantec trust network (STN) certification practice statement, Version 3.8.19, March 19, 2015, available at https://www.symantec.com/content/en/us/about/media/repository/stn-cps.pdf, last accessed July 1, 2015.

Tan, W.-K. and Tan, Y.-J., Critical factors influencing the successful development of e-micropayment program, in *Proceedings of the International Conference on Management and Service Science* (*MASS'09*), Wuhan/Beijing, September 20–22, 2009, pp. 1–4.

Taylor, P., View from the top, Hans Vestberg, Chief executive, Ericsson, *Financial Times*, October 7, 2013, p. 20.

Tenenbaum, J. M., Medich, C., Schiffman, A. M., and Wong, W. T., CommerceNet: Spontaneous electronic commerce on the Internet, in *COMPCON'95*, San Francisco, CA, March 5–9, 1995, pp. 38–43.

Thierauf, R. J., *Electronic Data Interchange in Finance and Accounting*, Quorum Books, New York, 1990.

Troulet-Lambert, O., L'évolution de l'EDI et l'EDI-ouvert: Les travaux du SC 30, *Télécom Interview*, 33, 41–46, 1997.

Tsai, C.-L., Chen, C.-J., and Zhuang, D.-J., Secure OTP and biometric verification scheme for mobile banking, in *Proceedings of the 2012 Third FTRA International Conference on Mobile, Ubiquitous, and Intelligent Computing* (*MUSIC*), Vancouver, British Columbia, Canada, June 26–28, 2012, pp. 138–141.

Tsukayama, H. and Halzack, S., Fraud raises question of whether Apple Pay is secure, *The Star-Ledger*, March 25, 2015, p. 12, Reprinted from *The Washington Post*.

Tung, B., *Kerberos: A Network Authentication System*, Addison Wesley Longman, Inc., Reading, MA, 1999.

Turner, S., Net effect, *Bank. Technol.*, 15(8), 49–52, 1998.

Tweddle, D., EDI in international trade: A customs view, in *The EDI Handbook*, Gikins, M. and Hitchcock, D., Eds., Blenheim Online, London, UK, 1988, pp. 139–146.

Tyson-Davies, R., The function of APACS, in *Electronic Banking and Security*, Welch, B., Ed., Blackwell Scientific, Oxford, UK, 1994, pp. 11–30.

Ugon, M., The microcomputer smart card: A multi-application device which secures telecommunications, in *EuroComm 88, Proceedings of the International Congress on Business, Public, and Home Communication*, Amsterdam, the Netherlands, Schuringa, T. M., Ed., Elsevier/North Holland, Amsterdam, the Netherlands, 1989, pp. 265–282.

UK IT Security Evaluation and Certification Scheme, Certification Report No. P165, Issue 1.0, March 2002, available at http://www.commoncriteriaportal.org/files/epfiles/CRP165.pdf, last accessed November 25, 2013.

UNCITRAL (United Nations Commission on International Trade Law), UNCITRAL model law on electronic commerce, 1996, available at http://www.uncitral.org/uncitral/en/uncitral_texts/electronic_commerce/1996Model.html, last accessed June 3, 2014.

Unique Identification Authority of India, Biometrics design standards for UID applications, Version 1.0, December 2009, available at http://www.uidai.gov.in/images/resource/Biometrics_Standards_Committee_report.pdf, last accessed June 1, 2015.

Untersinger, M., Les applications sécurisées sortent de l'anonymat, *Le Monde*, November 25, 2014, p. 11.

Van Damme, G. and Wouters, K., Practical Experiences with NFC security on Mobile phones, in *Proceedings of the Fifth Workshop on RFID Security 2009* (*RFID SEC09*), Leuven, Belgium, available at http://www.cosic.esat.kuleuven.be/publications/article-1288.pdf, last accessed December 22, 2013.

Van Herreweghen, E. and Wille, U., Risks and potentials of using EMV for Internet payments, in *Proceedings of the USENIX Workshop* on *Smartcard Technology*, Chicago, IL, 1999, pp. 163–173.

Van Hove, L., The New York City Smart Card trial in perspective: A research note, *Int. J. Electron. Commerce*, 5(2), 119–131, 2001.

Van Oorschot, P. and Wiener, M., Parallel collision search with applications to hash functions and discrete logarithms, in *Proceedings of the Second ACM Conference on Computer and Communications Security*, Fairfax, VA, November 2–4, 1994, pp. 210–128.

Van Someren, N., Odlyzko, A., Rivest, R., Jones, T., and Goldie-Scot, D., Panel: Does anyone really need micropayments? in *Proceedings of the Seventh International Conference on Financial Cryptography* (*FC2003*), Guadeloupe, French West Indies, January 27–30, 2003, Lecture Notes in Computer Science, Vol. 2742, Wright, R. N., Ed., Springer-Verlag, Berlin, Germany, 2003, p. 69.

Van Thanh, D., Security issues in mobile ecommerce, in *Proceedings of the 11th International Workshop on Database and Expert Systems Applications*, Washington, DC, 2000, pp. 412–425. Also in *Proceedings of the First International Conference on Electronic Commerce and Web Technologies*, London, UK, September 4–6, 2000, Lecture Notes in Computer Science, Vol. 1875, Bauknecht, K., Madria, S. J., and Pernul, G., Eds., Springer-Verlag, Berlin, Germany, 2000, pp. 467–476.

Varoufakis, Y., Bitcoin and the dangerous fantasy of 'apolitical' money, April 22, 2013, available at http://yanisvaroufakis.eu/2013/04/22/bitcoin-and-the-dangerous-fantasy-of-apolitical-money, last accessed April 3, 2015.

Vaudenay, S., Security flaws induced by CBC padding—Applications to SSL, IPSec, WTLS,…, in *CRYPTO 2002*, Lecture Notes in Computer Science, Vol. 2332, Knudsen, L. R., Ed., Springer-Verlag, Berlin, Germany, 2002, pp. 534–545.

Vigna, P. and Casey, M. J., *The Age of Cryptocurrency. How Bitcoin and Digital Money are Challenging the Global Economic Order*, St. Martin's Press, New York, 2015.

Visa, 3-D Secure™—System overview V. 1.0.2, September 26; 3-D Secure™—Protocol specifications, core functions, V. 1.02, July 16; 3-D Secure™—Functional requirements: Access control server, V. 1.02, July 16; 3-D Secure™—Functional requirements: Merchant server plug-in V.1.02, July 16, 2002.

Von Reischach, F., Karpischeck, S., Michahelles, F., and Adelmann, R., Evaluation of 1D barcode scanning on mobile phones, in *Proceedings of 2010 Internet of Things (IOT), IoT for a Green Planet*, Tokyo, Japan, November 29–December 1, 2010.

W3C, XML encryption syntax and processing, W3C recommendation, December 10, 2002, available at http://www.w3.org/TR/2002/REC-xmlenc-core-20021210, last accessed August 16, 2015.

W3C, XML key management specification (XKMS 2.0), Version 2.0, W3C Recommendation, June 28, 2005, available at http://www.w3.org/TR/2005/REC-xkms2-20050628, last accessed August 16, 2015.

W3C, XML signature syntax and processing, W3C recommendation, 2nd ed., June 10, 2008, available at http://www.w3.org/TR/2008/REC-xmldsig-core-20080610, last accessed August 16, 2015.

Wagner, D. and Schneier, B., Analysis of the SSL 3.0 protocol, in *Proceedings of the Second USENIX Workshop Electronic Commerce*, Oakland, CA, 1996, pp. 29–40.

Wakefield, J., Yahoo malware enslaves PCs to Bitcoin mining, *BBC News*, January 8, 2014, available at www.bbc.co.uk/news/technology-25653664, last accessed April 24, 2015.

Waldmeir, P., Freedom under attack: Civil liberties in the workplace are the latest casualties of the U.S. reaction to terrorism, *Financial Times*, December 13, 2001, p. 10.

Walker, R., 1992: Maintaining the UK's competitive edge in EDI, in *The EDI Handbook*, Gikins, M. and Hitchcock, D., Eds., Blenheim Online, London, UK, 1988, pp. 3–10.

Warrell, H., We know what you're learning, *Financial Times*, Weekend Edition, Life & Arts, July 25–26, 2015, p. 15.

Warren, J., Bitmessage: A peer-to-peer message authentication and delivery system, November 27, 2012, available at https://bitmessage.org/bitmessage.pdf, last accessed March 29, 2015.

Waters, R., Grand theft data, *Financial Times*, April 30/May 1, 2011, p. 5.

Watson, C., Fiumara, G., Tabassi, E., Cheng, S. L., Flanagan, P., and Salamon, W., Fingerprint Vendor Technology Evaluation, NIST Internal/Interagency Report NISTIR 8034, National Institute of Standards and Technology, Gaithersburg, MD, December 18, 2014.

Wayner, P., *Digital Cash: Commerce on the Net*, 2nd ed., Academic Press Professional, New York, 1997.

Westland, J. C., Kwok, M., Shu, J., Kwok, T., and Ho, H., Customer and merchant acceptance of electronic cash: Evidence from Mondex in Hong Kong, *Int. J. Electron. Commerce*, 2(4), 5–26, 1998.

Wiener, M. J., Performance comparison of public-key cryptosystems, *CryptoBytes*, 4(1), 1–5, 1998.

Wildes, R. P., Iris recognition: An emerging biometric technology, *Proc. IEEE*, 85(9), 1348–1363, 1997.

Williams, D., *Pro PayPal Ecommerce*, Apress, Berkeley, CA, 2007.

Wireless Application Forum, Wireless Application Protocol—Public Key Infrastructure Definition WPKI, WAP-217-WPKI, Version 24, April 2001, available at http://technical.openmobilealliance.org/Technical/Release_Program/docs/WPKI/WAP-217-WPKI-20010424-a.pdf, last accessed January 28, 2016.

Xing, L., Bai, X., Li, T., Wang, X., Chen, K., and Liao, X., Unauthorized cross-app resource access on MAC OS X and iOS, May 26, 2015, available at http://www.netfast.com/wp-content/uploads/2015/06/Apple-Zero-Day-Threat-Research.pdf, last accessed July 14, 2015.

Yang, J.-C., Biometrics verification techniques combing with digital signature for multimodal biometrics payment system, in *Proceedings of the 2010 International Congress on Management of e-Commerce and e-Government*, Chengdu, China, October 23–24, 2010, pp. 405–410.

Zaba, S., E-commerce payment protocols: Requirements and analysis, in *Proceedings of the Ninth IEEE Workshop on Computer Security Foundations (CSFW'96)*, Kenmare, Ireland, June 10–12, 1996, pp. 78–80.

Zambito, T., Brains behind $30M counterfeit scheme pleads guilty, *The Star-Ledger*, September 26, 2014, p. 6.

Zelle, D., Security protocol for the mTAN procedure, Presentation at *IT Security for the Next Generation*, International Round, Delft University of Technology, Delft, the Netherlands, May 11–13, 2012, available at http://www.kaspersky.com/images/Abstracts_Final%20Cup.pdf, last accessed February 2, 2015.

Zorpette, G. The end of cash, *IEEE Spectr.*, 49(6), 27–29, 2012.

*Websites**

General

http://www.ecb.europa.eu/paym/retpaym/html/index.en.html—Site of the Single Euro Payments Area (SEPA) project.

https://www.treasury.gov/Pages/default.aspx—Site of the U.S. Treasury Department.

http://www.bis.org—Site of the Bank of International Settlements, which contain data on the various transactions with the different scriptural means of payments for the G-10 countries.

http://www.databreachtoday.co—Updated reports on data breaches: news, events, preparations, training, jobs, etc.

http://krebsonsecurity.com—Site of Brian Krebs, an ex-reporter for *The Washington Post* from 1995 to 2009, on computer and Internet security.

http://www.disa.org/apps/acroglos—Glossary of acronyms of EDI.

Standards

http://ansi.org/—ANSI (American National Standards Institute) standards.

http://www.iata.org—IATA (International Air Transport Association) standards.

http://www.ebxml.org—ebXML standards.

http://www.unece.org/cefact/edifact/welcome.html—EDIFACT standards.

http://www.ietf.org—IETF standards.

http://www.rsasecurity.com—Information on the PKCS industry standards.

http://www.w3.org—W3C standards for electronic payments.

Encryption

http://www.emc.com/emc-plus/rsa-labs/historical/questions-and-answers.htm—Site with an excellent FAQ on the theory and practice of cryptographic systems.

https://www.schneier.com/cryptography/—Site that gives information on the practice of cryptography.

http://csrc.nist.gov/groups/ST/crypto_apps_infra/pki/index.html—Site that gives information on NIST's PKI program.

http://www.freeswan.org—Site that has open-source implementations for Linux of IPSec, IKE, and other protocols.

Kerberos

http://web.mit.edu/kerberos/www/index.html—The official Web page for Kerberos.

https://cybersafe.com/—Site for a commercial version of the free version of Kerberos (Heidmal). The free version was written by Johan Danielsson and Assar Westerlund from the Swedish Institute of Computer Science. It is not restricted by the U.S. laws on exporting encryption.

Certification

https://wiki.mozilla.org/CA:Problematic_Practices—A page will collect some controversial practices by certification authorities that have caused their applications for approval to be delayed.

Biometrics

http://biosecure.it-sudparis.eu/AB/—Site of the BioSecure Foundation (Association BioSecure) whose database includes reference systems (baseline algorithms) and assessment protocols for a variety of biometrics. This is a joint effort of 11 European institutions and comprises voice samples, fingerprints, faces, iris, online signatures, fingerprints, hand shapes, and iris scans.

https://www.ibia.org/—This is the site of the International Biometric and Identity Association (IBIA), a trade association that represents the industrial players in biometric technology and identity management. It was first established in Washington, DC, in 1998 as the International Biometric Industry Association.

https://biolab.csr.unibo.it/FVCOnGoing/UI/Form/Home.aspx—FVC-onGoing is a web-based online evaluation of fingerprint recognition algorithms. It tracks advanced fingerprint recognition technologies. It evolved out the series of international Fingerprint Verification Competitions (FVC) organized in 2000, 2002, 2004, and 2006. The competitions were organized by the Biometric System Laboratory, University of Bologna, the Pattern Recognition and Image Processing Laboratory of Michigan State University, the Biometric Test Center of San Jose State University, and the Biometric Recognition Group—ATVS, from the Universidad Autonoma de Madrid.

* All sites were accessed on January 31, 2016.

http://www.ee.surrey.ac.uk/CVSSP/banca/icba2004/csuresults.html—Results of the face verification competition organized during the 2004 International Conference on Biometric Authentication (ICBA). Since 2006, the conference became the International Conference on Biometrics (ICB).

SPAM Control

https://www.virusbtn.com/conference/vb2016/index—The Virus Bulletin is a security information portal, testing and certification body with information about the latest developments in malware and an annual international conference. Since 2008, the site hosts the Spammers' Compendium that tracks spamming methods.

EDIFACT

http://lexmercatoria.org—Site that publishes UNCITRAL Model Law on Electronic Commerce of 1996.

http://www.unece.orgcefact—Site of the UN/ECE (United Nations Economic Commission for Europe), which hosts the documents of the CEFACT including those that relate to ebXML.

http://uddi.xml.org—Site of the organization responsible for establishing the directory of services available on the web.

SSL/TLS/WTLS

http://www.openssl.org—Site of the project OpenSSL.

http://ftp.openbsd.org/pub/OpenBSD/LibreSSL—Site of LibreSSL, a free and open version of the SSL/TLS protocol forked from OpenSSL.

http://www.wapforum.org—Site of WAP Forum that contains the WAP specifications, including those of WTLS.

http://technical.openmobilealliance.org/Technical/use-agreement?fp=tech%2Faffiliates%2Fwap%2Fwap-261-wtls-20010406-a.pdf&rp=228—The latest version of the WAP protocol, Version 06 published in April 2001 (Wireless Application Protocol WAP-261-WTLS-20010406-a).

SET

http://www.maithean.com/docs/set_bk1.pdf; http://www.maithean.com/docs/set_bk3.pdf—Books 1 and 2 of the SET specifications can be downloaded this site.

http://ccc.cs.lakeheadu.ca/set/set_bk2.pdf—Book 2 of the SET specifications can be downloaded this site.

http://www.redbooks.ibm.com/redbooks/pdfs/sg244978.pdf—A document describing the use of SET for credit card payments on the web (June 1997).

Magnetic Stripe Cards

www.magtek.com/documentation/public/99800004-1.08.pdf—Summary of standards and specifications of magnetic stripe cards.

http://www.hackerscatalog.com/Services/TECH_Notes/twentysix.html—One of the many sites offering equipment to read and write magnetic tracks of a stripe card.

Purses

http://www.globalplatform.org—Site of the GlobalPlatform organization.

http://ftp.jrc.es/EURdoc/eur19922en.pdf—The Electronic Payment Systems Observatory (EPSO), issues 1–8, July 2000–September 2001. (Use Firefox; Safari cannot load data from the site)

Mobile Payments

System	Site
Mobito	https://www.o2.cz/osobni/en/203289-penezni_sluzby/292675-mobito.html
Mpass	www.mpass.de
Pay2Me	https://www.mobistar.be/fr/options-et-services/recharger-et-payer/m-banxafe/pay2me
Paybox	http://www1.paybox.com
PayPass (terminal requirements)	https://www.paypass.com/pdf/public_documents/FINAL_PayPass_Terminal_2007_v1.pdf
PayPass (implementation guide)	https://www.paypass.com/PP_Imp_Guides/PPCR_Manual.pdf

Smart (Microprocessor) Cards

http://www.visa.com, http://www.mastercard.com—Site that publishes the EMV specifications on the use of smart cards for payment system.

http://www.opencdp.org/ocf/—The OpenCard Framework was originally defined by the OpenCard Consortium, mainly driven by IBM and Gemplus. The work ended with version 1.2 of the specification and a reference implementation by IBM.

http://www.pcscworkgroup.com—Site that contains the PC/SC specifications for integrating smart cards and smart card readers with PCs under Windows.

http://www.oracle.com/technetwork/java/index.html—Site that gives the JavaCard specifications.

http://pcsclite.alioth.debian.org/musclecard.com/info.html—Site of MUSCLE, a movement to adapt integrated circuit cards to the Linux environment.

http://www.seis.se—Site of SEIS (Secured Electronic Information in Society), a nonprofit Swedish association that publishes the specifications of the electronic identity card that was later adopted as a Swedish Standard.

PayPal

https://www.paypal-marketing.com/paypal/html/hosted/emarketing/partner/directory—Solutions that are pre-integrated with PayPal.

http://railscasts.com/episodes/143-paypal-security—This is a text version of a video by Ryan Bates on PayPal security.

https://www.paypal.com/webapps/mpp/paypal-safety-and-security—A PayPal list of scams that target sellers, tactics for placing fraudulent orders, how to detect suspicious or unauthorized transaction, etc.

http://www.paypalsucks.com—This site contains customer's report of problems with eBay and PayPal, particularly current fraud events.

Electronic and Virtual Checks

http://www.eccho.org—ECCHO is a U.S. not-for-profit clearinghouse with a membership of more than 3000 financial institution members. It establishes rules and standardized mechanisms for electronic check exchanges. In addition, it provides the services of education and advocacy. It certifies National Check Professionals (NCPs).

http://fsroundtable.org/bits—BITS is the technology policy division of Financial Services Roundtable (FSR). It absorbed the Financial Services Technology Consortium (FSTC) in June 2009.

http://www.it-aac.org/itaachomepage.html—Site of the IT Acquisition Advisory Council in which the FSTC is a participant

http://www.echeck.org—Site that gives the Echeck project.

http://www.netcheque.org—Site presenting the prototype of the NetCheque system.

http://www.ofx.net/—OFX (Open Financial Exchange) is a unified specification for the electronic exchange of financial data between financial institutions, businesses, and consumers via the Internet.

Bitcoin and Cryptocurrencies

https://en.bitcoin.it/wiki/Protocol_documentation—Protocol documentation.

http://www.righto.com/2014/02/bitcoin-mining-hard-way-algorithms.html—This article explains Bitcoin mining in details, right down to the hex data and network traffic, in particular how a transaction gets mined into a block.

https://www.khanacademy.org/economics-finance-domain/core-finance/money-and-banking/bitcoin/v/bitcoin-transaction-block-chains—An online tutorial on Bitcoin offered by the Khan Academy.

http://bitcoincharts.com—Bitcoincharts is a site that provides financial and technical data related to the Bitcoin network.

https://blockexplorer.com/—Bitcoin Block Explorer is an open-source web tool that allows you to view information about the blocks, addresses, and transactions created by Bitcoin. The source code is on GitHub.

http://blockexplorer.com/q/getblockcount—Provides the current block count.

http://2.bp.blogspot.com/-DaJcdsyqQSs/UsiTXNHP-0I/AAAAAAAATC0/kiFRowh-J18/s1600/blockchain.png—Documentation of the Bitcoin blockchain from the Bitcoin-QT application due to John T. Ratcliff

http://www.coindesk.com/—This is a site that offers the latest news and lists the prices on the various exchanges (CoinDesk, Bitstamp, Bitfinex, BTC-e, LakeBTC, OKCoin).

http://www.hashcash.org—The hashcash is a proof-of-work algorithm, used as a denial-of-service countermeasure technique in a number of systems and in the Bitcoin mining function. The site provides the source code that includes a library form and works with a cryptographic hash, such as SHA1, SHA256, or SHA-3.

https://alloscomp.com/bitcoin/calculator—This site offers a calculator to estimate the expected earning in bitcoins and U.S. $ for a given hash rate and at the current difficulty.

Cryptocurrencies

Bitmessage	https://bitmessage.org/wiki/Main_Page
Bitnotar	https://github.com/bitcoinaustria/bitnotar
BitShares	https://bitshares.org
Chronobit	https://github.com/goblin/chronobit
CoinSpark	http://coinspark.org
Devcoin	http://devcoin.org
Dash (Darkcoin)	http://en.wiki.dashninja.pl/wiki/DarkSend
Dogecoin	http://www.reddit.com/r/dogeducation/wiki/dogecoin
Everledger (previously Blocktrace)	http://www.everledger.io
Litecoin	https://litecoin.org
Nxt	http://nxt.org
Paycoin	https://paycoin.com
Ripple	https://ripple.com/
Stellar	https://www.stellar.org
Zerocash	http://zerocash-project.org
Zerocoin	http://zerocoin.org

https://99bitcoins.com/how-to-use-darkcoin-wallet-darksend—Instructions on the installation and use of the Darkcoin Wallet, Darksend, and Dash (Darkcoin).

https://en.bitcoin.it/wiki/Bitcoind—Information regarding the first full-node implementation of Bitcoin.

Digital Estate Planning

http://www.thedigitalbeyond.com

Labeling Organizations

http://www.webtrust.org/item64428.aspx—Site of the WebTrust program that was offered jointly by the American Institute of Certified Public Accountants and the Canadian Institute of Chartered Accountants.

http://www.truste.org—Site of TRUSTe, an organization run by the Electronic Frontier Foundation and CommerceNet, which awards its privacy seal to sites that comply with TRUSTe oversight and conflict resolution procedures. TRUSTe also offers a special seal for sites addressed toward children to indicate their compliance with the COPPA requirements.

Index

D

E

F

N

Q

R

S

V